W9-BMR-266

Formulas/Equations

Distance Formula   If $P_1 = (x_1, y_1)$ and $P_2 = (x_2, y_2)$, the distance from P_1 to P_2 is

$$d(P_1, P_2) = \sqrt{(x_2 - x_1)^2 + (y_2 - y_1)^2}$$

Equation of a Circle   The equation of a circle of radius r with center at (h, k) is

$$(x - h)^2 + (y - k)^2 = r^2$$

Slope Formula   The slope m of the line containing the points $P_1 = (x_1, y_1)$ and $P_2 = (x_2, y_2)$ is

$$m = \frac{y_2 - y_1}{x_2 - x_1} \qquad \text{if } x_1 \neq x_2$$

$$m \text{ is undefined} \qquad \text{if } x_1 = x_2$$

Point–Slope Equation of a Line   The equation of a line with slope m containing the point (x_1, y_1) is

$$y - y_1 = m(x - x_1)$$

Slope–Intercept Equation of a Line   The equation of a line with slope m and y-intercept b is

$$y = mx + b$$

Quadratic Formula   The solutions of the equation $ax^2 + bx + c = 0, a \neq 0$, are

$$x = \frac{-b \pm \sqrt{b^2 - 4ac}}{2a}$$

If $b^2 - 4ac > 0$, there are two real unequal solutions.
If $b^2 - 4ac = 0$, there is a repeated real solution.
If $b^2 - 4ac < 0$, there are two complex solutions.

Geometry Formulas

Circle

r = Radius, A = Area, C = Circumference
$$A = \pi r^2 \qquad C = 2\pi r$$

Triangle

b = Base, h = Altitude (Height), A = Area
$$A = \tfrac{1}{2}bh$$

Rectangle

l = Length, w = Width, A = Area, P = Perimeter
$$A = lw \qquad P = 2l + 2w$$

Rectangular Box

l = Length, w = Width, h = Height, V = Volume
$$V = lwh$$

Sphere

r = Radius, V = Volume, S = Surface area
$$V = \tfrac{4}{3}\pi r^3 \qquad S = 4\pi r^2$$

Precalculus

Graphing and Data Analysis

MICHAEL SULLIVAN
Chicago State University

MICHAEL SULLIVAN, III
South Suburban College

Prentice Hall, Upper Saddle River, NJ 07458

Library of Congress Cataloging-in-Publication Data
Sullivan, Michael
    Precalculus: graphing and data analysis/
  Michael Sullivan, Michael Sullivan, III.
      p.   cm.
    Includes index.
    ISBN 0-13-778499-6
    1. Functions.   I. Sullivan, Michael.   II. Title.
  QA331.3.S928   1998
  510—dc21   97-23427
                                                          CIP

Senior Acquisitions Editor: *Sally Denlow*
Marketing Manager: *Patrice Lumumba Jones*
Production Editor: *Robert C. Walters*
Managing Editor: *Linda Mihatov Behrens*
Executive Managing Editor: *Kathleen Schiaparelli*
Assistant Vice President of Production and Manufacturing: *David W. Riccardi*
Supplements Editor/Editorial Assistant: *April Thrower*
Marketing Assistant: *Patrick Murphy*
Creative Director: *Paula Maylahn*
Art Director: *Amy Rosen*
Art Manager: *Gus Vibal*
Interior Design: *Elm Street Publishing Services, Inc.*
Cover Design: *Jeanette Jacobs*
Photo Editor: *Lori Morris-Nantz*
Photo Research: *Beth Boyd*
Manufacturing Buyer: *Alan Fischer*
Manufacturing Manager: *Trudy Pisciotti*
Cover Photo: SUNFLOWER in *Symmetry and Chaos,* by Michael Field and Martin
Golubitsky, OUP, 1992.

 © 1998 by Prentice-Hall, Inc.
Simon & Schuster/A Viacom Company
Upper Saddle River, NJ 07458

All rights reserved. No part of this book may be reproduced, in any form or by any means,
without permission in writing from the publisher.

Printed in the United States of America

10 9 8 7 6 5 4 3 2 1

ISBN 0-13-778499-6

Prentice-Hall International (UK) Limited, *London*
Prentice-Hall of Australia, Pty. Limited, *Sydney*
Prentice-Hall Canada Inc. *Toronto*
Prentice-Hall Hispanoamericana, S.A., *Mexico*
Prentice-Hall of India Private Limited, *New Delhi*
Prentice-Hall of Japan, Inc., *Tokyo*
Simon & Schuster Asia Pte. Ltd., *Singapore*
Editoria Prentice-Hall do Brasil, Ltda., *Rio de Janeiro*

For Our Students. . .
Past and Present

CONTENTS

CHAPTER 8 POLAR COORDINATES; VECTORS 591

CHAPTER 9 ANALYTIC GEOMETRY 665

CHAPTER 10 SYSTEMS OF EQUATIONS AND INEQUALITIES 745

CHAPTER 11 SEQUENCES; INDUCTION; COUNTING; PROBABILITY 853

Why We Wrote This Book

Times continue to change! In 1987, *Precalculus* was written, a traditional text that developed all the skills a student would need to succeed in college level mathematics courses. This text, now in its 4th edition, remains traditional in its approach. As technology developed, various philosophies regarding the use of incorporating graphing utilities into the curriculum emerged. These philosophies, as they relate to Precalculus, lead to different views regarding the use of technology in the classroom. In 1996, we wrote *Precalculus Enhanced with Graphing Utilities* in response to the call for a text to fully incorporate technology while maintaining strict mathematical content. Through the use of technology, the text allows students to recognize patterns, visualize, explore, and foreshadow concepts from areas such as Calculus. This book, *Precalculus: Graphing and Data Analysis,* was written to address the point of view that technology should not only be used to graph functions and solve equations, but also should be used to analyze data, draw scatter diagrams, and find curves of best fit. Thus, this text incorporates concepts never before discussed in a traditional Precalculus course while still maintaining strict mathematical content. Students are not simply "pushing buttons" on their calculators, but must answer thoughtful questions that force them to utilize mathematics.

The standards developed by MAA, AMATYC, and NCTM encourage modeling and connecting mathematics to various disciplines. This text introduces mathematical theory through the use of models developed in other disciplines. The text also expects students to develop their own models either from data or situations and proceed to analyze these models mathematically.

To this end, we begin with a discussion of bivariate data, representing it in scatter diagrams and determining whether the relationship between the variables is linear or nonlinear. Using a graphing utility, LINear REGression techniques are used to find and interpret curves of best fit for linearly related data. As the mathematics evolves, we use a graphing utility to find curves of best fit for quadratic, power, cubic, exponential, logarithmic, logistic, and sinusoidal curves of best fit. Finally, the ideas of the limit and the derivative of a function are introduced and explored. In each instance the data is used not only to find curves of best fit, but also to analyze the functions found using concepts such as local minima and maxima. By comparing the results of their analysis to the actual data, students are able to judge the reasonableness of their answer.

The data itself is truly real, drawn from sources such as the Statistical Abstract of the United States, the U.S. Census Bureau, the U.S. Department of Justice, the Bureau of Labor Statistics, the New York and NASDAQ Stock Exchanges, and so on. Each chapter begins with a popular issue, such as domestic and global use of oil, asking students to analyze and interpret data from web site locations. This data is dynamic, changing on a regular basis. The data is widely based, drawn from disciplines such as physics, biology, chemistry, economics, finance, psychology, and statistics.

Notwithstanding the emphasis on data analysis, we fully utilize the graphers ability to promote visualization, exploration, and foreshadowing of concepts. The technology is interwoven with the mathematics. Doing mathematics by hand using traditional methods and doing mathematics using technology are often combined, such as using the domain of a function to set a viewing window. Many examples are solved both graphically and algebraically. We encourage students to distinguish between solving a problem using the full power of the technology vs. recognizing when the technology is limited or the simple nature of the problem makes a solution by hand the better choice.

About This Book

Content   The content of this book contains the topics found in a traditional Precalculus text, except possibly for the chapter devoted to the limit and the derivative of a function. In addition, it contains topics unique to a technology approach: sections on data analysis and curves of best fit appear in chapters dealing with linear functions (LINear REGression), quadratic functions (QUADratic REGression), power functions (POWer REGression), polynomial functions (CUBic REGression), logarithmic and exponential functions (LOGarithmic REGression, EXPonential REGression, LOGISTIC REGression), and trigonometric functions (SINusoidal REGression). Also, many examples and exercises found in this text are ones that cannot be handled using traditional methods.

Organization   In recognition of the fact that some students may not be fully prepared for this course, a specially designed Appendix has been included to provide necessary review. The Appendix is referenced at the beginning of the chapter and at the appropriate place within the chapter.

Chapter 1 begins with a discussion of rectangular coordinates, and their use in graphing data in a scatter diagram. Graphing is done early in this chapter (1.2), with emphasis on the graphs of certain key equations. A section on modelling linearly related data appears here. Also included is a discussion of solving equations and inequalities, utilizing both an algebraic and graphing approach. The solution of polynomial and rational inequalities is postponed to Chapter 3, so the power of graphing polynomial and rational functions can be used.

Chapter 2 develops the important concept of a function from a discrete and a continuous point of view, emphasizing tables, graphs, and properties and including discussions on average rate of change, local maxima and minima, and increasing and decreasing functions. The last section focuses on the construction of functions in applications.

Chapter 3 discusses graphs and properties of polynomial and rational functions. Here, data resulting in scatter diagrams lead to curves of best fit that are quadratic, power, or cubic are analyzed. Complex numbers and quadratic equations with a negative discriminant provide motivation for the Fundamental Theorem of Algebra. Many exercises here ask students to draw scatter diagrams, find the appropriate curve of best fit, and analyze the resulting function using techniques introduced in the chapter.

Chapter 4, Exponential and Logarithmic Functions, places emphasis on graphs and properties, utilizing data with scatter diagrams that lead to curves of best fit that are exponential, logarithmic, or logistic.

Chapter 5, Trigonometric Functions, uses the unit circle approach to define the six trigonometric functions and develop properties of these functions. The graphs of the trigonometric functions are then discussed. The chapter concludes with the inverse trigonometric functions.

Chapter 6, Analytic Trigonometry, deals with trigonometric identities and trigonometric equations. A graphing utility is used to solve trigonometric equations for which no algebraic or trigonometric method is available.

Chapter 7, Applications of Trigonometric Functions, deals both with early applications of trigonometry (requiring solving right or oblique triangles) and with modern applications, such as simple harmonic motion, damped motion, and sinusoidal curve fitting.

Chapter 8, Polar Coordinates; Vectors, introduces polar coordinates, with an emphasis on graphing. Included is a discussion of Demoivre's Theorem. Independent of the preceding topics, sections on vectors in the plane, vectors in space, and the dot product are also provided.

Chapter 9, Analytic Geometry, discusses conics, including rotation of axes and polar forms. Parametric equations are also discussed, including an example of motion simulation.

Chapter 10, Systems of Equations and Inequalities, discusses both graphing and algebraic approaches. Many examples and exercises here can only be worked using technology.

Chapter 11, Sequences; Induction; Counting; Probability, is mostly traditional since the use of technology usually does not lead to more efficient methods.

Chapter 12, A Preview of Calculus: the Limit and the Derivative of a Function, introduces limits first through the use of Tables and then by graphing. Algebraic techniques for calculating limits follow. One-sided limits and continuity preceed the discussion of the derivative. Applications to velocity of falling objects and instantaneous rates of change are provided.

Instructor Supplements

Instructor Resource Manual
written by Michael Sullivan, Michael Sullivan III and Katy Murphy
ISBN: 0-13-685389-7
Contains complete step-by-step worked out solutions to all the even numbered exercises in the textbook. Also included are strategies for using the "Mission Possible" collaborative learning projects found in each chapter.

Written Test Item File
ISBN: 0-13-685355-2
Features questions that are found within the computerized test bank in a format that can be used for selecting specific problems for testing.

TestPro Computerized Test Generator
MAC ISBN: 0-13-685272-6
IBM ISBN: 0-13-685264-5
Allows instructors to generate tests from algorithms keyed to the text by chapter, section, and learning objective. Instructors select from thousands of test questions and hundreds of algorithms which generate different but equivalent questions. A user-friendly expression-building toolbar, editing and graphing capabilities are included. Customization toolbars allow for customized headers and layout options which provide instructors with the ability to add or delete workspace or add columns to conserve paper.

Tutorial Videos
ISBN: 0-13-685116-9

A new tutorial videotape series, created exclusively to accompany this text, include 15 minute segments for each section of the text. Each segment uses both traditional and graphical ways of solving mathematical problems. These videos provide an alternative process which can add to your students' success in this course. Included with each set of tapes is a permission letter to duplicate the tapes for your mathematics lab or library.

New York Times Supplement

A free newspaper supplement from Prentice Hall and the New York Times which includes interesting and relevant articles on mathematics in the world around us. Great for getting students to talk and write about mathematics. This supplement is updated each fall. To request free copies for all of your students, contact your Prentice Hall representative.

Prentice Hall Companion Website

Located at *www.prenhall.com/sullivan* this website offers you and your students additional information, projects and exercises to enrich the learning experience using this text. The Internet "Excursions" found in the beginning of each chapter have duplicate pages on the website with the links to the appropriate data and information-gathering sites described in the projects.

Acknowledgments

Textbooks are written by authors, but evolve from an idea into final form through the efforts of many people. Special thanks to Don Dellen, who first suggested this book and the other books in this series. Don's extensive contributions to publishing and mathematics are well known; we will all miss him dearly.

There are many people we would like to thank for their input, encouragement, patience and support. They have our deepest thanks and appreciation. We apologize for any omissions . . .

James Africh, *College of DuPage*
Steve Agronsky, *Cal Poly State University*
Dave Anderson, *South Suburban College*
Joby Milo Anthony, *University of Central Florida*
James E. Arnold, *University of Wisconsin-Milwaukee*
Agnes Azzolino, *Middlesex County College*
Wilson P. Banks, *Illinois State University*
Dale R. Bedgood, *East Texas State University*
Beth Beno, *South Suburban College*
William H. Beyer, *University of Akron*
Richelle Blair, *Lakeland Community College*
Trudy Bratten, *Grossmont College*
William J. Cable, *University of Wisconsin-Stevens Point*

Lois Calamia, *Brookdale Community College*
Roger Carlsen, *Moraine Valley Community College*
John Collado, *South Suburban College*
Denise Corbett, *East Carolina University*
Theodore C. Coskey, *South Seattle Community College*
John Davenport, *East Texas State University*
Duane E. Deal, *Ball State University*
Vivian Dennis, *Eastfield College*
Karen R. Dougan, *University of Florida*
Louise Dyson, *Clark College*
Paul D. East, *Lexington Community College*
Don Edmondson, *University of Texas-Austin*

Christopher Ennis, *University of Minnesota*

Garret J. Etgen, *University of Houston*

W. A. Ferguson, *University of Illinois-Urbana/Champaign*

Iris B. Fetts, *Clemson University*

Mason Flake, *student at Edison Community College*

Merle Friel, *Humboldt State University*

Richard A. Fritz, *Moraine Valley Community College*

Carolyn Funk, *South Suburban College*

Dewey Furness, *Ricke College*

Dawit Getachew, *Chicago State University*

Wayne Gibson, *Rancho Santiago College*

Joan Goliday, *Sante Fe Community College*

Frederic Gooding, *Goucher College*

Ken Gurganus, *University of North Carolina*

James E. Hall, *University of Wisconsin-Madison*

Judy Hall, *West Virginia University*

Edward R. Hancock, *DeVry Institute of Technology*

Brother Herron, *Brother Rice High School*

Kim Hughes, *California State College-San Bernardino*

Ron Jamison, *Brigham Young University*

Richard A. Jensen, *Manatee Community College*

Sandra G. Johnson, *St. Cloud State University*

Moana H. Karsteter, *Tallahassee Community College*

Arthur Kaufman, *College of Staten Island*

Thomas Kearns, *North Kentucky University*

Teddy Koukounas, *SUNY at Old Westbury*

Keith Kuchar, *Manatee Community College*

Tor Kwembe, *Chicago State University*

Linda J. Kyle, *Tarrant Country Jr. College*

H. E. Lacey, *Texas A & M University*

Christopher Lattin, *Oakton Community College*

Adele LeGere, *Oakton Community College*

Stanley Lukawecki, *Clemson University*

Virginia McCarthy, *Iowa State University*

James McCollow, *DeVry Institute of Technology*

Laurence Maher, *North Texas State University*

Jay A. Malmstrom, *Oklahoma City Community College*

James Maxwell, *Oklahoma State University-Stillwater*

Carolyn Meitler, *Concordia University*

Eldon Miller, *University of Mississippi*

James Miller, *West Virginia University*

Michael Miller, *Iowa State University*

Kathleen Miranda, *SUNY at Old Westbury*

Jane Murphy, *Middlesex Community College*

Bill Naegele, *South Suburban College*

James Nymann, *University of Texas-El Paso*

Sharon O'Donnell, *Chicago State University*

Seth F. Oppenheimer, *Mississippi State University*

E. James Peake, *Iowa State University*

Thomas Radin, *San Joaquin Delta College*

Ken A. Rager, *Metropolitan State College*

Elsi Reinhardt, *Truckee Meadows Community College*

Jane Ringwald, *Iowa State University*

Stephen Rodi, *Austin Community College*

Howard L. Rolf, *Baylor University*

Edward Rozema, *University of Tennessee at Chattanooga*

Dennis C. Runde, *Manatee Community College*

John Sanders, *Chicago State University*

Susan Sandmeyer, *Jamestown Community College*

A.K. Shamma, *University of West Florida*

Martin Sherry, *Lower Columbia College*

Anita Sikes, *Delgado Community College*

Timothy Sipka, *Alma College*

John Spellman, *Southwest Texas State University*

Becky Stamper, *Western Kentucky University*

Neil Stephens, *Hinsdale South High School*

Diane Tesar, *South Suburban College*

Tommy Thompson, *Brookhaven College*

Richard J. Tondra, *Iowa State University*

Marvel Townsend, *University of Florida*

Jim Trudnowski, *Carroll College*

Richard G. Vinson, *University of Southern Alabama*

Mary Voxman, *University of Idaho*

Darlene Whitkenack, *Northern Illinois University*

Chris Wilson, *West Virginia University*

Carlton Woods, *Auburn University*

George Zazi, *Chicago State University*

Recognition and thanks are due particularly to the following individuals for their valuable assistance in the preparation of this edition: Jerome Grant for his support and commitment; Sally Denlow, for her genuine interest and insightful direction; Bob Walters for his organizational skill as production supervisor; Patrice Lumumba Jones for his innovative marketing efforts; Ray Mullaney for his specific editorial comments; the entire Prentice Hall sales staff for their confidence; and to Katy Murphy for checking the answers to all the exercises.

Michael Sullivan
Michael Sullivan, III

As you begin your study of Precalculus, you might feel overwhelmed by the number of theorems, definitions, procedures and equations that confront you. You may even wonder whether you can learn all this material in a single course. For many of you, this may be your last mathematics course, while for others, just the first in a series of many. Don't worry—either way, this text was written with you in mind.

This text was designed to help you—the student, master the terminology and basic concepts of Precalculus. These aims have helped to shape every aspect of the book. Many learning aids are built into the format of the text to make your study of this material easier and more rewarding. This book is meant to be a "machine for learning," one that can help you to focus your efforts and get the most from the time and energy you invest.

This book requires that you have access to a graphing utility: a graphing calculator or a computer software package that has a graphing component. Be sure you have some familiarity with the device you are using before the course begins.

Here are some hints we give our students at the beginning of the course:

1. Take advantage of the feature PREPARING FOR THIS CHAPTER. At the beginning of each chapter, we have prepared a list of topics to review. Be sure to take the time to do this. It will help you proceed quicker and more confidently through the chapter.

2. Read the material in the book before the lecture. Knowing what to expect and what is in the book, you can take fewer notes and spend more time listening and understanding the lecture.

3. After each lecture, rewrite your notes as you re-read the book, jotting down any additional facts that seem helpful. Be sure to do the Now Work Problem x as you proceed through a section. After completing a section, be sure to do the assigned problems. Answers to the Odd ones are in the back of the book.

4. If you are confused about something, visit your instructor during office hours immediately, before you fall behind. Bring your attempted solutions to problems with you to show your instructor where you are having trouble.

5. To prepare for an exam, review your notes. Then proceed through the Chapter Review. It contains a capsule summary of all the important material of the chapter. If you are uncertain of any concept, go back into the chapter and study it further. Be sure to do the Review Exercises for practice.

Remember the two "golden rules" of Precalculus:

1. DON'T GET BEHIND! The course moves too fast, and it's hard to catch up.
2. WORK LOTS OF PROBLEMS. Everyone needs to practice, and problems show where you need more work. If you can't solve the homework problems without help, you won't be able to do them on exams.

We encourage you to examine the following overview for some hints on how to use this text.

Best Wishes!

Michael Sullivan
Michael Sullivan, III

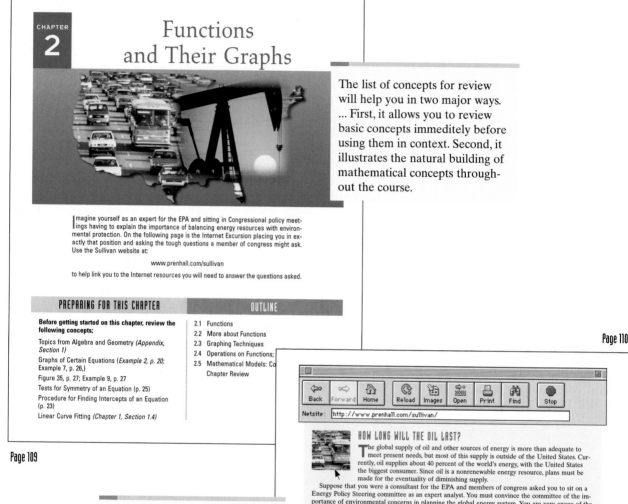

The list of concepts for review will help you in two major ways. ... First, it allows you to review basic concepts immediately before using them in context. Second, it illustrates the natural building of mathematical concepts throughout the course.

CHAPTER 2

Functions and Their Graphs

Imagine yourself as an expert for the EPA and sitting in Congressional policy meetings having to explain the importance of balancing energy resources with environmental protection. On the following page is the Internet Excursion placing you in exactly that position and asking the tough questions a member of congress might ask. Use the Sullivan website at:

www.prenhall.com/sullivan

to help link you to the Internet resources you will need to answer the questions asked.

PREPARING FOR THIS CHAPTER

Before getting started on this chapter, review the following concepts;

Topics from Algebra and Geometry *(Appendix, Section 1)*

Graphs of Certain Equations *(Example 2, p. 20; Example 7, p. 26,)*

Figure 35, p. 27; Example 9, p. 27

Tests for Symmetry of an Equation (p. 25)

Procedure for Finding Intercepts of an Equation (p. 23)

Linear Curve Fitting *(Chapter 1, Section 1.4)*

OUTLINE

2.1 Functions
2.2 More about Functions
2.3 Graphing Techniques
2.4 Operations on Functions;
2.5 Mathematical Models: Co
    Chapter Review

Each chapter begins by placing you in an intriguing and relevant situation where you will need to collect information from the Internet, analyze data according to material available within the chapter, and answer thought-provoking questions. All the appropriate links can be found on the "live" pages at:
http://www.prenhall.com/sullivan

Netsite: http://www.prenhall.com/sullivan/

HOW LONG WILL THE OIL LAST?

The global supply of oil and other sources of energy is more than adequate to meet present needs, but most of this supply is outside of the United States. Currently, oil supplies about 40 percent of the world's energy, with the United States the biggest consumer. Since oil is a nonrenewable energy resource, plans must be made for the eventuality of diminishing supply.

Suppose that you were a consultant for the EPA and members of congress asked you to sit on a Energy Policy Steering committee as an expert analyst. You must convince the committee of the importance of environmental concerns in planning the global energy system. You are very aware of the fact that present modes of energy use and production threaten serious environmental deterioration. Your plan is to create a set of what-if functions that will allow your committee to model many possible scenarios for their policy guidelines.

1. $E(t)$, the first function that you create, will allow you to model United States oil consumption over a period of time. Make a scatterplot using the EPA Data on United States energy consumption, per capita from 1950–1990. Since oil provides 40 percent of the energy consumed, you adjust your figures to get oil consumption per capita. Use the LINear REGression tool on your graphing utility to find the linear function $E(t)$ of best fit for modeling the data. Does this seem like a good model?

2. The second function, $P(t)$, models population. From the United States Census Data make a scatterplot, and then use a QUADratic REGression to model the data. Next, let $O(t) = E(t) \cdot P(t)$. What does $O(t)$ model? Graph $O(t)$. Compare to the actual figures from BP Petroleum.

3. Using the function $O(t)$, estimate the area under the graph for the years 1997 until 2000. To do this make a trapezoid, and find its area. Now estimate the areas under $O(t)$ using 10-year intervals. For what year does the total of all the trapezoid areas equal the remaining United States oil reserve?

4. Could this sort of analysis be used to predict when the world's supply of oil will run out? Can your committee think of any methods that might improve your predictions?

Perhaps the most central idea in mathematics is the notion of a *function.* This important chapter deals with what a function is, how to graph functions, how to perform operations on functions, and how functions are used in applications.

The word *function* apparently was introduced by René Descartes in 1637. For him, a function simply meant any positive integral power of a variable *x.* Gottfried Wilheim von Leibniz (1646–1716), who always emphasized the geometric side of mathematics, used the word function to denote any quantity associated with a curve, such as the coordinates of a point on the curve. Leonhard Euler (1707–1783) employed the word to mean any equation or formula involving variables and constants. His idea of a function is similar to the one most often used today in courses that precede calculus. Later, the use of functions in investigating heat flow equations led to a very broad definition, due to Lejeune Dirichlet (1805–1859), which describes a function as a rule or correspondence between two sets. It is his definition that we use here.

Page 109

Page 110

2.1 FUNCTIONS

1. Determine Whether a Relation Represents a Function
2. Identify the Graph of a Function
3. Find the Value of a Function
4. Find the Domain of a Function
5. Obtain Information from/about the Graph of a Function

In Section 1.1, we said a **relation** is a correspondence between two variables, say x and y, and can be written as a set of ordered pairs (x, y). When relations are written as ordered pairs (x, y), we say x is related to y. Often, we are interested in specifying the type of relation (such as an equation) that might exist between the two variables. For example, when variables appear to be linearly related (based on the scatter diagram), we find a line of best fit. In Section 1.4, we found that fertilizer and crop yield appeared to be linearly related and the line of best fit was $y = 0.717x + 4.786$, where x represents the amount of fertilizer and y represents the crop yield. Thus, if we

Begin each section by reading the learning objectives. These objectives will help you organize your studies and prepare for class.

Page 111

Increase your understanding, visualize, discover, explore, and solve problems using a graphing utility.

EXAMPLE 1   Vertical Shift Down

Use the graph of $f(x) = x^2$ to obtain the graph of $h(x) = x^2 - 4$.

Solution   Table 5 lists some points on the graphs of $f = Y_1$ and $h = Y_2$. The graph of h is identical to that of f, except that it is shifted down 4 units. See Figure 38.

FIGURE 38

Horizontal Shifts

EXPLORATION   On the same screen, graph each of the following functions:
$$Y_1 = x^2$$
$$Y_2 = (x - 1)^2$$
$$Y_3 = (x - 3)^2$$
$$Y_4 = (x + 2)^2$$
What do you observe?

Page 156

Page 155

Now work Problem 31.

Vertical and horizontal shifts are sometimes combined.

EXAMPLE 2   Combining Vertical and Horizontal Shifts

Graph the function $f(x) = (x - 1)^3 + 3$.

FIGURE 40

Solution   We graph f in steps. First, we note that the rule for f is basically a cube function. Thus, we begin with the graph of $y = x^3$, shown in blue in Figure 40. Next, to get the graph of $y = (x - 1)^3$, we shift the graph of $y = x^3$ horizontally 1 unit to the right. See the graph shown in red in Figure 40. Finally, to get the graph of $y = (x - 1)^3 + 3$, we shift the graph of $y = (x - 1)^3$ vertically up 3 units. See the graph shown in green in Figure 40. Note the points that have been plotted on each graph. Using key points such as these can be helpful in keeping track of just what is taking place.

In Example 2, if the vertical shift had been done first, followed by the horizontal shift, the final graph would have been the same. (Try it for yourself.)

Now work Problem 43.

Compressions and Stretches

EXPLORATION   On the same screen, graph each of the following functions:

FIGURE 41
$$Y_1 = |x|$$
$$Y_2 = 2|x|$$
$$Y_3 = 3|x|$$
$$Y_4 = \tfrac{1}{2}|x|$$

Result   Figure 41 illustrates the graphs. You should have observed the following pattern. The graphs of $Y_2 = 2|x|$ and $Y_3 = 3|x|$ can be obtained from the graph of $Y_1 = |x|$ by multiplying each y-coordinate of $Y_1 = |x|$ by factors of 2 and 3, respectively. This is sometimes referred to as a vertical *stretch* using factors of 2 and 3.

The graph of $Y_4 = \tfrac{1}{2}|x|$ can be obtained from the graph of $Y_1 = |x|$ by multiplying each y-coordinate by $\tfrac{1}{2}$. This is sometimes referred to as a vertical *compression* using a factor of $\tfrac{1}{2}$.

"Now Work"
Many examples end with "Now Work Problems." The problems suggested here are similar to the corresponding examples and provide a great way to check your understanding as you work through the chapter. The solutions to all "Now Work" problems can be found in the back of the text as well as in the student solutions manual.

OVERVIEW

stretching, the shape of the graph of a quadratic function will look like one of the parabolas in Figure 5.

E X A M P L E 1 Graphing a Quadratic Function

Graph the function $f(x) = 2x^2 + 8x + 5$. Find the vertex and axis of symmetry.

Graphing Solution Before graphing, notice the leading coefficient, 2, is positive and, therefore, the graph will open up and the vertex will be the lowest point. Now graph f. See Figure 6. We observe the graph does in fact open up. To estimate the vertex of the parabola, we use the MINIMUM command. The vertex is $(-2, -3)$. Therefore, the axis of symmetry is the line $x = -2$.

FIGURE 6

Algebraic Solution We begin by completing the square* on the right side:

$$f(x) = 2x^2 + 8x + 5$$
$$= 2(x^2 + 4x) + 5 \qquad \text{Factor out the 2 from } 2x^2 + 8x.$$
$$= 2(x^2 + 4x + 4) + 5 - 8 \qquad \text{Complete the square of } 2(x^2 + 4x). \text{ Notice that (2)}$$
$$\text{the factor of 2 requires that 8 be added and}$$
$$= 2(x + 2)^2 - 3 \qquad \text{subtracted.}$$

The graph of f can be obtained in three stages, as shown in Figure 7. Now compare this graph to the graph in Figure 5(a). The graph of $f(x) = 2x^2 + 8x + 5$ is a parabola that opens up and has its vertex (lowest point) at $(-2, -3)$. Its axis of symmetry is the line $x = -2$.

FIGURE 7

(a) $y = 2x^2$ (b) $y = 2(x + 2)^2$ (c) $y = 2(x$

*Refer to the Appendix, Section 2, for a review.

Procedures, both algebraic and technical, are clearly expressed throughout the text.

Page 49

"Mission Possible"
Learn to "think beyond the box." Collaborate with fellow students to solve these extended mathematical situations based on relevant "real-world" data.

MISSION POSSIBLE

Predicting The Future of Olympic Track Events

Some people think that women athletes are beginning to "catch up" to men in the Olympic Track Events. In fact, researchers at UCLA published data in 1992 that seemed to indicate that within 65 years the top men and women runners would be able to compete on an equal basis. Other researchers disagreed. Here are some data on the 200-meter run with winning times for the Olympics in the year given.

	Men's Time in Seconds	Women's Time in Seconds
1948	21.1	24.4
1952	20.7	23.7
1956	20.6	23.4
1960	20.5	24.0
1964	20.3	23.0
1968	19.83	22.5
1972	20.00	22.40
1976	20.23	22.37
1980	20.19	22.03
1984	19.80	21.81
1988	19.75	21.34
1992	19.73	21.72

(You may notice that the introduction of better timing devices meant more accurate measures starting in 1968 for the men and 1972 for the women.)

1. To begin your investigation of the UCLA conjecture, make a graph of these data. To get the kind of accuracy you need, you should use graph paper. You will need to use only the first quadrant. On your x-axis place the Olympic years from 1948 to 2048, counting by 4's. On your y-axis place the numbers from 15 to 25, which represent the seconds; if you're using graph paper, allow about four squares per one second. Use X's to represent the men's times on the graph and O's to represent the women's times. These should form what we call a *scatter plot*.

2. Using a ruler or straightedge, draw a line that represents roughly the slope and direction indicated by the men's scores. Do the same for the women's scores. Write a sentence or two explaining why you think your line is a good representation of the scatter plot.

3. Next find the equation for each line, using your graph to estimate the slope and y-intercept for each. Try to be as accurate as possible, remembering your units and using the points where the line crosses intersections of the grid. The slope will be in seconds per year.

4. Do your two lines appear to cross? In what year do they cross? If you solve the two equations algebraically, do you get the same answer?

5. Some graphing calculators enable you do this problem in a statistics mode. They will plot the individual points that you type in and find a line that represents the data. Use your graphing calculator to find out whether its equation matches yours.

6. Make a group decision about whether you think that women's times will "catch up" to men's times in the future. Write two to three sentences explaining why you believe they will or will not.

Logistic Curve Fitting

Logistic growth models can be used to model situations where the value of the dependent variable is limited. Many real-world situations conform to this scenario. For example, the population of the human race is limited by the availability of natural resources such as food, shelter, and so forth. When the value of the dependent variable is limited, a logistic growth model is often appropriate.

EXAMPLE 3   Fitting a Curve to a Logistic Growth Model

The data on the left, obtained from R. Pearl ("The Growth of Population," *Quarterly Review of Biology* 2 (1927): 532–548) represents the amount of yeast biomass after *t* hours in a culture.

Time (in hours)	Yeast Biomass	Time (in hours)	Yeast Biomass
0	9.6	10	513.3
1	18.3	11	559.7
2	29.0	12	594.8
3	47.2	13	629.4
4	71.1	14	640.8
5	119.1	15	651.1
6	174.6	16	655.9
7	257.3	17	659.6
8	350.7	18	661.8
9	441.0		

(a) Using a graphing utility, draw a scatter diagram of the data with time as the independent variable.
(b) Using a graphing utility, fit a logistic growth model to the data.
(c) Using a graphing utility, graph the function found in (b) on the scatter diagram.
(d) What is the predicted carrying capacity of the culture?
(e) Use the function found in (b) to predict the population of the culture at *t* = 19 hours.

Solution   (a) See Figure 54.
(b) A graphing utility fits a logistic growth model of the form $y = \dfrac{c}{1 + ae^{-bx}}$ by using the LOGISTIC regression option. See Figure 55. The logistic growth function of best fit to the data is

$$y = \frac{663.0}{1 + 71.6e^{-0.5470x}}$$

where *y* is the amount of yeast biomass in the culture and *x* is the time.
(c) See Figure 56.
(d) Based on the logistic growth function found in (b), the carrying capacity of the culture is 663.
(e) Using the logistic growth function found in (b), the predicted amount of yeast biomass at *t* = 19 is

$$y = \frac{663.0}{1 + 71.6e^{-0.5470(19)}} = 661.5$$

FIGURE 54

FIGURE 55

```
Logistic
y=c/(1+ae^(-bx))
a=71.57629487
b=.5469947267
c=663.0219908
```

FIGURE 56

Now work **Problem 13.**

Choosing the Best Model

We have discussed seven different models thus far that can be used to explain the relation between two variables, *x* and *y* (see Table 8).

TABLE 8		
Model	Functional Form	Section
1. Linear	$y = ax + b$	1.4
2. Quadratic	$y = ax^2 + bx + c$	3.1
3. Power	$y = ax^b$	3.2
4. Cubic	$y = ax^3 + bx^2 + cx + d$	3.3
5. Exponential	$y = ab^x$	4.8
6. Logarithmic	$y = a + b \ln x$	4.8
7. Logistic	$y = \dfrac{c}{1 + ae^{-bx}}$	4.8

How can we be certain the *model* we used is the "best" *model* to explain this relation? For example, why did we use an exponential model in Example 1 of this section?

Unfortunately, there is no such thing as *the* correct model. Modeling is not only a science but also an art form. Selecting an appropriate model requires experience and skill in the field in which you are modeling. For ex-

Many examples and exercises connect real-world situations to mathematical concepts.

Learning to work with models is a skill that transfers to many disciplines.

OVERVIEW

7.5 EXERCISES

In Problems 1–10, determine the amplitude and period of each function without graphing.

1. $y = 2 \sin x$ **2.** $y = 3 \cos x$ **3.** $y = -4 \cos 2x$ **4.** $y = -\sin \frac{1}{2} x$

5. $y = 6 \sin \pi x$ **6.** $y = -3 \cos 3x$ **7.** $y = -\frac{1}{2} \cos \frac{3}{2} x$ **8.** $y = \frac{4}{3} \sin \frac{2}{3} x$

9. $y = \frac{5}{3} \sin(-\frac{2\pi}{3} x)$ **10.** $y = \frac{9}{5} \cos(-\frac{3\pi}{2} x)$

In Problems 11–20, match the given function to one of the graphs (A)–(J).

(A)   (B)   (C)

Page 568

Each end-of-section exercise set begins with visual and concept-based problems, starting you out with the basics of the section.

87. Explain in your own words how you would use your calculator to find the value of $\cot^{-1} 10$.

88. Consult three books on calculus and write down the definition in each of $y = \sec^{-1} x$ and $y = \csc^{-1} x$. Compare these with the definitions given in this book.

Challenge Yourself! Critical Thinking and Writing Questions really get you thinking.

Page 468

Section 2.1 Functions   **129**

74. Advertising and Sales Revenue A marketing firm wishes to find a function that relates the sales S of a product and A, the amount spent on advertising the product. The data are obtained from past experience. Advertising and sales are measured in thousands of dollars.

Advertising Expenditures, A	Sales, S
20	335
22	339
22.5	338
24	343
24	341
27	350
28.3	351

(a) Does the relation defined by the set of ordered pairs (A, S) represent a function?
(b) Using a graphing utility, draw a scatter diagram of the data.
(c) Using a graphing utility, find the line of best fit relating advertising expenditures and sales.
(d) Interpret the slope.
(e) Express the relationship found in (c) using function notation.
(f) What is the domain of the function?
(g) Predict sales if advertising expenditures are $25,000.

75. Distance and Time Using the data below, find a function that relates the distance, s, in miles, driven by a Ford Taurus, and t, the time the Taurus has been driven.
(a) Does the relation defined by the set of ordered pairs (t, s) represent a function?
(b) Using a graphing utility, draw a scatter diagram of the data.

Time (Hours), t	Distance (Miles), s
0	0
1	30
2	55
3	83
4	100
5	150
6	210
7	260
8	300

(c) Using a graphing utility, find the line of best fit relating time and distance.
(d) Interpret the slope.
(e) Express the relationship found in (c) using function notation.
(f) What is the domain of the function?
(g) Predict the distance the car is driven after 11 hours.

76. High School vs. College GPA An administrator at Southern Illinois University wants to find a function that relates a student's college grade point average G to the high school grade point average, x. She randomly selects 8 students and obtains the following data:

High School GPA, x	College GPA, G
2.73	2.43
2.92	2.97
3.45	3.63
3.78	3.81
2.56	2.83
2.98	2.81
3.67	3.45
3.10	2.93

(a) Does the relation defined by the set of ordered pairs (x, G) represent a function?
(b) Using a graphing utility, draw a scatter diagram of the data.
(c) Using a graphing utility, find the line of best fit relating high school GPA and college GPA.
(d) Interpret the slope.
(e) Express the relationship found in (c) using function notation.
(f) What is the domain of the function?
(g) Predict a student's college GPA if her high school GPA is 3.23.

77. Effect of Gravity on Earth If a rock falls from a height of 20 meters on Earth, the height H (in meters) after x seconds is approximately

$$H(x) = 20 - 4.9x^2$$

(a) Use a graphing utility to graph the function H.
(b) What is the height of the rock when $x = 1$ second? $x = 1.1$ seconds? $x = 1.2$ seconds? $x = 1.3$ seconds?
(c) When is the height of the rock 15 meters? When is it 10 meters? When is it 5 meters?
(d) When does the rock strike the ground?

Construct your own mathematical models from data and experience how the math you learn in class applies to the world around you.

Page 129

CHAPTER REVIEW

THINGS TO KNOW

Definitions

Angle in standard position	Vertex is at the origin; initial side is along the positive x-axis
Degree (1°)	$1° = \frac{1}{360}$ revolution
Radian (1)	The measure of a central angle whose rays subtend an arc whose length is the radius of the circle
Unit circle	Center at origin; radius = 1 Equation: $x^2 + y^2 = 1$
Trigonometric functions	$P = (a, b)$ is the point on the unit circle corresponding to $\theta = t$ radians;

$$\sin t = \sin \theta = b \qquad \cos t = \cos \theta = a$$

$$\tan t = \tan \theta = \frac{b}{a}, \quad a \neq 0 \qquad \cot t = \cot \theta = \frac{a}{b}, \quad b \neq 0$$

$$\csc t = \csc \theta = \frac{1}{b}, \quad b \neq 0 \qquad \sec t = \sec \theta = \frac{1}{a}, \quad a \neq 0$$

Properties of the Trigonometric Functions

$y = \sin x$	Domain: $-\infty < x < \infty$ Range: $-1 \leq y \leq 1$ Periodic: period $= 2\pi$ (360°) Odd function
$y = \cos x$	Domain: $-\infty < x < \infty$ Range: $-1 \leq y \leq 1$ Periodic: period $= 2\pi$ (360°) Even function
$y = \tan x$	Domain: $-\infty < x < \infty$, except odd multiples of $\pi/2$ (90°) Range: $-\infty < y < \infty$ Periodic: period $= \pi$ (180°) Odd function

The chapter review is a great place to check your understanding of the chapter material. Start with "Things to Know." Check your understanding of the concepts listed there. Next, make sure you know "How To" solve the items within that section. "Fill in the Blanks" determines your comfort with vocabulary. "True False" is a stickler for knowing definitions. If you are uncertain of any concept, go back into the chapter and study it further. Be sure to do the Review Exercises for practice. These reviews are for your success in this course—Make good use of them.

HOW TO

Convert an angle from radian measure to degree measure

Convert an angle from degree measure to radian measure

Find the value of each of the remaining trigonometric functions if the value of one function and the quadrant of the angle are given

Use the theorem on cofunctions of complementary angles

Graph the trigonometric functions, including transformations

Use reference angles to find the value of a trigonometric function

Use a calculator to find the value of a trigonometric function

Find the exact value of certain inverse trigonometric functions

Use a calculator to find the approximate values of inverse trigonometric functions

FILL-IN-THE-BLANK ITEMS

1. Two rays drawn with a common vertex form a(n) _____. One of the rays is called the _____ _____; the other is called the _____ _____.
2. In the formula $s = r\theta$ for measuring the length s of arc along a circle of radius r, the angle θ must be measured in _____.
3. 180 degrees = _____ radians.
4. Two acute angles whose sum is a right angle are called _____.
5. The sine and _____ functions are cofunctions.
6. An angle is in _____ _____ if its vertex is at the origin and its initial side coincides with the positive x-axis.
7. The reference angle of 135° is _____.
8. The sine, cosine, cosecant, and secant functions have period _____; the tangent and cotangent functions have period _____.

TRUE/FALSE ITEMS

T F **1.** In the formula $s = r\theta$, r is the radius of a circle and s is the arc subtended by a central angle θ, where θ is measured in degrees.

T F **2.** $|\sin \theta| \leq 1$

T F **3.** $1 + \tan^2 \theta = \csc^2 \theta$

T F **4.** The only even trigonometric functions are the cosine and secant functions.

T F **5.** $\tan 62° = \cot 38°$

T F **6.** $\sin 182° = \cos 2°$

 phs of $y = \tan x$, $y = \cot x$, $y = \sec x$, and $y = \csc x$ have infinitely many vertical asymptotes.

REVIEW EXERCISES

In Problems 1–4, convert each angle in degrees to radians. Express your answer as a multiple of π.

1. 135°   **2.** 210°   **3.** 18°   **4.** 15°

In Problems 5–8, convert each angle in radians to degrees.

5. $3\pi/4$ **6.** $2\pi/3$ **7.** $-5\pi/2$ **8.** $-3\pi/2$

In Problems 9–30, find the exact value of each expression. Do not use a calculator.

9. $\tan \dfrac{\pi}{4} - \sin \dfrac{\pi}{6}$

10. $\cos \dfrac{\pi}{3} + \sin \dfrac{\pi}{2}$

11. $3 \sin 45° - 4 \tan \dfrac{\pi}{6}$

12. $4 \cos 60° + 3 \tan \dfrac{\pi}{3}$

13. $6 \cos \dfrac{3\pi}{4} + 2 \tan\left(-\dfrac{\pi}{3}\right)$

14. $3 \sin \dfrac{2\pi}{3} - 4 \cos \dfrac{5\pi}{2}$

Use these text supplements as your resources. Ask your instructor how to obtain these items to complement your learning style.

Algebra Review
Written by Michael Sullivan
ISBN: 0-13-590621-0
This paperback supplement offers four chapters of Intermediate Algebra Review with solutions to the odd exercises in the back of the supplement. Written by the author to ease the students into Precalculus courses, it offers detailed review material of basic concepts.

Student Solutions Manual
Written by Michael Sullivan, Michael Sullivan III, and Katy Sullivan Murphy
ISBN: 0-13-698714-1
Contains complete step-by-step worked out solutions to all the odd numbered exercises in the textbook. Follows the exact procedures of the text—no magical steps.

Tutorial Videos
ISBN: 0-13-685116-9
A new tutorial videotape series which has been created to accompany this specific test, includes 15 minute segments for each section of the text. Each segment uses both traditional and technological ways of solving the mathematical problems. Entertaining and educational, these videos provide an alternative process which can add to your success in this course. Ask your professor where the set is located for your school.

New York Times Supplement
A free newspaper from Prentice Hall and the New York Times which includes interesting and relevant articles on mathematics in the world around us. This supplement is revised each fall.

Life on the Internet—Mathematics
Guides you through the complexity of the Internet, offering navigation strategies, practice exercises and lists of mathematical resources. This supplement is updated each year to reflect the ongoing changes on the Internet.

Prentice Hall Companion Website
Located at *www.prenhall.com/sullivan* this website offers you additional information, projects and exercises to enrich the learning experience using this text. The Internet "Excursions" found in the beginning of each chapter have duplicate pages on the website with the links to the appropriate data and information gathering sites described in the projects.

LIST OF APPLICATIONS

Chapter 1	Sir Isaac Newton	Erich Lessing/Art Resource, NY
Chapter 2	Oil well at sunset near Bakersfield, California 405 and Sunset Blvd.	Dan Suzio/Photo Researchers, Inc. Joe Sohm/Photo Researchers, Inc.
Chapter 3	SETI Telescope	Jet Propulsion Laboratory/ NASA Headquarters
Chapter 4	Electron micrograph of the mite Varroae jacobeoni that occurs as an ectoparasite on the honeybee	Andrew Syred/Science Photo Library/Photo Researchers, Inc.
Chapter 5	Polyhedra	Prof. George W. Hart
Chapter 6	A Prism	George Dodson/Uniphoto
Chapter 7	Annapolis Tidal Generating Station	Nova Scotia Power
Chapter 8	Golden Gate Bridge	Brian Parker/Tom Stack & Associates
Chapter 9	Spirograph (drawing)	Kathy Enfalice
Chapter 10	Three-ring German Enigma Cypher Machine	Science Museum/Science & Society Picture Library
Chapter 11	Missey Glove—Mountain Bike Racer	Carl Yarborough/Carl Yarborough Photography
Chapter 12	Earth from space showing North and South America	NASA/GFSC/Tom Stack and Associates

Graphs and Equations

On the page following each chapter opener is an Internet Excursion. These Excursions will place you in intriguing situations where information from various Internet resources are needed in order to answer thought provoking questions. All of the links and even more information on these Excursions can be found on your Sullivan website.
The URL address is:

www.prenhall.com/sullivan

Once at the website, find the appropriate text and explore the student's pages of the site. You will find an interactive version of the Excursion as well as many other useful and helpful materials.

Turn the page and learn all about Isaac Newton and his mathematical discoveries. . . .

PREPARING FOR THIS CHAPTER

Before getting started on this chapter, review the following concepts:

Topics from Algebra and Geometry
(Appendix, Section 1)

Completing the Square; the Quadratic Formula
(Appendix, Section 2)

OUTLINE

THE CATALOG OF EQUATIONS

Trinity College, 1667. You have just begun your college work and one of your teachers is a strange fellow named *Isaac Newton*. His ideas are difficult to understand and very radical. Yet, when you listen to him speak, his passion makes a profound impression on you. His curious experiments and clever devices attract you to him.

After class one day, he asks you to help him with his Catalog of Equations. He quickly explains about the coordinate geometry of Descartes, and how algebra and geometry are now joined together. He assigns to you the equation:

$$(-2x + x^2 + y^2)^2 = x^2 + y^2$$

He gives you a manuscript for *The Witch*, and asks you to do a similar work for your assigned curve. As you walk away, you feel a great sense of honor in being selected to do research for Isaac Newton.

1. Have you ever thought about such a Catalog of Equations? Take a quick tour of the famous curves at the *MacTutor Website*. Notice the cartesian formulas (x and y) for the curves. How would you organize a catalog for the classification of algebraic curves?
2. Today a student does not hear much about a Catalog of Equations. Why do you suppose that is?
3. From looking at your equation, can you use any of the symmetry principles to get some idea of its graph? Can you graph this equation on your calculator? Are you able to graph it by making a chart of x and y values?
4. Visit the *Visual Dictionary of Plane Curves Website* and read about catacaustic curves. Can you determine the catacaustic of your curve with a lightsource on the edge? How could you graph it?
5. Tie together your explorations into a 'manuscript.'

Information often in the form of data, is the basis for making informed decisions. In this chapter we discuss some of the ways that data can be represented using rectangular coordinates. The idea of using a system of rectangular coordinates dates back to ancient times, when such a system was used for surveying and city planning. Apollonius of Perga, in 200 B.C., used a form of rectangular coordinates in his work on conics, although this use does not stand out as clearly as it does in modern treatments. Sporadic use of rectangular coordinates continued until the 1600's. By that time, algebra had developed sufficiently so that René Descartes (1596–1650) and Pierre de Fermat (1601–1665) could take the crucial step, which was the use of rectangular coordinates to translate geometry problems into algebra problems, and vice versa. This step was supremely important for two reasons. First, it allowed both geometers and algebraists to gain critical new insights into their subjects, which previously had been regarded as separate but now were seen to be connected in many important ways. Second, the insights gained made possible the development of calculus, which greatly enlarged the number of areas in which mathematics could be applied and made possible a much deeper understanding of these areas. With the advent of technology, in particular, graphing utilities, we are now able not only to visualize the dual roles of algebra and geometry, but we are also able to solve many problems that before this technology required advanced methods.

1.1 RECTANGULAR COORDINATES; GRAPHING UTILITIES; DATA IN ORDERED PAIRS

> **1** Distance Formula
> **2** Midpoint Formula
> **3** Construct and Interpret Line Graphs
> **4** Draw and Interpret Scatter Diagrams

We locate a point on the real number line by assigning it a single real number, called the *coordinate of the point.* For work in a two-dimensional plane, we locate points by using two numbers.

We begin with two real number lines located in the same plane: one horizontal and the other vertical. We call the horizontal line the **x-axis,** the vertical line the **y-axis,** and the point of intersection the **origin O.** We assign coordinates to every point on these number lines as shown in Figure 1, using a convenient scale. In mathematics, we usually use the same scale on each axis; in applications, a different scale is often used on each axis.

The origin O has a value of 0 on both the x-axis and the y-axis. We follow the usual convention that points on the x-axis to the right of O are associated with positive real numbers, and those to the left of O are associated with negative numbers. Those on the y-axis above O are associated with positive real numbers, and those below O are associated with negative real numbers. In Figure 1, the x-axis and y-axis are labeled as x and y, respectively, and we have used an arrow at the end of each axis to denote the positive direction.

The coordinate system described here is called a **rectangular,** or **Cartesian* coordinate system.** The plane formed by the x-axis and y-axis is sometimes called the **xy-plane,** and the x-axis and y-axis are referred to as the **coordinate axes.**

Any point P in the xy-plane can then be located by using an **ordered pair** (x, y) of real numbers. Let x denote the signed distance of P from the y-axis (*signed* in the sense that, if P is to the right of the y-axis, then $x > 0$, and if P is to the left of the y-axis, then $x < 0$); and let y denote the signed distance of P from the x-axis. The ordered pair (x, y), also called the **coordinates** of P, then gives us enough information to locate the point P in the plane.

For example, to locate the point whose coordinates are $(-3, 1)$, go 3 units along the x-axis to the left of O and then go straight up 1 unit. We **plot** this point by placing a dot at this location. See Figure 2, in which the points with coordinates $(-3, 1)$, $(-2, -3)$, $(3, -2)$, and $(3, 2)$ are plotted.

The origin has coordinates $(0, 0)$. Any point on the x-axis has coordinates of the form $(x, 0)$, and any point on the y-axis has coordinates of the form $(0, y)$.

If (x, y) are the coordinates of a point P, then x is called the **x-coordinate,** or **abscissa,** of P, and y is the **y-coordinate,** or **ordinate,** of P. We identify the point P by its coordinates (x, y) by writing $P = (x, y)$. Usually, we will simply say "the point (x, y)" rather than "the point whose coordinates are (x, y)."

FIGURE 1

FIGURE 2

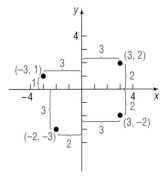

---

*Named after René Descartes (1596–1650), a French mathematician, philosopher, and theologian.

FIGURE 3

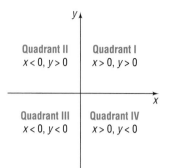

The coordinate axes divide the *xy*-plane into four sections, called **quadrants,** as shown in Figure 3. In quadrant I, both the *x*-coordinate and the *y*-coordinate of all points are positive; in quadrant II, *x* is negative and *y* is positive; in quadrant III, both *x* and *y* are negative; and in quadrant IV, *x* is positive and *y* is negative. Points on the coordinate axes belong to no quadrant.

Graphing Utilities

All graphing utilities, that is, all graphing calculators and all computer software graphing packages, graph equations by plotting points on a screen. The screen itself actually consists of small rectangles, called **pixels.** The more pixels the screen has, the better the resolution. Most graphing calculators have 48 pixels per square inch; most computer screens have 32 to 108 pixels per square inch. When a point to be plotted lies inside a pixel, the pixel is turned on (lights up). Thus, the graph of an equation is a collection of pixels. Figure 4 shows how the graph of $y = 2x$ looks on a TI-83 graphing calculator.

FIGURE 4
$y = 2x$

The screen of a graphing utility will display the coordinate axes of a rectangular coordinate system. However, you must set the scale on each axis. You must also include the smallest and largest values of *x* and *y* that you want included in the graph. This is called **setting the RANGE** and it gives the **viewing rectangle** or **window.**

Figure 5 illustrates a typical viewing rectangle.

To select the viewing rectangle, we must give values to the following expressions:

FIGURE 5

*X*min: the smallest value of *x*

*X*max: the largest value of *x*

*X*scl: the number of units per tick mark on the *x*-axis

*Y*min: the smallest value of *y*

*Y*max: the largest value of *y*

*Y*scl: the number of units per tick mark on the *y*-axis

Figure 6 illustrates these settings for a typical screen.

FIGURE 6

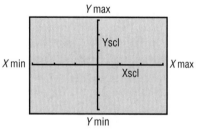

If the scale used on each axis is known, we can determine the minimum and maximum values of *x* and *y* shown on the screen by counting the tick marks. Look again at Figure 6. For a scale of 1 on each axis, the minimum and maximum values of *x* are −3 and 3, respectively; the minimum and maximum values of *y* are also −3 and 3. If the scale is 2 on each axis, then the minimum and maximum values of *x* are −6 and 6, respectively; the minimum and maximum values of *y* are −6 and 6, respectively.

Conversely, if we know the minimum and maximum values of *x* and *y*, we can determine the scales being used by counting the tick marks displayed. We shall follow the practice of showing the minimum and maximum values of *x* and *y* in our illustrations so that you will know how the RANGE was set. See Figure 7.

FIGURE 7

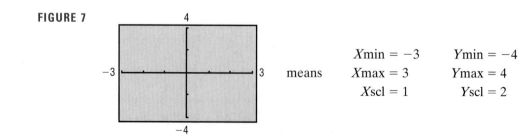

means $X\text{min} = -3$ $Y\text{min} = -4$
$X\text{max} = 3$ $Y\text{max} = 4$
$X\text{scl} = 1$ $Y\text{scl} = 2$

E X A M P L E   1   Finding the Coordinates of a Point Shown on a Graphing Utility Screen

Find the coordinates of the point shown in Figure 8.

FIGURE 8

Solution   First we note that the range setting used in Figure 8 is

$$X\text{min} = -3 \qquad X\text{max} = 3 \qquad X\text{scl} = 1$$
$$Y\text{min} = -4 \qquad Y\text{max} = 4 \qquad Y\text{scl} = 2$$

The point shown is 2 tick units to the left on the horizontal axis (scale = 1) and 1 tick up on the vertical (scale = 2). Thus, the coordinates of the point shown are $(-2, 2)$.

Now work Problems 5 and 15.

Distance between Points

1   If the same units of measurement, such as inches, centimeters, and so on, are used for both the x-axis and the y-axis, then all distances in the xy-plane can be measured using this unit of measurement.

E X A M P L E   2   Finding the Distance between Two Points

Find the distance d between the points $(1, 3)$ and $(5, 6)$.

Solution   First we plot the points $(1, 3)$ and $(5, 6)$ as shown in Figure 9(a). Then we draw a horizontal line from $(1, 3)$ to $(5, 3)$ and a vertical line from $(5, 3)$ to $(5, 6)$, forming a right triangle, as in Figure 9(b). One leg of the triangle is of length 4 and the other is of length 3. By the Pythagorean Theorem (see the Appendix, Section 1), the square of the distance d we seek is

$$d^2 = 4^2 + 3^2 = 16 + 9 = 25$$
$$d = 5$$

FIGURE 9

(a)   (b)

The **Distance Formula** provides a straightforward method for computing the distance between two points.

Theorem   Distance Formula

The distance between two points $P_1 = (x_1, y_1)$ and $P_2 = (x_2, y_2)$, denoted by $d(P_1, P_2)$, is

$$\sqrt{d(P_1, \sqrt{} P_2)} = \sqrt{(x_2 - x_1)^2 + (y_2 - y_1)^2} \qquad (1)$$

That is, to compute the distance between two points, find the difference of the x-coordinates, square it, and add this to the square of the difference of the y-coordinates. The square root of this sum is the distance. See Figure 10.

FIGURE 10

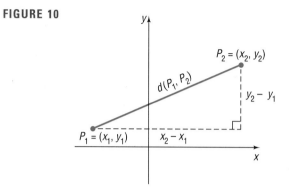

Proof of the Distance Formula   Let (x_1, y_1) denote the coordinates of point P_1, and let (x_2, y_2) denote the coordinates of point P_2. Assume that the line joining P_1 and P_2 is neither horizontal nor vertical. Refer to Figure 11(a). The coordinates of P_3 are (x_2, y_1). The horizontal distance from P_1 to P_3 is the absolute value of the difference of the x-coordinates, $|x_2 - x_1|$. The vertical distance from P_3 to P_2 is the absolute value of the difference of the y-coordinates, $|y_2 - y_1|$. See Figure 11(b). The distance $d(P_1, P_2)$ that we seek is the length of the hypotenuse of the right triangle, so, by the Pythagorean Theorem, it follows that

$$[d(P_1, P_2)]^2 = |x_2 - x_1|^2 + |y_2 - y_1|^2$$
$$= (x_2 - x_1)^2 + (y_2 - y_1)^2$$
$$d(P_1, P_2) = \sqrt{(x_2 - x_1)^2 + (y_2 - y_1)^2}$$

FIGURE 11

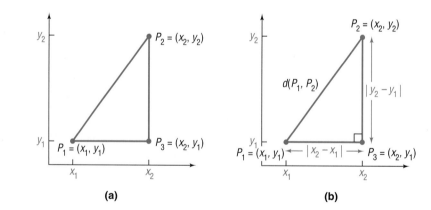

(a)   (b)

Now, if the line joining P_1 and P_2 is horizontal, then the y-coordinate of P_1 equals the y-coordinate of P_2; that is, $y_1 = y_2$. Refer to Figure 12(a). In this case, the distance formula (1) still works, because, for $y_1 = y_2$, it reduces to

$$d(P_1, P_2) = \sqrt{(x_2 - x_1)^2 + 0^2} = \sqrt{(x_2 - x_1)^2} = |x_2 - x_1|$$

A similar argument holds if the line joining P_1 and P_2 is vertical. See Figure 12(b). Thus, the distance formula is valid in all cases.

FIGURE 12

(a)   (b)

 Now work Problem 25.

The distance between two points $P_1 = (x_1, y_1)$ and $P_2 = (x_2, y_2)$ is never a negative number. Furthermore, the distance between two points is 0 only when the points are identical, that is, when $x_1 = x_2$ and $y_1 = y_2$. Also, because $(x_2 - x_1)^2 = (x_1 - x_2)^2$ and $(y_2 - y_1)^2 = (y_1 - y_2)^2$, it makes no difference whether the distance is computed from P_1 to P_2 or from P_2 to P_1; that is, $d(P_1, P_2) = d(P_2, P_1)$.

The introduction to this chapter mentioned that rectangular coordinates enable us to translate geometry problems into algebra problems, and vice versa. The next example shows how algebra (the distance formula) can be used to solve geometry problems.

E X A M P L E   3   Using Algebra to Solve Geometry Problems

Consider the three points $A = (-2, 1)$, $B = (2, 3)$, and $C = (3, 1)$.

(a)   Plot each point and form the triangle ABC.
(b)   Find the length of each side of the triangle.
(c)   Verify that the triangle is a right triangle.
(d)   Find the area of the triangle.

Solution   (a)   Points A, B, C and triangle ABC are plotted in Figure 13.

FIGURE 13

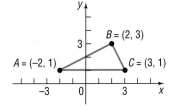

(b) $d(A, B) = \sqrt{[2 - (-2)]^2 + (3 - 1)^2} = \sqrt{16 + 4} = \sqrt{20} = 2\sqrt{5}$

$d(B, C) = \sqrt{(3 - 2)^2 + (1 - 3)^2} = \sqrt{1 + 4} = \sqrt{5}$

$d(A, C) = \sqrt{[3 - (-2)]^2 + (1 - 1)^2} = \sqrt{5^2 + 0^2} = 5$

(c)   To show that the triangle is a right triangle, we need to show that the sum of the squares of the lengths of two of the sides equals the square of the length of the third side. (Why is this sufficient?) Looking at Figure 13, it seems reasonable to conjecture that the right angle is at vertex B. Thus, we shall check to see whether

$$[d(A, B)]^2 + [d(B, C)]^2 = [d(A, C)]^2$$

We find that

$$[d(A, B)]^2 + [d(B, C)]^2 = (2\sqrt{5})^2 + (\sqrt{5})^2$$
$$= 20 + 5 = 25 = [d(A, C)]^2$$

so it follows from the Converse of the Pythagorean Theorem that triangle ABC is a right triangle.

(d) Because the right angle is at B, the sides AB and BC form the base and altitude of the triangle. Its area is therefore

$$\text{Area} = \frac{1}{2}(\text{Base})(\text{Altitude}) = \frac{1}{2}(2\sqrt{5})(\sqrt{5}) = 5 \text{ square units}$$ ▬

 Now work Problem 43.

Midpoint Formula

FIGURE 14

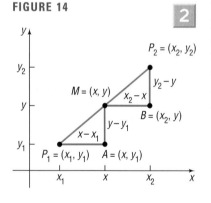

We now derive a formula for the coordinates of the **midpoint of a line segment.** Let $P_1 = (x_1, y_1)$ and $P_2 = (x_2, y_2)$ be the endpoints of a line segment, and let $M = (x, y)$ be the point on the line segment that is the same distance from P_1 as it is from P_2. See Figure 14. The triangles P_1AM and MBP_2 are congruent.* [Do you see why? Angle AP_1M = Angle BMP_2.† Angle P_1MA = Angle MP_2B, and $d(P_1, M) = d(M, P_2)$ are given. Thus, we have Angle–Side–Angle.] Hence, corresponding sides are equal in length. That is,

$$x - x_1 = x_2 - x \quad \text{and} \quad y - y_1 = y_2 - y$$
$$2x = x_1 + x_2 \qquad\qquad 2y = y_1 + y_2$$
$$x = \frac{x_1 + x_2}{2} \qquad\qquad y = \frac{y_1 + y_2}{2}$$

Theorem   Midpoint Formula

The midpoint (x, y) of the line segment from $P_1 = (x_1, y_1)$ to $P_2 = (x_2, y_2)$ is

$$(x, y) = \left(\frac{x_1 + x_2}{2}, \frac{y_1 + y_2}{2} \right) \qquad (2)$$

▬

---

*The following statement is a postulate from geometry. Two triangles are congruent if their sides are the same length (SSS), or if two sides and the included angle are the same (SAS), or if two angles and the included side are the same (ASA).

†Another postulate from geometry states that the transversal $\overline{P_1P_2}$ forms equal corresponding angles with the parallel lines $\overline{P_1A}$ and $\overline{MB}$.

Thus, to find the midpoint of a line segment, we average the x-coordinates and the y-coordinates of the endpoints.

E X A M P L E   4   Finding the Midpoint of a Line Segment

Find the midpoint of a line segment from $P_1 = (-5, 3)$ to $P_2 = (3, 1)$. Plot the points P_1 and P_2 and their midpoint. Check your answer.

Solution   We apply the midpoint formula (2) using $x_1 = -5$, $x_2 = 3$, $y_1 = 3$, and $y_2 = 1$. Then the coordinates (x, y) of the midpoint M are

$$x = \frac{x_1 + x_2}{2} = \frac{-5 + 3}{2} = -1 \quad \text{and} \quad y = \frac{y_1 + y_2}{2} = \frac{3 + 1}{2} = 2$$

That is, $M = (-1, 2)$. See Figure 15.

FIGURE 15

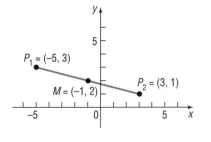

Check: Because M is the midpoint, we check the answer by verifying that $d(P_1, M) = d(M, P_2)$:

$$d(P_1, M) = \sqrt{[-1 - (-5)]^2 + (2 - 3)^2} = \sqrt{16 + 1} = \sqrt{17}$$
$$d(M, P_2) = \sqrt{[3 - (-1)]^2 + (1 - 2)^2} = \sqrt{16 + 1} = \sqrt{17}$$

 Now work Problem 53.

Analysis of Data Using Ordered Pairs

Time Series Data   Often, we are interested in observing trends in data over time in order to determine any pattern that may exist. Data in which we measure a variable at different points in time is called **time series data.** For example, the price of a stock at the end of each year for the past 10 years is time series data. Time series data may be represented graphically using line graphs.

E X A M P L E   5   Drawing a Line Graph from Time Series Data

The data in Table 1 represent the closing price of Walt Disney Co. stock at the end of each month from 1/95 through 12/95.

TABLE 1			
Date	**Closing Price**	**Date**	**Closing Price**
1/95	$50\frac{7}{8}$	7/95	$58\frac{5}{8}$
2/95	$53\frac{3}{8}$	8/95	$56\frac{1}{4}$
3/95	$53\frac{1}{2}$	9/95	$57\frac{3}{8}$
4/95	$55\frac{3}{8}$	10/95	$57\frac{5}{8}$
5/95	$55\frac{1}{2}$	11/95	$60\frac{1}{8}$
6/95	$55\frac{1}{2}$	12/95	$58\frac{7}{8}$

Source: New York Stock Exchange.

(a) Draw a line graph of the data in Table 1 by hand.
(b) Draw a line graph using a graphing utility.

(c) Using the line graph, at the end of which month was Disney stock highest?

(d) Using the line graph, at the end of which month was Disney stock lowest?

Solution   (a) A line graph of the data in Table 1 is obtained by plotting the date as the x-coordinate and the corresponding closing price as the y-coordinate. A line is then drawn connecting consecutive points. See Figure 16.

FIGURE 16

FIGURE 17

(b) To draw a line graph using a graphing utility, we enter the date numerically by letting 1 represent January, 1995, 2 represent February, 1995, and so on. We then enter the corresponding closing price of Disney stock. Finally, we set the appropriate viewing window by letting Xmin $= 0$, Xmax $= 13$, Ymin $= 50$, Ymax $= 61$. The graph is shown in Figure 17.

(c) Using the line graph, Disney stock was highest at the end of November, 1995.

(d) Using the line graph, Disney stock was lowest at the end of January, 1995.

Now work Problem 65.

Scatter Diagrams   A **relation** is a correspondence between two variables, say x and y, and can be written as a set of ordered pairs (x, y). When relations are written as ordered pairs, we say x is related to y. Often, we are interested in specifying the type of relation (such as an equation) that might exist between two quantitative variables. The first step in finding this relation is to plot the ordered pairs in a Cartesian coordinate system. The resulting graph is called a **scatter diagram.**

E X A M P L E   6   Drawing a Scatter Diagram

The data in Table 2 represent the total number of accidents for all U.S. General Aviation Flying and the corresponding total number of fatalities as a result of the accidents for the years 1984–1993. Draw a scatter diagram of the data, treating the number of accidents as *x* and the number of fatalities as *y*.

TABLE 2		
Total Number of Accidents, *x*	**Total Fatalities, *y***	**(*x, y*)**
3016	1042	(3016, 1042)
2738	955	(2738, 955)
2582	967	(2582, 967)
2494	838	(2494, 838)
2386	800	(2386, 800)
2230	768	(2230, 768)
2214	766	(2214, 766)
2170	781	(2170, 781)
2074	862	(2074, 862)
2022	715	(2022, 715)

Source: National Transportation Safety Board.

(a) Draw a scatter diagram by hand.

(b) Use a graphing utility to draw a scatter diagram.

 (c) Describe what happens to the total number of fatalities as the number of accidents increases.

Solution (a) To draw a scatter diagram by hand, we plot the ordered pairs listed in Table 2, with total number of accidents as the *x*-coordinate and total number of fatalities as the *y*-coordinate. See Figure 18. Notice that the points in a scatter diagram are not connected.

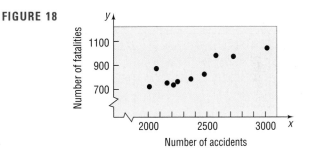

FIGURE 18

(b) Figure 19 shows a scatter diagram of the data using a graphing utility.

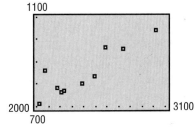

FIGURE 19

(c) From the scatter diagrams, as the number of accidents increases, the number of fatalities increases.

 Now work Problem 69.

1.1 | EXERCISES

In Problems 1 and 2, plot each point in the xy-plane. Tell in which quadrant or on what coordinate axis each point lies.

1. (a) $A = (-3, 2)$ (b) $B = (6, 0)$ (c) $C = (-2, -2)$

     (d) $D = (6, 5)$ (e) $E = (0, -3)$ (f) $F = (6, -3)$

2. (a) $A = (1, 4)$ (b) $B = (-3, -4)$ (c) $C = (-3, 4)$

     (d) $D = (4, 1)$ (e) $E = (0, 1)$ (f) $F = (-3, 0)$

3. Plot the points $(2, 0), (2, -3), (2, 4), (2, 1)$, and $(2, -1)$. Describe the set of all points of the form $(2, y)$, where y is a real number.

4. Plot the points $(0, 3), (1, 3), (-2, 3), (5, 3)$ and $(-4, 3)$. Describe the set of all points of the form $(x, 3)$, where x is a real number.

In Problems 5–8, determine the coordinates of the points shown. Tell in which quadrant each point lies. Assume all coordinates are integers.

5.   **6.**

7.   **8.**

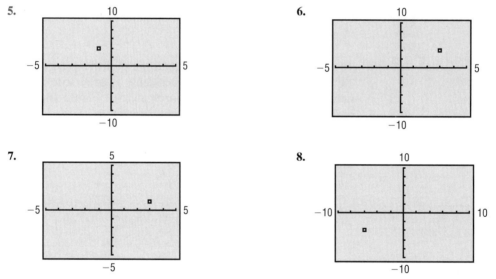

In Problems 9–14, select a RANGE setting so that each of the given points will lie within the viewing rectangle.

9. $(-10, 5), (3, -2), (4, -1)$ **10.** $(5, 0), (6, 8), (-2, -3)$

11. $(40, 20), (-20, -80), (10, 40)$ **12.** $(-80, 60), (20, -30), (-20, -40)$

13. $(0, 0), (100, 5), (5, 150)$ **14.** $(0, -1), (100, 50), (-10, 30)$

In Problems 15–24, determine the RANGE setting used for each viewing rectangle.

15.   **16.**

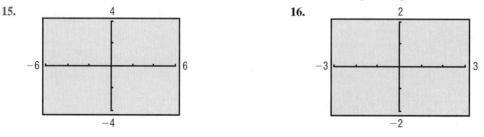

17. 18.

19. 20.

21. 22.

23. 24.

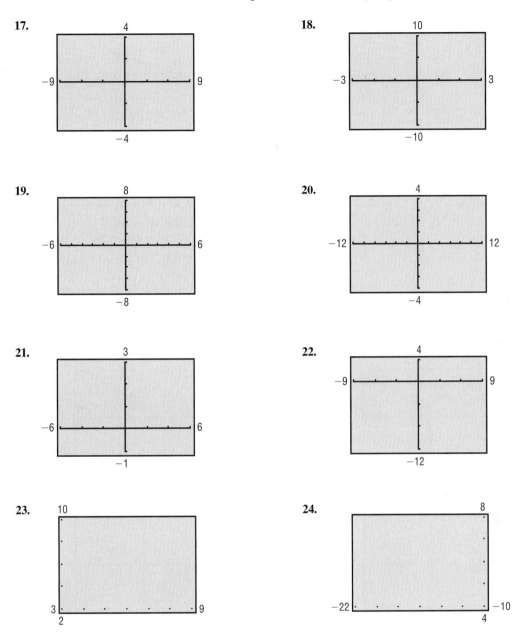

In Problems 25–38, find the distance $d(P_1, P_2)$ between the points P_1 and P_2.

25. 26.

27.

28.

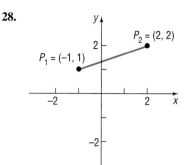

29. $P_1 = (3, -4)$; $P_2 = (5, 4)$

30. $P_1 = (-1, 0)$; $P_2 = (2, 4)$

31. $P_1 = (-3, 2)$; $P_2 = (6, 0)$

32. $P_1 = (2, -3)$; $P_2 = (4, 2)$

33. $P_1 = (4, -3)$; $P_2 = (6, 4)$

34. $P_1 = (-4, -3)$; $P_2 = (6, 2)$

35. $P_1 = (-0.2, 0.3)$; $P_2 = (2.3, 1.1)$

36. $P_1 = (1.2, 2.3)$; $P_2 = (-0.3, 1.1)$

37. $P_1 = (a, b)$; $P_2 = (0, 0)$

38. $P_1 = (a, a)$; $P_2 = (0, 0)$

In Problems 39–42, find the length of the line segment. Assume the endpoints have integer coordinates.

39.

40.

41.

42.

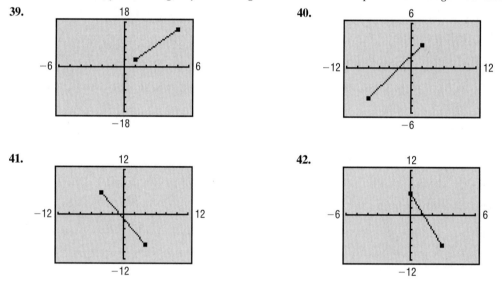

In Problems 43–48, plot each point and form the triangle ABC. Verify that the triangle is a right triangle. Find its area.

43. $A = (-2, 5)$; $B = (1, 3)$; $C = (-1, 0)$

44. $A = (-2, 5)$; $B = (12, 3)$; $C = (10, -11)$

45. $A = (-5, 3)$; $B = (6, 0)$; $C = (5, 5)$

46. $A = (-6, 3)$; $B = (3, -5)$; $C = (-1, 5)$

47. $A = (4, -3)$; $B = (0, -3)$; $C = (4, 2)$

48. $A = (4, -3)$; $B = (4, 1)$; $C = (2, 1)$

49. Find all points having an x-coordinate of 2 whose distance from the point $(-2, -1)$ is 5.

50. Find all points having a y-coordinate of -3 whose distance from the point $(1, 2)$ is 13.

51. Find all points on the x-axis that are 5 units from the point $(4, -3)$.

52. Find all points on the y-axis that are 5 units from the point $(4, 4)$.

In Problems 53–62, find the midpoint of the line segment joining the points P_1 and P_2.

53. $P_1 = (5, -4); \quad P_2 = (3, 2)$

54. $P_1 = (-1, 0); \quad P_2 = (2, 4)$

55. $P_1 = (-3, 2); \quad P_2 = (6, 0)$

56. $P_1 = (2, -3); \quad P_2 = (4, 2)$

57. $P_1 = (4, -3); \quad P_2 = (6, 1)$

58. $P_1 = (-4, -3); \quad P_2 = (2, 2)$

59. $P_1 = (-0.2, 0.3); \quad P_2 = (2.3, 1.1)$

60. $P_1 = (1.2, 2.3); \quad P_2 = (-0.3, 1.1)$

61. $P_1 = (a, b); \quad P_2 = (0, 0)$

62. $P_1 = (a, a); \quad P_2 = (0, 0)$

63. Mortgage Rates The following time series graph represents the 30 year mortgage rate based on a loan origination fee of 3 points (a point equals 1% of the loan amount) for September, 1994 to November, 1996.

(a) When were 30 year mortgage rates at their highest?

(b) When were 30 year mortgage rates at their lowest?

(c) What were 30 year mortgages in July, 1996?

64. Adjustable Rate Mortgages The following time series graph represents the 1 year adjustable rate mortgage (ARM) based on a loan origination fee of 3 points (a point equals 1% of the loan amount) for September, 1994 to November, 1996.

(a) When were 1 year ARMs at their highest?

(b) When were 1 year ARMs at their lowest?

(c) What were 1 year ARMs in July, 1996?

65. Price of Motorola Stock The following data represent the closing price of Motorola, Inc., stock at the end of each month in 1995.

Month	Closing Price
January, 1995	$59\frac{1}{4}$
February, 1995	$57\frac{1}{2}$
March, 1995	$54\frac{5}{8}$
April, 1995	$56\frac{7}{8}$
May, 1995	$59\frac{7}{8}$
June, 1995	$67\frac{1}{8}$
July, 1995	$76\frac{1}{2}$
August, 1995	$74\frac{5}{8}$
September, 1995	$76\frac{3}{8}$
October, 1995	$65\frac{3}{4}$
November, 1995	$61\frac{1}{2}$
December, 1995	57

Source: New York Stock Exchange.

(a) By hand, draw a line graph of the data.

(b) Using a graphing utility, draw a line graph of the data.

(c) At the end of which month was Motorola stock highest?

(d) At the end of which month was Motorola stock lowest?

66. Price of Microsoft Corp. Stock The following data represent the closing price of Microsoft Corporation stock at the end of each month in 1995.

Month	Closing Price
January, 1995	$29\frac{11}{16}$
February, 1995	$31\frac{1}{2}$
March, 1995	$35\frac{9}{16}$
April, 1995	$40\frac{7}{8}$
May, 1995	$42\frac{3}{8}$
June, 1995	$45\frac{3}{16}$
July, 1995	$45\frac{1}{4}$
August, 1995	$46\frac{1}{4}$
September, 1995	$45\frac{1}{4}$
October, 1995	50
November, 1995	$43\frac{9}{16}$
December, 1995	$43\frac{7}{8}$

Source: National Association of Security Dealers Automated Quotation.

(a) By hand, draw a line graph of the data.
(b) Using a graphing utility, draw a line graph of the data.
(c) Determine the month in which Microsoft stock was highest.
(d) Determine the month in which Microsoft stock was lowest.

67. **Fuel Consumption** The following data represent average fuel consumption per automobile for the years 1984 to 1993.

Year	Average Fuel Consumption
1984	536
1985	525
1986	526
1987	514
1988	509
1989	509
1990	502
1991	496
1992	512
1993	513

Source: U.S. Federal Highway Administration.

(a) By hand, draw a line graph of the data.
(b) Using a graphing utility, draw a line graph of the data.

(c) Determine the year in which fuel consumption was highest.
(d) Determine the year in which fuel consumption was lowest.
(e) What is the trend in the data? In other words, as time passes, what is happening to average fuel consumption?

68. **Gross Domestic Product** The following data represent per capita gross domestic product (GDP) for the United States for the years 1985 to 1994 in constant 1987 dollars (dollars adjusted for inflation). GDP is the total value of all goods and services produced in the United States.

Year	Per Capita Gross Domestic Product
1985	17,944
1986	18,299
1987	18,694
1988	19,252
1989	19,556
1990	19,593
1991	19,263
1992	19,490
1993	19,879
1994	20,476

Source: U.S. Bureau of Economic Analysis.

(a) By hand, draw a line graph of the data.
(b) Using a graphing utility, draw a line graph of the data.
(c) Determine the year in which per capita GDP was highest.
(d) Determine the year in which per capita GDP was lowest.
(e) What is the trend in the data? In other words, as time passes, what is happening to per capita GDP?

69. **Disposable Personal Income** The following data represent per capita disposable personal income and per capita personal consumption expenditures for the years 1985 to 1994 (in 1994 dollars adjusted for inflation). Treat per capita disposable income as x and per capita personal consumption as y.

Per Capita Disposable Income, x	Per Capita Personal Consumption, y
13,258	12,013
13,552	12,336
13,545	12,568
13,890	12,903
14,005	13,025
14,101	13,063
14,003	12,860
14,279	13,110
14,341	13,361
14,696	13,711

Source: U.S. Bureau of Economic Analysis.

(a) By hand, draw a scatter diagram.
(b) Using a graphing utility, draw a scatter diagram.
(c) What happens to per capita personal consumption as per capita disposable income increases?

70. Natural Gas versus Coal   The data below represent the amount of energy provided by natural gas and coal measured in trillions of BTU's (British Thermal Units) for New England states. Let the amount of energy provided by natural gas be x and the amount of energy provided by coal be y.

State	Natural Gas, x	Coal, y
Maine	5	22
New Hampshire	17	35
Vermont	8	1
Massachusetts	306	111
Rhode Island	79	0
Connecticut	114	22

Source: U.S. Energy Information Administration.

(a) By hand, draw a scatter diagram.
(b) Using a graphing utility, draw a scatter diagram.
(c) What happens to the amount of energy provided by coal when the amount of energy provided by natural gas increases?

71. Fuel Consumption   The following data represent the average fuel consumption per car (gallons) and the pollutant emissions of carbon monoxide (thousands of tons) for 1983 to 1993. Let the average fuel consumption per vehicle be represented by x and the pollutant emissions of carbon monoxide be represented by y.

Average Fuel Consumption Per Car, x	Pollutant Emissions of Carbon Monoxide, y
553	115,334
536	114,262
525	112,072
526	108,070
514	105,117
509	106,100
509	100,806
502	103,753
496	99,898
512	96,368
513	97,208

Source: Statistical Abstract of the United States, 1995.

(a) By hand, draw a scatter diagram.
(b) Using a graphing utility, draw a scatter diagram.
(c) What happens to the level of carbon monoxide as the average fuel consumption per car increases?

72. Hourly Pay and Productivity   The data below represent the average hourly earnings of production workers and productivity (output per hour) for the years 1986 to 1995. Let productivity be represented by x and average hourly earnings be represented by y.

Productivity, x	Average Hourly Earnings, y
94.2	8.76
94.1	8.98
94.6	9.28
95.3	9.66
96.1	10.01
96.7	10.32
100	10.57
100.2	10.83
100.7	11.12
100.8	11.44

Source: Bureau of Labor Statistics.

(a) By hand, draw a scatter diagram.
(b) Using a graphing utility, draw a scatter diagram.
(c) What happens to the average hourly earnings as productivity increases?

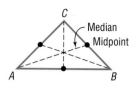

73. The **medians** of a triangle are the line segments from each vertex to the midpoint of the opposite side (see the figure on the right). Find the lengths of the medians of the triangle with vertices at $A = (0, 0)$, $B = (0, 6)$, and $C = (8, 0)$.

74. An **equilateral triangle** is one in which all three sides are of equal length. If two vertices of an equilateral triangle are $(0, 4)$ and $(0, 0)$, find the third vertex. How many of these triangles are possible?

*In Problems 75–78, find the length of each side of the triangle determined by the three points P_1, P_2, and P_3. State whether the triangle is an isosceles triangle, a right triangle, neither of these, or both. (An **isosceles triangle** is one in which at least two of the sides are of equal length.)*

75. $P_1 = (2, 1)$; $P_2 = (-4, 1)$; $P_3 = (-4, -3)$

76. $P_1 = (-1, 4)$; $P_2 = (6, 2)$; $P_3 = (4, -5)$

77. $P_1 = (-2, -1)$; $P_2 = (0, 7)$; $P_3 = (3, 2)$

78. $P_1 = (7, 2)$; $P_2 = (-4, 0)$; $P_3 = (4, 6)$

79. If r is a real number, prove that the point $P = (x, y)$ that divides the line segment from $P_1 = (x_1, y_1)$ to $P_2 = (x_2, y_2)$ in the ratio r, that is,

$$\frac{d(P_1, P)}{d(P_1, P_2)} = r$$

has coordinates

$$x = x_1 + r(x_2 - x_1) \text{ and } y = y_1 + r(y_2 - y_1)$$

[**Hint:** Use similar triangles.*]

85. Baseball A major league baseball "diamond" is actually a square, 90 feet on a side (see the figure). What is the distance directly from home plate to second base (the diagonal of the square)?

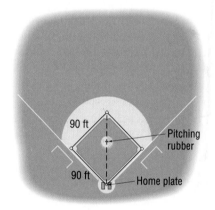

Problems 80–84 use the result of Problem 79.

80. Verify that the midpoint of a line segment divides the line segment from $P_1 = (x_1, y_1)$ to $P_2 = (x_2, y_2)$ in the ratio $r = \frac{1}{2}$.

81. What point P divides the line segment from P_1 to P_2 in the ratio $r = 1$?

82. What point P divides the line segment from P_1 to P_2 in the ratio $r = 0$?

83. Find the point P on the line joining $P_1 = (2, 4)$ and $P_2 = (5, 6)$ that is twice as far from P_1 as P_2 is from P_1 and that lies on the same side of P_1 as P_2 does.

84. Find the point P on the line joining $P_1 = (0, 4)$ and $P_2 = (-4, 1)$ that is three times as far from P_1 as P_2 is from P_1.

86. Little League Baseball The layout of a Little League playing field is a square, 60 feet on a side.† How far is it directly from home plate to second base (the diagonal of the square)?

*Two triangles are **similar** if they have the same angles. In similar triangles, corresponding sides are in proportion.

†*Source: Little League Baseball, Official Regulations and Playing Rules, 1991.*

87. Baseball Refer to Problem 85. Overlay a rectangular coordinate system on a major league baseball diamond so that the origin is at home plate, the positive x-axis lies in the direction from home plate to first base, and the positive y-axis lies in the direction from home plate to third base.

(a) What are the coordinates of first base, second base, and third base? Use feet as the unit of measurement.

(b) If the right fielder is located at (310, 15), how far is he from second base?

(c) If the center fielder is located at (300, 300), how far is she from third base?

88. Little League Baseball Refer to Problem 86. Overlay a rectangular coordinate system on a Little League baseball diamond so that the origin is at home plate, the positive x-axis lies in the direction from home plate to first base, and the positive y-axis lies in the direction from home plate to third base.

(a) What are the coordinates of first base, second base, and third base? Use feet as the unit of measurement.

(b) If the right fielder is located at (180, 20), how far is she from second base?

(c) If the center fielder is located at (220, 220), how far is he from third base?

89. Geometry Find the midpoint of each diagonal of a square with side of length s. Draw the conclusion that the diagonals of a square intersect at their midpoints.

[**Hint:** Use (0, 0), (0, s), (s, 0), and (s, s) as the vertices of the square.]

90. Geometry Verify that the points (0, 0), (a, 0), and $(a/2, \sqrt{3}a/2)$ are the vertices of an equilateral triangle. Then show that the midpoints of the three sides are the vertices of a second equilateral triangle. (See Problem 74.)

91. A Dodge Intrepid and a Mack truck leave an intersection at the same time. The Intrepid heads east at an average speed of 30 miles per hour, while the truck heads south at an average speed of 40 miles per hour. Find an expression for their distance apart d (in miles) at the end of t hours.

92. A hot air balloon, headed due east at an average speed of 15 miles per hour and at a constant altitude of 100 feet, passes over an intersection (see the figure). Find an expression for its distance d (measured in feet) from the intersection t seconds later.

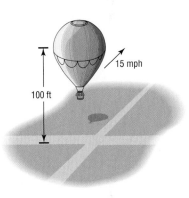

1.2 GRAPHS OF EQUATIONS; CIRCLES

1   Graph Equations by Hand and by Using a Graphing Utility

2   Find Intercepts Algebraically

3   Find Intercepts Using a Graphing Utility

4   Test an Equation for Symmetry with Respect to the (a) x-axis, (b) y-axis, and (c) Origin

5   Write the Standard Form of the Equation of a Circle

6   Graph a Circle by Hand and by Using a Graphing Utility; Identify the Center and Radius of a Circle

7   Write the General Form of the Equation of a Circle

1   The **graph of an equation** in two variables x and y consists of the set of points in the xy-plane whose coordinates (x, y) satisfy the equation. Let's look at some examples.

E X A M P L E   1   Graphing an Equation by Hand

Graph the equation: $y = 2x + 5$

FIGURE 20
$y = 2x + 5$

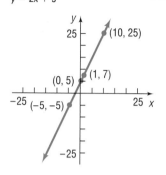

Solution   We want to find all points (x, y) that satisfy the equation. To locate some of these points (and thus get an idea of the pattern of the graph), we assign some numbers to x and find corresponding values for y:

If	Then	Point on Graph
$x = 0$	$y = 2(0) + 5 = 5$	$(0, 5)$
$x = 1$	$y = 2(1) + 5 = 7$	$(1, 7)$
$x = -5$	$y = 2(-5) + 5 = -5$	$(-5, -5)$
$x = 10$	$y = 2(10) + 5 = 25$	$(10, 25)$

By plotting these points and then connecting them, we obtain the graph of the equation (a straight line), as shown in Figure 20. ∎

E X A M P L E   2   Graphing an Equation by Hand

Graph the equation: $y = x^2$

Solution   Table 3 provides several points on the graph. In Figure 21 we plot these points and connect them with a smooth curve to obtain the graph (a *parabola*).

TABLE 3		
x	**$y = x^2$**	**(x, y)**
-4	16	$(-4, 16)$
-3	9	$(-3, 9)$
-2	4	$(-2, 4)$
-1	1	$(-1, 1)$
0	0	$(0, 0)$
1	1	$(1, 1)$
2	4	$(2, 4)$
3	9	$(3, 9)$
4	16	$(4, 16)$

FIGURE 21
$y = x^2$

Now work Problem 27.

The graphs of the equations shown in Figures 20 and 21 do not show all points. For example, in Figure 20, the point $(20, 45)$ is a part of the graph of $y = 2x + 5$, but it is not shown. Since the graph of $y = 2x + 5$ could be extended out as far as we please, we use arrows to indicate that the pattern shown continues. Thus, it is important when illustrating a graph to present enough of the graph so that any viewer of the illustration will "see" the rest of it as an obvious continuation of what is actually there. This is referred to as a **complete graph.**

So, one way to obtain a complete graph of an equation is to plot a sufficient number of points on the graph until a pattern becomes evident. Then these points are connected with a smooth curve following the suggested pattern. But how many points are sufficient? Sometimes knowledge about the equation tells us. For example, we will learn in the next section that, if an equation is of the form $y = mx + b$, then its graph is a line. In this case, two points would suffice to obtain the graph.

One purpose of this book is to investigate the properties of equations in order to decide whether a graph is complete. Sometimes we shall graph equations by hand by plotting a sufficient number of points on the graph until a pattern becomes evident; then we connect these points with a smooth curve following the suggested pattern. Shortly, we shall investigate various techniques that will enable us to graph an equation by hand without plotting so many points. Other times, we shall graph equations using a graphing utility.

From Examples 1 and 2, we see that a graph can be obtained by plotting points in a rectangle coordinate system and connecting them. Graphing utilities perform these same steps when graphing an equation. For example, the TI-83 determines 95 evenly spaced input values,* uses the equation to determine the output values, plots these points on the screen, and, finally (if in the connected mode), draws a line between consecutive points.

Most graphing utilities require the following steps in order to obtain the graph of an equation:

Steps for Graphing Equations Using a Graphing Utility

STEP 1: Solve the equation for y in terms of x.
STEP 2: Get into the graphing mode of your graphing utility. The screen will usually display $Y_1 =$, prompting you to enter the expression involving x that you found in STEP 1. (Consult your manual for the correct way to enter the expression; for example, $y = x^2$ might be entered as $x\text{^}2$ or as $x*x$ or as $x \, x^Y \, 2$.)
STEP 3: Select the viewing rectangle. Without prior knowledge about the behavior of the graph of the equation, it is common to select the **standard viewing rectangle** initially. The viewing rectangle is then adjusted based on the graph that appears. In this text, the standard viewing rectangle will be:

Xmin $= -10$
Xmax $= 10$
Xscl $= 1$
Ymin $= -10$
Ymax $= 10$
Yscl $= 1$

STEP 4: Execute.
STEP 5: Adjust the viewing rectangle until a complete graph is obtained.

E X A M P L E   3   Graphing an Equation on a Graphing Utility

Graph the equation: $6x^2 + 3y = 36$

Solution   STEP 1: We solve for y in terms of x.

$$6x^2 + 3y = 36$$
$$3y = -6x^2 + 36 \qquad \text{Subtract } 6x^2 \text{ from both sides of the equation.}$$
$$y = -2x^2 + 12 \qquad \text{Divide both sides of the equation by 3.}$$

*These input values depend on the values of Xmin and Xmax. For example, if Xmin $= -10$ and Xmax $= 10$, then the first input value will be -10 and the next input value will be $-10 + (10 - (-10))/94 = -9.7872$, and so on.

STEP 2: From the graphing mode, enter the expression $-2x^2 + 12$ after the prompt $Y_1 = $.

STEP 3: Set the viewing rectangle to the standard viewing rectangle.

STEP 4: Execute. The screen should look like Figure 22.

FIGURE 22

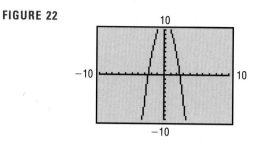

STEP 5: The graph of $y = -2x^2 + 12$ is not complete. The value of Ymax must be increased so the top portion of the graph is visible. After increasing the value of Ymax to 12, we obtain the graph in Figure 23. The graph is now complete.

FIGURE 23

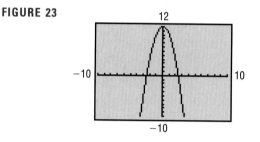

Look again at Figure 23. Although a complete graph is shown, the graph might be improved by adjusting the values of Xmin and Xmax. Figure 24 shows the graph of $y = -2x^2 + 12$ using Xmin $= -4$ and Xmax $= 4$. Do you think this is a better choice for the viewing rectangle?

FIGURE 24

FIGURE 25

Two techniques that reduce the number of points required to graph an equation by hand involve finding *intercepts* and checking for *symmetry*.

The points, if any, at which a graph crosses or touches the coordinate axes are called the **intercepts.** The x-coordinate of a point at which the graph crosses or touches the x-axis is an **x-intercept,** and the y-coordinate of a point at which the graph crosses or touches the y-axis is a **y-intercept.** For example, the graph in Figure 25 has three x-intercepts, -3, $\frac{3}{2}$, and 4.5, and three y-intercepts, -3.5, $-\frac{4}{3}$, and 3.

Procedure for Finding
Intercepts Algebraically

1. To find the x-intercept(s), if any, of the graph of an equation, let $y = 0$ in the equation and solve for x.
2. To find the y-intercept(s), if any, of the graph of an equation, let $x = 0$ in the equation and solve for y.

E X A M P L E   4

Finding Intercepts from an Equation

Find the x-intercept(s) and the y-intercept(s) of the graph of $y = x^2 - 4$.

Solution   To find the x-intercepts, we let $y = 0$ and obtain the equation

$$x^2 - 4 = 0$$

The equation has two solutions, -2 and 2. Thus, the x-intercepts are -2 and 2.
   To find the y-intercept(s), we let $x = 0$ and obtain the equation $y = -4$. The only y-intercept is -4.   ∎

Now work Problem 13(a).

Four tools that can be used to locate intercepts using a graphing utility are TRACE, BOX, VALUE, and ROOT (or ZERO).
   Most graphing utilities allow you to move from point to point along the graph, displaying on the screen the coordinates of each point. This feature is called TRACE.

E X A M P L E   5

Using TRACE to Locate Intercepts

Graph the equation $y = x^3 - 8$. Use TRACE to locate the intercepts.

Solution   Figure 26(a) shows the graph of $y = x^3 - 8$.
   Activate the TRACE feature. As you move the cursor along the graph, you will see the coordinates of each point displayed. Just before you get to the x-axis, the display will look like the one in Figure 26(b). (Due to differences in graphing utilities, your display may be slightly different than the one shown here.)

FIGURE 26

(a)   (b)   (c)

FIGURE 27

In Figure 26(b), the negative value of the y-coordinate indicates we are still below the x-axis. The next position of the cursor is shown in Figure 26(c).
   The positive value of the y-coordinate indicates that we are now above the x-axis. This means that between these two points the x-axis was crossed. The x-intercept lies between 1.9148936 and 2.0212766.
   Using TRACE we find that the y-intercept is -8. See Figure 27.   ∎

Most graphing utilities have a BOX feature that allows you to box in a specific part of the graph of an equation.

EXAMPLE 6

Using BOX

Graph the equation $y = x^3 - 8$ and use BOX to improve on the approximation found in Example 5 for the x-intercept.

Solution Using the viewing rectangle in Figure 26(a), graph the equation. Activate the BOX feature. (With some graphing utilities, this requires positioning the cursor at one corner of the box and then tracing out the sides of the box to the diagonal corner.) See Figure 28.

Once executed, the box becomes the viewing rectangle. The result is shown in Figure 29.

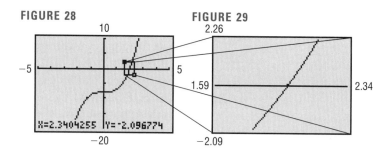

Now you can TRACE to get the approximations shown in Figure 30.

FIGURE 30

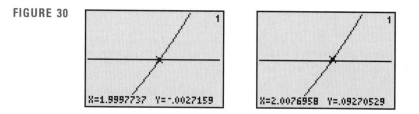

Now we know that the x-intercept lies between 1.9997737 and 2.0076958.

Most graphing utilities have an eVALUEate feature that, given a value of x, determines the value of y for an equation. We can use this feature to evaluate an equation at $x = 0$ to determine the y-intercept. Using VALUE, we find that the y-intercept of $y = x^3 - 8$ is -8.

Most graphing utilities also have a ROOT (or ZERO) feature that can be used to determine the x-intercept of an equation. Using ROOT on a TI-82, we find that the x-intercept of $y = x^3 - 8$ is 2. Consult your owner's manual to determine the appropriate keystrokes for these features.

4 Another useful tool for graphing equations by hand involves *symmetry*, particularly symmetry with respect to the x-axis, the y-axis, and the origin.

A graph is said to be **symmetric with respect to the *x*-axis** if, for every point (x, y) on the graph, the point $(x, -y)$ is also on the graph.

> A graph is said to be **symmetric with respect to the *y*-axis** if, for every point (*x*, *y*) on the graph, the point (−*x*, *y*) is also on the graph.

> A graph is said to be **symmetric with respect to the origin** if, for every point (*x*, *y*) on the graph, the point (−*x*, −*y*) is also on the graph.

Figure 31 illustrates the definition. Notice that, when a graph is symmetric with respect to the *x*-axis, the part of the graph above the *x*-axis is a reflection of the part below it, and vice versa. When a graph is symmetric with respect to the *y*-axis, the part of the graph to the right of the *y*-axis is a reflection of the part to the left of it, and vice versa. Notice that symmetry with respect to the origin may be viewed in two ways:

1. As a reflection about the *y*-axis, followed by a reflection about the *x*-axis.
2. As a projection along a line through the origin so that the distances from the origin are equal.

FIGURE 31

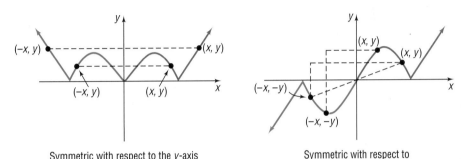

| Symmetric with respect to the *x*-axis | Symmetric with respect to the *y*-axis | Symmetric with respect to the origin |

 Now work Problem 13(b).

When the graph of an equation is symmetric, the number of points that you need to plot in order to see the pattern is reduced. For example, if the graph of an equation is symmetric with respect to the *y*-axis, then, once points to the right of the *y*-axis are plotted, an equal number of points on the graph can be obtained by reflecting them about the *y*-axis. Thus, before we graph an equation, we first want to determine whether it has any symmetry. The following tests are used for this purpose.

Tests for Symmetry

> To test the graph of an equation for symmetry with respect to the:
>
> **x-axis** Replace *y* by −*y* in the equation. If an equivalent equation results, the graph of the equation is symmetric with respect to the *x*-axis.
>
> **y-axis** Replace *x* by −*x* in the equation. If an equivalent equation results, the graph of the equation is symmetric with respect to the *y*-axis.
>
> **Origin** Replace *x* by −*x* and *y* by −*y* in the equation. If an equivalent equation results, the graph of the equation is symmetric with respect to the origin.

E X A M P L E   7

Graphing an Equation by Finding Intercepts and Checking for Symmetry

Graph the equation $y = x^3$ by hand. Find any intercepts and check for symmetry first.

Solution   First, we seek the intercepts. When $x = 0$, then $y = 0$; and when $y = 0$, then $x = 0$. Thus, the origin $(0, 0)$ is the only intercept. Now we test for symmetry:

x-Axis: Replace y by $-y$. Since the result, $-y = x^3$, is not equivalent to $y = x^3$, the graph is not symmetric with respect to the *x*-axis.

y-Axis: Replace x by $-x$. Since the result, $y = -x^3$, is not equivalent to $y = x^3$, the graph is not symmetric with respect to the *y*-axis.

Origin: Replace x by $-x$ and y by $-y$. Since the result, $-y = -x^3$, is equivalent to $y = x^3$, the graph is symmetric with respect to the origin.

To graph by hand, we use the equation to obtain several points on the graph. Because of the symmetry, we only need to locate points on the graph for which $x \geq 0$. See Table 4. Points on the graph could also be obtained using the TABLE feature on a graphing utility. (Check your manual to see if your graphing utility has this feature.) Using a TI-83, we obtain Table 5. Figure 32 shows the graph.

TABLE 4		
x	**$y = x^3$**	**(x, y)**
0	0	(0, 0)
1	1	(1, 1)
2	8	(2, 8)
3	27	(3, 27)

TABLE 5

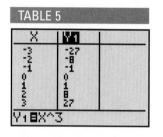

FIGURE 32
Symmetry with respect to the origin

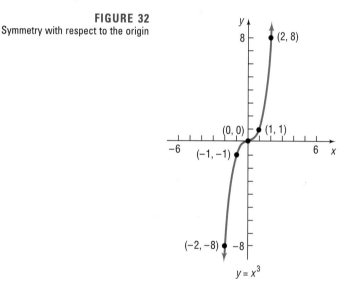

Now work Problem 91.

E X A M P L E   8

Graphing an Equation

Graph the equation $x = y^2$. Find any intercepts and check for symmetry first.

Solution   The lone intercept is $(0, 0)$. The graph is symmetric with respect to the *x*-axis. (Do you see why? Replace y by $-y$.) Figure 33 shows the graph drawn by hand.

To graph this equation using a graphing utility, we must write the equation in the form $y =$ expression involving x. So to graph $x = y^2$, we must solve for y:

FIGURE 33
Symmetry with respect to the *x*-axis

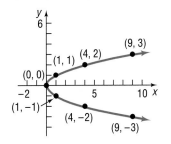

$$x = y^2$$
$$y^2 = x$$
$$y = \pm\sqrt{x}$$

Thus, to graph $x = y^2$, we need to graph both $Y_1 = \sqrt{x}$ and $Y_2 = -\sqrt{x}$ on the same screen. Figure 34 shows the result. Table 6 shows various values of y for a given value of x when $Y_1 = \sqrt{x}$ and $Y_2 = -\sqrt{x}$. Notice that when $x < 0$ we get an error. Can you explain why?

FIGURE 34

Look again at either Figure 33 or 34. If we restrict y so that $y \geq 0$, the equation $x = y^2$, $y \geq 0$, may be written equivalently as $y = \sqrt{x}$. The portion of the graph of $x = y^2$ in quadrant I is therefore the graph of $y = \sqrt{x}$. See Figure 35.

FIGURE 35
$y = \sqrt{x}$

Now work Problem 89.

E X A M P L E   9   Graphing the Equation $y = 1/x$

Consider the equation: $y = 1/x$

(a) Graph this equation using a graphing utility. Set the viewing rectangle as

Range
$$X\text{min} = -3$$
$$X\text{max} = 3$$
$$X\text{scl}\ = 1$$
$$Y\text{min} = -4$$
$$Y\text{max} = 4$$
$$Y\text{scl}\ = 1$$

(b) Use algebra to find any intercepts and test for symmetry.
(c) Draw the graph by hand.

Solution

FIGURE 36

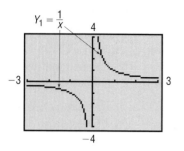

(a) Figure 36 illustrates the graph.

We infer from the graph that there are no intercepts; we may also infer that symmetry with respect to the origin is a possibility. The TRACE feature on a graphing utility can provide further evidence of symmetry with respect to the origin. Using TRACE, we observe that for any ordered pair (x, y) the ordered pair $(-x, -y)$ is also a point on the graph. For example, the points $(0.95744681, 1.0444444)$ and $(-0.95744681, -1.0444444)$ both lie on the graph.

(b) We check for intercepts first. If we let $x = 0$, we obtain a 0 denominator, which is not allowed. Hence, there is no y-intercept. If we let $y = 0$, we get the equation $1/x = 0$, which has no solution. Hence, there is no x-intercept. Thus, the graph of $y = 1/x$ does not cross the coordinate axes.

Next we check for symmetry:

x-Axis: Replacing y by $-y$ yields $-y = 1/x$, which is not equivalent to $y = 1/x$.

y-Axis: Replacing x by $-x$ yields $y = -1/x$, which is not equivalent to $y = 1/x$.

Origin: Replacing x by $-x$ and y by $-y$ yields $-y = -1/x$, which is equivalent to $y = 1/x$.

The graph is symmetric with respect to the origin. This confirms the inferences drawn in part (a) of the solution.

(c) We can use the equation to obtain some points on the graph. Because of symmetry, we need only find points (x, y) for which x is positive. Also, from the equation $y = 1/x$ we infer that if x is a large and positive number, then $y = 1/x$ is a positive number close to 0. We also infer that if x is a positive number close to 0 then $y = 1/x$ is a large and positive number. Armed with this information, we can graph the equation. Table 7 and Figure 37 illustrate some of these points and the graph of $y = 1/x$. Observe how the absence of intercepts and the existence of symmetry with respect to the origin were utilized.

TABLE 7

x	$y = 1/x$	(x, y)
$\frac{1}{10}$	10	$(\frac{1}{10}, 10)$
$\frac{1}{3}$	3	$(\frac{1}{3}, 3)$
$\frac{1}{2}$	2	$(\frac{1}{2}, 2)$
1	1	$(1, 1)$
2	$\frac{1}{2}$	$(2, \frac{1}{2})$
3	$\frac{1}{3}$	$(3, \frac{1}{3})$
10	$\frac{1}{10}$	$(10, \frac{1}{10})$

FIGURE 37

Square Screens

To get an undistorted view of a graph, the same scale must be used on each axis. However, most graphing utilities have a rectangular screen. Because of this, using the same RANGE for both x and y will result in a distorted view.

FIGURE 38

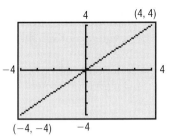

For example, Figure 38 shows the graph of the line $y = x$ connecting the points $(-4, -4)$ and $(4, 4)$.

We expect the line to bisect the first and third quadrants, but it doesn't. We need to adjust the selections for Xmin, Xmax, Ymin, and Ymax so that a **square screen** results. On most graphing utilities, this is accomplished by setting the ratio of x to y at 3:2.* In other words,

$$2(X\text{max} - X\text{min}) = 3(Y\text{max} - Y\text{min})$$

Figure 39 shows the graph of the line $y = x$ on a square screen using a TI-83. Notice that the line now bisects the first and third quadrants. Compare this illustration to Figure 38.

FIGURE 39

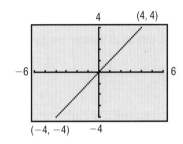

Circles

One advantage of a coordinate system is that it enables us to translate a geometric statement into an algebraic statement, and vice versa. Consider, for example, the following geometric statement that defines a circle.

> A **circle** is a set of points in the xy-plane that are a fixed distance r from a fixed point (h, k). The fixed distance r is called the **radius,** and the fixed point (h, k) is called the **center** of the circle.

FIGURE 40

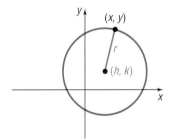

Figure 40 shows the graph of a circle. Is there an equation having this graph? If so, what is the equation? To find the equation, we let (x, y) represent the coordinates of any point on a circle with radius r and center (h, k). Then the distance between the points (x, y) and (h, k) must always equal r. That is, by the distance formula,

$$\sqrt{(x - h)^2 + (y - k)^2} = r$$

or, equivalently,

$$(x - h)^2 + (y - k)^2 = r^2$$

The **standard form of an equation of a circle** with radius r and center (h, k) is

$$(x - h)^2 + (y - k)^2 = r^2 \qquad (1)$$

*Some graphing utilities have a built-in function that automatically squares the screen. For example the TI-85 has a ZSQR function that does this. Some graphing utilities require a ratio other than 3:2 to square the screen. For example, the HP48G requires the ratio of x to y to be 1:2 for a square screen. Consult your manual.

25.

26.

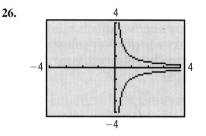

In Problems 27–32, tell whether the given points are on the graph of the equation.

27. Equation: $y = x^4 + \sqrt{x}$
Points: $(0, 0)$; $(1, 1)$; $(-1, 0)$

28. Equation: $y = x^3 - 2\sqrt{x}$
Points: $(0, 0)$; $(1, 1)$; $(1, -1)$

29. Equation: $y^2 = x^2 + 4$
Points: $(0, 2)$; $(2, 0)$; $(-2, 0)$

30. Equation: $y^3 = x + 1$
Points: $(1, 2)$; $(0, 1)$; $(-1, 0)$

31. Equation: $x^2 + y^2 = 4$
Points: $(0, 2)$; $(-2, 2)$; $(\sqrt{2}, \sqrt{2})$

32. Equation: $x^2 + 4y^2 = 4$
Points: $(0, 1)$; $(2, 0)$; $(2, \frac{1}{2})$

33. If $(a, 2)$ is a point on the graph of $y = 5x + 4$, what is a?

34. If $(2, b)$ is a point on the graph of $y = x^2 + 3x$, what is b?

35. If (a, b) is a point on the graph of $2x + 3y = 6$, write an equation that relates a to b.

36. If $(2, 0)$ and $(0, 5)$ are points on the graph of $y = mx + b$, what are m and b?

In Problems 37–40, find the center and radius of each circle. Write the standard form of the equation.

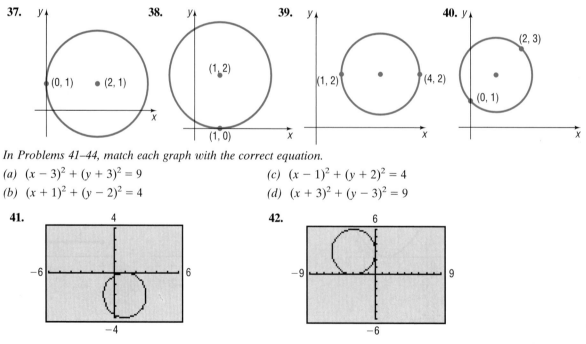

In Problems 41–44, match each graph with the correct equation.

(a) $(x - 3)^2 + (y + 3)^2 = 9$

(c) $(x - 1)^2 + (y + 2)^2 = 4$

(b) $(x + 1)^2 + (y - 2)^2 = 4$

(d) $(x + 3)^2 + (y - 3)^2 = 9$

41.

42.

43.

44.

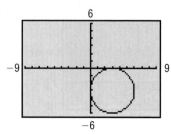

In Problems 45–48, find the standard form of the equation of each circle.

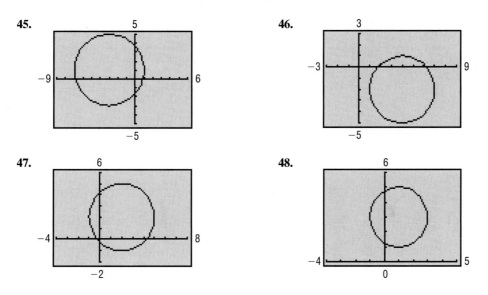

45.

46.

47.

48.

In Problems 49–58, write the standard form of the equation and the general form of the equation of each circle of radius r and center (h, k). By hand, graph each circle.

49. $r = 1; (h, k) = (1, -1)$

50. $r = 2; (h, k) = (-2, 1)$

51. $r = 2; (h, k) = (0, 2)$

52. $r = 3; (h, k) = (1, 0)$

53. $r = 5; (h, k) = (4, -3)$

54. $r = 4; (h, k) = (2, -3)$

55. $r = 2; (h, k) = (0, 0)$

56. $r = 3; (h, k) = (0, 0)$

57. $r = \frac{1}{2}; (h, k) = (\frac{1}{2}, 0)$

58. $r = \frac{1}{2}; (h, k) = (0, -\frac{1}{2})$

In Problems 59–68, find the center (h, k) and radius r of each circle. By hand, graph each circle.

59. $x^2 + y^2 = 4$

60. $x^2 + (y - 1)^2 = 1$

61. $(x - 3)^2 + y^2 = 4$

62. $(x + 1)^2 + (y - 1)^2 = 2$

63. $x^2 + y^2 + 4x - 4y - 1 = 0$

64. $x^2 + y^2 - 6x + 2y + 9 = 0$

65. $x^2 + y^2 - x + 2y + 1 = 0$

66. $x^2 + y^2 + x + y - \frac{1}{2} = 0$

67. $2x^2 + 2y^2 - 12x + 8y - 24 = 0$

68. $2x^2 + 2y^2 + 8x + 7 = 0$

In Problems 69–74, find the general form of the equation of each circle.

69. Center at the origin and containing the point $(-2, 3)$

70. Center $(-3, 1)$ and tangent to the y-axis

71. Center $(1, 0)$ and containing the point $(-3, 2)$

72. With endpoints of a diameter at $(1, 4)$ and $(-3, 2)$

73. Center $(2, 3)$ and tangent to the x-axis

74. With endpoints of a diameter at $(4, 3)$ and $(0, 1)$

In Problems 75–78, use the graph on the right.

75. Draw the graph to make it symmetric with respect to the x-axis.

76. Draw the graph to make it symmetric with respect to the y-axis.

77. Draw the graph to make it symmetric with respect to the origin.

78. Draw the graph to make it symmetric with respect to the x-axis, y-axis, and origin.

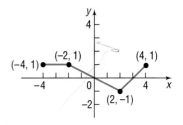

In Problems 79–82, use the graph on the right.

79. Draw the graph to make it symmetric with respect to the x-axis.

80. Draw the graph to make it symmetric with respect to the y-axis.

81. Draw the graph to make it symmetric with respect to the origin.

82. Draw the graph to make it symmetric with respect to the x-axis, y-axis, and origin.

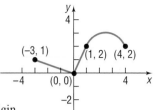

In Problems 83–96, list the intercepts and test for symmetry. Graph each equation using a graphing utility.

83. $x^2 = y$

84. $y^2 = x + 1$

85. $y = 3x$

86. $y = -5x$

87. $x^2 + y - 9 = 0$

88. $y^2 - x - 4 = 0$

89. $9x^2 + 4y^2 = 36$

90. $4x^2 + y^2 = 4$

91. $y = x^3 - 27$

92. $y = x^4 - 1$

93. $y = x^2 - 3x - 4$

94. $y = x^2 + 4$

95. $y = \dfrac{x}{x^2 + 9}$

96. $y = \dfrac{x^2 - 4}{x}$

In Problem 97, you may use a graphing utility, but it is not required.

97. (a) Graph $y = \sqrt{x^2}$, $y = x$, $y = |x|$, and $y = (\sqrt{x})^2$, noting which graphs are the same.

(b) Explain why the graphs of $y = \sqrt{x^2}$ and $y = |x|$ are the same.

(c) Explain why the graphs of $y = x$ and $y = (\sqrt{x})^2$ are not the same.

(d) Explain why the graphs of $y = \sqrt{x^2}$ and $y = x$ are not the same.

98. Make up an equation with the intercepts $(2, 0)$, $(4, 0)$, and $(0, 1)$. Compare your equation with a friend's equation. Comment on any similarities.

99. An equation is being tested for symmetry with respect to the x-axis, the y-axis, and the origin. Explain why, if two of these symmetries are present, then the remaining one must also be present.

100. Draw a graph that contains the points $(-2, -1)$, $(0, 1)$, $(1, 3)$, and $(3, 5)$. Compare your graph with those of other students. Are most of the graphs almost straight lines? How many are "curved"? Discuss the various ways these points might be connected.

1.3 LINES

1	Calculate and Interpret the Slope of a Line
2	Graph Lines by Hand and by Using a Graphing Utility
3	Find the Equation of Horizontal and Vertical Lines
4	Use the Point–Slope Form of a Line
5	Write the Equation of a Line in General Form
6	Write the Equation of a Line in Slope–Intercept Form
7	Identify the Slope and y-intercept of a Line from a Linear Equation
8	Define Parallel and Perpendicular Lines
9	Find Equations of Parallel Lines
10	Find Equations of Perpendicular Lines

In this section we study a certain type of equation that contains two variables, called a *linear equation,* and its graph, a *line.*

Slope of a Line

FIGURE 45

Consider the staircase illustrated in Figure 45. Each step contains exactly the same horizontal **run** and the same vertical **rise.** The ratio of the rise to the run, called the *slope,* is a numerical measure of the steepness of the staircase. For example, if the run is increased and the rise remains the same, the staircase becomes less steep. If the run is kept the same, but the rise is increased, the staircase becomes more steep. This important characteristic of a line is best defined using rectangular coordinates.

Let $P = (x_1, y_1)$ and $Q = (x_2, y_2)$ be two distinct points with $x_1 \neq x_2$. The **slope** m of the nonvertical line L containing P and Q is defined by the formula

$$m = \frac{y_2 - y_1}{x_2 - x_1} \qquad x_1 \neq x_2 \tag{1}$$

If $x_1 = x_2$, L is a **vertical line** and the slope m of L is **undefined** (since this results in division by 0).

Figure 46(a) provides an illustration of the slope of a nonvertical line; Figure 46(b) illustrates a vertical line.

FIGURE 46

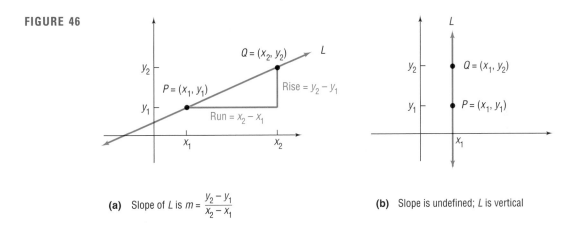

(a) Slope of L is $m = \dfrac{y_2 - y_1}{x_2 - x_1}$

(b) Slope is undefined; L is vertical

As Figure 46(a) illustrates, the slope m of a nonvertical line may be viewed as

$$m = \frac{y_2 - y_1}{x_2 - x_1} = \frac{\text{Rise}}{\text{Run}}$$

We can also express the slope m of a nonvertical line as

$$m = \frac{y_2 - y_1}{x_2 - x_1} = \frac{\text{Change in } y}{\text{Change in } x} = \frac{\Delta y}{\Delta x}$$

That is, the slope m of a nonvertical line L measures the amount y changes as x changes from x_1 to x_2. This is called the **average rate of change** of y with respect to x.

Two comments about computing the slope of a nonvertical line may prove helpful:

1. Any two distinct points on the line can be used to compute the slope of the line. (See Figure 47 for justification.)

FIGURE 47
Triangles *ABC* and *PQR* are similar (equal angles). Hence, ratios of corresponding sides are proportional. Thus:

Slope using *P* and $Q = \dfrac{y_2 - y_1}{x_2 - x_1}$

$= $ Slope using *A* and $B = \dfrac{d(B, C)}{d(A, C)}$

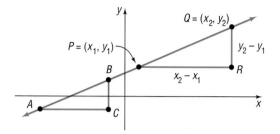

2. The slope of a line may be computed from $P = (x_1, y_1)$ to $Q = (x_2, y_2)$ or from Q to P because

$$\frac{y_2 - y_1}{x_2 - x_1} = \frac{y_1 - y_2}{x_1 - x_2}$$

E X A M P L E 1

Finding and Interpreting the Slope of a Line Joining Two Points

The slope m of the line joining the points $(1, 2)$ and $(5, -3)$ may be computed as

$$m = \frac{-3 - 2}{5 - 1} = \frac{-5}{4} \quad \text{or as} \quad m = \frac{2 - (-3)}{1 - 5} = \frac{5}{-4} = \frac{-5}{4}$$

For every 4 unit change in x, y will change by -5 units. Thus, if x increases by 4 units, then y will decrease by 5 units. ■

Now work Problem 7.

FIGURE 48

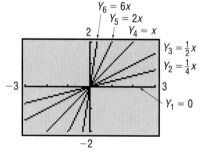

Seeing the Concept On the same square screen, graph the following equations:

$Y_1 = 0$ Slope of line is 0.

$Y_2 = \frac{1}{4}x$ Slope of line is $\frac{1}{4}$.

$Y_3 = \frac{1}{2}x$ Slope of line is $\frac{1}{2}$.

$Y_4 = x$ Slope of line is 1.

$Y_5 = 2x$ Slope of line is 2.

$Y_6 = 6x$ Slope of line is 6.

See Figure 48. ■

Seeing the Concept On the same square screen, graph the following equations:

$Y_1 = 0$ Slope of line is 0.

$Y_2 = -\frac{1}{4}x$ Slope of line is $-\frac{1}{4}$.

FIGURE 49

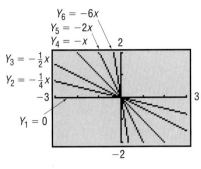

$Y_3 = -\frac{1}{2}x$ Slope of line is $-\frac{1}{2}$.

$Y_4 = -x$ Slope of line is -1.

$Y_5 = -2x$ Slope of line is -2.

$Y_6 = -6x$ Slope of line is -6.

See Figure 49.

Figures 48 and 49 illustrate the following facts:

1. When the slope of a line is positive, the line slants upward from left to right.
2. When the slope of a line is negative, the line slants downward from left to right.
3. When the slope is 0, the line is horizontal.

Figures 48 and 49 also illustrate that the closer the line is to the vertical position the greater the magnitude of the slope.

2   The next example illustrates how the slope of a line can be used to graph the line.

E X A M P L E   2   **Graphing a Line Given a Point and a Slope**

Draw a graph of the line that passes through the point $(3, 2)$ and has a slope of:

(a) $\frac{3}{4}$ (b) $-\frac{4}{5}$

Solution   (a) Slope = Rise/Run. The fact that the slope is $\frac{3}{4}$ means that for every horizontal movement (run) of 4 units to the right there will be a vertical movement (rise) of 3 units. If we start at the given point $(3, 2)$ and move 4 units to the right and 3 units up, we reach the point $(7, 5)$. By drawing the line through this point and the point $(3, 2)$, we have the graph. See Figure 50.

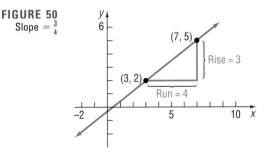

FIGURE 50
Slope $= \frac{3}{4}$

(b) The fact that the slope is

$$-\frac{4}{5} = \frac{-4}{5} = \frac{\text{Rise}}{\text{Run}}$$

means that for every horizontal movement of 5 units to the right there will be a corresponding vertical movement of -4 units (a downward movement). If we start at the given point $(3, 2)$ and move 5 units to the right and then 4 units down, we arrive at the point $(8, -2)$. By drawing the line through these points, we have the graph. See Figure 51.

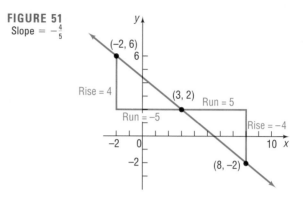

FIGURE 51
Slope $= -\frac{4}{5}$

Alternatively, we can set

$$-\frac{4}{5} = \frac{4}{-5} = \frac{\text{Rise}}{\text{Run}}$$

so that for every horizontal movement of -5 units (a movement to the left) there will be a corresponding vertical movement of 4 units (upward). This approach brings us to the point $(-2, 6)$, which is also on the graph shown in Figure 51. ■

Now work Problem 15.

Equations of Lines

Now that we have discussed the slope of a line, we are ready to derive equations of lines. As we shall see, there are several forms of the equation of a line. Let's start with an example.

E X A M P L E 3

Graphing a Line

Graph the equation: $x = 3$

Solution Using a graphing utility, we need to express the equation in the form $y = $ expression in x. But $x = 3$ cannot be put into this form. To overcome this problem, we must utilize a function in our graphing utility that allows vertical lines to be drawn. Consult your manual to determine the methodology required to draw vertical lines. Figure 52(a) shows the graph you should obtain.

To graph $x = 3$ by hand, we recall that we are looking for all points (x, y) in the plane for which $x = 3$. Thus, no matter what y-coordinate is used, the corresponding x-coordinate always equals 3. Consequently, the graph of the equation $x = 3$ is a vertical line with x-intercept 3 and undefined slope. See Figure 52(b).

FIGURE 52
$x = 3$

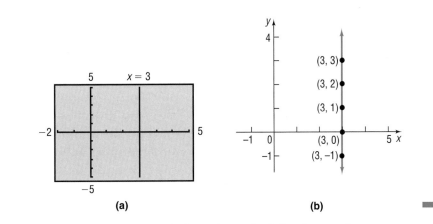

(a)   (b)

As suggested by Example 3, we have the following result:

Theorem   Equation of a Vertical Line

A vertical line is given by an equation of the form

$$x = a$$

where a is the x-intercept.

Now let L be a nonvertical line with slope m and containing the point (x_1, y_1). See Figure 53. For any other point (x, y) on L, we have

$$m = \frac{y - y_1}{x - x_1} \quad \text{or} \quad y - y_1 = m(x - x_1)$$

FIGURE 53

Theorem   Point–Slope Form of an Equation of a Line

An equation of a nonvertical line of slope m that passes through the point (x_1, y_1) is

$$y - y_1 = m(x - x_1) \tag{2}$$

E X A M P L E   4

Using the Point–Slope Form of a Line

An equation of the line with slope 4 and passing through the point $(1, 2)$ can be found by using the point–slope form with $m = 4$, $x_1 = 1$, and $y_1 = 2$:

$$y - y_1 = m(x - x_1)$$
$$y - 2 = 4(x - 1)$$
$$y = 4x - 2$$

See Figure 54.

FIGURE 54
$y = 4x - 2$

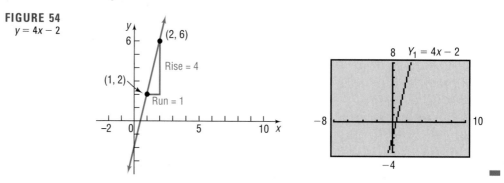

E X A M P L E   5

Finding the Equation of a Horizontal Line

Find an equation of the horizontal line passing through the point $(3, 2)$.

Solution
The slope of a horizontal line is 0. To get an equation, we use the point–slope form with $m = 0$, $x_1 = 3$, and $y_1 = 2$:

$$y - y_1 = m(x - x_1)$$
$$y - 2 = 0 \cdot (x - 3)$$
$$y - 2 = 0$$
$$y = 2$$

See Figure 55 for the graph.

FIGURE 55
$y = 2$

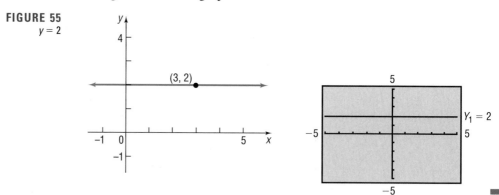

As suggested by Example 5, we have the following result:

Theorem   Equation of a Horizontal Line

A horizontal line is given by an equation of the form

$$y = b$$

where b is the y-intercept.

5

An equation of the line with slope $-\frac{1}{3}$ passing through the point $(2, 4)$ is

$$y - 4 = -\frac{1}{3}(x - 2)$$

Another form of the equation of this line can be obtained by multiplying both sides of the point–slope equation by 3 and collecting terms:

$$3(y - 4) = 3\left(-\frac{1}{3}\right)(x - 2) \quad \text{Multiply by 3.}$$
$$3y - 12 = -1(x - 2)$$
$$3y - 12 = -x + 2$$
$$x + 3y - 14 = 0$$

This is called the *general form*.

General Form of an Equation of a Line
The equation of a line L is in **general form** when it is written as

$$Ax + By + C = 0 \qquad (3)$$

where A, B, and C are three real numbers and A and B are not both 0.

Now work Problem 39.

Every line has an equation that is equivalent to an equation written in general form. For example, a vertical line whose equation is

$$x = a$$

can be written in the general form

$$1 \cdot x + 0 \cdot y - a = 0 \qquad A = 1, B = 0, C = -a$$

A horizontal line whose equation is

$$y = b$$

can be written in the general form

$$0 \cdot x + 1 \cdot y - b = 0 \qquad A = 0, B = 1, C = -b$$

Lines that are neither vertical nor horizontal have general equations of the form

$$Ax + By + C = 0 \qquad A \neq 0, B \neq 0$$

Because the equation of every line can be written in general form, any equation equivalent to (3) is called a **linear equation.**

The next example illustrates one way of graphing a linear equation.

E X A M P L E   6

Finding the Intercepts of a Line

Find the intercepts of the line $2x + 3y - 6 = 0$. Graph this line.

Solution

To find the point at which the graph crosses the x-axis, that is, to find the x-intercept, we need to find the number x for which $y = 0$. Thus, we let $y = 0$ to get

$$2x + 3(0) - 6 = 0$$
$$2x - 6 = 0$$
$$x = 3$$

FIGURE 56
$2x + 3y - 6 = 0$

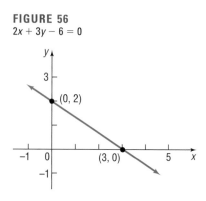

The x-intercept is 3. To find the y-intercept, we let $x = 0$ and solve for y:

$$2(0) + 3y - 6 = 0$$
$$3y - 6 = 0$$
$$y = 2$$

The y-intercept is 2.

To graph the line by hand, we use the intercepts. Since the x-intercept is 3 and the y-intercept is 2, we know two points on the line: $(3, 0)$ and $(0, 2)$. Because two points determine a unique line, we do not need further information to graph the line. See Figure 56.

To graph the line using a graphing utility, we need to solve for y.

$$2x + 3y - 6 = 0$$
$$3y = -2x + 6$$
$$y = -\tfrac{2}{3}x + 2$$

See Figure 57 for the graph.

FIGURE 57

$4 \quad Y_1 = -\tfrac{2}{3}x + 2$

6

Another useful equation of a line is obtained when the slope m and y-intercept b are known. In this event, we know both the slope m of the line and a point $(0, b)$ on the line; thus, we may use the point–slope form, equation (2), to obtain the following equation:

$$y - b = m(x - 0) \quad \text{or} \quad y = mx + b$$

Theorem Slope–Intercept Form of an Equation of a Line

An equation of a line L with slope m and y-intercept b is

$$y = mx + b \tag{4}$$

FIGURE 58
$y = mx + 2$

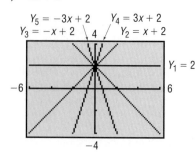

$Y_5 = -3x + 2 \quad Y_4 = 3x + 2$
$Y_3 = -x + 2 \quad Y_2 = x + 2$
$Y_1 = 2$

Seeing the Concept To see the role the slope m plays, graph the following lines on the same square screen.

$$Y_1 = 2$$
$$Y_2 = x + 2$$
$$Y_3 = -x + 2$$
$$Y_4 = 3x + 2$$
$$Y_5 = -3x + 2$$

See Figure 58. What do you conclude about the lines $y = mx + 2$?

FIGURE 59
$y = 2x + b$

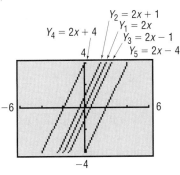

Seeing the Concept To see the role of the y-intercept b, graph the following lines on the same square screen.

$$Y_1 = 2x$$
$$Y_2 = 2x + 1$$
$$Y_3 = 2x - 1$$
$$Y_4 = 2x + 4$$
$$Y_5 = 2x - 4$$

See Figure 59. What do you conclude about the lines $y = 2x + b$?

When the equation of a line is written in slope–intercept form, it is easy to find the slope m and y-intercept b of the line. For example, suppose that the equation of a line is

$$y = -2x + 3$$

Compare it to $y = mx + b$

The slope of this line is -2 and its y-intercept is 3.
Let's look at another example.

E X A M P L E   7

Finding the Slope and y-Intercept of a Line

[7]

Find the slope m and y-intercept b of the line $2x + 4y - 8 = 0$. Graph the line.

Solution To obtain the slope and y-intercept, we transform the equation into its slope–intercept form. Thus, we need to solve for y:

$$2x + 4y - 8 = 0$$
$$4y = -2x + 8$$
$$y = -\frac{1}{2}x + 2$$

The coefficient of x, $-\frac{1}{2}$, is the slope, and the y-intercept is 2.

Figure 60 shows the graph using a graphing utility.

FIGURE 60

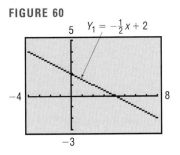

We can graph the line by hand in two ways:

1. Use the fact that the y-intercept is 2 and the slope is $-\frac{1}{2}$. Then, starting at the point $(0, 2)$, go to the right 2 units and then down 1 unit to the point $(2, 1)$. See Figure 61 on page 46.

Or:

2. Locate the intercepts. Because the y-intercept is 2, we know one intercept is $(0, 2)$. To obtain the x-intercept, let $y = 0$ and solve for x. When $y = 0$, we have

$$2x + 4 \cdot 0 - 8 = 0$$
$$2x - 8 = 0$$
$$x = 4$$

Thus, the intercepts are $(4, 0)$ and $(0, 2)$. See Figure 62.

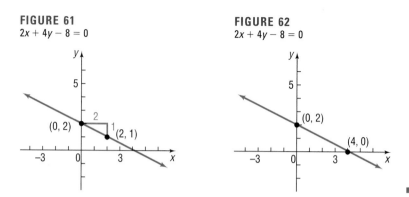

FIGURE 61
$2x + 4y - 8 = 0$

FIGURE 62
$2x + 4y - 8 = 0$

Now work **Problem 55.**

Parallel and Perpendicular Lines

FIGURE 63
The lines are parallel if and only if their slopes are equal.

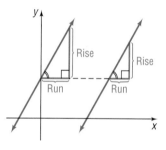

8 When two lines (in the plane) have no points in common, they are said to be **parallel.** Look at Figure 63. There we have drawn two lines and have constructed two right triangles by drawing sides parallel to the coordinate axes. These lines are parallel if and only if the right triangles are similar. (Do you see why? Two angles are equal.) And the triangles are similar if and only if the ratios of corresponding sides are equal.

This suggests the following result:

> **Theorem**
>
> Two distinct nonvertical lines are parallel if and only if their slopes are equal.

The use of the words "if and only if" in the preceding theorem means that actually two statements are being made, one the converse of the other.

If two distinct nonvertical lines are parallel, then their slopes are equal.

If two distinct nonvertical lines have equal slopes, then they are parallel.

Now work **Problem 27.**

E X A M P L E   8   **Showing That Two Lines Are Parallel**

Show that the lines given by the following equations are parallel:

$$L:\quad 2x + 3y - 6 = 0 \qquad M:\quad 4x + 6y = 0$$

Solution   To determine whether these lines have equal slopes, we write each equation in slope–intercept form:

$$L:\quad 2x + 3y - 6 = 0 \qquad\qquad M:\quad 4x + 6y = 0$$
$$3y = -2x + 6 \qquad\qquad\qquad 6y = -4x$$
$$y = -\tfrac{2}{3}x + 2 \qquad\qquad\qquad y = -\tfrac{2}{3}x$$
$$\text{Slope} = -\tfrac{2}{3} \qquad\qquad\qquad \text{Slope} = -\tfrac{2}{3}$$

Because these lines have the same slope, $-\frac{2}{3}$, but different y-intercepts, the lines are parallel. See Figure 64.

FIGURE 64
Parallel lines

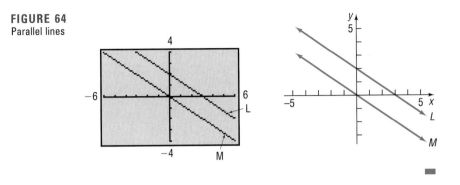

E X A M P L E   9

Finding a Line That Is Parallel to a Given Line

9

Find an equation for the line that contains the point $(2, -3)$ and is parallel to the line $2x + y - 6 = 0$.

Solution

The slope of the line we seek equals the slope of the line $2x + y - 6 = 0$, since the two lines are to be parallel. Thus, we begin by writing the equation of the line $2x + y - 6 = 0$ in slope–intercept form:

$$2x + y - 6 = 0$$
$$y = -2x + 6$$

The slope is -2. Since the line we seek contains the point $(2, -3)$, we use the point–slope form to obtain

$$y + 3 = -2(x - 2)$$
$$2x + y - 1 = 0 \qquad \text{General form}$$
$$y = -2x + 1 \qquad \text{Slope–intercept form}$$

This line is parallel to the line $2x + y - 6 = 0$ and contains the point $(2, -3)$. See Figure 65.

FIGURE 65

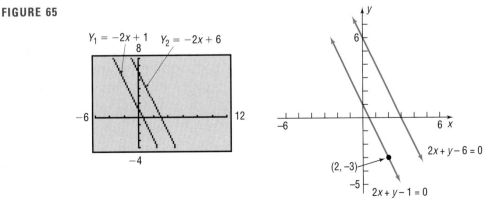

$Y_1 = -2x + 1 \qquad Y_2 = -2x + 6$

$2x + y - 6 = 0$

$(2, -3)$

$2x + y - 1 = 0$

Now work Problem 43.

FIGURE 66
Perpendicular lines

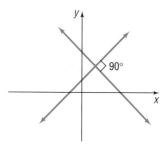

When two lines intersect at a right angle (90°), they are said to be **perpendicular.** See Figure 66.

The following result gives a condition, in terms of their slopes, for two lines to be perpendicular:

> ### Theorem
>
> Two nonvertical lines are perpendicular if and only if the product of their slopes is −1.

Here, we shall prove the "only if" part of the statement:

> If two nonvertical lines are perpendicular, then the product of their slopes is −1.

In Problem 102, you are asked to prove the "if" part of the theorem; that is:

> If two nonvertical lines have slopes whose product is −1, then the lines are perpendicular.

FIGURE 67

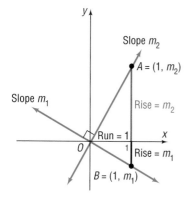

Proof Let m_1 and m_2 denote the slopes of the two lines. There is no loss in generality (that is, neither the angle nor the slopes are affected) if we situate the lines so that they meet at the origin. See Figure 67. The point $A = (1, m_2)$ is on the line having slope m_2, and the point $B = (1, m_1)$ is on the line having slope m_1. (Do you see why this must be true?)

Suppose that the lines are perpendicular. Then triangle OAB is a right triangle. As a result of the Pythagorean Theorem, it follows that

$$[d(O, A)]^2 + [d(O, B)]^2 = [d(A, B)]^2 \tag{5}$$

By the distance formula, we can write each of these distances as

$$[d(O, A)]^2 = (1 - 0)^2 + (m_2 - 0)^2 = 1 + m_2^2$$
$$[d(O, B)]^2 = (1 - 0)^2 + (m_1 - 0)^2 = 1 + m_1^2$$
$$[d(A, B)]^2 = (1 - 1)^2 + (m_2 - m_1)^2 = m_2^2 - 2m_1m_2 + m_1^2$$

Using these facts in equation (5), we get

$$(1 + m_2^2) + (1 + m_1^2) = m_2^2 - 2m_1m_2 + m_1^2$$

which, upon simplification, can be written as

$$m_1m_2 = -1$$

Thus, if the lines are perpendicular, the product of their slopes is −1. ∎

You may find it easier to remember the condition for two nonvertical lines to be perpendicular by observing that the equality $m_1m_2 = -1$ means that m_1 and m_2 are negative reciprocals of each other; that is, either $m_1 = -1/m_2$ or $m_2 = -1/m_1$.

E X A M P L E   10   Finding the Slope of a Line Perpendicular to Another Line

If a line has slope $\frac{3}{2}$, any line having slope $-\frac{2}{3}$ is perpendicular to it. ∎

MISSION POSSIBLE

Predicting The Future of Olympic Track Events

Some people think that women athletes are beginning to "catch up" to men in the Olympic Track Events. In fact, researchers at UCLA published data in 1992 that seemed to indicate that within 65 years the top men and women runners would be able to compete on an equal basis. Other researchers disagreed. Here are some data on the 200-meter run with winning times for the Olympics in the year given.

	Men's Time in Seconds	Women's Time in Seconds
1948	21.1	24.4
1952	20.7	23.7
1956	20.6	23.4
1960	20.5	24.0
1964	20.3	23.0
1968	19.83	22.5
1972	20.00	22.40
1976	20.23	22.37
1980	20.19	22.03
1984	19.80	21.81
1988	19.75	21.34
1992	19.73	21.72

(You may notice that the introduction of better timing devices meant more accurate measures starting in 1968 for the men and 1972 for the women.)

1.   To begin your investigation of the UCLA conjecture, make a graph of these data. To get the kind of accuracy you need, you should use graph paper. You will need to use only the first quadrant. On your x-axis place the Olympic years from 1948 to 2048, counting by 4's. On your y-axis place the numbers from 15 to 25, which represent the seconds; if you're using graph paper, allow about four squares per one second. Use X's to represent the men's times on the graph and O's to represent the women's times. These should form what we call a *scatter plot*.

2.   Using a ruler or straightedge, draw a line that represents roughly the slope and direction indicated by the men's scores. Do the same for the women's scores. Write a sentence or two explaining why you think your line is a good representation of the scatter plot.

3.   Next find the equation for each line, using your graph to estimate the slope and y-intercept for each. Try to be as accurate as possible, remembering your units and using the points where the line crosses intersections of the grid. The slope will be in seconds per year.

4.   Do your two lines appear to cross? In what year do they cross? If you solve the two equations algebraically, do you get the same answer?

5.   Some graphing calculators enable you do this problem in a statistics mode. They will plot the individual points that you type in and find a line that represents the data. Use your graphing calculator to find out whether its equation matches yours.

6.   Make a group decision about whether you think that women's times will "catch up" to men's times in the future. Write two to three sentences explaining why you believe they will or will not.

In Problems 31–54, find an equation for the line with the given properties. Express your answer using either the general form or the slope–intercept form of the equation of a line, whichever you prefer.

31. Slope = 3; passing through $(-2, 3)$

32. Slope = 2; passing through $(4, -3)$

33. Slope = $-\frac{2}{3}$; passing through $(1, -1)$

34. Slope = $\frac{1}{2}$; passing through $(3, 1)$

35. Passing through $(1, 3)$ and $(-1, 2)$

36. Passing through $(-3, 4)$ and $(2, 5)$

37. Slope = -3; y-intercept = 3

38. Slope = -2; y-intercept = -2

39. x-intercept = 2; y-intercept = -1

40. x-intercept = -4; y-intercept = 4

41. Slope undefined; passing through $(2, 4)$

42. Slope undefined; passing through $(3, 8)$

43. Parallel to the line $y = 2x$; passing through $(-1, 2)$

44. Parallel to the line $y = -3x$; passing through $(-1, 2)$

45. Parallel to the line $2x - y + 2 = 0$; passing through $(0, 0)$

46. Parallel to the line $x - 2y + 5 = 0$; passing through $(0, 0)$

47. Parallel to the line $x = 5$; passing through $(4, 2)$

48. Parallel to the line $y = 5$; passing through $(4, 2)$

49. Perpendicular to the line $y = \frac{1}{2}x + 4$; passing through $(1, -2)$

50. Perpendicular to the line $y = 2x - 3$; passing through $(1, -2)$

51. Perpendicular to the line $2x + y - 2 = 0$; passing through $(-3, 0)$

52. Perpendicular to the line $x - 2y + 5 = 0$; passing through $(0, 4)$

53. Perpendicular to the line $x = 8$; passing through $(3, 4)$

54. Perpendicular to the line $y = 8$; passing through $(3, 4)$

In Problems 55–74, find the slope and y-intercept of each line. Graph the line by hand. Check your graph using a graphing utility.

55. $y = 2x + 3$

56. $y = -3x + 4$

57. $\frac{1}{2}y = x - 1$

58. $\frac{1}{3}x + y = 2$

59. $y = \frac{1}{2}x + 2$

60. $y = 2x + \frac{1}{2}$

61. $x + 2y = 4$

62. $-x + 3y = 6$

63. $2x - 3y = 6$

64. $3x + 2y = 6$

65. $x + y = 1$

66. $x - y = 2$

67. $x = -4$

68. $y = -1$

69. $y = 5$

70. $x = 2$

71. $y - x = 0$

72. $x + y = 0$

73. $2y - 3x = 0$

74. $3x + 2y = 0$

75. Find an equation of the x-axis.

76. Find an equation of the y-axis.

In Problems 77–80, match each graph with the correct equation:

(a) $y = x$ (b) $y = 2x$ (c) $y = x/2$ (d) $y = 4x$

77.

78.

79.

80.

In Problems 81–86, write an equation of each line. Express your answer using either the general form or the slope–intercept form of the equation of a line, whichever you prefer.

81.

82.

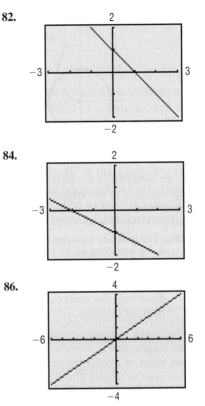

83.

84.

85.

86.

87. Measuring Temperature The relationship between Celsius (°C) and Fahrenheit (°F) degrees of measuring temperature is linear. Find an equation relating °C and °F if 0°C corresponds to 32°F and 100°C corresponds to 212°F. Use the equation to find the Celsius measure of 70°F.

88. Measuring Temperature The Kelvin (K) scale for measuring temperature is obtained by adding 273 to the Celsius temperature.
(a) Write an equation relating K and °C.
(b) Write an equation relating K and °F (see Problem 87).

89. Business: Computing Profit Each Sunday, a newspaper agency sells x copies of certain newspaper for $1.00 per copy. The cost to the agency of each newspaper is $0.50. The agency pays a fixed cost for storage, delivery, and so on, of $100 per Sunday.
(a) Write an equation that relates the profit P, in dollars, to the number x of copies sold. Graph this equation.
(b) What is the profit to the agency if 1000 copies are sold?
(c) What is the profit to the agency if 5000 copies are sold?

90. Business: Computing Profit Repeat Problem 89 if the cost to the agency is $0.45 per copy and the fixed cost is $125 per Sunday.

91. Cost of Electricity In 1991, Commonwealth Edison Company supplied electricity in the summer months to residential customers for a monthly customer charge of $9.06 plus 10.819¢ per kilowatt-hour supplied in the month.* Write an equation that relates the monthly charge C, in dollars, to the number x of kilowatt-hours used in the month. Graph this equation. What is the monthly charge for using 300 kilowatt-hours? For using 900 kilowatt-hours?

92. Show that an equation for a line with nonzero x- and y-intercepts can be written as

$$\frac{x}{a} + \frac{y}{b} = 1$$

where a is the x-intercept and b is the y-intercept. This is called the **intercept form** of the equation of a line.

*Source: Commonwealth Edison Co., Chicago, Illinois, 1991.

E X A M P L E   1   Distinguishing between Linear and Nonlinear Relations

Determine whether the relation between the two variables shown in each scatter diagram of Figure 70 is linear or nonlinear.

FIGURE 70

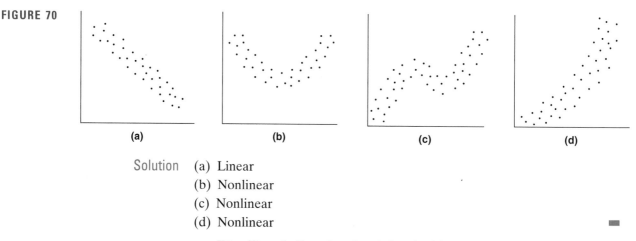

(a)   (b)   (c)   (d)

Solution   (a) Linear
(b) Nonlinear
(c) Nonlinear
(d) Nonlinear   ▬

We will study linearly related data in this section. Fitting an equation to nonlinear data will be discussed in later chapters.

Now work Problem 1.

Suppose that the scatter diagram of a set of data appears to be linearly related. One way to obtain an equation for such data is to draw a line by hand through two points on the scatter diagram and use the two points to obtain the equation of the line.

E X A M P L E   2   Finding an Equation for Linearly Related Data

A farmer collected the following data, which shows crop yields for various amounts of fertilizer used.

Plot	Fertilizer, x (Pounds/100ft²)	Yield, y (Bushels)
1	0	4
2	0	6
3	5	10
4	5	7
5	10	12
6	10	10
7	15	15
8	15	17
9	20	18
10	20	21
11	25	23
12	25	22

(a) Using a graphing utility, draw a scatter diagram of the data.

(b) Draw a line through two points on the scatter diagram and find an equation of the line.

(c) Graph the line on the scatter diagram.

Solution
(a) It is reasonable to expect that a relation exists between the amount of fertilizer used and crop yield. To draw a scatter diagram, we plot points, using fertilizer as the x-coordinate and yield as the y-coordinate. See Figure 71. From the scatter diagram, it appears that a linear relation exists between the amount of fertilizer used and yield.

FIGURE 71

(b) Select two points in the scatter diagram, say $(0, 4)$ and $(25, 23)$. (You should select your own two points and complete the solution.) The slope of the line joining the points $(0, 4)$ and $(25, 23)$ is

$$m = \frac{23 - 4}{25 - 0} = \frac{19}{25} = 0.76$$

FIGURE 72

The equation of the line with slope 0.76 and passing through $(0, 4)$ is found using the point–slope form with $m = 0.76$, $x_1 = 0$, and $y_1 = 4$:

$$y - y_1 = m(x - x_1)$$
$$y - 4 = 0.76(x - 0)$$
$$y = 0.76x + 4$$

(c) Figure 72 shows the scatter diagram with the graph of the line found in part (b). ∎

2

The line obtained in Example 2 depends on the selection of points, which will vary from person to person. So the line we found is probably different from the line you found. Although the line we found in Example 2 appears to "fit" the data well, there may be a line that "fits it better." Do you think your line "fits" the data better? Is there a line of "best fit"? As it turns out, there is a method for finding the line that best fits linearly related data.* This line is called the **line of best fit.**

Lines of Best Fit

E X A M P L E   3

Finding the Line of Best Fit

Use the data from Example 2 to find the line of best fit using a graphing utility.

Solution
Graphing utilities contain built-in programs that find the linear equation of "best fit" for a collection of points in a scatter diagram. (Look in your owner's manual under Linear Regression or Line of Best Fit for details on how to execute the program.) Upon executing the LINear REGression program, we obtain the results shown in Figure 73. The output the utility provides shows us the equation, $y = ax + b$, where a is the slope of the line and b is the y-intercept. The line of best fit that relates fertilizer and yield is

FIGURE 73

$$\text{Yield} = 0.717(\text{fertilizer}) + 4.79$$

*We shall not discuss in this book the underlying mathematics of lines of best fit. Most books in statistics and many in linear algebra discuss this topic.

FIGURE 74

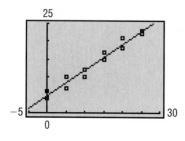

Figure 74 shows the graph of the line of best fit, along with the scatter diagram.

Does the line of best fit appear to be a good "fit"? In other words, does the line appear to accurately describe the relation between yield and fertilizer?

And just how good is this line of best fit? How accurately does this line describe the relationship between yield and fertilizer? The answers are given by what is called the *correlation coefficient.*

 Now work Problem 7.

Correlation Coefficients

Look again at Figure 73. The last line of output is $r = 0.98$. This number, called the **correlation coefficient, r,** $-1 < r < 1$, is a measure of the strength of the *linear relation* that exists between two variables. The closer $|r|$ is to 1, the more perfect the linear relationship is. If r is close to 0, there is little or no *linear* relationship between the variables. A negative value of r, $r < 0$, indicates that as x increases y decreases (negative slope); a positive value of r, $r > 0$, indicates that as x increases, y does also (positive slope). Thus, the data given in Example 3, having a correlation coefficient of 0.98, are indicative of a strong linear relationship with a positive slope.

Figure 75, on page 59, illustrates a variety of scatter diagrams and the relations that they suggest.

Prediction and Accuracy

3 Once an equation has been obtained, the equation can be used to predict values for the dependent variable for a given independent variable.

E X A M P L E   4   Prediction

Use the results of Example 3 to estimate the yield if the farmer uses 17 pounds of fertilizer/100 square feet.

Solution   To determine the yield, we substitute the value $x = 17$ pounds into the equation found in Example 3.

$$\text{Yield} = 0.717(17) + 4.79 \approx 17 \text{ bushels}$$

So, 17 pounds of fertilizer per 100 square feet will, on average, produce a yield of 17 bushels. Notice that we rounded our prediction to the nearest whole number. This is because our data are measured to the nearest whole number as well. Our predictions cannot be more precise than our data.

E X A M P L E   5   Interpreting the Slope

Use the equation of the line found in Example 3 and interpret the slope.

Solution   The slope of the line is 0.717. This can be interpreted as follows: For every 1 pound per square foot increase in fertilizer, the yield is expected to increase by 0.717 bushels.

FIGURE 75

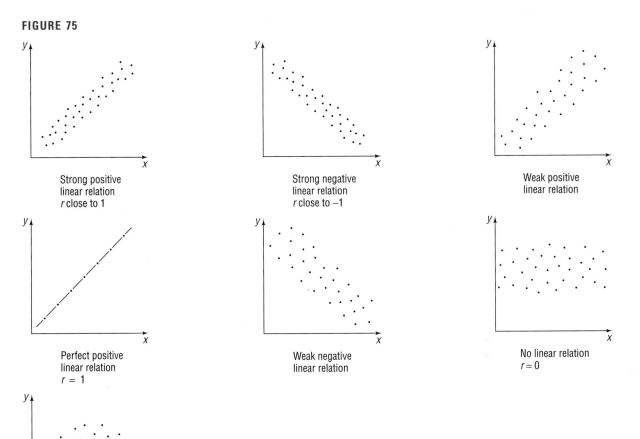

It is important that predictions are made within the scope of the model. That is, we can only make predictions regarding this model for $0 \le$ fertilizer ≤ 25, since this is the range for which we have observable data. Unless we collect additional data for $x > 25$, we can't make predictions for $x > 25$. The reason should be clear. It is not clear whether adding more fertilizer will continue to increase crop yield. In fact, it is conceivable that adding more fertilizer may actually reduce crop yield. ▬

1.4   EXERCISES

In Problems 1–6, examine the scatter diagram and determine whether the type of relation, if any, that may exist is linear or nonlinear.

1.

2.

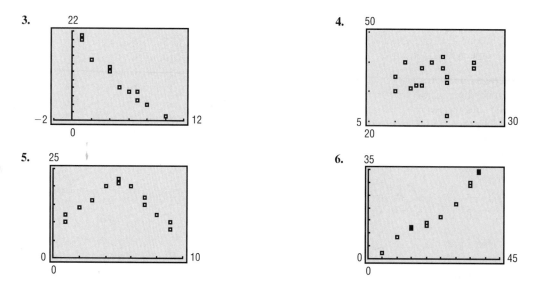

3. [scatter diagram, y-axis 22 to 0, x-axis −2 to 12]

4. [scatter diagram, y-axis 50 to 5, x-axis 20 to 30]

5. [scatter diagram, y-axis 25 to 0, x-axis 0 to 10]

6. [scatter diagram, y-axis 35 to 0, x-axis 0 to 45]

In Problems 7–14, (a) use a graphing utility to draw a scatter diagram; (b) select two points from the scatter diagram and find the equation of the line through the points selected; (c) use a graphing utility to graph the line found in (b) on the scatter diagram; (d) use a graphing utility to find the line of best fit; and (e) use a graphing utility to graph the line of best fit on the scatter diagram.

7.

x	3	4	5	6	7	8	9
y	4	6	7	10	12	14	16

8.

x	3	5	7	9	11	13
y	0	2	3	6	9	11

9.

x	−2	−1	0	1	2
y	−4	0	1	4	5

10.

x	−2	−1	0	1	2
y	7	6	3	2	0

11.

x	20	30	40	50	60
y	100	95	91	83	70

12.

x	5	10	15	20	25
y	2	4	7	11	18

13.

x	−20	−17	−15	−14	−10
y	100	120	118	130	140

14.

x	−30	−27	−25	−20	−14
y	10	12	13	13	18

15. **Consumption and Disposable Income** An economist wishes to estimate a line that relates personal consumption expenditures (C) and disposable income (I). Both C and I are in thousands of dollars. She interviews eight heads of households with families of size 4 and obtains the following data:

C (000)	I (000)
16	20
18	20
13	18
21	27
27	36
26	37
36	45
39	50

Let I represent the independent variable and C the dependent variable.
(a) Use a graphing utility to draw a scatter diagram.
(b) Use a graphing utility to find the line of best fit to the data.
(c) Interpret the slope. The slope of this line is called the **marginal propensity to consume.**
(d) Predict the consumption of a family whose disposable income is $42 thousand.

16. **Marginal Propensity to Save** The same economist as the one in Problem 15 wants to estimate a line that relates savings (S) and disposable income (I). Let $S = I − C$ be the dependent variable and I the independent variable. The slope of this line is called the **marginal propensity to save.**
(a) Use a graphing utility to draw a scatter diagram.
(b) Use a graphing utility to find the line of best fit to the data.

(c) Interpret the slope.

(d) Predict the savings of a family whose income is $42 thousand.

17. Average Speed of a Car Brian wanted to determine the relation that might exist between speed and miles per gallon of an automobile. Let X be the average speed of a car on the highway measured in miles per hour and let Y represent the miles per gallon of the automobile. The following data are collected:

X	50	55	55	60	60	62	65	65
Y	28	26	25	22	20	20	17	15

(a) Use a graphing utility to draw a scatter diagram.

(b) Use a graphing utility to find the line of best fit to the data.

(c) Interpret the slope.

(d) Predict the miles per gallon of a car traveling 61 miles per hour.

18. Height versus Weight A doctor wished to determine whether a relation exists between the height of a female and her weight. She obtained the heights and weights of 10 females aged 18 to 24. Let height be the independent variable, X, measured in inches, and weight be the dependent variable, Y, measured in pounds.

X (height)	60	61	62	62	64	65	65	67	68	68
Y (weight)	105	110	115	120	120	125	130	135	135	145

(a) Use a graphing utility to draw a scatter diagram.

(b) Use a graphing utility to find the line of best fit to the data.

(c) Interpret the slope.

(d) Predict the weight of a female aged 18 to 24 whose height is 66 inches.

19. Sales Data versus Income The following data represent sales and net income before taxes (both are in billions of dollars) for all manufacturing firms within the United States for 1980–1989. Treat sales as the independent variable and net income before taxes as the dependent variable.

(a) Use a graphing utility to draw a scatter diagram.

(b) Use a graphing utility to find the line of best fit to the data.

(c) Interpret the slope.

(d) Predict the net income before taxes of manufacturing firms in 1990 if sales are $2456.4 billion.

Year	Sales	Net Income Before Taxes
1980	1912.8	92.6
1981	2144.7	101.3
1982	2039.4	70.9
1983	2114.3	85.8
1984	2335.0	107.6
1985	2331.4	87.6
1986	2220.9	83.1
1987	2378.2	115.6
1988	2596.2	154.6
1989	2745.1	136.3

Source: Economic Report of the President, February 1995.

20. Employment and the Labor Force The following data represent the civilian labor force (people aged 16 years and older, excluding those serving in the military) and the number of employed people in the United States for the years 1981–1991. Treat the size of the labor force as the independent variable and the number employed as the dependent variable. Both the size of the labor force and the number employed are measured in thousands of people.

Year	Civilian Labor Force	Number Employed
1981	108,670	100,397
1982	110,204	99,526
1983	111,550	100,834
1984	113,544	105,005
1985	115,461	107,150
1986	117,834	109,597
1987	119,865	112,440
1988	121,669	114,968
1989	123,869	117,342
1990	124,787	117,914
1991	125,303	116,877

Source: Business Statistics, 1963–1991, U.S. Department of Commerce, Economics and Statistics Administration, Bureau of Economic Analysis, June 1992.

(a) Use a graphing utility to draw a scatter diagram.

(b) Use a graphing utility to find the line of best fit to the data.

    (c) Interpret the slope.
    (d) Predict the number of employed if the civilian labor force is 122,340 thousand people.

21. Disposable Personal Income Use the data from Problem 69, Section 1.1.
    (a) With a graphing utility, find the line of best fit, letting per capita disposable income represent the independent variable.
    (b) Interpret the slope.
    (c) Predict per capita personal consumption if per capita disposable income is $14,989.

22. Natural Gas versus Coal Use the data from Problem 70, Section 1.1.
    (a) With a graphing utility, find the line of best fit, letting the amount of natural gas represent the independent variable.
    (b) Interpret the slope.
    (c) Predict the amount of energy provided by coal if the amount of energy provided by natural gas is 67 trillion BTU.

23. Fuel Consumption Use the data from Problem 71, Section 1.1.
    (a) With a graphing utility, find the line of best fit, letting the average fuel consumption per car represent the independent variable.
    (b) Interpret the slope.
    (c) Predict the pollutant emissions of carbon monoxide if the average fuel consumption per car is 505 gallons.

24. Hourly Pay and Productivity Use the data from Problem 72, Section 1.1.
    (a) With a graphing utility, find the line of best fit, letting productivity represent the independent variable.
    (b) Interpret the slope.
    (c) Predict average hourly earnings if productivity is 101.3.

1.5 SOLVING EQUATIONS

  1 Solve Equations Algebraically
  2 Solve Equations Using a Graphing Utility
  3 Solve Quadratic Equations by Graphing and by Factoring
  4 Solve Quadratic Equations by Graphing and by the Quadratic Formula

1 An **equation in one variable** is a statement in which two expressions, at least one containing the variable, are equal. The expressions are called the **sides** of the equation. Since an equation is a statement, it may or may not be true, depending on the value of the variable. Unless otherwise restricted, the admissible values of the variable are those in the domain of the variable. Those admissible values of the variable, if any, that result in a true statement are called **solutions,** or **roots,** of the equation. To **solve an equation** means to find all the solutions of the equation.

For example, the following are all equations in one variable, x:

$$x + 5 = 9 \qquad x^2 + 5x = 2x - 2 \qquad \frac{x^2 - 4}{x + 1} = 0 \qquad x^2 + 9 = 5$$

The first of these statements, $x + 5 = 9$, is true when $x = 4$ and false for any other choice of x. Thus, 4 is a solution of the equation $x + 5 = 9$. We also say that 4 **satisfies** the equation $x + 5 = 9$, because, when x is replaced by 4, a true statement results.

Sometimes an equation will have more than one solution. For example, the equation

$$\frac{x^2 - 4}{x + 1} = 0$$

has either $x = -2$ or $x = 2$ as a solution.

Sometimes we will write the solutions of an equation in set notation. This set is called the **solution set** of the equation. For example, the solution set of the equation $x^2 - 9 = 0$ is $\{-3, 3\}$.

Unless indicated otherwise, we will limit ourselves to real solutions. Some equations have no real solution. For example, $x^2 + 9 = 5$ has no real solution, because there is no real number whose square when added to 9 equals 5.

An equation that is satisfied for every choice of the variable for which both sides are defined is called an **identity.** For example, the equation

$$3x + 5 = x + 3 + 2x + 2$$

is an identity, because this statement is true for any real number x.

Two or more equations that have precisely the same solutions are called **equivalent equations.** For example, all the following equations are equivalent, because each has only the solution $x = 5$.

$$2x + 3 = 13$$
$$2x = 10$$
$$x = 5$$

These three equations illustrate one method for solving many types of equations: Replace the original equation by an equivalent equation, and continue until an equation with an obvious solution, such as $x = 5$, is reached. The question, though, is: "How do I obtain an equivalent equation?" In general, there are five ways to do so.

Procedures That Result in Equivalent Equations

1. Interchange the two sides of the equation:

   Replace $3 = x$ by $x = 3$

2. Simplify the sides of the equation by combining like terms, eliminating parentheses, and so on:

   Replace $(x + 2) + 6 = 2x + (x + 1)$
   by $x + 8 = 3x + 1$

3. Add or subtract the same expression on both sides of the equation:

   Replace $3x - 5 = 4$
   by $(3x - 5) + 5 = 4 + 5$

4. Multiply or divide both sides of the equation by the same nonzero expression:

   Replace $\dfrac{3x}{x-1} = \dfrac{6}{x-1}$ $x \neq 1$

   by $\dfrac{3x}{x-1} \cdot (x-1) = \dfrac{6}{x-1} \cdot (x-1)$

5. If one side of the equation is 0 and the other side can be factored, then we may use the zero-product law* and set each factor equal to 0:

   Replace $x(x - 3) = 0$
   by $x = 0$ or $x - 3 = 0$

*The zero-product law states that if $ab = 0$ then $a = 0$, or $b = 0$, or both equal 0.

Whenever it is possible to solve an equation in your head, do so. For example:

The solution of $2x = 8$ is $x = 4$.
The solution of $3x - 15 = 0$ is $x = 5$.

Often, though, some rearrangement is necessary.

E X A M P L E 1

Solving an Equation

Solve the equation: $(x + 1)(2x) = (x + 1)(2)$

Solution   We begin by collecting all terms on the left side:

$$(x + 1)(2x) = (x + 1)(2)$$
$$(x + 1)(2x) - (x + 1)(2) = 0$$
$$(x + 1)(2x - 2) = 0 \quad \text{Factor.}$$
$$x + 1 = 0 \quad \text{or} \quad 2x - 2 = 0 \quad \text{Apply the zero-product law.}$$
$$x = -1 \qquad\qquad 2x = 2$$
$$x = 1$$

The solution set is $\{-1, 1\}$.

E X A M P L E 2

Solving an Equation

Solve the equation: $\dfrac{3x}{x - 1} + 2 = \dfrac{3}{x - 1}$

Solution   First, we note that the domain of the variable is $\{x \mid x \neq 1\}$. Since the two quotients in the equation have the same denominator, $x - 1$, we can simplify by multiplying both sides by $x - 1$. The resulting equation is equivalent to the original equation, since we are multiplying by $x - 1$, which is not 0 (remember, $x \neq 1$).

$$\frac{3x}{x - 1} + 2 = \frac{3}{x - 1}$$

$$\left(\frac{3x}{x - 1} + 2\right) \cdot (x - 1) = \frac{3}{x - 1} \cdot (x - 1) \qquad \begin{array}{l}\text{Multiply both sides by } x - 1;\\ \text{cancel on the right.}\end{array}$$

$$\frac{3x}{x - 1} \cdot (x - 1) + 2 \cdot (x - 1) = 3 \qquad \begin{array}{l}\text{Use the distributive property on}\\ \text{the left side; cancel on the left.}\end{array}$$

$$3x + (2x - 2) = 3 \qquad \text{Simplify.}$$
$$5x - 2 = 3$$
$$5x = 5 \qquad \text{Add 2 to each side.}$$
$$x = 1 \qquad \text{Divide both sides by 5.}$$

The solution appears to be 1. But recall that $x = 1$ is not in the domain of the variable. Thus, the equation has no solution.

Now work Problem 27.

Steps for Solving Equations

> STEP 1: List any restrictions on the domain of the variable.
> STEP 2: Simplify the equation by replacing the original equation by a succession of equivalent equations following the procedures listed earlier.
> STEP 3: If the result of STEP 2 is a product of factors equal to 0, use the zero-product law and set each factor equal to 0 (procedure 5).
> STEP 4: Check your solution(s).

E X A M P L E   3

Solving an Equation

Solve the equation: $x^3 = 25x$

Solution

We first rearrange the equation to get 0 on the right side:

$$x^3 = 25x$$
$$x^3 - 25x = 0$$

We notice that x is a factor of each term on the left:

$$x(x^2 - 25) = 0$$
$$x(x + 5)(x - 5) = 0 \qquad \text{Difference of two squares.}$$
$$x = 0 \quad \text{or} \quad x + 5 = 0 \quad \text{or} \quad x - 5 = 0 \qquad \text{Set each factor equal to 0.}$$
$$x = -5 \quad \text{or} \qquad x = 5 \qquad \text{Solve.}$$

The solution set is $\{-5, 0, 5\}$.

Now work Problem 29.

Because there are two points whose distance from the origin is 5 units, -5 and 5, the equation $|x| = 5$ will have the solution set $\{-5, 5\}$.

E X A M P L E   4

Solving an Equation Involving Absolute Value

Solve the equation: $|x + 4| = 13$

Solution

There are two possibilities:

$$x + 4 = 13 \quad \text{or} \quad x + 4 = -13$$
$$x = 9 \qquad\qquad x = -17$$

Thus, the solution set is $\{-17, 9\}$.

Now work Problem 47.

Solving Equations Using a Graphing Utility

In solving the equation $2x = 6$, we are able to obtain an exact solution, $x = 3$. To solve the equation $x^3 = 25x$ required some additional work (see Example 3), but, again, exact solutions were obtained.

We shall see as we proceed through this book that some equations can be solved using algebraic techniques that result in exact solutions being obtained. Whenever algebraic techniques exist, we shall discuss them and use them to obtain exact solutions.

For many equations, though, there are no algebraic techniques that lead to a solution. For such equations, a graphing utility can often be used to investigate possible solutions. When a graphing utility is used to solve an equation, usually *approximate* solutions are obtained. Unless otherwise stated, we shall follow the practice of giving approximate solutions as decimals *correct to two decimal places.*

Let's look at an example that explains this terminology.

EXAMPLE 5   Writing a Number Correct to n Decimal Places

Writing the Number	1/7	$\sqrt{2}$	π
Correct to 1 decimal place	0.1	1.4	3.1
Correct to 2 decimal places	0.14	1.41	3.14
Correct to 3 decimal places	0.142	1.414	3.141
Correct to 4 decimal places	0.1428	1.4142	3.1415
Correct to 5 decimal places	0.14285	1.41421	3.14159
Correct to 6 decimal places	0.142857	1.414213	3.141592

As the example illustrates, **correct to n decimal places** means the decimal that results from truncation after the nth decimal.

We learned in Section 1.2 that the x-intercepts, if any, of an equation in two variables x and y are found by letting $y = 0$ and solving the equation for x. Thus, solving an equation for x is equivalent to finding the x-intercepts of the graph of the corresponding equation. It is this relationship between the solutions of an equation and the graph of the corresponding equation that enables us to use a graphing utility to solve equations.

EXAMPLE 6   Using a Graphing Utility to Approximate Solutions of an Equation

Find the smaller of the two solutions of the equation $x^2 - 6x + 7 = 0$. Express the answer correct to two decimal places.

Solution   The solutions of the equation $x^2 - 6x + 7 = 0$ are the same as the x-intercepts of the graph of $Y_1 = x^2 - 6x + 7$. We begin by graphing the equation using a scale of 1 on each axis.

Figure 76 shows the graph.

FIGURE 76

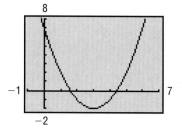

The smaller of the two x-intercepts (solutions of the equation) lies between 1 and 2. (Remember that Xscl = 1.) Thus, correct to 0 decimal places, the smaller solution of the equation is $x = 1$.

Next, we BOX the graph from approximately $x = 1$ to $x = 2$ and from $y = -1$ to $y = 1$. See Figure 77. Adjust the RANGE setting so that Xscl = 0.1 and Yscl = 0.1. Graph Y_1 again. See Figure 78.

FIGURE 77   **FIGURE 78**

The x-intercept lies between 1.5 and 1.6. (Remember that the BOX was from approximately $x = 1$ to $x = 2$ and Xscl = 0.1.) Thus, correct to one decimal place, the smaller solution of the equation is $x = 1.5$.

Next, BOX again from $x = 1.5$ to $x = 1.6$ and from $y = -0.1$ to $y = 0.1$. See Figure 79. Adjust the RANGE so that Xscl = 0.01 and Yscl = 0.01 and graph Y_1. See Figure 80.

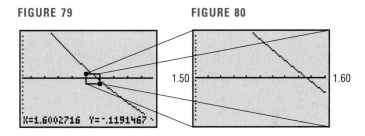

FIGURE 79   **FIGURE 80**

The x-intercept lies between 1.58 and 1.59. Remember that the BOX was from $x = 1.5$ to $x = 1.6$ with Xscl = 0.01. Thus, correct to two decimal places, the smaller of the two solutions is $x = 1.58$.   ▬

The steps to follow for approximating solutions of equations correct to any desired number of decimals are given next.

Steps for Approximating Solutions of Equations

STEP 1: Write the equation in the form {expression in x} = 0.
STEP 2: Graph Y_1 = {expression in x}, using Xscl = 1.
STEP 3: BOX the graph so that the x-intercept is within the BOX. Adjust Xscl and Yscl to one-tenth of their current value and graph Y_1 again.
STEP 4: If additional accuracy is desired, repeat STEP 3.

In these steps, each repetition of Step 3 gives one more decimal place of accuracy. Thus, when Xscl = 0.01, the x-intercept will be known correct to two decimal places. Let's look at another example.

E X A M P L E 7

Use a Graphing Utility to Approximate the Solution(s) of the Equation

$$x^3 - 1.5x^2 + x - 1.2 = 0$$

Solution

STEP 1: The equation is already in the proper form.
STEP 2: Graph the equation

$$Y_1 = x^3 - 1.5x^2 + x - 1.2$$

FIGURE 81

Figure 81 shows the graph. Notice that the scale is 1 on each axis. Also notice that the x-intercept lies between 1 and 2.

STEP 3: Figure 82 shows the box chosen. Notice that its width is about 1 unit and its height is about 2 units. Change Xscl = 0.1, Yscl = 0.1 and graph.

Figure 83 shows the result. Xscl and Yscl have both been changed to 0.1. From Figure 83 we see that the x-intercept is near 1.4.

FIGURE 82   **FIGURE 83**

STEP 4: We repeat Step 3, using a BOX from approximately $x = 1.38$ to $x = 1.41$ and changing Xscl = 0.01 and Yscl = 0.01. This gives us Figure 84. The x-intercept is 1.39, correct to two decimal places.

FIGURE 84

FIGURE 85

If your graphing utility has a ZERO or ROOT command, you can use it to find the solution of an equation. Using the ZERO command on a TI-83, the solution of the equation $x^3 - 1.5x^2 + x - 1.2 = 0$ (Example 7) is given as $x = 1.3985052$, which is correct to six decimal places. See Figure 85.

Now work Problem 113.

Note: If the x-intercept that you are trying to approximate appears to fall on a tick mark, then you should evaluate Y_1 at that tick mark. If the value is zero, then there is an exact solution. If the value is not zero, then you know whether the graph is above or below the x-axis at the tick mark by noting the sign at the tick mark.

Quadratic Equations

Quadratic equations are equations such as

$$2x^2 + x + 8 = 0 \qquad 3x^2 - 5x + 6 = 0 \qquad x^2 - 9 = 0$$

A general definition is given next.

A **quadratic equation** is an equation equivalent to one of the form

$$ax^2 + bx + c = 0 \tag{1}$$

where a, b, and c are real numbers and $a \neq 0$.

A quadratic equation written in the form $ax^2 + bx + c = 0$ is said to be in **standard form.**

Sometimes, a quadratic equation is called a **second-degree equation,** because the left side is a polynomial of degree 2. We shall discuss two ways of solving quadratic equations algebraically: by factoring and by use of the quadratic formula. We shall also solve quadratic equations using a graphing utility.

When a quadratic equation is written in standard form, $ax^2 + bx + c = 0$, it may be possible to factor the expression on the left side as the product of two first-degree polynomials. Then, by setting each factor equal to 0 and solving the resulting linear equations, we obtain the solutions of the quadratic equation.

Let's look at an example.

E X A M P L E 8   Solving a Quadratic Equation by Graphing and by Factoring

Solve the equation: $x^2 = 12 - x$

Solution   We put the equation in standard form by adding $x - 12$ to each side:

$$x^2 = 12 - x$$
$$x^2 + x - 12 = 0$$

FIGURE 86   Graphing Solution   Figure 86 shows the graph of the equation

$$Y_1 = x^2 + x - 12$$

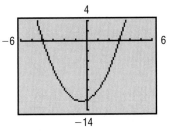

From the graph it appears that there are two x-intercepts, one near -4, the other near 3. We conjecture that the equation has two x-intercepts: -4 and 3. This is verified by eVALUEating Y_1 at -4 and at 3. Since the VALUE of Y_1 at both -4 and 3 is zero, the solution set is $\{-4, 3\}$.

Algebraic Solution The left side of the equation in standard form,

$$x^2 + x - 12 = 0$$

may now be factored as

$$(x + 4)(x - 3) = 0$$

so that

$$x + 4 = 0 \quad \text{or} \quad x - 3 = 0$$
$$x = -4 \qquad\qquad x = 3$$

The solution set is $\{-4, 3\}$.

When the left side factors into two linear equations with the same solution, the quadratic equation is said to have a **repeated solution.** We also call this solution a **root of multiplicity 2,** or a **double root.**

E X A M P L E   9   Solving a Quadratic Equation by Graphing and by Factoring

Solve the equation: $x^2 - 6x + 9 = 0$

Graphing Solution Figure 87 shows the graph of the equation

$$Y_1 = x^2 - 6x + 9$$

From the graph it appears that there is one x-intercept, 3. Since the VALUE of Y_1 at $x = 3$ is zero, the equation has only the repeated solution 3.

FIGURE 87

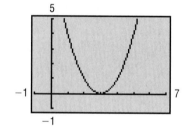

Algebraic Solution This equation is already in standard form, and the left side can be factored:

$$x^2 - 6x + 9 = 0$$
$$(x - 3)(x - 3) = 0$$

so

$$x = 3 \quad \text{or} \quad x = 3$$

This equation has only the repeated solution 3.

Now work Problems 69 and 75.

Quadratic equations can also be solved by using the *quadratic formula.* See the Appendix, Section 2, for the derivation.

Theorem   Quadratic Formula

If $b^2 - 4ac \geq 0$, the real solution(s) of the quadratic equation

$$ax^2 + bx + c = 0 \qquad a \neq 0 \qquad (2)$$

is (are) given by the **quadratic formula:**

$$x = \frac{-b \pm \sqrt{b^2 - 4ac}}{2a} \qquad (3)$$

The quantity $b^2 - 4ac$ is called the **discriminant** of the quadratic equation, because its value tells us whether the equation has real solutions. In fact, it also tells us how many solutions to expect.

Discriminant of a Quadratic Equation

For a quadratic equation $ax^2 + bx + c = 0$,

1. If $b^2 - 4ac > 0$, there are two unequal real solutions.
2. If $b^2 - 4ac = 0$, there is a repeated real solution, a root of multiplicity two.
3. If $b^2 - 4ac < 0$, there is no real solution.*

Thus, when asked to find the real solutions, if any, of a quadratic equation, always evaluate the discriminant first to see how many real solutions there are.

E X A M P L E   10   Solving a Quadratic Equation by Graphing and by Using the Quadratic Formula

Find the real solutions, if any, of the equation: $3x^2 - 5x + 1 = 0$

Solution   The equation is in standard form, so we compare it to $ax^2 + bx + c = 0$ to find a, b, and c:

$$3x^2 - 5x + 1 = 0$$
$$ax^2 + bx + c = 0$$

With $a = 3$, $b = -5$, and $c = 1$, we evaluate the discriminant $b^2 - 4ac$:

$$b^2 - 4ac = (-5)^2 - 4(3)(1) = 25 - 12 = 13$$

Since $b^2 - 4ac > 0$, there are two real solutions.

*We shall consider quadratic equations whose discriminant is negative in Chapter 3.

FIGURE 88

Graphing Solution   Figure 88 shows the graph of the equation

$$Y_1 = 3x^2 - 5x + 1$$

As expected, we see that there are two x-intercepts: one between 0 and 1, the other between 1 and 2. The solutions to the equation are 0.23 and 1.43 correct to two decimal places.

Algebraic Solution   We use the quadratic formula with $a = 3$, $b = -5$, $c = 1$, and $b^2 - 4ac = 13$.

$$x = \frac{-b \pm \sqrt{b^2 - 4ac}}{2a} = \frac{5 \pm \sqrt{13}}{6}$$

The solution set is $\{(5 - \sqrt{13})/6, (5 + \sqrt{13})/6\}$. These solutions are exact. ∎

 Now work Problem 93.

E X A M P L E   11   Solving Quadratic Equations by Graphing and by Using the Quadratic Formula

Find the real solutions, if any, of the equation:

$$\tfrac{25}{2}x^2 - 30x + 18 = 0$$

Solution   The equation is given in standard form. However, to simplify the arithmetic, we clear the fractions by multiplying both sides of the equation by 2:

$$\tfrac{25}{2}x^2 - 30x + 18 = 0$$
$$25x^2 - 60x + 36 = 0 \quad \text{Clear fractions.}$$
$$ax^2 + bx + c = 0 \quad \text{Compare to standard form.}$$

With $a = 25$, $b = -60$, and $c = 36$, we evaluate the discriminant:

$$b^2 - 4ac = (-60)^2 - 4(25)(36) = 3600 - 3600 = 0$$

The equation has a repeated solution.

Graphing Solution   Figure 89 shows the graph of the equation

$$Y_1 = \tfrac{25}{2}x^2 - 30x + 18$$

From the graph, we see that the one x-intercept is between 1 and 2. The solution to the equation is 1.2.

FIGURE 89

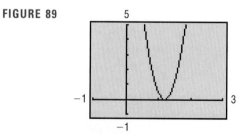

Algebraic Solution   We use $a = 25$, $b = -60$, $c = 36$, and $b^2 - 4ac = 0$ to solve

$$25x^2 - 60x + 36 = 0$$

Using the quadratic formula, we find

$$x = \frac{-b \pm \sqrt{b^2 - 4ac}}{2a} = \frac{60 \pm \sqrt{0}}{50} = \frac{60}{50} = \frac{6}{5}$$

The repeated solution is $\frac{6}{5}$, which is exact.

E X A M P L E   12   **Solving Quadratic Equations by Graphing and by Using the Quadratic Formula**

Find the real solutions, if any, of the equation:

$$3x^2 + 2 = 4x$$

Solution   The equation, as given, is not in standard form.

$$3x^2 + 2 = 4x$$
$$3x^2 - 4x + 2 = 0 \quad \text{Put in standard form.}$$
$$ax^2 + bx + c = 0 \quad \text{Compare to standard form.}$$

With $a = 3$, $b = -4$, and $c = 2$, we find

$$b^2 - 4ac = 16 - 24 = -8$$

Since $b^2 - 4ac < 0$, the equation has no real solution.

Graphing Solution   We use the standard form of the equation and graph the equation

$$Y_1 = 3x^2 - 4x + 2$$

See Figure 90. We see that there are no x-intercepts, so the equation has no real solution, as expected.

FIGURE 90

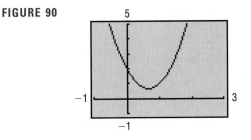

Sometimes a given equation can be transformed into a quadratic equation so that it can be solved using the quadratic formula.

E X A M P L E   13   **Solving Quadratic Equations by Graphing and by Using the Quadratic Formula**

Find the real solutions, if any, of the equation: $9 + \dfrac{3}{x} - \dfrac{2}{x^2} = 0$, $x \neq 0$

Graphing Solution Figure 91 shows the graph of the equation

$$Y_1 = 9 + \frac{3}{x} - \frac{2}{x^2}$$

From the graph, we conjecture that there are two x-intercepts, one between -1 and 0, the other between 0 and 1. The solutions to the equation are -0.66 and 0.33 correct to two decimal places.

FIGURE 91

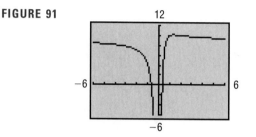

Comment: The graph of the preceding equation on most graphing utilities will appear ragged near $x = 0$. This is caused because the equation is not defined for $x = 0$. This problem can be easily overcome. Simply multiply both sides of the equation by x^2 and obtain $Y_2 = 9x^2 + 3x - 2$. The resulting equation will have the same x-intercepts as the original equation. See Figure 92.

FIGURE 92

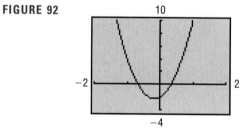

Algebraic Solution In its present form, the equation

$$9 + \frac{3}{x} - \frac{2}{x^2} = 0$$

is not a quadratic equation. However, it can be transformed into one by multiplying each side by x^2. The result is

$$9x^2 + 3x - 2 = 0$$

Although we multiplied each side by x^2, we know that $x^2 \neq 0$ (do you see why?), so this quadratic equation is equivalent to the original equation.

Using $a = 9$, $b = 3$, and $c = -2$, the discriminant is

$$b^2 - 4ac = 9 + 72 = 81$$

Since $b^2 - 4ac > 0$, the new equation has two real solutions:

$$x = \frac{-b \pm \sqrt{b^2 - 4ac}}{2a} = \frac{-3 \pm \sqrt{81}}{18} = \frac{-3 \pm 9}{18}$$

$$x = \frac{-3 + 9}{18} = \frac{6}{18} = \frac{1}{3} \quad \text{or} \quad x = \frac{-3 - 9}{18} = \frac{-12}{18} = \frac{-2}{3}$$

The solution set is $\{-\frac{2}{3}, \frac{1}{3}\}$. These are the exact solutions.

SUMMARY

Procedure for Solving a
Quadratic Equation
Algebraically

To solve a quadratic equation, first put it in standard form:

$$ax^2 + bx + c = 0$$

Then:

STEP 1: Identify a, b, and c.

STEP 2: Evaluate the discriminant $b^2 - 4ac$.

STEP 3: (a) If the discriminant is negative, the equation has no real solution.

(b) If the discriminant is nonnegative, determine whether the left side can be factored. If you can easily spot factors, use the factoring method to solve the equation. Otherwise, use the quadratic formula.

1.5 EXERCISES

In Problems 1–8, mentally solve each equation.

1. $3x = 21$

2. $3x = -24$

3. $5x + 15 = 0$

4. $3x + 18 = 0$

5. $2x - 3 = 5$

6. $3x + 4 = -8$

7. $\frac{1}{3}x = \frac{5}{12}$

8. $\frac{2}{3}x = \frac{9}{2}$

In Problems 9–80, solve each equation algebraically.

9. $6 - x = 2x + 9$

10. $3 - 2x = 2 - x$

11. $2(3 + 2x) = 3(x - 4)$

12. $3(2 - x) = 2x - 1$

13. $8x - (2x + 1) = 3x - 10$

14. $5 - (2x - 1) = 10$

15. $\frac{1}{2}x - 4 = \frac{3}{4}x$

16. $1 - \frac{1}{2}x = 5$

17. $0.9t = 0.4 + 0.1t$

18. $0.9t = 1 + t$

19. $\frac{2}{y} + \frac{4}{y} = 3$

20. $\frac{4}{y} - 5 = \frac{5}{2y}$

21. $(x + 7)(x - 1) = (x + 1)^2$

22. $(x + 2)(x - 3) = (x - 3)^2$

23. $x(2x - 3) = (2x + 1)(x - 4)$

24. $x(1 + 2x) = (2x - 1)(x - 2)$

25. $z(z^2 + 1) = 3 + z^3$

26. $w(4 - w^2) = 8 - w^3$

27. $\frac{x}{x - 3} + 3 = \frac{3}{x - 3}$

28. $\frac{3x}{x + 2} = \frac{-6}{x + 2} - 2$

29. $x^2 = 9x$

30. $x^3 = x^2$

31. $t^3 - 9t^2 = 0$

32. $4z^3 - 8z^2 = 0$

33. $\frac{2x}{x^2 - 4} = \frac{4}{x^2 - 4} - \frac{1}{x + 2}$

34. $\frac{x}{x^2 - 9} + \frac{1}{x + 3} = \frac{3}{x^2 - 9}$

35. $\frac{x}{x + 2} = \frac{1}{2}$

36. $\frac{3x}{x - 1} = 2$

37. $\frac{3}{2x - 3} = \frac{2}{x + 5}$

38. $\frac{-2}{x + 4} = \frac{-3}{x + 1}$

39. $(x + 2)(3x) = (x + 2)(6)$

40. $(x - 5)(2x) = (x - 5)(4)$

41. $\frac{6t + 7}{4t - 1} = \frac{3t + 8}{2t - 4}$

42. $\frac{8w + 5}{10w - 7} = \frac{4w - 3}{5w + 7}$

43. $\frac{2}{x - 2} = \frac{3}{x + 5} + \frac{10}{(x + 5)(x - 2)}$

44. $\frac{1}{2x + 3} + \frac{1}{x - 1} = \frac{1}{(2x + 3)(x - 1)}$

45. $|2x| = 6$

46. $|3x| = 12$

47. $|2x + 3| = 5$

48. $|3x - 1| = 2$

49. $|1 - 4t| = 5$

50. $|1 - 2z| = 3$

51. $|-2x| = 8$

52. $|-x| = 1$

53. $|-2|x = 4$

54. $|3|x = 9$

55. $\frac{2}{3}|x| = 8$

56. $\frac{3}{4}|x| = 9$

57. $\left|\frac{x}{3} + \frac{2}{5}\right| = 2$

58. $\left|\frac{x}{2} - \frac{1}{3}\right| = 1$

59. $|x - 2| = -\frac{1}{2}$

60. $|2 - x| = -1$

61. $|x^2 - 4| = 0$

62. $|x^2 - 9| = 0$

63. $|x^2 - 2x| = 3$

64. $|x^2 + x| = 12$

65. $|x^2 + x - 1| = 1$

66. $|x^2 + 3x - 2| = 2$

67. $x^2 = 4x$

68. $x^2 = -8x$

69. $z^2 + 4z - 12 = 0$

70. $v^2 + 7v + 12 = 0$

71. $2x^2 - 5x - 3 = 0$

72. $3x^2 + 5x + 2 = 0$

73. $x(x - 7) + 12 = 0$

74. $x(x + 1) = 12$

75. $4x^2 + 9 = 12x$

76. $25x^2 + 16 = 40x$

77. $6x - 5 = \dfrac{6}{x}$

78. $x + \dfrac{12}{x} = 7$

79. $\dfrac{4(x - 2)}{x - 3} + \dfrac{3}{x} = \dfrac{-3}{x(x - 3)}$

80. $\dfrac{5}{x + 4} = 4 + \dfrac{3}{x - 2}$

In Problems 81–86, solve each equation. The letters a, b, and c are constants.

81. $ax - b = c, \quad a \neq 0$

82. $1 - ax = b, \quad a \neq 0$

83. $\dfrac{x}{a} + \dfrac{x}{b} = c, \quad a \neq 0, b \neq 0, \quad a \neq -b$

84. $\dfrac{a}{x} + \dfrac{b}{x} = c, \quad c \neq 0$

85. $\dfrac{1}{x - a} + \dfrac{1}{x + a} = \dfrac{2}{x - 1}$

86. $\dfrac{b + c}{x + a} = \dfrac{b - c}{x - a}, \quad c \neq 0, \quad a \neq 0$

In Problems 87–92, use the discriminant to determine whether each quadratic equation has two unequal real solutions, a repeated real solution, or no real solution without solving the equation. Use a graphing utility to verify your result.

87. $2x^2 - 6x + 7 = 0$

88. $x^2 + 4x + 7 = 0$

89. $9x^2 - 30x + 25 = 0$

90. $25x^2 - 20x + 4 = 0$

91. $3x^2 + 5x - 2 = 0$

92. $2x^2 - 3x - 4 = 0$

In Problems 93–100, use a graphing utility to approximate the real solutions, if any, of each equation correct to two decimal places. Then use the quadratic formula to obtain exact solutions. Compare the two results. Experiment further to see how close you can make the graphing utility solution to the exact solution.

93. $x^2 - 4x + 2 = 0$

94. $x^2 + 4x + 2 = 0$

95. $x^2 + \sqrt{3}x - 3 = 0$

96. $x^2 + \sqrt{2}x - 2 = 0$

97. $\pi x^2 - x - \pi = 0$

98. $\pi x^2 + \pi x - 2 = 0$

99. $3x^2 + 8\pi x + \sqrt{29} = 0$

100. $\pi x^2 - 15\sqrt{2}x + 20 = 0$

In Problems 101–112, use a graphing utility to approximate the real solutions, if any, of each equation correct to two decimal points. Then use any algebraic method that you wish to obtain exact solutions. Compare the two results. Experiment further to see how close you can make the graphing utility solution to the exact solution.

101. $x^2 - 7 = 0$

102. $x^2 - 8 = 0$

103. $16x^2 - 8x + 1 = 0$

104. $9x^2 - 6x + 1 = 0$

105. $10x^2 - 19x - 15 = 0$

106. $6x^2 + 7x - 20 = 0$

107. $2 + z = 6z^2$

108. $2 = y + 6y^2$

109. $x^2 + \sqrt{2}x = \frac{1}{2}$

110. $\frac{1}{2}x^2 = \sqrt{2}x + 1$

111. $x^2 + x = 4$

112. $x^2 + x = 1$

In Problems 113–120, use a graphing utility to approximate the positive solutions for each equation. Express the answer correct to two decimal places.

113. $x^3 + 3.2x^2 - 16.83x - 5.31 = 0$

114. $x^3 + 3.2x^2 - 7.25x - 6.3 = 0$

115. $x^4 - 1.4x^3 - 33.71x^2 + 23.94x + 292.41 = 0$

116. $x^4 + 1.2x^3 - 7.46x^2 - 4.692x + 15.2881 = 0$

117. $\pi x^3 - (8.88\pi + 1)x^2 - (42.066\pi - 8.88)x + 42.066 = 0$

118. $\pi x^3 - (5.63\pi + 2)x^2 - (108.392\pi - 11.26)x + 216.784 = 0$

119. $x^3 + 19.5x^2 - 1021x + 1000.5 = 0$

120. $x^3 + 14.2x^2 - 4.8x - 12.4 = 0$

Problems 121–126 list some formulas that occur in applications. Solve each formula for the indicated variable.

121. Electricity $\dfrac{1}{R} = \dfrac{1}{R_1} + \dfrac{1}{R_2}$ for R

122. Finance $A = P(1 + rt)$ for r

123. Mechanics $F = \dfrac{mv^2}{R}$ for R

124. Chemistry $PV = nRT$ for T

125. Mathematics $S = \dfrac{a}{1 - r}$ for r

126. Mechanics $v = -gt + v_0$ for t

127. Which of the following pairs of equations are equivalent? Explain.
(a) $x^2 = 9$; $x = 3$
(b) $x = \sqrt{9}$; $x = 3$
(c) $(x - 1)(x - 2) = (x - 1)^2$; $x - 2 = x - 1$

128. The equation

$$\frac{5}{x + 3} + 3 = \frac{8 + x}{x + 3}$$

has no solution, yet when we go through the process of solving it we obtain $x = -3$. Write a brief paragraph to explain what causes this to happen.

129. Make up an equation that has no solution and give it to a fellow student to solve. Ask the fellow student to write a critique of your equation.

130. Describe three ways you might solve a quadratic equation. State your preferred method; explain why you chose it.

131. Explain the benefits of evaluating the discriminant of a quadratic equation before attempting to solve it.

132. Make up three quadratic equations: one having two distinct solutions, one having no real solution, and one having exactly one real solution.

133. The word *quadratic* seems to imply four (*quad*), yet a quadratic equation is an equation that involves a polynomial of degree 2. Investigate the origin of the term *quadratic* as it is used in the expression *quadratic equation*. Write a brief essay on your findings.

1.6 SETTING UP EQUATIONS: APPLICATIONS

1. Translating Verbal Descriptions into Mathematical Expressions
2. Setting Up Applied Problems
3. Solving Interest Problems
4. Solving Mixture Problems
5. Solving Uniform Motion Problems
6. Solving Constant Rate Jobs Problems

The previous section provides the tools for solving equations. But, unfortunately, applied problems do not come in the form "Solve the equation. . . ." Instead, they are narratives that supply information—hopefully, enough to answer the question that inevitably arises. Thus, to solve applied problems we must be able to translate the verbal description into the language of mathematics. We do this by using symbols (usually letters of the alphabet) to represent unknown quantities and then finding relationships (such as equations) that involve these symbols. The process of doing this is called **mathematical modeling.**

in the form of a loan or by a bank from an individual in the form of a savings account) is called the **principal.** The **rate of interest,** expressed as a percent, is the amount charged for the use of the principal for a given period of time, usually on a yearly (that is, per annum) basis.

Simple Interest Formula

If a principal of P dollars is borrowed for a period of t years at a per annum interest rate r, expressed as a decimal, the interest I charged is

$$I = Prt \tag{1}$$

Interest charged according to formula (1) is called **simple interest.**

E X A M P L E   3

Financial Planning

Judy, an investor with $70,000, decides to place part of her money in corporate bonds paying 12% per year and the rest in a Certificate of Deposit paying 8% per year. If she wishes to obtain an overall return of 9% per year, how much should she place in each investment?

Solution   STEP 1: The question is asking for two dollar amounts: the principal to invest in the corporate bonds and the principal to invest in the Certificate of Deposit.

STEP 2: We let x represent the amount (in dollars) to be invested in the bonds. Then $70,000 - x$ is the amount that will be invested in the certificate. (Do you see why?)

STEP 3: We set up a table:

	Principal $	Rate	Time yr	Interest $
Bonds	x	12% = 0.12	1	0.12x
Certificate	70,000 − x	8% = 0.08	1	0.08(70,000 − x)
Total	70,000	9% = 0.09	1	0.09(70,000) = 6300

Since the total interest from the investments is to equal $0.09(70,000) = 6300$, we must have the equation

$$0.12x + 0.08(70,000 - x) = 6300$$

(Note that the units are consistent: the unit is dollars on each side.)

STEP 4:
$$0.12x + 5600 - 0.08x = 6300$$
$$0.04x = 700$$
$$x = 17,500$$

Thus, Judy should place $17,500 in the bonds and place $70,000 - $17,500 = $52,500 in the certificate.

STEP 5: The interest on the bonds after 1 year is $0.12(\$17,500) = \$2100$; the interest on the certificate after 1 year is $0.08(\$52,500) = \$4200$. The total annual interest is $6300, the required amount. ■

Now work Problem 23.

Mixture Problems

The next example is a type usually referred to as a **mixture problem.**

E X A M P L E   4   Chemistry: Mixing Acids

In a chemistry laboratory the concentration of one solution is 10% hydrochloric acid (HCl) and that of a second solution is 60% HCl. How many milliliters (mL) of each should be mixed to obtain 50 mL of a 30% HCl solution?

Solution   Let x represent the number of milliliters of the 10% solution. Then $50 - x$ equals the number of milliliters of the 60% solution. See Figure 94.

FIGURE 94

Based on the figure, we form a table:

	Amount mL	Concentration of HCl	Amount of Pure Acid mL
10% HCl	x	10% = 0.10	$0.10x$
60% HCl	$50 - x$	60% = 0.60	$0.60(50 - x)$
30% HCl	50	30% = 0.30	$0.30(50) = 15$

The amount of HCl in the 30% solution (15 milliliters) must equal the sum of the amounts of HCl found in the 10% solution and the 60% solution. Thus, we have the equation

$$0.10x + 0.60(50 - x) = 15$$
$$0.10x + 30 - 0.60x = 15$$
$$-0.50x = -15$$
$$x = 30 \text{ milliliters}$$

Thus, 30 milliliters of the 10% acid solution, when mixed with 20 milliliters of the 60% acid solution, yields 50 milliliters of a 30% acid solution.

Check: To check this answer, we note that there are $0.10(30) = 3$ milliliters of acid in the 10% solution and $0.60(20) = 12$ milliliters of acid in the 60%

solution. The 50 milliliter mixture therefore contains 15 milliliters of acid, for an acid concentration of $\frac{15}{50} = 0.30 = 30\%$. ∎

Now work Problem 53.

Uniform Motion

The next example deals with moving objects.

If an object moves at an average velocity v, the distance s covered in time t is given by the formula

$$s = vt \qquad\qquad (2)$$

That is, Distance = Velocity · Time. Objects that are moving in accordance with formula (2) are said to be in **uniform motion.**

E X A M P L E   5   Physics: Uniform Motion

Greg, who is a long-distance runner, runs at an average velocity of 8 miles per hour. Two hours after Greg leaves your house, you leave in your Buick Reatta and follow the same route. If your average velocity is 40 miles per hour, how long will it be before you catch up to Greg? How far will each of you be from your house?

Solution   Refer to Figure 95. We use t to represent the time (in hours) that it takes the Reatta to catch up with Greg. When this occurs, the total time elapsed for the runner is $t + 2$ hours.

FIGURE 95

Set up the following table:

	Velocity mi/hr	Time hr	Distance mi
Greg	8	$t + 2$	$8(t + 2)$
Reatta	40	t	$40t$

Since the distance traveled is the same, we are led to the following equation:

$$8(t + 2) = 40t$$
$$8t + 16 = 40t$$
$$32t = 16$$
$$t = \frac{1}{2} \text{ hour}$$

It will take you $\frac{1}{2}$ hour to catch up to Greg. Each of you will have gone 20 miles.

Check: In 2.5 hours, Greg travels a distance of $(2.5)(8) = 20$ miles. In $\frac{1}{2}$ hour, the Reatta travels a distance of $(\frac{1}{2})(40) = 20$ miles. ▬

E X A M P L E   6

Physics: Uniform Motion

A motorboat heads upstream a distance of 24 miles on the Illinois River, whose current is running at 3 miles per hour. The trip up and back takes 6 hours. Assuming that the motorboat maintained a constant speed relative to the water, what was its speed?

Solution

See Figure 96. We use v to represent the constant speed of the motorboat relative to the water. Then the true speed going upstream is $v - 3$ miles per hour, and the true speed going downstream is $v + 3$ miles per hour. Since Distance = Velocity × Time, then Time = Distance/Velocity. We set up a table.

FIGURE 96

	Velocity mi/hr	Distance mi	Time = Distance/Velocity hr
Upstream	$v - 3$	24	$\dfrac{24}{v - 3}$
Downstream	$v + 3$	24	$\dfrac{24}{v + 3}$

Since the total time up and back is 6 hours, we have

$$\frac{24}{v - 3} + \frac{24}{v + 3} = 6$$

$$\frac{24(v + 3) + 24(v - 3)}{(v - 3)(v + 3)} = 6$$

$$\frac{48v}{v^2 - 9} = 6$$

$$48v = 6(v^2 - 9)$$

$$6v^2 - 48v - 54 = 0$$

$$v^2 - 8v - 9 = 0$$

$$(v - 9)(v + 1) = 0$$

$$v = 9 \quad \text{or} \quad v = -1$$

We discard the solution $v = -1$ mile per hour, so the speed of the motorboat relative to the water is 9 miles per hour. ▬

Now work Problem 51.

Constant Rate Jobs

This example involves jobs that are performed at a **constant rate.** Our assumption is that if a job can be done in t units of time then $1/t$ of the job is done in 1 unit of time.

E X A M P L E   7

Working Together to Do a Job

At 10 AM Danny is asked by his father to weed the garden. From past experience Danny knows this will take him 4 hours working alone. His older brother, Mike, when it is his turn to do the job, requires 6 hours. Since Mike wants to go golfing with Danny and has a reservation for 1 PM, he agrees to help Danny. Assuming no gain or loss for efficiency, when will they finish if they work together? Can they make the golf date?

Solution

In 1 hour, Danny does $\frac{1}{4}$ of the job, and in 1 hour, Mike does $\frac{1}{6}$ of the job. Let t be the time (in hours) it takes them to do the job together. In 1 hour, then, $1/t$ of the job is completed. We reason as follows:

$$\left(\begin{array}{c}\text{Part done by Danny}\\\text{in 1 hour}\end{array}\right) + \left(\begin{array}{c}\text{Part done by Mike}\\\text{in 1 hour}\end{array}\right) = \left(\begin{array}{c}\text{Part done together}\\\text{in 1 hour}\end{array}\right)$$

Now we set up the table in the margin. From the table,

	Hours to Do Job	Part of Job Done in 1 Hour
Danny	4	$\frac{1}{4}$
Mike	6	$\frac{1}{6}$
Together	t	$\frac{1}{t}$

$$\frac{1}{4} + \frac{1}{6} = \frac{1}{t}$$

$$\frac{3+2}{12} = \frac{1}{t}$$

$$t = \frac{12}{5}$$

Working together, the job can be done in $\frac{12}{5}$ hours, or 2 hours, 24 minutes. They should make the golf date, since they will finish at 12:24 PM.   ▄

Now work Problem 39.

Other Applied Problems

The next two examples illustrate problems that you will probably see again in a slightly different form if you study calculus.

E X A M P L E   8

Preview of a Calculus Problem

From each corner of a square piece of sheet metal, remove a square of side 9 centimeters. Turn up the edges to form an open box. If the box is to hold 144 cubic centimeters, what should be the dimensions of the piece of sheet metal?

Solution

We use Figure 97 as a guide. We have labeled by x the length of a side of the square piece of sheet metal. The box will be of height 9 centimeters and its square base will have $x - 18$ as the length of a side. The volume (Length × Width × Height) of the box is therefore

$$9(x - 18)(x - 18) = 9(x - 18)^2$$

FIGURE 97

Volume = $9(x-18)(x-18)$

Since the volume of the box is to be 144 cubic centimeters, we have

$$9(x-18)^2 = 144$$
$$(x-18)^2 = 16$$
$$x-18 = \pm 4$$
$$x = 18 \pm 4$$
$$x = 22 \quad \text{or} \quad x = 14$$

We discard the solution $x = 14$ (do you see why?) and conclude that the sheet metal should be 22 centimeters by 22 centimeters.

Check: If we begin with a piece of sheet metal 22 centimeters by 22 centimeters, cut out a 9 centimeter square from each corner, and fold up the edges, we get a box whose dimensions are 9 by 4 by 4. The volume of the box is $9 \times 4 \times 4 = 144$ cubic centimeters, as required. ∎

  Now work Problem 31.

EXAMPLE 9   Preview of a Calculus Problem

A piece of wire 8 feet in length is to be cut into two pieces. Each piece will then be bent into a square. Where should the cut in the wire be made if the sum of the areas of these squares is to be 2 square feet?

FIGURE 98

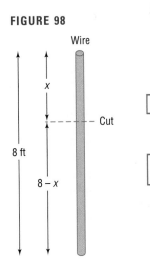

Solution   We use Figure 98 as a guide. We have labeled by x the length of one of the pieces of wire after it has been cut. The remaining piece will be of length $8 - x$. If each length is bent into a square, then one of the squares has a side of length $x/4$ and the other a side of length $(8 - x)/4$. Since the sum of the areas of these two squares is 2, we have the equation

$$\left(\frac{x}{4}\right)^2 + \left(\frac{8-x}{4}\right)^2 = 2$$
$$\frac{x^2}{16} + \frac{64 - 16x + x^2}{16} = 2$$
$$2x^2 - 16x + 64 = 32$$
$$2x^2 - 16x + 32 = 0 \quad \text{Put in standard form.}$$
$$x^2 - 8x + 16 = 0 \quad \text{Divide by 2.}$$
$$(x - 4)^2 = 0 \quad \text{Factor.}$$
$$x = 4$$

Since $x = 4, 8 - x = 4$, and the original piece of wire should be cut into two pieces, each of length 4 feet.

Check: If the length of each piece of wire is 4 feet, then each piece can be formed into a square whose side is 1 foot. The area of each square is then 1 square foot, so the sum of the areas is 2 square feet, as required. ∎

1.6 EXERCISES

In Problems 1–10, translate each sentence into a mathematical equation. Be sure to identify the meaning of all symbols.

1. **Geometry** The area of a circle is the product of the number π times the square of the radius.

2. **Geometry** The circumference of a circle is the product of the number π times twice the radius.

3. **Geometry** The area of a square is the square of the length of a side.

4. **Geometry** The perimeter of a square is four times the length of a side.

5. **Physics** Force equals the product of mass times acceleration.

6. **Physics** Pressure is force per unit area.

7. **Physics** Work equals force times distance.

8. **Physics** Kinetic energy is one-half the product of the mass times the square of the velocity.

9. **Business** The total variable cost of manufacturing x dishwashers is $150 per dishwasher times the number of dishwashers manufactured.

10. **Business** The total revenue derived from selling x dishwashers is $250 per dishwasher times the number of dishwashers sold.

11. **Finance** A total of $20,000 is to be invested, some in bonds and some in Certificates of Deposit (CDs). If the amount invested in bonds is to exceed that in CDs by $2000, how much will be invested in each type of instrument?

12. **Finance** A total of $10,000 is to be divided up between Yani and Diane, with Diane to receive $2000 less than Yani. How much will each receive?

13. **Finance** An inheritance of $900,000 is to be divided among David, Paige, and Dan in the following manner: Paige is to receive $\frac{3}{4}$ of what David gets, while Dan gets $\frac{1}{2}$ of what David gets. How much does each receive?

14. **Sharing the Cost of a Pizza** Carole and Canter agree to share the cost of an $18 pizza based on how much each ate. If Carole ate $\frac{2}{3}$ the amount Canter ate, how much should each pay?

Carole's portion

Canter's portion

15. **Computing Hourly Wages** Laura who is paid time-and-a-half for hours worked in excess of 40 hours, had gross weekly wages of $442 for 48 hours worked. What is her regular hourly rate?

16. **Computing Hourly Wages** Amy is paid time-and-a-half for hours worked in excess of 40 hours and double-time for hours worked on Sunday. If Amy had gross weekly wages of $342 for working 50 hours, 4 of which were on Sunday, what is her regular hourly rate?

17. **Football** In an NFL football game, the Bears scored a total of 41 points, including one safety (2 points) and two field goals (3 points each). After scoring a touchdown (6 points), a team is given the chance to score 1 or 2 extra points. The bears, in trying to score 1 extra point after each touchdown, missed 2 extra points. How many touchdowns did the Bears get?

18. **Basketball** In a basketball game, the Bulls scored a total of 103 points and made three times as many field goals (2 points each) as free throws (1 point each). They also made eleven 3 point baskets. How many field goals did they have?

19. **Geometry** The perimeter of a rectangle is 60 feet. Find its length and width if the length is 8 feet longer than the width.

20. **Geometry** The perimeter of a rectangle is 42 meters. Find its length and width if the length is twice the width.

21. **Enclosing a Garden** Scott has 46 feet of fencing to be used to enclose a rectangular garden that has a border 2 feet wide surrounding it (see the figure).

(a) If the length of the garden is to be twice its width, what will be the dimensions of the garden?
(b) What is the area of the garden?
(c) If the length and width of the garden were to be the same, what would be the dimensions of the garden?
(d) What would be the area of the square garden?

22. **Chemistry: Sugar Molecules** A sugar molecule has twice as many atoms of hydrogen as it does oxygen and one more atom of carbon than oxygen. If a sugar molecule has a total of 45 atoms, how many are oxygen? How many are hydrogen?

23. **Financial Planning** Betsy, a recent retiree, requires $6000 per year in extra income. She has $50,000 to invest and can invest in B-rated bonds paying 15% per year or in a Certificate of Deposit (CD) paying 7% per year. How much money should be invested in each to realize exactly $6000 in interest per year?

24. **Financial Planning** After 2 years, Betsy (see Problem 23) finds she now will require $7000 per year. Assuming that the remaining information is the same, how should the money be reinvested?

25. **Banking** A bank loaned out $12,000, part of it at the rate of 8% per year and the rest at the rate of 18% per year. If the interest received totaled $1000, how much was loaned at 8%?

26. **Banking** Wendy, a loan officer at a bank, has $1,000,000 to lend and is required to obtain an average return of 18% per year. If she can lend at the rate of 19% or at the rate of 16%, how much can she lend at the 16% rate and still meet her requirement?

27. **Dimensions of a Window** The area of the opening of a rectangular window is to be 143 square feet. If the length is to be 2 feet more than the width, what are the dimensions?

28. **Dimensions of a Window** The area of a rectangular window is to be 306 square centimeters. If the length exceeds the width by 1 centimeter, what are the dimensions?

29. **Geometry** Find the dimensions of a rectangle whose perimeter is 26 meters and whose area is 40 square meters.

30. **Watering a Field** An adjustable water sprinkler that sprays water in a circular pattern is placed at the center of a square field whose area is 1250 square feet (see the figure). What is the shortest radius setting that can be used if the field is to be completely enclosed within the circle?

31. **Constructing a Box** An open box is to be constructed from a square piece of sheet metal by removing a square of side 1 foot from each corner and turning up the edges. If the box is to hold 4 cubic feet, what should be the dimensions of the sheet metal?

32. **Constructing a Box** Rework Problem 31 if the piece of sheet metal is a rectangle whose length is twice its width.

33. **Physics** A ball is thrown vertically upward from the top of a building 96 feet tall with an initial velocity of 80 feet per second. The distance s (in feet) of the ball from the ground after t seconds is $s = 96 + 80t - 16t^2$.
(a) After how many seconds does the ball strike the ground?
(b) After how many seconds will the ball pass the top of the building on its way down?

34. **Constructing a Coffee Can** A 39 ounce can of Hills Bros.® coffee requires 188.5 square inches of aluminum. If its height is 7 inches, what is its radius? (The surface area A of a right circular

cylinder is $A = 2\pi r^2 + 2\pi rh$, where r is the radius and h is the height.)

7 in.

35. Business: Discount Pricing A builder of tract homes reduced the price of a model by 15%. If the new price is $125,000, what was its original price? How much can be saved by purchasing the model?

36. Business: Discount Pricing A car dealer, at a year-end clearance, reduces the list price of last year's models by 15%. If a certain four-door model has a discounted price of $8000, what was its list price? How much can be saved by purchasing last year's model?

37. Business: Marking Up the Price of Books A college book store marks up the price it pays the publisher for a book by 25%. If the selling price of a book is $56.00, how much did the book store pay for the book?

38. Personal Finance: Cost of a Car The suggested list price of a new car is $12,000. The dealer's cost is 85% of list. How much will you pay if the dealer is willing to accept $100 over cost for the car?

39. Working Together on a Job Craig can deliver his newspapers in 30 minutes. It takes Jim 20 minutes to do the same route. How long would it take them to deliver the newspapers if they work together?

40. Working Together on a Job Ray, by himself, can paint four rooms in 10 hours. If he hires Chris to help, they can do the same job together in 6 hours. If he lets Chris work alone, how long will it take for Chris to paint the four rooms?

41. Computing Grades Going into the final exam, which will count as two tests, Brooke has test scores of 80, 83, 71, 61, and 95. What score does Brooke need on the final in order to have an average score of 80?

42. Computing Grades Going into the final exam, which will count as two-thirds of the final grade, Mike has test scores of 86, 80, 84, and 90. What score does Mike need on the final in order to earn a B, which requires an average score of 80? What does he need to earn an A, which requires an average of 90?

43. Football A tight end can run the 100 yard dash in 12 seconds. A defensive back can do it in 10

seconds. The tight end catches a pass at his own 20 yard line with the defensive back at the 15 yard line. (See the figure.) If no other players are nearby, at what yard line will the defensive back catch up to the tight end?

44. Computing Business Expenses Jane Manning, an outside saleswoman, uses her car for both business and pleasure. Last year, she traveled 30,000 miles, using 900 gallons of gasoline. Her car gets 40 miles per gallon on the highway and 25 in the city. She can deduct all highway travel, but no city travel, on her taxes. How many miles should Jane Manning be allowed as a business expense?

45. Constructing a Border around a Garden A landscaper, who just completed a rectangular flower garden measuring 6 feet by 10 feet, orders 1 cubic yard of premixed cement, all of which is to be used to create a border of uniform width around the garden. If the border is to have a depth of 3 inches, how wide will the border be? (1 cubic yard = 27 cubic feet)

6 ft

10 ft

46. Physics An object is propelled vertically upward with an initial velocity of 20 meters per second. The distance s (in meters) of the object from the ground after t seconds is $s = -4.9t^2 + 20t$.
(a) When will the object be 15 meters above the ground?
(b) When will it strike the ground?
(c) Will the object reach a height of 100 meters?
(d) What is the maximum height?

47. Reducing the Size of a Candy Bar A jumbo chocolate bar with a rectangular shape measures 12 centimeters in length, 7 centimeters in width, and 3 centimeters in thickness. Due to escalating costs of cocoa, management decides to reduce the volume of the bar by 10%. To accomplish this reduction, management decides the new bar should have the same 3 centimeter thickness, but the length and width each should be reduced an equal number of centimeters. What should be the dimensions of the new candy bar?

48. Reducing the Size of a Candy Bar Rework Problem 47 if the reduction is to be 20%.

49. Constructing a Border around a Pool A pool in the shape of a circle measures 10 feet across. One cubic yard of concrete is to be used to create a circular border of uniform width around the pool. If the border is to have a depth of 3 inches, how wide will the border be? (1 cubic yard = 27 cubic feet)

← 10 ft →

50. Constructing a Border around a Pool Rework Problem 49 if the depth of the border is 4 inches.

51. Physics: Uniform Motion A motorboat can maintain a constant speed of 16 miles per hour relative to the water. The boat makes a trip upstream to a certain point in 20 minutes; the return trip takes 15 minutes. What is the speed of the current? (See the figure.)

52. Purity of Gold The purity of gold is measured in karats, with pure gold being 24 karats. Other purities of gold are expressed as proportional parts of pure gold. Thus, 18 karat gold is $\frac{18}{24}$, or 75% pure gold; 12 karat gold is $\frac{12}{24}$, or 50% pure gold; and so on. How much 12 karat gold should be mixed with pure gold to obtain 60 grams of 16 karat gold?

53. Business: Mixing Nuts A nut store normally sells cashews for $4.00 per pound and peanuts for $1.50 per pound. But at the end of the month, the peanuts had not sold well, so in order to sell 60 pounds of peanuts the manager decided to mix the 60 pounds of peanuts with some cashews and sell the mixture for $2.50 per pound. How many pounds of cashews should be mixed with the peanuts to ensure no change in the profit?

54. Business: Pricing Candy A candy store sells boxes of candy containing caramels and cremes. Each box sells for $9.50 and holds 30 pieces of candy (all pieces are the same size). If the caramels cost $0.25 to produce and the cremes cost $0.45 to produce, how many of each should be in a box for no loss or gain to occur?

55. Running a Race Mike can run the mile in 6 minutes, and Dan can run the mile in 9 minutes. If Mike gives Dan a head start of 1 minute, how far from the start will Mike pass Dan? (See the figure.) How long does it take?

56. Physics: Uniform Motion A motorboat heads upstream on a river that has a current of 3 miles per hour. The trip upstream takes 5 hours, while the return trip takes 2.5 hours. What is the speed of the motorboat? (Assume that the motorboat maintains a constant speed relative to the water.)

57. Rescue at Sea A ship that is in danger of sinking radios the Coast Guard for assistance. When the rescue craft leaves the Coast Guard station, the ship is 60 miles away and heading directly toward the station. If the average speed of the ship is 10 miles per hour and the average speed of the rescue craft is 20 miles per hour, how long will it take for the rescue craft to reach the ship? (See the figure.)

58. Physics: Uniform Motion Two cars enter the Florida Turnpike at Commercial Boulevard at 8:00 AM, each heading for Wildwood. One car's average speed is 10 miles per hour more than the other's. The faster car arrives at Wildwood at 11:00 AM, $\frac{1}{2}$ hour before the other car. What is the average speed of each car? How far did each travel?

59. Emptying Oil Tankers An oil tanker can be emptied by the main pump in 4 hours. An auxiliary pump can empty the tanker in 9 hours. If the main pump is started at 9 AM, when should the auxiliary pump be started so that the tanker is emptied by noon?

60. Cement Mix A 20 pound bag of Economy brand cement mix contains 25% cement and 75% sand. How much pure cement must be added to produce a cement mix that is 40% cement?

61. Emptying a Tub A bathroom tub will fill in 15 minutes with both faucets open and the stopper in place. With both faucets closed and the stopper removed, the tub will empty in 20 minutes. How long will it take for the tub to fill if both faucets are open and the stopper is removed?

62. Range of an Airplane An air rescue plane averages 300 miles per hour in still air. It carries enough fuel for 5 hours of flying time. If, upon takeoff, it encounters a wind of 30 miles per hour, how far can it fly and return safely? (Assume that the wind remains constant.)

63. Home Equity Loans Suppose that you obtain a home equity loan of $100,000 that requires only a monthly interest payment at 10% per annum, with the principal due after 5 years. You decide to invest part of the loan in a 5 year CD that pays 9% compounded and paid monthly and part in a B+ rated bond due in 5 years that pays 12% compounded and paid monthly. What is the most you can invest in the CD to ensure that the monthly home equity loan payment is made?

64. Comparing Olympic Heroes In the 1984 Olympics, Carl Lewis of the United States won the gold medal in the 100 meter race with a time of 9.99 seconds. In the 1896 Olympics, Thomas Burke, also of the United States, won the gold medal in the 100 meter race in 12.0 seconds. If they ran in the same race repeating their respective times, by how many meters would Lewis beat Burke?

65. Computing Average Speed In going from Chicago to Atlanta, a car averages 45 miles per hour, and in going from Atlanta to Miami, it averages 55 miles per hour. If Atlanta is halfway between Chicago and Miami, what is the average speed from Chicago to Miami? Discuss an intuitive solution. Write a paragraph defending your intuitive solution. Then solve the problem algebraically. Is your intuitive solution the same as the algebraic one? If not, find the flaw.

66. Speed of a Plane On a recent flight from Phoenix to Kansas City, a distance of 919 nautical miles, the plane arrived 20 minutes early. On leaving the aircraft, I asked the captain, "What was our tail wind?" He replied, "I don't know, but our ground speed was 550 knots." How can you determine if enough information is provided to find the tail wind? If possible, find the tail wind. (1 knot = 1 nautical mile per hour)

67. Critical Thinking You are the manager of a clothing store and have just purchased 100 dress shirts for $20.00 each. After 1 month of selling the shirts at the regular price, you plan to have a sale giving 40% off the original selling price. However, you still want to make a profit of $4 on each shirt at the sale price. What should you price the shirts at initially to ensure this? If, instead of 40% off at the sale, you give 50% off, by how much is your profit reduced?

68. Critical Thinking Make up a word problem that requires solving a linear equation as part of its solution. Exchange problems with a friend. Write a critique of your friend's problem.

69. Critical Thinking Without solving, explain what is wrong with the following mixture problem: How many liters of 25% ethanol should be added to 20 liters of 48% ethanol to obtain a solution of 58% ethanol? Now go through an algebraic solution. What happens?

1.7 INEQUALITIES

1 Use Properties of Inequalities

2 Solve Inequalities Algebraically

3 Solve Inequalities Using a Graphing Utility

An **inequality in one variable** is a statement involving two expressions, at least one containing the variable, separated by one of the inequality symbols $<, \leq, >$, or $\geq$. To **solve an inequality** means to find all values of the variable for which the statement is true. These values are called **solutions** of the inequality.

For example, the following are all inequalities involving one variable, x:

$$x + 5 < 8 \qquad 2x - 3 \geq 4 \qquad x^2 - 1 \leq 3 \qquad \frac{x + 1}{x - 2} > 0$$

In working with inequalities, we will need to know certain properties that they obey.

We begin with the **trichotomy property,** which states that either two numbers are equal or one of them is less than the other.

For any pair of real numbers a and b,

Trichotomy Property

$$a < b \quad \text{or} \quad a = b \quad \text{or} \quad b < a$$

If $b = 0$, the trichotomy property states that, for any real number a,

$$a < 0 \quad \text{or} \quad a = 0 \quad \text{or} \quad a > 0$$

That is, any real number is negative or 0 or positive, a fact we have already noted.

For any real number a, we have

Nonnegative Property

$$a^2 \geq 0$$

In the following properties, a, b, and c are real numbers.

Transitive Property of Inequalities

If $a < b$ and $b < c$, then $a < c$.
If $a > b$ and $b > c$, then $a > c$.

Addition Property of Inequalities

If $a < b$, then $a + c < b + c$.
If $a > b$, then $a + c > b + c$.

The **addition property** states that the sense, or direction, of an inequality remains unchanged if the same number is added to each side.

Multiplication Properties of Inequalities

> If $a < b$ and if $c > 0$, then $ac < bc$.
> If $a < b$ and if $c < 0$, then $ac > bc$.
> If $a > b$ and if $c > 0$, then $ac > bc$.
> If $a > b$ and if $c < 0$, then $ac < bc$.

The **multiplication properties** state that the sense, or direction, of an inequality *remains the same* if each side is multiplied by a *positive* real number, while the direction is *reversed* if each side is multiplied by a *negative* real number.

The **reciprocal property** states that the reciprocal of a positive real number is positive and the reciprocal of a negative real number is negative.

Reciprocal Property for Inequalities

> If $a > 0$, then $\dfrac{1}{a} > 0$.
>
> If $a < 0$, then $\dfrac{1}{a} < 0$

Solving Inequalities

Two inequalities having exactly the same solution set are called **equivalent inequalities.**

As with equations, one method for solving an inequality is to replace it by a series of equivalent inequalities, until an inequality with an obvious solution, such as $x < 3$, is obtained. We obtain equivalent inequalities by applying some of the same operations as those used to find equivalent equations. The addition property and the multiplication properties form the basis for the following procedures.

Procedures That Leave the
Inequality Symbol Unchanged

1. Simplify both sides of the inequality by combining like terms and eliminating parentheses:

    Replace $(x + 2) + 6 > 2x + (x + 1)$
    by $x + 8 > 3x + 1$

2. Add or subtract the same expression on both sides of the inequality:

    Replace $3x - 5 < 4$
    by $(3x - 5) + 5 < 4 + 5$

3. Multiply or divide both sides by the inequality by the same *positive* expression:

    Replace $4x > 16$ by $\dfrac{4x}{4} > \dfrac{16}{4}$

Procedures That Reverse the Sense or Direction of the Inequality Symbol

1. Interchange the two sides of the inequality:

   Replace $\quad 3 < x \quad$ by $\quad x > 3$

2. Multiply or divide both sides of the inequality by the same *negative* expression:

   Replace $\quad -2x > 6 \quad$ by $\quad \dfrac{-2x}{-2} < \dfrac{6}{-2}$

To solve an inequality using a graphing utility, we follow these steps:

3

Steps for Solving Inequalities Graphically

STEP 1: Write the inequality in one of the following forms:

$$Y_1 < Y_2, \qquad Y_1 > Y_2, \qquad Y_1 \leq Y_2, \qquad Y_1 \geq Y_2$$

where Y_1, Y_2 are expressions involving the variable x.

STEP 2: Graph Y_1 and Y_2 on the same screen.

If the inequality is of the form $Y_1 < Y_2$, determine on what interval Y_1 is below Y_2.

If the inequality is of the form $Y_1 > Y_2$, determine on what interval Y_1 is above Y_2.

If the inequality is not strict, include the x-coordinates of the points of intersection in the solution.

E X A M P L E   1   Solving an Inequality

Solve the inequality $4x + 7 \geq 2x - 3$, and graph the solution set.

Graphing Solution   We graph $Y_1 = 4x + 7$ and $Y_2 = 2x - 3$ on the same screen. See Figure 99. Using the INTERSECT command, we find that Y_1 and Y_2 intersect at $x = -5$. The graph of Y_1 is above that of Y_2, $Y_1 > Y_2$, to the right of the point of intersection. Since the inequality is not strict, the solution set is $\{x | x \geq -5\}$ or, using interval notation, $[-5, \infty)$.

FIGURE 99

Algebraic Solution

$4x + 7 \geq 2x - 3$	
$4x + 7 - 7 \geq 2x - 3 - 7$	Subtract 7 from both sides.
$4x \geq 2x - 10$	Simplify.
$4x - 2x \geq 2x - 10 - 2x$	Subtract 2x from both sides.
$2x \geq -10$	Simplify.
$\dfrac{2x}{2} \geq \dfrac{-10}{2}$	Divide both sides by 2. (The sense of the inequality symbol is unchanged.)
$x \geq -5$	Simplify.

FIGURE 100
$-5 \leq x < \infty$ or $[-5, \infty)$

The solution set is $\{x | -5 \leq x < \infty\}$ or, using interval notation, all numbers in the interval $[-5, \infty)$.

See Figure 100 for the graph.

■

 Now work Problem 13.

E X A M P L E 2

Solving Combined Inequalities

Solve the inequality $-5 < 3x - 2 < 1$ and draw a graph to illustrate the solution.

Graphing Solution

To solve a combined inequality, we graph each part: $Y_1 = -5$, $Y_2 = 3x - 2$, and $Y_3 = 1$. We seek the values of x for which the graph of Y_2 is between the graphs of Y_1 and Y_3. See Figure 101. The point of intersection of Y_1 and Y_2 is $(-1, 5)$, and the point of intersection of Y_2 and Y_3 is $(1, 1)$. The inequality is true for all values of x between these two intersection points. Since the inequality is strict, the solution set is $\{x | -1 < x < 1\}$ or, using interval notation, $(-1, 1)$.

FIGURE 101

Algebraic Solution

Recall that the inequality

$$-5 < 3x - 2 < 1$$

is equivalent to the two inequalities

$$-5 < 3x - 2 \quad \text{and} \quad 3x - 2 < 1$$

We will solve each of these inequalities separately. For the first inequality,

$-5 < 3x - 2$	
$-5 + 2 < 3x - 2 + 2$	Add 2 to both sides.
$-3 < 3x$	Simplify.
$\dfrac{-3}{3} < \dfrac{3x}{3}$	Divide both sides by 3.
$-1 < x$	Simplify.

The second inequality is solved as follows:

$3x - 2 < 1$	
$3x - 2 + 2 < 1 + 2$	Add 2 to both sides.
$3x < 3$	Simplify.
$\dfrac{3x}{3} < \dfrac{3}{3}$	Divide both sides by 3.
$x < 1$	Simplify.

The solution set of the original pair of inequalities consists of all x for which

FIGURE 102
$-1 < x < 1$ or $(-1, 1)$

$$-1 < x \quad \text{and} \quad x < 1$$

This may be written more compactly as $\{x \mid -1 < x < 1\}$. In interval notation, the solution is $(-1, 1)$. See Figure 102 for the graph. ∎

We observe in the preceding process that the two inequalities we solved required exactly the same steps. A shortcut to solving the original inequality algebraically is to deal with the two inequalities at the same time, as follows:

$$
\begin{array}{lll}
-5 < & 3x - 2 < 1 & \\
-5 + 2 < 3x - 2 + 2 < 1 + 2 & \quad \text{Add 2 to each part.} \\
-3 < & 3x & < 3 \quad \text{Simplify.} \\
\dfrac{-3}{3} < & \dfrac{3x}{3} & < \dfrac{3}{3} \quad \text{Divide each part by 3.} \\
-1 < & x & < 1 \quad \text{Simplify.}
\end{array}
$$

Let's look at an inequality involving absolute value.

E X A M P L E   3   Solving an Inequality Involving Absolute Value

Solve the inequality: $|x| < 4$

Graphing Solution   We graph $Y_1 = |x|$ and $Y_2 = 4$ on the same screen. See Figure 103. Using the INTERSECT command (twice), we find that Y_1 and Y_2 intersect at $x = -4$ and at $x = 4$. The graph of Y_1 is below that of Y_2, $Y_1 < Y_2$, between the points of intersection. Since the inequality is strict, the solution set is $\{x \mid -4 < x < 4\}$ or, using interval notation, $(-4, 4)$.

FIGURE 103

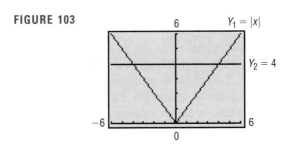

Algebraic Solution   We are looking for all points whose coordinate x is a distance less than 4 units from the origin. See Figure 104 for an illustration. Because any x between -4 and 4 satisfies the condition $|x| < 4$, the solution set consists of all numbers x for which $-4 < x < 4$; that is, all x in $(-4, 4)$. ∎

FIGURE 104
$-4 < x < 4$ or $(-4, 4)$

Algebraic Solution

$|2x - 5| > 3$ This follows the form of statement (3); the expression
                  $u = 2x - 5$ is inside the absolute value bars.

$2x - 5 < -3$	or	$2x - 5 > 3$	Apply statement (3).
$2x - 5 + 5 < -3 + 5$	or	$2x - 5 + 5 > 3 + 5$	Add 5 to each part.
$2x < 2$	or	$2x > 8$	Simplify.
$\dfrac{2x}{2} < \dfrac{2}{2}$	or	$\dfrac{2x}{2} > \dfrac{8}{2}$	Divide each part by 2.
$x < 1$	or	$x > 4$	Simplify.

FIGURE 110
$-\infty < x < 1$ or $4 < x < \infty$;
$(-\infty, 1)$ or $(4, \infty)$

-2 -1 0 1 2 3 4 5 6 7

The solution set is $\{x | -\infty < x < 1$ or $4 < x < \infty\}$, that is, all x in $(-\infty, 1)$ or $(4, \infty)$. See Figure 110.   ■

Warning: A common error to be avoided is to attempt to write the solution $x < 1$ or $x > 4$ as $1 > x > 4$, which is incorrect, since there are no numbers x for which $1 > x$ *and* $x > 4$. Another common error is to "mix" the symbols and write $1 < x > 4$, which, of course, makes no sense.

1.7   EXERCISES

In Problems 1–4, an inequality is given. Write the equivalent inequality obtained by:
(a) Adding −3 to each side of the given inequality.
(b) Subtracting 5 from each side of the given inequality.
(c) Multiplying each side of the given inequality by 3.
(d) Multiplying each side of the given inequality by −2.

1. $3 < 5$ **2.** $2 > 1$ **3.** $2x + 1 < 2$ **4.** $1 - 2x > 5$

In Problems 5–12, fill in the blank with the correct inequality symbol.

5. If $x < 5$, then $x - 5$ _____ 0.   **6.** If $x < -4$, then $x + 4$ _____ 0.

7. If $x > -4$, then $x + 4$ _____ 0.   **8.** If $x > 6$, then $x - 6$ _____ 0.

9. If $x > -4$, then $3x$ _____ -12.   **10.** If $x < 3$, then $2x$ _____ 6.

11. If $x < 6$, then $-2x$ _____ -12.   **12.** If $x > -2$, then $-4x$ _____ 8.

In Problems 13–68, solve each inequality (a) graphically and (b) algebraically. Graph the solution set.

13. $x + 1 < 5$ **14.** $x - 6 < 1$ **15.** $1 - 2x \leq 3$

16. $2 - 3x \leq 5$ **17.** $3x - 7 > 2$ **18.** $2x + 5 > 1$

19. $3x - 1 \geq 3 + x$ **20.** $2x - 2 \geq 3 + x$ **21.** $-2(x + 3) < 8$

22. $-3(1 - x) < 12$ **23.** $4 - 3(1 - x) \leq 3$ **24.** $8 - 4(2 - x) \leq -2x$

25. $\frac{1}{2}(x - 4) > x + 8$ **26.** $3x + 4 > \frac{1}{3}(x - 2)$ **27.** $\dfrac{x}{2} \geq 1 - \dfrac{x}{4}$

28. $\dfrac{x}{3} \geq 2 + \dfrac{x}{6}$ **29.** $0 \leq 2x - 6 \leq 4$ **30.** $4 \leq 2x + 2 \leq 10$

31. $-5 \leq 4 - 3x \leq 2$ **32.** $-3 \leq 3 - 2x \leq 9$ **33.** $-3 < \dfrac{2x - 1}{4} < 0$

34. $0 < \dfrac{3x + 2}{2} < 4$ **35.** $1 < 1 - \frac{1}{2}x < 4$ **36.** $0 < 1 - \frac{1}{3}x < 1$

37. $(x + 2)(x - 3) > (x - 1)(x + 1)$ **38.** $(x - 1)(x + 1) > (x - 3)(x + 4)$ **39.** $x(4x + 3) \leq (2x + 1)^2$

40. $x(9x - 5) \le (3x - 1)^2$

41. $\dfrac{1}{2} \le \dfrac{x + 1}{3} < \dfrac{3}{4}$

42. $\dfrac{1}{3} < \dfrac{x + 1}{2} \le \dfrac{2}{3}$

43. $(4x + 2)^{-1} < 0$

44. $(2x - 1)^{-1} > 0$

45. $0 < \dfrac{1}{x} < \dfrac{1}{5}$

46. $0 < \dfrac{1}{x} < \dfrac{1}{3}$

47. $0 < (2x - 4)^{-1} < \tfrac{1}{2}$

48. $0 < (3x + 6)^{-1} < \tfrac{1}{3}$

49. $|2x| < 8$

50. $|3x| < 15$

51. $|3x| > 12$

52. $|2x| > 6$

53. $|x - 2| < 1$

54. $|x + 4| < 2$

55. $|3t - 2| \le 4$

56. $|2u + 5| \le 7$

57. $|x - 3| \ge 2$

58. $|x + 4| \ge 2$

59. $|1 - 4x| < 5$

60. $|1 - 2x| < 3$

61. $|1 - 2x| > 3$

62. $|2 - 3x| > 1$

63. $|x + 3| > 0$

64. $|3 - x| > 0$

65. $|x + 2| > -3$

66. $|2 - x| > -2$

67. $|2x - 1| < 0.02$

68. $|3x - 2| < 0.02$

69. Express the fact that x differs from 2 by less than $\frac{1}{2}$ as an inequality involving an absolute value. Solve for x.

70. Express the fact that x differs from -1 by less than 1 as an inequality involving an absolute value. Solve for x.

71. Express the fact that x differs from -3 by more than 2 as an inequality involving an absolute value. Solve for x.

72. Express the fact that x differs from 2 by more than 3 as an inequality involving an absolute value. Solve for x.

73. A young adult may be defined as someone older than 21, but less than 30 years of age. Express this statement using inequalities.

74. Middle-aged may be defined as being 40 or more and less than 60. Express the statement using inequalities.

75. **Body Temperature** "Normal" human body temperature is 98.6°F. If a temperature x that differs from normal by at least 1.5° is considered unhealthy, write the condition for an unhealthy temperature x as an inequality involving an absolute value, and solve for x.

76. **Household Voltage** In the United States, normal household voltage is 115 volts. However, it is not uncommon for actual voltage to differ from normal voltage by at most 5 volts. Express this situation as an inequality involving an absolute value. Use x as the actual voltage and solve for x.

77. **Life Expectancy** Metropolitan Life Insurance Co. reported that an average 25-year-old male in 1996 could expect to live at least 48.4 more years, and an average 25-year-old female in 1996 could expect to live at least 54.7 more years.

(a) To what age can an average 25-year-old male expect to live? Express your answer as an inequality.

(b) To what age can an average 25-year-old female expect to live? Express your answer as an inequality.

(c) Who can expect to live longer, a male or a female? By how many years?

78. **General Chemistry** For a certain ideal gas, the volume V (in cubic centimeters) equals 20 times the temperature T (in degrees Celsius). If the temperature varies from 80° to 120°C, inclusive, what is the corresponding range of the volume of the gas?

79. **Real Estate** A real estate agent agrees to sell a large apartment complex according to the following commission schedule: $45,000 plus 25% of the selling price in excess of $900,000. Assuming that the complex will sell at some price between $900,000 and $1,100,000, inclusive, over what range does the agent's commission vary? How does the commission vary as a percent of selling price?

80. **Sales Commission** A used car salesperson is paid a commission of $25 plus 40% of the selling price in excess of owner's cost. The owner claims that used cars typically sell for at least owner's cost plus $70 and at most owner's cost plus $300. For each sale made, over what range can the salesperson expect the commission to vary?

81. **Federal Tax Withholding** The percentage method of withholding for federal income tax (1995)* states that a single person whose weekly wages, after subtracting withholding allowances, are over $476, but not over $999, shall have $63.90 plus 28% of the excess over $476 withheld. Over what range does the amount withheld vary if the weekly wages vary from $500 to $550, inclusive?

82. **Federal Tax Withholding** Rework Problem 81 if the weekly wages vary from $600 to $700, inclusive.

83. **Electricity Rates** Commonwealth Edison Company's summer charge for electricity is 10.819¢ per kilowatt-hour.† In addition, each monthly bill contains a customer charge of $9.06. If last summer's bills ranged from a low of $82.14 to a high of $279.63, over what range did usage vary (in kilowatt-hours)?

84. **Water Bills** The Village of Oak Lawn charges homeowners $21.60 per quarter year plus $1.70 per 1000 gallons for water usage in excess of 12,000 gallons.‡ In 1995, one homeowner's quarterly bill ranged from a high of $65.75 to a low of $28.40. Over what range did water usage vary?

85. **Markup of a New Car** The markup over dealer's cost of a new car ranges from 12% to 18%. If the sticker price is $8800, over what range will the dealer's cost vary?

86. **IQ Tests** A standard intelligence test has an average score of 100. According to statistical theory, of the people who take the test, the 2.5% with the highest scores will have scores of more than 1.96 σ above the average, where σ (sigma, a number called the *standard deviation*) depends on the nature of the test. If $\sigma = 12$ for this test and there is (in principle) no upper limit to the score possible on the test, write the interval of possible test scores of the people in the top 2.5%.

Source: Employer's Tax Guide. Department of the Treasury, Internal Revenue Service, 1995.

†*Source:* Commonwealth Edison Co., Chicago, Illinois, 1991.

‡*Source:* Village of Oak Lawn, Illinois, 1995.

87. **Computing Grades** In your Economics 101 class, you have scores of 68, 82, 87, and 89 on the first four of five tests. To get a grade of B, the average of the first five test scores must be greater than or equal to 80 and less than 90. Solve an inequality to find the range of the score you need on the last test to get a B.

What do I need to get a B?

88. **Computing Grades** Repeat Problem 87 if the fifth test counts double.

89. A car that averages 25 miles per gallon has a tank that holds 20 gallons of gasoline. After a trip that covered at least 300 miles, the car ran out of gasoline. What is the range of the amount of gasoline (in gallons) that was in the tank at the start of the trip?

90. Repeat Problem 89 if the same car runs out of gasoline after a trip of no more than 250 miles.

91. **Physics** A ball is thrown vertically upward with an initial velocity of 80 feet per second. The distance s (in feet) of the ball from the ground after t seconds is $s = 80t - 16t^2$. For what time interval is the ball more than 96 feet above the ground? (See the figure.)

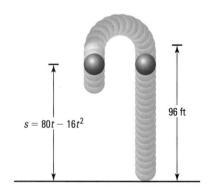

$s = 80t - 16t^2$

96 ft

92. **Physics** Rework Problem 91 to find when the ball is less than 64 feet above the ground.

93. **Business** The monthly revenue achieved by selling x wristwatches is figured to be $x(40 - 0.2x)$ dollars. The wholesale cost of each watch is $28. How many watches must be sold each month to achieve a profit (revenue − cost) of at least $100?

94. Business The monthly revenue achieved by selling x boxes of candy is figured to be $x(5 - 0.05x)$ dollars. The wholesale cost of each box of candy is $1.50. How many boxes must be sold each month to achieve a profit of at least $60?

95. Arithmetic Mean If $a < b$, show that $a < (a + b)/2 < b$. The number $(a + b)/2$ is called the **arithmetic mean** of a and b.

96. Refer to Problem 95. Show that the arithmetic mean of a and b is equidistant from a and b.

97. Geometric Mean If $0 < a < b$, show that $a < \sqrt{ab} < b$. The number $\sqrt{ab}$ is called the **geometric mean** of a and b.

98. Refer to Problems 95 and 97. Show that the geometric mean of a and b is less than the arithmetic mean of a and b.

99. Harmonic Mean For $0 < a < b$, let h be defined by

$$\frac{1}{h} = \frac{1}{2}\left(\frac{1}{a} + \frac{1}{b}\right)$$

Show that $a < h < b$. The number h is called the **harmonic mean** of a and b.

100. Refer to Problems 95, 97, and 99. Show that the harmonic mean of a and b equals the geometric mean squared divided by the arithmetic mean.

101. Make up an inequality that has no solution. Make up one that has exactly one solution.

102. The inequality $x^2 + 1 < -5$ has no solution. Explain why.

103. Do you prefer to use inequality notation or interval notation to express the solution to an inequality? Give your reasons. Are there particular circumstances when you prefer one to the other? Cite examples.

104. How would you explain to a fellow student the underlying reason for the multiplication property for inequalities (page 92); that is, the sense or direction of an inequality remains the same if each side is multiplied by a positive real number, while the direction is reversed if each side is multiplied by a negative real number.

CHAPTER REVIEW

THINGS TO KNOW

Formulas

Distance formula	$d = \sqrt{(x_2 - x_1)^2 + (y_2 - y_1)^2}$
Midpoint formula	$(x, y) = \left(\dfrac{x_1 + x_2}{2}, \dfrac{y_1 + y_2}{2}\right)$
Slope	$m = \dfrac{y_2 - y_1}{x_2 - x_1}$, if $x_1 \neq x_2$; undefined if $x_1 = x_2$
Parallel lines	Equal slopes $(m_1 = m_2)$
Perpendicular lines	Product of slopes is -1 $(m_1 \cdot m_2 = -1)$

Equations

Vertical line	$x = a$
Horizontal line	$y = b$
Point–slope form of the equation of a line	$y - y_1 = m(x - x_1)$; m is the slope of the line, (x_1, y_1) is a point on the line
General form of the equation of a line	$Ax + By + C = 0$, A, B not both 0
Slope–intercept form of the equation of a line	$y = mx + b$; m is the slope of the line, b is the y-intercept
Standard form of the equation of a circle	$(x - h)^2 + (y - k)^2 = r^2$; r is the radius of the circle, (h, k) is the center of the circle
General form of the equation of a circle	$x^2 + y^2 + ax + by + c = 0$
Equation of the unit circle	$x^2 + y^2 = 1$

87. **Physics: Uniform Motion** A man is walking at an average speed of 4 miles per hour alongside a railroad track. A freight train, going in the same direction at an average speed of 30 miles per hour, requires 5 seconds to pass the man. How long is the freight train? Give your answer in feet.

30 mi/hr

Length of train — 4 mi/hr —

$t = 0$ ← 5 sec → $t = 5$

88. One formula stating the relationship between the length l and width w of a rectangle of "pleasing proportion" is $l^2 = w(l + w)$. How should a 4 foot by 8 foot sheet of plasterboard be cut so that the result is a rectangle of "pleasing proportion" with a width of 4 feet?

89. **Business: Determining the Cost of a Charter** A group of 20 senior citizens can charter a bus for a 1 day excursion trip for $15 per person. The charter company agrees to reduce the price of each ticket by 10¢ for each additional passenger in excess of 20 who goes on the trip, up to a maximum of 44 passengers (the capacity of the bus). If the final bill from the charter company was $482.40, how many seniors went on the trip, and how much did each pay?

90. Show that points $A = (3, 4)$, $B = (1, 1)$, and $C = (-2, 3)$ are the vertices of an isosceles triangle.

91. Show that the points $A = (-2, 0)$, $B = (-4, 4)$, and $C = (8, 5)$ are the vertices of a right triangle in two ways:
(a) By using the converse of the Pythagorean Theorem
(b) By using the slopes of the lines joining the vertices

92. Show that the points $A = (2, 5)$, $B = (6, 1)$, and $C = (8, -1)$ lie on a line by using slopes.

93. Show that the points $A = (1, 5)$, $B = (2, 4)$, and $C = (-3, 5)$ lie on a circle with center $(-1, 2)$. What is the radius of this circle?

94. The endpoints of the diameter of a circle are $(-3, 2)$ and $(5, -6)$. Find the center and radius of the circle. Write the general equation of this circle.

95. Find two numbers y such that the distance from $(-3, 2)$ to $(5, y)$ is 10.

96. **Evening Up a Race** In a 100 meter race, George crosses the finish line 5 meters ahead of Bob. To even things up, George suggests to Bob that they race again, this time with George lining up 5 meters behind the start.
(a) Assuming that George and Bob run at the same pace as before, does the second race end in a tie?
(b) If not, who wins?
(c) By how many meters does he win?
(d) How far back should George start so that the race ends in a tie?

After running the race a second time, Bob, to even things up, suggests to George that he (Bob) line up 5 meters in front of the start.
(e) Assuming again that they run at the same pace as in the first race, does the third race result in a tie?
(f) If not, who wins?
(g) By how many meters?
(h) How far up should Bob start so that the race ends in a tie?

97. Explain the differences among the following three problems. Are there any similarities in their solution?

(a) Write the expression as a single quotient:
$$\frac{x}{x - 2} + \frac{x}{x^2 - 4}$$
(b) Solve: $\dfrac{x}{x - 2} + \dfrac{x}{x^2 - 4} = 0$
(c) Solve: $\dfrac{x}{x - 2} + \dfrac{x}{x^2 - 4} < 0$

98. Make up four problems that you might be asked to do given the two points $(-3, 4)$ and $(6, 1)$. Each problem should involve a different concept. Be sure that your directions are clearly stated.

99. Describe each of the following graphs. Give justification.

(a) $x = 0$ (b) $y = 0$ (c) $x + y = 0$
(d) $xy = 0$ (e) $x^2 + y^2 = 0$

Functions
and Their Graphs

Imagine yourself as an expert for the EPA and sitting in Congressional policy meetings having to explain the importance of balancing energy resources with environmental protection. On the following page is the Internet Excursion placing you in exactly that position and asking the tough questions a member of congress might ask. Use the Sullivan website at:

www.prenhall.com/sullivan

to help link you to the Internet resources you will need to answer the questions asked.

PREPARING FOR THIS CHAPTER

Before getting started on this chapter, review the following concepts;

Topics from Algebra and Geometry *(Appendix, Section 1)*

Graphs of Certain Equations *(Example 2, p. 20; Example 7, p. 26,)*

Figure 35, p. 27; Example 9, p. 27

Tests for Symmetry of an Equation (p. 25)

Procedure for Finding Intercepts of an Equation (p. 23)

Linear Curve Fitting *(Chapter 1, Section 1.4)*

OUTLINE

Netsite: http://www.prenhall.com/sullivan/

HOW LONG WILL THE OIL LAST?

The global supply of oil and other sources of energy is more than adequate to meet present needs, but most of this supply is outside of the United States. Currently, oil supplies about 40 percent of the world's energy, with the United States the biggest consumer. Since oil is a nonrenewable energy resource, plans must be made for the eventuality of diminishing supply.

Suppose that you were a consultant for the EPA and members of congress asked you to sit on a Energy Policy Steering committee as an expert analyst. You must convince the committee of the importance of environmental concerns in planning the global energy system. You are very aware of the fact that present modes of energy use and production threaten serious environmental deterioration. Your plan is to create a set of what-if functions that will allow your committee to model many possible scenarios for their policy guidelines.

1. $E(t)$, the first function that you create, will allow you to model United States oil consumption over a period of time. Make a scatterplot using the EPA Data on United States energy consumption, per capita from 1950–1990. Since oil provides 40 percent of the energy consumed, you adjust your figures to get oil consumption per capita. Use the LINear REGression tool on your graphing utility to find the linear function $E(t)$ of best fit for modeling the data. Does this seem like a good model?

2. The second function, $P(t)$, models population. From the United States Census Data make a scatterplot, and then use a QUADratic REGression to model the data. Next, let $O(t) = E(t) \cdot P(t)$. What does $O(t)$ model? Graph $O(t)$. Compare to the actual figures from BP Petroleum.

3. Using the function $O(t)$, estimate the area under the graph for the years 1997 until 2000. To do this make a trapezoid, and find its area. Now estimate the areas under $O(t)$ using 10-year intervals. For what year does the total of all the trapezoid areas equal the remaining United States oil reserve?

4. Could this sort of analysis be used to predict when the world's supply of oil will run out? Can your committee think of any methods that might improve your predictions?

Perhaps the most central idea in mathematics is the notion of a *function*. This important chapter deals with what a function is, how to graph functions, how to perform operations on functions, and how functions are used in applications.

The word *function* apparently was introduced by René Descartes in 1637. For him, a function simply meant any positive integral power of a variable *x*. Gottfried Wilheim von Leibniz (1646–1716), who always emphasized the geometric side of mathematics, used the word function to denote any quantity associated with a curve, such as the coordinates of a point on the curve. Leonhard Euler (1707–1783) employed the word to mean any equation or formula involving variables and constants. His idea of a function is similar to the one most often used today in courses that precede calculus. Later, the use of functions in investigating heat flow equations led to a very broad definition, due to Lejeune Dirichlet (1805–1859), which describes a function as a rule or correspondence between two sets. It is his definition that we use here.

2.1   FUNCTIONS

1   Determine Whether a Relation Represents a Function
2   Identify the Graph of a Function
3   Find the Value of a Function
4   Find the Domain of a Function
5   Obtain Information from/about the Graph of a Function

In Section 1.1, we said a **relation** is a correspondence between two variables, say x and y, and can be written as a set of ordered pairs (x, y). When relations are written as ordered pairs (x, y), we say x is related to y. Often, we are interested in specifying the type of relation (such as an equation) that might exist between the two variables. For example, when variables appear to be linearly related (based on the scatter diagram), we find a line of best fit. In Section 1.4, we found that fertilizer and crop yield appeared to be linearly related and the line of best fit was $y = 0.717x + 4.786$, where x represents the amount of fertilizer and y represents the crop yield. Thus, if we know how much fertilizer has been used, we can predict the yield by using the rule $y = 0.717x + 4.786$. This rule is an example of a *function*.

As another example, suppose an icicle falls off a building from a height of 64 feet above the ground. According to a law of physics, the distance s (in feet) of the icicle from the ground after t seconds is given (approximately) by the formula $s = 64 - 16t^2$. When $t = 0$ seconds, the icicle is $s = 64$ feet above the ground. After 1 second, the icicle is $s = 64 - 16(1)^2 = 48$ feet above the ground. After 2 seconds, the icicle strikes the ground. The formula $s = 64 - 16t^2$ provides a way of finding the distance s when the time t ($0 \leq t \leq 2$) is prescribed. There is a correspondence between each time t in the interval $0 \leq t \leq 2$ and the distance s. We say that the distance s is a *function* of the time t because

1.   There is a correspondence between the set of times and the set of distances.
2.   There is exactly one distance s obtained for a prescribed time t in the interval $0 \leq t \leq 2$.

Let's now look at the definition of a function.

Definition of Function

Let X and Y be two nonempty sets of real numbers.* A **function** from X into Y is a rule or a correspondence that associates with each element of X a unique element of Y.

---

*The two sets X and Y can also be sets of complex numbers, (discussed in Section 3.6), and then we have defined a complex function. In the broad definition (due to Lejeune Dirichlet), X and Y can be any two sets.

FIGURE 1

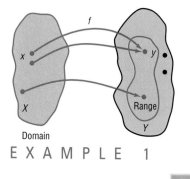

Domain

The set X is called the **domain** of the function. For each element x in X, the corresponding element y in Y is called the **value** of the function at x, or the **image** of x. The set of all images of the elements of the domain is called the **range** of the function. See Figure 1.

Since there may be some elements in Y that are not the image of some x in X, it follows that the range of a function may be a subset of Y, as shown in Figure 1.

E X A M P L E 1

Determining Whether a Relation Represents a Function

Determine whether the following relations represent functions.

(a) For this relation, the domain represents the employees of Sara's Pre-Owned Car Mart and the range represents their base salary.

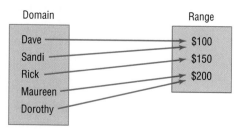

(b) For this relation, the domain represents the employees of Sara's Pre-Owned Car Mart and the range represents their phone number(s).

Domain	Range
Dave	555 – 2345
Sandi	549 – 9402
Rick	930 – 3956
Maureen	555 – 8294
Dorothy	839 – 9013

Solution (a) The relation is a function because each element in the domain corresponds to a unique element in the range. Notice that more than one element in the domain can correspond to the same element in the range.

(b) The relation is not a function because each element in the domain does not correspond to a unique element in the range. Maureen has two telephone numbers; therefore, if Maureen is chosen from the domain, a unique telephone number cannot be assigned to her. ∎

If the relation is a function, then an element in the domain allows for unique identification of the corresponding element in the range. However, if the relation is not a function, then some element in the domain does not allow for unique identification of the corresponding value in the range. For example, if Maureen is chosen from the domain in Example 1(a), she is identified with a unique base salary of $200, so base salary is a function of the person selected. However, if Maureen is chosen from the domain in Example 1(b), she is not identified with a unique telephone number (we don't know whether to call her at 555-8294 or 930-3956), so telephone number is not a function of the person selected.

We may think of a function as a set of ordered pairs (x, y), in which no two distinct pairs have the same first element. Therefore, a function is a specific type of relation. The set of all first elements x is the domain of the function, and the set of all second elements y is its range. Thus, there is associated with each element x in the domain a unique element y in the range.

E X A M P L E   2

Determining Whether a Relation Represents a Function

Determine whether each relation represents a function.

(a) $\{(1, 4), (2, 5), (3, 6), (4, 7)\}$
(b) $\{(1, 4), (2, 4), (3, 5), (6, 10)\}$
(c) $\{(-3, 9), (-2, 4), (0, 0), (1, 1), (-3, 8)\}$

Solution
(a) This relation is a function because there are no distinct ordered pairs with the same first element.
(b) This relation is a function because there are no distinct ordered pairs with the same first element.
(c) This relation is not a function because there is a first element, -3, that corresponds to two different second elements, 9 and 8. ▬

In Example 2(b), notice that 1 and 2 in the domain each have the same image in the range. This does not violate the definition of a function—two different first elements can have the same second element. A violation of the definition occurs when two ordered pairs have the same first element and different second elements, as in Example 2(c).

The rule (or correspondence) referred to in the definition of a function is most often given as an equation in two variables, usually denoted x and y.

E X A M P L E   3

Example of a Function

Consider the function defined by the equation

$$y = 2x - 5 \qquad 1 \leq x \leq 6$$

The domain $1 \leq x \leq 6$ specifies that the number x is restricted to the real numbers from 1 to 6, inclusive. The rule $y = 2x - 5$ specifies that the number x is to be multiplied by 2 and then 5 is to be subtracted from the result to get y. For example, the value of the function at $x = \frac{3}{2}$ (that is, the image of $x = \frac{3}{2}$) is $y = 2 \cdot \frac{3}{2} - 5 = -2$. ▬

Now work Problems 1 and 5.

The Graph of a Function

In applications, a graph often demonstrates more clearly the relationship between two variables than, say, an equation or table would. For example, Table 1 shows the price per share of McDonald's Corp. stock at the end of each week from Sept. 6, 1996 to Dec. 13, 1996.

When we use functions in applications, the domain may be restricted by physical or geometric considerations. For example, the domain of the function f defined by $f(x) = x^2$ is the set of all real numbers. However, if f is used as the rule for obtaining the area of a square when the length x of a side is known, then we must restrict the domain of f to the positive real numbers, since the length of a side can never be 0 or negative.

Independent Variable; Dependent Variable

Consider a function $y = f(x)$. The variable x is called the **independent variable,** because it can be assigned any of the permissible numbers from the domain. The variable y is called the **dependent variable,** because its value depends on x.

Any symbol can be used to represent the independent and dependent variables. For example, if f is the *cube function,* then f can be defined by $f(x) = x^3$ or $f(t) = t^3$ or $f(z) = z^3$. All three rules are identical: Each tells us to cube the independent variable. In practice, the symbols used for the independent and dependent variables are based on common usage.

In applications, it is often apparent from the data which variable is the independent variable and which is the dependent variable. For example, with time-series data, the independent variable is time. Since we can only obtain one value of a variable at any point in time, time-series data always represents a function. With other applications, it is helpful to think of the independent variable as the variable that determines the value of the dependent variable.

E X A M P L E   9   Construction Cost

Sally, a builder of homes, is interested in finding a function that relates the cost C of building a house and x, the number of square feet of the house so she can easily provide estimates to clients for various sized homes. The data she obtained from constructing homes last year is listed in Table 2.

TABLE 2	
Square Feet, x	Cost, C
1500	165,000
1600	176,300
1700	183,000
1700	189,500
1830	201,700
1970	217,000
2050	220,000
2100	237,400

(a) Does the relation defined by the set of ordered pairs (x, C) represent a function?

(b) Using a graphing utility, draw a scatter diagram of the data.

(c) Using a graphing utility, find the line of best fit relating square feet and cost.

(d) Interpret the slope.

(e) Express the relationship found in (c) using function notation.

(f) What is the domain of the function?

(g) What is the cost to build a 2000 square foot house?

 (h) Give some reasons to explain why the cost of building a 1700 square foot house might cost $183,000 in one case, and $189,500 in another. Does this also explain why, perhaps, $C(2000) > C(2050)$?

Solution (a) No, because the first element 1700 has, corresponding to it, two second elements, 183,000 and 189,500.

(b) See Figure 7.

FIGURE 7

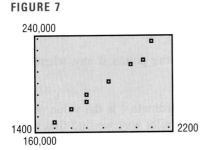

(c) The line of best fit is $y = 112.09x - 3734.02$.

(d) The cost per square foot to build a house is approximately $112.09. Thus, for each additional square foot of house being built, the cost increases by $112.09.

(e) Using function notation, $C(x) = 112.09x - 3734.02$.

(f) The domain is $\{x | x > 0\}$ since a house cannot have 0 or negative square feet.

(g) The cost to build a 2000 square foot house is

$$C(2000) = 112.09(2000) - 3734.02 \approx \$220,446$$

(h) One explanation is that the fixtures placed in one house are more expensive than those placed in the other house. ▬

Observe in the solution to Example 9 that we used the symbol C in two ways: It is used to name the function, and is used to symbolize the dependent variable. This double use is common in applications and should not cause any difficulty.

Refer back to Example 9. Notice the data in Table 2 does not represent a function since the same first element is paired with two different second elements ((1700, 183000), (1700, 189500)). However, when we find the line of best fit for this data, we find a function that relates the data. Thus, **curve fitting** is a process whereby we find a functional relationship between two or more variables even though the data, itself, may not represent a function.

E X A M P L E   10   Area of a Circle

Express the area of a circle as a function of its radius.

Solution We know that the formula for the area A of a circle of radius r is $A = \pi r^2$. If we use r to represent the independent variable and A to represent the dependent variable, the function expressing this relationship is

$$A(r) = \pi r^2$$

In this setting, the domain is $\{r | r > 0\}$. (Do you see why?) ▬

 Now work Problems 73 and 81.

Examples 9 and 10 demonstrate that some functions are determined from data while others are based on geometric, physical, or other relations.

The distance d_2 from the landing point to town is $12 - x$, and the time t_2 it takes to cover this distance at an average walking speed of 5 miles per hour obey the equation

$$d_2 = 5t_2$$

Thus,

$$t_2 = \frac{d_2}{5} = \frac{12 - x}{5}$$

The total time T of the trip is $t_1 + t_2$. Thus,

$$T(x) = \frac{\sqrt{4 + x^2}}{3} + \frac{12 - x}{5}$$

(b) Since x equals the distance from P to where the boat lands, it follows that the domain of T is $0 \le x \le 12$.

(c) If the boat is landed 4 miles from P, then $x = 4$. The time T the trip takes is

$$T(4) = \frac{\sqrt{20}}{3} + \frac{8}{5} \approx 3.09 \text{ hours}$$

FIGURE 10

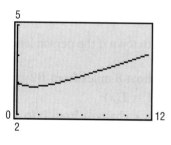

(d) If the boat is landed 8 miles from P, then $x = 8$. The time T the trip takes is

$$T(8) = \frac{\sqrt{68}}{3} + \frac{4}{5} \approx 3.55 \text{ hours}$$

(e) See Figure 10.

(f) As x varies from 0 to 12, the time T varies from about 2.93 to about 4.05.

(g) Using the TRACE function, for x approximately 1.53 miles, the time T is least, about 2.93 hours. See Figure 11(a).

Note: Most graphing utilities have a function minimum command that determines the minimum value of a function for a specified domain. Consult your manual. Using this command, we find that x is approximately 1.50 miles with the time T about 2.93 hours. See Figure 11(b).

FIGURE 11

(a) Using TRACE   **(b)** Using minimum on a TI 83

Now work Problem 83.

SUMMARY

We list here some of the important vocabulary introduced in this section, with a brief description of each term.

Function   A rule or correspondence between two sets of real numbers so that each number x in the first set, the domain, has corresponding to it exactly one number y in the second set.

A set of ordered pairs (x, y) or $(x, f(x))$ in which no two distinct pairs have the same first element.

The range is the set of *y* values of the function for the *x* values in the domain.

A function *f* may be defined implicitly by an equation involving *x* and *y* or explicitly by writing $y = f(x)$.

Unspecified domain   If a function *f* is defined by an equation and no domain is specified, then the domain will be taken to be the largest set of real numbers for which the rule defines a real number.

Function notation $y = f(x)$

f is a symbol for the rule that defines the function.

x is the independent variable.

y is the dependent variable.

$f(x)$ is the value of the function at *x*, or the image of *x*.

Graph of a function   The collection of points (x, y) that satisfies the equation $y = f(x)$.

A collection of points is the graph of a function provided vertical lines intersect the graph in at most one point (vertical-line test).

2.1 EXERCISES

In Problems 1–12, determine whether each relation represents a function.

1.

2.

3.

4.

5. $\{(2, 6), (-3, 6), (4, 9), (1, 10)\}$

6. $\{(-2, 5), (-1, 3), (3, 7), (4, 12)\}$

7. $\{(1, 3), (2, 3), (3, 3), (4, 3)\}$

8. $\{(0, -2), (1, 3), (2, 3), (3, 7)\}$

9. $\{(-2, 4), (-2, 6), (0, 3), (3, 7)\}$

10. $\{(-4, 4), (-3, 3), (-2, 2), (-1, 1), (-4, 0)\}$

11. $\{(-2, 4), (-1, 1), (0, 0), (1, 1)\}$

12. $\{(-2, 16), (-1, 4), (0, 3), (1, 4)\}$

In Problems 13–20, find the following values for each function:
(a) $f(0)$; (b) $f(1)$; (c) $f(-1)$; (d) $f(-x)$; (e) $-f(x)$; (f) $f(x + 1)$

13. $f(x) = -3x^2 + 2x - 4$

14. $f(x) = 2x^2 + x - 1$

15. $f(x) = \dfrac{x}{x^2 + 1}$

16. $f(x) = \dfrac{x^2 - 1}{x + 4}$

17. $f(x) = |x| + 4$

18. $f(x) = \sqrt{x^2 + x}$

19. $f(x) = \dfrac{2x + 1}{3x - 5}$

20. $f(x) = 1 - \dfrac{1}{(x + 2)^2}$

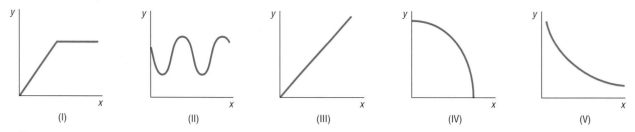

(I)   (II)   (III)   (IV)   (V)

64. Match each of the following functions with the graph that best describes the situation.
    (a) The temperature of a bowl of soup as a function of time.
    (b) The number of hours of daylight during the year.
    (c) The population of Florida as a function of time.
    (d) The distance of a car traveling at a constant velocity as a function of time.
    (e) The height of a golf ball hit with a 7-iron as a function of time.

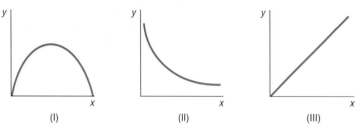

(I)   (II)   (III)   (IV)   (V)

65. Consider the following scenario: Barbara decides to take a walk. She leaves home, walks 2 blocks in 5 minutes at a constant speed, and realizes she forgot to lock the door. So, Barbara runs home in 1 minute. While at her doorstep it takes her 1 minute to find her keys and lock the door. Barbara walks 5 blocks in 15 minutes, then decides to jog home. It takes her 7 minutes to get home. Draw a graph of Barbara's distance from home (in blocks) as a function of time.

66. Consider the following scenario: Jayne enjoys riding her bicycle through the woods. At the forest preserve, she gets on her bicycle and rides up a 2,000-foot hill in 10 minutes. She then travels down the hill in 3 minutes. The next 5,000 feet is level terrain and she covers the distance in 20 minutes. She rests for 15 minutes. Jayne then travels 10,000 feet in 30 minutes. Draw a graph of Jayne's distance traveled (in feet) as a function a of time.

67. If $f(x) = 2x^3 + Ax^2 + 4x - 5$ and $f(2) = 5$, what is the value of A?

68. If $f(x) = 3x^2 - Bx + 4$ and $f(-1) = 12$, what is the value of B?

69. If $f(x) = (3x + 8)/(2x - A)$ and $f(0) = 2$, what is the value of A?

70. If $f(x) = (2x - B)/(3x + 4)$ and $f(2) = \frac{1}{2}$, what is the value of B?

71. If $f(x) = (2x - A)/(x - 3)$ and $f(4) = 0$, what is the value of A? Where is f not defined?

72. If $f(x) = (x - B)/(x - A)$, $f(2) = 0$, and $f(1)$ is undefined, what are the values of A and B?

73. **Demand for Jeans**   The marketing manager at Levi-Strauss wishes to find a function that relates the demand D of men's jeans and p, the price of the jeans. The following data was obtained based on a price history of the jeans.

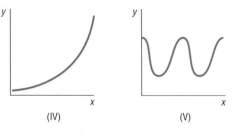

Price ($/Pair), p	Demand (Pairs of Jeans Sold per Day), D
20	60
22	57
23	56
23	53
27	52
29	49
30	44

(a) Does the relation defined by the set of ordered pairs (p, D) represent a function?

(b) Using a graphing utility, draw a scatter diagram of the data.

(c) Using a graphing utility, find the line of best fit relating price and quantity demanded.

(d) Interpret the slope.

(e) Express the relationship found in (c) using function notation.

(f) What is the domain of the function?

(g) How many jeans will be demanded if the price is $28 a pair?

74. Advertising and Sales Revenue A marketing firm wishes to find a function that relates the sales S of a product and A, the amount spent on advertising the product. The data are obtained from past experience. Advertising and sales are measured in thousands of dollars.

Advertising Expenditures, A	Sales, S
20	335
22	339
22.5	338
24	343
24	341
27	350
28.3	351

(a) Does the relation defined by the set of ordered pairs (A, S) represent a function?
(b) Using a graphing utility, draw a scatter diagram of the data.
(c) Using a graphing utility, find the line of best fit relating advertising expenditures and sales.
(d) Interpret the slope.
(e) Express the relationship found in (c) using function notation.
(f) What is the domain of the function?
(g) Predict sales if advertising expenditures are $25,000.

75. Distance and Time Using the data below, find a function that relates the distance, s, in miles, driven by a Ford Taurus, and t, the time the Taurus has been driven.

(a) Does the relation defined by the set of ordered pairs (t, s) represent a function?
(b) Using a graphing utility, draw a scatter diagram of the data.

Time (Hours), t	Distance (Miles), s
0	0
1	30
2	55
3	83
4	100
5	150
6	210
7	260
8	300

(c) Using a graphing utility, find the line of best fit relating time and distance.
(d) Interpret the slope.
(e) Express the relationship found in (c) using function notation.
(f) What is the domain of the function?
(g) Predict the distance the car is driven after 11 hours.

76. High School vs. College GPA An administrator at Southern Illinois University wants to find a function that relates a student's college grade point average G to the high school grade point average, x. She randomly selects 8 students and obtains the following data:

High School GPA, x	College GPA, G
2.73	2.43
2.92	2.97
3.45	3.63
3.78	3.81
2.56	2.83
2.98	2.81
3.67	3.45
3.10	2.93

(a) Does the relation defined by the set of ordered pairs (x, G) represent a function?
(b) Using a graphing utility, draw a scatter diagram of the data.
(c) Using a graphing utility, find the line of best fit relating high school GPA and college GPA.
(d) Interpret the slope.
(e) Express the relationship found in (c) using function notation.
(f) What is the domain of the function?
(g) Predict a student's college GPA if her high school GPA is 3.23.

77. Effect of Gravity on Earth If a rock falls from a height of 20 meters on Earth, the height H (in meters) after x seconds is approximately

$$H(x) = 20 - 4.9x^2$$

(a) Use a graphing utility to graph the function H.
(b) What is the height of the rock when $x = 1$ second? $x = 1.1$ seconds? $x = 1.2$ seconds? $x = 1.3$ seconds?
(c) When is the height of the rock 15 meters? When is it 10 meters? When is it 5 meters?
(d) When does the rock strike the ground?

2.2 MORE ABOUT FUNCTIONS

1. Find the Average Rate of Change of a Function
2. Use a Graphing Utility to Determine Where a Function is Increasing and Decreasing
3. Use a Graphing Utility to Locate Local Maxima and Minima
4. Determine Even or Odd Functions from a Graph
5. Identify Even or Odd Functions from the Equation
6. Graph Certain Important Functions
7. Graph Piecewise-Defined Functions

Average Rate of Change

1. In Section 1.3 we said the slope of a straight line could be interpreted as the average rate of change. Often, we are interested in the rate at which functions change. To find the average rate of change of a function between any two points on its graph, we calculate the slope of the line containing the two points.

E X A M P L E   1   Finding the Average Rate of Change of a Function

Suppose you drop a ball from a cliff 1000 feet high. You measure the distance s the ball has fallen after time t using a motion detector and obtain the data in Table 3.

(a) Draw a scatter diagram of the data by hand treating time as the independent variable.
(b) Draw a line from the point $(0, 0)$ to $(2, 64)$.
(c) Find the average rate of change of the ball between 0 and 2 seconds, that is, find the slope of the line in (b).
(d) Interpret the average rate of change found in (c).
(e) Draw a line from the point $(5, 400)$ to $(7, 784)$.
(f) Find the average rate of change of the ball between 5 and 7 seconds.
(g) Interpret the average rate of change found in (f).
(h) What is happening to the average rate of change as time passes?

TABLE 3	
Time, t (Seconds)	Distance, s (Feet)
0	0
1	16
2	64
3	144
4	256
5	400
6	576
7	784

Solution

(a) We plot the ordered pairs $(0, 0)$, $(1, 16)$, and so on, using rectangular coordinates. See Figure 12.
(b) Draw a line from $(0, 0)$ to $(2, 64)$. See Figure 13.
(c) The average rate of change is found by computing the slope of the line formed by joining the points $(0, 0)$ and $(2, 64)$.

$$\text{Average Rate of Change} = \frac{\Delta s}{\Delta t} = \frac{64 - 0}{2 - 0} = \frac{64}{2} = 32 \text{ ft/sec}$$

(d) Since the change in distance divided by the change in time represents a speed, we would interpret the slope of the line as an average speed. The average speed of the ball between 0 and 2 seconds is 32 feet per second.
(e) Draw a line from $(5, 400)$ to $(7, 784)$. See Figure 14.

FIGURE 12

FIGURE 13

FIGURE 14

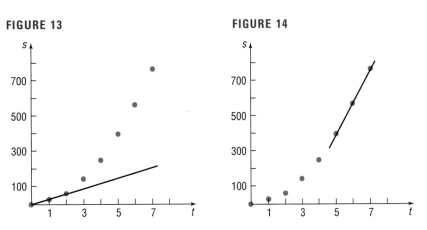

(f) The average rate of change of the ball between 5 and 7 seconds is

$$\text{Average Rate of Change} = \frac{\Delta s}{\Delta t} = \frac{784 - 400}{7 - 5} = \frac{384}{2} = 192 \ \text{ft/sec}$$

(g) The average speed of the ball between 5 and 7 seconds is 192 feet per second.

(h) The average speed of the ball is increasing as time passes since the ball is accelerating due to the effect of gravity. ▬

If we know the function f that relates the time t to the distance s the ball has fallen, $s = f(t)$, then the average speed of the ball between 0 and 2 seconds is

$$\frac{f(2) - f(0)}{2 - 0} = \frac{64 - 0}{2} = 32 \ \text{ft/sec}$$

The average speed of the ball between 5 and 7 seconds is

$$\frac{f(7) - f(5)}{7 - 5} = \frac{784 - 400}{2} = 192 \ \text{ft/sec}$$

Expressions like these occur frequently in calculus.

If c is in the domain of a function $y = f(x)$, the **average rate of change of f** between c and x is defined as

$$\text{average rate of change} = \frac{\Delta y}{\Delta x} = \frac{f(x) - f(c)}{x - c} \qquad x \neq c \qquad (1)$$

This expression is also called the **difference quotient** of f at c.

The average rate of change of a function has an important geometric interpretation. Look at the graph of $y = f(x)$ in Figure 15. We have labeled two points on the graph: $(c, f(c))$ and $(x, f(x))$. The slope of the line containing these two points is

$$\frac{f(x) - f(c)}{x - c}$$

This line is called a **secant line.** Thus, the average rate of change of a function equals the slope of a secant line containing two points on its graph.

FIGURE 15

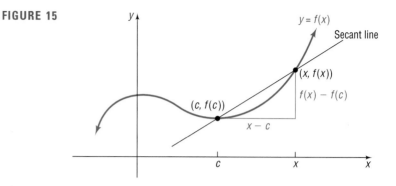

E X A M P L E 2

Finding the Average Rate of Change

The function $s(t) = -4.9t^2 + 100t + 5$ gives the height s (in meters) of a bullet fired straight up as a function of time t (in seconds).

(a) Find the average rate of change of the height of the bullet between 2 and t seconds.

(b) Using the result found in (a), find the average rate of change of the height of the bullet between 2 and 3 seconds.

Solution (a) From expression (1), we seek

$$\frac{\Delta s}{\Delta t} = \frac{s(t) - s(2)}{t - 2}, \qquad t \neq 2$$

We begin by finding $s(2)$:

$$s(2) = -4.9(2)^2 + 100(2) + 5 = -19.6 + 200 + 5 = 185.4$$

Then the average rate of change of s between 2 and t seconds is

$$\frac{s(t) - s(2)}{t - 2} = \frac{-4.9t^2 + 100t + 5 - 185.4}{t - 2} = \frac{-4.9t^2 + 100t - 180.4}{t - 2}$$

$$= \frac{(-4.9t + 90.2)(t - 2)}{t - 2} = -4.9t + 90.2$$

(b) The average rate of change between 2 and 3 seconds is found by letting $t = 3$ in the expression found in (a). Thus, the average rate of change of the bullet between 2 and 3 seconds is $-4.9(3) + 90.2 = 75.5$ meters per second. ■

Now work Problem 31.

Increasing and Decreasing Functions

Consider the graph given in Figure 16. If you look from left to right along the graph of the function, you will notice that parts of the graph are rising, parts are falling, and parts are horizontal. In such cases, the function is described as *increasing, decreasing,* and *constant,* respectively.

FIGURE 16

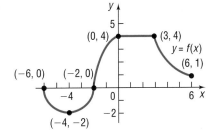

E X A M P L E   3

Determining Where a Function Is Increasing, Decreasing, or Constant

Where is the function in Figure 16 increasing? Where is it decreasing? Where is it constant?

Solution

To answer the question of where a function is increasing, where it is decreasing, and where it is constant, we use inequalities involving the independent variable x or we use open intervals* of x-coordinates. The graph in Figure 16 is rising (increasing) from the point $(-4, -2)$ to the point $(0, 4)$, so we conclude that it is increasing on the open interval $(-4, 0)$ (or for $-4 < x < 0$). The graph is falling (decreasing) from the point $(-6, 0)$ to the point $(-4, -2)$ and from the point $(3, 4)$ to the point $(6, 1)$. We conclude that the graph is decreasing on the open intervals $(-6, -4)$ and $(3, 6)$ (or for $-6 < x < -4$ and $3 < x < 6$). The graph is constant on the open interval $(0, 3)$ (or for $0 < x < 3$). ▬

More precise definitions follow.

> A function f is **increasing** on an open interval I if, for any choice of x_1 and x_2 in I, with $x_1 < x_2$, we have $f(x_1) < f(x_2)$.

> A function f is **decreasing** on an open interval I if, for any choice of x_1 and x_2 in I, with $x_1 < x_2$, we have $f(x_1) > f(x_2)$.

> A function f is **constant** on an open interval I if, for all choices of x in I, the values $f(x)$ are equal.

Thus, the graph of an increasing function goes up from left to right, the graph of a decreasing function goes down from left to right, and the graph of a constant function remains at a fixed height. Figure 17 illustrates the definitions.

*The open interval (a, b) consists of all real numbers x for which $a < x < b$. Refer to Section 1 of the Appendix, if necessary.

FIGURE 17

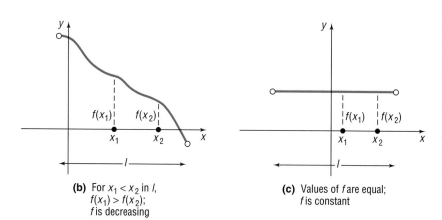

(a) For $x_1 < x_2$ in I,
$f(x_1) < f(x_2)$;
f is increasing

(b) For $x_1 < x_2$ in I,
$f(x_1) > f(x_2)$;
f is decreasing

(c) Values of f are equal;
f is constant

E X A M P L E   4   Using a Graphing Utility to Determine Where a Function Is Increasing and Decreasing

 Use a graphing utility to graph the function $f(x) = x^3 - 2x^2 + 3$ for $-1 < x < 2$. Determine where f is increasing and where it is decreasing.

Solution

FIGURE 18

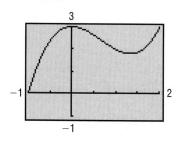

Figure 18 shows the graph of f. Use TRACE, beginning at the point $(-1, 0)$. As we proceed along the graph, the values of y increase until we cross the y-axis (where $y = 3$). We infer that f is increasing on the interval $(-1, 0)$.

As we continue along the graph to the right of the y-axis, the values of f decrease until we reach $x \approx 1.33$ (where $y \approx 1.815$). We infer that f is decreasing on the interval $(0, 1.33)$.

As we continue, the values of y increase from $x = 1.33$ to $x = 2$. We infer that f is increasing on the interval $(1.33, 2)$.   ▬

Local Maximum; Local Minimum

When the graph of a function is increasing to the left of $x = c$ and decreasing to the right of $x = c$, then at c the value of f is largest. This value is called a *local maximum* of f.

When the graph of a function is decreasing to the left of $x = c$ and is increasing to the right of $x = c$, then at c the value of f is the smallest. This value is called a *local minimum* of f.

Thus, if f has a local maximum at c, then the value of f at c is greater than the values of f near c. If f has a local minimum at c, then the value of f at c is less than the values of f near c. The word *local* is used to suggest that it is only near c that the value $f(c)$ is largest or smallest. See Figure 19.

FIGURE 19
f has local maximum at x_1 and x_3; f has a
local minimum at x_2.

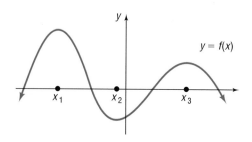

A function *f* has a **local maximum at *c*** if there is an interval *I* containing *c* so that, for all *x* in *I*, $f(x) < f(c)$. We call $f(c)$ a **local maximum of *f*.**

A function *f* has a **local minimum at *c*** if there is an interval *I* containing *c* so that, for all *x* in *I*, $f(x) > f(c)$. We call $f(c)$ a **local minimum of *f*.**

For example, in Figure 18, *f* has a local maximum at 0 and a local minimum at 1.33. The local maximum is $f(0) = 3$; the local minimum is $f(1.33) = 1.815$.

To locate the exact value at which a function *f* has a local maximum or a local minimum usually requires calculus. However, a graphing utility may be used to approximate these values.

E X A M P L E   5   Using a Graphing Utility to Locate Local Maxima and Minima

Use a graphing utility to graph $f(x) = 6x^3 - 12x + 5$ for $-2 < x < 2$. Determine where *f* has a local maximum and where *f* has a local minimum.

Solution   Figure 20 shows the graph of *f* on the interval $(-2, 2)$.

Using TRACE, we find the local maximum to occur at approximately -0.80 and the local minimum to occur at 0.80, correct to two decimal places. The local maximum is 11.53 and the local minimum is -1.53, correct to two decimal places. See Figures 20(a) and (b).

FIGURE 20

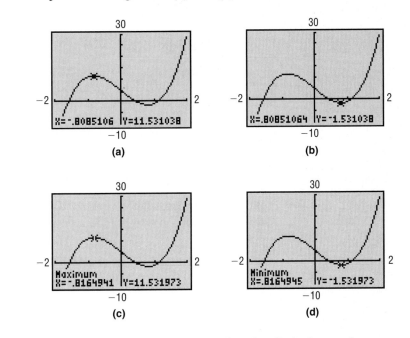

(a)

(b)

(c)

(d)

Most graphing utilities have a function that finds the maximum or minimum point of a graph within a given domain. Using this function, we find the maximum occurs at $(-0.81, 11.53)$ and the minimum occurs at $(0.81, -1.53)$, correct to two decimal places, for $-2 < x < 2$, as seen in Figures 20(c) and (d).

Now work Problem 95.

Even and Odd Functions

A function *f* is **even** if for every number *x* in its domain the number −*x* is also in the domain and

$$f(-x) = f(x)$$

In other words, if $y = f(x)$, then *f* is even if and only if whenever the point (x, y) is on the graph of *f*, then the point $(-x, y)$ is also on the graph.

A function *f* is **odd** if for every number *x* in its domain the number −*x* is also in the domain and

$$f(-x) = -f(x)$$

In other words, if $y = f(x)$, then *f* is odd if and only if whenever the point (x, y) is on the graph of *f*, then the point $(-x, -y)$ is also on the graph.

Refer to Section 1.2, where the tests for symmetry are listed. The following results are then evident.

Theorem

A function is even if and only if its graph is symmetric with respect to the *y*-axis. A function is odd if and only if its graph is symmetric with respect to the origin.

E X A M P L E   6

Determining Even and Odd Functions from the Graph

4

Determine whether each graph given in Figure 21 is the graph of an even function, an odd function, or a function that is neither even nor odd.

FIGURE 21

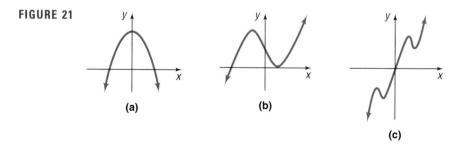

(a)   (b)

(c)

Solution   The graph in Figure 21(a) is that of an even function, because the graph is symmetric with respect to the *y*-axis. The function whose graph is given in

Figure 21(b) is neither even nor odd, because the graph is neither symmetric with respect to the y-axis nor symmetric with respect to the origin. The function whose graph is given in Figure 21(c) is odd, because its graph is symmetric with respect to the origin. ■

Now work Problem 9.

A graphing utility can be used to conjecture whether a function is even, odd, or neither. As stated, a function is even if $f(-x) = f(x)$. This condition implies that when an even function contains the point (x, y) it must also contain the point $(-x, y)$. Therefore, if TRACE indicates that both the point (x, y) and the point $(-x, y)$ are on the graph for every x, then we would conjecture that the function is even.*

In addition, a function is odd if $f(-x) = -f(x)$. This condition implies that an odd function contains the points $(-x, -y)$ and (x, y). TRACE could be used in the same way to conjecture that the function is odd.*

In the next example, we show how to verify whether a function is even, odd, or neither.

5

E X A M P L E   7

Identifying Even and Odd Functions

Determine whether each of the following functions is even, odd, or neither. Then determine whether the graph is symmetric with respect to the y-axis or with respect to the origin.

(a) $f(x) = x^2 - 5$ (b) $g(x) = x^3 - 1$

(c) $h(x) = 5x^3 - x$ (d) $F(x) = |x|$

Solution (a) Graph the function. Use TRACE to determine different pairs of points (x, y) and $(-x, y)$. For example, the point $(1.957447, -1.168402)$ is on the graph. See Figure 22(a). Now move the cursor to -1.957447 and determine the corresponding y-coordinate. See Figure 22(b). Since $(1.957447, -1.168402)$ and $(-1.957447, -1.168402)$ both lie on the graph, we have evidence that the function is even. Repeating this procedure yields similar results. Therefore, we conjecture that the function is even.

FIGURE 22

(a)   (b)

To algebraically verify the conjecture, we replace x by $-x$ in $f(x) = x^2 - 5$. Then

$$f(-x) = (-x)^2 - 5 = x^2 - 5 = f(x)$$

Since $f(-x) = f(x)$, we conclude that f is an even function, and the graph is symmetric with respect to the y-axis.

*−Xmin and Xmax must be equal for this to work.

(b) Graph the function. Using TRACE, the point (1.787234, 4.7087929) is on the graph. See Figure 23(a). Now move the cursor to -1.787234 and determine the corresponding y-coordinate, -6.708793. See Figure 23(b). We conjecture that the function is neither even nor odd since (x, y) does not equal $(-x, y)$ (so it is not even) and (x, y) also does not equal $(-x, -y)$ (so it is not odd).

FIGURE 23

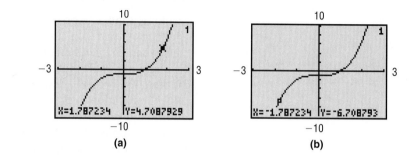

(a)   (b)

To algebraically verify the conjecture, we replace x by $-x$. Then

$$g(-x) = (-x)^3 - 1 = -x^3 - 1$$

Since $g(-x) \neq g(x)$ and $g(-x) \neq -g(x)$, we conclude that g is neither even nor odd. The graph is not symmetric with respect to the y-axis nor with respect to the origin.

(c) Graph the function. Using the TRACE, the point (1.0212766, 4.3047109) is on the graph. See Figure 24(a). We move the cursor to -1.021277 and determine the corresponding y-coordinate, -4.304711. See Figure 24(b). Since (x, y) equals $(-x, -y)$, correct to five decimal places, we have evidence that the function is odd. An examination of other pairs of points leads to the conjecture that the function is odd.

FIGURE 24

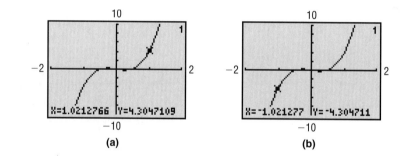

(a)   (b)

To algebraically verify the conjecture, we replace x by $-x$ in $h(x) = 5x^3 - x$. Then

$$h(-x) = 5(-x)^3 - (-x) = -5x^3 + x = -(5x^3 - x) = -h(x)$$

Since $h(-x) = -h(x)$, h is an odd function, and the graph of h is symmetric with respect to the origin.

(d) Graph the function. Using TRACE, the point $(-5.744681, 5.7446809)$ is on the graph. See Figure 25(a). We move the cursor to 5.7446809 and determine the corresponding y-coordinate, 5.7446809. See Figure 25(b). Since (x, y) equals $(-x, y)$, we have evidence that the function is even. Repeating this procedure yields similar results. We conjecture that the function is even.

FIGURE 25

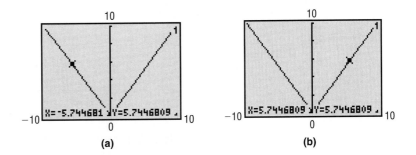

(a)   (b)

To algebraically verify the conjecture, we replace x by $-x$ in $F(x) = |x|$. Then

$$F(-x) = |-x| = |x| = F(x)$$

Since $F(-x) = F(x)$, F is an even function, and the graph of F is symmetric with respect to the y-axis. ▬

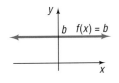 **Now work Problem 41.**

Library of Functions

We now give names to some of the functions we have encountered. In going through this list, pay special attention to the characteristics of each function, particularly to the shape of each graph. Knowing these graphs will lay the foundation for later graphing techniques.

Linear Functions

$f(x) = mx + b$ m and b are real numbers

The domain of a **linear function** f consists of all real numbers. The graph of this function is a nonvertical straight line with slope m and y-intercept b. A linear function is increasing if $m > 0$, decreasing if $m < 0$, and constant if $m = 0$.

Constant Function

$f(x) = b$ b a real number

FIGURE 26

See Figure 26.

A **constant function** is a special linear function ($m = 0$). Its domain is the set of all real numbers; its range is the set consisting of a single number b. Its graph is a horizontal line whose y-intercept is b. The constant function is an even function whose graph is constant over its domain.

Identity Function

$$f(x) = x$$

See Figure 27.

FIGURE 27

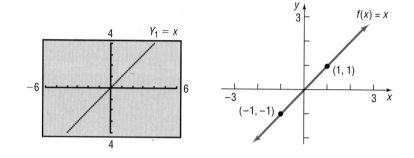

The **identity function** is also a special linear function. Its domain and its range are the set of all real numbers. Its graph is a line whose slope is $m = 1$ and whose y-intercept is 0. The line consists of all points for which the x-coordinate equals the y-coordinate. The identity function is an odd function that is increasing over its domain. Note that the graph bisects quadrants I and III.

Square Function

$$f(x) = x^2$$

See Figure 28.

FIGURE 28

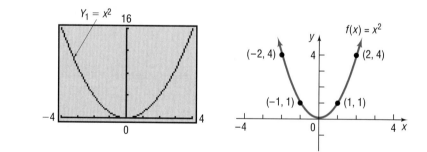

The domain of the **square function** f is the set of all real numbers; its range is the set of nonnegative real numbers. The graph of this function is a parabola, whose intercept is at $(0, 0)$. The square function is an even function that is descreasing on the interval $(-\infty, 0)$ and increasing on the interval $(0, \infty)$.

Cube Function

$$f(x) = x^3$$

See Figure 29.

FIGURE 29

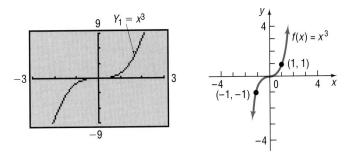

The domain and range of the **cube function** are the set of all real numbers. The intercept of the graph is at $(0, 0)$. The cube function is odd and is increasing on the interval $(-\infty, \infty)$.

Square Root Function

$$f(x) = \sqrt{x}$$

See Figure 30.

FIGURE 30

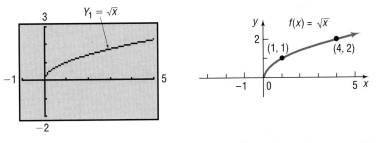

The domain and range of the **square root function** are the set of nonnegative real numbers. The intercept of the graph is at $(0, 0)$. The square root function is neither even nor odd and is increasing on the interval $(0, \infty)$.

Reciprocal Function

$$f(x) = \frac{1}{x}$$

Refer to Example 9, p. 27, for a discussion of the equation $y = 1/x$. See Figure 31.

FIGURE 31

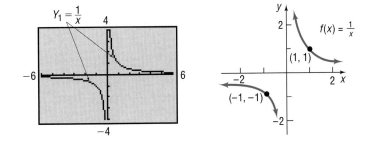

The domain and range of the **reciprocal function** are the set of all nonzero real numbers. The graph has no intercepts. The reciprocal function is decreasing on the intervals $(-\infty, 0)$ and $(0, \infty)$ and is an odd function.

Absolute Value Function

$$f(x) = |x|$$

See Figure 32.

FIGURE 32

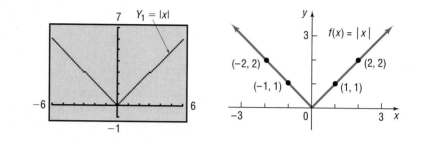

The domain of the **absolute value function** is the set of all real numbers; its range is the set of nonnegative real numbers. The intercept of the graph is at $(0, 0)$. If $x \geq 0$, then $f(x) = x$ and the graph of f is part of the line $y = x$; if $x < 0$, then $f(x) = -x$ and the graph of f is part of the line $y = -x$. The absolute value function is an even function; it is decreasing on the interval $(-\infty, 0)$ and increasing on the interval $(0, \infty)$.

Comment: If your utility has no built-in absolute value function, you can still graph $f(x) = |x|$ by using the fact that $|x| = \sqrt{(x^2)}$.

The notation $\text{int}(x)$ stands for the largest integer less than or equal to x. For example,

$$\text{int}(1) = 1 \qquad \text{int}(2.5) = 2 \qquad \text{int}(\tfrac{1}{2}) = 0 \qquad \text{int}(\tfrac{-3}{4}) = -1 \qquad \text{int}(\pi) = 3$$

This type of correspondence occurs frequently enough in mathematics that we give it a name.

Greatest-Integer Function

$$f(x) = \text{int}(x) = \text{Greatest integer less than or equal to } x$$

We obtain the graph of $f(x) = \text{int}(x)$ by plotting several points. See Table 4. For values of x, $-1 \leq x < 0$, the value of $f(x) = \text{int}(x)$ is -1; for values of x, $0 \leq x < 1$, the value of f is 0. See Figure 33 for the graph.

TABLE 4	
X	Y1
-1	-1
-.75	-1
-.5	-1
-.25	-1
0	0
.25	0
.5	0
Y1∎int(X)	

FIGURE 33

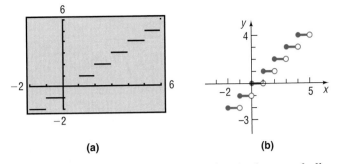

(a)   (b)

The domain of the **greatest integer function** is the set of all real numbers; its range is the set of integers. The y-intercept of the graph is at 0. The x-intercepts lie in the interval $[0, 1)$. The greatest-integer function is neither even nor odd. It is constant on every interval of the form $[k, k + 1)$, for k an integer. In Figure 33(b), we use a solid dot to indicate, for example, that at $x = 1$ the value of f is $f(1) = 1$; we use an open circle to illustrate that the function does not assume the value of 0 at $x = 1$.

From the graph of the greatest-integer function, we can see why it is also called a **step function.** At $x = 0, x = \pm 1, x = \pm 2$, and so on, this function exhibits what is called a *discontinuity;* that is, at integer values, the graph suddenly "steps" from one value to another without taking on any of the intermediate values. For example, to the immediate left of $x = 3$, the y-coordinates are 2, and to the immediate right of $x = 3$, the y-coordinates are 3.

Comment: When graphing a function, you can choose either the **connected mode,** in which points plotted on the screen are connected, making the graph appear without any breaks, or the **dot mode,** in which only the points plotted appear. When graphing the greatest integer function with a graphing utility, it is necessary to be in the **dot mode.** This is to prevent the utility from "connecting the dots" when $f(x)$ changes from one integer value to the next. See Figure 34.

FIGURE 34
$y = \text{int}(x)$

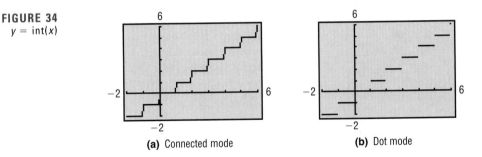

(a) Connected mode   (b) Dot mode

The functions we have discussed so far are basic. Whenever you encounter one of them, you should see a mental picture of its graph. For example, if you encounter the function $f(x) = x^2$, you should see in your mind's eye a picture like Figure 28.

Now work Problems 1–8.

Piecewise Defined Functions

7

Sometimes, a function is defined by a rule consisting of two or more equations. The choice of which equation to use depends on the value of the independent variable x. For example, the absolute value function $f(x) = |x|$ is actually defined by two equations: $f(x) = x$ if $x \geq 0$ and $f(x) = -x$ if $x < 0$. For convenience, we generally combine these equations into one expression as

$$f(x) = |x| = \begin{cases} x & \text{if } x \geq 0 \\ -x & \text{if } x < 0 \end{cases}$$

When functions are defined by more than one equation, they are called **piecewise defined** functions.

Let's look at another example of a piecewise defined function.

EXAMPLE 8

Analyzing a Piecewise Defined Function

For the following function f,

$$f(x) = \begin{cases} -x + 1 & \text{if } -1 \leq x < 1 \\ 2 & \text{if } x = 1 \\ x^2 & \text{if } x > 1 \end{cases}$$

(a) Find $f(0)$, $f(1)$, and $f(2)$ (b) Determine the domain of f.
(c) Graph f. (d) Use the graph to find the range of f.

Solution (a) To find $f(0)$, we observe that when $x = 0$ the equation for f is given by $f(x) = -x + 1$. So we have

$$f(0) = -0 + 1 = 1$$

When $x = 1$, the equation for f is $f(x) = 2$. Thus

$$f(1) = 2$$

When $x = 2$, the equation for f is $f(x) = x^2$. So

$$f(2) = 2^2 = 4$$

(b) To find the domain of f, we look at its definition. We conclude that the domain of f is $\{x | x \geq -1\}$, or $[-1, \infty)$.

(c) On a graphing utility, the procedure for graphing a piecewise defined function varies depending on the particular utility. In general, you need to enter each piece as a function with a restricted domain.

To graph the middle piece [the point $(1, 2)$], use the $\boxed{\text{STAT} \atop \text{PLOT}}$ function. See Figure 35(a).

In graphing piecewise defined functions, it is usually better to be in the dot mode, since such functions may have breaks (discontinuities).

To graph f on paper, we graph "each piece." Thus, we first graph the line $y = -x + 1$ and keep only the part for which $-1 \leq x < 1$. Then we plot the point $(1, 2)$, because when $x = 1$, $f(x) = 2$. Finally, we graph the parabola $y = x^2$ and keep only the part for which $x > 1$. See Figure 35(b).

FIGURE 35

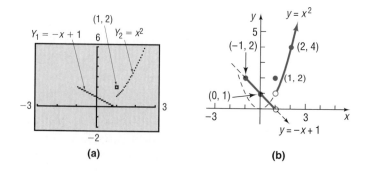

(a)   (b)

(d) From the graph, we conclude that the range of f is $\{y|y > 0\}$, or $(0, \infty)$.

Now work Problem 69.

E X A M P L E   9

Cost of Electricity

In the winter, Commonwealth Edison Company supplies electricity to residences for a monthly customer charge of \$8.91 plus 10.494¢ per kilowatt-hour (kWhr) for the first 400 kWhr supplied in the month, and 7.91¢ per kWhr for all usage over 400 kWhr in the month.*

(a) What is the charge for using 300 kWhr in a month?
(b) What is the charge for using 700 kWhr in a month?
(c) If C is the monthly charge for x kWhr, express C as a function of x.

Solution   (a) For 300 kWhr, the charge is \$8.91 plus 10.494¢ = \$0.10494 per kWhr. Thus,

$$\text{Charge} = \$8.91 + \$0.10494(300) = \$40.39$$

(b) For 700 kWhr, the charge is \$8.91 plus 10.494¢ for the first 400 kWhr plus 7.91¢ for the 300 kWhr in excess of 400. Thus,

$$\text{Charge} = \$8.91 + \$0.10494(400) + \$0.0791(300) = \$74.62$$

(c) If $0 \le x \le 400$, the monthly charge C (in dollars) can be found by multiplying x times \$0.10494 and adding the monthly customer charge of \$8.91. Thus, if $0 \le x \le 400$, then $C(x) = 0.10494x + 8.91$. For $x > 400$, the charge is $0.10494(400) + 8.91 + 0.0791(x - 400)$, since $x - 400$ equals the usage in excess of 400 kWhr, which costs \$0.0791 per kWhr. Thus, if $x > 400$, then

$$
\begin{aligned}
C(x) &= 0.10494(400) + 8.91 + 0.0791(x - 400) \\
&= 50.89 + 0.0791(x - 400) \\
&= 0.0791x + 19.25
\end{aligned}
$$

The rule for computing C follows two equations:

$$
C(x) = \begin{cases} 0.10494x + 8.91 & \text{if } 0 \le x \le 400 \\ 0.0791x + 19.25 & \text{if } x > 400 \end{cases}
$$

See Figure 36 for the graph.

FIGURE 36

Source: Commonwealth Edison Co., Chicago, Illinois, 1997.

2.2 EXERCISES

In Problems 1–8, match each graph to the function listed whose graph most resembles the one given.

A. *Constant function*
B. *Linear function*
C. *Square function*
D. *Cube function*
E. *Square root function*
F. *Reciprocal function*
G. *Absolute value function*
H. *Greatest-integer function*

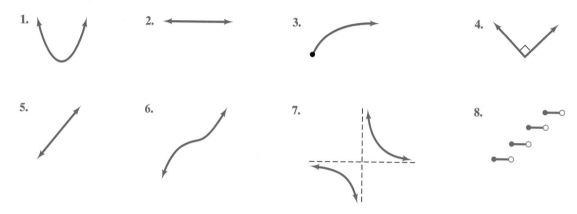

1.　　　　2.　　　　3.　　　　4.

5.　　　　6.　　　　7.　　　　8.

In Problems 9–24, the graph of a function is given. Use the graph to find

(a) *Its domain and range*

(b) *The intervals on which it is increasing, decreasing, or constant*

(c) *Whether it is even, odd, or neither*

(d) *The intercepts, if any*

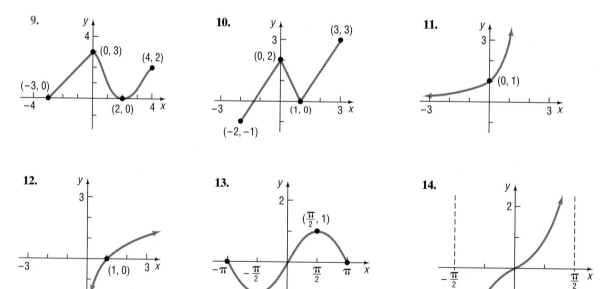

9.

10.

11.

12.

13.

14.

15.

16.

17.

18.

19.

20.

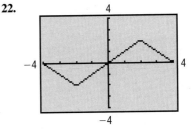

In Problems 21–24, assume that the entire graph is shown.

21.

22.

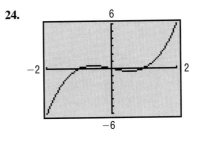

23.

24.

25. If

$$f(x) = \begin{cases} x^2 & \text{if } x < 0 \\ 2 & \text{if } x = 0 \\ 2x + 1 & \text{if } x > 0 \end{cases}$$

find: (a) $f(-2)$ (b) $f(0)$ (c) $f(2)$

26. If

$$f(x) = \begin{cases} x^3 & \text{if } x < 0 \\ 3x + 2 & \text{if } x \geq 0 \end{cases}$$

find: (a) $f(-1)$ (b) $f(0)$ (c) $f(1)$

27. If $f(x) = \text{int}(2x)$, find: (a) $f(1.2)$ (b) $f(1.6)$ (c) $f(-1.8)$

28. If $f(x) = \text{int}(x/2)$, find: (a) $f(1.2)$ (b) $f(1.6)$ (c) $f(-1.8)$

In Problems 29–40, find the average rate of change of f between 1 and x,

$$\frac{f(x) - f(1)}{x - 1} \qquad x \neq 1$$

for each function f. Be sure to simplify.

29. $f(x) = 3x$

30. $f(x) = -2x$

31. $f(x) = 1 - 3x$

32. $f(x) = x^2 + 1$

33. $f(x) = 3x^2 - 2x$

34. $f(x) = 4x - 2x^2$

35. $f(x) = x^3 - x$

36. $f(x) = x^3 + x$

37. $f(x) = \dfrac{2}{x + 1}$

38. $f(x) = \dfrac{1}{x^2}$

39. $f(x) = \sqrt{x}$

40. $f(x) = \sqrt{x + 3}$

In Problems 41–52, tell whether each function is even, odd, or neither without drawing a graph. Graph each function and use TRACE to verify your results.

41. $f(x) = 4x^3$

42. $f(x) = 2x^4 - x^2$

43. $g(x) = 2x^2 - 5$

44. $h(x) = 3x^3 + 2$

45. $F(x) = \sqrt[3]{x}$

46. $G(x) = \sqrt{x}$

47. $f(x) = x + |x|$

48. $f(x) = \sqrt[3]{2x^2 + 1}$

49. $g(x) = \dfrac{1}{x^2}$

50. $h(x) = \dfrac{x}{x^2 - 1}$

51. $h(x) = \dfrac{x^3}{3x^2 - 9}$

52. $F(x) = \dfrac{x}{|x|}$

53. How many x-intercepts can a function defined on an interval have if it is increasing on that interval? Explain.

54. How many y-intercepts can a function have? Explain.

In Problems 55–80:

(a) Find the domain of each function.

(b) Locate any intercepts.

(c) Graph each function by hand.

(d) Based on the graph, find the range.

(e) Verify your results using a graphing utility.

55. $f(x) = 3x - 3$

56. $f(x) = 4 - 2x$

57. $g(x) = x^2 - 4$

58. $g(x) = x^2 + 4$

59. $h(x) = -x^2$

60. $F(x) = 2x^2$

61. $f(x) = \sqrt{x - 2}$

62. $g(x) = \sqrt{x + 2}$

63. $h(x) = \sqrt{2 - x}$

64. $F(x) = -\sqrt{x}$

65. $f(x) = |x| + 3$

66. $g(x) = |x + 3|$

67. $h(x) = -|x|$

68. $F(x) = |3 - x|$

69. $f(x) = \begin{cases} 2x & \text{if } x \neq 0 \\ 0 & \text{if } x = 0 \end{cases}$

70. $f(x) = \begin{cases} 3x & \text{if } x \neq 0 \\ 4 & \text{if } x = 0 \end{cases}$

71. $f(x) = \begin{cases} 1 + x & \text{if } x < 0 \\ x^2 & \text{if } x \geq 0 \end{cases}$

72. $f(x) = \begin{cases} 1/x & \text{if } x < 0 \\ \sqrt{x} & \text{if } x \geq 0 \end{cases}$

73. $f(x) = \begin{cases} |x| & \text{if } -2 \leq x < 0 \\ 1 & \text{if } x = 0 \\ x^3 & \text{if } x > 0 \end{cases}$

74. $f(x) = \begin{cases} 3 + x & \text{if } -3 \leq x < 0 \\ 3 & \text{if } x = 0 \\ \sqrt{x} & \text{if } x > 0 \end{cases}$

75. $g(x) = \begin{cases} 1 & \text{if } x \text{ is an integer} \\ -1 & \text{if } x \text{ is not an integer} \end{cases}$

76. $g(x) = \begin{cases} x & \text{if } x \geq 1 \\ 1 & \text{if } x < 1 \end{cases}$

77. $h(x) = 2\text{int}(x)$

78. $f(x) = \text{int}(2x)$

79. $F(x) = \begin{cases} 4 - x^2 & \text{if } |x| \leq 2 \\ x^2 - 4 & \text{if } |x| > 2 \end{cases}$

80. $G(x) = |x^2 - 4|$

Problems 81–84 require the following definition. Secant Line: The slope of the secant line containing the two points $(x, f(x))$ and $(x + h, f(x + h))$ on the graph of a function $y = f(x)$ may be given as

$$\frac{f(x + h) - f(x)}{h}$$

In Problems 81–84, express the slope of the secant line of each function in terms of x and h. Be sure to simplify your answer.

81. $f(x) = 2x + 5$

82. $f(x) = -3x + 2$

83. $f(x) = x^2 + 2x$

84. $f(x) = 1/x$

In Problems 85–88, the graph of a piecewise-defined function is given. Write a definition for each function.

85.

86.

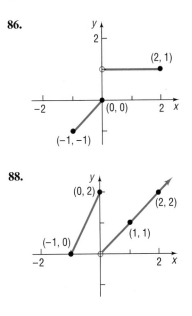

87.

graph with points (−1, 1), (0, 0), (1, 1), (2, 0)

88.

graph with points (0, 2), (−1, 0), (1, 1), (2, 2)

89. Growth of Bacteria The following data represents the population of an unknown bacteria.

Time (Days)	Population
0	50
1	153
2	234
3	357
4	547
5	839
6	1280

(a) Draw a scatter diagram of the data by hand treating the population as the independent variable.
(b) Draw a line from the point (0, 50) to (1, 153) on the scatter diagram found in (a).
(c) Find the average rate of change of the population between 0 and 1 days.
(d) Interpret the average rate of change found in (c).
(e) Draw a line from the point (5, 839) to (6, 1280) on the scatter diagram found in (a).
(f) Find the average rate of change of the population between 5 and 6 days.
(g) Interpret the average rate of change found in (f).
(h) What is happening to average rate of change of population as time passes?

90. Cost of Manufacturing Bikes The following data represents the monthly cost of producing bicycles at Tunney's Bicycle Shop.

Number of Bicycles, x	Total Cost of Production, C (Dollars)
0	24,000
25	27,750
60	31,500
102	35,250
150	39,000
190	42,750
223	46,500
249	50,250

(a) Draw a scatter diagram of the data by hand treating the number of bicycles produced as the independent variable.
(b) Draw a line from the point (0, 24000) to (25, 27750) on the scatter diagram found in (a).
(c) Find the average rate of change of the cost between 0 and 25 bicycles.
(d) Interpret the average rate of change found in (c). In economics, this is called **marginal cost.**
(e) Draw a line from the point (190, 42750) to (223, 46500) on the scatter diagram found in (a).
(f) Find the average rate of change of the cost between 190 and 223 bicycles.
(g) Interpret the average rate of change found in (f).

91. Revenue from Selling Bikes The following data represents the total revenue that would be received from selling x bicycles at Tunney's Bicycle Shop.

Number of Bicycles, x	Total Revenue, R (Dollars)
0	0
25	28,000
60	45,000
102	53,400
150	59,160
190	62,360
223	64,835
249	66,525

(a) Draw a scatter diagram of the data by hand treating the number of bicycles produced as the independent variable.
(b) Draw a line from the point $(0, 0)$ to $(25, 28000)$ on the scatter diagram found in (a).
(c) Find the average rate of change of revenue between 0 and 25 bicycles.
(d) Interpret the average rate of change found in (c). In economics, this is called **marginal revenue.**
(e) Draw a line from the point $(190, 62360)$ to $(223, 64835)$ on the scatter diagram found in (a).
(f) Find the average rate of change of revenue between 190 and 223 bicycles.
(g) Interpret the average rate of change found in (f).
(h) What is happening to marginal revenue as the number of bicycles sold increases?
(i) Refer to Problem 90(d). A firm's profit will increase as long as marginal revenue is greater than marginal cost. At what level of output does Tunney's Bicycle Shop's profits stop increasing?

92. Cost of Tuition The following data represents the average cost of tuition and required fees (in dollars) at public four-year colleges.

Year	Average Cost of Tuition
1990	2,035
1991	2,159
1992	2,410
1993	2,604
1994	2,822

Source: U.S. National Center for Education Statistics.

(a) Draw a scatter diagram of the data by hand treating the year as the independent variable.
(b) Draw a line from the point $(1990, 2035)$ to $(1992, 2410)$ on the scatter diagram found in (a).
(c) Find the average rate of change of the cost of tuition and required fees between 1990 and 1992.
(d) Interpret the average rate of change found in (c).
(e) Draw a line from the point $(1992, 2410)$ to $(1994, 2822)$ on the scatter diagram found in (a).
(f) Find the average rate of change of the cost of tuition and required fees between 1992 and 1994.
(g) Interpret the average rate of change found in (f).
(h) What is happening to the average rate of change for the cost of tuition and required fees as time passes?

In Problems 93 and 94, decide whether each function is even. Given a reason.

93. $f(x) = \begin{cases} x^2 + 4 & \text{if } x \neq 2 \\ 6 & \text{if } x = 2 \end{cases}$

94. $f(x) = \begin{cases} x^2 + 4 & \text{if } x \neq 2 \\ 5 & \text{if } x = 2 \end{cases}$

In Problems 95–102, use a graphing utility to graph each function over the indicated interval. Determine where the function is increasing and where it is decreasing. Approximate any local maxima and local minima.

95. $f(x) = x^3 - 3x + 2 \quad (-2, 2)$

96. $f(x) = x^3 - 3x^2 + 5 \quad (-1, 3)$

97. $f(x) = x^5 - x^3 \quad (-2, 2)$

98. $f(x) = x^4 - x^2 \quad (-2, 2)$

99. $f(x) = -0.2x^3 - 0.6x^2 + 4x - 6 \quad (-6, 4)$

100. $f(x) = -0.4x^3 + 0.6x^2 + 3x - 2 \quad (-4, 5)$

101. $f(x) = 0.25x^4 + 0.3x^3 - 0.9x^2 + 3 \quad (-3, 2)$

102. $f(x) = -0.4x^4 - 0.5x^3 + 0.8x^2 - 2 \quad (-3, 2)$

103. Maximizing the Volume of a Box   An open box with a square base is to be made from a square piece of cardboard 24 inches on a side by cutting out a square from each corner and turning up the sides. (See the illustration.) The volume V of the box as a function of the length x of the side of the square cut from each corner is

$$V(x) = x(24 - 2x)^2$$

Graph V and determine where V is largest.

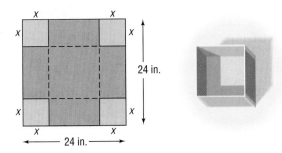

104. Minimizing the Material Needed to Make a Box   An open box with a square base is required to have a volume of 10 cubic feet. The amount A of material used to make such a box as a function of the length x of a side of the square base is

$$A(x) = x^2 + 40/x$$

Graph A and determine where A is smallest.

105. Maximum Height of a Ball   The height s of a ball (in feet) thrown with an initial velocity of 80 feet per second from an initial height of 6 feet is given as a function of the time t (in seconds) by

$$s(t) = -16t^2 + 80t + 6$$

(a) Graph s.
(b) Determine the time at which height is maximum.
(c) What is the maximum height?

106. Minimum Average Cost   The average cost of producing x riding lawn mowers per hour is given by

$$A(x) = 0.3x^2 + 21x - 251 + \frac{2500}{x}$$

(a) Graph A.
(b) Determine the number of riding lawn mowers to produce in order to minimize average cost.
(c) What is the minimum average cost?

107. Graph $y = x^2$. Then on the same screen graph $y = x^2 + 2$, followed by $y = x^2 + 4$, followed by $y = x^2 - 2$. What pattern do you observe? Can you predict the graph of $y = x^2 - 4$? Of $y = x^2 + 5$?

108. Graph $y = x^2$. Then on the same screen graph $y = (x - 2)^2$, followed by $y = (x - 4)^2$, followed by $y = (x + 2)^2$. What pattern do you observe? Can you predict the graph of $y = (x + 4)^2$? Of $y = (x - 5)^2$?

109. Graph $y = |x|$. Then on the same screen graph $y = 2|x|$, followed by $y = 4|x|$, followed by $y = \frac{1}{2}|x|$. What pattern do you observe? Can you predict the graph of $y = \frac{1}{4}|x|$? Of $y = 5|x|$?

110. Graph $y = x^2$. Then on the same screen graph $y = -x^2$. What pattern do you observe? Now try $y = |x|$ and $y = -|x|$. What do you conclude?

111. Graph $y = \sqrt{x}$. Then on the same screen graph $y = \sqrt{-x}$. What pattern do you observe? Now try $y = 2x + 1$ and $y = 2(-x) + 1$. What do you conclude?

112. Graph $y = x^3$. Then on the same screen graph $y = (x - 1)^3 + 2$. Could you have predicted the result?

113. Graph $y = x^2$, $y = x^4$, and $y = x^6$ on the same screen. What do you notice is the same about each graph? What do you notice that is different?

114. Graph $y = x^3$, $y = x^5$, and $y = x^7$ on the same screen. What do you notice is the same about each graph? What do you notice that is different?

115. Cost of Natural Gas   On December 13, 1996, the Peoples Gas Company had the following rate schedule* for natural gas usage in single-family residences:

Monthly service charge	$9.00
Per therm service charge,	
1st 50 therms	$0.36375/therm
Over 50 therms	$0.11445/therm
Gas charge	$0.3256/therm

(a) What is the charge for using 50 therms in a month?
(b) What is the charge for using 500 therms in a month?
(c) Construct a function that relates the monthly charge C for x therms of gas.
(d) Graph this function.

116. Cost of Natural Gas   On December 19, 1996, Northern Illinois Gas Company had the following rate schedule† for natural gas usage in single-family residences:

Monthly customer charge	$6.00
Distribution charge,	
1st 20 therms	$0.2012/therm
Next 30 therms	$0.1117/therm
Over 50 therms	$0.0374/therm
Gas supply charge	$0.3113/therm

*Source: The Peoples Gas Company, Chicago, Illinois.
†Source: Northern Illinois Gas Company, Naperville, Illinois.

(a) What is the charge for using 40 therms in a month?

(b) What is the charge for using 202 therms in a month?

(c) Construct a function that gives the monthly charge C for x therms of gas.

(d) Graph this function.

117. Let f denote any function with the property that, whenever x is in its domain, then so is $-x$. Define the functions $E(x)$ and $O(x)$ to be

$$E(x) = \frac{1}{2}[f(x) + f(-x)]$$

$$O(x) = \frac{1}{2}[f(x) - f(-x)]$$

(a) Show that $E(x)$ is an even function.

(b) Show that $O(x)$ is an odd function.

(c) Show that $f(x) = E(x) + O(x)$.

(d) Draw the conclusion that any such function f can be written as the sum of an even function and an odd function.

118. Let f and g be two functions defined on the same interval $[a, b]$. Suppose that we define two functions min (f, g) and max (f, g) as follows:

$$\min(f, g)(x) = \begin{cases} f(x) & \text{if } f(x) \le g(x) \\ g(x) & \text{if } f(x) > g(x) \end{cases}$$

$$\max(f, g)(x) = \begin{cases} g(x) & \text{if } f(x) \le g(x) \\ f(x) & \text{if } f(x) > g(x) \end{cases}$$

Show that

$$\min(f, g)(x) = \frac{f(x) + g(x)}{2} - \frac{|f(x) - g(x)|}{2}$$

Develop a similar formula for $\max(f, g)$.

119. Consider the equation

$$y = \begin{cases} 1 & \text{if } x \text{ is rational} \\ 0 & \text{if } x \text{ is irrational} \end{cases}$$

Is this a function? What is its domain? What is its range? What is its y-intercept, if any? What are its x-intercepts, if any? Is it even, odd, or neither? How would you describe its graph?

120. Define some functions that pass through $(0, 0)$ and $(1, 1)$ and are increasing for $x \ge 0$. Begin your list with $y = \sqrt{x}, y = x,$ and $y = x^2$. Can you propose a general result about such functions?

121. Can you think of a function that is both even and odd?

2.3 GRAPHING TECHNIQUES

1. Graph Functions Using Horizontal and Vertical Shifts
2. Graph Functions Using Compressions and Stretches
3. Graph Functions Using Reflections about the x-Axis or y-Axis

At this stage, if you were asked to graph any of the functions defined by $y = x, y = x^2, y = x^3, y = \sqrt{x}, y = |x|,$ or $y = 1/x$, your response should be, "Yes, I recognize these functions and know the general shapes of their graphs." (If this is not your answer, review the previous section and Figures 27 through 32.)

Sometimes, we are asked to graph a function that is "almost" like one we already know how to graph. In this section, we look at some of these functions and develop techniques for graphing them. Collectively, these techniques are referred to as **transformations.**

EXPLORATION   **Vertical Shifts**

1   On the same screen, graph each of the following functions:

$$Y_1 = x^2$$
$$Y_2 = x^2 + 1$$
$$Y_3 = x^2 + 2$$
$$Y_4 = x^2 - 1$$
$$Y_5 = x^2 - 2$$

What do you observe?

Result Figure 37 illustrates the graphs. You should have observed a general pattern. With $Y_1 = x^2$ on the screen, the graph of $Y_2 = x^2 + 1$ is identical to that of $Y_1 = x^2$, except that it is shifted vertically up 1 unit. Similarly, $Y_3 = x^2 + 2$ is identical to that of $Y_1 = x^2$, except that it is shifted vertically up 2 units. The graph of $Y_4 = x^2 - 1$ is identical to that of $Y_1 = x^2$, except that it is shifted vertically down 1 unit.

FIGURE 37

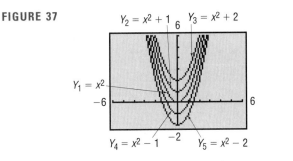

We are led to the following conclusion:

> If a real number c is added to the right side of a function $y = f(x)$, the graph of the new function $y = f(x) + c$ is the graph of f **shifted vertically** up (if $c > 0$) or down (if $c < 0$).

Let's look at an example.

E X A M P L E   1   Vertical Shift Down

Use the graph of $f(x) = x^2$ to obtain the graph of $h(x) = x^2 - 4$.

Solution Table 5 lists some points on the graphs of $f = Y_1$ and $h = Y_2$. The graph of h is identical to that of f, except that it is shifted down 4 units. See Figure 38.

FIGURE 38

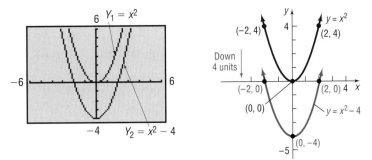

Horizontal Shifts

E X P L O R A T I O N   On the same screen, graph each of the following functions:

$$Y_1 = x^2$$
$$Y_2 = (x - 1)^2$$
$$Y_3 = (x - 3)^2$$
$$Y_4 = (x + 2)^2$$

What do you observe?

Result

FIGURE 39

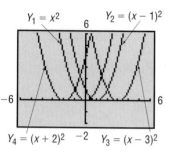

$Y_1 = x^2$ $Y_2 = (x-1)^2$

$Y_4 = (x+2)^2$ $Y_3 = (x-3)^2$

Figure 39 illustrates the graphs.

You should have observed the following pattern. With the graph of $Y_1 = x^2$ on the screen, the graph of $Y_2 = (x-1)^2$ is identical to that of $Y = x^2$, except it is shifted horizontally to the right 1 unit. Similarly, the graph of $Y_3 = (x-3)^2$ is identical to that of $Y_1 = x^2$, except it is shifted horizontally to the right 3 units. Finally, the graph of $Y_4 = (x+2)^2$ is identical to that of $Y_1 = x^2$, except it is shifted horizontally to the left 2 units. ▬

We are led to the following conclusion.

> If a real number c is added to the argument x of a function f, the graph of the new function $g(x) = f(x + c)$ is the graph of f **shifted horizontally** left (if $c > 0$) or right (if $c < 0$).

Now work Problem 31.

Vertical and horizontal shifts are sometimes combined.

E X A M P L E   2

FIGURE 40

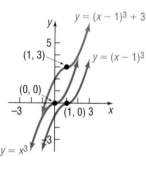

$y = (x-1)^3 + 3$

$(1, 3)$

$y = (x-1)^3$

$(0, 0)$

$(1, 0)$ 3

$y = x^3$

Combining Vertical and Horizontal Shifts

Graph the function $f(x) = (x - 1)^3 + 3$.

Solution

We graph f in steps. First, we note that the rule for f is basically a cube function. Thus, we begin with the graph of $y = x^3$, shown in blue in Figure 40. Next, to get the graph of $y = (x - 1)^3$, we shift the graph of $y = x^3$ horizontally 1 unit to the right. See the graph shown in red in Figure 40. Finally, to get the graph of $y = (x - 1)^3 + 3$, we shift the graph of $y = (x - 1)^3$ vertically up 3 units. See the graph shown in green in Figure 40. Note the points that have been plotted on each graph. Using key points such as these can be helpful in keeping track of just what is taking place. ▬

In Example 2, if the vertical shift had been done first, followed by the horizontal shift, the final graph would have been the same. (Try it for yourself.)

Now work Problem 43.

Compressions and Stretches

E X P L O R A T I O N

FIGURE 41

2

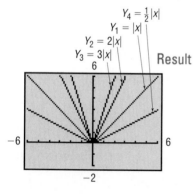

$Y_4 = \frac{1}{2}|x|$

$Y_1 = |x|$

$Y_2 = 2|x|$

$Y_3 = 3|x|$

On the same screen, graph each of the following functions:

$$Y_1 = |x|$$
$$Y_2 = 2|x|$$
$$Y_3 = 3|x|$$
$$Y_4 = \frac{1}{2}|x|$$

Result

Figure 41 illustrates the graphs. You should have observed the following pattern. The graphs of $Y_2 = 2|x|$ and $Y_3 = 3|x|$ can be obtained from the graph of $Y_1 = |x|$ by multiplying each y-coordinate of $Y_1 = |x|$ by factors of 2 and 3, respectively. This is sometimes referred to as a vertical *stretch* using factors of 2 and 3.

The graph of $Y_4 = \frac{1}{2}|x|$ can be obtained from the graph of $Y_1 = |x|$ by multiplying each y-coordinate by $\frac{1}{2}$. This is sometimes referred to as a vertical *compression* using a factor of $\frac{1}{2}$.

Look at Tables 6 and 7, where $Y_1 = |x|$, $Y_3 = 3|x|$, and $Y_4 = \frac{1}{2}|x|$. Notice that the values for Y_3 in Table 6 are three times the values of Y_1. Therefore, the graph of Y_3 will be vertically *stretched* by a factor of 3. Likewise, the values of Y_4 in Table 7 are half the values of Y_1. Therefore, the graph of Y_4 will be vertically *compressed* by a factor of $\frac{1}{2}$.

> When the right side of a function $y = f(x)$ is multiplied by a positive number k, the graph of the new function $y = kf(x)$ is a **vertically compressed** (if $0 < k < 1$) or **stretched** (if $k > 1$) version of the graph of $y = f(x)$.

Now work Problem 33.

If the argument x of a function $y = f(x)$ is multiplied by a positive number k, the graph of the new function $y = f(kx)$ is also a compressed or stretched version of the graph of $y = f(x)$, but, in this case, it occurs horizontally rather than vertically. To see why, we look at the following Exploration.

E X P L O R A T I O N   On the same screen, graph each of the following functions:

$$Y_1 = f(x) = x^2$$
$$Y_2 = f(2x) = (2x)^2$$
$$Y_3 = f\left(\frac{1}{2}x\right) = \left(\frac{1}{2}x\right)^2$$

Result   You should have obtained the graphs shown in Figure 42. The graph of $Y_2 = (2x)^2$ is the graph of $Y_1 = x^2$ compressed horizontally. Look at Table 8(a). Notice $(1, 1)$, $(2, 4)$, $(4, 16)$, and $(16, 256)$ are points on the graph of $Y_1 = x^2$. Also, $(0.5, 1)$, $(1, 4)$, $(2, 16)$, and $(8, 256)$ are points on the graph of $Y_2 = (2x)^2$. Thus, the graph of $Y_2 = (2x)^2$ is obtained by multiplying the x-coordinate of each point on the graph of $Y_1 = x^2$ by 1/2. The graph of $Y_3 = \left(\frac{1}{2}x\right)^2$ is the graph of $Y_1 = x^2$ stretched horizontally. Look at Table 8(b). Notice $(0.5, 0.25)$, $(1, 1)$, $(2, 4)$, $(4, 16)$, are points on the graph of $Y_1 = x^2$. Also, $(1, 0.25)$, $(2, 1)$, $(4, 4)$, $(8, 16)$ are points on the graph of $Y_3 = \left(\frac{1}{2}x\right)^2$. Thus, the graph of $Y_3 = \left(\frac{1}{2}x\right)^2$ is obtained by multiplying the x-coordinate of each point on the graph of $Y_1 = x^2$ by a factor of 2.

FIGURE 42

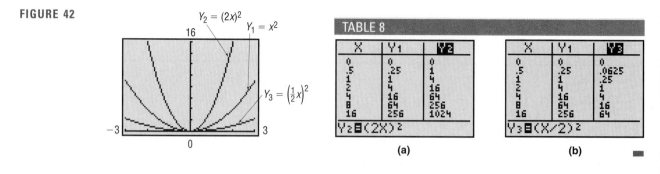

TABLE 8

(a)   (b)

E X A M P L E   3   Graphing Using Stretches and Compressions

The graph of $y = f(x)$ is given in Figure 43. Use this graph to find the graphs of

(a) $y = 3f(x)$ (b) $y = f(3x)$

FIGURE 43

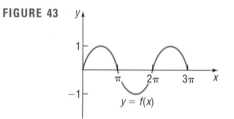

$y = f(x)$

Solution   (a) The graph of $y = 3f(x)$ is obtained by multiplying each y-coordinate of
$y = f(x)$ by a factor of 3. See Figure 44(a).

(b) The graph of $y = f(3x)$ is obtained from the graph of $y = f(x)$ by multiplying each x-coordinate of $y = f(x)$ by a factor of $\frac{1}{3}$. See Figure 44(b).

FIGURE 44

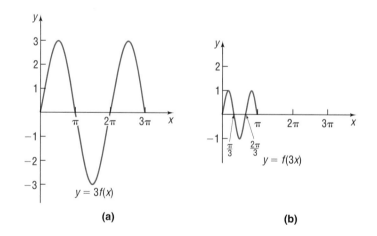

(a)   (b)

Reflections about the x-Axis and the y-Axis

E X P L O R A T I O N   Reflection about the x-Axis

3

(a) Graph $Y_1 = x^2$ followed by $Y_2 = -x^2$.

(b) Graph $Y_1 = |x|$ followed by $Y_2 = -|x|$.

(c) Graph $Y_1 = x^2 - 4$ followed by $Y_2 = -(x^2 - 4) = -x^2 + 4$.

Result See Tables 9(a), (b), and (c) and Figures 45(a), (b), and (c). In each instance, the second graph is the reflection about the x-axis of the first graph.

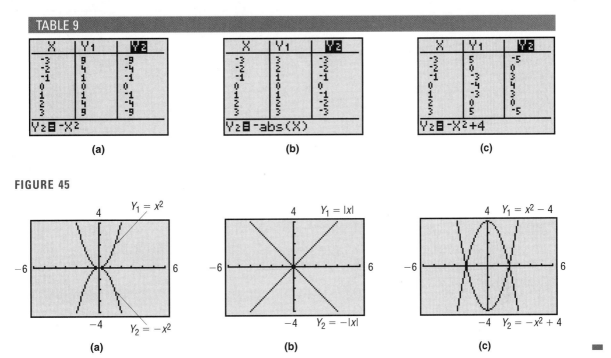

TABLE 9

(a)   (b)   (c)

FIGURE 45

(a)   (b)   (c)

When the right side of the equation $y = f(x)$ is multiplied by -1, the graph of the new function $y = -f(x)$ is the **reflection about the x-axis** of the graph of the function $y = f(x)$.

Now work Problem 39.

E X P L O R A T I O N   (a) Graph $Y_1 = \sqrt{x}$ followed by $Y_2 = \sqrt{-x}$
(b) Graph $Y_1 = x + 1$ followed by $Y_2 = -x + 1$.
(c) Graph $Y_1 = x^4 + x$ followed by $Y_2 = (-x)^4 + (-x) = x^4 - x$.

Result See Tables 10(a), (b), and (c) and Figures 46(a), (b), and (c). In each instance, the second graph is the reflection about the y-axis of the first graph.

TABLE 10

(a)   (b)   (c)

FIGURE 46

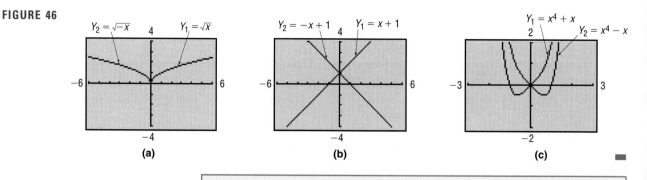

(a)   (b)   (c)

When the graph of the function $y = f(x)$ is known, the graph of the new function $y = f(-x)$ is the **reflection about the y-axis** of the graph of the function $y = f(x)$.

SUMMARY OF GRAPHING TECHNIQUES

Table 11 summarizes the graphing procedures we have just discussed.

TABLE 11

To Graph:	Draw the Graph of f and:
Vertical shifts	
$\quad y = f(x) + c, \quad c > 0$	Raise the graph of f by c units.
$\quad y = f(x) - c, \quad c > 0$	Lower the graph of f by c units.
Horizontal shifts	
$\quad y = f(x + c), \quad c > 0$	Shift the graph of f to the left c units.
$\quad y = f(x - c), \quad c > 0$	Shift the graph of f to the right c units.
Compressing or stretching	
$\quad y = kf(x), \quad k > 0$	Multiply each y value of $y = f(x)$ by k.
$\quad y = f(kx), \quad k > 0$	Multiply each x coordinate of $y = f(x)$ by $\dfrac{1}{k}$.
Reflection about the x-axis	
$\quad y = -f(x)$	Reflect the graph of f about the x-axis.
Reflection about the y-axis	
$\quad y = f(-x)$	Reflect the graph of f about the y-axis.

The examples that follow combine some of the procedures outlined in this section to get the required graph.

E X A M P L E   4   Combining Graphing Procedures

Graph the function $f(x) = \dfrac{3}{x - 2} + 1$.

Solution   We use the following steps to obtain the graph of f:

STEP 1: $y = \dfrac{1}{x}$ Reciprocal function.

STEP 2: $y = \dfrac{3}{x}$ Vertical stretch of the graph of $y = \dfrac{1}{x}$ by a factor of 3; multiply by 3.

STEP 3: $y = \dfrac{3}{x - 2}$ Horizontal shift to the right 2 units; replace x by $x - 2$.

STEP 4: $y = \dfrac{3}{x - 2} + 1$ Vertical shift up 1 unit; add 1.

See Figure 47.

FIGURE 47

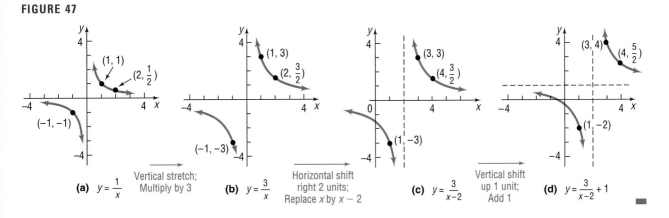

(a) $y = \dfrac{1}{x}$ Vertical stretch; Multiply by 3   **(b)** $y = \dfrac{3}{x}$ Horizontal shift right 2 units; Replace x by $x - 2$ **(c)** $y = \dfrac{3}{x-2}$ Vertical shift up 1 unit; Add 1   **(d)** $y = \dfrac{3}{x-2} + 1$

There are other orderings of the steps shown in Example 4 that would also result in the graph of f. For example, try this one:

STEP 1: $y = \dfrac{1}{x}$ Reciprocal function.

STEP 2: $y = \dfrac{1}{x - 2}$ Horizontal shift to the right 2 units; replace x by $x - 2$.

STEP 3: $y = \dfrac{3}{x - 2}$ Vertical stretch of the graph of $y = \dfrac{1}{x - 2}$ by factor of 3; multiply by 3.

STEP 4: $y = \dfrac{3}{x - 2} + 1$ Vertical shift up 1 unit; add 1.

 Now work Problem 45.

E X A M P L E 5   Combining Graphing Procedures

Graph the function $f(x) = \sqrt{1 - x} + 2$.

Solution   We use the following steps to get the graph of $y = \sqrt{1 - x} + 2$:

STEP 1: $y = \sqrt{x}$ Square root function.
STEP 2: $y = \sqrt{x + 1}$ Horizontal shift left 1 unit; replace x by $x + 1$.
STEP 3: $y = \sqrt{-x + 1} = \sqrt{1 - x}$ Reflect about y-axis; replace x by $-x$.
STEP 4: $y = \sqrt{1 - x} + 2$ Vertical shift up 2 units; add 2.

See Figure 48.

FIGURE 48

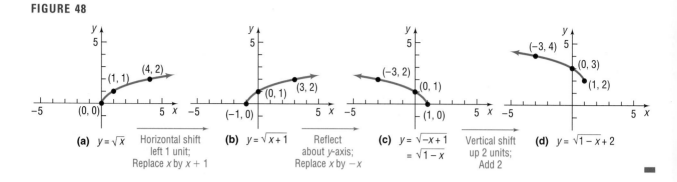

(a) $y = \sqrt{x}$ Horizontal shift left 1 unit; Replace x by $x + 1$

(b) $y = \sqrt{x + 1}$ Reflect about y-axis; Replace x by $-x$

(c) $y = \sqrt{-x + 1}$ $= \sqrt{1 - x}$ Vertical shift up 2 units; Add 2

(d) $y = \sqrt{1 - x} + 2$

2.3 | EXERCISES

In Problems 1–12, match each graph to one of the following functions:

A. $y = x^2 + 2$ B. $y = -x^2 + 2$ C. $y = |x| + 2$ D. $y = -|x| + 2$

E. $y = (x - 2)^2$ F. $y = -(x + 2)^2$ G. $y = |x - 2|$ H. $y = -|x + 2|$

I. $y = 2x^2$ J. $y = -2x^2$ K. $y = 2|x|$ L. $y = -2|x|$

1.

2.

3.

4.

5.

6.

7.

8.

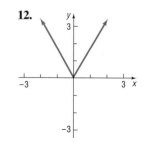

9.

10.

11.

12.

In Problems 13–16, match each graph to one of the following functions:

A. $y = x^3$ B. $y = (x + 2)^3$ C. $y = -2x^3$ D. $y = x^3 + 2$

13.   **14.**

15.   **16.**

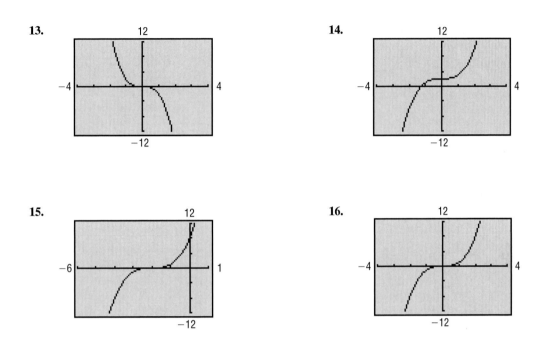

In Problems 17–24, write the function whose graph is the graph of $y = x^3$, but is

17. Shifted to the right 4 units   **18.** Shifted to the left 4 units

19. Shifted up 4 units   **20.** Shifted down 4 units

21. Reflected about the y-axis   **22.** Reflected about the x-axis

23. Vertically stretched by a factor of 4   **24.** Horizontally stretched by a factor of 4

In Problems 25–54, graph each function by hand using the techniques of shifting, compressing, stretching, and/or reflecting. Start with the graph of the basic function (for example, $y = x^2$) and show all stages. Verify your answer by using a graphing utility.

25. $f(x) = x^2 - 1$ **26.** $f(x) = x^2 + 4$ **27.** $g(x) = x^3 + 1$ **28.** $g(x) = x^3 - 1$

29. $h(x) = \sqrt{x - 2}$ **30.** $h(x) = \sqrt{x + 1}$ **31.** $f(x) = (x - 1)^3$ **32.** $f(x) = (x + 2)^3$

33. $g(x) = 4\sqrt{x}$ **34.** $g(x) = \frac{1}{2}\sqrt{x}$ **35.** $h(x) = \dfrac{1}{2x}$ **36.** $h(x) = \dfrac{4}{x}$

37. $f(x) = -|x|$ **38.** $f(x) = -\sqrt{x}$ **39.** $g(x) = -\dfrac{1}{x}$ **40.** $g(x) = -x^3$

41. $h(x) = \text{int}(-x)$ **42.** $h(x) = \dfrac{1}{-x}$ **43.** $f(x) = (x + 1)^2 - 3$ **44.** $f(x) = (x - 2)^2 + 1$

45. $g(x) = \sqrt{x - 2} + 1$ **46.** $g(x) = |x + 1| - 3$ **47.** $h(x) = \sqrt{-x} - 2$ **48.** $h(x) = \dfrac{4}{x} + 2$

49. $f(x) = (x + 1)^3 - 1$ **50.** $f(x) = 4\sqrt{x - 1}$ **51.** $g(x) = 2|1 - x|$ **52.** $g(x) = 4\sqrt{2 - x}$

53. $h(x) = 2\,\text{int}(x - 1)$ **54.** $h(x) = -x^3 + 2$

In Problems 55–60, the graph of a function f is illustrated. Use the graph of f as the first step toward graphing each of the following functions:

(a) $F(x) = f(x) + 3$ (b) $G(x) = f(x + 2)$ (c) $P(x) = -f(x)$

(d) $Q(x) = \frac{1}{2}f(x)$ (e) $g(x) = f(-x)$ (f) $h(x) = 3f(x)$

55.

56.

57.

58.

59.

60.

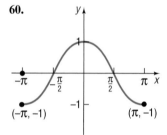

61. Exploration
  (a) Use a graphing utility to graph $y = x + 1$ and $y = |x + 1|$.
  (b) Graph $y = 4 - x^2$ and $y = |4 - x^2|$.
  (c) Graph $y = x^3 + x$ and $y = |x^3 + x|$.
  (d) What do you conclude about the relationship between the graphs of $y = f(x)$ and $y = |f(x)|$?

63. The graph of a function f is illustrated in the figure.
  (a) Draw the entire graph of $y = |f(x)|$.
  (b) Draw the entire graph of $y = f(|x|)$.

62. Exploration
  (a) Use a graphing utility to graph $y = x + 1$ and $y = |x| + 1$.
  (b) Graph $y = 4 - x^2$ and $y = 4 - |x|^2$.
  (c) Graph $y = x^3 + x$ and $y = |x|^3 + |x|$.
  (d) What do you conclude about the relationship between the graphs of $y = f(x)$ and $y = f(|x|)$?

64. The graph of a function f is illustrated in the figure.
  (a) Draw the entire graph of $y = |f(x)|$.
  (b) Draw the entire graph of $y = f(|x|)$.

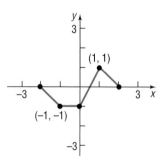

In Problems 65–70, complete the square of each quadratic expression. Then graph each function by hand using the technique of shifting. Verify your results using a graphing utility. (If necessary, refer to Appendix 2 to review Completing the Square.)

65. $f(x) = x^2 + 2x$

66. $f(x) = x^2 - 6x$

67. $f(x) = x^2 - 8x + 1$

68. $f(x) = x^2 + 4x + 2$

69. $f(x) = x^2 + x + 1$

70. $f(x) = x^2 - x + 1$

71. The equation $y = (x - c)^2$ defines a *family of parabolas*, one parabola for each value of c. On one set of coordinate axes, graph the members of the family for $c = 0$, $c = 3$, $c = -2$.

72. Repeat Problem 71 for the family of parabolas $y = x^2 + c$.

73. **Temperature Measurements** The relationship between the Celsius (°C) and Fahrenheit (°F) scales for measuring temperature is given by the equation

$$F = \frac{9}{5}C + 32$$

The relationship between the Celsius (°C) and Kelvin (K) scales is $K = C + 273$. Graph the equation $F = \frac{9}{5}C + 32$ using degrees Fahrenheit on the y-axis and degrees Celsius on the x-axis. Use the techniques introduced in this section to obtain the graph showing the relationship between Kelvin and Fahrenheit temperatures.

74. **Period of a Pendulum** The period T (in seconds) of a simple pendulum is a function of its length l (in feet) defined by the equation

$$T = 2\pi\sqrt{\frac{l}{g}}$$

where $g \approx 32.2$ feet per second per second is the acceleration of gravity.

(a) Use a graphing utility to graph the function $T = T(l)$.

(b) Now graph the functions $T = T(l + 1)$, $T = T(l + 2)$, and $T = T(l + 3)$.

(c) Discuss how adding to the length l changes the period T.

(d) Now graph the functions $T = T(2l)$, $T = T(3l)$, and $T = T(4l)$.

(e) Discuss how multiplying the length l by factors of 2, 3, and 4 changes the period T.

75. **Cigarette Company Profits** The daily profits of a cigarette company from selling x packs of cigarettes are given by

$$p(x) = -0.05x^2 + 100x - 2000$$

The government wishes to impose a tax on cigarettes (sometimes called a **sin tax**) that gives the company the option of either paying a flat tax of \$10,000 per day or a tax of 10% on profits. As Chief Financial Officer (CFO) of the company, you need to decide which tax is the better option for the company.

(a) On the same viewing rectangle graph $Y_1 = p(x) - 10,000$ and $Y_2 = (1 - 0.10)p(x)$.

(b) Based on the graph which option would you select? Why?

(c) Using the terminology learned in this section, describe each graph in terms of the graph of $p(x)$.

(d) Suppose the government offered the option of a flat tax option of \$4,800 or a tax of 10% on profits. Which option would you select? Why?

2.4 OPERATIONS ON FUNCTIONS; COMPOSITE FUNCTIONS

> **1** Form the Sum, Difference, Product, and Quotient of Two Functions
>
> **2** Form the Composite Function and Find Its Domain

In this section, we introduce some operations on functions. We shall see that functions, like numbers, can be added, subtracted, multiplied, and divided. For example, if $f(x) = x^2 + 9$ and $g(x) = 3x + 5$, then

$$f(x) + g(x) = (x^2 + 9) + (3x + 5) = x^2 + 3x + 14$$

The new function $y = x^2 + 3x + 14$ is called the *sum function* $f + g$. Similarly,

$$f(x) \cdot g(x) = (x^2 + 9)(3x + 5) = 3x^3 + 5x^2 + 27x + 45$$

The new function $y = 3x^3 + 5x^2 + 27x + 45$ is called the *product function f · g*. The general definitions are given next.

If f and g are functions:
Their **sum f + g** is the function defined by

$$(f + g)(x) = f(x) + g(x)$$

The domain of $f + g$ consists of the numbers x that are in the domain of f and in the domain of g.

Their **difference f − g** is the function defined by

$$(f - g)(x) = f(x) - g(x)$$

The domain of $f - g$ consists of the numbers x that are in the domain of f and in the domain of g.

Their **product f · g** is the function defined by

$$(f \cdot g)(x) = f(x) \cdot g(x)$$

The domain of $f \cdot g$ consists of the numbers x that are in the domain of f and in the domain of g.

Their **quotient f/g** is the function defined by

$$\left(\frac{f}{g}\right)(x) = \frac{f(x)}{g(x)}, \quad g(x) \neq 0$$

The domain of f/g consists of the numbers x for which $g(x) \neq 0$ that are in the domain of f and in the domain of g.

EXAMPLE 1

Operations on Functions

Let f and g be two functions defined as
$$f(x) = \sqrt{x + 2} \quad \text{and} \quad g(x) = \sqrt{x - 3}$$
Find the following, and determine the domain in each case:
(a) $(f + g)(x)$ (b) $(f - g)(x)$ (c) $(f \cdot g)(x)$ (d) $(f/g)(x)$

Solution

(a) $(f + g)(x) = f(x) + g(x) = \sqrt{x + 2} + \sqrt{x - 3}$
(b) $(f - g)(x) = f(x) - g(x) = \sqrt{x + 2} - \sqrt{x - 3}$
(c) $(f \cdot g)(x) = f(x) \cdot g(x) = (\sqrt{x + 2})(\sqrt{x - 3}) = \sqrt{(x + 2)(x - 3)}$
(d) $\left(\frac{f}{g}\right)(x) = \frac{f(x)}{g(x)} = \frac{\sqrt{x + 2}}{\sqrt{x - 3}} = \sqrt{\frac{x + 2}{x - 3}}$

The domain of f consists of all numbers x for which $x \geq -2$; the domain of g consists of all numbers x for which $x \geq 3$. The numbers x common to both these domains are those for which $x \geq 3$. As a result, the numbers x for which $x \geq 3$ comprise the domain of the sum function $f + g$, the difference function $f - g$, and the product function $f \cdot g$. For the quotient function f/g, we must exclude from this set the number 3, because the denominator, g, has the value 0 when $x = 3$. Thus, the domain of f/g consists of all x for which $x > 3$. ■

Now work Problem 1.

In calculus, it is sometimes helpful to view a complicated function as the sum, difference, product, or quotient of simpler functions. For example,

$F(x) = x^2 + \sqrt{x}$ is the sum of $f(x) = x^2$ and $g(x) = \sqrt{x}$.
$H(x) = (x^2 - 1)/(x^2 + 1)$ is the quotient of $f(x) = x^2 - 1$ and
$g(x) = x^2 + 1$.

Composite Functions

Consider the function $y = (2x + 3)^2$. If we write $y = f(u) = u^2$ and $u = g(x) = 2x + 3$, then, by a substitution process, we can obtain the original function: $y = f(u) = f(g(x)) = (2x + 3)^2$. This process is called **composition.**

In general, suppose that f and g are two functions, and suppose that x is a number in the domain of g. By evaluating g at x, we get $g(x)$. If $g(x)$ is in the domain of f, then we may evaluate f at $g(x)$ and thereby obtain the expression $f(g(x))$. The correspondence from x to $f(g(x))$ is called a *composite function $f \circ g$.*

Look carefully at Figure 49. Only those x's in the domain of g for which $g(x)$ is in the domain of f can be in the domain of $f \circ g$. The reason is that if $g(x)$ is not in the domain of f, then $f(g(x))$ is not defined. Thus, the domain of $f \circ g$ is a subset of the domain of g; the range of $f \circ g$ is a subset of the range of f.

FIGURE 49

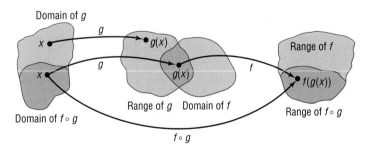

Domain of g

g

x • → • $g(x)$

g

Range of f

x • $g(x)$

f

$f(g(x))$

Domain of $f \circ g$

Range of g Domain of f

Range of $f \circ g$

$f \circ g$

Given two functions f and g, the **composite function,** denoted by $f \circ g$ (read as "f composed with g"), is defined by

$$(f \circ g)(x) = f(g(x))$$

The domain of $f \circ g$ is the set of all numbers x in the domain of g such that $g(x)$ is in the domain of f.

Figure 50 provides a second illustration of the definition. Notice that the "inside" function g in $f(g(x))$ is done first.

FIGURE 50

INPUT x ⟶ g $\xrightarrow{g(x)}$ f ⟶   OUTPUT $f(g(x))$

Let's look at some examples.

E X A M P L E 2

Evaluating a Composite Function

Suppose that $f(x) = 2x^2 - 3$ and $g(x) = 4x$. Find

(a) $(f \circ g)(1)$ (b) $(g \circ f)(1)$ (c) $(f \circ f)(-2)$ (d) $(g \circ g)(-1)$

Solution   (a) $(f \circ g)(1) = f(g(1)) = f(4) = 2 \cdot 16 - 3 = 29$

$$g(x) = 4x \qquad f(x) = 2x^2 - 3$$
$$g(1) = 4$$

(b) $(g \circ f)(1) = g(f(1)) = g(-1) = 4 \cdot (-1) = -4$

$$f(x) = 2x^2 - 3 \quad g(x) = 4x$$
$$f(1) = -1$$

(c) $(f \circ f)(-2) = f(f(-2)) = f(5) = 2 \cdot 25 - 3 = 47$

$$f(-2) = 5$$

(d) $(g \circ g)(-1) = g(g(-1)) = g(-4) = 4 \cdot (-4) = -16$

$$g(-1) = -4$$

■

Now work Problem 21.

E X A M P L E 3

Finding a Composite Function

Suppose that $f(x) = \sqrt{x}$ and $g(x) = \dfrac{1}{x - 1}$. Find the following composite functions, and then find the domain of each composite function.

(a) $f \circ g$ (b) $g \circ f$ (c) $f \circ f$ (d) $g \circ g$

Solution   The domain of f is $\{x \mid x \geq 0\}$ and the domain of g is $\{x \mid x \neq 1\}$.

(a) $(f \circ g)(x) = f(g(x)) = f\left(\dfrac{1}{x - 1}\right) = \sqrt{\dfrac{1}{x - 1}} = \dfrac{1}{\sqrt{x - 1}}$

The domain of $f \circ g$ consists of those x in the domain of g for which $g(x) = 1/(x - 1) \geq 0$, or equivalently, $x > 1$. Thus, the domain of $f \circ g$ is $\{x \mid x > 1\}$.

(b) $(g \circ f)(x) = g(f(x)) = g(\sqrt{x}) = \dfrac{1}{\sqrt{x} - 1}$

The domain of $g \circ f$ consists of those x in the domain of f for which $f(x) = \sqrt{x} \neq 1$, or equivalently, $x \neq 1$. Thus, the domain of $g \circ f$ is $\{x \mid x \geq 0, x \neq 1\}$.

(c) $(f \circ f)(x) = f(f(x)) = f(\sqrt{x}) = \sqrt{\sqrt{x}} = \sqrt[4]{x}$

The domain of $f \circ f$ consists of those x in the domain of f for which $f(x) = \sqrt{x} \geq 0$, or equivalently, $x \geq 0$. Thus, the domain of $f \circ f$ is $\{x \mid x \geq 0\}$.

(d) $(g \circ g)(x) = g(g(x)) = g\left(\dfrac{1}{x-1}\right) = \dfrac{1}{\dfrac{1}{x-1} - 1} = \dfrac{x-1}{1-(x-1)} = -\dfrac{x-1}{x-2}$

The domain of $g \circ g$ consists of those x in the domain of g for which $g(x) = 1/(x-1) \neq 1$, or equivalently, $x \neq 2$. Thus, the domain of $g \circ g$ is $\{x | x \neq 1, x \neq 2\}$.

Check: In Example 3(d), see for yourself! Determine what happens if $x = 2$ for the composite $(g \circ g)(x)$.   ▬

Now work Problem 35.

Examples 3(a) and 3(b) illustrate that, in general, $f \circ g \neq g \circ f$. However, sometimes $f \circ g$ does equal $g \circ f$, as shown in the next example.

E X A M P L E   4   Showing Two Composite Functions Equal

If $f(x) = 3x - 4$ and $g(x) = \frac{1}{3}(x + 4)$, show that $(f \circ g)(x) = (g \circ f)(x) = x$ for every x.

Solution $(f \circ g)(x) = f(g(x))$

$\qquad\qquad = f\left(\dfrac{x+4}{3}\right) \qquad\qquad g(x) = \dfrac{1}{3}(x+4) = \dfrac{x+4}{3}.$

$\qquad\qquad = 3\left(\dfrac{x+4}{3}\right) - 4 \qquad$ Substitute $g(x)$ into the rule for
$\qquad\qquad\qquad\qquad\qquad\qquad\qquad$ f, $f(x) = 3x - 4.$

$\qquad\qquad = x + 4 - 4 = x$

$\quad (g \circ f)(x) = g(f(x))$

$\qquad\qquad = g(3x - 4) \qquad\qquad f(x) = 3x - 4.$

$\qquad\qquad = \frac{1}{3}[(3x - 4) + 4] \qquad$ Substitute $f(x)$ into the rule for g,
$\qquad\qquad\qquad\qquad\qquad\qquad\qquad$ $g(x) = \frac{1}{3}(x+4).$

$\qquad\qquad = \frac{1}{3}(3x) = x$

Thus, $(f \circ g)(x) = (g \circ f)(x) = x.$ ▬

In Section 4.1, we shall see that there is an important relationship between functions f and g for which $(f \circ g)(x) = (g \circ f)(x) = x.$

Now work Problem 45.

Calculus Application

Some techniques in calculus require that we be able to determine the components of a composite function. For example, the function $H(x) = \sqrt{x + 1}$ is the composition of the functions f and g, where $f(x) = \sqrt{x}$ and $g(x) = x + 1$, because $H(x) = (f \circ g)(x) = f(g(x)) = f(x + 1) = \sqrt{x + 1}$.

E X A M P L E   5   Finding the Components of a Composite Function

Find functions f and g such that $f \circ g = H$ if $H(x) = (x^2 + 1)^{50}$.

Solution   The function H takes $x^2 + 1$ and raises it to the power 50. A natural way to decompose H is to raise the function $g(x) = x^2 + 1$ to the power 50. Thus,

FIGURE 53

FIGURE 54

Solution (a) Figure 53 illustrates the graph. The distance d from P to O is

$$d = \sqrt{x^2 + y^2}$$

Since P is a point on the graph of $y = x^2 - 1$, we have

$$d(x) = \sqrt{x^2 + (x^2 - 1)^2} = \sqrt{x^4 - x^2 + 1}$$

Thus, we have expressed the distance d as a function of x.

(b) If $x = 0$, the distance d is

$$d(0) = \sqrt{1} = 1$$

(c) If $x = 1$, the distance d is

$$d(1) = \sqrt{1 - 1 + 1} = 1$$

(d) If $x = \sqrt{2}/2$, the distance d is

$$d\left(\frac{\sqrt{2}}{2}\right) = \sqrt{\left(\frac{\sqrt{2}}{2}\right)^4 - \left(\frac{\sqrt{2}}{2}\right)^2 + 1} = \sqrt{\frac{1}{4} - \frac{1}{2} + 1} = \frac{\sqrt{3}}{2}$$

(e) Figure 54 shows the graph of $Y_1 = \sqrt{x^4 - x^2 + 1}$.

Using the minimum function on a graphing utility, we find that, when $x \approx 0.70$ the value of d is smallest ($d \approx 0.86$) correct to two decimal places. ▬

Now work Problem 13.

E X A M P L E   4   Filling a Swimming Pool

A rectangular swimming pool 20 meters long and 10 meters wide is 4 meters deep at one end and 1 meter deep at the other. Figure 55 illustrates a cross-sectional view of the pool. Water is being pumped into the pool at the deep end.

(a) Find a function that expresses the volume V of water in the pool as a function of the height x of the water at the deep end.

(b) Find the volume when the height is 1 meter.

(c) Find the volume when the height is 2 meters.

(d) Use a graphing utility to graph the function $V = V(x)$. At what height is the volume 20 cubic meters? 100 cubic meters?

FIGURE 55

Solution (a) Let L denote the distance (in meters) measured at water level from the deep end to the short end. Notice that L and x form the sides of a triangle that is similar to the triangle whose sides are 20 meters by 3 meters. Thus, L and x are related by the equation

$$\frac{L}{x} = \frac{20}{3} \quad \text{or} \quad L = \frac{20x}{3} \qquad 0 \le x \le 3$$

Consulting for the Silver Satellite & Cable TV Co.

Your team works for the Silver Satellite & Cable TV Company in the Research & Development Department. You've been asked to come up with a formula to determine the cost of running cable from a connection box to a new cable household. The first example you are working with involves the Frick family who own a rural home with a driveway two miles long extending to the house from a nearby highway. The nearest connection box is along the highway, but 5 miles from the driveway.

It costs the company $10 per mile to install cable along the highway and $14 per mile to install cable off the highway. Because the Frick house is surrounded by farmland which they own, it would be possible to run the cable overland to the house directly from the connection box or from any point between the connection box and the driveway.

1. Draw a sketch of this problem situation, assuming the highway is a straight road and the driveway is also a straight road perpendicular to the highway. Include two or more possible routes for the cable.

2. Suppose x represents the distance in miles the cable runs along the highway from the connection box before turning off toward the house. Express the total cost of installation as a function of x. (You may choose to answer #3 before #2 if you would like to examine concrete instances before creating the function.)

3. Make a table of the possible integral values of x and the corresponding cost in each instance. Is there one choice which appears to cost the least?

4. If you charge the Fricks $80 for installation, would you be willing to let them choose which way the cable would go? Explain.

5. Using a graphing calculator, graph the function from #2 and determine if there is a nonintegral choice for x that would make the installation cost even cheaper. Use BOX or TRACE (or MINIMUM) to determine the least possible cost.

6. Before proceeding further with the installation, you check the local regulations for cable companies and find that there is a pending state legislation that says the cable cannot turn off the highway more than 0.5 miles from the Fricks' driveway. If this legislation passes, what will be the ultimate cost of installing the Fricks' cable?

7. If the cable company wishes to install cable in 5,000 homes in this area, and assuming the figures for the Frick installation are typical, how much will the new legislation cost the company over all if they cannot use the cheapest installation cost, but instead have to follow the new state regulations?

26. A circle of radius r is inscribed in a square (see the figure).

(a) Express the area A of the square as a function of the radius r of the circle.

(b) Express the perimeter p of the square as a function of r.

27. A wire 10 meters long is to be cut into two pieces. One piece will be shaped as a square, and the other piece will be shaped as a circle (see the figure).

(a) Express the total area A enclosed by the pieces of wire as a function of the length x of a side of the square.

(b) What is the domain of A?

(c) Graph $A = A(x)$. For what value of x is A smallest?

28. A wire 10 meters long is to be cut into two pieces. One piece will be shaped as an equilateral triangle, and the other piece will be shaped as a circle.

(a) Express the total area A enclosed by the pieces of wire as a function of the length x of a side of the equilateral triangle.

(b) What is the domain of A?

(c) Graph $A = A(x)$. For what value of x is A smallest?

29. A semicircle of radius r is inscribed in a rectangle so that the diameter of the semicircle is the length of the rectangle (see the figure).

(a) Express the area A of the rectangle as a function of the radius r of the semicircle.

(b) Express the perimeter p of the rectangle as a function of r.

30. An equilateral triangle is inscribed in a circle of radius r. See the figure. Express the circumference C of the circle as a function of the length x of the side of the triangle.

[**Hint:** First show that $r^2 = x^2/3$.]

31. An equilateral triangle is inscribed in a circle of radius r. See the figure. Express the area A within the circle, but outside the triangle, as a function of the length x of a side of the triangle.

32. **Cost of Transporting Goods** A trucking company transports goods between Chicago and New York, a distance of 960 miles. The company's policy is to charge, for each pound, $0.50 per mile for the first 100 miles, $0.40 per mile for the next 300 miles, $0.25 per mile for the next 400 miles, and no charge for the remaining 160 miles.

(a) Graph the relationship between the cost of transportation in dollars and mileage over the entire 960 mile route.

(b) Find the cost as a function of mileage for hauls between 100 and 400 miles from Chicago.

(c) Find the cost as a function of mileage for hauls between 400 and 800 miles from Chicago.

33. **Car Rental Costs** An economy car rented in Florida from National Car Rental® on a weekly basis costs $95 per week.* Extra days cost $24 per day until the day rate exceeds the weekly rate, in which case the weekly rate applies. Find the cost C of renting an economy car as a piecewise-defined function of the number x of days used, where $7 \le x \le 14$. Graph this function. [**Note:** Any part of a day counts as a full day.]

34. Rework Problem 33 for a luxury car, which costs $219 on a weekly basis with extra days at $45 per day.

35. Water is poured into a container in the shape of a right circular cone with radius 4 feet and height 16 feet (see the figure). Express the volume V of water in the cone as a function of the height h of the water.

[**Hint:** The volume V of a cone of radius r and height h is $V = \frac{1}{3}\pi r^2 h$.]

*Source: National Car Rental®, 1995.

36. Federal Income Tax Two 1997 Tax Rate Schedules are given in the accompanying table. If x equals the amount on Form 1040, line 37, and y equals the tax due, construct a function f for each schedule.

1997 TAX RATE SCHEDULES

SCHEDULE X— USE IF YOUR FILING STATUS IS SINGLE						SCHEDULE Y-1—USE IF YOUR FILING STATUS IS MARRIED FILING JOINTLY OR QUALIFYING WIDOW(ER)					
If the amount on Form 1040, line 37, is: Over—	But not over—	Enter on Form 1040, line 38			of the amount over—	If the amount on Form 1040, line 37, is: Over—	But not over—	Enter on Form 1040, line 38			of the amount over—
$ 0	$ 24,650	$ 0	+	15%	$ 0	$ 0	$ 41,200	$ 0	+	15%	$ 0
24,650	59,750	3,698	+	28	24,650	41,200	99,600	6,180	+	28	41,200
59,750	124,650	13,526	+	31	59,750	99,600	151,750	22,532	+	31	99,600
124,650	271,050	33,645	+	36	124,650	151,750	271,050	38,699	+	36	151,750
271,050		86,349	+	39.6	271,050	271,050		81,647	+	39.6	271,050

37. Linear Curve Fitting The manager of a clothing store collected the following prices and quantity demanded for shirts.

Price ($/Shirt)	Quantity Purchased
15	1000
17	920
18	890
19	870
21	800
23	720
24	670
25	620
26	580

(a) Use your graphing utility to find the line of best fit that relates price and quantity purchased. Treat price as the independent variable and quantity as the dependent variable.
(b) Determine the revenue function $R(p)$, where p is the price.
(c) Graph $R(p)$ using a graphing utility.
(d) What is the revenue maximizing price?
(e) What quantity will maximize revenue? What is the maximum revenue?

38. Linear Curve Fitting The manager of a clothing store collected the following prices and quantity demanded for pants.

Price ($/Pair)	Quantity Purchased
29	1000
31	985
32	960
34	920
36	875
37	845
39	790
40	765
41	740
43	695

(a) Use your graphing utility to find the line of best fit that relates price and quantity purchased. Treat price as the independent variable and quantity as the dependent variable.
(b) Determine the revenue function $R(p)$, where p is the price.
(c) Graph $R(p)$ using a graphing utility.
(d) What is the revenue maximizing price?
(e) What quantity will maximize revenue? What is the maximum revenue?

CHAPTER REVIEW

THINGS TO KNOW

Function

A rule or correspondence between two sets of real numbers so that each number x in the first set, the domain, has corresponding to it exactly one number y in the second set. The range is the set of y values of the function for the x values in the domain.

x is the independent variable; y is the dependent variable.

A function f may be defined implicitly by an equation involving x and y or explicitly by writing $y = f(x)$.

A function can also be characterized as a set of ordered pairs (x, y) or $(x, f(x))$ in which no two distinct pairs have the same first element.

Function notation

$y = f(x)$

f is a symbol for the function or rule that defines the function.

x is the argument, or independent variable.

y is the dependent variable.

$f(x)$ is the value of the function at x, or the image of x.

Domain

If unspecified, the domain of a function f is the largest set of real numbers for which the rule defines a real number.

Vertical line test

A set of points in the plane is the graph of a function if and only if every vertical line intersects the graph in at most one point.

Local maximum

A function f has a local maximum at c if there is an interval I containing c so that for all x in I, $f(x) < f(c)$.

Local minimum

A function f has a local minimum at c if there is an interval I containing c so that for all x in I, $f(x) > f(c)$.

Even function f

$f(-x) = f(x)$ for every x in the domain ($-x$ must also be in the domain).

Odd function f

$f(-x) = -f(x)$ for every x in the domain ($-x$ must also be in the domain).

LIBRARY OF FUNCTIONS

Linear function

$f(x) = mx + b$

Graph is a straight line with slope m and y-intercept b.

Constant function

$f(x) = b$

Graph is a horizontal line with y-intercept b (see Figure 26).

Identity function

$f(x) = x$

Graph is a straight line with slope 1 and y-intercept 0 (see Figure 27).

Square function

$f(x) = x^2$

Graph is a parabola with intercept at $(0, 0)$ (see Figure 28).

Cube function

$f(x) = x^3$ See Figure 29.

Square root function

$f(x) = \sqrt{x}$ See Figure 30.

Reciprocal function

$f(x) = 1/x$ See Figure 31.

Absolute value function

$f(x) = |x|$ See Figure 32.

HOW TO

Determine whether a relation represents a function

Find the domain and range of a function from its graph

Find the domain of a function given its equation

Determine whether a function is even or odd without graphing it

Graph certain functions by shifting, compressing, stretching, and/or reflecting (see Table 11)

Use a graphing utility to determine where the graph of a function is increasing or decreasing

Use a graphing utility to find the local maxima and local minima of a function

Find the composite of two functions

Construct functions in applications, including piecewise-defined functions

FILL-IN-THE-BLANK ITEMS

1. If f is a function defined by the equation $y = f(x)$, then x is called the _____ variable and y is the _____ variable.

2. A set of points in the xy-plane is the graph of a function if and only if no _____ line contains more than one point of the set.

3. A(n) _____ function f is one for which $f(-x) = f(x)$ for every x in the domain of f; a(n) _____ function f is one for which $f(-x) = -f(x)$ for every x in the domain of f.

4. Suppose that the graph of a function f is known. Then the graph of $y = f(x - 2)$ may be obtained by a(n) _____ shift of the graph of f to the _____ a distance of 2 units.

5. If $f(x) = x + 1$ and $g(x) = x^3$, then _____ $= (x + 1)^3$.

TRUE/FALSE ITEMS

T F **1.** Vertical lines intersect the graph of a function in no more than one point.

T F **2.** The y-intercept of the graph of the function $y = f(x)$ whose domain is all real numbers is $f(0)$.

T F **3.** Even functions have graphs that are symmetric with respect to the origin.

T F **4.** The graph of $y = f(-x)$ is the reflection about the y-axis of the graph of $y = f(x)$.

T F **5.** $f(g(x)) = f(x) \cdot g(x)$

REVIEW EXERCISES

1. Given that f is a linear function, $f(4) = -5$, and $f(0) = 3$, write the equation that defines f.

3. A function f is defined by

$$f(x) = \frac{Ax + 5}{6x - 2}$$

If $f(1) = 4$, find A.

2. Given that g is a linear function with slope $= -4$ and $g(-2) = 2$, write the equation that defines g.

4. A function g is defined by

$$g(x) = \frac{A}{x} + \frac{8}{x^2}$$

If $g(-1) = 0$, find A.

66. Speed as a Function of Time The speed v (in feet per second) of a car is a linear function of the time t (in seconds) for $10 \leq t \leq 30$. If after each second the speed of the car has increased by 5 feet per second, and if after 20 seconds the speed is 80 feet per second, how fast is the car going after 30 seconds? Find the function $v = v(t)$.

67. Strength of a Beam The strength of a rectangular wooden beam is proportional to the product of the width and the cube of its depth (see the figure). If the beam is to be cut from a log in the shape of a cylinder of radius 3 feet, express the strength S of the beam as a function of the width x. What is the domain of S?

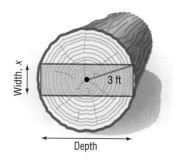

Width, x

3 ft

Depth

68. Productivity vs. Earnings The data below represents the average hourly earnings and productivity (output per hour) of production workers for the years 1986–1995. Let productivity x be the independent variable and average hourly earnings y be the dependent variable.

Productivity	Average Hourly Earnings
94.2	8.76
94.1	8.98
94.6	9.28
95.3	9.66
96.1	10.01
96.7	10.32
100	10.57
100.2	10.83
100.7	11.12
100.8	11.44

Source: Bureau of Labor Statistics.

(a) Draw a scatter diagram of the data by hand.
(b) Draw a line from the point (94.2, 8.76) to (96.7, 10.32) on the scatter diagram found in (a).
(c) Find the average rate of change of hourly earnings for productivity between 94.2 and 96.7.
(d) Interpret the average rate of change found in (c).
(e) Draw a line from the point (96.7, 10.32) to (100.8, 11.44) on the scatter diagram found in (a).
(f) Find the average rate of change of hourly earnings for productivity between 96.7 and 100.8.
(g) Interpret the average rate of change found in (f).
(h) What is happening to the average rate of change of hourly earnings as productivity increases?

69. Speed of a Parachutist The following data represents the distance a parachutist has fallen over time, ignoring air resistance.

Time (Seconds)	Distance (Meters)
0	0
5	112.5
10	490
15	1102.5
20	1960

(a) Draw a scatter diagram of the data by hand treating time as the independent variable.
(b) Draw a line from the point (0, 0) to (5, 112.5).
(c) Find the average rate of change of distance between 0 and 5 seconds.
(d) Interpret the average rate of change found in (c).
(e) Draw a line from the points (15, 1102.5) to (20, 1960).
(f) Find the average rate of change of distance between 15 and 20 seconds.
(g) Interpret the average rate of change found in (f).
(h) What is happening to the average rate of change of distance as time passes?

Polynomial and Rational Functions

Technology, such as radio telescopes and satellite dishes, have their foundations in the mathematical concepts of parabolas and the quadratic formula. On the following page, you have been hired as a high school physics teacher and need to introduce your class to Galileo's *Analysis of Motion*. Use the Sullivan website at:

www.prenhall.com/sullivan

to help link you to the Internet resources you will need to prepare your lesson.

PREPARING FOR THIS CHAPTER

Before getting started on this chapter, review the following concepts:

Completing the Square; the Quadratic Formula *(Appendix, Section 2)*

Linear Curve Fitting *(Section 1.4)*

Library of Functions (pp. 141–145)

Synthetic Division *(Appendix, Section 3)*

Inequalities *(Section 1.7)*

OUTLINE

Netsite: http://www.prenhall.com/sullivan/

THE PARABOLA

Quadratic functions, like all polynomials, are formed using only the basic rules of arithmetic: addition, subtraction, multiplication, and division. The graph of a quadratic function is called a **parabola.** The parabola has properties that make it useful for satellite dishes, car headlights, radio telescopes, and reflecting telescopes, including *the liquid mirror telescope.* Although the Greeks knew about the parabola as a conic section, it was *Galileo* who first observed that a quadratic function can be used to describe the motion of a falling body.

Suppose that you are a high school physics teacher and you want to put together an Internet lesson for your class. You would like to begin with a short lecture on Galileo's *Analysis of Motion.* His analysis was a major advance in science. The quadratic function was like his sword for fighting against the bondage of the dark ages. You plan to duplicate *his experiments* with in-class hands-on activity.

1. As you start to prepare your Internet lesson, you search the net for "parabolas." You are overwhelmed with the number of links. How might you handle this information overload?
2. How many different kinds of applications of the parabola did you find on the web? Did you find any links that refer to suspension bridges?
3. For the math side you want to make a list of important definitions, and find some good illustrations of the essential features of parabolas. You want to try to find interactive sites. Did you find any?
4. Given the physical description of the *largest radio telescope on the planet* at Arecibo, Puerto Rico, approximate the dish with a quadratic function. Then approximate it with a circle. Graph both functions with your calculator. Can you determine the amount of difference between the two graphs? What advantages are there in making the radio telescope circular?
5. Suppose you perform Galileo's *parabola experiment.* Model the data with a quadratic function. How well does the graph of your quadratic function fit the data? How well does your *data* fit the ideal law?

In Chapter 2, we graphed linear functions $f(x) = ax + b$, $a \neq 0$; the square function $f(x) = x^2$; and the cube function $f(x) = x^3$. Each of these functions belongs to the class of functions called *polynomial functions,* which we discuss further in this chapter. We will also discuss *rational functions,* which are ratios of polynomial functions. In this chapter, we place special emphasis on the graphs of polynomial and rational functions. This emphasis will demonstrate the importance of evaluating polynomials and solving polynomial equations (Section 3.5). Section 3.6 introduces complex numbers and the role they play in solving quadratic equations with a negative discriminant. This paves the way for the Fundamental Theorem of Algebra (Section 3.7). Finally, polynomial and rational inequalities are discussed in Section 3.8.

3.1 QUADRATIC FUNCTIONS; CURVE FITTING

1 Graph a Quadratic Function by Hand and by Using a Graphing Utility
2 Identify the Vertex and Axis of Symmetry of a Quadratic Function
3 Determine the Maximum or Minimum Value of a Quadratic Function
4 Use a Graphing Utility to Find the Quadratic Function of Best Fit

A **quadratic function** is a function of the form

$$f(x) = ax^2 + bx + c \qquad (1)$$

where *a, b,* and *c* are real numbers and $a \neq 0$. The domain of a quadratic function consists of all real numbers.

Many applications require a knowledge of quadratic functions. For example, suppose Texas Instruments collects the data shown in Table 1 that relates the number *c* of calculators sold at the price *p* per calculator. Since the price of a product determines the quantity that will be purchased, we treat price as the independent variable.

TABLE 1	
Price per Calculator, *p* (Dollars)	**Number of Calculators, *c***
60	11,100
65	10,115
70	9,652
75	8,731
80	8,087
85	7,205
90	6,439

Using the LINear REGression option on a graphing utility, we find the relation between the number *c* of calculators and the price *p* per calculator is

$$c = 20208.43 - 152.63p$$

Then the revenue *R* derived from selling *c* calculators at the price *p* per calculator is

$$
\begin{aligned}
R &= cp \\
&= (20208.43 - 152.63p)p \\
&= -152.63p^2 + 20208.43p
\end{aligned}
$$

FIGURE 1

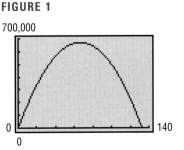

So, the revenue *R* is expressed as a quadratic function of the price *p*. Figure 1 illustrates the graph of this revenue function. Since both *c* and *p* must be nonnegative, we only show the graph of *R* in the first quadrant.

A second situation in which a quadratic function appears involves the motion of a projectile. Based on Newton's second law of motion (force equals mass times acceleration, $F = ma$), it can be shown that, ignoring air resistance, the path of a projectile propelled upward at an inclination to the horizontal is the graph of a quadratic function. See Figure 2 for an illustration.

FIGURE 2
Path of a cannonball

Graphing Quadratic Functions

We know how to graph quadratic functions of the form $f(x) = ax^2$, $a \neq 0$, based on prior discussions. Figure 3 shows the graph of three functions of the form $f(x) = ax^2$, $a > 0$, for $a = 1$, $a = \frac{1}{2}$, and $a = 3$. Notice that the larger the value of a, the "narrower" the graph is and the smaller the value of a the "wider" the graph is.

FIGURE 3

FIGURE 4

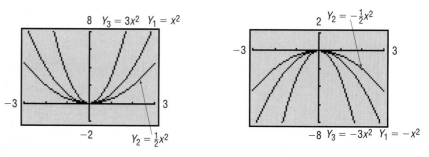

Figure 4 shows the graphs of $f(x) = ax^2$ for $a < 0$. Notice these graphs are reflections about the x-axis of the graphs in Figure 3. Based on the results of these two figures, we can draw some general conclusions about the graph of $f(x) = ax^2$. First, as $|a|$ increases, the graph becomes "narrower" and as $|a|$ gets closer to zero, the graph gets "wider." Second, if a is positive, then the graph opens "up" and if a is negative, then the graph opens "down."

The graphs in Figures 3 and 4 are typical of the graphs of all quadratic functions, which we call **parabolas.*** Refer to Figure 5, where two parabolas are pictured. The one on the left **opens up** and has a lowest point; the one on the right **opens down** and has a highest point. The lowest or highest point of a parabola is called the **vertex.** The vertical line passing through the vertex in each parabola in Figure 5 is called the **axis of symmetry** (usually abbreviated to **axis**) of the parabola. Because the parabola is symmetric about its axis, the axis of symmetry of a parabola can be used to advantage in graphing the parabola by hand.

The parabolas shown in Figure 5 are the graphs of a quadratic function $f(x) = ax^2 + bx + c$, $a \neq 0$. Notice that the coordinate axes are not included in the figure. Depending on the values of a, b, and c, the axes could be placed anywhere. The important fact is that, except possibly for compression or

FIGURE 5
Graphs of a quadratic function,
$f(x) = ax^2 + bx + c$, $a \neq 0$

Axis of symmetry

Vertex is highest point

Vertex is lowest point

Axis of symmetry

(a) Opens up

(b) Opens down

---

*We shall study parabolas using a geometric definition later in this book.

stretching, the shape of the graph of a quadratic function will look like one of the parabolas in Figure 5.

E X A M P L E   1

Graphing Solution

Graphing a Quadratic Function

Graph the function $f(x) = 2x^2 + 8x + 5$. Find the vertex and axis of symmetry.

Before graphing, notice the leading coefficient, 2, is positive and, therefore, the graph will open up and the vertex will be the lowest point. Now graph f. See Figure 6. We observe the graph does in fact open up. To estimate the vertex of the parabola, we use the MINIMUM command. The vertex is $(-2, -3)$. Therefore, the axis of symmetry is the line $x = -2$.

FIGURE 6

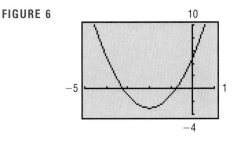

Algebraic Solution

We begin by completing the square* on the right side:

$$f(x) = 2x^2 + 8x + 5$$
$$= 2(x^2 + 4x) + 5 \qquad \text{Factor out the 2 from } 2x^2 + 8x.$$
$$= 2(x^2 + 4x + 4) + 5 - 8 \qquad \text{Complete the square of } 2(x^2 + 4x). \qquad (2)$$
$$ \qquad \text{Notice that the factor of 2 requires}$$
$$= 2(x + 2)^2 - 3 \qquad \text{that 8 be added and subtracted.}$$

The graph of f can be obtained in three stages, as shown in Figure 7. Now compare this graph to the graph in Figure 5(a). The graph of $f(x) = 2x^2 + 8x + 5$ is a parabola that opens up and has its vertex (lowest point) at $(-2, -3)$. Its axis of symmetry is the line $x = -2$.

FIGURE 7

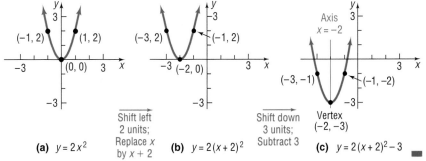

(a) $y = 2x^2$ Shift left 2 units; Replace x by $x + 2$ (b) $y = 2(x + 2)^2$ Shift down 3 units; Subtract 3   (c) $y = 2(x + 2)^2 - 3$

*Refer to the Appendix, Section 2, for a review.

The method used in the Algebraic Solution to Example 1 can be used to graph any quadratic function $f(x) = ax^2 + bx + c, a \neq 0$, as follows:

$$f(x) = ax^2 + bx + c$$

$$= a\left(x^2 + \frac{b}{a}x\right) + c \qquad \text{Factor out } a \text{ from } ax^2 + bx.$$

$$= a\left(x^2 + \frac{b}{ax} + \frac{b^2}{4a^2}\right) + c - a\left(\frac{b^2}{4a^2}\right) \qquad \begin{array}{l}\text{Complete the square by adding and} \\ \text{subtracting } a(b^2/4a^2). \text{ Look closely at} \\ \text{this step!}\end{array}$$

$$= a\left(x + \frac{b}{2a}\right)^2 + c - \frac{b^2}{4a}$$

$$= a\left(x + \frac{b}{2a}\right)^2 + \frac{4ac - b^2}{4a}$$

Based on the above results, we conclude:

If $h = -b/2a$ and $k = (4ac - b^2)/4a$, then
$$f(x) = ax^2 + bx + c = a(x - h)^2 + k \qquad (3)$$

The graph of f is the parabola $y = ax^2$ shifted horizontally h units and vertically k units. As a result, the vertex is at (h, k), and the graph opens up if $a > 0$ and down if $a < 0$. The axis is the vertical line $x = h$.

For example, compare equation (3) with equation (2) of Example 1.

$$f(x) = 2(x + 2)^2 - 3$$
$$f(x) = a(x - h)^2 + k$$

We conclude that $a = 2$, so the graph opens up. Also, we find that $h = -2$ and $k = -3$, so its vertex is at $(-2, -3)$.

It is not required to complete the square to obtain the vertex. In almost every case, it is easier to obtain the vertex of a quadratic function f by remembering that its x-coordinate is $h = -b/2a$. The y-coordinate can then be found by evaluating f at $-b/2a$.

These results are summarized next:

Characteristics of the Graph of a Quadratic Function

$$f(x) = ax^2 + bx + c$$

$$\text{Vertex} = \left(\frac{-b}{2a}, f\left(\frac{-b}{2a}\right)\right) \quad \text{Axis of Symmetry: The line } x = \frac{-b}{2a} \quad (4)$$

E X A M P L E   2   **Locating the Vertex without Graphing**

Without graphing, locate the vertex and axis of the parabola defined by $f(x) = -3x^2 + 6x + 1$. Does it open up or down?

Solution   For this quadratic function, $a = -3, b = 6$, and $c = 1$. The x-coordinate of the vertex is

$$\frac{-b}{2a} = \frac{-6}{-6} = 1$$

The y-coordinate of the vertex is therefore

$$f\left(\frac{-b}{2a}\right) = f(1) = -3 + 6 + 1 = 4$$

The vertex is located at the point $(1, 4)$. The axis of symmetry is the line $x = 1$. Finally, because $a = -3 < 0$, the parabola opens down. ■

The information we gathered in Example 2, together with the location of the intercepts, usually provides enough information to graph $f(x) = ax^2 + bx + c$, $a \neq 0$, by hand. The y-intercept is the value of f at $x = 0$, that is, $f(0) = c$. The x-intercepts, if there are any, are found by solving the equation

$$f(x) = ax^2 + bx + c = 0$$

This equation has two, one, or no real solutions, depending on whether the discriminant $b^2 - 4ac$ is positive, 0, or negative. Thus, it has corresponding x-intercepts, as follows:

The x-intercepts of a Quadratic Function

> 1. If the discriminant $b^2 - 4ac > 0$, the graph of $f(x) = ax^2 + bx + c$ has two distinct x-intercepts and so will cross the x-axis in two places.
> 2. If the discriminant $b^2 - 4ac = 0$, the graph of $f(x) = ax^2 + bx + c$ has one x-intercept and touches the x-axis at its vertex.
> 3. If the discriminant $b^2 - 4ac < 0$, the graph of $f(x) = ax^2 + bx + c$ has no x-intercept and so will not cross or touch the x-axis.

Figure 8 illustrates these possibilities for parabolas that open up.

FIGURE 8
$f(x) = ax^2 + bx + c, \ a > 0$

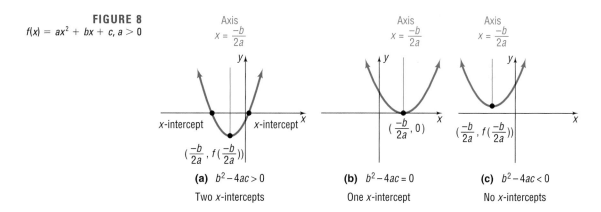

(a) $b^2 - 4ac > 0$
Two x-intercepts

(b) $b^2 - 4ac = 0$
One x-intercept

(c) $b^2 - 4ac < 0$
No x-intercepts

E X A M P L E   3

Graphing a Quadratic Function By Hand Using Its Vertex, Axis, and Intercepts

Use the information from Example 2 and the locations of the intercepts to graph $f(x) = -3x^2 + 6x + 1$.

Solution   In Example 2, we found the vertex to be at $(1, 4)$ and the axis of symmetry to be $x = 1$. The y-intercept is found by letting $x = 0$. Thus, the y-intercept is $f(0) = 1$. The x-intercepts are found by letting $f(x) = 0$. This results in the equation

$$-3x^2 + 6x + 1 = 0$$

The discriminant $b^2 - 4ac = (6)^2 - 4(-3)(1) = 36 + 12 = 48 > 0$, so the equation has two real solutions and the graph has two x-intercepts. Using the quadratic formula, we find

$$x = \frac{-b + \sqrt{b^2 - 4ac}}{2a} = \frac{-6 + \sqrt{48}}{-6} = \frac{-6 + 4\sqrt{3}}{-6} \approx -0.15$$

and

$$x = \frac{-b - \sqrt{b^2 - 4ac}}{2a} = \frac{-6 - \sqrt{48}}{-6} = \frac{-6 - 4\sqrt{3}}{-6} \approx 2.15$$

The x-intercepts are approximately -0.15 and 2.15.

The graph is illustrated in Figure 9. Notice how we used the y-intercept and the axis of symmetry, $x = 1$, to obtain the additional point $(2, 1)$ on the graph. ■

FIGURE 9
$f(x) = -3x^2 + 6x + 1$

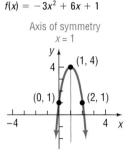

EXAMPLE 4 Analyzing the Graph of a Quadratic Function

Graph $f(x) = x^2 - 6x + 9$. Determine whether the graph opens up or down. Find its vertex, axis of symmetry, y-intercept, and x-intercepts, if any.

Solution Figure 10 shows the graph of $f(x) = x^2 - 6x + 9$.

From the graph, we conjecture there is one y-intercept near 9 and one x-intercept near 3. The parabola opens up.

Now we confirm these conclusions using algebra. The parabola opens up since $a = 1 > 0$. Since $a = 1$, $b = -6$, and $c = 9$, the x-coordinate of the vertex is

$$\frac{-b}{2a} = \frac{-(-6)}{2(1)} = 3$$

The y-coordinate of the vertex is

$$f(3) = (3)^2 - 6(3) + 9 = 0$$

So the vertex is at $(3, 0)$. The axis of symmetry is the line $x = 3$. The y-intercept is $f(0) = 9$. Since the vertex $(3, 0)$ lies on the x-axis, the graph touches the x-axis at the x-intercept. ■

FIGURE 10

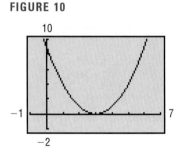

Now work Problem 25.

EXAMPLE 5 Analyzing the Graph of a Quadratic Function

Graph $f(x) = 2x^2 + x + 1$. Determine whether the graph opens up or down. Find its vertex, axis of symmetry, y-intercept, and x-intercepts, if any.

Solution Graph the function. See Figure 11. From the graph we can see the graph opens up. We also observe the graph has no x-intercepts and one y-intercept 1.

Now we confirm these conclusions. Since $a = 2$, $b = 1$, and $c = 1$, the x-coordinate of the vertex is

$$\frac{-b}{2a} = -\frac{1}{4}$$

FIGURE 11

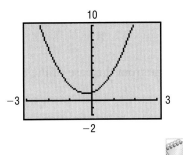

The y-coordinate of the vertex is

$$f\left(-\frac{1}{4}\right) = 2\left(\frac{1}{16}\right) + \left(-\frac{1}{4}\right) + 1 = \frac{7}{8}$$

So the vertex is at $\left(-\frac{1}{4}, \frac{7}{8}\right)$. The axis of symmetry is the line $x = -\frac{1}{4}$. The y-intercept is $f(0) = 1$. Since the discriminant $b^2 - 4ac = (1)^2 - 4(2)(1) = -7 < 0$, the equation $2x^2 + x + 1 = 0$ has no real solutions and therefore the graph has no x-intercepts, confirming the conclusions reached earlier. ■

Now work Problem 31.

Applications

We have already seen that the graph of a quadratic function $f(x) = ax^2 + bx + c$ is a parabola with vertex at $(-b/2a, f(-b/2a))$. This vertex is the highest point on the graph if $a < 0$ and the lowest point on the graph if $a > 0$. If the vertex is the highest point ($a < 0$), then $f(-b/2a)$ is the **maximum value** of f. If the vertex is the lowest point ($a > 0$), then $f(-b/2a)$ is the **minimum value** of f. These ideas give rise to many applications.

E X A M P L E   6   Maximizing Revenue

The marketing department at Texas Instruments has found that, when certain calculators are sold at a price of p dollars per unit, the revenue R (in dollars) as a function of the price p is

$$R(p) = -152.63p^2 + 20208.43p$$

What unit price should be established in order to maximize revenue? If this price is charged, what is the maximum revenue?

Solution   The revenue R is

$$R(p) = -152.63p^2 + 20208.43p = ap^2 + bp + c$$

The function R is a quadratic function with $a = -152.63$, $b = 20,208.43$, and $c = 0$. Because $a < 0$, the vertex is the highest point of the parabola. The revenue R is therefore a maximum when the price p is

FIGURE 12
$R(p) = -152.63p^2 + 20208.43p$

700,000

0   140
0

$$p = \frac{-b}{2a} = \frac{-20208.43}{2(-152.63)} = \frac{-20208.43}{-305.26} = \$66.20$$

The maximum revenue R is

$$R(66.20) = -152.63(66.20)^2 + 20208.43(66.20) \approx \$668,906.25$$

See Figure 12 for an illustration. ■

E X A M P L E   7   Analyzing the Motion of a Projectile

A projectile is fired from a cliff 500 feet above the water at an inclination of 45° to the horizontal, with a muzzle velocity of 400 feet per second. In

3.1 EXERCISES

In Problems 1–8, match each graph to one of the following functions:

A. $y = x^2 - 1$ B. $y = -x^2 - 1$ C. $y = x^2 - 2x + 1$ D. $y = x^2 + 2x + 1$

E. $y = x^2 - 2x + 2$ F. $y = x^2 + 2x$ G. $y = x^2 - 2x$ H. $y = x^2 + 2x + 2$

1.

2.

3.

4.

5.

6.

7.

8.

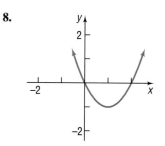

In Problems 9–12, match each graph to one of the following functions.

A. $y = (1/3)x^2 + 2$ B. $y = 3x^2 + 2$
C. $y = x^2 + 5x + 1$ D. $y = -x^2 + 5x + 1$

9.   **10.**

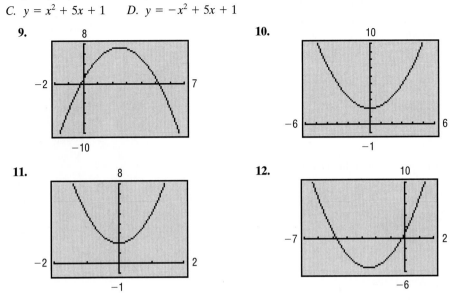

11.   **12.**

In Problems 13–42, graph each quadratic function by hand by determining whether its graph opens up or down and by finding its vertex, axis of symmetry, y-intercept, and x-intercepts, if any. Verify your results using a graphing utility.

13. $f(x) = \dfrac{1}{4}x^2$

14. $f(x) = 2x^2$

15. $f(x) = \dfrac{1}{4}x^2 - 2$

16. $f(x) = 2x^2 - 3$

17. $f(x) = \dfrac{1}{4}x^2 + 2$

18. $f(x) = 2x^2 + 4$

19. $f(x) = -\dfrac{1}{4}x^2 + 1$

20. $f(x) = -2x^2 - 2$

21. $f(x) = x^2 + 4x + 2$

22. $f(x) = x^2 - 6x - 1$

23. $f(x) = 2x^2 - 4x + 1$

24. $f(x) = 3x^2 + 6x$

25. $f(x) = -x^2 - 2x$

26. $f(x) = -2x^2 + 6x + 2$

27. $f(x) = \dfrac{1}{2}x^2 + x - 1$

28. $f(x) = \dfrac{2}{3}x^2 + \dfrac{4}{3}x - 1$

29. $f(x) = x^2 + 2x - 8$

30. $f(x) = x^2 - 2x - 3$

31. $f(x) = -x^2 - 3x + 4$

32. $f(x) = -x^2 + x + 2$

33. $f(x) = x^2 + 2x + 1$

34. $f(x) = -x^2 + 4x - 4$

35. $f(x) = 2x^2 - x + 2$

36. $f(x) = 4x^2 - 2x + 1$

37. $f(x) = -2x^2 + 2x - 3$

38. $f(x) = -3x^2 + 3x - 2$

39. $f(x) = 3x^2 + 6x + 2$

40. $f(x) = 2x^2 + 5x + 3$

41. $f(x) = -4x^2 - 6x + 2$

42. $f(x) = 3x^2 - 8x + 2$

In Problems 43–48, determine without graphing, whether the given quadratic function has a maximum value or a minimum value and then find the value. Verify your results using a graphing utility.

43. $f(x) = 2x^2 + 12x - 3$

44. $f(x) = 4x^2 - 8x + 3$

45. $f(x) = -x^2 + 10x - 4$

46. $f(x) = -2x^2 + 8x + 3$

47. $f(x) = -3x^2 + 12x + 1$

48. $f(x) = 4x^2 - 4x$

49. On one set of coordinate axes, graph the family of parabolas $f(x) = x^2 + 2x + c$ for $c = -3$, $c = 0$ and $c = 1$. Describe the characteristics of a member of this family.

50. On one set of coordinate axes, graph the family of parabolas $f(x) = x^2 + cx + 1$ for $c = -4$, $c = 0$, and $c = 4$. Describe the general characteristics of this family.

51. Graph $y = x^2 + 1$. Then graph $y = x^2 + x + 1$, followed by $y = x^2 + 2x + 1$, followed by $y = x^2 + 3x + 1$. What is happening? Do you see any pattern?

52. Graph $y = x^2 + x + 1$. Then graph $y = 2x^2 + x + 1$, followed by $y = 3x^2 + x + 1$, followed by $y = 4x^2 + x + 1$. What is happening? Do you see any pattern?

53. Maximizing Revenue   Suppose that the manufacturer of a gas clothes dryer has found that, when the unit price is p dollars, the revenue R (in dollars) is

$$R = -4p^2 + 4000p$$

What unit price should be established for the dryer to maximize revenue? What is the maximum revenue?

54. Maximizing Revenue   The John Deere company has found that the revenue from sales of heavy-duty tractors is a function of the unit price p it charges. If the revenue R is

$$R = -\frac{1}{2}p^2 + 1900p$$

what unit price p should be charged to maximize revenue? What is the maximum revenue?

55. Rectangles with Fixed Perimeter   What is the largest rectangular area that can be enclosed with 400 feet of fencing? What are the dimensions of the rectangle?

56. Rectangles with Fixed Perimeter   What are the dimensions of a rectangle of a fixed perimeter P that result in the largest area?

57. Enclosing the Most Area with a Fence   A farmer with 4000 meters of fencing wants to enclose a rectangular plot that borders on a river. If the farmer does not fence the side along the river, what is the largest area that can be enclosed? (See the figure.)

4000 − 2x

58. Enclosing the Most Area with a Fence   A farmer with 2000 meters of fencing wants to enclose a rectangular plot that borders on a straight highway. If the farmer does not fence the side along the highway, what is the largest area that can be enclosed?

59. Enclosing the Most Area with a Fence   A farmer with 10,000 meters of fencing wants to enclose a rectangular field and then divide it into two plots with a fence parallel to one of the sides (see the figure). What is the largest area that can be enclosed?

60. Enclosing the Most Area with a Fence   A farmer with 10,000 meters of fencing wants to enclose a rectangular field and then divide it into three plots with two fences parallel to one of the sides. What is the largest area that can be enclosed?

61. Analyzing the Motion of a Projectile   A projectile is fired from a cliff 200 feet above the water at an inclination of 45° to the horizontal, with a muzzle velocity of 50 feet per second. The height h of the projectile above the water is given by

$$h(x) = \frac{-32x^2}{(50)^2} + x + 200$$

where x is the horizontal distance of the projectile from the base of the cliff.
(a) Using a graphing utility, graph the function $h, 0 \le x \le 200$.
(b) Find the maximum height of the projectile.
(c) How far from the base of the cliff will the projectile strike the water?
(d) Use a graphing utility to verify the solutions found in (b) and (c).
(e) When the height of the projectile is 100 feet above the water, how far is it from the cliff?

62. Analyzing the Motion of a Projectile   A projectile is fired at an inclination of 45° to the horizontal, with a muzzle velocity of 100 feet per second. The height h of the projectile is given by

$$h(x) = \frac{-32x^2}{(100)^2} + x$$

where x is the horizontal distance of the projectile from the firing point.

(a) Using a graphing utility, graph the function h, $0 \le x \le 350$.

(b) Find the maximum height of the projectile.

(c) How far from the firing point will the projectile strike the ground?

(d) Use a graphing utility to verify the results obtained in parts (b) and (c).

(e) When the height of the projectile is 50 feet above the ground, how far has it traveled horizontally?

63. **Navigation** The *U.S.S. Independence* maintains a constant speed of 10 knots heading due north. At 4:00 P.M. the ship's radar detects a destroyer 100 nautical miles due east of the carrier. If the destroyer is heading due west at 20 knots, when will the two ships be the closest? (1 knot = 1 nautical mile per hour)

64. **Air Traffic Control** An air traffic controller sees two aircraft flying at the same altitude on his screen. One, a Piper Cub, is headed due west at 150 miles per hour. The other, a Lear jet, is 15 miles due north of the Piper and is headed due south at 400 miles per hour. How close will the two aircraft come to each other?

65. **Suspension Bridge** A suspension bridge with weight uniformly distributed along its length has twin towers that extend 75 meters above the road surface and are 400 meters apart. The cables are parabolic in shape and are suspended from the tops of the towers. The cables touch the road surface at the center of the bridge. Find the height of the cables at a point 100 meters from the center. (Assume that the road is level.)

66. **Architecture** A parabolic arch has a span of 120 feet and a maximum height of 25 feet. Choose suitable rectangular coordinate axes and find the equation of the parabola. Then calculate the height of the arch at points 10 feet, 20 feet, and 40 feet from the center.

67. **Constructing Rain Gutters** A rain gutter is to be made of aluminum sheets that are 12 inches wide by turning up the edges 90°. What depth

will provide maximum cross-sectional area and hence allow the most water to flow?

68. **Navigation** At 4 P.M. the *Ecstasy* leaves the Port of Miami heading due east at a constant speed of 15 knots. At the same time, a pleasure boat located 100 nautical miles northeast of the Port of Miami is headed due south at a constant speed of 12 knots. When are the two ships closest? How close do they get to each other? (Express your answer in nautical miles; 1 knot = 1 nautical mile per hour.)

69. **Norman Windows** A Norman window has the shape of a rectangle surmounted by a semicircle of diameter equal to the width of the rectangle (see the figure). If the perimeter of the window is 20 feet, what dimensions will admit the most light (maximize the area)?

[**Hint:** Circumference of circle $= 2\pi r$; Area of circle $= \pi r^2$, where r is the radius of the circle.]

70. **Constructing a Stadium** A track and field playing area is in the shape of a rectangle with semicircles at each end (see the figure). The inside perimeter of the track is to be 1500 meters.

What should the dimensions of the rectangle be so that the area of the rectangle is a maximum?

71. **Architecture** A special window has the shape of a rectangle surmounted by an equilateral triangle (see the figure). If the perimeter of the window is 16 feet, what dimensions will admit the most light?

    [**Hint:** Area of an equilateral triangle = $(\sqrt{3}/4)x^2$, where x is the length of a side of the triangle.]

72. **Analyzing the Motion of a Projectile** A projectile is fired at an inclination of 45° to the horizontal with an initial velocity of v_0 feet per second. If the starting point is the origin, the x-axis is horizontal, and the y-axis is vertical, then the height y (in feet) after a horizontal distance x has been traversed is approximately

    $$y = \frac{-32x^2}{v_0^2} + x$$

    (a) Find the maximum height in terms of the initial velocity v_0.
    (b) If the initial velocity is doubled, what happens to the maximum height?
    (c) Assuming the ground is flat, how far from the starting point will the projectile land if the initial velocity is 64 feet per second?

73. **Charter Club Memberships** A charter flight club charges its members $400 per year. But, for each new member in excess of 60, the charge for every member is reduced by $5. What number of members leads to a maximum revenue?

74. **Car Rentals** A car rental agency has 24 identical cars. The owner of the agency finds that all the cars can be rented at a price of $10 per day. However, for each $2 increase in rental, one of the cars is not rented. What should be charged to maximize income?

75. **Life Cycle Hypothesis** An individual's income varies with his/her age. Below is the median income I of individuals of different age groups within the United States for 1995. For each class, let the class midpoint represent the independent variable, x. For the class "65 years and older," we will assume the class midpoint is 69.5.

Age	Class Midpoint, x	Median Income, I
15–24 years	19.5	$20,979
25–34 years	29.5	$34,701
35–44 years	39.5	$43,465
45–54 years	49.5	$48,058
55–64 years	59.5	$38,077
65 years and older	69.5	$19,096

Source: U.S. Census Bureau.

(a) Using a graphing utility, draw a scatter diagram of the data. Comment on the type of relation that may exist between the two variables.
(b) Find the quadratic function of best fit.
(c) Using a graphing utility, draw the quadratic function of best fit on your scatter diagram.
(d) Using the function found in (b), determine the age at which an individual can expect to earn the most income.
(e) Predict the peak income earned.
(f) Compare your result in (d) and (e) to the data and comment.

76. **Advertising** A small manufacturing firm collected the data below on advertising expenditures (in thousands of dollars) and total revenue (in thousands of dollars).

Advertising	Total Revenue
20	$6101
22	$6222
25	$6350
25	$6378
27	$6453
28	$6423
29	$6360
31	$6231

(a) Using a graphing utility, draw a scatter diagram of the data. Comment on the type of relation that may exist between the two variables.

(b) Find the quadratic function of best fit.
(c) Using a graphing utility, draw the quadratic function of best fit on your scatter diagram.
(d) Using the function found in (b), determine the optimal level of advertising for this firm.
(e) Determine the revenue the firm can expect if it uses the optimal level of advertising.
(f) Compare your result in (d) and (e) to the data and comment.

77. Fuel Consumption The data below represents the average fuel consumption C by cars (in billions of gallons) for the years 1980–1993.

Year, t	Average Fuel Consumption, C
1980	71.9
1981	71.0
1982	70.1
1983	69.9
1984	68.7
1985	69.3
1986	71.4
1987	70.6
1988	71.9
1989	72.7
1990	72.0
1991	70.7
1992	73.9
1993	75.1

Source: U.S. Federal Highway Administration.

(a) Using a graphing utility, draw a scatter diagram of the data. Comment on the type of relation that may exist between the two variables.
(b) Find the quadratic function of best fit.
(c) Using a graphing utility, draw the quadratic function of best fit on your scatter diagram.
(d) Using the function found in (b), determine the year in which average fuel consumption was lowest.
(e) Compare your result in (d) to the data and comment.
(f) Use the function found in (b) to predict the average fuel consumption for 1994.

78. Miles Per Gallon An engineer collects data showing the speed s of a Ford Taurus and its average miles per gallon, M. See the table.

Speed, s	Miles per Gallon, M
30	18
35	20
40	23
40	25
45	25
50	28
55	30
60	29
65	26
65	25
70	25

(a) Using a graphing utility, draw a scatter diagram of the data. Comment on the type of relation that may exist between the two variables.
(b) Find the quadratic function of best fit.
(c) Using a graphing utility, draw the quadratic function of best fit on your scatter diagram.
(d) Using the function found in (b), determine the speed that maximizes miles per gallon.
(e) Compare your result in (d) to the data.
(f) Using the function found in (b), predict the miles per gallon of the car if you travel an average of 63 miles per hour.

79. Height of a Ball A physicist throws a ball at an inclination of 45° to the horizontal. The data below represents the height of the ball h at the instant it has traveled x feet horizontally.

Distance, x	Height, h
20	25
40	40
60	55
80	65
100	71
120	77
140	77
160	75
180	71
200	64

(a) Using a graphing utility, draw a scatter diagram of the data. Comment on the type of relation that may exist between the two variables.

(b) Find the quadratic function of best fit.
(c) Using a graphing utility, draw the quadratic function of best fit on your scatter diagram.
(d) Using the function found in (b), how far will the ball travel before it reaches its maximum height?
(e) Using the function found in (b), determine the maximum height of the ball.
(f) Compare your result in (d) and (e) to the data.
(g) Determine the horizontal distance the ball will travel based on the function found in (b).

80. **Enrollment in Public Schools** The data below represents the enrollment E in all public schools (both elementary and high school) for the academic years 1980–1981 to 1988–1989. Let 1 represent the academic year 1980–1981, 2 the academic year 1981–1982, and so on.

Year, t	Enrollment, E
1	41.5
2	40.8
3	40.1
4	39.6
5	39.1
6	39.1
7	39.6
8	39.8
9	40.1

(a) Using a graphing utility, draw a scatter diagram of the data. Comment on the type of relation that may exist between the two variables.
(b) Find the quadratic function of best fit.
(c) Using a graphing utility, draw the quadratic function of best fit on your scatter diagram.
(d) Using the function found in (b), determine when enrollment was lowest.
(e) Compare your result in (d) to the data.
(f) Using the function found in (b), predict the enrollment for 1991–1992.

81. **Chemical Reactions** A self-catalytic chemical reaction results in the formation of a compound that causes the formation ratio to increase. If the reaction rate V is given by

$$V(x) = kx(a - x) \qquad 0 \le x \le a$$

where k is a positive constant, a is the initial amount of the compound, and x is the variable

amount of the compound, for what value of x is the reaction rate a maximum?

82. A rectangle has one vertex on the line $y = 10 - x$, $x > 0$, another at the origin, one on the positive x-axis, and one on the positive y-axis. Find the largest area A that can be enclosed by the rectangle.

83. **Calculus: Simpson's Rule** The figure shows the graph of $y = ax^2 + bx + c$. Suppose the points $(-h, y_0)$, $(0, y_1)$, and (h, y_2) are on the graph. It can be shown that the area enclosed by the parabola, the x-axis and the lines $x = -h$ and $x = h$ is

$$\text{Area} = \frac{h}{3}(2ah^2 + 6c)$$

Show that this area may also be given by

$$\text{Area} = \frac{h}{3}(y_0 + 4y_1 + y_2)$$

84. Let $f(x) = ax^2 + bx + c$, where a, b, and c are odd integers. If x is an integer, show that $f(x)$ must be an odd integer.
[**Hint:** x is either an even integer or and odd integer.]

85. Make up a quadratic function that opens down and has only one x-intercept. Compare yours with others in the class. What are the similarities? What are the differences?

86. **CBL Experiment** As a ball bounces up and down, the maximum height the ball attains continually decreases from one bounce to the next. For a given bounce, plotting the height of the ball against time results in a parabola. The motion of the ball will be analyzed. (Activity 9, Real-World Math with the CBL System, 1994.)

87. **CBL Experiment** A cart is pushed up a ramp and allowed to return down the ramp. Plotting the distance the cart is from a motion detector against time results in the graph of a parabola. The parabola will be analyzed to determine the characteristics of the cart's motion. (Activity 8, Real-World Math with the CBL System, 1994.)

3.2 POWER FUNCTIONS; CURVE FITTING

1 Determine the Coefficient and Degree of a Monomial
2 Graph Power Functions in Stages
3 Find the Power Function of Best Fit from Data

1 A **monomial in one variable** is the product of a constant times a variable raised to a nonnegative integer power. Thus, a monomial is of the form ax^n where a is a constant, x is a variable, and $n \geq 0$ is an integer. The constant a is called the coefficient of the monomial. If $a \neq 0$, then n is called the degree of the monomial.

EXAMPLE 1 Identifying the Coefficient and Degree of a Monomial

Determine the coefficient and degree of the following monomials.

(a) $-4x^3$ (b) $10x$ (c) 7

Solution (a) The coefficient is -4 and the degree is 3.
(b) The coefficient is 10 and the degree is 1.
(c) The coefficient is 7 and the degree is 0 since $7 = 7x^0$. ∎

We now discuss *power functions*, functions whose rule is given by a monomial.

Power Functions

A **power function of degree n** is a function of the form

$$f(x) = ax^n \tag{1}$$

where a is a real number, $a \neq 0$, and $n > 0$ is an integer.

The graph of a power function of degree 1, $f(x) = ax$, is a straight line, with slope a, that passes through the origin. The graph of a power function of degree 2, $f(x) = ax^2$, is a parabola, with vertex at the origin, that opens up if $a > 0$ and down if $a < 0$.

If we know how to graph a power function of the form $f(x) = x^n$, then a compression or stretch and, perhaps, a reflection about the x-axis will enable us to obtain the graph of $g(x) = ax^n$. Consequently, we shall concentrate on graphing power functions of the form $f(x) = x^n$.

We begin with power functions of even degree of the form $f(x) = x^n, n \geq 2$ and n *even*. Using your graphing utility and a RANGE of $-2 \leq x \leq 2$, $-4 \leq y \leq 16$, graph the function $f(x) = x^4$. Can you conjecture the domain of this function from the graph? The range? On the same screen, graph $f(x) = x^8$. Now, also on the same screen, graph $f(x) = x^{12}$. What do you notice about the graphs as the magnitude of the exponent increases? Repeat the procedure given above for a RANGE of $-1 \leq x \leq 1; 0 \leq y \leq 1$. What do you notice? See Figures 21(a) and (b).

FIGURE 21

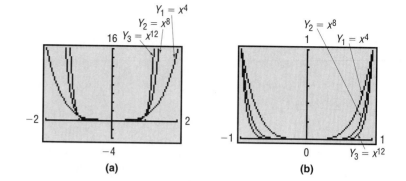

(a)   (b)

From the graphs, we can conjecture a few characteristics of even power functions. First, even power functions are symmetric with respect to the y-axis. Second, the domain is the set of all real numbers and the range is the set of nonnegative real numbers. Third, the graph always contains the points $(0, 0)$, $(1, 1)$, and $(-1, 1)$. Fourth, as the exponent increases in magnitude, the graph increases very rapidly as x increases, but for x near the origin the graph tends to flatten out and lie closer to the origin.

Note that for large n, it appears the graph coincides with the x-axis near the origin, but it does not; the graph actually touches the x-axis only at the origin. See Table 2, where $Y_1 = x^8$ and $Y_2 = x^{12}$. In Table 2, the expression 1E-8 means $1 \cdot 10^{-8}$. Also, for large n, it may appear that for $x < -1$ or for $x > 1$, the graph is vertical, but it is not; it is only increasing very rapidly. If you use TRACE along one of the graphs these distinctions would be clear.

Now we consider power functions of odd degree of the form $f(x) = x^n$, n odd. Using your graphing utility and a RANGE of $-2 \le x \le 2$, $-16 \le y \le 16$, graph the function $f(x) = x^3$. You have seen this graph several times before. Based on the graph, can you conjecture the domain and range of this function? On the same screen, graph $f(x) = x^7$ and $f(x) = x^{11}$. What do you notice about the graphs as the magnitude of the exponent increases? Repeat the procedure given above for a RANGE of $-1 \le x \le 1$; $-1 \le y \le 1$. What do you notice? The graphs on your screen should look like Figures 22(a) and (b).

TABLE 2

X	Y1	Y2
-1	1	1
1	1	1
.5	.00391	2.4E-4
.1	1E-8	1E-12
.01	1E-16	1E-24
.001	1E-24	1E-36
0	0	0

Y1☰X^8

FIGURE 22

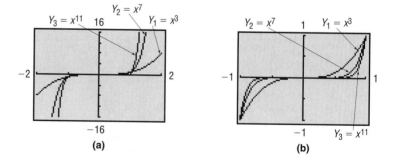

(a)   (b)

From the graphs in Figure 22(a) and (b), we can conjecture a few characteristics of odd power functions. First, odd power functions are symmetric with respect to the origin. Second, the domain and range of odd power functions are the set of all real numbers. Third, the graph of an odd power function always contains the points $(0, 0)$, $(1, 1)$, and $(-1, -1)$. Fourth, as the exponent increases in magnitude, the graphs become more vertical when $x > 1$ or $x < -1$ and the graphs tend to flatten out and lie closer to the x-axis when

x is near the origin. Note that it appears the graph coincides with the x-axis near the origin, but it does not; the graph actually touches the x-axis only at the origin. Also it appears that as x increases, the graph is vertical, but it is not, it is increasing very rapidly. TRACE along the graphs to verify these distinctions.

2 The methods of shifting, compression, stretching, and reflection studied in Section 2.3, when used with the facts just presented, will enable us to graph and analyze a variety of polynomials.

E X A M P L E   2   Graphing In Stages

Using a graphing utility, show each stage to obtain the graph of $f(x) = 1 - x^5$.

Solution   STAGE 1: Graph $y = x^5$. See Figure 23 and Table 3.

FIGURE 23

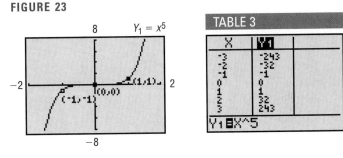

STAGE 2: Graph $y = -x^5$ (Reflection about y-axis). See Figure 24 and Table 4.

FIGURE 24

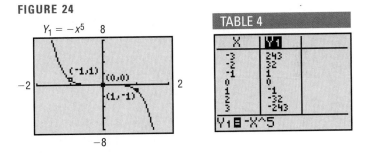

STAGE 3: Graph $y = 1 - x^5$ (Shift up 1 unit). See Figure 25 and Table 5.

FIGURE 25

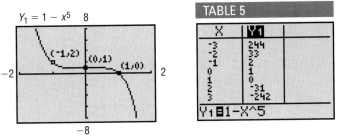

By following key points as we shift, reflect, and so forth, we determine the intercepts of f are $(0, 1)$ and $(1, 0)$.   ▬

E X A M P L E 3   Graphing In Stages

Using a graphing utility, show each stage to obtain the graph of $f(x) = \frac{1}{2}(x - 1)^4$.

Solution   STAGE 1: Graph $y = x^4$. See Figure 26 and Table 6.

FIGURE 26

STAGE 2: Graph $y = \frac{1}{2}x^4$ (compression). See Figure 27 and Table 7.

FIGURE 27

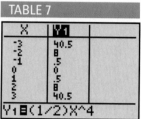

STAGE 3: Graph $y = \frac{1}{2}(x - 1)^4$ (Shift right 1 unit). See Figure 28 and Table 8.

FIGURE 28

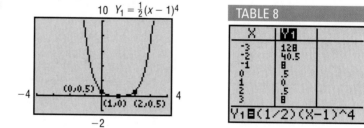

The intercepts of f are $(0, \frac{1}{2})$ and $(1, 0)$.

    Now work Problem 7.

Curve Fitting

In many applications, the relation that exists between two variables x and y follows a function of the form

$$y = ax^b$$

where $a > 0$ and $b > 0$ are two real numbers. This function is called a **generalized power function.** Figure 29 shows scatter diagrams that typically follow a generalized power function. Below each scatter diagram are restrictions on the values of the parameters, a and b, and the independent and dependent variables x and y.

FIGURE 29

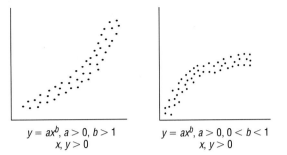

$$y = ax^b, a > 0, b > 1$$
$$x, y > 0$$

$$y = ax^b, a > 0, 0 < b < 1$$
$$x, y > 0$$

E X A M P L E   4

Fitting a Curve to a Generalized Power Function

The *period* of a pendulum is the time required for one oscillation; the pendulum is usually referred to as *simple* when the angle made to the vertical is less than 5°. An experiment is conducted in which simple pendulums are constructed with different lengths, l, and the corresponding period T is recorded. The data below is collected:

Length l (in feet)	1	2	3	4	5	6	7
Period T (in seconds)	1.10	1.55	1.89	2.24	2.51	2.76	2.91

(a) Using a graphing utility, draw a scatter diagram of the data with length as the independent variable and the period as the dependent variable.

(b) Using a graphing utility, fit a generalized power function to the data.

(c) Graph the generalized power function found in (b) on the scatter diagram.

(d) If the length of a simple pendulum is known to be 2.3 feet, what is the period of the pendulum?

Solution (a) After entering the data into the graphing utility, we obtain the scatter diagram shown in Figure 30.

FIGURE 30 3.5

(b) A graphing utility fits the data in Figure 30 to a generalized power function of the form $y = ax^b$ by using the PoWeR REGression option. See Figure 31.

FIGURE 31

The period T of a simple pendulum is related to its length l by the power function

$$T = 1.0948l^{0.5098}$$

For generalized power functions, a value for r is given, since such functions can be expressed as linear functions using logarithms. (See Chapter 4). Notice that the value of r is close to 1, indicating a good fit.*

(c) Figure 32 shows the graph of the power function on the scatter diagram.

FIGURE 32

(d) The period T of a pendulum whose length l is 2.3 feet is

$$T = (1.0948)2.3^{0.5098} = 1.674 \text{ seconds}$$

3.2 EXERCISES

For Problems 1–6, determine the coefficient and degree of each monomial.

1. $8x^5$ **2.** $-12x^3$ **3.** 4 **4.** $10x^5$ **5.** $-7x$ **6.** 5

In Problems 7–14, using a graphing utility, show the steps required to graph each function.

7. $f(x) = (x + 1)^4$ **8.** $f(x) = x^4 + 2$ **9.** $f(x) = \dfrac{1}{2}x^4$

10. $f(x) = -x^4$ **11.** $f(x) = 2(x + 1)^4 + 1$ **12.** $f(x) = 3 - (x + 2)^4$

13. $f(x) = -\dfrac{1}{2}(x - 2)^4 - 1$ **14.** $f(x) = 1 - 2(x + 1)^4$

15. CBL Experiment Cathy is conducting an experiment to measure the relation between a light bulb's intensity and the distance from the light source. She measures a 100-watt light bulb's intensity 1 meter from the bulb and at 0.1-meter intervals up to 2 meters from the bulb and obtains the following data.

---

*In physics, it is proven that $T = \dfrac{2\pi}{\sqrt{32}} \sqrt{l}$, provided friction is ignored. Since $\sqrt{l} = l^{1/2} = l^{0.5}$

and $\dfrac{2\pi}{\sqrt{32}} \approx 1.1107$, the power function obtained is reasonably close to what is expected.

Distance (Meters)	Intensity
1.0	0.29645
1.1	0.25215
1.2	0.20547
1.3	0.17462
1.4	0.15342
1.5	0.13521
1.6	0.11450
1.7	0.10243
1.8	0.09231
1.9	0.08321
2.0	0.07342

(a) Using your graphing utility, draw a scatter diagram with distance as the independent variable and intensity as the dependent variable.

(b) Using your graphing utility, fit a generalized power function to the data.

(c) Graph the function found in (b) on the scatter diagram.

(d) What will the intensity of a 100-watt light bulb be if you stand 2.3 meters away?

(e) It is known that intensity I is inversely proportional to the square of the distance x, so that $I = \dfrac{a}{x^2}$. How close is the estimate of the exponent?

16. **CBL Experiment** Cathy repeats the experiment from Problem 15, but this time uses a 40-watt light bulb and obtains the following data.

Distance (Meters)	Intensity
1.0	0.0972
1.1	0.0804
1.2	0.0674
1.3	0.0572
1.4	0.0495
1.5	0.0433
1.6	0.0384
1.7	0.0339
1.8	0.0294
1.9	0.0268
2.0	0.0224

(a) Using your graphing utility, draw a scatter diagram with distance as the independent

variable and intensity as the dependent variable.

(b) Using your graphing utility, fit a generalized power function to the data.

(c) Graph the function found in (b) on the scatter diagram.

(d) What will the intensity of a 40-watt light bulb be if you stand 2.3 meters away?

17. **CBL Experiment** David is conducting an experiment in order to estimate the acceleration of an object due to gravity. David takes a ball and drops it from different heights and records the time it takes for the ball to hit the ground. Using an optic laser connected to a stop watch in order to determine the time, he collects the following data:

Time (Seconds)	Distance (Feet)
1.003	16
1.365	30
1.769	50
2.093	70
2.238	80

(a) Draw a scatter diagram using time as the independent variable and distance as the dependent variable.

(b) Fit a generalized power function to the data.

(c) Graph the function found in (b) on the scatter diagram.

(d) Predict how long it will take an object to fall 100 feet.

(e) Physics theory states the distance an object falls is directly proportional to the time squared. Rewrite the model found in (b) so it is of the form $s = \frac{1}{2}gt^2$, where s is distance, g is the acceleration due to gravity, and t is time. What is David's estimate of g? (It is known that the acceleration due to gravity is approximately 32 feet/sec^2.)

18. **CBL Experiment** Paul, David's friend, doesn't believe David's estimate of acceleration due to gravity is correct. Paul repeats the experiment described in Problem 17 and obtains the following data:

Time (Seconds)	Distance (Feet)
0.7907	10
1.1160	20
1.4760	35
1.6780	45
1.9380	60

(a) Draw a scatter diagram using time as the dependent variable and distance as the independent variable.
(b) Fit a generalized power function to the data.
(c) Graph the function found in (b) on the scatter diagram.
(d) Predict how long it will take an object to fall 100 feet. Why do you think this estimate is different than the one from Problem 17?

(e) Physics theory states the distance an object falls is directly proportional to the time squared. Rewrite the model found in (b) so it is of the form $s = \frac{1}{2}gt^2$, where s is distance, g is the acceleration due to gravity, and t is time. What is Paul's estimate of g? [It is known that the acceleration due to gravity is approximately 32 feet/sec^2.]

3.3 POLYNOMIAL FUNCTIONS; CURVE FITTING

1 Identify Polynomials and Their Degree
2 Identify the Zeros of a Polynomial and Their Multiplicity
3 Analyze the Graph of a Polynomial
4 Find the Cubic Function of Best Fit from Data

A **polynomial function** is a function of the form

$$f(x) = a_n x^n + a_{n-1} x^{n-1} + \cdots + a_1 x + a_0 \qquad (1)$$

where $a_n, a_{n-1}, \ldots, a_1, a_0$ are real numbers and n is a nonnegative integer. The domain consists of all real numbers.

1   Thus, a polynomial function is a function whose rule is given by a polynomial in one variable.*

E X A M P L E   1   Identifying Polynomial Functions

Determine which of the following are polynomial functions. For those that are, state the degree; for those that are not, tell why not.

(a) $f(x) = 2 - 3x^4$ (b) $g(x) = \sqrt{x}$ (c) $h(x) = \dfrac{x^2 - 2}{x^3 - 1}$

(d) $F(x) = 0$ (e) $G(x) = 8$

Solution   (a) f is a polynomial function of degree 4.
(b) g is not a polynomial function. The variable x is raised to the $\frac{1}{2}$ power, which is not a nonnegative integer.
(c) h is not a polynomial function. It is the ratio of two polynomials, and the polynomial in the denominator is of positive degree.
(d) F is the zero polynomial function; it is not assigned a degree.
(e) G is a nonzero constant function, a polynomial function of degree 0. ▬

Now work Problems 1 and 5.

---

*A review of polynomials may be found in the Appendix, Section 1.

We have already discussed in detail polynomial functions of degrees 0, 1, and 2. See Table 9 for a summary of the characteristics of the graphs of these polynomial functions.

TABLE 9

Degree	Form	Name	Graph
No degree	$f(x) = 0$	Zero function	The x-axis
0	$f(x) = a_0, \quad a_0 \neq 1$	Constant function	Horizontal line with y-intercept a_0
1	$f(x) = a_1 x + a_0, \quad a_1 \neq 0$	Linear function	Nonvertical, nonhorizontal line with slope a_1 and y-intercept a_0
2	$f(x) = a_2 x^2 + a_1 x + a_0, \quad a_2 \neq 0$	Quadratic function	Parabola: Graph opens up if $a_2 > 0$; graph opens down if $a_2 < 0$

Graphing Polynomials

FIGURE 33

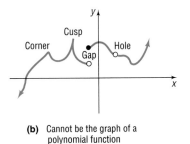

(a) Graph of a polynomial function: smooth, continuous

(b) Cannot be the graph of a polynomial function

To graph most polynomial functions of degree 3 or higher by hand requires techniques beyond the scope of this text. If you take a course in calculus you will learn that the graph of every polynomial function is both smooth and continuous. By **smooth,** we mean that the graph contains no sharp corners or cusps; by **continuous,** we mean that the graph has no gaps or holes and can be drawn without lifting pencil from paper. See Figures 33(a) and 33(b).

Figure 34 shows the graph of a polynomial function with four x-intercepts. Notice that at the x-intercepts the graph must either cross the x-axis or touch the x-axis. Consequently, between consecutive x-intercepts the graph is either above the x-axis or below the x-axis. We will make use of this characteristic of the graph of a polynomial shortly.

FIGURE 34

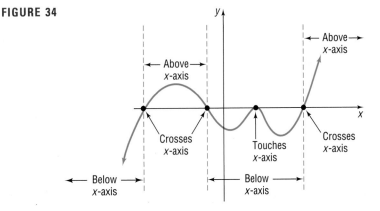

If a polynomial function f is factored completely, it is easy to solve the equation $f(x) = 0$ and locate the x-intercepts of the graph. For example, if $f(x) = (x - 1)^2(x + 3)$, then the solutions of the equation

$$f(x) = (x - 1)^2(x + 3) = 0$$

are easily identified as 1 and -3. Based on this result, we make the following observations:

If f is a polynomial function and r is a real number for which $f(r) = 0$, then r is called a (real) **zero of f,** or **root of f.**

If r is a (real) zero of f, then

(a) r is an x-intercept of the graph of f.
(b) $(x - r)$ is a factor of f.

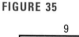

If the same factor $x - r$ occurs more than once, then r is called a **repeated,** or **multiple, zero of f.** More precisely, we have the following definition.

If $(x - r)^m$ is a factor of a polynomial f and $(x - r)^{m+1}$ is not a factor of f, then r is called a **zero of multiplicity m of f.**

E X A M P L E   2   Identifying Zeros and Their Multiplicities

For the polynomial

$$f(x) = 5(x - 2)(x + 3)^2\left(x - \frac{1}{2}\right)^4$$

2 is a zero of multiplicity 1
-3 is a zero of multiplicity 2
$\frac{1}{2}$ is a zero of multiplicity 4

∎

Suppose it is possible to factor completely a polynomial function and, as a result, locate all the x-intercepts of its graph (the real zeros of the function). The following example illustrates the role the multiplicity of the x-intercept plays.

E X A M P L E   3   Investigating the Role of Multiplicity

For the polynomial $f(x) = x^2(x - 2)$.

(a) Find the x- and y-intercepts of the graph.
(b) Using a graphing utility, graph the polynomial.
(c) For each x-intercept, determine whether it is of odd or even multiplicity.

Solution   (a) The y-intercept is $f(0) = 0^2(0 - 2) = 0$. The x-intercepts satisfy the equation

$$f(x) = x^2(x - 2) = 0$$

from which we find

$$x^2 = 0 \quad \text{or} \quad x - 2 = 0$$
$$x = 0 \qquad\qquad x = 2$$

The x-intercepts are 0 and 2.

(b) See Figure 35.

FIGURE 35

(c) We can see from the factored form of f that 0 is a zero or root of multiplicity 2 and 2 is a zero or root of multiplicity 1, so 0 is of even multiplicity and 2 is of odd multiplicity.

Notice from the graph in (b) that the graph of the polynomial at the root 0 just touches the x-axis so that the sign of $f(x)$ is the same on each side of the root. At the root 2, the graph of the polynomial crosses the x-axis so that the sign of $f(x)$ changes from one side of the root to the other. This observation leads us to the following result:

If r is a Zero of Even Multiplicity

Sign of $f(x)$ does not change from one side to the other side of r.	Graph **touches** x-axis at r.

If r is a Zero of Odd Multiplicity

Sign of $f(x)$ changes from one side to the other side of r.	Graph **crosses** x-axis at r.

Now work Problem 11.

Look again at Figure 35. We can use a graphing utility to determine the graph's minimum point in the interval $0 < x < 2$. After utilizing BOX (or MINIMUM if your utility has this feature), we find the graph's minimum point in the interval $0 < x < 2$ is $(1.33, -1.18)$ correct to two decimal places. In calculus, a method is presented for locating minimum and maximum points. There, such points are called *local minima* or *local maxima*. *Local minima* and *local maxima* are points where the graph changes direction (i.e., changes from an increasing function to a decreasing function or vice versa). We call these points **turning points.**

Look again at Figure 35. The graph of $f(x) = x^2(x - 2) = x^3 - 2x^2$, a polynomial of degree 3, has two turning points.

Exploration Graph $Y_1 = x^3$, $Y_2 = x^3 - x$, and $Y_3 = x^3 + 3x^2 + 4$. How many turning points do you see? How does the number of turning points relate to the degree? Graph $Y_1 = x^4$, $Y_2 = x^4 - \frac{4}{3}x^3$, and $Y_3 = x^4 - 2x^2$. How many turning points do you see? How does the number of turning points compare to the degree?

The following theorem from calculus supplies the answer.

> **Theorem**
>
> If f is a polynomial function of degree n, then f has at most $n - 1$ turning points.

One last remark about Figure 35. Notice that the graph of $f(x) = x^2(x - 2)$ looks somewhat like the graph of $y = x^3$. In fact, for very large values of x, either positive or negative, there is little difference.

Exploration Consider the functions Y_1 and Y_2 given on page 220 in (a), (b), (c). Graph Y_1 and Y_2 on the same viewing window. TRACE for large positive and large negative values of x. What do you notice about the graphs of Y_1 and Y_2 as x becomes very large and positive or very large and negative?

(a) $Y_1 = x^2(x - 2)$; $Y_2 = x^3$
(b) $Y_1 = x^4 - 3x^3 + 7x - 3$; $Y_2 = x^4$
(c) $Y_1 = -2x^3 + 4x^2 - 8x + 10$; $Y_2 = -2x^3$

Theorem

For large values of x, either positive or negative, the graph of the polynomial

$$f(x) = a_n x^n + a_{n-1} x^{n-1} + \cdots + a_1 x + a_0$$

resembles the graph of the power function

$$y = a_n x^n$$

The following display summarizes some features of the graph of a polynomial function.

SUMMARY: GRAPH OF A POLYNOMIAL FUNCTION

$f(x) = a_n x^n + a_{n-1} x^{n-1} + \cdots + a_1 x + a_0, a_n \neq 0$

Degree of the polynomial f: n
Maximum number of turning points: $n - 1$
At zero of even multiplicity: graph of f touches x-axis
At zero of odd multiplicity: graph of f crosses x-axis
Between zeros, graph of f is either above the x-axis or below it.
For large x, graph of f behaves like graph of $y = a_n x^n$.

E X A M P L E   4

Analyzing the Graph of a Polynomial Function

For the polynomial $f(x) = x^4 - 3x^3 - 4x^2$,

(a) Using a graphing utility, graph f.
(b) Find the x- and y-intercepts.
(c) Determine whether each x-intercept is of odd or even multiplicity.
(d) Find the power function that the graph of f resembles for large values of x.
(e) Determine the number of turning points on the graph of f.
(f) Determine the *local maxima* and *local minima*, if any exist, correct to two decimal places.

Solution

(a) See Figure 36.

(b) The y-intercept is $f(0) = 0$. Factoring, we find $f(x) = x^4 - 3x^3 - 4x^2 = x^2(x - 4)(x + 1)$. We find the x-intercepts by solving the equation

$$f(x) = x^2(x - 4)(x + 1) = 0$$

So

$$x^2 = 0 \quad \text{or} \quad x - 4 = 0 \quad \text{or} \quad x + 1 = 0$$
$$x = 0 \qquad\qquad x = 4 \qquad\qquad x = -1$$

The x-intercepts are $-1, 0$, and 4.

(c) The intercept 0 is a zero of even multiplicity, 2, so the graph of f will touch the x-axis at 0; 4 and -1 are zeros of odd multiplicity, 1, so the graph of f will cross the x-axis at 4 and -1. This means the graph will cross the x-axis at -1 and 4, and touch the x-axis at 0. Look again at Figure 36. The graph is below the x-axis for $-1 < x < 0$!

(d) The graph of f behaves like $f(x) = x^4$ for large x.

(e) Since f is of degree 4, the graph can have at most three turning points. From the graph, we see the graph has three turning points: one between -1 and 0, one at $(0, 0)$, and one between 2 and 4.

(f) Correct to two decimal places, the *local maximum* is $(0, 0)$ and the *local minima* are $(-0.68, -0.69)$ and $(2.93, -36.10)$. ▬

FIGURE 36

E X A M P L E   5

Analyzing the Graph of a Polynomial Function

Follow the instructions of Example 4 for the polynomial below:

$$f(x) = x^3 + 2.48x^2 - 4.3155x + 1.484406$$

Solution

(a) See Figure 37.

(b) The y-intercept is $f(0) = 1.484406$. In Example 4 we could easily factor $f(x)$ to find the x-intercepts. However, it is not readily apparent how $f(x)$ factors in this example. Therefore, we use a graphing utility and find the x-intercepts to be -3.74 and 0.63.

(c) The x-intercept -3.74 is of odd multiplicity since the graph of f crosses the x-axis at -3.74; the x-intercept 0.63 is of even multiplicity since the graph of f touches the x-axis at 0.63.

(d) The graph behaves like $f(x) = x^3$ for large x.

(e) Since f is of degree 3, the graph can have at most two turning points. From the graph we see the graph has two turning points: one between -3 and -2; the other at $(0.63, 0)$.

(f) Correct to two decimal places, the *local maximum* is $(-2.28, 12.36)$ and the *local minimum* is $(0.63, 0)$. ▬

FIGURE 37

Now work Problem 23.

Curve Fitting

In Section 1.4, we discussed finding a line of best fit from data. In Section 3.1, we discussed finding a quadratic function of best fit. It is possible to find

functions of best fit for any polynomial of degree n. However, most statisticians recommend finding polynomials of best fit of degree no higher than 3.*
Data that follows a cubic relation should look like Figure 38(a) or (b).

FIGURE 38

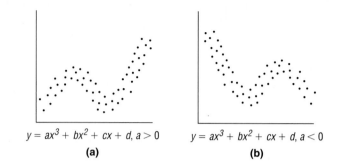

$$y = ax^3 + bx^2 + cx + d, a > 0$$
(a)

$$y = ax^3 + bx^2 + cx + d, a < 0$$
(b)

E X A M P L E 6

Finding a Cubic Function of Best Fit

The data in Table 10 represents the number of barrels of refined oil products imported into the United States for 1977–1992, where 1 represents 1977, 2 represents 1978, and so on.

TABLE 10	
Year, *x*	Refined Oil Products (1000 Barrels per Day), *R*
1977, 1	2193
1978, 2	2008
1979, 3	1937
1980, 4	1646
1981, 5	1599
1982, 6	1625
1983, 7	1722
1984, 8	2011
1985, 9	1866
1986, 10	2045
1987, 11	2004
1988, 12	2295
1989, 13	2217
1990, 14	2123
1991, 15	1844
1992, 16	1805

Source: U.S. Energy Information Administration, *Monthly Energy Review*, February, 1995.

*Two points determine a unique linear function. Three noncollinear points determine a unique quadratic function. Four points determine a unique cubic function, and n points will determine a unique polynomial of degree $n - 1$. Therefore, higher degree polynomials will always "fit" data at least as well as lower degree polynomials. However, higher degree polynomials yield highly erratic predictions. Since the ultimate goal of curve fitting is not necessarily to find the model that best fits the data, but instead to explain relationships between two or more variables, polynomial models of degree three or less are usually used.

(a) Draw a scatter diagram of the data. Comment on the type of relation that may exist between the two variables.

(b) Find the cubic function of best fit.

(c) Draw the cubic function of best fit on your scatter diagram.

(d) Use the function found in (b) to predict the number of barrels of refined oil products imported in 1993 ($x = 17$).

 (e) Do you think the function found in (b) will be useful in predicting the number of barrels of refined oil products for the year 1999? Why?

Solution   (a) Figure 39 shows the scatter diagram.

FIGURE 39

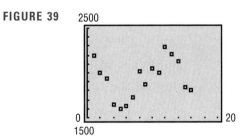

A cubic relation with $a < 0$ may exist between the two variables.

(b) Upon executing the CUBIC REGression program, we obtain the results shown in Figure 40. The output the utility provides shows us the equation $y = ax^3 + bx^2 + cx + d$. The cubic function of best fit is

$$R(x) = -2.4117x^3 + 63.4297x^2 - 453.7614x + 2648.0027,$$

where x represents the year and R represents the number of barrels of refined oil products imported.

FIGURE 40

(c) Figure 41 shows the graph of the cubic function of best fit on the scatter diagram. The function seems to fit the data well.

FIGURE 41

(d) We evaluate the function $R(x)$ for $x = 17$:

$$R(17) = -2.4117(17)^3 + 63.4297(17)^2 - 453.7614(17) + 2648.0027 \approx 1417$$

So, we predict 1417 barrels of refined oil products will be imported into the United States in 1993.

(e) The graph of the function will continually decrease for years beyond 1993, indicating the number of barrels of refined oil will continue to decline. This is not a likely scenario. So, the function probably will not be useful in predicting the number of barrels of refined oil products imported into the United States in 1999. ▬

3.3 EXERCISES

In Problems 1–10, determine which functions are polynomial functions. For those that are, state the degree. For those that are not, tell why not.

1. $f(x) = 4x + x^3$

2. $f(x) = 5x^2 + 4x^4$

3. $g(x) = \dfrac{1 - x^2}{2}$

4. $h(x) = 3 - \dfrac{1}{2}x$

5. $f(x) = 1 - \dfrac{1}{x}$

6. $f(x) = x(x - 1)$

7. $g(x) = x^{3/2} - x^2 + 2$

8. $h(x) = \sqrt{x}(\sqrt{x} - 1)$

9. $F(x) = 5x^4 - \pi x^3 + \dfrac{1}{2}$

10. $F(x) = \dfrac{x^2 - 5}{x^3}$

In Problems 11–20, for each polynomial function, list each real zero and its multiplicity. Determine whether the graph crosses or touches the x-axis at each x-intercept.

11. $f(x) = 3(x - 7)(x + 3)^2$

12. $f(x) = 4(x + 4)(x + 3)^3$

13. $f(x) = 4(x^2 + 1)(x - 2)^3$

14. $f(x) = 2(x - 3)(x + 4)^3$

15. $f(x) = -2\left(x + \dfrac{1}{2}\right)^2 (x^2 + 4)^2$

16. $f(x) = \left(x - \dfrac{1}{3}\right)^2 (x - 1)^3$

17. $f(x) = (x - 5)^3(x + 4)^2$

18. $f(x) = (x + \sqrt{3})^2(x - 2)^4$

19. $f(x) = 3(x^2 + 8)(x^2 + 9)^2$

20. $f(x) = -2(x^2 + 3)^3$

In Problems 21–56, for each polynomial function f:

(a) Using a graphing utility, graph f.

(b) Find the x- and y-intercepts.

(c) Determine whether each x-intercept is of odd or even multiplicity.

(d) Find the power function that the graph of f resembles for large values of x.

(e) Determine the number of turning points on the graph of f.

(f) Determine the local maxima and local minima, if any exist, correct to two decimal places.

21. $f(x) = (x - 1)^2$

22. $f(x) = (x - 2)^3$

23. $f(x) = x^2(x - 3)$

24. $f(x) = x(x + 2)^2$

25. $f(x) = 6x^3(x + 4)$

26. $f(x) = 5x(x - 1)^3$

27. $f(x) = -4x^2(x + 2)$

28. $f(x) = -\frac{1}{2}x^3(x + 4)$

29. $f(x) = x(x - 2)(x + 4)$

30. $f(x) = x(x + 4)(x - 3)$

31. $f(x) = 4x - x^3$

32. $f(x) = x - x^3$

33. $f(x) = x^2(x - 2)(x + 2)$

34. $f(x) = x^2(x - 3)(x + 4)$

35. $f(x) = x^2(x - 2)^2$

36. $f(x) = x^3(x - 3)$

37. $f(x) = x^2(x - 3)(x + 1)$

38. $f(x) = x^2(x - 3)(x - 1)$

39. $f(x) = x(x + 2)(x - 4)(x - 6)$

40. $f(x) = x(x - 2)(x + 2)(x + 4)$

41. $f(x) = x^2(x - 2)(x^2 + 3)$

42. $f(x) = x^2(x^2 + 1)(x + 4)$

43. $f(x) = x^3 + 0.2x^2 - 1.5876x - 0.31752$

44. $f(x) = x^3 - 0.8x^2 - 4.6656x + 3.73248$

45. $f(x) = x^3 + 2.56x^2 - 3.31x + 0.89$

46. $f(x) = x^3 - 2.91x^2 - 7.668x - 3.8151$

47. $f(x) = x^4 - 2.5x^2 + 0.5625$

48. $f(x) = x^4 - 18.5x^2 + 50.2619$

49. $f(x) = x^4 + 0.65x^3 - 16.6319x^2 + 14.209335x - 3.1264785$

50. $f(x) = x^4 + 3.45x^3 - 11.6639x^2 - 5.864241x - 0.69257738$

51. $y = \pi x^3 + \sqrt{2}x^2 - x - 2$

52. $y = -2x^3 + \pi x^2 + \sqrt{3}x + 1$

53. $y = 2x^4 - \pi x^3 + \sqrt{5}x - 4$

54. $y = -1.2x^4 + 0.5x^2 - \sqrt{3}x + 2$

55. $y = -2x^5 - \sqrt{2}x^2 - x - \sqrt{2}$

56. $y = \pi x^5 + \pi x^4 + \sqrt{3}x + 1$

57. Consult the illustration. Which of the following polynomial functions might have this graph? (More than one answer may be possible.)

58. Consult the illustration. Which of the following polynomial functions might have this graph? (More than one answer may be possible.)

(a) $y = -4x(x - 1)(x - 2)$
(b) $y = x^2(x - 1)^2(x - 2)$
(c) $y = 3x(x - 1)(x - 2)$
(d) $y = x(x - 1)^2(x - 2)^2$
(e) $y = x^3(x - 1)(x - 2)$
(f) $y = -x(1 - x)(x - 2)$

(a) $y = 2x^3(x - 1)(x - 2)^2$
(b) $y = x^2(x - 1)(x - 2)$
(c) $y = x^3(x - 1)^2(x - 2)$
(d) $y = x^2(x - 1)^2(x - 2)^2$
(e) $y = 5x(x - 1)^2(x - 2)$
(f) $y = -2x(x - 1)^2(2 - x)$

(c) Of radius 4 feet?

(d) Of radius 5 feet?

(e) Graph $A = A(r)$. For what value of r is A smallest?

99. Consult the illustration. Make up a rational function that might have this graph. Compare yours with a friend's. What similarities do you see? What differences?

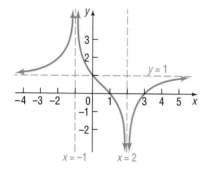

100. Can the graph of a rational function have both a horizontal and an oblique asymptote? Explain.

101. Write a few paragraphs that provide a general strategy for graphing a rational function. Be sure to mention the following: proper, improper, intercepts, and asymptotes.

102. Make up a rational function that has the following characteristics: crosses the x-axis at 3; touches the x-axis at -2; one vertical asymptote, $x = 1$; and one horizontal asymptote, $y = 2$. Give your rational function to a fellow classmate and ask for a written critique of your rational function.

103. Make up a rational function that has $y = 2x + 1$ as an oblique asymptote. Explain the methodology you used.

3.5   THE REAL ZEROS OF A POLYNOMIAL FUNCTION

> **1** Utilize the Remainder and Factor Theorems
>
> **2** Use Descartes' Rule of Signs to Determine the Number of Positive and the Number of Negative Real Zeros of a Polynomial Function
>
> **3** Use the Rational Zeros Theorem to List the Potential Rational Zeros of a Polynomial Function
>
> **4** Find the Real Zeros of a Polynomial Function
>
> **5** Solve Polynomial Equations
>
> **6** Utilize the Upper and Lower Bounds Test

Remainder and Factor Theorems

1 Recall that when we divide one polynomial (the dividend) by another (the divisor) we obtain a quotient polynomial and a remainder, the remainder being either the zero polynomial or a polynomial whose degree is less than the degree of the divisor. To check our work, we verify that

$$(\text{Divisor})(\text{Quotient}) + \text{Remainder} = \text{Dividend}$$

This checking routine is the basis for a famous theorem called the **division algorithm*** **for polynomials,** which we now state without proof.

---

*A systematic process in which certain steps are repeated a finite number of times is called an **algorithm.** Thus, long division is an algorithm.

M I S S I O N   P O S S I B L E

Responding to a Challenge by the Cardassians

While playing with virtual reality in your school's computer lab, you are suddenly faced with some very real and very nasty-looking Cardassians who have burst into your school through your computer network. As usual, they are very scornful of the intelligence of Earthlings, and when you try to protest, they throw this challenge at you. "Find a rational function defined by this graph," they say, "and we promise to support your reputation for intellectual achievement in every intergalactic community this side of the Macklin Nebula. Otherwise, consider yourselves the laughing stock of the universe." And they departed back through the computer screen, laughing at what they perceived to be their extreme cleverness.

As you look at the computer screen now, all you can see is this graph:

You decide to tackle this as a team project to save the reputation of all the humans on this planet! The Cardassians mentioned "rational function," so you know the solution will have the form

$$R(x) = \frac{a_n x^n + a_{n-1} x^{n-1} + \cdots + a_0}{b_m x^m + b_{m-1} x^{m-1} + \cdots + b_0}$$

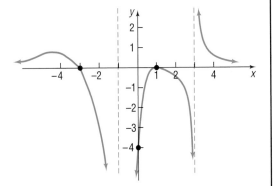

Sometimes it is easier to think of it in factored form.

1. You notice some things right away. There are two vertical asymptotes. How do they fit into the solution? Put them in.
2. There are two x-intercepts. Where do they show up in the solution? Put them in.
3. Use a graphing utility to check what you have so far. Does it look like the original graph? Notice that 1 and -3 appear to be the only real zeros. What can you tell about the multiplicity of each one? Do you need to change the multiplicity of either zero in your solution?
4. How can you account for the fact that the graph of the function goes to $-\infty$ on both sides of one vertical asymptote and at the other vertical asymptote it goes to $-\infty$ on one side and to ∞ on the other? Can you change the denominator in some way to account for this?
5. Is the information you have so far consistent with the horizontal asymptote $y = 0$? If not, how should you adjust your function so that it is consistent?
6. Is the information you have so far consistent with the y-intercept? If not, how should you adjust your function so that it is consistent?
7. Are you finished? Have you checked your solution on a graphing utility? Are you satisfied you have it right?
8. Are there other functions whose graph might look like the one given? If so, what common characteristics do they have? Will the Cardassians be more impressed if more than one correct answer is sent?
9. Compare your group's solution to those of other groups before typing the solution into the computer. Then send all the correct solutions off to cardass@slime.uni.

negative zeros. There is an x-intercept near 1 (where the graph crosses the x-axis) and another near 2 (which may be a root of even multiplicity, since the graph appears to touch the x-axis near 2).

Since 1 appears to be a zero and 1 is a potential rational zero, we use synthetic division to determine if $x - 1$ is a factor of f:

$$
\begin{array}{r|rrrrr}
1) & 1 & -5 & 12 & -24 & 32 & -16 \\
   &   & 1 & -4 & 8 & -16 & 16 \\
\hline
   & 1 & -4 & 8 & -16 & 16 & 0
\end{array}
$$

Since the remainder is 0, we know 1 is a zero and $x - 1$ is a factor of f. Thus, 1 is a solution to the equation.

We now work with the first depressed equation of f:

$$x^4 - 4x^3 + 8x^2 - 16x + 16 = 0$$

We check the potential rational zero 2 next using synthetic division.

$$
\begin{array}{r|rrrrr}
2) & 1 & -4 & 8 & -16 & 16 \\
   &   & 2 & -4 & 8 & -16 \\
\hline
   & 1 & -2 & 4 & -8 & 0
\end{array}
$$

Since the remainder is 0, we conclude 2 is a zero and $x - 2$ is a factor of f. Thus, 2 is a solution to the equation.

We work with the second depressed equation of f: $x^3 - 2x^2 + 4x - 8 = 0$. Earlier, we conjectured 2 may be a root of even multiplicity, so we check 2 again.

$$
\begin{array}{r|rrrr}
2) & 1 & -2 & 4 & -8 \\
   &   & 2 & 0 & 8 \\
\hline
   & 1 & 0 & 4 & 0
\end{array}
$$

Since the remainder is 0, 2 is a zero and $x - 2$ is a factor of f. Thus, $x = 2$ is a zero of multiplicity 2.

The third depressed equation is $x^2 + 4 = 0$. This equation has no real solutions.

Thus, the real solutions are 1 and 2 (the latter being a repeated solution). ∎

Based on the solution to Example 7, the polynomial function f can be factored as follows:

$$f(x) = x^5 - 5x^4 + 12x^3 - 24x^2 + 32x - 16 = (x - 1)(x - 2)^2(x^2 + 4)$$

Now work Problem 61.

E X A M P L E   8   Finding the Zeros of a Polynomial

Use Descartes' Rule of Signs and the Rational Zeros Theorem to find the real zeros of the polynomial function

$$g(x) = x^5 - 7x^4 + 8x^3 - 20x^2 + 20x - 2$$

Express any irrational zeros correct to two decimal places.

Solution   STEP 1: There are at most five zeros.
STEP 2: There are five, three, or one positive real zeros. Also, because

$$g(-x) = -x^5 - 7x^4 - 8x^3 - 20x^2 - 20x - 2$$

there are no negative real zeros.

FIGURE 61

STEP 3: The potential rational zeros of g are 1 and 2.
STEP 4: Because there are no negative real zeros, we graph f only for $x \geq -1$. See Figure 61. The graph has the characteristics we expect: it behaves like $y = x^5$ for large x and has y-intercept -2. The information obtained in steps 1 and 2 is confirmed. ZOOMing-IN near $(\frac{1}{2}, 1)$, we see the graph has three x-intercepts: one between 0 and 1, one near 1, and one near 6. We test the potential rational zero 1 using synthetic division.

The remainder is 0 so 1 is a zero and $x - 1$ is a factor of g.
STEP 5: The first depressed equation of g

$$x^4 - 6x^3 + 2x^2 - 18x + 2 = 0$$

has only 1 and 2 as potential rational zeros. Neither of these is, in fact, a zero. So the remaining zeros of g are not rational. Correct to two decimal places, these irrational zeros are: 0.10 and 6.14. ▬

In Example 7, the quadratic factor $x^2 + 4$ that appears in the factored form of $f(x)$ is called *irreducible*, because the polynomial $x^2 + 4$ cannot be factored over the real numbers. In general, we say that a quadratic factor $ax^2 + bx + c$ is **irreducible** if it cannot be factored over the real numbers, that is, if it is prime over the real numbers.

Refer back to the polynomial function f of Example 6. We found that f has five real zeros, so, by the Factor Theorem, its factored form will contain five linear factors. The polynomial function g of Example 8 has three real zeros, so its factored form will contain three linear factors and one irreducible quadratic factor. The next result tells us what to expect when we factor a polynomial.

Theorem

Every polynomial function (with real coefficients) can be uniquely factored into a product of linear factors and/or irreducible quadratic factors.

▬

We shall prove this result in Section 3.7, and, in fact, we shall draw several additional conclusions about the zeros of a polynomial function. One conclusion is worth noting now. If a polynomial (with real coefficients) is of odd degree, then it must contain at least one linear factor. (Do you see why?) Therefore, it will have at least one real zero.

Corollary

A polynomial function (with real coefficients) of odd degree has at least one real zero.

▬

▬

One of the challenges in using a graphing utility is to set the viewing window so that a complete graph is obtained. The next theorem is a tool which can be used to find bounds on the zeros. This will assure the function does not have any zeros above or below these bounds.

Upper and Lower Bounds

The search for the zeros of a polynomial function can be reduced somewhat if upper and lower bounds to the zeros can be found. A number M is an **upper bound** to the zeros of a polynomial f if no zero of f exceeds M. The number m is a **lower bound** if no zero of f is less than m.

Thus, if m is a lower bound and M is an upper bound to the zeros of a polynomial f, then

$$m \leq \text{Any zero of } f \leq M$$

One immediate advantage of knowing the values of a lower bound m and an upper bound M is that, for polynomials with integer coefficients, it may allow you to eliminate some potential rational zeros, that is, any that lie outside the interval $[m, M]$. The next result tells us how to locate lower and upper bounds.

> ### Theorem   Bounds on Zeros
>
> Let f denote a polynomial function whose leading coefficient is positive.
>
> If $M > 0$ is a real number and if the third row in the process of synthetic division of f by $x - M$ contains only numbers that are positive or 0, then M is an upper bound to the zeros of f.
>
> If $m \leq 0$ is a real number and if the third row in the process of synthetic division of f by $x - m$ contains numbers that are alternately positive (or 0) and negative (or 0), then m is a lower bound to the zeros of f.

Proof (Outline)   We shall give only an outline of the proof of the first part of the theorem. Suppose that M is a positive real number, and the third row in the process of synthetic division of the polynomial f by $x - M$ contains only numbers that are positive or 0. Then there is a quotient q and a remainder R so that

$$f(x) = (x - M)q(x) + R$$

where the coefficients of $q(x)$ are positive or 0 and the remainder $R \geq 0$. Then, for any $x > M$, we must have $x - M > 0$, $q(x) > 0$, and $R \geq 0$, so that $f(x) > 0$. That is, there is no zero of f larger than M.

We can use the Upper and Lower Bounds Test to determine if the RANGE we have selected for our graph is large enough to show all the real zeros of the function, ensuring no zeros lie off the viewing rectangle. Consider Example 9.

E X A M P L E 9

Using the Upper and Lower Bounds Test

Use the Upper and Lower Bounds Test on the polynomial function $f(x) = x^5 - 5x^4 + 12x^3 - 24x^2 + 32x - 16$ to verify there are no zeros larger than 2 and no zeros smaller than 1.

Solution   To get an upper bound to the zeros, the usual practice is to start with the value of the next highest integer larger than the largest zero found and continue with consecutive integers until the third row of the process of synthetic division yields only numbers that are positive or 0.

We begin with the graph of f. See Figure 62.

The largest zero we see is near 2. The next highest integer, 3, is where we begin. We use synthetic division to divide f by $x - 3$. If the third row yields only numbers that are positive or zero, then 3 is an upper bound.

FIGURE 62

$$
\begin{array}{r|rrrrr}
3) & 1 & -5 & 12 & -24 & 32 & -16 \\
  &   & 3 & -6 & 18 & -18 & 42 \\
\hline
  & 1 & -2 & 6 & -6 & 14 & 26 \\
\end{array}
$$

Since the third row yields some negative numbers, 3 is not an upper bound. We will try 4, the next consecutive integer.

$$
\begin{array}{r|rrrrr}
4) & 1 & -5 & 12 & -24 & 32 & -16 \\
  &   & 4 & -4 & 32 & 32 & 256 \\
\hline
  & 1 & -1 & 8 & 8 & 64 & 240 \\
\end{array}
$$

Since the third row yields some negative numbers, 4 is not an upper bound. We will try 5.

$$
\begin{array}{r|rrrrr}
5) & 1 & -5 & 12 & -24 & 32 & -16 \\
  &   & 5 & 0 & 60 & 180 & 1060 \\
\hline
  & 1 & 0 & 12 & 36 & 212 & 1044 \\
\end{array}
$$

The last row yields only numbers that are positive or 0; therefore, 5 is an upper bound. We are certain there are no zeros larger than 5. This means 5 is a good choice for Xmax when we next graph f.

To get a lower bound to the zeros, we start with the next integer smaller than the smallest zero we see.

Look again at Figure 62. The smallest zero we see is near 1, so we begin by testing 0.

$$
\begin{array}{r|rrrrr}
0) & 1 & -5 & 12 & -24 & 32 & -16 \\
  &   & 0 & 0 & 0 & 0 & 0 \\
\hline
  & 1 & -5 & 12 & -24 & 32 & -16 \\
\end{array}
$$

Since the third row of the process of synthetic division yields numbers that alternate in sign, 0 is a lower bound; we are certain there are no zeros less than 0.*

This means 0 is a good choice for Xmin.   ■

Now work Problem 35.

*__Note:__ In determining lower bounds, a 0 in the bottom row following a nonzero entry may be counted as positive or negative, as needed. If the next entry is also a 0, it must be counted opposite to the way the preceding 0 was counted (i.e., if the third row yields two consecutive zeros and the first 0 was counted as positive, the next zero would count as negative).

Steps for Finding the Zeros
of a Polynomial

STEP 1: Use the degree of the polynomial to determine the maximum number of zeros.

STEP 2: Use Descartes' Rule of Signs to determine the possible number of positive real zeros and negative real zeros.

STEP 3: If the polynomial has integer coefficients, use the Rational Zeros Theorem to identify those rational numbers that are potential zeros.

STEP 4: (a) Using a graphing utility, graph the polynomial.
(b) Conjecture possible zeros based on the graph and the list of potential rational zeros.
(c) Test each potential rational zero.
(d) Approximate any irrational zeros correct to two decimal places.

STEP 5: Use the Upper and Lower Bounds test to be sure no zeros exist off the viewing window.

Let's work one more example that shows the STEPS listed above.

EXAMPLE   10

Finding the Zeros of a Polynomial

Find all the real zeros of the polynomial function

$$f(x) = x^5 - 1.8x^4 - 17.79x^3 + 31.672x^2 + 37.95x - 8.712$$

Solution

STEP 1: There are at most 5 zeros.

STEP 2: By Descartes' Rule of Signs there are three or one positive real zero(s). Also because

$$f(-x) = -x^5 - 1.8x^4 + 17.79x^3 + 31.672x^2 - 37.95x - 8.712$$

there are two or no real negative zeros.

STEP 3: Since there are noninteger coefficients, the Rational Zeros Theorem does not apply.

STEP 4: See Figure 63 for the graph of f. We see f appears to have four x-intercepts: one near -4, one near -1, one between 0 and 1, and one near 3. The x-intercept near 3 might be a zero of even multiplicity since the graph seems to touch the x-axis at the point. From step 2, we learned the function would have two or no negative real zeros and three or one positive real zeros. Thus, we know there are no other negative real zeros off the viewing window. However, we may only be seeing two positive real zeros and Descartes' Rule of Signs says there are three.

We use the factor theorem to determine if -4, and -1 are zeros. Since $f(-4) = f(-1) = 0$, we know -4 and -1 are the two negative zeros. Correct to two decimal places, the remaining zeros are 0.20 and 3.30.

STEP 5: We will check to see if 4 is an upper bound to the zeros.

FIGURE 63

$$
\begin{array}{r|rrrrrr}
4) & 1 & -1.8 & -17.79 & 31.672 & 37.95 & -8.712 \\
  &   & 4 & 8.8 & & & \\
\hline
  & 1 & 2.2 & -8.99 & & & \\
\end{array}
$$

We stop the process since the Upper Bound Theorem fails. We try 5 next.

$$
\begin{array}{r|rrrrr}
5) & 1 & -1.8 & -17.79 & 31.672 & 37.95 & -8.712 \\
  &   & 5 & 16 \\
\hline
  & 1 & 3.2 & -1.79
\end{array}
$$

Again we stop the process and try 6.

$$
\begin{array}{r|rrrrr}
6) & 1 & -1.8 & -17.79 & 31.672 & 37.95 & -8.712 \\
  &   & 6 & 25.2 & 44.46 & 456.792 & 2968.452 \\
\hline
  & 1 & 4.2 & 7.41 & 76.132 & 494.742 & 2959.74
\end{array}
$$

So, 6 is an upper bound to the zeros. Therefore, either 3.30 is a root of multiplicity 2 or there are two distinct zeros, both of which equal 3.30 correct to two decimal places.

Note: Notice f has 4 turning points. Thus, we know there are no zeros off the viewing rectangle. (Do you know why?) ■

HISTORICAL FEATURE Formulas for the solution of third- and fourth-degree polynomial equations exist, and, while not very practical, they do have an interesting history.

In the 1500s in Italy, mathematical contests were a popular pastime, and persons possessing methods for solving problems kept them secret. (Solutions that were published were already common knowledge.) Niccolo of Brescia (1499–1557), commonly referred to as Tartaglia ("the stammerer"), had the secret for solving cubic (third-degree) equations, which gave him a decided advantage in the contests. Girolamo Cardano (1501–1576) found out that Tartaglia had the secret, and, being interested in cubics, he requested it from Tartaglia. The reluctant Tartaglia hesitated for some time, but finally, swearing Cardano to secrecy with midnight oaths by candlelight, told him the secret. Cardano then published the solution in his book *Ars Magna* (1545), giving Tartaglia the credit but rather compromising the secrecy. Tartaglia exploded into bitter recriminations, and each wrote pamphlets that reflected on the other's mathematics, moral character, and ancestry. Tartaglia's method is discussed in Problems 85–89 in Exercise 3.5.

The quartic (fourth-degree) equation was solved by Cardano's student Lodovico Ferrari, and this solution also was included, with credit and this time with permission, in the *Ars Magna*.

Attempts were made to solve the fifth-degree equation in similar ways, all of which failed. In the early 1800s, P. Ruffini, Niels Abel, and Evariste Galois all found ways to show that it is not possible to solve fifth-degree equations by formula, but the proofs required the introduction of new methods. Galois' methods eventually developed into a large part of modern algebra. ■

3.5 EXERCISES

In Problems 1–10, use the Factor Theorem to determine whether $x - c$ is a factor of $f(x)$.

1. $f(x) = 4x^3 - 3x^2 - 8x + 4$; $c = 2$

2. $f(x) = -4x^3 + 5x^2 + 8$; $c = -3$

3. $f(x) = 3x^4 - 6x^3 - 5x + 10$; $c = 2$

4. $f(x) = 4x^4 - 15x^2 - 4$; $c = 2$

5. $f(x) = 3x^6 + 82x^3 + 27$; $c = -3$

6. $f(x) = 2x^6 - 18x^4 + x^2 - 9$; $c = -3$

7. $f(x) = 4x^6 - 64x^4 + x^2 - 15$; $c = -4$

8. $f(x) = x^6 - 16x^4 + x^2 - 16$; $c = -4$

9. $f(x) = 2x^4 - x^3 + 2x - 1$; $c = \dfrac{1}{2}$

10. $f(x) = 3x^4 + x^3 - 3x + 1$; $c = -\dfrac{1}{3}$

In Problems 11–22, tell the maximum number of zeros each polynomial function may have. Then, use Descartes' Rule of Signs to determine how many positive and how many negative zeros each polynomial function may have. Do not attempt to find the zeros.

11. $f(x) = -4x^7 + x^3 - x^2 + 2$
12. $f(x) = 5x^4 + 2x^2 - 6x - 5$
13. $f(x) = 2x^6 - 3x^2 - x + 1$

14. $f(x) = -3x^5 + 4x^4 + 2$
15. $f(x) = 3x^3 - 2x^2 + x + 2$
16. $f(x) = -x^3 - x^2 + x + 1$

17. $f(x) = -x^4 + x^2 - 1$
18. $f(x) = x^4 + 5x^3 - 2$
19. $f(x) = x^5 + x^4 + x^2 + x + 1$

20. $f(x) = x^5 - x^4 + x^3 - x^2 + x - 1$
21. $f(x) = x^6 - 1$
22. $f(x) = x^6 + 1$

In Problems 23–34, list the potential rational zeros of each polynomial function. Do not attempt to find the zeros.

23. $f(x) = 3x^4 - 3x^3 + x^2 - x + 1$
24. $f(x) = x^5 - x^4 + 2x^2 + 3$
25. $f(x) = x^5 - 6x^2 + 9x - 3$

26. $f(x) = 2x^5 - x^4 - x^2 + 1$
27. $f(x) = -4x^3 - x^2 + x + 2$
28. $f(x) = 6x^4 - x^2 + 2$

29. $f(x) = 3x^4 - x^2 + 2$
30. $f(x) = -4x^3 + x^2 + x + 2$
31. $f(x) = 2x^5 - x^3 + 2x^2 + 4$

32. $f(x) = 3x^5 - x^2 + 2x + 3$
33. $f(x) = 6x^4 + 2x^3 - x^2 + 2$
34. $f(x) = -6x^3 - x^2 + x + 3$

In Problems 35–40, find integer-valued upper and lower bounds to the zeros of each polynomial function.

35. $f(x) = 2x^3 + x^2 - 1$
36. $f(x) = 3x^3 - 2x^2 + x + 4$

37. $f(x) = x^3 - 5x^2 - 11x + 11$
38. $f(x) = 2x^3 - x^2 - 11x - 6$

39. $f(x) = x^4 + 3x^3 - 5x^2 + 9$
40. $f(x) = 4x^4 - 12x^3 + 27x^2 - 54x + 81$

In Problems 41–60, find the rational zeros of f. List any irrational zeros correct to two decimal places.

41. $f(x) = x^3 + 2x^2 - 5x - 6$
42. $f(x) = x^3 + 8x^2 + 11x - 20$

43. $f(x) = 2x^3 - x^2 + 2x - 1$
44. $f(x) = 2x^3 + x^2 + 2x + 1$

45. $f(x) = x^4 + x^2 - 2$
46. $f(x) = x^4 - 3x^2 - 4$

47. $f(x) = 4x^4 + 7x^2 - 2$
48. $f(x) = 4x^4 + 15x^2 - 4$

49. $f(x) = x^4 + x^3 - 3x^2 - x + 2$
50. $f(x) = x^4 - x^3 - 6x^2 + 4x + 8$

51. $f(x) = 4x^5 - 8x^4 - x + 2$
52. $f(x) = 4x^5 + 12x^4 - x - 3$

53. $f(x) = x^3 + 3.2x^2 - 16.83x - 5.31$
54. $f(x) = x^3 + 3.2x^2 - 7.25x - 6.3$

55. $f(x) = x^4 - 1.4x^3 - 33.71x^2 + 23.94x + 292.41$

56. $f(x) = x^4 + 1.2x^3 - 7.46x^2 - 4.692x + 15.2881$

57. $f(x) = \pi x^3 - (8.88\pi + 1)x^2 - (42.066\pi - 8.88)x + 42.066$

58. $f(x) = \pi x^3 - (5.62\pi + 2)x^2 - (108.392\pi - 11.26)x + 216.784$

59. $f(x) = x^3 + 19.5x^2 - 1021x + 1000.5$

60. $f(x) = x^3 + 42.2x^2 - 664.8x + 1490.4$

In Problems 61–70, solve each equation in the real number system.

61. $x^4 - x^3 + 2x^2 - 4x - 8 = 0$
62. $2x^3 + 3x^2 + 2x + 3 = 0$

63. $3x^3 + 4x^2 - 7x + 2 = 0$
64. $2x^3 - 3x^2 - 3x - 5 = 0$

65. $3x^3 - x^2 - 15x + 5 = 0$
66. $2x^3 - 11x^2 + 10x + 8 = 0$

67. $x^4 + 4x^3 + 2x^2 - x + 6 = 0$
68. $x^4 - 2x^3 + 10x^2 - 18x + 9 = 0$

69. $x^3 - \frac{2}{3}x^2 + \frac{8}{3}x + 1 = 0$
70. $x^3 - \frac{2}{3}x^2 + 3x - 2 = 0$

71. Average Cost of Manufacturing In Problem 93 of Section 3.4, you found the average cost function for manufacturing Chevy Cavaliers. Use the methods learned in this section to determine how many Cavaliers can be manufactured if the average cost is $10,500. In other words, solve the equation $\overline{C}(x) = 10.5$.

72. Average Cost of Printing In Problem 94 of Section 3.4, you found the average cost function for printing textbooks. Use the methods learned in this section to determine how many textbooks can be printed if the average cost is $8.90. In other words, solve the equation $\overline{C}(x) = 8.9$.

73. Find k such that $f(x) = x^3 - kx^2 + kx + 2$ has the factor $x - 2$.

74. Find k such that $f(x) = x^4 - kx^3 + kx^2 + 1$ has the factor $x + 2$.

75. What is the remainder when $f(x) = 2x^{20} - 8x^{10} + x - 2$ is divided by $x - 1$?

76. What is the remainder when $f(x) = -3x^{17} + x^9 - x^5 + 2x$ is divided by $x + 1$?

77. Is $\frac{1}{3}$ a zero of $f(x) = 2x^3 + 3x^2 - 6x + 7$? Explain.

78. Is $\frac{1}{3}$ a zero of $f(x) = 4x^3 - 5x^2 - 3x + 1$? Explain.

79. Is $\frac{3}{5}$ a zero of $f(x) = 2x^6 - 5x^4 + x^3 - x + 1$? Explain.

80. Is $\frac{2}{3}$ a zero of $f(x) = x^7 + 6x^5 - x^4 + x + 2$? Explain.

81. What is the length of the edge of a cube if, after a slice 1 inch thick is cut from one side, the volume remaining is 294 cubic inches?

82. What is the length of the edge of a cube if its volume could be doubled by an increase of 6 centimeters in one edge, an increase of 12 centimeters in a second edge, and a decrease of 4 centimeters in the third edge?

83. Let $f(x)$ be a polynomial function whose coefficients are integers. Suppose that r is a real zero of f and that the leading coefficient of f is 1. Use the Rational Zeros Theorem to show that r is either an integer or an irrational number.

84. Prove the Rational Zeros Theorem.

[**Hint:** Let p/q, where p and q have no common factors except 1 and -1, be a solution of the polynomial $f(x) = a_n x^n + a_{n-1} x^{n-1} + \cdots + a_1 x + a_0$, whose coefficients are all integers. Show that $a_n p^n + a_{n-1} p^{n-1} q + \cdots + a_1 p q^{n-1} + a_0 q^n = 0$. Now, because p is a factor of the first n terms of this equation, p must also be a factor of the term $a_0 q^n$. Since p is not a factor of q (why?), p must be a factor of a_0. Similarly, q must be a factor of a_n.]

In Problems 85–89, develop the Tartaglia–Cardano solution of the cubic equation and show why it is not altogether practical.

85. Show that the general cubic equation $y^3 + by^2 + cy + d = 0$ can be transformed into an equation of the form $x^3 + px + q = 0$ by using the substitution $y = x - b/3$.

86. In the equation $x^3 + px + q = 0$, replace x by $H + K$. Let $3HK = -p$, and show that $H^3 + K^3 = -q$.
[**Hint:** $3H^2 K + 3HK^2 = 3HKx$.]

87. Based on Problem 86, we have the two equations
$$3HK = -p \quad \text{and} \quad H^3 + K^3 = -q$$
Solve for K in $3HK = -p$ and substitute into $H^3 + K^3 = -q$. Then show that
$$H = \sqrt[3]{\frac{-q}{2} + \sqrt{\frac{q^2}{4} + \frac{p^3}{27}}}$$
[**Hint:** Look for an equation that is quadratic in form.]

88. Use the solution for H from Problem 87 and the equation $H^3 + K^3 = -q$ to show that

$$K = \sqrt[3]{\frac{-q}{2} - \sqrt{\frac{q^2}{4} + \frac{p^3}{27}}}$$

89. Use the results from Problems 86–88 to show that the solution of $x^3 + px + q = 0$ is

$$x = \sqrt[3]{\frac{-q}{2} + \sqrt{\frac{q^2}{4} + \frac{p^3}{27}}} + \sqrt[3]{\frac{-q}{2} - \sqrt{\frac{q^2}{4} + \frac{p^3}{27}}}$$

90. Use the result of Problem 89 to solve the equation $x^3 - 6x - 9 = 0$.

91. Use the result of Problem 89 to solve the equation $x^3 + 3x - 14 = 0$.

92. Use the methods of this chapter to solve the equation $x^3 + 3x - 14 = 0$.

93. A function f has the property that $f(2 + x) = f(2 - x)$ for all x. If f has exactly four real zeros, find their sum.

3.6 COMPLEX NUMBERS; QUADRATIC EQUATIONS WITH A NEGATIVE DISCRIMINANT

1 Add, Subtract, Multiply, and Divide Complex Numbers

2 Solve Quadratic Equations with a Negative Discriminant

One property of a real number is that its square is nonnegative. For example, there is no real number x for which

$$x^2 = -1$$

To remedy this situation, we introduce a number called the **imaginary unit,** which we denote by *i* and whose square is -1. Thus,

$$i^2 = -1$$

This should not surprise you. If our universe were to consist only of integers, there would be no number x for which $2x = 1$. This unfortunate circumstance was remedied by introducing numbers such as $\frac{1}{2}$ and $\frac{2}{3}$, the *rational numbers.* If our universe were to consist only of rational numbers, there would be no number x whose square equals 2. That is, there would be no number x for which $x^2 = 2$. To remedy this, we introduced numbers such as $\sqrt{2}$ and $\sqrt[3]{5}$, the *irrational numbers.* The *real numbers,* you will recall, consist of the rational numbers and the irrational numbers. Now, if our universe were to consist only of real numbers, then there would be no number x whose square is -1. To remedy this, we introduce a number i, whose square is -1.

In the progression outlined, each time we encountered a situation that was unsuitable, we introduced a new number system to remedy this situation. And each new number system contained the earlier number system as a subset. The number system that results from introducing the number i is called the **complex number system.**

Complex numbers are numbers of the form **a + bi,** where *a* and *b* are real numbers. The real number *a* is called the **real part** of the number a + bi; the real number *b* is called the **imaginary part** of a + bi.

For example, the complex number $-5 + 6i$ has the real part -5 and the imaginary part 6.

When a complex number is written in the form $a + bi$, where a and b are real numbers, we say it is in **standard form.** However, if the imaginary part of a complex number is negative, such as in the complex number $3 + (-2)i$, we agree to write it instead in the form $3 - 2i$.

Also, the complex number $a + 0i$ is usually written merely as a. This serves to remind us that the real numbers are a subset of the complex numbers. The complex number $0 + bi$ is usually written as bi. Sometimes the complex number bi is called a **pure imaginary number.**

Equality, addition, subtraction, and multiplication of complex numbers are defined so as to preserve the familiar rules of algebra for real numbers. Thus, two complex numbers are equal if and only if their real parts are equal and their imaginary parts are equal. That is,

Equality of Complex Numbers

$$a + bi = c + di \quad \text{if and only if} \quad a = c \text{ and } b = d \qquad (1)$$

Sum of Complex Numbers

Two complex numbers are added by forming the complex number whose real part is the sum of the real parts and whose imaginary part is the sum of the imaginary parts. That is,

$$(a + bi) + (c + di) = (a + c) + (b + d)i \qquad (2)$$

To subtract two complex numbers, we follow the rule

Difference of Complex Numbers

$$(a + bi) - (c + di) = (a - c) + (b - d)i \qquad (3)$$

E X A M P L E 1

Adding and Subtracting Complex Numbers

(a) $(3 + 5i) + (-2 + 3i) = [3 + (-2)] + (5 + 3)i = 1 + 8i$
(b) $(6 + 4i) - (3 + 6i) = (6 - 3) + (4 - 6)i = 3 + (-2)i = 3 - 2i$ ∎

 Now work Problem 5.

Products of complex numbers are calculated as illustrated in Example 2.

E X A M P L E 2

Multiplying Complex Numbers

$$(5 + 3i) \cdot (2 + 7i) = 5 \cdot (2 + 7i) + 3i(2 + 7i) = 10 + 35i + 6i + 21i^2$$
$$\uparrow \qquad\qquad\qquad\qquad \uparrow$$
$$\text{Distributive property} \qquad\qquad \text{Distributive property}$$
$$= 10 + 41i + 21(-1)$$
$$\uparrow$$
$$i^2 = -1$$
$$= -11 + 41i \qquad ∎$$

Based on the procedure of Example 2, we define the **product** of two complex numbers by the formula

Product of Complex Numbers

$$(a + bi) \cdot (c + di) = (ac - bd) + (ad + bc)i \qquad (4)$$

Do not bother to memorize formula (4). Instead, whenever it is necessary to multiply two complex numbers, follow the usual rules for multiplying two binomials, as in Example 2, remembering that $i^2 = -1$. For example,

$$(2i)(2i) = 4i^2 = -4$$
$$(2 + i)(1 - i) = 2 - 2i + i - i^2 = 3 - i$$

 Now work Problem 11.

Algebraic properties for addition and multiplication, such as the commutative, associative, and distributive properties, hold for complex numbers. Of these, the property that every nonzero complex number has a multiplicative inverse, or reciprocal, requires a closer look.

Conjugates

If $z = a + bi$ is a complex number, then its **conjugate**, denoted by $\bar{z}$, is defined as

$$\bar{z} = \overline{a + bi} = a - bi$$

For example, $\overline{2 + 3i} = 2 - 3i$ and $\overline{-6 - 2i} = -6 + 2i$.

E X A M P L E 3

Multiplying a Complex Number by Its Conjugate

Find the product of the complex number $z = 3 + 4i$ and its conjugate $\bar{z}$.

Solution   Since $\bar{z} = 3 - 4i$, we have

$$z\bar{z} = (3 + 4i)(3 - 4i) = 9 + 12i - 12i - 16i^2 = 9 + 16 = 25 \qquad \blacksquare$$

The result obtained in Example 3 has an important generalization:

Theorem

The product of a complex number and its conjugate is a nonnegative real number. Thus, if $z = a + bi$, then

$$z\bar{z} = a^2 + b^2 \qquad (5)$$

Proof   If $z = a + bi$, then

$$z\bar{z} = (a + bi)(a - bi) = a^2 - (bi)^2 = a^2 - b^2i^2 = a^2 + b^2 \qquad \blacksquare$$

To express the reciprocal of a nonzero complex number z in standard form, multiply the numerator and denominator by its conjugate $\bar{z}$. Thus, if $z = a + bi$ is a nonzero complex number, then

$$\frac{1}{a + bi} = \frac{1}{z} = \frac{1}{z} \cdot \frac{\bar{z}}{\bar{z}} = \frac{\bar{z}}{z\bar{z}} = \frac{a - bi}{(a + bi)(a - bi)}$$

$$= \frac{a - bi}{a^2 + b^2}$$

$\uparrow$
Use (5).

$$= \frac{a}{a^2 + b^2} - \frac{b}{a^2 + b^2}i$$

E X A M P L E   4   **Writing the Reciprocal of a Complex Number in Standard Form**

Write $\dfrac{1}{3 + 4i}$ in standard form $a + bi$; that is, find the reciprocal of $3 + 4i$.

Solution   The idea is to multiply the numerator and denominator by the conjugate of $3 + 4i$, that is, the complex number $3 - 4i$. The result is

$$\frac{1}{3 + 4i} = \frac{1}{3 + 4i} \cdot \frac{3 - 4i}{3 - 4i} = \frac{3 - 4i}{9 + 16} = \frac{3}{25} - \frac{4}{25}i \qquad \blacksquare$$

To express the quotient of two complex numbers in standard form, we multiply the numerator and denominator of the quotient by the conjugate of the denominator.

E X A M P L E   5   **Writing the Quotient of Complex Numbers in Standard Form**

Write each of the following in standard form:

(a) $\dfrac{1 + 4i}{5 - 12i}$ (b) $\dfrac{2 - 3i}{4 - 3i}$

Solution (a) $\dfrac{1 + 4i}{5 - 12i} = \dfrac{1 + 4i}{5 - 12i} \cdot \dfrac{5 + 12i}{5 + 12i} = \dfrac{5 + 20i + 12i + 48i^2}{25 + 144}$

$\qquad\qquad = \dfrac{-43 + 32i}{169} = \dfrac{-43}{169} + \dfrac{32}{169}i$

(b) $\dfrac{2 - 3i}{4 - 3i} = \dfrac{2 - 3i}{4 - 3i} \cdot \dfrac{4 + 3i}{4 + 3i} = \dfrac{8 - 12i + 6i - 9i^2}{16 + 9} = \dfrac{17 - 6i}{25} = \dfrac{17}{25} - \dfrac{6}{25}i$

Now work Problem 19.

E X A M P L E   6   Writing Other Expressions in Standard Form

If $z = 2 - 3i$ and $w = 5 + 2i$, write each of the following expressions in standard form:

(a) $\dfrac{z}{w}$ (b) $\overline{z + w}$ (c) $z + \overline{z}$

Solution (a) $\dfrac{z}{w} = \dfrac{z \cdot \overline{w}}{w \cdot \overline{w}} = \dfrac{(2 - 3i)(5 - 2i)}{(5 + 2i)(5 - 2i)} = \dfrac{10 - 15i - 4i + 6i^2}{25 + 4}$

$\qquad\qquad = \dfrac{4 - 19i}{29} = \dfrac{4}{29} - \dfrac{19}{29}i$

(b) $\overline{z + w} = \overline{(2 - 3i) + (5 + 2i)} = \overline{7 - i} = 7 + i$

(c) $z + \overline{z} = (2 - 3i) + (2 + 3i) = 4$

The conjugate of a complex number has certain general properties that we shall find useful later.

For a real number $a = a + 0i$, the conjugate is $\overline{a} = \overline{a + 0i} = a - 0i = a$. That is,

> **Theorem**
>
> The conjugate of a real number is the real number itself.

Other properties of the conjugate that are direct consequences of the definition are given next. In each statement, z and w represent complex numbers.

> **Theorem**
>
> The conjugate of the conjugate of a complex number is the complex number itself:
>
> $$\overline{(\overline{z})} = z \qquad\qquad (6)$$
>
> The conjugate of the sum of two complex numbers equals the sum of their conjugates:
>
> $$\overline{z + w} = \overline{z} + \overline{w} \qquad\qquad (7)$$

We take the conjugate of both sides to get

$$\overline{a_n r^n + a_{n-1} r^{n-1} + \cdots + a_1 r + a_0} = \overline{0}$$

$$\overline{a_n r^n} + \overline{a_{n-1} r^{n-1}} + \cdots + \overline{a_1 r} + \overline{a_0} = \overline{0} \qquad \text{The conjugate of a sum equals the sum of the conjugates (see Section 3.6).}$$

$$\overline{a_n}(\overline{r})^n + \overline{a_{n-1}}(\overline{r})^{n-1} + \cdots + \overline{a_1}\,\overline{r} + \overline{a_0} = \overline{0} \qquad \text{The conjugate of a product equals the product of the conjugates.}$$

$$a_n(\overline{r})^n + a_{n-1}(\overline{r})^{n-1} + \cdots + a_1\overline{r} + a_0 = 0 \qquad \text{The conjugate of a real number equals the real number.}$$

This last equation states that $f(\overline{r}) = 0$; that is, $\overline{r}$ is a zero of f. ∎

The value of this result should be clear. Once we know that, say, $3 + 4i$ is a zero of a polynomial with real coefficients, then we know that $3 - 4i$ is also a zero. This result has an important corollary.

Corollary

A complex polynomial f of odd degree with real coefficients has at least one real zero.

∎

Proof Because complex zeros occur as conjugate pairs in a complex polynomial with real coefficients, there will always be an even number of zeros that are not real numbers. Consequently, since f is of odd degree, one of its zeros has to be a real number. ∎

For example, the polynomial $f(x) = x^5 - 3x^4 + 4x^3 - 5$ has at least one zero that is a real number, since f is of degree 5 (odd) and has real coefficients.

E X A M P L E 1   Using the Conjugate Pairs Theorem

A polynomial f of degree 5 whose coefficients are real numbers has the zeros 1, $5i$, and $1 + i$. Find the remaining two zeros.

Solution   Since complex zeros appear as conjugate pairs, it follows that $-5i$, the conjugate of $5i$, and $1 - i$, the conjugate of $1 + i$, are the two remaining zeros. ∎

Now work Problem 1.

E X A M P L E 2   Finding a Polynomial Function Whose Zeros Are Given

(a) Find a polynomial f of degree 4 whose coefficients are real numbers and has the zeros 1, 1, and $-4 + i$.

(b) Graph the polynomial found in (a) to verify your result.

Solution   (a) Since $-4 + i$ is a zero, by the Conjugate Pairs Theorem, $-4 - i$ must also be a zero of f. Because of the Factor Theorem, if $f(c) = 0$, then $x - c$ is a factor of $f(x)$. So, we can now write f as

$$f(x) = a(x - 1)(x - 1)[x - (-4 + i)][x - (-4 - i)]$$

where a is any real number. If we let $a = 1$, we obtain

$$
\begin{aligned}
f(x) &= (x - 1)(x - 1)[x - (-4 + i)][x - (-4 - i)] \\
&= (x^2 - 2x + 1)[x^2 - (-4 + i)x - (-4 - i)x + (-4 + i)(-4 - i)] \\
&= (x^2 - 2x + 1)(x^2 + 4x - ix + 4x + ix + 16 + 4i - 4i - i^2) \\
&= (x^2 - 2x + 1)(x^2 + 8x + 17) \\
&= (x^4 + 8x^3 + 17x^2 - 2x^3 - 16x^2 - 34x + x^2 + 8x + 17) \\
&= x^4 + 6x^3 + 2x^2 - 26x + 17
\end{aligned}
$$

FIGURE 64

(b) A quick analysis of the polynomial f tells us what to expect:

> At most 3 turning points.
> For large x, the graph will behave like $y = x^4$.
> A repeated real zero at 1 so that the graph will touch the x-axis at 1.
> The only x-intercept is at 1.

Figure 64 shows the complete graph. (Do you see why? The graph has exactly 3 turning points.) ▬

Exploration   Graph the function found in Example 2 for $a = 2$ and $a = -1$. Does the value of a affect the zeros of f? How does the value of a affect the graph of f? ▬

Now we can prove the theorem we conjectured earlier in Section 3.5.

Theorem

Every polynomial function with real coefficients can be uniquely factored over the real numbers into a product of linear factors and/or irreducible quadratic factors.

▬

Proof   Every complex polynomial f of degree n has exactly n zeros and can be factored into a product of n linear factors. If its coefficients are real, then those zeros that are complex numbers will always occur as conjugate pairs. As a result, if $r = a + bi$ is a complex zero, then so is $\bar{r} = a - bi$. Consequently, when the linear factors $x - r$ and $x - \bar{r}$ of f are multiplied, we have

$$(x - r)(x - \bar{r}) = x^2 - (r + \bar{r})x + r\bar{r} = x^2 - 2ax + a^2 + b^2$$

This second-degree polynomial has real coefficients and is irreducible (over the real numbers). Thus, the factors of f are either linear or irreducible quadratic factors. ▬

E X A M P L E   3

Finding the Complex Zeros of a Polynomial

(a) It is known that $2 + i$ is a zero of $f(x) = x^4 - 8x^3 + 64x - 105$. Find the remaining zeros.
(b) Graph f to verify your results.

Solution (a) Since $2 + i$ is a zero of f and f has real coefficients, then $2 - i$ must also be a zero of f. (Complex zeros appear as conjugate pairs in a polynomial with real coefficients.) So, $x - (2 + i)$ and $x - (2 - i)$ are factors of f. Therefore, their product,

$$[x - (2 + i)][x - (2 - i)]$$
$$= x^2 - (2 - i)x - x(2 + i) + (2 + i)(2 - i) = x^2 - 4x + 5$$

is also a factor of f. We use long division to obtain the other factor:

$$
\begin{array}{r}
x^2 - 4x - 21 \\
x^2 - 4x + 5 \overline{) x^4 - 8x^3 + 64x - 105} \\
\underline{x^4 - 4x^3 + 5x^2} \\
-4x^3 - 5x^2 + 64x \\
\underline{-4x^3 + 16x^2 - 20x} \\
-21x^2 + 84x - 105 \\
\underline{-21x^2 + 84x - 105} \\
0
\end{array}
$$

FIGURE 65

So, $f(x) = (x^2 - 4x + 5)(x^2 - 4x - 21) = (x^2 - 4x + 5)(x - 7)(x + 3)$. By the Factor Theorem, the zeros of f are $-3, 7, 2 + i$, and $2 - i$.

(b) A quick analysis of the polynomial f tells us what to expect:

At most 3 turning points.

Two x-intercepts: at -3 and 7.

For large x, the graph will behave like $y = x^4$.

Figure 65 shows the complete graph. (Do you see why? The graph has exactly 3 turning points.)

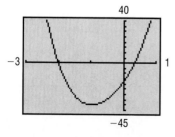

Now work Problem 17.

The steps on page 262 in Section 3.5 used to find the real zeros of a polynomial function can also be used to find the complex zeros of a polynomial.

E X A M P L E   4   Finding the Complex Zeros of a Polynomial

Find the complex zeros of the polynomial function
$f(x) = 3x^4 + 5x^3 + 25x^2 + 45x - 18$.

STEP 1: The degree of f is 4. So, f will have 4 complex zeros.
STEP 2: Descartes' Rule of Signs provides information about the real zeros. For this polynomial, there is one positive real zero. Because

$$f(-x) = 3x^4 - 5x^3 + 25x^2 - 45x - 18$$

there are three, or one, negative real zero(s).

STEP 3: The Rational Zeros Theorem provides information about the potential rational zeros of polynomials with integer coefficients. For this polynomial (which has integer coefficients), the potential rational zeros are

FIGURE 66

$$\pm \frac{1}{3}, \pm \frac{2}{3}, \pm 1, \pm 2, \pm 3, \pm 6, \pm 9, \pm 18$$

STEP 4: Figure 66 shows the graph of f. The graph has the characteristics we expect of this polynomial of degree 4: It behaves like $y = 3x^4$

for large x and has y-intercept -18. There are x-intercepts near -2 and between 0 and 1. Since $f(-2) = 0$, we know -2 is a zero of f. We use synthetic division to factor f.

$$
\begin{array}{r|rrrr}
-2)3 & 5 & 25 & 45 & -18 \\
 & -6 & 2 & -54 & 18 \\
\hline
3 & -1 & 27 & -9 & 0
\end{array}
$$

So, $f(x) = (x + 2)(3x^3 - x^2 + 27x - 9)$. From the graph of f and the list of potential rational zeros, it appears 1/3 may also be a zero of f. We use synthetic division on the depressed equation of f to confirm this.

$$
\begin{array}{r|rrrr}
\tfrac{1}{3})3 & -1 & 27 & -9 \\
 & 1 & 0 & 9 \\
\hline
3 & 0 & 27 & 0
\end{array}
$$

Thus, 1/3 is a zero of f and $x - 1/3$ is therefore a factor. Using the bottom row of the synthetic division, we find

$$
f(x) = (x + 2)\left(x - \frac{1}{3}\right)(3x^2 + 27) = 3(x + 2)\left(x - \frac{1}{3}\right)(x^2 + 9).
$$

The factor $x^2 + 9$ does not have any real zeros; its complex zeros are $\pm 3i$. Thus, the complex zeros of $f(x) = 3x^4 + 5x^3 + 25x^2 + 45x - 18$ are $-2, 1/3, 3i, -3i$. ∎

3.7 EXERCISES

In Problems 1–10, information is given about a complex polynomial f(x) whose coefficients are real numbers. Find the remaining zeros of f.

1. Degree 3; zeros: $3, 4 - i$
2. Degree 3; zeros: $4, 3 + i$
3. Degree 4; zeros: $i, 1 + i$
4. Degree 4; zeros: $1, 2, 2 + i$
5. Degree 5; zeros: $1, i, 2i$
6. Degree 5; zeros: $0, 1, 2, i$
7. Degree 4; zeros: $i, 2, -2$
8. Degree 4; zeros: $2 - i, -i$
9. Degree 6; zeros: $2, 2 + i, -3 -i, 0$
10. Degree 6; zeros: $i, 3 - 2i, -2 + i$

In Problems 11–16, form a polynomial f(x) with real coefficients having the given degree and zeros.

11. Degree 4; zeros: $3 + 2i$; 4, multiplicity 2
12. Degree 4; zeros: $i, 1 + 2i$
13. Degree 5; zeros: 2, multiplicity 1; $1 - i$; $1 + i$
14. Degree 6; zeros: $i, 4 - i$; $2 + i$
15. Degree 4; zeros: 3, multiplicity 2; $-i$
16. Degree 5: zeros: 1, multiplicity 3; $1 + i$

In Problems 17–24, use the given zero to find the remaining zeros of each function. Graph the function to verify your results.

17. $f(x) = x^3 - 4x^2 + 4x - 16$; zero: $2i$
18. $g(x) = x^3 + 3x^2 + 25x + 75$; zero: $-5i$
19. $f(x) = 2x^4 + 5x^3 + 5x^2 + 20x - 12$; zero: $-2i$
20. $h(x) = 3x^4 + 5x^3 + 25x^2 + 45x - 18$; zero: $3i$
21. $h(x) = x^4 - 9x^3 + 21x^2 + 21x - 130$; zero: $3 - 2i$
22. $f(x) = x^4 - 7x^3 + 14x^2 - 38x - 60$; zero: $1 + 3i$
23. $h(x) = 3x^5 + 2x^4 + 15x^3 + 10x^2 - 528x - 352$; zero: $-4i$
24. $g(x) = 2x^5 - 3x^4 - 5x^3 - 15x^2 - 207x + 108$; zero: $3i$

E X A M P L E   2

Solving a Polynomial Inequality

Solve the inequality $x^4 > x$, and graph the solution set.

Graphing Solution   We graph $Y_1 = x^4$ and $Y_2 = x$ on the same screen. See Figure 69. Using the INTERSECT command, we find that Y_1 and Y_2 intersect at $x = 0$ and at $x = 1$. The graph of Y_1 is above that of Y_2, $Y_1 > Y_2$, to the left of $x = 0$ and to the right of $x = 1$. Since the inequality is strict, the solution set is of $\{x | x < 0$ or $x > 1\}$ or, using interval notation, $(-\infty, 0)$ or $(1, \infty)$.

Algebraic Solution   STEP 1: Rearrange the inequality so that 0 is on the right side.

FIGURE 69

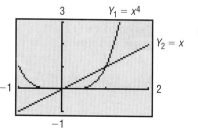

$$x^4 > x$$
$$x^4 - x > 0 \quad \text{Subtract } x \text{ from both sides of the inequality.}$$

STEP 2: Find the critical numbers of $f(x) = x^4 - x$ by solving $x^4 - x = 0$.

$$x^4 - x = 0$$
$$x(x^3 - 1) = 0 \qquad \text{Factor out } x.$$
$$x(x - 1)(x^2 + x + 1) = 0 \qquad \text{Factor difference of two cubes.}$$
$$x = 0 \quad \text{or} \quad x - 1 = 0 \quad \text{or} \quad x^2 + x + 1 = 0 \qquad \text{Set each factor equal to zero and solve.}$$
$$x = 0 \quad \text{or} \quad x = 1$$

The equation $x^2 + x + 1 = 0$ has no real solutions. (Do you see why?)

STEP 3: We use the critical numbers to separate the real number line into three intervals

$$-\infty < x < 0 \qquad 0 < x < 1 \qquad 1 < x < \infty$$

STEP 4: We select a test number in each interval found in step 3 and evaluate $f(x) = x^4 - x$ at each test number to determine if $f(x)$ is positive or negative. See Table 17.

TABLE 17			
Interval	Test Number	$f(x)$	Positive/Negative
$-\infty < x < 0$	-1	$f(-1) = (-1)^4 - (-1) = 2$	Positive
$0 < x < 1$	$1/2$	$f(1/2) = -7/16$	Negative
$1 < x < \infty$	2	$f(2) = 14$	Positive

Since we want to know where $f(x)$ is positive, we conclude that the solutions are numbers x for which $-\infty < x < 0$ or $1 < x < \infty$. Because the original inequality is strict, numbers x that satisfy the equation $x^4 = x$ are not solutions. Thus, the solution set of the given inequality is $\{x | -\infty < x < 0$ or $1 < x < \infty\}$ or using interval notation, $(-\infty, 0)$ or $(1, \infty)$.

Figure 70 shows the graph of the solution set.

FIGURE 70
$x < 0$ or $x > 1$

Now work Problem 17.

Let's solve a rational inequality.

2

E X A M P L E   3

Solving a Rational Inequality

Solve the inequality $\dfrac{4x + 5}{x + 2} \geq 3$, and graph the solution set.

Graphing Solution   We first note that the domain of the variable consists of all real numbers except -2. We graph $Y_1 = (4x + 5)/(x + 2)$ and $Y_2 = 3$ on the same screen. See Figure 71. Using the INTERSECT command, we find that Y_1 and Y_2 intersect at $x = 1$. The graph of Y_1 is above that of Y_2, $Y_1 > Y_2$, to the left of $x = -2$ and to the right of $x = 1$. Since the inequality is not strict, the solution set is $\{x \mid -\infty < x < -2 \text{ or } 1 \leq x < \infty\}$. Using interval notation, the solution set is $(-\infty, -2)$ or $[1, \infty)$.

Algebraic Solution   **STEP 1:**   We first note that the domain of the variable consists of all real numbers except -2. We rearrange terms so that 0 is on the right side:

FIGURE 71

$Y_1 = \frac{4x+5}{x+2}$ 10

$Y_1 = \frac{4x+5}{x+2}$

$Y_2 = 3$

-6 6

-1

$$\frac{4x + 5}{x + 2} - 3 \geq 0 \qquad \text{Subtract 3 from both sides of the inequality.}$$

STEP 2:   Find the critical numbers of $f(x) = \dfrac{4x + 5}{x + 2} - 3$. To find the critical numbers we must express f as a quotient.

$$f(x) = \frac{4x + 5}{x + 2} - 3 \qquad \text{Least Common Denominator: } x + 2.$$

$$f(x) = \frac{4x + 5}{x + 2} - 3 \cdot \frac{x + 2}{x + 2} \qquad \text{Multiply } -3 \text{ by } \frac{x + 2}{x + 2}.$$

$$f(x) = \frac{4x + 5 - 3x - 6}{x + 2} \qquad \text{Distribute } -3.$$

$$f(x) = \frac{x - 1}{x + 2} \qquad \text{Collect like terms.}$$

The critical numbers are -2 and 1.

STEP 3:   We use the critical numbers to separate the real number line into three intervals:

$$-\infty < x < -2 \qquad -2 < x < 1 \qquad 1 < x < \infty$$

STEP 4:   We select a test number in each interval found in step 3 and evaluate $f(x) = \dfrac{4x + 5}{x + 2} - 3$ at each test number to determine if $f(x)$ is positive or negative. See Table 18.

TABLE 18			
Interval	**Test Number**	***f*(x)**	**Positive/Negative**
$-\infty < x < -2$	-3	$f(-3) = 4$	Positive
$-2 < x < 1$	0	$f(0) = -1/2$	Negative
$1 < x < \infty$	2	$f(2) = 1/4$	Positive

We want to know where $f(x)$ is positive. We conclude that the solutions are all x such that $-\infty < x < -2$ or $1 < x < \infty$. Because the original inequality is not strict, numbers x that satisfy the equation $\dfrac{x-1}{x+2} = 0$ are also solutions of the inequality. Since $\dfrac{x-1}{x+2} = 0$ only if $x = 1$, we conclude that the solution set is $\{x | -\infty < x < -2 \text{ or } 1 \le x < \infty\}$ or using interval notation $(-\infty, -2)$ or $[1, \infty)$.

Figure 72 shows the graph of the solution set.　■

FIGURE 72
$-\infty < x < -2$ or $1 \le x < \infty$
$(-\infty, -2)$ or $[1, \infty)$

![number line from -5 to 4 with open circle at -2 and closed circle at 1](number line)
$-5\ -4\ -3\ -2\ -1\ \ 0\ \ 1\ \ 2\ \ 3\ \ 4$

Now work Problem 37.

E X A M P L E   4

Minimum Sales Requirements

Tami is considering leaving her $30,000 a year job and buying a cookie company. According to the financial records of the firm, the relationship between pounds of cookies sold and profit is exhibited by Table 19.

TABLE 19	
Pounds of Cookies (in Hundreds), x	**Profit, P**
0	−20,000
50	−5990
75	412
120	10,932
200	26,583
270	36,948
340	44,381
420	49,638
525	49,225
610	44,381
700	34,220

(a) Draw a scatter diagram of the data in Table 19 with the pounds of cookies sold as the independent variable.

(b) Find the quadratic function of best fit using a graphing utility.

(c) Using the function found in (b), determine the number of pounds of cookies Tami must sell in order for the profits to exceed $30,000 a year and therefore make it worthwhile for her to quit her job.

(d) Using the function found in (b), determine the number of pounds of cookies Tami should sell in order to maximize profits.

(e) Using the function found in (b), determine the maximum profit Tami can expect to earn.

Solution   (a) Figure 73 shows the scatter diagram.

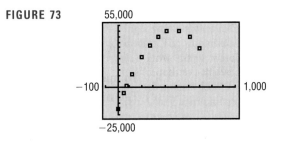

FIGURE 73

(b) Using a graphing utility, the quadratic function of best fit is

$$p(x) = -0.31x^2 + 295.86x - 20042.52$$

See Figure 74.

FIGURE 74

(c) We want profit to exceed \$30,000; therefore, we want to solve the inequality

$$-0.31x^2 + 295.86x - 20042.52 > 30000$$

We graph

$$Y_1 = p(x) = -0.31x^2 + 295.86x - 20042.52 \quad \text{and} \quad Y_2 = 30000$$

on the same screen. See Figure 75.

FIGURE 75

Using the INTERSECT command, we find that Y_1 and Y_2 intersect at $x = 220$ and at $x = 735$. The graph of Y_1 is above that of Y_2, $Y_1 > Y_2$, between the points of intersection. Since the inequality is strict, the solution set is $\{x | 220 < x < 735\}$.

BEE PARASITIC MITE SYNDROME [BPMS]

The Bee Crisis. If you've noticed a lack of bees buzzing around your home, you're not alone. In recent years, there has been a *serious loss* of bee colonies in the United States. The cause is from two parasitic mites: Acarapis woodi and Varroa jacobsoni, which first showed up in 1987. Both mites now infest bees in every state except Hawaii. There is no official wild bee census, but estimates go as high as 90 percent of the wild bee colonies have been wiped out! Scientists expect that all the bee colonies in the United States will be exposed to mites.

Varroa mites are a particularly serious threat to the United States honeybee industry. In 1986, more than 211,000 beekeepers produced about 200 million pounds of honey valued at $103.1 million, according to the National Honey Board. To agriculture, the mite problem poses an even bigger threat. Crops valued at $20 billion, ranging from blueberries in Maine to almonds in California, depend on bees for pollination each year. Serious economic losses could result if Varroa mites were to devastate the United States bee population.

As a research entomologist, you want to find alternative methods of control other than the use of pesticides. Your research into Varroa resistant hybrids may offer some hope. You need to create some mathematical models that reflect possible outcomes of your research. These models will help you to determine a focus for your genetic work.

1. What are some of the *difficulties* in fighting the Varroa mite? What are likely results unless some control measures are used?
2. Pesticides are recommended to curb the Varroa epidemic. This is the main defense currently in use. How effective are the pesticides? Does the short-term gain result in any long-term costs?
3. If your research in bee hybridization allows you to breed a mite-resistant honey bee such that only one-third of your bees die from Varroa per year, would this be considered a success?
4. If you study the graphs of bee-mite interaction, you will notice some fairly well defined *logistic curves*. You search the Internet for references to logistic growth models. Using a graphing utility, determine a logistic growth function of best fit that models the situation.
5. Collect and assemble the data on *honey production* for the 1990's. Can the Varroa infestation be seen in this data?
6. Given the results of questions 4 and 5, could you justify removing the use of pesticides for the treatment of a colony?

Until now, our study of functions has concentrated primarily on polynomial and rational functions. These functions belong to the class of **algebraic functions,** that is, functions that can be expressed in terms of sums, differences, products, quotients, powers, or roots of polynomials. Functions that are not algebraic are termed **transcendental** (they transcend, or go beyond, algebraic functions).

In this chapter, we study two transcendental functions: the *exponential* and *logarithmic functions.* These functions occur frequently in a wide variety of applications, such as biology, chemistry, economics, and psychology.

The chapter begins with a discussion of inverse functions.

4.1 ONE-TO-ONE FUNCTIONS; INVERSE FUNCTIONS

1. Determine Whether a Function Is One-to-One
2. Obtain the Graph of the Inverse Function from the Graph of the Function
3. Find an Inverse Function

In Section 2.1, we said a function f can be thought of as a machine that receives as input a number, say x, from the domain, manipulates it, and outputs the value $f(x)$. The **inverse** of f receives as input a number $f(x)$, manipulates it, and outputs the value x. The inverse of f reverses the process of f. If the function f is the set of ordered pairs (x, y), then the inverse of f is the set of ordered pairs (y, x).

E X A M P L E   1

Finding the Inverse of a Function

Find the inverse of the following functions.

(a) Let the domain of the function represent the employees of Yolanda's Preowned Car Mart and let the range represent their base salaries.

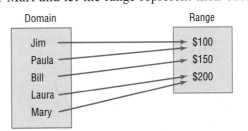

(b) Let the domain of the function represent the employees of Yolanda's Preowned Car Mart and let the range represent their spouse's names.

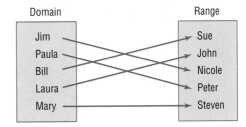

(c) $\{(-3, -27), (-2, -8), (-1, -1), (0, 0), (1, 1), (2, 8), (3, 27)\}$
(d) $\{(-3, 9), (-2, 4), (-1, 1), (0, 0), (1, 1), (2, 4), (3, 9)\}$

Solution
(a) The elements in the domain represent inputs to the function and the elements in the range represent the outputs. To find the inverse, interchange the elements in the domain with the elements in the range. For example, the function receives as input Bill and outputs $150. So, the inverse receives as input $150 and outputs Bill.

(b)

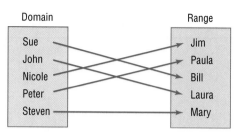

Domain   Range

(c) The inverse of the function is found by interchanging the entries in each ordered pair. So the inverse of the given function is $\{(-27, -3),$ $(-8, -2), (-1, -1), (0, 0), (1, 1), (8, 2), (27, 3)\}$.

(d) The inverse of the given function is $\{(9, -3), (4, -2), (1, -1), (0, 0),$ $(1, 1), (4, 2), (9, 3)\}$. ∎

We notice the inverses found in Examples 1(b) and (c) represent functions since each element in the domain corresponds to a unique element in the range. The inverses found in Examples 1(a) and (d) do not represent functions since each element in the domain does not correspond to a unique element in the range. Compare the function in Example 1(c) with the function in Example 1(d). For the function in Example 1(c), to every unique x-coordinate there corresponds a unique y-coordinate; for the function in Example 1(d), every unique x-coordinate does not correspond to a unique y-coordinate. Functions where unique x-coordinates correspond to unique y-coordinates are called *one-to-one* functions. So, in order for the inverse of a function, f, to be a function itself, f must be one-to-one.

> A function f is said to be **one-to-one** if, for any choice of numbers x_1 and x_2, $x_1 \neq x_2$, in the domain of f, then $f(x_1) \neq f(x_2)$.

In other words, if f is a one-to-one function, then for each x in the domain of f, there is exactly one y in the range, and no y in the range is the image of more than one x in the domain. See Figure 1.

FIGURE 1

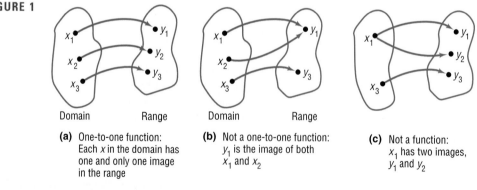

Domain   Range   Domain   Range

(a) One-to-one function: Each x in the domain has one and only one image in the range

(b) Not a one-to-one function: y_1 is the image of both x_1 and x_2

(c) Not a function: x_1 has two images, y_1 and y_2

Now work Problem 1.

If the graph of a function f is known, there is a simple test, called the **horizontal line test,** to determine whether f is one-to-one.

FIGURE 2
$f(x_1) = f(x_2) = h$, but $x_1 \ne x_2$; f is not a one-to-one function.

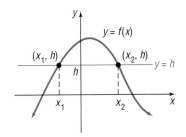

Theorem Horizontal Line Test

If horizontal lines intersect the graph of a function f in at most one point, then f is one-to-one.

The reason this test works can be seen in Figure 2, where the horizontal line $y = h$ intersects the graph at two distinct points, (x_1, h) and (x_2, h), with the same second element. Thus, f is not one-to-one.

E X A M P L E   2

Using the Horizontal Line Test

For each given function, use the graph to determine whether the function is one-to-one.

(a) $f(x) = x^2$ (b) $g(x) = x^3$

Solution   (a) Figure 3(a) illustrates the horizontal line test for $f(x) = x^2$. The horizontal line $y = 1$ meets the graph of f twice, at $(1, 1)$ and at $(-1, 1)$, so f is not one-to-one.

(b) Figure 3(b) illustrates the horizontal line test for $g(x) = x^3$. Because each horizontal line will intersect the graph of g exactly once, it follows that g is one-to-one.

FIGURE 3

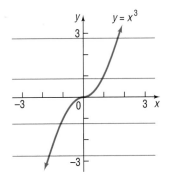

(a) A horizontal line intersects the graph twice; thus, f is not one-to-one

(b) Horizontal lines intersect the graph exactly once; thus, g is one-to-one

Now work Problem 9.

Let's look more closely at the one-to-one function $g(x) = x^3$. This function is an increasing function. Because an increasing (or decreasing) function will always have different y values for unequal x values, it follows that a function that is increasing (or decreasing) on its domain is also a one-to-one function.

> ### Theorem
>
> An increasing (decreasing) function is a one-to-one function.

Inverse of a Function $y = f(x)$

We mentioned in Chapter 2 that a function $y = f(x)$ can be thought of as a rule that tells us to do something to the argument x. For example, the function $f(x) = 2x$ multiplies the argument by 2. An *inverse function* of f undoes whatever f does. For example, the function $g(x) = \frac{1}{2}x$, which divides the argument by 2, is an inverse of $f(x) = 2x$. See Figure 4.

FIGURE 4
$f(x) = 2x;\ g(x) = \frac{1}{2}x$

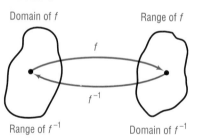

To put it another way, if we think of f as an input/output machine that processes an input x into $f(x)$, then the inverse function reverses this process, taking $f(x)$ back to x.

For a function $y = f(x)$ to have an inverse *function*, f must be one-to-one. Then for each x in its domain there is exactly one y in its range; furthermore, to each y in the range, there corresponds exactly one x in the domain. The correspondence from the range of f onto the domain of f is, therefore, also a function. It is this function that is the *inverse of f*. A definition is given next.

> Let f denote a one-to-one function $y = f(x)$. The **inverse of f,** denoted by f^{-1}, is a function such that $f^{-1}(f(x)) = x$ for every x in the domain of f and $f(f^{-1}(x)) = x$ for every x in the domain of f^{-1}.

Warning: Be careful! The -1 used in f^{-1} is not an exponent. Thus, f^{-1} does *not* mean the reciprocal of f; it means the inverse of f.

FIGURE 5

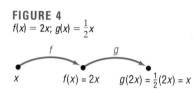

Domain of f Range of f

f

f^{-1}

Range of f^{-1} Domain of f^{-1}

Figure 5 illustrates the definition.

Two facts are now apparent about a function f and its inverse f^{-1}.

> $$\text{Domain of } f = \text{Range of } f^{-1} \qquad \text{Range of } f = \text{Domain of } f^{-1}$$

Look again at Figure 5 to visualize the relationship. If we start with x, apply f, and then apply f^{-1}, we get x back again. If we start with x, apply f^{-1}, and then apply f, we get the number x back again. To put it simply, what f does, f^{-1} undoes, and vice versa:

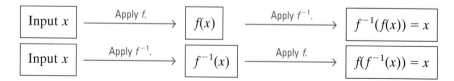

| Input x | $\xrightarrow{\text{Apply } f.}$ | $f(x)$ | $\xrightarrow{\text{Apply } f^{-1}.}$ | $f^{-1}(f(x)) = x$ |
| Input x | $\xrightarrow{\text{Apply } f^{-1}.}$ | $f^{-1}(x)$ | $\xrightarrow{\text{Apply } f.}$ | $f(f^{-1}(x)) = x$ |

In other words,

> $$f^{-1}(f(x)) = x \quad \text{and} \quad f(f^{-1}(x)) = x$$

The preceding conditions can be used to verify that a function is, in fact, the inverse of f, as Example 3 demonstrates.

E X A M P L E   3   Verifying Inverse Functions

(a) We verify that the inverse of $g(x) = x^3$ is $g^{-1}(x) = \sqrt[3]{x}$ by showing that
$$g^{-1}(g(x)) = g^{-1}(x^3) = \sqrt[3]{x^3} = x$$
and
$$g(g^{-1}(x)) = g(\sqrt[3]{x}) = (\sqrt[3]{x})^3 = x$$

(b) We verify that the inverse of $h(x) = 3x$ is $h^{-1}(x) = \frac{1}{3}x$ by showing that
$$h^{-1}(h(x)) = h^{-1}(3x) = \frac{1}{3}(3x) = x$$
and
$$h(h^{-1}(x)) = h(\frac{1}{3}x) = 3(\frac{1}{3}x) = x$$

(c) We verify that the inverse of $f(x) = 2x + 3$ is $f^{-1}(x) = \frac{1}{2}(x - 3)$ by showing that
$$f^{-1}(f(x)) = f^{-1}(2x + 3) = \frac{1}{2}[(2x + 3) - 3] = \frac{1}{2}(2x) = x$$
and
$$f(f^{-1}(x)) = f(\frac{1}{2}(x - 3)) = 2[\frac{1}{2}(x - 3)] + 3 = (x - 3) + 3 = x \quad \blacksquare$$

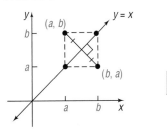

Now work Problem 21.

Exploration Simultaneously graph $Y_1 = x$, $Y_2 = x^3$, and $Y_3 = \sqrt[3]{x}$ on a square screen, using the viewing rectangle $-3 \le x \le 3, -2 \le y \le 2$. What do you observe about the graphs of $Y_2 = f(x) = x^3$, its inverse $Y_3 = f^{-1}(x) = \sqrt[3]{x}$, and the line $Y_1 = x$?

Do you see the symmetry of the graph of f and its inverse with respect to the line $Y_1 = x$? ∎

For the functions in Example 3(c), we list points on the graph of $f = Y_1$ and on the graph of $f^{-1} = Y_2$ in Table 1.

FIGURE 6

TABLE 1

X	Y1
-5	-7
-4	-5
-3	-3
-2	-1
-1	1
0	3
1	5

Y₁■2X+3

X	Y2
-7	-5
-5	-4
-3	-3
-1	-2
1	-1
3	0
5	1

Y₂■(1/2)(X-3)

FIGURE 7

We notice that whenever (a, b) is on the graph of f then (b, a) is on the graph of f^{-1}. Figure 6 shows these points plotted. Also shown there is the graph of $y = x$, which you should observe is a line of symmetry of the points.

Geometric Interpretation

Suppose that (a, b) is a point on the graph of the one-to-one function f defined by $y = f(x)$. Then $b = f(a)$. This means that $a = f^{-1}(b)$, so (b, a) is a point on the graph of the inverse function f^{-1}. The relationship between the point (a, b) on f and the point (b, a) on f^{-1} is shown in Figure 7. The line joining (a, b) and (b, a) is perpendicular to the line $y = x$ and is bisected by the line $y = x$. (Do you see why?) It follows that the point (b, a) on f^{-1} is the reflection about the line $y = x$ of the point (a, b) on f.

> ### Theorem
>
> The graph of a function f and the graph of its inverse f^{-1} are symmetric with respect to the line $y = x$.

◾

Figure 8 illustrates this result. Notice that, once the graph of f is known, the graph of f^{-1} may be obtained by folding the paper along the line $y = x$.

FIGURE 8

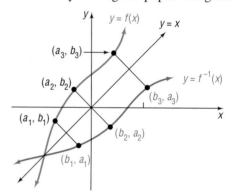

E X A M P L E   4

Graphing the Inverse Function

The graph in Figure 9(a) is that of a one-to-one function $y = f(x)$. Draw the graph of its inverse.

Solution   We begin by adding the graph of $y = x$ to Figure 9(a). Since the points $(-2, -1), (-1, 0)$, and $(2, 1)$ are on the graph of f, we know that the points $(-1, -2), (0, -1)$, and $(1, 2)$ must be on the graph of f^{-1}. Keeping in mind that the graph of f^{-1} is the reflection about the line $y = x$ of the graph of f, we can draw f^{-1}. See Figure 9(b).

FIGURE 9

(a)   (b)   ◾

Now work Problem 15.

Finding the Inverse Function

The fact that the graph of a one-to-one function f and its inverse are symmetric with respect to the line $y = x$ tells us more. It says that we can obtain

f^{-1} by interchanging the roles of x and y in f. Look again at Figure 8. If f is defined by the equation

$$y = f(x)$$

then f^{-1} is defined by the equation

$$x = f(y)$$

The equation $x = f(y)$ defines f^{-1} *implicitly*. If we can solve this equation for y, we will have the *explicit* form of f^{-1}, that is,

$$y = f^{-1}(x)$$

Let's use this procedure to find the inverse of $f(x) = 2x + 3$. (Since f is a linear function and is increasing, we know that f is one-to-one.)

E X A M P L E   5   Finding the Inverse Function

Find the inverse of $f(x) = 2x + 3$. Also find the domain and range of f and f^{-1}. Graph f and f^{-1} on the same coordinate axes.

Solution   In the equation $y = 2x + 3$, interchange the variables x and y. The result,

$$x = 2y + 3$$

is an equation that defines the inverse f^{-1} implicitly. Solving for y, we obtain

$$2y + 3 = x$$
$$2y = x - 3$$
$$y = \tfrac{1}{2}(x - 3)$$

The explicit form of the inverse f^{-1} is therefore

$$f^{-1}(x) = \tfrac{1}{2}(x - 3)$$

which we verified in Example 3(c).

Then we find

$$\text{Domain } f = \text{Range } f^{-1} = (-\infty, \infty)$$
$$\text{Range } f = \text{Domain } f^{-1} = (-\infty, \infty)$$

The graphs of $Y_1 = f(x) = 2x + 3$ and its inverse $Y_2 = f^{-1}(x) = \tfrac{1}{2}(x - 3)$ are shown in Figure 10. Note the symmetry of the graphs with respect to the line $Y_3 = x$.

FIGURE 10

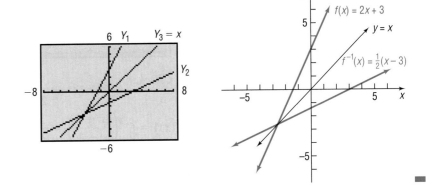

We outline next the steps to follow for finding the inverse of a one-to-one function.

In Problems 19–24, the graph of an exponential function is given. Match each graph to one of the following functions:

A. $y = 4^x$ B. $y = 4^{-x}$ C. $y = 4^{x-1}$ D. $y = 4^x - 1$ E. $y = -4^x$ F. $y = 1 - 4^x$

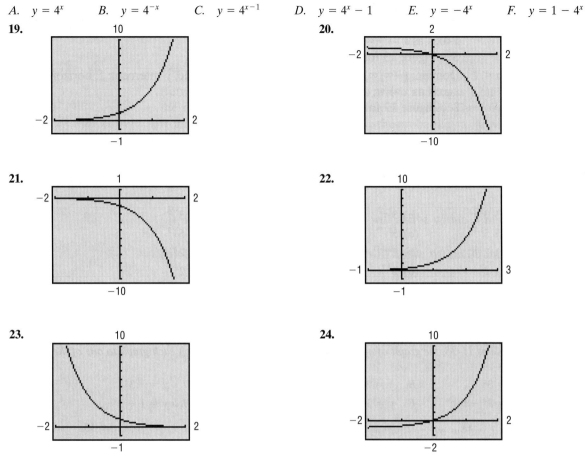

19.

20.

21.

22.

23.

24.

In Problems 25–32, using a graphing utility, show the stages required to graph each function. Determine the domain, range and horizontal asymptote of each function.

25. $f(x) = e^{-x}$
26. $f(x) = -e^x$
27. $f(x) = e^{x+2}$
28. $f(x) = e^x - 1$
29. $f(x) = 5 - e^{-x}$
30. $f(x) = 9 - 3e^{-x}$
31. $f(x) = 2 - e^{-x/2}$
32. $f(x) = 7 - 3e^{-2x}$

33. If $4^x = 7$, what does 4^{-2x} equal?
34. If $2^x = 3$, what does 4^{-x} equal?
35. If $3^{-x} = 2$, what does 3^{2x} equal?
36. If $5^{-x} = 3$, what does 5^{3x} equal?

37. **Optics** If a single pane of glass obliterates 3% of the light passing through it, then the percent p of light that passes through n successive panes is given approximately by the equation

$$p = 100e^{-0.03n}$$

(a) What percent of light will pass through 10 panes?
(b) What percent of light will pass through 25 panes?

38. **Atmospheric Pressure** The atmospheric pressure p on a balloon or plane decreases with increasing height. This pressure, measured in millimeters of mercury, is related to the number of kilometers h above sea level by the formula

$$p = 760e^{-0.145h}$$

(a) Find the atmospheric pressure at a height of 2 kilometers (over a mile).
(b) What is it at a height of 10 kilometers (over 30,000 feet)?

39. **Space Satellites** The number of watts w provided by a space satellite's power supply over a period of d days is given by the formula

$$w = 50e^{-0.004d}$$

(a) How much power will be available after 30 days?

(b) How much power will be available after 1 year (365 days)?

40. **Healing of Wounds** The normal healing of wounds can be modeled by an exponential function. If A_0 represents the original area of the wound and if A equals the area of the wound after n days, then the formula

$A = A_0 e^{-0.35n}$

describes the area of a wound on the nth day following an injury when no infection is present to retard the healing. Suppose a wound initially had an area of 100 square centimeters.

(a) If healing is taking place, how large should the area of the wound be after 3 days?

(b) How large should it be after 10 days?

41. **Drug Medication** The formula

$D = 5e^{-0.4h}$

can be used to find the number of milligrams D of a certain drug that is in a patient's bloodstream h hours after the drug has been administered. How many milligrams will be present after 1 hour? After 6 hours?

42. **Spreading of Rumors** A model for the number of people N in a college community who have heard a certain rumor is

$N = P(1 - e^{-0.15d})$

where P is the total population of the community and d is the number of days that have elapsed since the rumor began. In a community of 1000 students, how many students will have heard the rumor after 3 days?

43. **Exponential Probability** Between 12:00 P.M. and 1:00 P.M. cars arrive at Citibank's drive-thru at the rate of 6 cars per hour (0.1 cars per minute). The following formula from statistics can be used to determine the probability a car will arrive within t minutes of 12:00 P.M.

$F(t) = 1 - e^{-0.1t}$

(a) Determine the probability a car will arrive within 10 minutes of 12:00 P.M. (That is, before 12:10 P.M.).

(b) Determine the probability a car will arrive within 40 minutes of 12:00 P.M. (before 12:40 P.M.).

(c) Graph F using your graphing utility.

(d) Using TRACE, determine how many minutes are needed for the probability to reach 50%?

(e) What value does F approach as t becomes unbounded in the positive direction?

44. **Exponential Probability** Between 5:00 P.M. and 6:00 P.M. cars arrive at Jiffy Lube at the rate of 9 cars per hour (0.15 cars per minute). The following formula from statistics can be used to determine the probability a car will arrive within t minutes of 5:00 P.M.:

$F(t) = 1 - e^{-0.15t}$

(a) Determine the probability a car will arrive within 15 minutes of 5:00 P.M. (That is, before 5:15 P.M.).

(b) Determine the probability a car will arrive within 30 minutes of 5:00 P.M. (before 5:30 P.M.).

(c) Graph F using your graphing utility.

(d) Using TRACE, determine how many minutes are needed for the probability to reach 60%?

(e) What value does F approach as t becomes unbounded in the positive direction?

45. **Response to TV Advertising** The percent R of viewers who respond to a television commercial for a new product after t days is found by using the formula

$R = 70 - 100e^{-0.2t}$

(a) What percent is expected to respond after 10 days?

(b) What percent has responded after 20 days?

(c) What is the highest percent of people expected to respond?

(d) Graph $R = 70 - 100e^{-0.2t}$, $t > 0$. TRACE and compare the values of R for $t = 10$ and $t = 20$ to the ones obtained in parts (a) and (b). How many days are required for R to exceed 40%?

46. **Profit** The annual profit P of a company due to the sales of a particular item after it has been on the market x years is determined to be

$P = \$100,000 - \$60,000(\tfrac{1}{2})^x$

(a) What is the profit after 5 years?

(b) What is the profit after 10 years?

(c) What is the most profit the company can expect from this product?

(d) Graph the profit function. TRACE and compare the values of P for $x = 5$ and $x = 10$ to the one obtained in parts (a) and (b). How many years does it take before a profit of \$65,000 is obtained?

47. Alternating Current in a RL Circuit The equation governing the amount of current I (in amperes) after time t (in seconds) in a single RL circuit consisting of a resistance R (in ohms), an inductance L (in henrys), and an electromotive force E (in volts) is

$$I = \frac{E}{R}[1 - e^{-(R/L)t}]$$

(a) If $E = 120$ volts, $R = 10$ ohms, and $L = 5$ henrys, how much current I_1 is available after 0.3 second? After 0.5 second? After 1 second?

(b) What is the maximum current?

(c) Graph this function $I = I_1(t)$, measuring I along the y-axis and t along the x-axis.

(d) If $E = 120$ volts, $R = 5$ ohms, and $L = 10$ henrys, how much current I_2 is available after 0.3 second? After 0.5 second? After 1 second?

(e) What is the maximum current?

(f) Graph this function $I = I_2(t)$ on the same viewing window as $I_1(t)$.

48. Alternating Current in a RC Circuit The equation governing the amount of current I (in milliamperes) after time t (in milliseconds) in a single RC circuit consisting of a resistance R (in ohms), a capacitance C (in microfarads), and an electromotive force E (in volts) is

$$I = \frac{E}{R}e^{-t/(RC)}$$

(a) If $E = 120$ volts, $R = 2000$ ohms, and $C = 1.0$ microfarad, how much current I_1 is available initially ($t = 0$)? After 1000 milliseconds? After 3000 milliseconds?

(b) What is the maximum current?

(c) Graph this function $I = I_1(t)$, measuring I along the y-axis and t along the x-axis.

(d) If $E = 120$ volts, $R = 1000$ ohms, and $C = 2.0$ microfarads, how much current I_2 is available initially? After 1000 milliseconds? After 3000 milliseconds?

(e) What is the maximum current?

(f) Graph this function $I = I_2(t)$ on the same viewing window as $I_1(t)$.

49. The Challenger Disaster* After the *Challenger* disaster in 1986, a study of the 23 launches that preceded the fatal flight was made. A mathematical model was developed involving the relationship between the Fahrenheit temperature x around the O-rings and the number y of eroded or leaky primary O-rings. The model stated that

$$y = \frac{6}{1 + e^{-(5.085 - 0.1156x)}}$$

where the number 6 indicates the 6 primary O-rings on the spacecraft.

(a) What is the predicted number of eroded or leaky primary O-rings at a temperature of 100°F?

(b) What is the predicted number of eroded or leaky primary O-rings at a temperature of 60°F?

(c) What is the predicted number of eroded or leaky primary O-rings at a temperature of 30°F?

*Linda Tappin, "Analyzing Data Relating to the *Challenger* Disaster," *Mathematics Teacher*, Vol. 87, No. 6, September 1994, pp. 423–426.

(d) Graph the equation and TRACE. At what temperature is the predicted number of eroded or leaky O-rings 1? 3? 5?

50. **Postage Stamps*** The cumulative number y of different postage stamps (regular and commemorative only) issued by the U.S. Post Office can be approximated (modeled) by the exponential function

$$y = 78e^{0.025x}$$

where x is the number of years since 1848.

(a) What is the predicted cumulative number of stamps that will have been issued by the year 1995? Check with the Postal Service and comment on the accuracy of using the function.

(b) What is the predicted cumulative number of stamps that will have been issued by the year 1998?

(c) The cumulative number of stamps actually issued by the United States was 2 in 1848, 88 in 1868, 218 in 1888, and 341 in 1908. What conclusion can you draw about using the given function as a model over the first few decades in which stamps were issued?

51. **Another Formula for e** Use a calculator to compute the value of

$$2 + \frac{1}{2!} + \frac{1}{3!} + \; \cdots \; + \frac{1}{n!}$$

for $n = 4$, 6, 8, and 10. Compare each result with e.

[**Hint:** $1! = 1, 2! = 2 \cdot 1, 3! = 3 \cdot 2 \cdot 1,$
$n! = n(n - 1) \cdot \ldots \cdot (3)(2)(1)$].

*David Kullman, "Patterns of Postage-stamp Production," *Mathematics Teacher*, Vol. 85, No. 3, March 1992, pp. 188–189.

52. **Another Formula for e** Use a calculator to compute the various values of the expression

$$2 + \cfrac{1}{1 + \cfrac{1}{2 + \cfrac{2}{3 + \cfrac{3}{4 + 4}}}}$$

etc.

Compare the values to e.

53. If $f(x) = a^x$, show that:
$$\frac{f(x + h) - f(x)}{h} = a^x \left(\frac{a^h - 1}{h} \right)$$

54. If $f(x) = a^x$, show that: $f(A + B) = f(A) \cdot f(B)$

55. If $f(x) = a^x$, show that: $f(-x) = \dfrac{1}{f(x)}$

56. If $f(x) = a^x$, show that: $f(\alpha x) = [f(x)]^\alpha$

Problems 57 and 58 provide definitions for two other transcendental functions.

57. The **hyperbolic sine function,** designated by $\sinh x$, is defined as

$$\sinh x = \frac{1}{2}(e^x - e^{-x})$$

(a) Show that $f(x) = \sinh x$ is an odd function.
(b) Graph $y = e^x$, $y = -e^{-x}$, and $f(x) = \sinh x$ on the same set of coordinate axes.
(c) How do you explain the relationship of the three graphs?

58. The **hyperbolic cosine function,** designated by $\cosh x$, is defined as

$$\cosh x = \frac{1}{2}(e^x + e^{-x})$$

(a) Show that $f(x) = \cosh x$ is an even function.
(b) Graph $y = e^x$, $y = e^{-x}$, and $f(x) = \cosh x$ on the same set of coordinate axes.
(c) Refer to Problem 57. Show that, for every x,
$$(\cosh x)^2 - (\sinh x)^2 = 1$$

59. **Historical Problem** Pierre de Fermat (1601–1665) conjectured that the function

$$f(x) = 2^{(2^x)} + 1$$

for $x = 1, 2, 3, \ldots$, would always have a value equal to a prime number. But Leonhard Euler (1707–1783) showed that this formula fails for $x = 5$. Use a calculator to determine the prime numbers produced by f for $x = 1, 2, 3, 4$. Then show that $f(5) = 641 \times 6{,}700{,}417$, which is not prime.

`60.` The bacteria in a 4-liter container double every minute. After 60 minutes the container is full. How long did it take to fill half the container?

61. Explain in your own words what the number e is. Provide at least two applications that require the use of this number.

62. Do you think there is a power function that increases more rapidly than an exponential function whose base is greater than 1? Explain.

4.3   LOGARITHMIC FUNCTIONS

> **1** Change Exponential Expressions to Logarithmic Expressions
> **2** Change Logarithmic Expressions to Exponential Expressions
> **3** Evaluate Logarithmic Functions
> **4** Determine the Domain of a Logarithmic Function
> **5** Graph Logarithmic Functions

Recall that a one-to-one function $y = f(x)$ has an inverse that is defined (implicitly) by the equation $x = f(y)$. In particular, the exponential function $y = f(x) = a^x$, $a > 0$, $a \neq 1$, is one-to-one and, hence, has an inverse that is defined implicitly by the equation

$$x = a^y \qquad a > 0, a \neq 1$$

This inverse is so important that it is given a name, the *logarithmic function*.

> The **logarithmic function to the base a**, where $a > 0$ and $a \neq 1$, is denoted by $y = \log_a x$ (read as "y is the logarithm to the base a of x") and is defined by
>
> $$y = \log_a x \qquad \text{if and only if} \qquad x = a^y$$

E X A M P L E   1   Relating Logarithms to Exponents

(a) If $y = \log_3 x$, then $x = 3^y$. Thus, $9 = 3^2$ is equivalent to $2 = \log_3 9$.
(b) If $y = \log_5 x$, then $x = 5^y$. Thus, $\frac{1}{5} = 5^{-1}$ is equivalent to $-1 = \log_5 \left(\frac{1}{5}\right)$.

E X A M P L E   2   Changing Exponential Expressions to Logarithmic Expressions

1 Change each exponential expression to an equivalent expression involving a logarithm.

(a) $1.2^3 = m$ (b) $e^b = 9$ (c) $a^4 = 24$

Solution   We use the fact that $y = \log_a x$ and $x = a^y$, $a > 0$, $a \neq 1$, are equivalent.

(a) If $1.2^3 = m$, then $3 = \log_{1.2} m$.
(b) If $e^b = 9$, then $b = \log_e 9$.
(c) If $a^4 = 24$, then $4 = \log_a 24$.

Now work Problem 1.

EXAMPLE 3

Changing Logarithmic Expressions to Exponential Expressions

Change each logarithmic expression to an equivalent expression involving an exponent.

(a) $\log_a 4 = 5$ (b) $\log_e b = -3$ (c) $\log_3 5 = c$

Solution

(a) If $\log_a 4 = 5$, then $a^5 = 4$.
(b) If $\log_e b = -3$, then $e^{-3} = b$.
(c) If $\log_3 5 = c$, then $3^c = 5$.

Now work Problem 13.

To find the exact value of a logarithm, we write the logarithm in exponential notation and use the following fact:

$$\text{If } a^u = a^v, \quad \text{then} \quad u = v. \tag{1}$$

The result (1) is a consequence of the fact that exponential functions are one-to-one.

EXAMPLE 4

Finding the Exact Value of a Logarithmic Function

Find the exact value of

(a) $\log_2 8$ (b) $\log_3 \frac{1}{3}$ (c) $\log_5 25$

Solution

(a) For $y = \log_2 8$, we have the equivalent exponential equation $2^y = 8 = 2^3$, so, by (1), $y = 3$. Thus, $\log_2 8 = 3$.
(b) For $y = \log_3 \frac{1}{3}$, we have $3^y = \frac{1}{3} = 3^{-1}$, so $y = -1$. Thus, $\log_3 \frac{1}{3} = -1$.
(c) For $y = \log_5 25$, we have $5^y = 25 = 5^2$, so $y = 2$. Thus, $\log_5 25 = 2$.

Now work Problem 25.

Domain of a Logarithmic Function

The logarithmic function $y = \log_a x$ has been defined as the inverse of the exponential function $y = a^x$. That is, if $f(x) = a^x$, then $f^{-1}(x) = \log_a x$. Based on the discussion given in Section 4.1 on inverse functions, we know that for a function f and its inverse f^{-1}

Domain f^{-1} = Range f and   Range f^{-1} = Domain f

Consequently, it follows that

Domain of logarithmic function = Range of exponential function = $(0, \infty)$
Range of logarithmic function = Domain of exponential function = $(-\infty, \infty)$

In the next box, we summarize some properties of the logarithmic function:

$$y = \log_a x \quad \text{(defining equation: } x = a^y)$$
$$\text{Domain: } 0 < x < \infty \quad \text{Range: } -\infty < y < \infty$$

Notice that the domain of a logarithmic function consists of the *positive* real numbers.

E X A M P L E   5

Finding the Domain of a Logarithmic Function

Find the domain of each logarithmic function:

(a) $F(x) = \log_2 (1 - x)$ (b) $g(x) = \log_5 \left(\dfrac{1 + x}{1 - x} \right)$ (c) $h(x) = \log_{1/2} |x|$

Solution (a) The domain of F consists of all x for which $(1 - x) > 0$; that is, all $x < 1$, or $(-\infty, 1)$.

(b) The domain of g is restricted to

$$\frac{1 + x}{1 - x} > 0$$

Solving this inequality, we find that the domain of g consists of all x between -1 and 1; that is, $-1 < x < 1$, or $(-1, 1)$.

(c) Since $|x| > 0$ provided $x \neq 0$, the domain of h consists of all nonzero real numbers. ∎

Now work Problem 39.

Graphs of Logarithmic Functions

Since exponential functions and logarithmic functions are inverses of each other, the graph of a logarithmic function $y = \log_a x$ is the reflection about the line $y = x$ of the graph of the exponential function $y = a^x$, as shown in Figure 25.

FIGURE 25

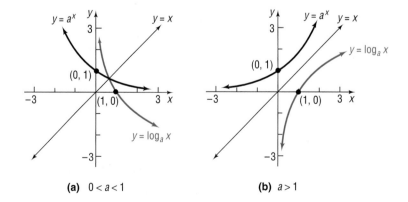

(a) $0 < a < 1$ **(b)** $a > 1$

> Facts about the Graph of a Logarithmic Function $f(x) = \log_a x$
>
> 1. The x-intercept of the graph is 1. There is no y-intercept.
> 2. The y-axis is a vertical asymptote of the graph.
> 3. A logarithmic function is decreasing if $0 < a < 1$ and increasing if $a > 1$.
> 4. The graph is smooth and continuous, with no corners or gaps.

If the base of a logarithmic function is the number e, then we have the **natural logarithm function.** This function occurs so frequently in applications that it is given a special symbol, **ln** (from the Latin, *logarithmus naturalis*). Thus,

> $$y = \ln x \quad \text{if and only if} \quad x = e^y$$

Since $y = \ln x$ and the exponential function $y = e^x$ are inverse functions, we can obtain the graph of $y = \ln x$ by reflecting the graph of $y = e^x$ about the line $y = x$. See Figure 26.

Table 6 displays other points on the graph of $f(x) = \ln x$. Notice for $x < 0$ we obtain an error message. Do you recall why?

FIGURE 26

EXAMPLE 6

Graphing Logarithmic Functions

Graph $f(x) = -\ln x$ by starting with the graph of $y = \ln x$. Determine the domain, range, and vertical asymptote of $f(x) = -\ln x$.

Solution

The graph of $f(x) = -\ln x$ is obtained by a reflection about the x-axis of the graph of $f(x) = \ln x$. See Figure 27.

FIGURE 27

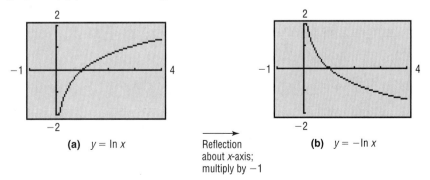

(a) $y = \ln x$

Reflection about x-axis; multiply by -1

(b) $y = -\ln x$

The domain of $f(x) = -\ln x$ is $(0, \infty)$, the range is $(-\infty, \infty)$, and the vertical asymptote is $x = 0$.

EXAMPLE 7 Graphing Logarithmic Functions

Graph $f(x) = \ln(x + 2)$. Determine the domain, range, and vertical asymptote of $f(x) = \ln(x + 2)$.

Solution The graph is obtained by applying a horizontal shift to the left 2 units, as shown in Figure 28.

FIGURE 28

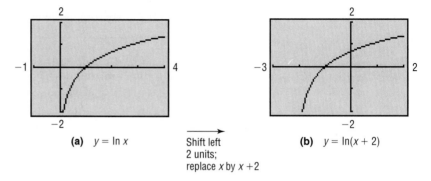

(a) $y = \ln x$ — Shift left 2 units; replace x by $x + 2$ — **(b)** $y = \ln(x + 2)$

The domain of $f(x) = \ln(x + 2)$ is $(-2, \infty)$, the range is $(-\infty, \infty)$, and the vertical asymptote is $x = -2$.

EXAMPLE 8 Graphing Logarithmic Functions

Graph $f(x) = \ln(1 - x)$. Determine the domain, range, and vertical asymptote of $f(x) = \ln(1 - x)$.

Solution To obtain the graph of $y = \ln(1 - x)$, we use the steps illustrated in Figure 29.

FIGURE 29

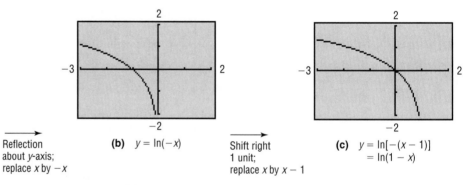

(a) $y = \ln x$ — Reflection about y-axis; replace x by $-x$ — **(b)** $y = \ln(-x)$ — Shift right 1 unit; replace x by $x - 1$ — **(c)** $y = \ln[-(x - 1)]$ $= \ln(1 - x)$

The domain of $f(x) = \ln(1 - x)$ is $(-\infty, 1)$, the range is $(-\infty, \infty)$, and the vertical asymptote is $x = 1$.

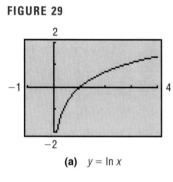

Now work Problem 67.

EXAMPLE 9 Alcohol and Driving

The concentration of alcohol in a person's blood is measurable. Recent medical research suggests that the risk R (given as a percent) of having an accident while driving a car can be modeled by the equation

$$R = 6e^{kx}$$

where x is the variable concentration of alcohol in the blood and k is a constant.

(a) Suppose that a concentration of alcohol in the blood of 0.04 results in a 10% risk ($R = 10$) of an accident. Find the constant k in the equation. Graph $R = 6e^{kx}$, using this value of k.

(b) Using this value of k, what is the risk if the concentration is 0.17?

(c) Using the same value of k, what concentration of alcohol corresponds to a risk of 100%?

(d) If the law asserts that anyone with a risk of having an accident of 20% or more should not have driving privileges, at what concentration of alcohol in the blood should a driver be arrested and charged with a DUI— Driving Under the Influence?

Solution   (a) For a concentration of alcohol in the blood of 0.04 and a risk of 10%, we let $x = 0.04$ and $R = 10$ in the equation and solve for k.

$$R = 6e^{kx}$$

$$10 = 6e^{k(0.04)}$$

$$\frac{10}{6} = e^{0.04k} \qquad \text{Change to a logarithmic expression.}$$

$$0.04k = \ln \frac{10}{6} = 0.5108256$$

$$k = 12.77$$

See Figure 30 for the graph of $R = 6e^{12.77x}$.

FIGURE 30

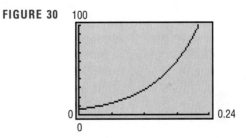

(b) Using $k = 12.77$ and $x = 0.17$ in the equation, we find the risk R to be

$$R = 6e^{kx} = 6e^{(12.77)(0.17)} = 52.6$$

For a concentration of alcohol in the blood of 0.17, the risk of an accident is about 52.6%. Verify this answer by TRACEing the graph.

(c) Using $k = 12.77$ and $R = 100$ in the equation, we find the concentration x of alcohol in the blood to be

$$R = 6e^{kx}$$

$$100 = 6e^{12.77x}$$

$$\frac{100}{6} = e^{12.77x} \qquad \text{Change to a logarithmic expression.}$$

$$12.77x = \ln \frac{100}{6} = 2.8134$$

$$x = 0.22$$

For a concentration of alcohol in the blood of 0.22, the risk of an accident is 100%. Verify this answer by TRACEing the graph.

(d) Using $k = 12.77$ and $R = 20$ in the equation, we find the concentration x of alcohol in the blood to be

$$R = 6e^{kx}$$
$$20 = 6e^{12.77x}$$
$$\frac{20}{6} = e^{12.77x}$$
$$12.77x = \ln\frac{20}{6} = 1.204$$
$$x = 0.094$$

A driver with a concentration of alcohol in the blood of 0.094 or more should be arrested and charged with DUI. Verify by TRACEing the graph. ▬

Note: Most states use 0.10 as the blood alcohol content at which a DUI citation is given. A few states use 0.08.

SUMMARY

Properties of the Logarithmic Function

$f(x) = \log_a x, \quad a > 1$ Domain: $(0, \infty)$; Range: $(-\infty, \infty)$; x-intercept: 1; y-intercept: none; vertical asymp
$(y = \log_a x$ means $x = a^y)$ tote: y-axis; increasing; one-to-one
    See Figure 25(b) for a typical graph.

$f(x) = \log_a x, \quad 0 < a < 1$ Domain: $(0, \infty)$; Range: $(-\infty, \infty)$; x-intercept: 1; y-intercept: none; vertical
$(y = \log_a x$ means $x = a^y)$ asymptote: y-axis; decreasing; one-to-one
    See Figure 25(a) for a typical graph.

4.3 EXERCISES

In Problems 1–12, change each exponential expression to an equivalent expression involving a logarithm.

1. $9 = 3^2$ **2.** $16 = 4^2$ **3.** $a^2 = 1.6$ **4.** $a^3 = 2.1$ **5.** $1.1^2 = M$ **6.** $2.2^3 = N$

7. $2^x = 7.2$ **8.** $3^x = 4.6$ **9.** $x^{\sqrt{2}} = \pi$ **10.** $x^\pi = e$ **11.** $e^x = 8$ **12.** $e^{2.2} = M$

In Problems 13–24, change each logarithmic expression to an equivalent expression involving an exponent.

13. $\log_2 8 = 3$ **14.** $\log_3(\frac{1}{9}) = -2$ **15.** $\log_a 3 = 6$ **16.** $\log_b 4 = 2$

17. $\log_3 2 = x$ **18.** $\log_2 6 = x$ **19.** $\log_2 M = 1.3$ **20.** $\log_3 N = 2.1$

21. $\log_{\sqrt{2}} \pi = x$ **22.** $\log_\pi x = \frac{1}{2}$ **23.** $\ln 4 = x$ **24.** $\ln x = 4$

In Problems 25–36, find the exact value of each logarithm without using a calculator.

25. $\log_2 1$ **26.** $\log_8 8$ **27.** $\log_5 25$ **28.** $\log_3(\frac{1}{9})$ **29.** $\log_{1/2} 16$ **30.** $\log_{1/3} 9$

31. $\log_{10}\sqrt{10}$ **32.** $\log_5\sqrt[3]{25}$ **33.** $\log_{\sqrt{2}} 4$ **34.** $\log_{\sqrt{3}} 9$ **35.** $\ln\sqrt{e}$ **36.** $\ln e^3$

In Problems 37–46, find the domain of each function.

37. $f(x) = \ln(3 - x)$ **38.** $g(x) = \ln(x^2 - 1)$ **39.** $F(x) = \log_2 x^2$ **40.** $H(x) = \log_5 x^3$

41. $h(x) = \log_{1/2}(x^2 - x - 6)$ **42.** $G(x) = \log_{1/2}\left(\frac{1}{x}\right)$ **43.** $f(x) = \frac{1}{\ln x}$ **44.** $g(x) = \ln(x - 5)$

45. $g(x) = \log_5\left(\frac{x + 1}{x}\right)$ **46.** $h(x) = \log_3\left(\frac{x^2}{x - 1}\right)$

In Problems 47–50, use a calculator to evaluate each expression. Round your answer to three decimal places.

47. $\ln \dfrac{5}{3}$

48. $\dfrac{\ln 5}{3}$

49. $\dfrac{\ln(10/3)}{0.04}$

50. $\dfrac{\ln(2/3)}{-0.1}$

51. Find a such that the graph of $f(x) = \log_a x$ contains the point $(2, 2)$.

52. Find a such that the graph of $f(x) = \log_a x$ contains the point $\left(\frac{1}{2}, -4\right)$.

In Problems 53–60, the graph of a logarithmic function is given. Match each graph to one of the following functions:

A. $y = \log_3 x$

B. $y = \log_3(-x)$

C. $y = -\log_3 x$

D. $y = -\log_3(-x)$

E. $y = \log_3 x - 1$

F. $y = \log_3(x - 1)$

G. $y = \log_3(1 - x)$

H. $y = 1 - \log_3 x$

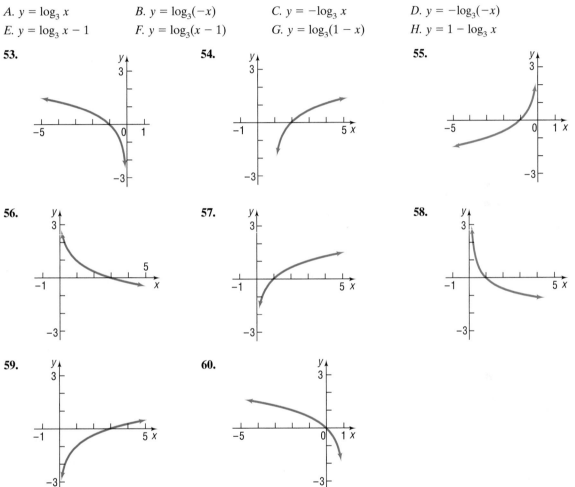

53. **54.** **55.**

56. **57.** **58.**

59. **60.**

In Problems 61–66, the graph of a logarithmic function is given. Match each graph to one of the following functions:

A. $y = \log_4 x$

B. $y = \log_4(-x)$

C. $y = \log_4(x - 1)$

D. $y = -\log_4 x$

E. $y = 1 - \log_4 x$

F. $y = -\log_4(-x)$

61. **62.**

63. 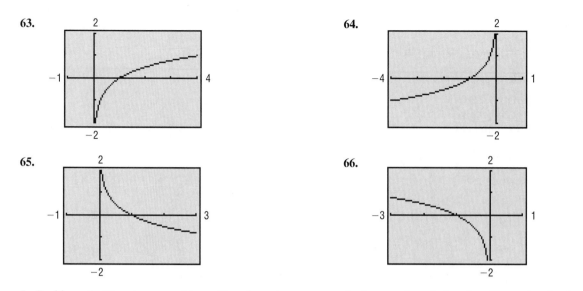 **64.**

65. **66.**

In Problems 67–76, using a graphing utility, show the stages required to graph each function. Determine the domain, range, and vertical asymptote of each function.

67. $f(x) = \ln(x + 4)$ **68.** $f(x) = \ln(x - 3)$ **69.** $f(x) = \ln(-x)$ **70.** $f(x) = -\ln(-x)$

71. $g(x) = \ln 2x$ **72.** $h(x) = \ln \frac{1}{2}x$ **73.** $f(x) = 3 \ln x$ **74.** $f(x) = -2 \ln x$

75. $g(x) = \ln(3 - x)$ **76.** $h(x) = \ln(4 - x)$

77. **Optics** If a single pane of glass obliterates 10% of the light passing through it, then the percent P of light that passes through n successive panes is given approximately by the equation

$$P = 100e^{-0.1n}$$

(a) How many panes are necessary to block at least 50% of the light?
(b) How many panes are necessary to block at least 75% of the light?

78. **Chemistry** The pH of a chemical solution is given by the formula

$$pH = -\log_{10} [H^+]$$

where $[H^+]$ is the concentration of hydrogen ions in moles per liter. Values of pH range from 0 (acidic) to 14 (alkaline).
(a) Find the pH of a 1 liter container of water with 0.0000001 mole of hydrogen ion.
(b) Find the hydrogen ion concentration of a mildly acidic solution with a pH of 4.2.

79. **Space Satellites** The number of watts w provided by a space satellite's power supply after d days is given by the formula

$$w = 50e^{-0.004d}$$

(a) How long will it take for the available power to drop to 30 watts?
(b) How long will it take for the available power to drop to only 5 watts?

80. **Healing of Wounds** The normal healing of wounds can be modeled by an exponential function. If A_0 represents the original area of the wound and if A equals the area of the wound after n days, then the formula

$$A = A_0 e^{-0.35n}$$

describes the area of a wound on the nth day following an injury when no infection is present to retard the healing. Suppose a wound initially had an area of 100 square centimeters.
(a) If healing is taking place, how many days should pass before the wound is one-half its original size?
(b) How long before the wound is 10% of its original size?

81. Drug Medication The formula

$$D = 5e^{-0.4h}$$

can be used to find the number of milligrams D of a certain drug that is in a patient's bloodstream h hours after the drug has been administered. When the number of milligrams reaches 2, the drug is to be administered again. What is the time between injections?

82. Spreading of Rumors A model for the number of people N in a college community who have heard a certain rumor is

$$N = P(1 - e^{-0.15d})$$

where P is the total population of the community and d is the number of days that have elapsed since the rumor began. In a community of 1000 students, how many days will elapse before 450 students have heard the rumor?

83. Current in a RL Circuit The equation governing the amount of current I (in amperes) after time t (in seconds) in a simple RL circuit consisting of a resistance R (in ohms), an inductance L (in henrys), and an electromotive force E (in volts) is

$$I = \frac{E}{R}[1 - e^{-(R/L)t}]$$

If $E = 12$ volts, $R = 10$ ohms, and $L = 5$ henrys, how long does it take to obtain a current of 0.5 ampere? Of 1.0 ampere? Graph the equation.

84. Learning Curve Psychologists sometimes use the function

$$L(t) = A(1 - e^{-kt})$$

to measure the amount L learned at time t. The number A represents the amount to be learned, and the number k measures the rate of learning. Suppose that a student has an amount A of 200 vocabulary words to learn. A psychologist determines that the student learned 20 vocabulary words after 5 minutes.
(a) Determine the rate of learning k.
(b) Approximately how many words will the student have learned after 10 minutes?
(c) After 15 minutes?
(d) How long does it take for the student to learn 180 words?

85. Alcohol and Driving The concentration of alcohol in a person's blood is measurable. Suppose the risk R (given as a percent) of having an accident while driving a car can be modeled by the equation

$$R = 3e^{kx}$$

where x is the variable concentration of alcohol in the blood and k is a constant.

(a) Suppose a concentration of alcohol in the blood of 0.06 results in a 10% risk ($R = 10$) of an accident. Find the constant k in the equation.
(b) Using this value of k, what is the risk if the concentration is 0.17?
(c) Using the same value of k, what concentration of alcohol corresponds to a risk of 100%?
(d) If the law asserts that anyone with a risk of having an accident of 15% or more should not have driving privileges, at what concentration of alcohol in the blood should a driver be arrested and charged with a DUI?
(e) Compare this situation with that of Example 9. If you were a lawmaker, which situation would you support? Give your reasons.

86. Is there any function of the form $y = x^\alpha$, $0 < \alpha < 1$, that increases more slowly than a logarithmic function whose base is greater than 1? Explain.

87. Constructing a Function Look back at Figure 14 in Section 4.2. Assuming that the points (1950, 20) and (1990, 50) are on the graph, find an exponential equation $y = Ae^{bt}$, that fits the data.

[**Hint:** Let $t = 0$ correspond to the year 1950. Then show that $A = 20$. Now find b.]

Is the projection of 102 million in 2015 confirmed by your model? Try to obtain similar data about the United States birthrate and construct a function to fit those data.

88. Critical Thinking In buying a new car, one consideration might be how well the price of the car holds up over time. Different makes of cars have different depreciation rates. One way to compute a depreciation rate for a car is given here. Suppose the current prices of a certain Mercedes automobile are as follows:

New	1 Year Old	2 Years Old	3 Years Old	4 Years Old	5 Years Old
$38,000	$36,600	$32,400	$28,750	$25,400	$21,200

Use the formula New = Old(e^{Rt}) to find R, the annual depreciation rate, for a specific time t. When might be the best time to trade the car in? Consult the NADA ("blue") book and compare two like models that you are interested in. Which has the better depreciation rate?

4.4 PROPERTIES OF LOGARITHMS

1 Write a Logarithmic Expression as a Sum/Difference of Logarithms
2 Write a Logarithmic Expression as a Single Logarithm
3 Evaluate Logarithms Whose Base Is Neither 10 nor e
4 Graph Logarithmic Functions Whose Base Is Neither 10 nor e

Logarithms have some very useful properties that can be derived directly from the definition and the laws of exponents.

EXAMPLE 1 Establishing Properties of Logarithms

(a) Show that $\log_a 1 = 0$. (b) Show that $\log_a a = 1$.

Solution (a) This fact was established when we graphed $y = \log_a x$ (see Figure 25). Algebraically, for $y = \log_a 1$, we have $a^y = 1 = a^0$, so $y = 0$.
(b) For $y = \log_a a$, we have $a^y = a = a^1$, so $y = 1$. ∎

To summarize

$$\log_a 1 = 0 \qquad \log_a a = 1$$

Theorem Properties of Logarithms

In the properties given next, M and a are positive real numbers, with $a \neq 1$, and r is any real number.
The number $\log_a M$ is the exponent to which a must be raised to obtain M. That is,

$$a^{\log_a M} = M \tag{1}$$

The logarithm to the base a of a raised to a power equals that power. That is,

$$\log_a a^r = r \tag{2}$$

Proof of Property (1) Let $x = \log_a M$. Change this logarithmic expression to the equivalent exponential expression:

$$a^x = M$$

But $x = \log_a M$, so

$$a^{\log_a M} = M$$ ▬

Proof of Property (2) Let $x = a^r$. Change this exponential expression to the equivalent logarithmic expression

$$\log_a x = r$$

But $x = a^r$, so

$$\log_a a^r = r$$ ▬

E X A M P L E 2

Using Properties (1) and (2)

(a) $2^{\log_2 \pi} = \pi$ (b) $\log_{0.2} 0.2^{-\sqrt{2}} = -\sqrt{2}$ (c) $\ln e^{kt} = kt$ ▬

Other useful properties of logarithms are given now.

Theorem   Properties of Logarithms

In the following properties, M, N, and a are positive real numbers, with $a \neq 1$, and r is any real number.

The Log of a Product Equals the Sum of the Logs

$$\log_a (MN) = \log_a M + \log_a N \tag{3}$$

The Log of a Quotient Equals the Difference of the Logs

$$\log_a\left(\frac{M}{N}\right) = \log_a M - \log_a N \tag{4}$$

$$\log_a\left(\frac{1}{N}\right) = -\log_a N \tag{5}$$

$$\log_a M^r = r \log_a M \tag{6}$$

▬

We shall derive properties (3) and (6) and leave the derivations of properties (4) and (5) as exercises (see Problems 63 and 64).

Proof of Property (3) Let $A = \log_a M$ and let $B = \log_a N$. These expressions are equivalent to the exponential expressions

$$a^A = M \quad \text{and} \quad a^B = N$$

Now

$$\log_a (MN) = \log_a (a^A a^B) = \log_a a^{A+B} \qquad \text{Law of exponents.}$$
$$= A + B \qquad \text{Property (2) of logarithms.}$$
$$= \log_a M + \log_a N \qquad \blacksquare$$

Proof of Property (6)   Let $A = \log_a M$. This expression is equivalent to

$$a^A = M$$

Now

$$\log_a M^r = \log_a (a^A)^r = \log_a a^{rA} \qquad \text{Law of exponents.}$$
$$= rA \qquad \text{Property (2) of logarithms.}$$
$$= r \log_a M \qquad \blacksquare$$

1   Logarithms can be used to transform products into sums, quotients into differences, and powers into factors. Such transformations prove useful in certain types of calculus problems.

E X A M P L E   3

Writing a Logarithmic Expression as a Sum of Logarithms

Write $\log_a(x\sqrt{x^2 + 1})$ as a sum of logarithms. Express all powers as factors.

Solution

$$\log_a(x\sqrt{x^2 + 1}) = \log_a x + \log_a \sqrt{x^2 + 1} \qquad \text{Property (3).}$$
$$= \log_a x + \log_a(x^2 + 1)^{1/2}$$
$$= \log_a x + \tfrac{1}{2}\log_a(x^2 + 1) \qquad \text{Property (6).} \quad \blacksquare$$

E X A M P L E   4

Writing a Logarithmic Expression as a Difference of Logarithms

Write

$$\log_a \frac{x^2}{(x - 1)^3}$$

as a difference of logarithms. Express all powers as factors.

Solution

$$\log_a \frac{x^2}{(x - 1)^3} = \log_a x^2 - \log_a(x - 1)^3 = 2 \log_a x - 3 \log_a(x - 1) \qquad \blacksquare$$
$$\uparrow \qquad\qquad\qquad\qquad \uparrow$$
$$\text{Property (4).} \qquad\qquad\quad \text{Property (6).}$$

Now work Problem 13.

E X A M P L E   5

Writing a Logarithmic Expression as a Sum and Difference of Logarithms

Write

$$\log_a \frac{x^3\sqrt{x^2 + 1}}{(x + 1)^4}$$

as a sum and difference of logarithms. Express all powers as factors.

Solution $$\log_a \frac{x^3\sqrt{x^2+1}}{(x+1)^4} = \log_a(x^3\sqrt{x^2+1}) - \log_a(x+1)^4$$
$$= \log_a x^3 + \log_a \sqrt{x^2+1} - \log_a(x+1)^4$$
$$= \log_a x^3 + \log_a(x^2+1)^{1/2} - \log_a(x+1)^4$$
$$= 3\log_a x + \tfrac{1}{2}\log_a(x^2+1) - 4\log_a(x+1)$$ ▬

2

Another use of properties (3) through (6) is to write sums and/or differences of logarithms with the same base as a single logarithm.

E X A M P L E   6

Writing Expressions as a Single Logarithm

Write each of the following as a single logarithm:

(a) $\log_a 7 + 4\log_a 3$
(b) $\tfrac{2}{3}\log_a 8 - \log_a(3^4 - 8)$
(c) $\log_a x + \log_a 9 + \log_a(x^2+1) - \log_a 5$

Solution   (a) $\log_a 7 + 4\log_a 3 = \log_a 7 + \log_a 3^4$ Property (6).
$$= \log_a 7 + \log_a 81$$
$$= \log_a(7 \cdot 81)$$ Property (3).
$$= \log_a 567$$

(b) $\tfrac{2}{3}\log_a 8 - \log_a(3^4 - 8) = \log_a 8^{2/3} - \log_a(81 - 8)$ Property (6).
$$= \log_a 4 - \log_a 73$$
$$= \log_a\left(\tfrac{4}{73}\right)$$ Property (4).

(c) $\log_a x + \log_a 9 + \log_a(x^2+1) - \log_a 5 = \log_a 9x + \log_a(x^2+1) - \log_a 5$
$$= \log_a[9x(x^2+1)] - \log_a 5$$
$$= \log_a\left[\frac{9x(x^2+1)}{5}\right]$$ ▬

Warning: A common error made by some students is to express the logarithm of a sum as the sum of logarithms:

$$\log_a(M+N) \quad \text{is not equal to} \quad \log_a M + \log_a N$$

Correct statement. $\log_a(MN) = \log_a M + \log_a N$ Property (3).

Another common error is to express the difference of logarithms as the quotient of logarithms:

$$\log_a M - \log_a N \quad \text{is not equal to} \quad \frac{\log_a M}{\log_a N}$$

Correct statement. $\log_a M - \log_a N = \log_a\left(\dfrac{M}{N}\right)$ Property (4).

Now work Problem 23.

or

$$\log_{\sqrt{2}} \sqrt{5} = \frac{\ln \sqrt{5}}{\ln \sqrt{2}} = \frac{\frac{1}{2} \ln 5}{\frac{1}{2} \ln 2} \approx \frac{1.6094}{0.6931} = 2.3219$$

Now work Problem 35.

4   We also use the Change-of-Base Formula to graph logarithmic functions whose base is neither 10 nor e.

E X A M P L E   10

Graphing a Logarithmic Function Whose Base Is Neither 10 Nor *e*

Use a graphing utility to graph: $y = \log_2 x$

FIGURE 31

Solution

Since graphing utilities only have logarithms with the base 10 or the base e, we need to use the Change-of-Base Formula to express $y = \log_2 x$ in terms of logarithms with base 10 or base e. We can graph either $y = \ln x/\ln 2$ or $y = \log x/\log 2$ to obtain the graph of $y = \log_2 x$. See Figure 31.

Check: Verify that $y = \ln x/\ln 2$ and $y = \log x/\log 2$ result in the same graph by graphing each one on the same viewing rectangle.

Now work Problem 43.

SUMMARY OF PROPERTIES OF LOGARITHMS

In the list that follows, $a > 0$, $a \neq 1$, and $b > 0$, $b \neq 1$; also, $M > 0$ and $N > 0$.

Definition	$y = \log_a x$ means $x = a^y$
Properties of logarithms	$\log_a 1 = 0$; $\log_a a = 1$
	$a^{\log_a M} = M$; $\log_a a^r = r$
	$\log_a(MN) = \log_a M + \log_a N$
	$\log_a\left(\dfrac{M}{N}\right) = \log_a M - \log_a N$
	$\log_a\left(\dfrac{1}{N}\right) = -\log_a N$
	$\log_a M^r = r \log_a M$
Change-of-Base Formula	$\log_a M = \dfrac{\log_b M}{\log_b a}$

HISTORICAL FEATURE   Logarithms were invented about 1590 by John Napier (1550–1617) and Jobst Bürgi (1552–1632), working independently. Napier, whose work had the greater influence, was a Scottish lord, a secretive man whose neighbors were inclined to believe him to be in league with the devil. His approach to logarithms was quite different from ours; it was based on the relationship between arithmetic and geometric sequences (see the chapter on induction and sequences), and not on the inverse function relationship of logarithms to exponential functions (described in Section 4.3). Napier's tables, published in 1614, listed what would now be called *natural logarithms* of sines and were rather difficult to use. A London professor, Henry Briggs, became interested in the tables and visited Napier. In their conversations, they developed the idea of common logarithms, and Briggs then converted Napier's tables into tables of common logarithms, which were published in 1617. Their importance for calculation was immediately recognized, and by 1650 they were being printed as far away as China. They remained an important calculation tool until the advent of the inexpensive handheld calculator about 1972, which has decreased their calculational, but not their theoretical, importance.

A side effect of the invention of logarithms was the popularization of the decimal system of notation for real numbers.

4.4 EXERCISES

In Problems 1–12, suppose ln 2 = a and ln 3 = b. Use properties of logarithms to write each logarithm in terms of a and b.

1. $\ln 6$

2. $\ln \frac{2}{3}$

3. $\ln 1.5$

4. $\ln 0.5$

5. $\ln 2e$

6. $\ln\left(\frac{3}{e}\right)$

7. $\ln 12$

8. $\ln 24$

9. $\ln \sqrt[5]{18}$

10. $\ln \sqrt[4]{48}$

11. $\log_2 3$

12. $\log_3 2$

In Problems 13–22, write each expression as a sum and/or difference of logarithms. Express powers as factors.

13. $\ln(x^2\sqrt{1-x})$

14. $\ln(x\sqrt{1+x^2})$

15. $\log_2\left(\frac{x^3}{x-3}\right)$

16. $\log_5\left(\frac{\sqrt[3]{x^2+1}}{x^2-1}\right)$

17. $\log\left[\frac{x(x+2)}{(x+3)^2}\right]$

18. $\log\frac{x^3\sqrt{x+1}}{(x-2)^2}$

19. $\ln\left[\frac{x^2-x-2}{(x+4)^2}\right]^{1/3}$

20. $\ln\left[\frac{(x-4)^2}{x^2-1}\right]^{2/3}$

21. $\ln\frac{5x\sqrt{1-3x}}{(x-4)^3}$

22. $\ln\left[\frac{5x^2\sqrt[3]{1-x}}{4(x+1)^2}\right]$

In Problems 23–32, write each expression as a single logarithm.

23. $3\log_5 u + 4\log_5 v$

24. $\log_3 u^2 - \log_3 v$

25. $\log_{1/2}\sqrt{x} - \log_{1/2} x^3$

26. $\log_2\left(\frac{1}{x}\right) + \log_2\left(\frac{1}{x^2}\right)$

27. $\ln\left(\frac{x}{x-1}\right) + \ln\left(\frac{x+1}{x}\right) - \ln(x^2-1)$

28. $\log\left(\frac{x^2+2x-3}{x^2-4}\right) - \log\left(\frac{x^2+7x+6}{x+2}\right)$

29. $8\log_2\sqrt{3x-2} - \log_2\left(\frac{4}{x}\right) + \log_2 4$

30. $21\log_3\sqrt[3]{x} + \log_3 9x^2 - \log_5 25$

31. $2\log_a 5x^3 - \frac{1}{2}\log_a(2x+3)$

32. $\frac{1}{3}\log(x^3+1) + \frac{1}{2}\log(x^2+1)$

33. Write the exponential model, $y = ab^x$, as a linear model.

34. Is the logarithmic model, $y = a + b\ln x$, linear? If so, what is the slope? What is the *y*-intercept?

In Problems 35–42, use the Change-of-Base Formula and a calculator to evaluate each logarithm. Round your answer to three decimal places.

35. $\log_3 21$ **36.** $\log_5 18$ **37.** $\log_{1/3} 71$ **38.** $\log_{1/2} 15$

39. $\log_{\sqrt{2}} 7$ **40.** $\log_{\sqrt{5}} 8$ **41.** $\log_\pi e$ **42.** $\log_\pi \sqrt{2}$

For Problems 43–48, graph each function using a graphing utility and the Change-of-Base Formula.

43. $y = \log_4 x$ **44.** $y = \log_5 x$ **45.** $y = \log_2(x + 2)$ **46.** $y = \log_4(x - 3)$

47. $y = \log_{x-1}(x + 1)$ **48.** $y = \log_{x+2}(x - 2)$

49. Show that: $\log_a(x + \sqrt{x^2 - 1}) + \log_a(x - \sqrt{x^2 - 1}) = 0$

50. Show that: $\log_a(\sqrt{x} + \sqrt{x - 1}) + \log_a(\sqrt{x} - \sqrt{x - 1}) = 0$

51. Show that: $\ln(1 + e^{2x}) = 2x + \ln(1 + e^{-2x})$

52. If $f(x) = \log_a x$, show that: $\dfrac{f(x + h) - f(x)}{h} = \log_a\left(1 + \dfrac{h}{x}\right)^{1/h}$, $h \neq 0$

53. If $f(x) = \log_a x$, show that: $-f(x) = \log_{1/a} x$ **54.** If $f(x) = \log_a x$, show that: $f(1/x) = -f(x)$

55. If $f(x) = \log_a x$, show that: $f(AB) = f(A) + f(B)$ **56.** If $f(x) = \log_a x$, show that: $f(x^\alpha) = \alpha f(x)$

In Problems 57–66, express y as a function of x. The constant C is a positive number.

57. $\ln y = \ln x + \ln C$ **58.** $\ln y = \ln(x + C)$ **59.** $\ln y = \ln x + \ln(x + 1) + \ln C$

60. $\ln y = 2\ln x - \ln(x + 1) + \ln C$ **61.** $\ln y = 3x + \ln C$ **62.** $\ln y = -2x + \ln C$

63. $\ln(y - 3) = -4x + \ln C$ **64.** $\ln(y + 4) = 5x + \ln C$

65. $3\ln y = \frac{1}{2}\ln(2x + 1) - \frac{1}{3}\ln(x + 4) + \ln C$ **66.** $2\ln y = -\frac{1}{2}\ln x + \frac{1}{3}\ln(x^2 + 1) + \ln C$

67. Find the value of $\log_2 3 \cdot \log_3 4 \cdot \log_4 5 \cdot \log_5 6 \cdot \log_6 7 \cdot \log_7 8$.

68. Find the value of $\log_2 4 \cdot \log_4 6 \cdot \log_6 8$.

69. Find the value of $\log_2 3 \cdot \log_3 4 \cdot \ldots \cdot \log_n(n + 1) \cdot \log_{n+1} 2$.

70. Find the value of $\log_2 2 \cdot \log_2 4 \cdot \ldots \cdot \log_2 2^n$.

71. Show that $\log_a(M/N) = \log_a M - \log_a N$, where a, M, and N are positive real numbers, with $a \neq 1$.

72. Show that $\log_a(1/N) = -\log_a N$, where a and N are positive real numbers, with $a \neq 1$.

73. Find the domain of $f(x) = \log_a x^2$ and the domain of $g(x) = 2\log_a x$. Since $\log_a x^2 = 2\log_a x$, how do you reconcile the fact that the domains are not equal? Write a brief explanation.

4.5 LOGARITHMIC AND EXPONENTIAL EQUATIONS

> **1** Solve Logarithmic Equations
> **2** Solve Exponential Equations
> **3** Solve Logarithmic and Exponential Equations Using a Graphing Utility

Logarithmic Equations

1 Equations that contain terms of the form $\log_a x$, where a is a positive real number, with $a \neq 1$, are often called **logarithmic equations.**

As before, our practice will be to solve equations, whenever possible, by finding exact solutions using algebraic methods. In such cases, we will also verify the solution obtained by using a graphing utility. When algebraic methods cannot be used, approximate solutions will be obtained using a graphing utility. The reader is encouraged to pay particular attention to the form of equations for which exact solutions are possible.

M I S S I O N   P O S S I B L E

McNewton's Coffee

Your team has been called in to solve a problem encountered by a fast food restaurant. They believe that their coffee should be brewed at 170° Fahrenheit. However, at that temperature it is too hot to drink, and a customer who accidentally spills the coffee on himself might receive third-degree burns.

What they need is a special container that will heat the water to 170°, brew the coffee at that temperature, then cool it quickly to a drinkable temperature, say 140°F, and hold it there, or at least keep it at or above 120°F for a reasonable period of time without further cooking. To cool down the coffee, three companies have submitted proposals with these specifications:

(a) The CentiKeeper Company has a container that will reduce the temperature of a liquid from 200°F to 100°F in 90 minutes by maintaining a constant temperature of 70°F.

(b) The TempControl Company has a container that will reduce the temperature of a liquid from 200°F to 110°F in 60 minutes by maintaining a constant temperature of 60°F.

(c) The Hot'n'Cold, Inc., has a container that will reduce the temperature of a liquid from 210°F to 90°F in 30 minutes by maintaining a constant temperature of 50°F.

Your job is to make a recommendation as to which container to purchase. For this you will need Newton's law of cooling which follows:

$$u(t) = T + (u_0 - T)e^{kt}, \quad k < 0$$

In this formula, T represents the temperature of the surrounding medium, u_0 is the initial temperature of the heated object, t is the length of time in minutes, k is a negative constant, and u represents the temperature at time t.

1. Use Newton's Law of Cooling to find the constant k of the formula for each container.
2. Use a graphing utility to graph each relation.
3. How long does it take each container to lower the coffee temperature from 170°F to 140°F?
4. How long will the coffee temperature remain between 120°F and 140°F?
5. On the basis of this information, which company should get the contract with McNewton's? What are your reasons?
6. Define "capital cost" and "operating cost." How might they affect your choice?

E X A M P L E   1

Solving a Logarithmic Equation

Solve: $\log_3(4x - 7) = 2$.

Solution   We can obtain an exact solution by changing the logarithm to exponential form.

$$\log_3(4x - 7) = 2$$
$$4x - 7 = 3^2$$
$$4x - 7 = 9$$
$$4x = 16$$
$$x = 4$$

We verify the solution by graphing $Y_1 = \log_3(4x - 7)$ and $Y_2 = 2$ to determine where the graphs intersect. See Figure 32. The graphs intersect at $x = 4$.

FIGURE 32

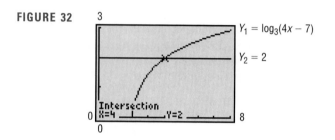

E X A M P L E   2

Solving a Logarithmic Equation

Solve: $2 \log_5 x = \log_5 9$.

Solution   Because each logarithm is to the same base, 5, we can obtain an exact solution as follows:

$$2 \log_5 x = \log_5 9$$
$$\log_5 x^2 = \log_5 9 \qquad \text{Property (6), Section 4.4.}$$
$$x^2 = 9 \qquad \text{Property (8), Section 4.4.}$$

$x = 3$ or $x = -3$ Recall that logarithms of negative numbers are not defined, so, in the expression $2 \log_5 x$, x must be positive. Therefore, -3 is extraneous and we discard it.

The equation has only one solution, 3.

You should verify for yourself that 3 is the only solution using a graphing utility.

Now work Problem 1.

E X A M P L E   3   Solving a Logarithmic Equation

Solve: $\log_4(x + 3) + \log_4(2 - x) = 1$.

Solution   To obtain an exact solution, we need to express the left side as a single logarithm. Then we will change the expression to exponential form.

$$\log_4(x + 3) + \log_4(2 - x) = 1$$
$$\log_4[(x + 3)(2 - x)] = 1 \qquad \text{Property (3), Section 4.4.}$$
$$(x + 3)(2 - x) = 4^1 = 4$$
$$-x^2 - x + 6 = 4$$
$$x^2 + x - 2 = 0$$
$$(x + 2)(x - 1) = 0$$
$$x = -2 \quad \text{or} \quad x = 1 \qquad \blacksquare$$

You should verify that both of these are solutions using a graphing utility.

 Now work Problem 11.

Care must be taken when solving logarithmic equations algebraically. Be sure to check each apparent solution in the original equation and discard any that are extraneous. In the expression $\log_a M$, remember that a and M are positive and $a \neq 1$.

Exponential Equations

2   Equations that involve terms of the form a^x, $a > 0$, $a \neq 1$, are often referred to as **exponential equations.** Such equations sometimes can be solved by appropriately applying the laws of exponents and equation (1), namely,

$$\text{If } a^u = a^v, \quad \text{then } u = v. \qquad (1)$$

To use equation (1), each side of the equality must be written with the same base.

E X A M P L E   4   Solving an Exponential Equation

Solve the equation: $3^{x+1} = 81$.

Solution   Since $81 = 3^4$, we can write the equation as

$$3^{x+1} = 81 = 3^4$$

Now we have the same base, 3, on each side, so we can apply (1) to obtain

$$x + 1 = 4$$
$$x = 3$$

We verify the solution by graphing $Y_1 = 3^{x+1}$ and $Y_2 = 81$ to determine where the graphs intersect. See Figure 33. The graphs intersect at $x = 3$.

FIGURE 33

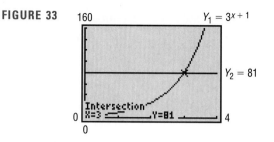

Now work Problem 19.

E X A M P L E   5

Solving an Exponential Equation

Solve the equation: $e^{-x^2} = (e^x)^2 \cdot \dfrac{1}{e^3}.$

Solution   We use some laws of exponents first to get the same base e on each side:

$$e^{-x^2} = (e^x)^2 \cdot \frac{1}{e^3} = e^{2x} \cdot e^{-3} = e^{2x-3}$$

Now apply (1) to get

$$-x^2 = 2x - 3$$
$$x^2 + 2x - 3 = 0$$
$$(x + 3)(x - 1) = 0$$
$$x = -3 \quad \text{or} \quad x = 1$$

You should verify these solutions using a graphing utility.

E X A M P L E   6

Solving an Exponential Equation

Solve the equation: $4^x - 2^x - 12 = 0.$

Solution   We note that $4^x = (2^2)^x = 2^{2x} = (2^x)^2$, so the equation is actually quadratic in form, and we can rewrite it as

$$(2^x)^2 - 2^x - 12 = 0$$

Now we can factor as usual:

$$(2^x - 4)(2^x + 3) = 0$$
$$2^x - 4 = 0 \quad \text{or} \quad 2^x + 3 = 0$$
$$2^x = 4 \qquad\qquad 2^x = -3$$

The equation on the left has the solution $x = 2$, since $2^x = 4 = 2^2$; the equation on the right has no solution, since $2^x > 0$ for all x. ∎

In each of the preceding three examples, we were able to write each exponential expression using the same base, obtaining exact solutions to the equation. When this is not possible, logarithms can sometimes be used to obtain the solution.

E X A M P L E 7   **Solving an Exponential Equation**

Solve for x: $2^x = 5$.

Solution   We write the exponential equation as the equivalent logarithmic equation:

$$2^x = 5$$

$$x = \log_2 5 = \underset{\uparrow}{\frac{\ln 5}{\ln 2}}$$

Change-of-Base Formula (10).

Alternatively, we can solve the equation $2^x = 5$ by taking the natural logarithm (or common logarithm) of each side. Taking the natural logarithm,

$$2^x = 5$$
$$\ln 2^x = \ln 5$$
$$x \ln 2 = \ln 5$$
$$x = \frac{\ln 5}{\ln 2}$$

Using a calculator, the solution, rounded to three decimal places, is

$$x = \frac{\ln 5}{\ln 2} \approx 2.322$$

∎

Now work Problem 33.

E X A M P L E 8   **Solving an Exponential Equation**

Solve for x: $8 \cdot 3^x = 5$.

Solution
$$8 \cdot 3^x = 5 \qquad \text{Isolate } 3^x \text{ on the left side.}$$
$$3^x = \tfrac{5}{8} \qquad \text{Proceed as in Example 7.}$$
$$x = \log_3\left(\tfrac{5}{8}\right) = \frac{\ln \tfrac{5}{8}}{\ln 3}$$

Using a calculator, the solution, rounded to three decimal places, is

$$x = \frac{\ln\left(\tfrac{5}{8}\right)}{\ln 3} \approx -0.428$$

∎

E X A M P L E   9

Solving an Exponential Equation

Solve for x: $5^{x-2} = 3^{3x+2}$.

Solution   Because the bases are different, we take the natural logarithm of each side and apply appropriate properties of logarithms. The result is an equation in x that we can solve.

$$5^{x-2} = 3^{3x+2}$$
$$\ln 5^{x-2} = \ln 3^{3x+2} \qquad \text{Property (7).}$$
$$(x - 2)\ln 5 = (3x + 2)\ln 3 \qquad \text{Property (6).}$$
$$(\ln 5)x - 2\ln 5 = (3\ln 3)x + 2\ln 3$$
$$(\ln 5 - 3\ln 3)x = 2\ln 3 + 2\ln 5$$
$$x = \frac{2(\ln 3 + \ln 5)}{\ln 5 - 3\ln 3} \approx -3.212$$

Now work Problem 41.

Graphing Utility Solutions

3 The techniques introduced in this section only apply to certain types of logarithmic and exponential equations. Solutions for other types are usually studied in calculus, using numerical methods. However, we can use a graphing utility to approximate the solution.

E X A M P L E   10

Solving Equations Using a Graphing Utility

Solve: $\log_3 x + \log_4 x = 4$.
Express the solution(s) correct to two decimal places.

Solution   The solution is found by graphing

$$Y_1 = \log_3 x + \log_4 x = \frac{\log x}{\log 3} + \frac{\log x}{\log 4} \text{ and } Y_2 = 4.$$

(Remember you must use the Change-of-Base Formula to graph Y_1). Y_1 is an increasing function (do you know why?) and so there is only one point of intersection for Y_1 and Y_2. Figure 34 shows the graphs of Y_1 and Y_2. The solution is 11.60 correct to two decimal places.

FIGURE 34

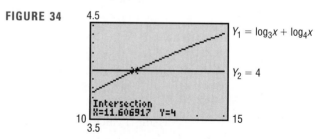

Can you discover an algebraic solution to Example 10?
[**Hint:** Factor $\log x$ from Y_1.]

E X A M P L E   11   Solving Equations Using a Graphing Utility

Solve: $x + e^x = 2$.
Express the solution(s) correct to two decimal places.

Solution   The solution is found by graphing $Y_1 = x + e^x$ and $Y_2 = 2$. Y_1 is an increasing function (do you know why?) and so there is only one point of intersection for Y_1 and Y_2. Figure 35 shows the graphs of Y_1 and Y_2. The solution is 0.44 correct to two decimal places.

FIGURE 35

4.5 EXERCISES

In Problems 1–58, solve each equation. Verify your solution using a graphing utility.

1. $\log_2(2x + 1) = 3$

2. $\log_3(3x - 2) = 2$

3. $\log_3(x^2 + 1) = 2$

4. $\log_5(x^2 + x + 4) = 2$

5. $\frac{1}{2}\log_3 x = 2\log_3 2$

6. $-2\log_4 x = \log_4 9$

7. $2\log_5 x = 3\log_5 4$

8. $3\log_2 x = -\log_2 27$

9. $3\log_2(x - 1) + \log_2 4 = 5$

10. $2\log_3(x + 4) - \log_3 9 = 2$

11. $\log x + \log(x + 15) = 2$

12. $\log_4 x + \log_4(x - 3) = 1$

13. $\log_x 4 = 2$

14. $\log_x\left(\frac{1}{8}\right) = 3$

15. $\log_3(x - 1)^2 = 2$

16. $\log_2(x + 4)^3 = 6$

17. $\log_{1/2}(3x + 1)^{1/3} = -2$

18. $\log_{1/3}(1 - 2x)^{1/2} = -1$

19. $2^{2x+1} = 4$

20. $5^{1-2x} = \frac{1}{5}$

21. $3^{x^3} = 9^x$

22. $4^{x^2} = 2^x$

23. $8^{x^2-2x} = \frac{1}{2}$

24. $9^{-x} = \frac{1}{3}$

25. $2^x \cdot 8^{-x} = 4^x$

26. $\left(\frac{1}{2}\right)^{1-x} = 4$

27. $2^{2x} + 2^x - 12 = 0$

28. $3^{2x} + 3^x - 2 = 0$

29. $3^{2x} + 3^{x+1} - 4 = 0$

30. $4^x - 2^x = 0$

31. $4^x = 8$

32. $9^{2x} = 27$

33. $2^x = 10$

34. $3^x = 14$

35. $8^{-x} = 1.2$

36. $2^{-x} = 1.5$

37. $3^{1-2x} = 4^x$

38. $2^{x+1} = 5^{1-2x}$

39. $\left(\frac{3}{5}\right)^x = 7^{1-x}$

40. $\left(\frac{4}{3}\right)^{1-x} = 5^x$

41. $1.2^x = (0.5)^{-x}$

42. $(0.3)^{1+x} = 1.7^{2x-1}$

43. $\pi^{1-x} = e^x$

44. $e^{x+3} = \pi^x$

45. $5(2^{3x}) = 8$

46. $0.3(4^{0.2x}) = 0.2$

47. $400e^{0.2x} = 600$

48. $500e^{0.3x} = 600$

49. $\log_a(x - 1) - \log_a(x + 6) = \log_a(x - 2) - \log_a(x + 3)$

50. $\log_a x + \log_a(x - 2) = \log_a(x + 4)$

51. $\log_{1/3}(x^2 + x) - \log_{1/3}(x^2 - x) = -1$

52. $\log_4(x^2 - 9) - \log_4(x + 3) = 3$

53. $\log_2 8^x = -3$

54. $\log_3 3^x = -1$

55. $\log_2(x^2 + 1) - \log_4 x^2 = 1$

[**Hint:** Change $\log_4 x^2$ to base 2.]

56. $\log_2(3x + 2) - \log_4 x = 3$

57. $\log_{16} x + \log_4 x + \log_2 x = 7$

58. $\log_9 x + 3\log_3 x = 14$

In Problems 59–74, use a graphing utility to solve each equation. Express your answer correct to two decimal places.

59. $\log_5 x + \log_3 x = 1$　　　　　　　　　　　　**60.** $\log_2 x + \log_6 x = 3$

61. $\log_5(x + 1) - \log_4(x - 2) = 1$　　　　　　　**62.** $\log_2(x - 1) - \log_6(x + 2) = 2$

63. $e^x = -x$　　　　　**64.** $e^{2x} = x + 2$　　　　　**65.** $e^x = x^2$　　　　　**66.** $e^x = x^3$

67. $\ln x = -x$　　　　**68.** $\ln 2x = -x + 2$　　　　**69.** $\ln x = x^3 - 1$　　　　**70.** $\ln x = -x^2$

71. $e^x + \ln x = 4$　　　**72.** $e^x - \ln x = 4$　　　　**73.** $e^{-x} = \ln x$　　　　**74.** $e^{-x} = -\ln x$

4.6　COMPOUND INTEREST

1　Determine the Future Value of a Lump Sum of Money
2　Calculate Effective Rates of Return
3　Determine the Present Value of a Lump Sum of Money
4　Determine the Time Required to Double or Triple a Lump Sum of Money

1　Interest is money paid for the use of money. The total amount borrowed (whether by an individual from a bank in the form of a loan or by a bank from an individual in the form of a savings account) is called the **principal.** The **rate of interest,** expressed as a percent, is the amount charged for the use of the principal for a given period of time, usually on a yearly (that is, per annum) basis.

If a principal of P dollars is borrowed for a period of t years at a per annum interest rate r, expressed as a decimal, the interest I charged is

Simple Interest Formula

$$I = Prt \qquad\qquad (1)$$

Interest charged according to formula (1) is called **simple interest.**

In working with problems involving interest we use the term **payment period** as follows:

Annually	Once per year	Monthly	12 times per year
Semiannually	Twice per year	Daily	365 times per year*
Quarterly	4 times per year		

When the interest due at the end of a payment period is added to the principal so that the interest computed at the end of the next payment period is based on this new principal amount (old principal + interest), the interest is said to have been **compounded.** Thus, **compound interest** is interest paid on previously earned interest.

EXAMPLE 1　　Computing Compound Interest

A credit union pays interest of 8% per annum compounded quarterly on a certain savings plan. If $1000 is deposited in such a plan and the interest is left to accumulate, how much is in the account after 1 year?

*Most banks use a 360 day "year." Why do you think they do?

Solution   We use the simple interest formula, $I = Prt$. The principal P is $1000 and the rate of interest is $8\% = 0.08$. After the first quarter of a year, the time t is $\frac{1}{4}$ year, so the interest earned is

$$I = Prt = (\$1000)(0.08)(\tfrac{1}{4}) = \$20$$

The new principal is $P + I = \$1000 + \$20 = \$1020$. At the end of the second quarter, the interest on this principal is

$$I = (\$1020)(0.08)(\tfrac{1}{4}) = \$20.40$$

At the end of the third quarter, the interest on the new principal of $1020 + $20.40 = $1040.40 is

$$I = (\$1040.40)(0.08)(\tfrac{1}{4}) = \$20.81$$

Finally, after the fourth quarter, the interest is

$$I = (\$1061.21)(0.08)(\tfrac{1}{4}) = \$21.22$$

Thus, after 1 year the account contains $1082.43. ▬

The pattern of the calculations performed in Example 1 leads to a general formula for compound interest. To fix our ideas, let P represent the principal to be invested at a per annum interest rate r, which is compounded n times per year. (For computing purposes, r is expressed as a decimal.) The interest earned after each compounding period is the principal times r/n. Thus, the amount A after one compounding period is

$$A = P + P\left(\frac{r}{n}\right) = P\left(1 + \frac{r}{n}\right)$$

After two compounding periods, the amount A, based on the new principal $P(1 + r/n)$, is

$$A = \underbrace{P\left(1 + \frac{r}{n}\right)}_{\text{New principal.}} + \underbrace{P\left(1 + \frac{r}{n}\right)\left(\frac{r}{n}\right)}_{\text{Interest on new principal.}} = P\left(1 + \frac{r}{n}\right)\left(1 + \frac{r}{n}\right) = P\left(1 + \frac{r}{n}\right)^2$$

After three compounding periods,

$$A = P\left(1 + \frac{r}{n}\right)^2 + P\left(1 + \frac{r}{n}\right)^2\left(\frac{r}{n}\right) = P\left(1 + \frac{r}{n}\right)^2\left(1 + \frac{r}{n}\right) = P\left(1 + \frac{r}{n}\right)^3$$

Continuing this way, after n compounding periods (1 year),

$$A = P\left(1 + \frac{r}{n}\right)^n$$

Because t years will contain $n \cdot t$ compounding periods, after t years we have

$$A = P\left(1 + \frac{r}{n}\right)^{nt}$$

> **Theorem**   **Compound Interest Formula**
>
> The amount A after t years due to a principal P invested at an annual interest rate r compounded n times per year is
>
> $$A = P\left(1 + \frac{r}{n}\right)^{nt} \qquad (2)$$

E X A M P L E   2

Comparing Investments Using Different Compounding Periods

Investing $1000 at an annual rate of 10% compounded annually, quarterly, monthly, and daily will yield the following amounts after 1 year:

Annual compounding: $A = P(1 + r)$
$$= (\$1000)(1 + 0.10) = \$1100.00$$

Quarterly compounding: $A = P\left(1 + \dfrac{r}{4}\right)^4$
$$= (\$1000)(1 + 0.025)^4 = \$1103.81$$

Monthly compounding: $A = P\left(1 + \dfrac{r}{12}\right)^{12}$
$$= (\$1000)(1 + 0.00833)^{12} = \$1104.71$$

Daily compounding: $A = P\left(1 + \dfrac{r}{365}\right)^{365}$
$$= (\$1000)(1 + 0.000274)^{365} = \$1105.16$$

 Now work Problem 1.

From Example 2, we can see that the effect of compounding more frequently is that the amount after 1 year is higher: $1000 compounded 4 times a year at 10% results in $1103.81; $1000 compounded 12 times a year at 10% results in $1104.71; and $1000 compounded 365 times a year at 10% results in $1105.16. This leads to the following question: What would happen to the amount after 1 year if the number of times the interest is compounded were increased without bound?

Let's find the answer. Suppose P is the principal, r is the per annum interest rate, and n is the number of times the interest is compounded each year. The amount after 1 year is

$$A = P\left(1 + \frac{r}{n}\right)^n$$

Now suppose that the number n of times the interest is compounded per year gets larger and larger; that is, suppose that $n \to \infty$. Then,

$$A = P\left(1 + \frac{r}{n}\right)^n = P\left[1 + \frac{1}{n/r}\right]^n = P\left(\left[1 + \frac{1}{n/r}\right]^{n/r}\right)^r = P\left[\left(1 + \frac{1}{h}\right)^h\right]^r \qquad (3)$$
$$\underset{h = \frac{n}{r}}{\uparrow}$$

In (3), as $n \to \infty$, then $h = n/r \to \infty$, and the expression in brackets equals e. [Refer to (2) on p. 317]. Thus, $A \to Pe^r$. Table 7 compares $(1 + r/n)^n$, for large values of n, to e^r for $r = 0.05$, $r = 0.10$, $r = 0.15$, and $r = 1$. The larger n gets, the closer $(1 + r/n)^n$ gets to e^r. Thus, no matter how frequent the compounding, the amount after 1 year has the definite ceiling Pe^r.

TABLE 7	$\left(1 + \dfrac{r}{n}\right)^n$			
	$n = 100$	$n = 1000$	$n = 10,000$	e^r
$r = 0.05$	1.0512579	1.05127	1.051271	1.0512711
$r = 0.10$	1.1051157	1.1051654	1.1051703	1.1051709
$r = 0.15$	1.1617037	1.1618212	1.1618329	1.1618342
$r = 1$	2.7048138	2.7169239	2.7181459	2.7182818

When interest is compounded so that the amount after 1 year is Pe^r, we say the interest is **compounded continuously.**

Theorem   Continuous Compounding

The amount A after t years due to a principal P invested at an annual interest rate r compounded continuously is

$$A = Pe^{rt} \qquad (4)$$

E X A M P L E   3   Using Continuous Compounding

The amount A that results from investing a principal P of $1000 at an annual rate r of 10% compounded continuously for a time t of 1 year is

$$A = \$1000e^{0.10} = (\$1000)(1.10517) = \$1105.17$$

Now work Problem 9.

2   The **effective rate of interest** is the equivalent annual simple rate of interest that would yield the same amount as compounding after 1 year. For example, based on Example 3, a principal of $1000 will result in $1105.17 at a rate of 10% compounded continuously. To get this same amount using a simple rate of interest would require that interest of $1105.17 − $1000.00 = $105.17 be earned on the principal. Since $105.17 is 10.517% of $1000, a simple rate of interest of 10.517% is needed to equal 10% compounded continuously. Thus, the effective rate of interest of 10% compounded continuously is 10.517%.

Based on the results of Examples 2 and 3, we find the following comparisons:

	Annual Rate	**Effective Rate**
Annual compounding	10%	10%
Quarterly compounding	10%	10.381%
Monthly compounding	10%	10.471%
Daily compounding	10%	10.516%
Continuous compounding	10%	10.517%

Now work Problem 21.

17. To get $80 after $3\frac{1}{4}$ years at 9% compounded continuously

18. To get $800 after $2\frac{1}{2}$ years at 8% compounded continuously

19. To get $400 after 1 year at 10% compounded continuously

20. To get $1000 after 1 year at 12% compounded continuously

21. Find the effective rate of interest for $5\frac{1}{4}$% compounded quarterly.

22. What interest rate compounded quarterly will give an effective interest rate of 7%?

23. What annual rate of interest is required to double an investment in 3 years? Verify your answer using a graphing utility.

24. What annual rate of interest is required to double an investment in 10 years? Verify your answer using a graphing utility.

In Problems 25–28, which of the two rates would yield the larger amount in 1 year?
[**Hint:** Start with a principal of $10,000 in each instance.]

25. 6% compounded quarterly or $6\frac{1}{4}$% compounded annually?

26. 9% compounded quarterly or $9\frac{1}{4}$% compounded annually?

27. 9% compounded monthly or 8.8% compounded daily?

28. 8% compounded semiannually or 7.9% compounded daily?

29. How long does it take for an investment to double in value if it is invested at 8% per annum compounded monthly? Compounded continuously? Verify your answer using a graphing utility.

30. How long does it take for an investment to double in value if it is invested at 10% per annum compounded monthly? Compounded continuously? Verify your answer using a graphing utility.

31. If Tanisha has $100 to invest at 8% per annum compounded monthly, how long will it be before she has $150? If the compounding is continuous, how long will it be?

32. If Angela has $100 to invest at 10% per annum compounded monthly, how long will it be before she has $175? If the compounding is continuous, how long will it be?

33. How many years will it take for an initial investment of $10,000 to grow to $25,000? Assume a rate of interest of 6% compounded continuously.

34. How many years will it take for an initial investment of $25,000 to grow to $80,000? Assume a rate of interest of 7% compounded continuously.

35. What will a $90,000 house cost 5 years from now if the inflation rate over that period averages 3% compounded annually?

36. Sears charges 1.25% per month on the unpaid balance for customers with charge accounts (interest is compounded monthly). A customer charges $200 and does not pay her bill for 6 months. What is the bill at that time?

37. Jerome will be buying a new car for $15,000 in 3 years. How much money should he ask his parents for now so that, if he invests it at 5% compounded continuously, he will have enough to buy the car?

38. John will require $3000 in 6 months to pay off a loan that has no prepayment privileges. If he has the $3000 now, how much of it should he save in an account paying 3% compounded monthly so that in 6 months he will have exactly $3000?

39. George is contemplating the purchase of 100 shares of a stock selling for $15 per share. The stock pays no dividends. The history of the stock indicates that is should grow at an annual rate of 15% per year. How much will the 100 shares of stock be worth in 5 years?

40. Tracy is contemplating the purchase of 100 shares of a stock selling for $15 per share. The stock pays no dividends. Her broker says the stock will be worth $20 per share in 2 years. What is the annual rate of return on this investment?

41. A business purchased for $650,000 in 1994 is sold in 1997 for $850,000. What is the annual rate of return for this investment?

42. Tanya has just inherited a diamond ring appraised at $5000. If diamonds have appreciated in value at an annual rate of 8%, what was the value of the ring 10 years ago when the ring was purchased?

43. Jim places $1000 in a bank account that pays 5.6% compounded continuously. After 1 year, will he have enough money to buy a computer system that costs $1060? If another bank will pay Jim 5.9% compounded monthly, is this a better deal?

44. On January 1, Kim places $1000 in a certificate of deposit that pays 6.8% compounded continuously and matures in 3 months. Then Kim places the $1000 and the interest in a passbook account that pays 5.25% compounded monthly. How much does Kim have in the passbook account on May 1?

45. Will invests $2000 in a bond trust that pays 9% interest compounded semiannually. His friend Henry invests $2000 in a Certificate of Deposit (CD) that pays $8\frac{1}{2}$% compounded continuously. Who has more money after 20 years, Will or Henry?

46. Suppose that April has access to an investment that will pay 10% interest compounded continuously. Which is better: To be given $1000 now so that she can take advantage of this investment opportunity or to be given $1325 after 3 years?

47. Colleen and Bill have just purchased a house for $150,000, with the seller holding a second mortgage of $50,000. They promise to pay the seller $50,000 plus all accrued interest 5 years from now. The seller offers them three interest options on the second mortgage:
(a) Simple interest at 12% per annum
(b) $11\frac{1}{2}$% interest compounded monthly
(c) $11\frac{1}{4}$% interest compounded continuously
Which option is best; that is, which results in the least interest on the loan?

48. The First National Bank advertises that it pays interest on saving accounts at the rate of 4.25% compounded daily. Find the effective rate if the bank uses (a) 360 days or (b) 365 days in determining the daily rate.

Problems 49–52 involve zero-coupon bonds. A zero-coupon bond is a bond that is sold now at a discount and will pay its face value at some time in the future when it matures; no interest payments are made.

49. A zero-coupon bond can be redeemed in 20 years for $10,000. How much should you be willing to pay for it now if you want a return of:
(a) 10% compounded monthly?
(b) 10% compounded continuously?

50. A child's grandparents are considering buying a $40,000 face value zero-coupon bond at birth so that she will have enough money for her college education 17 years later. If they want a rate of return of 8% compounded annually, what should they pay for the bond?

51. How much should a $10,000 face value zero-coupon bond, maturing in 10 years, be sold for now if its rate of return is to be 8% compounded annually?

52. If Pat pays $12,485.52 for a $25,000 face value zero-coupon bond that matures in 8 years, what is his annual rate of return?

 53. Explain in your own words what the term *compound interest* means. What does *continuous compounding* mean?

54. Explain in your own words the meaning of present value.

55. Write a program that will calculate the amount after *n* years if a principal *P* is invested at *r*% per annum compounded quarterly. Use it to verify your answers to Problems 1 and 3.

56. Write a program that will calculate the principal needed now to get the amount *A* in *n* years at *r*% per annum compounded daily. Use it to verify your answer to Problem 13.

57. Write a program that will calculate the annual rate of interest required to double an investment in *n* years. Use it to verify your answers to Problems 23 and 24.

58. Write a program that will calculate the number of months required for an initial investment of *x* dollars to grow to *y* dollars at *r*% per annum compounded continuously. Use it to verify your answer to Problems 33 and 34.

59. Time to Double or Triple an Investment The formula

$$y = \frac{\ln m}{n \ln\left(1 + \dfrac{r}{n}\right)}$$

can be used to find the number of years *y* required to multiply an investment *m* times when *r* is the per annum interest rate compounded *n* times a year.
(a) How many years will it take to double the value of an IRA that compounds annually at the rate of 12%?
(b) How many years will it take to triple the value of a savings account that compounds quarterly at an annual rate of 6%?
(c) Give a derivation of this formula.

60. Time to Reach an Investment Goal The formula

$$y = \frac{\ln A - \ln P}{r}$$

can be used to find the number of years *y* required for an investment *P* to grow to a value

A when compounded continuously at an annual rate *r*.

(a) How long will it take to increase an initial investment of $1000 to $8000 at an annual rate of 10%?

(b) What annual rate is required to increase the value of a $2000 IRA to $30,000 in 35 years?

(c) Give a derivation of this formula.

61. Critical Thinking You have just contracted to buy a house and will seek financing in the amount of $100,000. You go to several banks. Bank 1 will lend you $100,000 at the rate of 8.75% amortized over 30 years with a loan origination fee of 1.75%. Bank 2 will lend you $100,000 at the rate of 8.375% amortized over 15 years with a loan origination fee of 1.5%. Bank 3 will lend you $100,000 at the rate of 9.125% amortized over 30 years with no loan origination fee. Bank 4 will lend you $100,000 at the rate of 8.625% amortized over 15 years with no loan origination fee. Which loan would you take? Why? Be sure to have sound reasons for your choice. If the amount of the monthly payment does not matter to you, which loan would you take? Again, have sound reasons for your choice. Use the information in the table to assist you. Compare your final decision with others in the class. Discuss.

	Monthly Payment	Loan Origination Fee
Bank 1	$786.70	$1,750.00
Bank 2	$977.42	$1,500.00
Bank 3	$813.63	$0.00
Bank 4	$990.68	$0.00

4.7 GROWTH AND DECAY

1 Find Equations of Populations That Obey the Law of Uninhibited Growth
2 Find Equations of Populations That Obey the Law of Decay
3 Use Newton's Law of Cooling
4 Use Logistic Growth Models

1 Many natural phenomena have been found to follow the law that an amount *A* varies with time *t* according to

$$A = A_0 e^{kt} \tag{1}$$

where A_0 is the original amount ($t = 0$) and $k \neq 0$ is a constant.

If $k > 0$, then equation (1) states that the amount *A* is increasing over time; if $k < 0$, the amount *A* is decreasing over time. In either case, when an amount *A* varies over time according to equation (1), it is said to follow the **exponential law** or the **law of uninhibited growth** ($k > 0$) **or decay** ($k < 0$). See Figure 40.

FIGURE 40

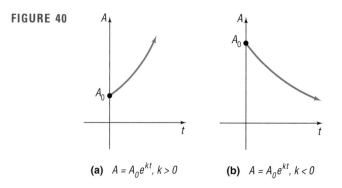

(a) $A = A_0 e^{kt}, k > 0$ (b) $A = A_0 e^{kt}, k < 0$

For example, we saw in Section 4.6 that continuously compounded interest follows the law of uninhibited growth. In this section we shall look at three additional phenomena that follow the exponential law.

Biology

Cell division is a universal process in the growth of living organisms such as amoebas, plants, human skin cells, and many others. Based on an ideal situation in which no cells die and no byproducts are produced, the number of cells present at a given time follows the law of uninhibited growth. Actually, however, after enough time has passed, growth at an exponential rate will cease due to the influence of factors such as lack of living space, dwindling food supply, and so on. The law of uninhibited growth accurately reflects only the early stages of the cell division process.

The cell division process begins with a culture containing N_0 cells. Each cell in the culture grows for a certain period of time and then divides into two identical cells. We assume that the time needed for each cell to divide in two is constant and does not change as the number of cells increases. These cells then grow, and eventually each divides in two, and so on.

A model that gives the number N of cells in the culture after a time t has passed (in the early stages of growth) is

---

Uninhibited Growth of Cells

$$N(t) = N_0 e^{kt} \qquad k > 0 \qquad\qquad (2)$$

---

where k is a positive constant.

In using equation (2) to model the growth of cells, we are using a function that yields positive real numbers, even though we are counting the number of cells, which must be an integer. This is a common practice in many applications.

E X A M P L E   1

Bacterial Growth

A colony of bacteria increases according to the law of uninhibited growth.

(a) If the number of bacteria doubles in 3 hours, how long will it take for the size of the colony to triple?

(b) Using a graphing utility, verify that the population doubles in 3 hours.

(c) Using a graphing utility, approximate the time it takes for the population to double a second time (i.e., increase four times). Does this answer seem reasonable?

Solution   (a) Using formula (2), the number N of cells at a time t is

$$N(t) = N_0 e^{kt}$$

where N_0 is the initial number of bacteria present and k is a positive number. We first seek the number k. The number of cells doubles in 3 hours; thus, we have

$$N(3) = 2N_0$$

But $N(3) = N_0 e^{k(3)}$, so

$$N_0 e^{k(3)} = 2N_0$$
$$e^{3k} = 2$$
$$3k = \ln 2 \quad \text{Write the exponential equation as a logarithm.}$$
$$k = \tfrac{1}{3} \ln 2 \approx \tfrac{1}{3}(0.6931) = 0.2310$$

FIGURE 41

Formula (2) for this growth process is therefore

$$N(t) = N_0 e^{0.2310t}$$

The time t needed for the size of the colony to triple requires that $N = 3N_0$. Thus, we substitute $3N_0$ for N to get

$$3N_0 = N_0 e^{0.2310t}$$
$$3 = e^{0.2310t}$$
$$0.2310t = \ln 3$$
$$t = \frac{1}{0.2310} \ln 3 \approx \frac{1.0986}{0.2310} = 4.756 \text{ hours}$$

FIGURE 42

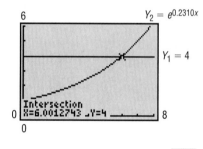

It will take about 4.756 hours for the size of the colony to triple.

(b) Figure 41 shows the graph of $Y_1 = 2$ and $Y_2 = e^{0.2310x}$, where x represents the time in hours. Using INTERSECT, it takes 3 hours for the population to double.

(c) Figure 42 shows the graph of $Y_1 = 4$ and $Y_2 = e^{0.2310x}$, where x represents the time in hours. Using INTERSECT, it takes 6 hours for the population to double a second time. ∎

Now work Problem 1.

Radioactive Decay

Radioactive materials follow the law of uninhibited decay. Thus, the amount A of a radioactive material present at time t is given by the model

Uninhibited Radioactive Decay

$$A = A_0 e^{kt} \qquad k < 0 \tag{3}$$

where A_0 is the original amount of radioactive material and k is a negative number.

All radioactive substances have a specific **half-life,** which is the time required for half of the radioactive substance to decay. In **carbon dating,** we use the fact that all living organisms contain two kinds of carbon, carbon-12 (a stable carbon) and carbon-14 (a radioactive carbon, with a half-life of 5600 years). While an organism is living, the ratio of carbon-12 to carbon-14 is constant. But when an organism dies, the original amount of carbon-12 present remains unchanged, whereas the amount of carbon-14 begins to decrease. This change in the amount of carbon-14 present relative to the amount of carbon-12 present makes it possible to calculate when an organism died.

EXAMPLE 2

Estimating the Age of Ancient Tools

Traces of burned wood found along with ancient stone tools in an archaeological dig in Chile were found to contain approximately 1.67% of the original amount of carbon-14.

(a) If the half-life of carbon-14 is 5600 years, approximately when was the tree cut and burned?
(b) Using a graphing utility, graph the relation between the percentage of carbon-14 remaining and time.
(c) Determine the time that elapses until half of the carbon-14 remains. This answer should equal the half-life of carbon-14.
(d) Verify the answer found in (a).

Solution
(a) Using equation (3), the amount A of carbon-14 present at time t is

$$A = A_0 e^{kt}$$

where A_0 is the original amount of carbon-14 present and k is a negative number. We first seek the number k. To find it, we use the fact that after 5600 years half of the original amount of carbon-14 remains. Thus,

$$\frac{1}{2} A_0 = A_0 e^{k(5600)}$$

$$\frac{1}{2} = e^{5600k}$$

$$5600k = \ln \frac{1}{2}$$

$$k = \frac{1}{5600} \ln \frac{1}{2} \approx -0.000124$$

Formula (3) therefore becomes

$$A = A_0 e^{-0.000124t}$$

If the amount A of carbon-14 now present is 1.67% of the original amount, it follows that

$$0.0167 A_0 = A_0 e^{-0.000124t}$$
$$0.0167 = e^{-0.000124t}$$
$$-0.000124t = \ln 0.0167$$

$$t = \frac{1}{-0.000124} \ln 0.0167 \approx 33{,}000 \text{ years}$$

FIGURE 43

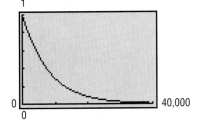

The tree was cut and burned about 33,000 years ago. Some archaeologists use this conclusion to argue that humans lived in the Americas 33,000 years ago, much earlier than is generally accepted.

(b) Figure 43 shows the graph of $y = e^{-0.000124x}$, where y is the fraction of carbon-14 present and x is the time.

(c) By graphing $Y_1 = 0.5$ and $Y_2 = e^{-0.000124x}$, where x is time, and using INTERSECT, we find it takes 5590 years until half the carbon-14 remains.

(d) By graphing $Y_1 = 0.0167$ and $Y_2 = e^{-0.000124x}$, where x is time, and using INTERSECT, we find it takes 33,003 years until 1.67% of the carbon-14 remains. ▬

Now work Problem 3.

Newton's Law of Cooling

3 **Newton's Law of Cooling*** states that the temperature of a heated object decreases exponentially over time toward the temperature of the surrounding medium. That is, the temperature u of a heated object at a given time t can be modeled by the function

Newton's Law of Cooling

$$u(t) = T + (u_0 - T)e^{kt} \qquad k < 0 \qquad (4)$$

where T is the constant temperature of the surrounding medium, u_0 is the initial temperature of the heated object, and k is a negative constant.

E X A M P L E 3

Using Newton's Law of Cooling

An object is heated to 100°C and is then allowed to cool in a room whose air temperature is 30°C.

(a) If the temperature of the object is 80°C after 5 minutes, when will its temperature be 50°C?

(b) Using a graphing utility, graph the relation found between the temperature y and time x.

(c) Using a graphing utility, verify that when $x = 18.6$, then $y = 50$.

(d) Using a graphing utility, determine the elapsed time before the object is 35°C.

(e) What do you notice about y, the temperature, as x, time, increases?

Solution (a) Using equation (4) with $T = 30$ and $u_0 = 100$, the temperature (in degrees Celsius) of the object at time t (in minutes) is

$$u(t) = 30 + (100 - 30)e^{kt} = 30 + 70e^{kt} \qquad (5)$$

*Named after Sir Isaac Newton (1642–1727), one of the cofounders of calculus.

where k is a negative constant. To find k, we use the fact that $u = 80$ when $t = 5$. Then

$$80 = 30 + 70e^{k(5)}$$
$$50 = 70e^{5k}$$
$$e^{5k} = \frac{50}{70}$$
$$5k = \ln \frac{5}{7}$$
$$k = \frac{1}{5} \ln \frac{5}{7} \approx -0.0673$$

Formula (5) therefore becomes

$$u = 30 + 70e^{-0.0673t}$$

Now, we want to find t when $u = 50°C$, so

$$50 = 30 + 70e^{-0.0673t}$$
$$20 = 70e^{-0.0673t}$$
$$e^{-0.0673t} = \frac{20}{70}$$
$$-0.0673t = \ln \frac{2}{7}$$
$$t = \frac{1}{-0.0673} \ln \frac{2}{7} \approx 18.6 \text{ minutes}$$

Thus, the temperature of the object will be 50°C after about 18.6 minutes.

(b) Figure 44 shows the graph of $y = 30 + 70e^{-0.0673x}$, where y is the temperature and x is the time.

(c) By graphing $Y_1 = 50$ and $Y_2 = 30 + 70e^{-0.0673x}$, where x is time, and using INTERSECT, we find it takes $x = 18.6$ minutes for the temperature to cool to 50°C.

(d) By graphing $Y_1 = 35$ and $Y_2 = 30 + 70e^{-0.0673x}$, where x is time, and using INTERSECT, we find it takes $x = 39.21$ minutes for the temperature to cool to 35°C.

(e) As x increases, the value of $e^{-0.0673x}$ approaches zero, so the value of y approaches 30°C.

FIGURE 44

100

0

0   50

Logistic Models

The exponential growth model $A = A_0e^{kt}, k > 0$ assumes uninhibited growth, meaning the value of the function grows without limit. Recall we stated that cell division could be modeled using this function assuming no cells die and no byproducts are produced. However, cell division would eventually be limited by factors such as living space, food supply, and so forth. The **logistic growth model** is an exponential function that can model situations where the growth of the dependent variable is limited. The logistic growth model is given by the function

Logistic Growth Model

$$P(t) = \frac{c}{1 + ae^{-bt}}$$

where a, b, and c are constants with $c > 0$ and $b > 0$.

The number c is called the **carrying capacity** because the value $P(t)$ approaches c as t approaches infinity; that is, $\lim_{t \to \infty} P(t) = c$. Thus, c represents the maximum value the function can attain.

E X A M P L E   4   Fruit Fly Population

Fruit flies are placed in a half-pint milk bottle with a banana (for food) and yeast plants (for food and to provide a stimulus to lay eggs). Suppose the fruit fly population after t days is given by

$$P(t) = \frac{230}{1 + 56.5e^{-0.37t}}$$

(a) Using a graphing utility, graph $P(t)$.

(b) What is the carrying capacity of the half-pint bottle? That is, what is $P(t)$ as $t \to \infty$?

(c) How many fruit flies were initially placed in the half-pint bottle?

(d) When will the population of fruit flies be 180?

FIGURE 45

Solution (a) See Figure 45.

(b) As $t \to \infty$, $e^{-0.37t} \to 0$, and $P(t) \to 230/1$. The carrying capacity of the half-pint bottle is 230 fruit flies.

(c) To find the initial number of fruit flies in the half-pint bottle, we evaluate $P(0)$:

$$P(0) = \frac{230}{1 + 56.5e^{-0.37(0)}}$$

$$= \frac{230}{1 + 56.5}$$

$$= 4$$

So, initially there were four fruit flies in the half-pint bottle.

(d) To determine when the population of fruit flies will be 180, we solve the equation

$$\frac{230}{1 + 56.5e^{-0.37t}} = 180$$

$$230 = 180(1 + 56.5e^{-0.37t})$$

$$1.2778 = 1 + 56.5e^{-0.37t}$$

$$0.2778 = 56.5e^{-0.37t}$$

$$0.0049 = e^{-0.37t}$$

$$\ln(0.0049) = -0.37t$$

$$t \approx 14.4 \text{ days}$$

It will take approximately 14.4 days for the population to reach 180 fruit flies.

We could have also solved this problem by graphing $Y_1 = \dfrac{230}{1 + 56.5e^{-0.37t}}$ and $Y_2 = 180$, using INTERSECT to find the solution shown in Figure 46.

FIGURE 46

Exploration   On the same viewing rectangle, graph $Y_1 = \dfrac{500}{1 + 24e^{-0.03t}}$ and $Y_2 = \dfrac{500}{1 + 24e^{-0.08t}}$. What effect does b have on the logistic growth function?

4.7 EXERCISES

1. **Growth of an Insect Population**   The size P of a certain insect population at time t (in days) obeys the equation $P = 500e^{0.02t}$.
   (a) After how many days will the population reach 1000? 2000?
   (b) Using a graphing utility, graph the relation between P and t. Verify your answers in (a) using INTERSECT.

2. **Growth of Bacteria**   The number N of bacteria present in a culture at time t (in hours) obeys the equation $N = 1000e^{0.01t}$.
   (a) After how many hours will the population equal 1500? 2000?
   (b) Using a graphing utility, graph the relation between N and t. Verify your answers in (a) using INTERSECT.

3. **Radioactive Decay**   Strontium-90 is a radioactive material that decays according to the equation $A = A_0 e^{-0.0244t}$, where A_0 is the initial amount present and A is the amount present at time t (in years). What is the half-life of strontium-90?

4. **Radioactive Decay**   Iodine-131 is a radioactive material that decays according to the equation $A = A_0 e^{-0.087t}$, where A_0 is the initial amount present and A is the amount present at time t (in days). What is the half-life of iodine-131?

5. (a) Use the information in Problem 3 to determine how long it takes for 100 grams of strontium-90 to decay to 10 grams.

   (b) Graph the relation $A = 100e^{-0.0244t}$ and verify your answer found in (a).

6. (a) Use the information in Problem 4 to determine how long it takes for 100 grams of iodine-131 to decay to 10 grams.
   (b) Graph the relation $A = 100e^{-0.087t}$ and verify your answer found in (a).

7. **Growth of a Colony of Mosquitoes**   The population of a colony of mosquitoes obeys the law of uninhibited growth. If there are 1000 mosquitoes initially, and there are 1800 after 1 day, what is the size of the colony after 3 days? How long is it until there are 10,000 mosquitoes?

8. **Bacterial Growth**   A culture of bacteria obeys the law of uninhibited growth. If 500 bacteria are present initially, and there are 800 after 1 hour, how many will be present in the culture after 5 hours? How long is it until there are 20,000 bacteria?

9. **Population Growth**   The population of a southern city follows the exponential law. If the population doubled in size over an 18-month period and the current population is 10,000, what will the population be 2 years from now?

10. **Population Growth**   The population of a midwestern city follows the exponential law. If the population decreased from 900,000 to 800,000 from 1993 to 1995, what will the population be in 1997?

11. **Radioactive Decay**   The half-life of radium is 1690 years. If 10 grams are present now, how much will be present in 50 years?

12. **Radioactive Decay** The half-life of radioactive potassium is 1.3 billion years. If 10 grams are present now, how much will be present in 100 years? In 1000 years?

13. **Estimating the Age of a Tree** A piece of charcoal is found to contain 30% of the carbon-14 it originally had.
    (a) When did the tree from which the charcoal came die? Use 5600 years as the half-life of carbon-14.
    (b) Using a graphing utility, graph the relation between the percentage of carbon-14 remaining and time.
    (c) Using INTERSECT, determine the time that elapses until half of the carbon-14 remains.
    (d) Verify the answer found in (a).

14. **Estimating the Age of a Fossil** A fossilized leaf contains 70% of its normal amount of carbon-14.
    (a) How old is the fossil?
    (b) Using a graphing utility, graph the relation between the percentage of carbon-14 remaining and time.
    (c) Using INTERSECT, determine the time that elapses until half of the carbon-14 remains.
    (d) Verify the answer found in (a).

15. **Cooling Time of a Pizza** A pizza baked at 450°F is removed from the oven at 5:00 P.M. into a room that is a constant 70°F. After 5 minutes, the pizza is at 300°F.
    (a) At what time can you begin eating the pizza if you want its temperature to be 135°F?
    (b) Using a graphing utility, graph the relation between temperature and time.
    (c) Using INTERSECT, determine the time that needs to elapse before the pizza is 160°F?
    (d) TRACE the function for large values of time. What do you notice about y, the temperature?

16. **Newton's Law of Cooling** A thermometer reading 72°F is placed in a refrigerator where the temperature is a constant 38°F. If the thermometer reads 60°F after 2 minutes, what will it read after 7 minutes?
    (a) How long will it take before the thermometer reads 39°F?
    (b) Using a graphing utility, graph the relation between temperature and time.
    (c) Using INTERSECT, determine the time needed to elapse before the thermometer reads 45°F.
    (d) TRACE the function for large values of time. What do you notice about y, the temperature?

17. **Newton's Law of Cooling** A thermometer reading 8°C is brought into a room with a constant temperature of 35°C.
    (a) If the thermometer reads 15°C after 3 minutes, what will it read after being in the room for 5 minutes? For 10 minutes?
    (b) Graph the relation between temperature and time. TRACE to verify your answers are correct.

[**Hint:** You need to construct a formula similar to equation (4).]

18. **Thawing Time of a Steak** A frozen steak has a temperature of 28°F. It is placed in a room with a constant temperature of 70°F. After 10 minutes, the temperature of the steak has risen to 35°F. What will the temperature of the steak be after 30 minutes? How long will it take the steak to thaw to a temperature of 45°F? [See the hint given for Problem 17.] Graph the relation between temperature and time. TRACE to verify your answer is correct.

19. **Decomposition of Salt in Water** Salt (NaCl) decomposes in water into sodium (NA^+) and chloride (Cl^-) ions according to the law of uninhibited decay. If the initial amount of salt is 25 kilograms and, after 10 hours, 15 kilograms of salt are left, how much salt is left after 1 day? How long does it take until $\frac{1}{2}$ kilogram of salt is left?

20. **Voltage of a Conductor** The voltage of a certain condenser decreases over time according to the law of uninhibited decay. If the initial voltage is 40 volts, and 2 seconds later it is 10 volts, what is the voltage after 5 seconds?

21. **Radioactivity from Chernobyl** After the release of radioactive material into the atmosphere from a nuclear power plant at Chernobyl (Ukraine) in 1986, the hay in Austria was contaminated by iodine-131 (half-life 8 years). If it is all right to feed the hay to cows when 10% of the iodine-131 remains, how long do the farmers need to wait to use this hay?

22. Pig Roasts The hotel Bora-Bora is having a pig roast. At noon, the chef put the pig in a large earthen oven. The pig's original temperature was 75°F. At 2:00 P.M., the chef checked the pig's temperature and was upset because it had reached only 100°F. If the oven's temperature remains a constant 325°F, at what time may the hotel serve its guests, assuming that pork is done when it reaches 175°F?

23. Proportion of the Population That Owns a VCR The logistic growth model

$$P(t) = \frac{0.9}{1 + 6e^{-0.32t}}$$

relates the proportion of U.S. households that own a VCR to the year. Let $t = 0$ represent 1984, $t = 1$ represent 1985, and so on.

(a) What proportion of U.S. households owned a VCR in 1984?

(b) Determine the maximum proportion of households that will own a VCR.

(c) Using a graphing utility, graph $P(t)$.

(d) When will 0.8 (80%) of U.S. households own a VCR?

24. Market Penetration of Intel's Coprocessor The logistic growth model

$$P(t) = \frac{0.90}{1 + 3.5e^{-0.339t}}$$

relates the proportion of new personal computers sold at Best Buy that have Intel's latest coprocessor t months after it has been introduced.

(a) What proportion of new personal computers sold at Best Buy will have Intel's latest coprocessor when it is first introduced (i.e., at $t = 0$)?

(b) Determine the maximum proportion of new personal computers sold at Best Buy that will have Intel's latest coprocessor.

(c) Using a graphing utility, graph $P(t)$.

(d) When will 0.75 (75%) of new personal computers sold at Best Buy will have Intel's latest coprocessor?

25. Population of a Bacteria Culture The logistic growth model

$$P(t) = \frac{1000}{1 + 32.33e^{-0.439t}}$$

represents the population of a bacteria after t hours.

(a) Using a graphing utility, graph $P(t)$.

(b) What is the carrying capacity of the environment?

(c) What was the initial amount of bacteria in the population?

(d) When will the amount of bacteria be 800?

26. Population of a Endangered Species Often environmentalists will capture an endangered species and transport the species to a controlled environment where the species can produce offspring and regenerate its population. Suppose 6 American Bald Eagles are captured and transported to Montana and set free. Based on experience, the environmentalists expect the population to grow according to the model

$$P(t) = \frac{500}{1 + 83.33e^{-0.162t}}$$

(a) Using a graphing utility, graph $P(t)$.

(b) What is the carrying capacity of the environment?

(c) What is the predicted population of the American Bald Eagle in 20 years?

(d) When will the population be 300?

4.8 EXPONENTIAL, LOGARITHMIC, AND LOGISTIC CURVE FITTING

1 Use a Graphing Utility to Obtain the Exponential Function of Best Fit

2 Use a Graphing Utility to Obtain the Logarithmic Function of Best Fit

3 Use a Graphing Utility to Obtain the Logistic Function of Best Fit

4 Determine the Model of Best Fit Based on "r"

In Section 1.4 we discussed how to find the linear equation of best fit ($y = ax + b$). In Chapter 3, we discussed how to find power equations

$(y = ax^b)$, quadratic equations $(y = ax^2 + bx + c)$, and cubic equations $(y = ax^3 + bx^2 + cx + d)$ of best fit.

In this section we will discuss how to use a graphing utility to find equations of best fit that describe the relation between two variables when the relation is thought to be exponential $(y = ab^x)$, logarithmic $(y = a + b \ln x)$, or logistic $\left(y = \dfrac{c}{1 + ae^{-bx}} \right)$. As before, we draw a scatter diagram of the data to help determine the appropriate model to use.

Figure 47 shows scatter diagrams that will typically be observed for the three models. Below each scatter diagram are any restrictions on the values of the parameters.

FIGURE 47

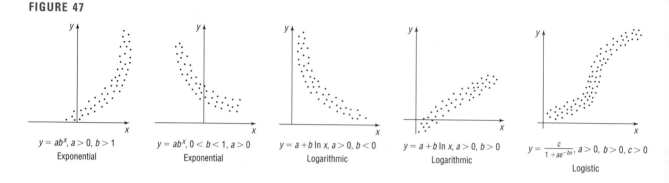

| $y = ab^x, a > 0, b > 1$ | $y = ab^x, 0 < b < 1, a > 0$ | $y = a + b \ln x, a > 0, b < 0$ | $y = a + b \ln x, a > 0, b > 0$ | $y = \dfrac{c}{1 + ae^{-bx}}, a > 0, b > 0, c > 0$ |
| Exponential | Exponential | Logarithmic | Logarithmic | Logistic |

Determining the appropriate model to use requires statistical techniques that are beyond the scope of this text. However, there are some rudimentary techniques we can use that will allow us to determine which model is most appropriate. Before we do this, however, we need to discuss how graphing utilities are used to fit curves to data.

Most graphing utilities have REGression options that fit data to a specific type of curve. Once the data has been entered and a scatter diagram obtained, the type of curve you want to fit to the data is selected. Then that REGression option is used to obtain the curve of "best fit" of the type selected.

The correlation coefficient, r, will appear only if the model can be written as a linear expression. Thus, r will appear for the linear, power, exponential, and logarithmic models since these models can be written as a linear expression (see Section 4.4). Remember, the closer $|r|$ is to 1, the better the fit.

Let's look at some examples.

Exponential Curve Fitting

1 As we saw in Section 4.7, many growth and decay models are exponential.

E X A M P L E 1   Fitting a Curve to an Exponential Model

Beth is interested in finding a function that explains the relation between the amount of an unknown radioactive material and time. She obtains 100 grams of the radioactive material and every day for 9 days measures the amount of radioactive material in the sample and obtains the following data:

Day	Weight (in Grams)
0	100
1	91.6
2	84.1
3	74.2
4	68.3
5	63.7
6	58.5
7	53.8
8	50.2
9	47.3

(a) Using a graphing utility, draw a scatter diagram.
(b) Using a graphing utility, fit an exponential model to the data.
(c) Express the curve in the form $A = A_0 e^{kt}$.
(d) Graph the exponential function found in (b) or (c) on the scatter diagram.
(e) Using the solution to (c), estimate the half-life of the unknown radioactive material.
(f) Predict how much of the material will remain after 30 days.

Solution

(a) After entering the data into the graphing utility, we obtain the scatter diagram shown in Figure 48.

FIGURE 48

(b) A graphing utility fits the data in Figure 48 to an exponential model of the form $y = ab^x$ by using the EXPonential REGression option.* See Figure 49.

The exponential function of best fit to the data is

$$y = 98.2463(0.9188)^x$$

Notice that $r = -0.997178162$, so $|r|$ is close to 1, indicating a good fit.

(c) To express $y = ab^x$ in the form $A = A_0 e^{kt}$, we proceed as follows.

FIGURE 49

*If your utility does not have such an option but does have a LINear REGression option, you can transform the exponential model to a linear model by the following technique (see Section 4.4):

$y = ab^x$	Exponential form.
$\ln y = \ln(ab^x)$	Take natural logs of each side.
$\ln y = \ln a + \ln b^x$	Property of logarithms.
$\ln y = \ln a + (\ln b)x$	Property of logarithms.
$Y = \ln a + (\ln b)x$	

Now apply the LINear REGression techniques discussed in Chapter 1. Be careful though. The dependent variable is now $Y = \ln y$, while the independent variable remains x. Once the line of best fit is obtained, you can find a and b and obtain the exponential curve $y = ab^x$.

$$ab^x = A_0 e^{kt}$$
$$a = A_0 \qquad b^x = e^{kt} = (e^k)^t$$
$$b = e^k$$

Thus,

$$A_0 = 98.2463 \text{ and } 0.9188 = e^k$$
$$k = -0.0846$$

FIGURE 50

As a result, $A = A_0 e^{kt} = 98.2463 e^{-0.0846t}$

(d) See Figure 50.

(e) We set $A = \frac{1}{2} A_0$ and solve the equation $\frac{1}{2} A_0 = A_0 e^{-0.08463t}$ for t. The result is that the half-life is $t = 8.2$ days.

(f) After $t = 30$ days, the amount A of the radioactive material left is

$$A = 98.2463 e^{-0.0846(30)} = 7.8 \text{ grams}$$

Now work Problem 1.

Logarithmic Curve Fitting

2

Many models do not follow uninhibited exponential growth but instead, as the independent variable increases, the growth of the dependent variable diminishes. Under these circumstances, a logarithmic model may be appropriate.

E X A M P L E   2   Fitting a Curve to a Logarithmic Model

Jodi, a meteorologist, is interested in finding a function that explains the relation between the height of a weather balloon (in kilometers) and the atmospheric pressure (measured in millimeters of mercury) on the balloon. She collects the data in the table below:

Atmospheric Pressure, p	Height, h
760	0
740	0.184
725	0.328
700	0.565
650	1.079
630	1.291
600	1.634
580	1.862
550	2.235

(a) Using a graphing utility, draw a scatter diagram of the data with atmospheric pressure as the independent variable.

(b) Using a graphing utility, fit a logarithmic model to the data.

(c) Draw the logarithmic function found in (b) on the scatter diagram.

(d) Use the function found in (b) to predict the height of the weather balloon if the atmospheric pressure is 560 millimeters of mercury.

Solution   (a) After entering the data into the graphing utility, we obtain the scatter diagram shown in Figure 51.

(b) A graphing utility fits the data in Figure 51 to a logarithmic model of the form $y = a + b \ln x$ by using the Logarithm REGression option. See Figure 52. The logarithmic function of best fit to the data is

$$h = 45.7863 - 6.9025 \ln p$$

where h is the height of the weather balloon and p is the atmospheric pressure. Notice that $|r|$ is close to 1, indicating a good fit.

(c) Figure 53 shows the graph of $h = 45.7863 - 6.9025 \ln p$ on the scatter diagram.

(d) Using the function found in (b), Jodi predicts the height of the weather balloon when the atmospheric pressure is 560 to be

$$h = 45.7863 - 6.9025 \ln 560$$
$$\approx 2.108 \text{ kilometers}$$

FIGURE 51

FIGURE 52

FIGURE 53

Now work Problem 11.

Logistic Curve Fitting

Logistic growth models can be used to model situations where the value of the dependent variable is limited. Many real-world situations conform to this scenario. For example, the population of the human race is limited by the availability of natural resources such as food, shelter, and so forth. When the value of the dependent variable is limited, a logistic growth model is often appropriate.

E X A M P L E   3

Fitting a Curve to a Logistic Growth Model

The data on the left, obtained from R. Pearl ("The Growth of Population," *Quarterly Review of Biology* 2 (1927): 532–548) represents the amount of yeast biomass after t hours in a culture.

Time (in hours)	Yeast Biomass	Time (in hours)	Yeast Biomass
0	9.6	10	513.3
1	18.3	11	559.7
2	29.0	12	594.8
3	47.2	13	629.4
4	71.1	14	640.8
5	119.1	15	651.1
6	174.6	16	655.9
7	257.3	17	659.6
8	350.7	18	661.8
9	441.0		

(a) Using a graphing utility, draw a scatter diagram of the data with time as the independent variable.

(b) Using a graphing utility, fit a logistic growth model to the data.

(c) Using a graphing utility, graph the function found in (b) on the scatter diagram.

(d) What is the predicted carrying capacity of the culture?

(e) Use the function found in (b) to predict the population of the culture at $t = 19$ hours.

exponential model, Figure 58(b) shows the results for a power model, and Figure 58(c) shows the results for a linear model.

FIGURE 58

(a)	(b)	(c)

By looking at the value of r, we conclude that the power model is superior to both the exponential model and the linear model. Thus, we conclude the relation between the distance a planet is from the sun (x) and its sidereal year (y) is $y = 0.00112x^{1.49938}$.*

We can predict the sidereal year of the newly discovered planet by substituting $x = 4200$ (4.2 billion = 4200 million) into the function just found:

$$y = 0.00112(4200)^{1.49938}$$
$$\approx 303.3 \text{ sidereal years}$$

■

Now work Problem 17.

4.8 EXERCISES

1. **Biology** A certain bacteria initially increases according to the law of uninhibited growth. A biologist collects the following data for this bacteria:

Time (Hours)	Population
0	1000
1	1415
2	2000
3	2828
4	4000
5	5656
6	8000

(a) Using a graphing utility, draw a scatter diagram.
(b) Using a graphing utility, fit an exponential curve to the data.

(c) Express the curve in the form $N = N_0 e^{kt}$.
(d) Graph the exponential function found in (b) or (c) on the scatter diagram.
(e) Using the solution to (b), predict the population at $t = 7$ hours.

2. **Biology** A colony of bacteria initially increases according to the law of uninhibited growth. A biologist collects the following data for this bacteria:

Time (Days)	Population
0	50
1	153
2	234
3	357
4	547
5	839
6	1280

*If we square both sides of this equation we obtain
$$y^2 \approx 0.0000013x^3$$
which implies the square of the sidereal year is proportional to the cube of the distance from the sun to the planet. This is Kepler's Third Law of Planetary Motion, named after Johann Kepler. Kepler discovered this relation in 1618 through experimentation without the aid of data analysis.

(a) Using a graphing utility, draw a scatter diagram.

(b) Using a graphing utility, fit an exponential curve to the data.

(c) Express the curve in the form $N = N_0 e^{kt}$.

(d) Graph the exponential function found in (b) or (c) on the scatter diagram.

(e) Using the solution to (b), predict the population at $t = 7$ days.

3. **Economics/Marketing** A store manager collected the following data regarding price and quantity demanded of shoes:

Price ($/Unit)	Quantity Demanded
79	10
67	20
54	30
46	40
38	50
31	60

(a) Using a graphing utility, draw a scatter diagram with quantity demanded on the x-axis and price on the y-axis.

(b) Using a graphing utility, fit an exponential curve to the data.

(c) Graph the exponential function found in (b) on the scatter diagram.

(d) Predict the quantity demanded if the price is $60.

4. **Economics/Marketing** A store manager collected the following data regarding price and quantity supplied of dresses:

Price ($/Unit)	Quantity Supplied
25	10
32	20
40	30
46	40
60	50
74	60

(a) Using a graphing utility, draw a scatter diagram with quantity supplied on the x-axis and price on the y-axis.

(b) Using a graphing utility, fit an exponential curve to the data.

(c) Graph the exponential function found in (b) on the scatter diagram.

(d) Predict the quantity supplied if the price is $45.

5. **Chemistry** A chemist has a 100-gram sample of a radioactive material. He records the amount of radioactive material every week for 6 weeks and obtains the following data:

Week	Weight (in Grams)
0	100.0
1	88.3
2	75.9
3	69.4
4	59.1
5	51.8
6	45.5

(a) Using a graphing utility, draw a scatter diagram with week as the independent variable.

(b) Using a graphing utility, fit an exponential curve to the data.

(c) Express the curve in the form $A = A_0 e^{kt}$.

(d) Graph the exponential function found in (b) or (c) on the scatter diagram.

(e) From the result found in (b), determine the half-life of the radioactive material.

(f) How much radioactive material will be left after 50 weeks?

6. **Chemistry** A chemist has a 1000-gram sample of a radioactive material. She records the amount of radioactive material remaining in the sample every day for a week and obtains the following data:

Day	Weight (in Grams)
0	1000.0
1	897.1
2	802.5
3	719.8
4	651.1
5	583.4
6	521.7
7	468.3

(a) Using a graphing utility, draw a scatter diagram with day as the independent variable.
(b) Using a graphing utility, fit an exponential curve to the data.
(c) Express the curve in the form $A = A_0 e^{kt}$.
(d) Graph the exponential function found in (b) or (c) on the scatter diagram.
(e) From the result found in (b), find the half-life of the radioactive material.
(f) How much radioactive material will be left after 20 days?

7. **Finance** The following data represents the amount of money an investor has in an investment account each year for 10 years. She wishes to determine the effective rate of return on her investment.

Year	Value of Account
1985	$10,000
1986	$10,573
1987	$11,260
1988	$11,733
1989	$12,424
1990	$13,269
1991	$13,968
1992	$14,823
1993	$15,297
1994	$16,539

(a) Using a graphing utility, draw a scatter diagram with time as the independent variable and the value of the account as the dependent variable.
(b) Using a graphing utility, fit an exponential curve to the data.
(c) Based on the answer to (b), what was the effective rate of return from this account over the past 10 years?
(d) If the investor plans on retiring in 2020, what will the predicted value of this account be?

8. **Finance** The following data shows the amount of money an investor has in an investment account each year for 7 years. He wishes to determine the effective rate of return on his investment.

Year	Value of Account
1988	$20,000
1989	$21,516
1990	$23,355
1991	$24,885
1992	$27,434
1993	$30,053
1994	$32,622

(a) Using a graphing utility, draw a scatter diagram with time as the independent variable and the value of the account as the dependent variable.
(b) Using a graphing utility, fit an exponential curve to the data.
(c) Based on the answer to (b), what was the effective rate of return from this account over the past 7 years?
(d) If the investor plans on retiring in 2020, what will the predicted value of this account be?

9. **CBL Experiment** The following data was collected by placing a temperature probe in a cup of hot water, removing the probe, and then recording temperature over time.

According to Newton's Law of Cooling, this data should follow an exponential model.
(a) Using a graphing utility, draw a scatter diagram for the data.
(b) Using a graphing utility, fit an exponential model to the data.
(c) Graph the exponential function found in (b) on the scatter diagram.
(d) Predict how long it will take for the water to reach a temperature of 110°F.

Time	Temperature (F°)
0	175.69
1	173.52
2	171.21
3	169.07
4	166.59
5	164.21
6	161.89
7	159.66
8	157.86
9	155.75
10	153.70
11	151.93
12	150.08
13	148.50
14	146.84
15	145.33
16	143.83
17	142.38
18	141.22
19	140.09
20	138.69
21	137.59
22	136.78
23	135.70
24	134.91
25	133.86

Time	Temperature (F°)
0	165.07
1	164.77
2	163.99
3	163.22
4	162.82
5	161.96
6	161.20
7	160.45
8	159.35
9	158.61
10	157.89
11	156.83
12	156.11
13	155.08
14	154.40
15	153.72

10. **CBL Experiment**   The following data was collected by placing a temperature probe in a portable heater, removing the probe, and then recording temperature over time.

According to Newton's Law of Cooling, this data should follow an exponential model.
(a) Using a graphing utility, draw a scatter diagram for the data.
(b) Using a graphing utility, fit an exponential model to the data.
(c) Graph the exponential function found in (b) on the scatter diagram.
(d) Predict how long it will take for the air blown out of the heater to reach a temperature of 110°F.

11. **Economics/Marketing**   The following data represents the price and quantity demanded in 1997 for IBM personal computers at Best Buy.
(a) Using a graphing utility, draw a scatter diagram of the data with price as the dependent variable.
(b) Using a graphing utility, fit a logarithmic model to the data.
(c) Using a graphing utility, draw the logarithmic function found in (b) on the scatter diagram.
(d) Use the function found in (b) to predict the number of IBM personal computers that would be demanded if the price were $1650.

Price ($/Computer)	Quantity Demanded
2300	152
2000	159
1700	164
1500	171
1300	176
1200	180
1000	189

Year	CPI
1984	103.2
1985	105.6
1986	109.0
1987	113.5
1988	118.2
1989	125.1
1990	132.4
1991	136.3
1992	137.9
1993	140.9

Source: Bureau of Labor Statistics.

24. **Economics** The following data is the level of the Consumer Price Index for electricity.

Year	CPI
1984	105.3
1985	108.9
1986	110.4
1987	110.0
1988	111.5
1989	114.7
1990	117.4
1991	121.8
1992	124.2
1993	126.7

Source: Bureau of Labor Statistics.

(a) Using a graphing utility, draw a scatter diagram of the data using time as the independent variable and the CPI as the dependent variable.
(b) Use an exponential model, logarithmic model, power model, and linear model to find the "best" model to describe the relation between time and the CPI.
(c) Use this model to predict the CPI for electricity for 1994.

25. **Economics** The following data represents the per capita Gross Domestic Product (GDP) in constant 1987 dollars (this removes the effects of inflation).

GDP

Year	Per Capita Gross Domestic Product (1987 Dollars)
1988	19,252
1989	19,556
1990	19,593
1991	19,238
1992	19,518
1993	19,888

Source: U.S. Bureau of Economic Analysis.

(a) Using a graphing utility, draw a scatter diagram of the data using time as the independent variable and the per capita GDP as the dependent variable.
(b) Use an exponential model, logarithmic model, power model, and linear model to find the "best" model to describe the relation between time and the per capita GDP.
(c) Use this model to predict per capita GDP for 1994.

26. **Economics** The following data represents the per capita Gross National Product (GNP) in constant 1987 dollars (this removes the effects of inflation).

GNP

Year	Per Capita Gross National Product (1987 Dollars)
1988	19,284
1989	19,615
1990	19,670
1991	19,290
1992	19,548
1993	19,897

Source: U.S. Bureau of Economic Analysis.

(a) Using a graphing utility, draw a scatter diagram of the data using time as the independent variable and the GNP as the dependent variable.
(b) Use an exponential model, logarithmic model, power model, and linear model to find the "best" model to describe the relation between time and the per capita GNP.
(c) Use this model to predict the per capita GNP for 1994.

4.9 LOGARITHMIC SCALES

1 Decibels (Loudness of Sound)
2 Richter Scale (Earthquake Magnitude)

Common logarithms often appear in the measurement of quantities, because they provide a way to scale down positive numbers that vary from very small to very large. For example, if a certain quantity can take on values from $0.0000000001 = 10^{-10}$ to $10,000,000,000 = 10^{10}$, the common logarithms of such numbers would be between -10 and 10, respectively.

Loudness of Sound

1 Our first application utilizes a logarithmic scale to measure the loudness of a sound. Physicists define the **intensity of a sound wave** as the amount of energy the wave transmits through a given area. For example, the least intense sound that a human ear can detect at a frequency of 100 Hertz is about 10^{-12} watt per square meter. The **loudness** $L(x)$, measured in **decibels** (named in honor of Alexander Graham Bell), of a sound of intensity x (measured in watts per square meter) is defined as

$$L(x) = 10 \log \frac{x}{I_0} \qquad (1)$$

where $I_0 = 10^{-12}$ watt per square meter is the least intense sound that a human ear can detect. If we let $x = I_0$ in equation (1), we get

$$L(I_0) = 10 \log \frac{I_0}{I_0} = 10 \log 1 = 0$$

Thus, at the threshold of human hearing, the loudness is 0 decibels. Figure 59 on page 384 gives the loudness of some common sounds.

Note that a decibel is not a linear unit like the meter. For example, a noise level of 10 decibels is 10 times as great as a noise level of 0 decibels. [If $L(x) = 10$, then $x = 10I_0$.] A noise level of 20 decibels is 100 times as great as a noise level of 0 decibels. [If $L(x) = 20$, then $x = 100I_0$.] A noise level of 30 decibels is 1000 times as great as a noise level of 0 decibels, and so on.

EXAMPLE 1   Finding the Intensity of a Sound

Use Figure 59 to find the intensity of the sound of a dripping faucet.

Solution   From Figure 59, we see that the loudness of the sound of a dripping faucet is 30 decibels. Thus, by equation (1), its intensity x may be found as follows:

$$30 = 10 \log\left(\frac{x}{I_0}\right)$$

$$3 = \log\left(\frac{x}{I_0}\right) \qquad \text{Divide by 10.}$$

$$\frac{x}{I_0} = 10^3 \qquad \text{Write in exponential form.}$$

$$x = 1000I_0$$

FIGURE 59
Loudness of common sounds (in decibels)

Decibels

140	Shotgun blast, jet 100 feet away at takeoff	Pain	
130	Motor test chamber	Human ear pain threshold	
120	Firecrackers, severe thunder, pneumatic jackhammer, hockey crowd	Uncomfortably loud	
110	Amplified rock music		
100	Textile loom, subway train, elevated train, farm tractor, power lawn mower, newspaper press	Loud	
90	Heavy city traffic, noisy factory		
80	Diesel truck going 40 mi/hr 50 feet away, crowded restaurant, garbage disposal, average factory, vacuum cleaner	Moderately loud	
70	Passenger car going 50 mi/hr 50 feet away		
60	Quiet typewriter, singing birds, window air conditioner, quiet automobile	Quiet	
50	Normal conversation, average office		
40	Household refrigerator, quiet office	Very quiet	
30	Average home, dripping faucet, whisper 5 feet away		
20	Light rainfall, rustle of leaves	Average person's threshold of hearing	
10	Whisper across room	Just audible	
0		Threshold for acute hearing	

where $I_0 = 10^{-12}$ watt per square meter. Thus, the intensity of the sound of a dripping faucet is 1000 times as great as a noise level of 0 decibels; that is, such a sound has an intensity of $1000 \cdot 10^{-12} = 10^{-9}$ watt per square meter. ■

Now work Problem 5.

E X A M P L E   2   Finding the Loudness of a Sound

Use Figure 59 to determine the loudness of a subway train if it is known that this sound is 10 times as intense as the sound due to heavy city traffic.

Solution   The sound due to heavy city traffic has a loudness of 90 decibels. Its intensity, therefore, is the value of x in the equation

$$90 = 10 \log\left(\frac{x}{I_0}\right)$$

A sound 10 times as intense as x has loudness $L(10x)$. Thus, the loudness of the subway train is

$$L(10x) = 10 \log\left(\frac{10x}{I_0}\right) \qquad \text{Replace } x \text{ by } 10x.$$

$$= 10 \log\left(10 \cdot \frac{x}{I_0}\right)$$

$$= 10\left[\log 10 + \log\left(\frac{x}{I_0}\right)\right] \qquad \text{Log of product = Sum of logs.}$$

$$= 10 \log 10 + 10 \log\left(\frac{x}{I_0}\right) \qquad \log 10 = 1.$$

$$= 10 + 90 = 100 \text{ decibels} \qquad \blacksquare$$

Magnitude of an Earthquake

Our second application uses a logarithmic scale to measure the magnitude of an earthquake.

The **Richter scale*** is one way of converting seismographic readings into numbers that provide an easy reference for measuring the magnitude M of an earthquake. All earthquakes are compared to a **zero-level earthquake** whose seismographic reading measures 0.001 millimeter at a distance of 100 kilometers from the epicenter. An earthquake whose seismographic reading measures x millimeters has **magnitude** $M(x)$ given by

$$M(x) = \log\left(\frac{x}{x_0}\right) \qquad\qquad (2)$$

where $x_0 = 10^{-3}$ is the reading of a zero-level earthquake the same distance from its epicenter.

EXAMPLE 3

Finding the Magnitude of an Earthquake

What is the magnitude of an earthquake whose seismographic reading is 0.1 millimeter at a distance of 100 kilometers from its epicenter?

Solution   If $x = 0.1$, the magnitude $M(x)$ of this earthquake is

$$M(0.1) = \log\left(\frac{x}{x_0}\right) = \log\left(\frac{0.1}{0.001}\right) = \log\left(\frac{10^{-1}}{10^{-3}}\right) = \log 10^2 = 2$$

This earthquake thus measures 2.0 on the Richter scale. $\blacksquare$

Now work Problem 7.

Based on formula (2), we define the **intensity of an earthquake** as the ratio of x to x_0. For example, the intensity of the earthquake described in Example 3 is $\frac{0.1}{0.001} = 10^2 = 100$. That is, it is 100 times as intense as a zero-level earthquake.

EXAMPLE 4

Comparing the Intensity of Two Earthquakes

The devastating San Francisco earthquake of 1906 measured 6.9 on the Richter scale. How did the intensity of that earthquake compare to the Papua New Guinea earthquake of 1988, which measured 6.7 on the Richter scale?

*Named after the American scientist, C.F. Richter, who devised it in 1935.

POLYHEDRA STAINED-GLASS LAMPS

Polyhedra, or many sided 3-D objects, have been known and studied for a long time. The Greek philosopher Plato studied the five regular sided polyhedra, now known as *platonic solids*. Many objects found in nature, such as crystals, some types of virus, and buckyballs, have the shape of polyhedra. Ever since *Buckminster Fuller* popularized the geodesic dome there has been a resurgence of interest in polyhedra.

Suppose that you are the owner and operator of a stained-glass business and a customer who lives in a geodesic dome wants to commission you to make her a stained-glass lamp in the shape of a polyhedron. As a master craftsman, building such a lamp will not pose much of a problem. Since you have no ready-made designs drawn up for the customer to choose from, you ask the customer if she would like to choose a pattern from the *World Wide Web Library of Polyhedra*. The virtual reality displays make a big impression on your client. She selects a *5cube* as her choice for a lamp.

1. The stained-glass lamp that you are about to build at first appears to be very complicated. However, as you continue to look at the 5cube, you begin to see an underlying structure. Can you see the underlying structure? Do you think you could build a 5cube?
2. You will be able to build the lamp by making six square sections. Each of these sections is made up only of triangles. How many different types of triangles are there? How many pieces will you need to cut for each of the different triangles?
3. Since the cost of a stained-glass lamp runs about $5 per piece of glass, (and each triangle will be a piece), what do you estimate is the cost of the lamp?
4. Next you have the challenge of figuring out how to make the patterns for cutting the glass. You plan to make a *paper model* of the lamp with an overall diameter of 20 inches. Can you use your calculator and a formula to determine the size of the cubes?
5. Once you have the side lengths of the cubes that compose the 5cube, what do you need to know in order to calculate the triangles? Can you determine these lengths by computation?

Trigonometry was developed by Greek astronomers who regarded the sky as the inside of a sphere, so it was natural that triangles on a sphere were investigated early (by Menelaus of Alexandria about A.D. 100) and that triangles in the plane were studied much later. The first book containing a systematic treatment of plane and spherical trigonometry was written by the Persian astronomer Nasîr ed-dîn (about A.D. 1250).

Regiomontanus (1436–1476) is the man most responsible for moving trigonometry from astronomy into mathematics. His work was improved by Copernicus (1473–1543) and Copernicus's student Rhaeticus (1514–1576). Rhaeticus's book was the first to define the six trigonometric functions as ratios of sides of triangles, although he did not give the functions their present names. Credit for this is due to Thomas Fincke (1583), but Fincke's notation was by no means universally accepted at the time. The notation was finally stabilized by the textbooks of Leonhard Euler (1707–1783).

Trigonometry has since evolved from its use by surveyors, navigators, and engineers to present applications involving ocean tides, the rise and fall of food supplies in certain ecologies, brain wave patterns, and many other phenomena.

There are two widely accepted approaches to the development of the trigonometric functions: One uses circles, especially the unit circle; the other uses right triangles. In this book, we develop the trigonometric functions using the unit circle. In Section 5.4, we shall show that right triangle trigonometry is a special case of the unit circle approach.

5.1 ANGLES AND THEIR MEASURE

1. Convert Between Degrees, Minutes, Seconds and Decimal Forms for Angles
2. Find the Arc Length of a Circle
3. Convert from Degrees to Radians
4. Convert from Radians to Degrees
5. Find the Linear Speed of an Object Traveling in Circular Motion

FIGURE 1

A **ray,** or **half-line,** is that portion of a line that starts at a point V on the line and extends indefinitely in one direction. The starting point V of a ray is called its **vertex.** See Figure 1.

If two rays are drawn with a common vertex, they form an **angle.** We call one of the rays of an angle the **initial side** and the other the **terminal side.** The angle that is formed is identified by showing the direction and amount of rotation from the initial side to the terminal side. If the rotation is in the counterclockwise direction, the angle is **positive;** if the rotation is clockwise, the angle is **negative.** See Figure 2. Lowercase Greek letters, such as α (alpha), β (beta), γ (gamma), θ (theta), and so on, will be used to denote angles. Notice in Figure 2(a) that the angle α is positive because the direction of the rotation from the initial side to the terminal side is counterclockwise. The angle β in Figure 2(b) is negative because the rotation is clockwise. The angle γ in Figure 2(c) is positive. Notice that the angle α in Figure 2(a) and the angle γ have the same initial side and the same terminal side. However, α and γ are unequal because the amount of rotation required to go from the initial side to the terminal side is greater for angle γ than for angle α.

FIGURE 2

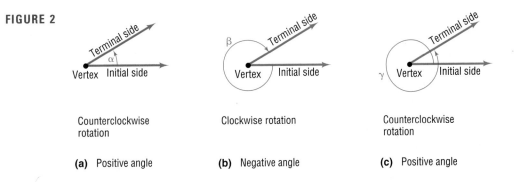

(a) Positive angle

(b) Negative angle

(c) Positive angle

An angle θ is said to be in **standard position** if its vertex is at the origin of a rectangular coordinate system and its initial side coincides with the positive x-axis. See Figure 3.

When an angle θ is in standard position, the terminal side either will lie in a quadrant, in which case we say θ **lies in that quadrant,** or it will lie

FIGURE 3

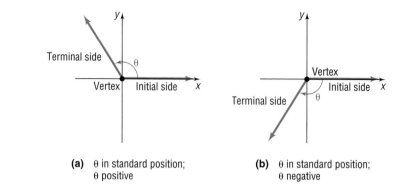

(a) θ in standard position;
θ positive

(b) θ in standard position;
θ negative

on the *x*-axis or the *y*-axis, in which case we say θ is a **quadrantal angle.** For example, the angle θ in Figure 4(a) lies in quadrant II, the angle θ in Figure 4(b) lies in quadrant IV, and the angle θ in Figure 4(c) is a quadrantal angle.

FIGURE 4

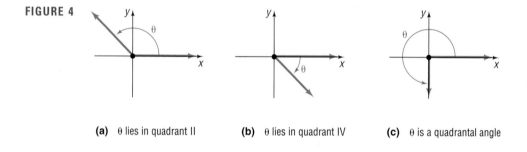

(a) θ lies in quadrant II

(b) θ lies in quadrant IV

(c) θ is a quadrantal angle

We measure angles by determining the amount of rotation needed for the initial side to become coincident with the terminal side. There are two commonly used measures for angles: *degrees* and *radians*.

Degrees

The angle formed by rotating the initial side exactly once in the counter-clockwise direction until it coincides with itself (1 revolution) is said to measure 360 degrees, abbreviated 360°. Thus **one degree, 1°,** is $\frac{1}{360}$ revolution. A **right angle** is an angle of 90°, or $\frac{1}{4}$ revolution; a **straight angle** is an angle of 180°, or $\frac{1}{2}$ revolution. See Figure 5. As Figure 5(b) shows, it is customary to indicate a right angle by using the symbol ⌐.

FIGURE 5

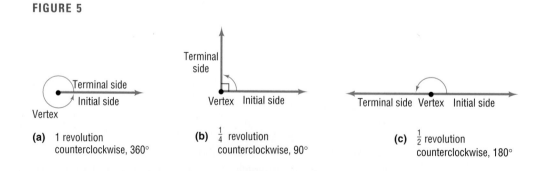

(a) 1 revolution
counterclockwise, 360°

(b) $\frac{1}{4}$ revolution
counterclockwise, 90°

(c) $\frac{1}{2}$ revolution
counterclockwise, 180°

E X A M P L E   1

Drawing an Angle

Draw each angle:

(a) 45°   (b) −90°   (c) 225°   (d) 405°

Solution

(a) An angle of 45° is $\frac{1}{2}$ of a right angle. See Figure 6.

(b) An angle of −90° is $\frac{1}{4}$ revolution in the clockwise direction. See Figure 7.

FIGURE 6

FIGURE 7

(c) An angle of 225° consists of a rotation through 180° followed by a rotation through 45°. See Figure 8.

(d) An angle of 405° consists of 1 revolution (360°) followed by a rotation through 45°. See Figure 9.

FIGURE 8

FIGURE 9

 Now work Problem 1.

Although subdivisions of a degree may be obtained by using decimals, we also may use the notion of *minutes* and *seconds*. **One minute,** denoted by **1′,** is defined as $\frac{1}{60}$ degree. **One second,** denoted by **1″,** is defined as $\frac{1}{60}$ minute or, equivalently, $\frac{1}{3600}$ degree. An angle of, say, 30 degrees, 40 minutes, 10 seconds is written compactly as 30°40′10″. To summarize:

> 1 counterclockwise revolution = 360°
> 60′ = 1°   60″ = 1′

(1)

Because calculators use decimals, it is sometimes necessary to convert from the degree, minute, second notation (D°M′S″) to a decimal form, and vice versa. Check your calculator; it should be capable of doing the conversion for you.

Before getting started, though, you must set the mode to degrees, because there are two common ways to measure angles: degree mode and radian mode. (We will define radians shortly.) Usually, a menu is used to change from one mode to another. Check your instruction manual to find out how your particular calculator works.

Note: Formulas must be consistent with regard to the units used. In equation (4), we write

$$s = r\theta$$

To see the units, however, we must go back to equation (3) and write

$$\frac{\theta \text{ radians}}{1 \text{ radian}} = \frac{s \text{ length units}}{r \text{ length units}}$$

$$s \text{ length units} = (r \text{ length units})\frac{\theta \text{ radians}}{1 \text{ radian}}$$

Since the radians cancel, we are left with

$$s \text{ length units} = (r \text{ length units})\theta \quad s = r\theta$$

where θ appears to be "dimensionless" but, in fact, is measured in radians. Thus, in using the formula $s = r\theta$, the dimension of radians for θ is usually omitted, and any convenient unit of length (such as inches or meters) may be used for s and r.

E X A M P L E   3

Finding the Length of Arc of a Circle

Find the length of the arc of a circle of radius 2 meters subtended by a central angle of 0.25 radian.

Solution   We use equation (4) with $r = 2$ meters and $\theta = 0.25$. The length s of the arc is

$$s = r\theta = 2(0.25) = 0.5 \text{ meter}$$ ▬

Now work Problem 33.

Relationship between Degrees and Radians

FIGURE 14
1 revolution = 2π radians

$s = 2\pi r$

1 revolution

r

Consider a circle of radius r. A central angle of 1 revolution will subtend an arc equal to the circumference of the circle (Figure 14). Because the circumference of a circle equals $2\pi r$, we use $s = 2\pi r$ in equation (4) to find that, for an angle θ of 1 revolution,

$$s = r\theta$$
$$2\pi r = r\theta$$
$$\theta = 2\pi \text{ radians}$$

Thus,

1 revolution = 2π radians	(5)

so that

$$360° = 2\pi \text{ radians}$$

or

$$180° = \pi \text{ radians} \tag{6}$$

Divide both sides of equation (6) by 180. Then

$$1 \text{ degree} = \frac{\pi}{180} \text{ radian}$$

Divide both sides of (6) by π. Then

$$\frac{180}{\pi} \text{ degrees} = 1 \text{ radian}$$

Thus, we have the following two conversion formulas:

$$1 \text{ degree} = \frac{\pi}{180} \text{ radian} \qquad 1 \text{ radian} = \frac{180}{\pi} \text{ degrees} \tag{7}$$

E X A M P L E   4

3

Converting from Degrees to Radians

Convert each angle in degrees to radians:

(a) 60°   (b) 150°   (c) −45°   (d) 90°

Solution   (a) $60° = 60 \cdot 1 \text{ degree} = 60 \cdot \frac{\pi}{180} \text{ radian} = \frac{\pi}{3} \text{ radians}$

(b) $150° = 150 \cdot \frac{\pi}{180} \text{ radian} = \frac{5\pi}{6} \text{ radians}$

(c) $-45° = -45 \cdot \frac{\pi}{180} \text{ radian} = -\frac{\pi}{4} \text{ radian}$

(d) $90° = 90 \cdot \frac{\pi}{180} \text{ radian} = \frac{\pi}{2} \text{ radians}$

Now work Problem 13.

Example 4 illustrates that angles that are fractions of a revolution are expressed in radian measure as fractional multiples of π, rather than as decimals. Thus, a right angle, as in Example 4(d), is left in the form $\pi/2$ radians, which is exact, rather than using the approximation $\pi/2 \approx 3.1416/2 = 1.5708$ radians.

E X A M P L E   5

4

Converting Radians to Degrees

Convert each angle in radians to degrees.

(a) $\frac{\pi}{6} \text{ radian}$ (b) $\frac{3\pi}{2} \text{ radians}$ (c) $-\frac{3\pi}{4} \text{ radians}$ (d) $\frac{7\pi}{3} \text{ radians}$

5.1 EXERCISES

In Problems 1–12, draw each angle.

1. 30°   **2.** 60°   **3.** 135°   **4.** −120°   **5.** 450°   **6.** 540°

7. $3\pi/4$ **8.** $4\pi/3$ **9.** $-\pi/6$ **10.** $-2\pi/3$ **11.** $16\pi/3$ **12.** $21\pi/4$

In Problems 13–22, convert each angle in degrees to radians. Express your answer as a multiple of π.

13. 30°   **14.** 120°   **15.** 240°   **16.** 330°   **17.** −60°

18. −30°   **19.** 180°   **20.** 270°   **21.** 135°   **22.** −225°

In Problems 23–32, convert each angle in radians to degrees.

23. $\pi/3$ **24.** $5\pi/6$ **25.** $-5\pi/4$ **26.** $-2\pi/3$ **27.** $\pi/2$

28. 4π **29.** $\pi/12$ **30.** $5\pi/12$ **31.** $2\pi/3$ **32.** $5\pi/4$

In Problems 33–40, s denotes the length of arc of a circle of radius r subtended by the central angle θ. Find the missing quantity.

33. $r = 10$ meters, $\theta = \frac{1}{2}$ radian, $s = ?$ **34.** $r = 6$ feet, $\theta = 2$ radians, $s = ?$

35. $\theta = \frac{1}{3}$ radian, $s = 2$ feet, $r = ?$ **36.** $\theta = \frac{1}{4}$ radian, $s = 6$ centimeters, $r = ?$

37. $r = 5$ miles, $s = 3$ miles, $\theta = ?$ **38.** $r = 6$ meters, $s = 8$ meters, $\theta = ?$

39. $r = 2$ inches, $\theta = 30°$, $s = ?$ **40.** $r = 3$ meters, $\theta = 120°$, $s = ?$

In Problems 41–48, convert each angle in degrees to radians. Express your answer in decimal form, rounded to two decimal places.

41. 17°   **42.** 73°   **43.** −40°   **44.** −51°   **45.** 125°   **46.** 200°   **47.** 340°   **48.** 350°

In Problems 49–56, convert each angle in radians to degrees. Express your answer in decimal form, rounded to two decimal places.

49. 3.14   **50.** π **51.** 10.25   **52.** 0.75   **53.** 2   **54.** 3   **55.** 6.32   **56.** $\sqrt{2}$

In Problems 57–62, convert each angle to a decimal in degrees. Round off your answer to two decimal places.

57. 40°10′25″   **58.** 61°42′21″   **59.** 1°2′3″   **60.** 73°40′40″   **61.** 9°9′9″   **62.** 98°22′45″

In Problems 63–68, convert each angle to D°M′S″ form. Round off your answer to the nearest second.

63. 40.32°   **64.** 61.24°   **65.** 18.255°   **66.** 29.411°   **67.** 19.99°   **68.** 44.01°

69. Minute Hand of a Clock The minute hand of a clock is 6 inches long. How far does the tip of the minute hand move in 15 minutes? How far does it move in 25 minutes?

70. Movement of a Pendulum A pendulum swings through an angle of 20° each second. If the pendulum is 40 inches long, how far does its tip move each second?

71. An object is traveling around a circle with a radius of 5 centimeters. If in 20 seconds a central angle of $\frac{1}{3}$ radian is swept out, what is the angular speed of the object? What is its linear speed?

72. An object is traveling around a circle with a radius of 2 meters. If in 20 seconds the object travels 5 meters, what is its angular speed? What is its linear speed?

73. Bicycle Wheels The diameter of each wheel of a bicycle is 26 inches. If you are traveling at a speed of 35 miles per hour on this bicycle, through how many revolutions per minute are the wheels turning?

74. Car Wheels The radius of each wheel of a car is 15″. If the wheels are turning at the rate of 3 revolutions per second, how fast is the car moving? Express your answer in inches per second and in miles per hour.

75. Windshield Wipers The windshield wiper of a car is 18 inches long. How many inches will the tip of the wiper trace out in $\frac{1}{3}$ revolution?

76. Windshield Wipers The windshield wiper of a car is 18 inches long. If it takes 1 second to trace out $\frac{1}{3}$ revolution, how fast is the tip of the wiper moving?

77. Speed of the Moon The mean distance of the Moon from Earth is 2.39×10^5 miles. Assuming that the orbit of the Moon around Earth is circular and that 1 revolution takes 27.3 days, find the linear speed of the Moon. Express your answer in miles per hour.

78. Speed of the Earth The mean distance of Earth from the Sun is 9.29×10^7 miles. Assuming that the orbit of Earth around the Sun is circular and that 1 revolution takes 365 days, find the linear speed of Earth. Express your answer in miles per hour.

79. Pulleys Two pulleys, one with radius 2 inches and the other with radius 8 inches, are connected by a belt. (See the figure.) If the 2 inch pulley is caused to rotate at 3 revolutions per minute, determine the revolutions per minute of the 8 inch pulley.

[**Hint:** The linear speeds of the pulleys, that is, the speed of the belt, are the same.]

80. Pulleys Two pulleys, one with radius r_1 and the other with radius r_2, are connected by a belt. The pulley with radius r_1 rotates at ω_1 revolutions per minute, whereas the pulley with radius r_2 rotates at ω_2 revolutions per minute. Show that $r_1/r_2 = \omega_2/\omega_1$.

81. Computing the Speed of River Current To approximate the speed of the current of a river, a circular paddle wheel with radius 4 feet is lowered into the water. If the current causes the wheel to rotate at a speed of 10 revolutions per minute, what is the speed of the current? Express your answer in miles per hour.

82. Spin Balancing Tires A spin balancer rotates the wheel of a car at 480 revolutions per minute. If the diameter of the wheel is 26 inches, what road speed is being tested? Express your answer in miles per hour. At how many revolutions per minute should the balancer be set to test a road speed of 80 miles per hour?

83. The Cable Cars of San Francisco At the Cable Car Museum you can see the four cable lines that are used to pull cable cars up and down the hills of San Francisco. Each cable travels at a speed of 9.55 mi/hr, caused by a rotating wheel whose diameter is 8.5 feet. How fast is the wheel rotating? Express your answer in revolutions per minute.

84. Difference in Time of Sun Rise Naples, Florida, is approximately 90 miles due west of Ft. Lauderdale. How much sooner would a person in Ft. Lauderdale first see the rising Sun than a person in Naples?

[**Hint:** Consult the figure (p. 406). When a person at Q sees the first rays of the Sun, a person at P is still in the dark. The person at P sees the first rays after Earth has rotated so that P is at the location Q.]

If $b \neq 0$, the **cosecant function** is defined as

$$\csc t = \frac{1}{b}$$

If $a \neq 0$, the **secant function** is defined as

$$\sec t = \frac{1}{a}$$

If $b \neq 0$, the **cotangent function** is defined as

$$\cot t = \frac{a}{b}$$

Notice in these definitions that if $a = 0$, that is, if the point $P = (0, b)$ is on the y-axis, then the tangent function and the secant function are undefined. Also, if $b = 0$, that is, if the point $P = (a, 0)$ is on the x-axis, then the cosecant function and the cotangent function are undefined.

Because we use the unit circle in these definitions of the trigonometric functions, they are also sometimes referred to as **circular functions.**

E X A M P L E   1

FIGURE 19
$\theta = t$ radians

1

Finding the Value of the Six Trigonometric Functions

Let t be a real number and let $P = (-\frac{1}{2}, \frac{\sqrt{3}}{2})$ be the point on the unit circle that corresponds to t. See Figure 19.
Then

$$\sin t = \frac{\sqrt{3}}{2} \qquad \cos t = -\frac{1}{2} \qquad \tan t = \frac{\frac{\sqrt{3}}{2}}{-\frac{1}{2}} = -\sqrt{3}$$

$$\csc t = \frac{1}{\frac{\sqrt{3}}{2}} = \frac{2\sqrt{3}}{3} \qquad \sec t = \frac{1}{-\frac{1}{2}} = -2 \qquad \cot t = \frac{-\frac{1}{2}}{\frac{\sqrt{3}}{2}} = \frac{-\sqrt{3}}{3}$$

Trigonometric Functions of Angles

FIGURE 20

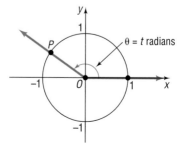

Let P be the point on the unit circle corresponding to the real number t. Then the angle θ, in standard position and measured in radians, whose terminal side is the ray from the origin through P is

$$\theta = t \text{ radians}$$

See Figure 20.

Thus, on the unit circle, the measure of the angle θ in radians equals the value of the real number t. As a result, we can say that

$$\sin t = \sin \theta$$

Real number $\theta = t$ radians

and so on. We now can define the trigonometric functions of the angle θ.

If $\theta = t$ radians, the **six trigonometric functions of the angle** θ are defined as

$$\sin \theta = \sin t \qquad \cos \theta = \cos t \qquad \tan \theta = \tan t$$
$$\csc \theta = \csc t \qquad \sec \theta = \sec t \qquad \cot \theta = \cot t$$

Even though the distinction between trigonometric functions of real numbers and trigonometric functions of angles is important, it is customary to refer to trigonometric functions of real numbers and trigonometric functions of angles collectively as *the trigonometric functions*. We shall follow this practice from now on.

If an angle θ is measured in degrees, we shall use the degree symbol when writing a trigonometric function of θ, as, for example, in $\sin 30°$ and $\tan 45°$. If an angle θ is measured in radians, then no symbol is used when writing a trigonometric function of θ, as, for example, in $\cos \pi$ and $\sec (\pi/3)$.

Finally, since the values of the trigonometric functions of an angle θ are determined by the coordinates of the point $P = (a, b)$ on the unit circle corresponding to θ, the units used to measure the angle θ are irrelevant. For example, it does not matter whether we write $\theta = \pi/2$ radians or $\theta = 90°$. The point on the unit circle corresponding to this angle is $P = (0, 1)$. Hence,

$$\sin \frac{\pi}{2} = \sin 90° = 1 \quad \text{and} \quad \cos \frac{\pi}{2} = \cos 90° = 0$$

Evaluating the Trigonometric Functions

To find the exact value of a trigonometric function of an angle θ requires that we locate the corresponding point P on the unit circle. In fact, though, any circle whose center is at the origin can be used.

Let θ be any nonquadrantal angle placed in standard position. Let $P^* = (a^*, b^*)$ be the point where the terminal side of θ intersects the unit circle. See Figure 21.

FIGURE 21

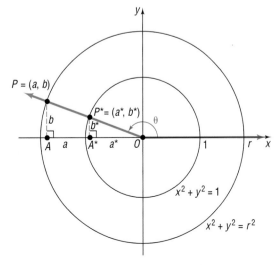

Let $P = (a, b)$ be any point on the terminal side of θ. Suppose that r is the distance from the origin to P. Then P is on the circle $x^2 + y^2 = r^2$. Refer again to Figure 21. Notice that the triangles OA^*P^* and OAP are similar; thus, ratios of corresponding sides are equal:

$$\frac{b^*}{1} = \frac{b}{r} \qquad \frac{a^*}{1} = \frac{a}{r} \qquad \frac{b^*}{a^*} = \frac{b}{a}$$

$$\frac{1}{b^*} = \frac{r}{b} \qquad \frac{1}{a^*} = \frac{r}{a} \qquad \frac{a^*}{b^*} = \frac{a}{b}$$

These results lead us to formulate the following theorem:

Theorem

For an angle θ in standard position, let $P = (a, b)$ be any point on the terminal side of θ. Let r equal the distance from the origin to P. Then

$$\sin \theta = \frac{b}{r} \qquad \cos \theta = \frac{a}{r} \qquad \tan \theta = \frac{b}{a}, \quad a \neq 0$$

$$\csc \theta = \frac{r}{b}, \quad b \neq 0 \qquad \sec \theta = \frac{r}{a}, \quad a \neq 0 \qquad \cot \theta = \frac{a}{b}, \quad b \neq 0$$

E X A M P L E   2

FIGURE 22

Finding the Exact Value of the Six Trigonometric Functions

Find the exact value of each of the six trigonometric functions of an angle θ if $(4, -3)$ is a point on its terminal side.

Solution   Figure 22 illustrates the situation for θ a positive angle. For the point $(a, b) = (4, -3)$, we have $a = 4$ and $b = -3$. Then $r = \sqrt{a^2 + b^2} = \sqrt{16 + 9} = 5$. Thus,

$$\sin \theta = \frac{b}{r} = -\frac{3}{5} \qquad \cos \theta = \frac{a}{r} = \frac{4}{5} \qquad \tan \theta = \frac{b}{a} = -\frac{3}{4}$$

$$\csc \theta = \frac{r}{b} = -\frac{5}{3} \qquad \sec \theta = \frac{r}{a} = \frac{5}{4} \qquad \cot \theta = \frac{a}{b} = -\frac{4}{3}$$

Now work Problem 1.

To evaluate the trigonometric functions of a given angle θ in standard position, we need to find the coordinates of any point on the terminal side of that angle. This is not always so easy to do. In the examples that follow, we will evaluate the trigonometric functions of certain angles for which this process is relatively easy. A calculator will be used to evaluate the trigonometric functions of most angles.

E X A M P L E   3

Finding the Exact Value of the Six Trigonometric Functions of Quadrantal Angles

Find the exact value of each of the trigonometric functions at:

(a) $\theta = 0 = 0°$ 　　　　(b) $\theta = \pi/2 = 90°$

(c) $\theta = \pi = 180°$ 　　　(d) $\theta = 3\pi/2 = 270°$

FIGURE 23
$\theta = 0 = 0°$

FIGURE 24
$\theta = \pi/2 = 90°$

FIGURE 25
$\theta = \pi = 180°$

FIGURE 26
$\theta = 3\pi/2 = 270°$

Solution (a) The point $P = (1, 0)$ is on the terminal side of $\theta = 0 = 0°$ and is a distance of 1 unit from the origin. See Figure 23. Thus,

$$\sin 0 = \sin 0° = \frac{0}{1} = 0 \qquad \cos 0 = \cos 0° = \frac{1}{1} = 1$$

$$\tan 0 = \tan 0° = \frac{0}{1} = 0 \qquad \sec 0 = \sec 0° = \frac{1}{1} = 1$$

Since the y-coordinate of P is 0, csc 0 and cot 0 are not defined.

(b) The point $P = (0, 1)$ is on the terminal side of $\theta = \pi/2 = 90°$ and is a distance of 1 unit from the origin. See Figure 24. Thus,

$$\sin \frac{\pi}{2} = \sin 90° = \frac{1}{1} = 1 \qquad \cos \frac{\pi}{2} = \cos 90° = \frac{0}{1} = 0$$

$$\csc \frac{\pi}{2} = \csc 90° = \frac{1}{1} = 1 \qquad \cot \frac{\pi}{2} = \cot 90° = \frac{0}{1} = 0$$

Since the x-coordinate of P is 0, $\tan(\pi/2)$ and $\sec(\pi/2)$ are not defined.

(c) The point $P = (-1, 0)$ is on the terminal side of $\theta = \pi = 180°$ and is a distance of 1 unit from the origin. See Figure 25. Thus,

$$\sin \pi = \sin 180° = \frac{0}{1} = 0 \qquad \cos \pi = \cos 180° = \frac{-1}{1} = -1$$

$$\tan \pi = \tan 180° = \frac{0}{1} = 0 \qquad \sec \pi = \sec 180° = \frac{1}{-1} = -1$$

Since the y-coordinate of P is 0, csc π and cot π are not defined.

(d) The point $P = (0, -1)$ is on the terminal side of $\theta = 3\pi/2 = 270°$ and is a distance of 1 unit from the origin. See Figure 26. Thus,

$$\sin \frac{3\pi}{2} = \sin 270° = \frac{-1}{1} = -1 \qquad \cos \frac{3\pi}{2} = \cos 270° = \frac{0}{1} = 0$$

$$\csc \frac{3\pi}{2} = \csc 270° = \frac{1}{-1} = -1 \qquad \cot \frac{3\pi}{2} = \cot 270° = \frac{0}{-1} = 0$$

Since the x-coordinate of P is 0, $\tan(3\pi/2)$ and $\sec(3\pi/2)$ are not defined. ∎

Note: The results obtained in Example 3 are the same whether we pick a point on the unit circle or a point on any circle whose center is at the origin. For example, $P = (0, 5)$ is a point on the terminal side of $\theta = \pi/2 = 90°$ and is a distance of 5 units from the origin. Using this point, $\sin \frac{\pi}{2} = \frac{5}{5} = 1$, $\cos \frac{\pi}{2} = \frac{0}{5} = 0$, and so on, as in Example 3(b).

Table 2 summarizes the values of the trigonometric functions found in Example 3.

TABLE 2	Quadrantal Angles						
θ (Radians)	θ (Degrees)	sin θ	cos θ	tan θ	csc θ	sec θ	cot θ
0	0°	0	1	0	Not defined	1	Not defined
$\pi/2$	90°	1	0	Not defined	1	Not defined	0
π	180°	0	-1	0	Not defined	-1	Not defined
$3\pi/2$	270°	-1	0	Not defined	-1	Not defined	0

Now work Problem 39.

SUMMARY

For the quadrantal angles and the angles 30°, 45°, 60° and their integral multiples, we can find the exact value of each trigonometric function by using the geometric features of these angles and symmetry.

Figure 36 shows points on the unit circle that are on the terminal sides of some angles that are integral multiples of $\pi/6$ (30°), $\pi/4$ (45°), and $\pi/3$ (60°).

FIGURE 36

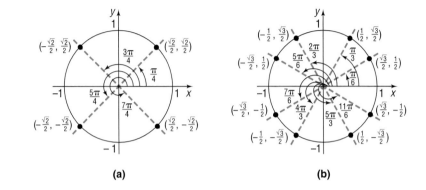

(a)   (b)

Notice the relationship between integral multiples of an angle and symmetry. For example, in Figure 36(a), the point on the unit circle corresponding to $7\pi/4$ is the reflection about the x-axis of the point corresponding to $\pi/4$. The point on the unit circle corresponding to $5\pi/4$ is the reflection about the origin of the point corresponding to $\pi/4$. Thus, using Figure 36(a), we see that

$$\sin\frac{7\pi}{4} = -\frac{\sqrt{2}}{2} \qquad \cos\left(-\frac{\pi}{4}\right) = \frac{\sqrt{2}}{2} \qquad \tan\frac{5\pi}{4} = \frac{-\sqrt{2}/2}{-\sqrt{2}/2} = 1$$

Using Figure 36(b), we see that

$$\sin\frac{11\pi}{6} = -\frac{1}{2} \qquad \cos\frac{7\pi}{6} = -\frac{\sqrt{3}}{2} \qquad \tan\frac{4\pi}{3} = \frac{-\sqrt{3}/2}{-\frac{1}{2}} = \sqrt{3}$$

For most other angles, besides the quadrantal angles and those listed in Figure 36, we can only approximate the value of each trigonometric function using a calculator.

HISTORICAL FEATURE The name *sine* for the sine function is due to a medieval confusion. The name comes from the Sanskrit word *jiva* (meaning chord), first used in India by Aryabhata the Elder (AD 510). He really meant half-chord, but abbreviated it. This was brought into Arabic as *jiba,* which was meaningless. Because the proper Arabic word *jaib* would be written the same way (short vowels are not written out in Arabic), *jiba* was pronounced as *jaib,* which meant bosom or hollow, and *jaib* remains as the Arabic word for sine to this day. Scholars translating the Arabic works into Latin found that the word *sinus* also meant bosom or hollow, and from *sinus* we get the word *sine.*

The name *tangent*, due to Thomas Fincke (1583), can be understood by looking at Figure 37. The line segment $\overline{DC}$ is tangent to the circle at C. If $d(O, B) = d(O, C) = 1$, then the length of the line segment $\overline{DC}$ is

FIGURE 37

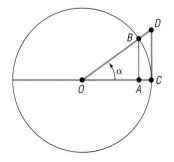

$$d(D, C) = \frac{d(D, C)}{1} = \frac{d(D, C)}{d(O, C)} = \tan \alpha$$

The old name for the tangent is *umbra versa* (meaning turned shadow), referring to the use of the tangent in solving height problems with shadows.

The names of the remaining functions came about as follows. If α and β are complementary angles, then $\cos \alpha = \sin \beta$. Because β is the complement of α, it was natural to write the cosine of α as *sin co* α. Probably for reasons involving ease of pronunciation, the *co* migrated to the front, and then cosine received a three-letter abbreviation to match sin, sec, and tan. The two other cofunctions were similarly treated, except that the long forms *cotan* and *cosec* survive to this day in some countries. ▬

5.2 EXERCISES

In Problems 1–10, a point on the terminal side of an angle θ is given. Find the exact value of each of the six trigonometric functions of θ.

1. $(-3, 4)$ **2.** $(5, -12)$ **3.** $(2, -3)$ **4.** $(-1, -2)$ **5.** $(-2, -2)$

6. $(1, -1)$ **7.** $(-3, -2)$ **8.** $(2, 2)$ **9.** $\left(\frac{1}{3}, -\frac{1}{4}\right)$ **10.** $(-0.3, -0.4)$

In Problems 11–30, find the exact value of each expression. Do not use a calculator.

11. $\sin 45° + \cos 60°$

12. $\sin 30° - \cos 45°$

13. $\sin 90° + \tan 45°$

14. $\cos 180° - \sin 180°$

15. $\sin 45° \cos 45°$

16. $\tan 45° \cos 30°$

17. $\csc 45° \tan 60°$

18. $\sec 30° \cot 45°$

19. $4 \sin 90° - 3 \tan 180°$

20. $5 \cos 90° - 8 \sin 270°$

21. $2 \sin \dfrac{\pi}{3} - 3 \tan \dfrac{\pi}{6}$

22. $2 \sin \dfrac{\pi}{4} + 3 \tan \dfrac{\pi}{4}$

23. $\sin \dfrac{\pi}{4} - \cos \dfrac{\pi}{4}$

24. $\tan \dfrac{\pi}{3} + \cos \dfrac{\pi}{3}$

25. $2 \sec \dfrac{\pi}{4} + 4 \cot \dfrac{\pi}{3}$

26. $3 \csc \dfrac{\pi}{3} + \cot \dfrac{\pi}{4}$

27. $\tan \pi - \cos 0$

28. $\sin \dfrac{3\pi}{2} + \tan \pi$

29. $\csc \dfrac{\pi}{2} + \cot \dfrac{\pi}{2}$

30. $\sec \pi - \csc \dfrac{\pi}{2}$

In Problems 31–46, find the exact value of each of the six trigonometric functions of the given angle. If any are not defined, say "not defined." Do not use a calculator.

31. $2\pi/3$ **32.** $3\pi/4$ **33.** $150°$ **34.** $330°$

35. $-\pi/6$ **36.** $-\pi/3$ **37.** $225°$ **38.** $210°$

39. $5\pi/2$ **40.** 3π **41.** $-180°$ **42.** $-270°$

43. $3\pi/2$ **44.** $-\pi$ **45.** $450°$ **46.** $-90°$

In Problems 47–70, use a calculator to find the approximate value of each expression rounded to two decimal places.

47. $\sin 28°$ **48.** $\cos 14°$ **49.** $\tan 21°$ **50.** $\sin 15°$

51. $\sec 41°$ **52.** $\csc 55°$ **53.** $\cot 70°$ **54.** $\tan 80°$

55. $\sin \dfrac{\pi}{10}$ **56.** $\cos \dfrac{\pi}{8}$ **57.** $\tan \dfrac{5\pi}{12}$ **58.** $\sin \dfrac{3\pi}{10}$

In Problems 67–78, find the exact value of each expression. Do not use a calculator.

67. $\sin(-\pi) + \cos 5\pi$

68. $\tan\left(-\dfrac{5\pi}{5}\right) - \cot\dfrac{7\pi}{2}$

69. $\sec(-\pi) + \csc\left(-\dfrac{\pi}{2}\right)$

70. $\tan(-6\pi) + \cos\dfrac{9\pi}{4}$

71. $\sin\left(-\dfrac{9\pi}{4}\right) - \tan\left(-\dfrac{9\pi}{4}\right)$

72. $\cos\left(-\dfrac{17\pi}{4}\right) - \sin\left(-\dfrac{3\pi}{2}\right)$

73. $\sin^2 40° + \cos^2 40°$

74. $\sec^2 18° - \tan^2 18°$

75. $\sin 80° \csc 80°$

76. $\tan 10° \cot 10°$

77. $\tan 40° - \dfrac{\sin 40°}{\cos 40°}$

78. $\cot 20° - \dfrac{\cos 20°}{\sin 20°}$

79. If $\sin\theta = 0.3$, find the value of
$\sin\theta + \sin(\theta + 2\pi) + \sin(\theta + 4\pi)$.

80. If $\cos\theta = 0.2$, find the value of
$\cos\theta + \cos(\theta + 2\pi) + \cos(\theta + 4\pi)$.

81. If $\tan\theta = 3$, find the value of
$\tan\theta + \tan(\theta + \pi) + \tan(\theta + 2\pi)$.

82. If $\cot\theta = -2$, find the value of
$\cot\theta + \cot(\theta - \pi) + \cot(\theta - 2\pi)$.

83. Calculating the Time of a Trip From a parking lot, you want to walk to a house on the ocean. The house is located 1500 feet down a paved path that parallels the ocean, which is 500 feet away. See the illustration. Along the path you can walk 300 feet per minute, but in the sand on the beach you can only walk 100 feet per minute.

The time T to get from the parking lot to the beachhouse can be expressed as a function of the angle θ shown in the illustration and is

$$T(\theta) = 5 - \dfrac{5}{3\tan\theta} + \dfrac{5}{\sin\theta}, \qquad 0 < \theta < \dfrac{\pi}{2}$$

Calculate the time T if you walk directly from the parking lot to the house.
[**Hint:** $\tan\theta = 500/1500$.]

and then jog directly back in the sand to get from one house to the other. See the illustration. The time T to get from one house to the other as a function of the angle θ shown in the illustration is

$$T(\theta) = 1 + \dfrac{2}{3\sin\theta} - \dfrac{1}{4\tan\theta}, \qquad 0 < \theta < \dfrac{\pi}{2}$$

(a) Calculate the time T for $\tan\theta = 1/4$.
(b) Describe the path taken.
(c) Explain why θ must be larger than 14°.

85. For what numbers θ is $f(\theta) = \tan\theta$ not defined?

86. For what numbers θ is $f(\theta) = \cot\theta$ not defined?

87. For what numbers θ is $f(\theta) = \sec\theta$ not defined?

88. For what numbers θ is $f(\theta) = \csc\theta$ not defined?

89. What is the value of $\sin k\pi$, where k is any integer?

90. What is the value of $\cos k\pi$, where k is any integer?

91. Show that the range of the tangent function is the set of all real numbers.

92. Show that the range of the cotangent function is the set of all real numbers.

93. Show that the period of $f(\theta) = \sin\theta$ is 2π.
[**Hint:** Assume that $0 < p < 2\pi$ exists so that $\sin(\theta + p) = \sin\theta$ for all θ. Let $\theta = 0$ to find p. Then let $\theta = \pi/2$ to obtain a contradiction.]

94. Show that the period of $f(\theta) = \cos\theta$ is 2π.

95. Show that the period of $f(\theta) = \sec\theta$ is 2π.

84. Calculating the Time of a Trip Two oceanfront homes are located 8 miles apart on a straight stretch of beach, each a distance of 1 mile from a paved road that parallels the ocean. Sally can jog 8 miles per hour along the paved road, but only 3 miles per hour in the sand that separates the road from the ocean. Because of a river directly between the two houses, it is necessary to jog in the sand to the road, continue on the road,

96. Show that the period of $f(\theta) = \csc\theta$ is 2π.

97. Show that the period of $f(\theta) = \tan\theta$ is π.

98. Show that the period of $f(\theta) = \cot\theta$ is π.

99. Prove the reciprocal identities given in formula (2).

100. Prove the quotient identities given in formula (3).

101. Establish the identity:
$$(\sin\theta\cos\phi)^2 + (\sin\theta\sin\phi)^2 + \cos^2\theta = 1.$$

102. Write down five characteristics of the tangent function. Explain the meaning of each.

103. Describe your understanding of the meaning of a periodic function.

5.4   RIGHT TRIANGLE TRIGONOMETRY

1 Find the Value of Trigonometric Functions of Acute Angles

2 Use the Complementary Angle Theorem

3 Find the Reference Angle

FIGURE 46

(a)

(b)

FIGURE 47

A triangle in which one angle is a right angle (90°) is called a **right triangle.** Recall that the side opposite the right angle is called the **hypotenuse,** and the remaining two sides are called the **legs** of the triangle. In Figure 46(a), we have labeled the hypotenuse as c, to indicate its length is c units, and, in a like manner, we have labeled the legs as a and b. Because the triangle is a right triangle, the Pythagorean Theorem tells us that

$$a^2 + b^2 = c^2$$

Now, suppose that θ is an **acute angle;** that is, $0° < \theta < 90°$ (if θ is measured in degrees) or $0 < \theta < \pi/2$ (if θ is measured in radians). Place θ in standard position, and let $P = (a, b)$ be any point except the origin O on the terminal side of θ. Form a right triangle by dropping the perpendicular from P to the x-axis, as shown in Figure 46(b).

By referring to the lengths of the sides of the triangle by the names hypotenuse (c), opposite (b), and adjacent (a), as indicated in Figure 47, we can express the trigonometric functions of θ as ratios of the sides of a right triangle:

$$\sin\theta = \frac{\text{Opposite}}{\text{Hypotenuse}} = \frac{b}{c} \qquad \cos\theta = \frac{\text{Adjacent}}{\text{Hypotenuse}} = \frac{a}{c}$$

$$\tan\theta = \frac{\text{Opposite}}{\text{Adjacent}} = \frac{b}{a} \qquad \csc\theta = \frac{\text{Hypotenuse}}{\text{Opposite}} = \frac{c}{b} \qquad (1)$$

$$\sec\theta = \frac{\text{Hypotenuse}}{\text{Adjacent}} = \frac{c}{a} \qquad \cot\theta = \frac{\text{Adjacent}}{\text{Opposite}} = \frac{a}{b}$$

Notice that each of the trigonometric functions of the acute angle θ is positive.

E X A M P L E   1

Finding the Value of Trigonometric Functions from a Right Triangle

Find the exact value of each of the six trigonometric functions of the angle θ in Figure 48.

75. Use a graphing utility set in radian mode to complete the following table. What can you conclude about the ratio $(\sin\theta)/\theta$ as θ approaches 0?

θ	0.5	0.4	0.2	0.1	0.01	0.001	0.0001	0.00001
$\sin\theta$								
$\dfrac{\sin\theta}{\theta}$								

76. Use a graphing utility set in radian mode to complete the following table. What can you conclude about the ratio $(\cos\theta - 1)/\theta$ as θ approaches 0?

θ	0.5	0.4	0.2	0.1	0.01	0.001	0.0001	0.00001
$\cos\theta - 1$								
$\dfrac{\cos\theta - 1}{\theta}$								

77. Suppose that the angle θ is a central angle of a circle of radius 1 (see the figure). Show that

(a) Angle $OAC = \dfrac{\theta}{2}$

(b) $|CD| = \sin\theta$ and $|OD| = \cos\theta$

(c) $\tan\dfrac{\theta}{2} = \dfrac{\sin\theta}{1 + \cos\theta}$

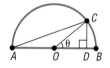

78. Show that the area of an isosceles triangle is $A = a^2 \sin\theta\cos\theta$, where a is the length of one of the two equal sides and θ is the measure of one of the two equal angles (see the figure).

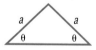

79. Let $n > 0$ be any real number, and let θ be any angle for which $0 < \theta < \pi/(1 + n)$. Then we can construct a triangle with the angles θ and $n\theta$ and included side of length 1 (do you see why?) and place it on the unit circle as illustrated. Now, drop the perpendicular from C to $D = (x, 0)$, and show that

$$x = \frac{\tan n\theta}{\tan\theta + \tan n\theta}$$

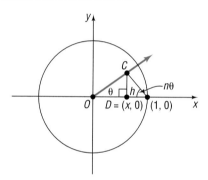

80. Refer to the figure below. The smaller circle, whose radius is a, is tangent to the larger circle, whose radius is b. The ray OA contains a diameter of each circle, and the ray OB is tangent to each circle. Show that

$$\cos\theta = \frac{\sqrt{ab}}{\dfrac{a + b}{2}}$$

(That is, $\cos\theta$ equals the ratio of the geometric mean of a and b to the arithmetic mean of a and b).

[**Hint:** First show that $\sin\theta = (b - a)/(b + a)$.]

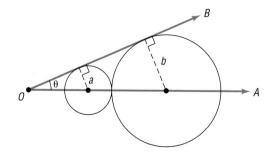

81. Refer to the figure. If $|OA| = 1$, show that
(a) Area $\triangle OAC = \frac{1}{2} \sin \alpha \cos \alpha$
(b) Area $\triangle OCB = \frac{1}{2}|OB|^2 \sin \beta \cos \beta$
(c) Area $\triangle OAB = \frac{1}{2}|OB| \sin(\alpha + \beta)$

(d) $|OB| = \dfrac{\cos \alpha}{\cos \beta}$

(e) $\sin(\alpha + \beta) = \sin \alpha \cos \beta + \cos \alpha \sin \beta$
[**Hint:** Area $\triangle OAB =$ Area $\triangle OAC +$ Area $\triangle OCB$.]

82. Refer to the accompanying figure, where a unit circle is drawn. The line DB is tangent to the circle.
(a) Express the area of $\triangle OBC$ in terms of $\sin \theta$ and $\cos \theta$.
(b) Express the area of $\triangle OBD$ in terms of $\sin \theta$ and $\cos \theta$.
(c) The area of the sector of the circle $\overset{\frown}{OBC}$, is $\frac{1}{2}\theta$, where θ is measured in radians. Use the results of parts (a) and (b) and the fact that

Area $\triangle OBC <$ Area $\overset{\frown}{OBC} <$ Area $\triangle OBD$

to show that

$$1 < \frac{\theta}{\sin \theta} < \frac{1}{\cos \theta}$$

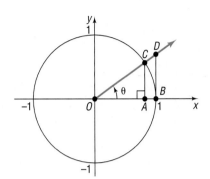

83. If $\cos \alpha = \tan \beta$ and $\cos \beta = \tan \alpha$, where α and β are acute angles, show that

$$\sin \alpha = \sin \beta = \sqrt{\frac{3 - \sqrt{5}}{2}}$$

84. If θ is an acute angle, explain why $\sec \theta > 1$.

85. If θ is an acute angle, explain why $0 < \sin \theta < 1$.

5.5 GRAPHS OF THE TRIGONOMETRIC FUNCTIONS

1. Graph Transformations of the Sine Function
2. Graph Transformations of the Cosine Function
3. Graph Transformations of the Tangent Function
4. Graph Transformations of the Cosecant, Secant, and Cotangent Functions

We have discussed properties of the trigonometric functions $f(\theta) = \sin \theta$, $f(\theta) = \cos \theta$, and so on. In this section, we shall use the traditional symbols x to represent the independent variable (or argument) and y for the dependent variable (or value at x) for each function. Thus, we write the six trigonometric functions as

$$
\begin{array}{ccc}
y = \sin x & y = \cos x & y = \tan x \\
y = \csc x & y = \sec x & y = \cot x
\end{array}
$$

Our purpose in this section is to graph each of these functions. Unless indicated otherwise, we shall use radian measure throughout for the independent variable x.

The Graph of $y = \sin x$

Since the sine function has period 2π, we need to graph $y = \sin x$ only on the interval $[0, 2\pi]$. The remainder of the graph will consist of repetitions of this portion of the graph.

We begin by constructing Table 7, which lists some points on the graph of $y = \sin x, 0 \le x \le 2\pi$. As the table shows, the graph of $y = \sin x, 0 \le x \le 2\pi$, begins at the origin. As x increases from 0 to $\pi/2$, the value of $y = \sin x$ increases from 0 to 1; as x increases from $\pi/2$ to π to $3\pi/2$, the value of y decreases from 1 to 0 to -1; as x increases from $3\pi/2$ to 2π, the value of y increases from -1 to 0. Based on this, we set the viewing rectangle as shown in Figure 62(a) and graph $y = \sin x$, $0 \le x \le 2\pi$. See Figure 62(b). Figure 62(c) shows the graph drawn by hand.

TABLE 7

FIGURE 62

(a)   (b)

(c)

The graph in Figure 62 is one period of the graph of $y = \sin x$. To obtain a more complete graph of $y = \sin x$, we repeat this period in each direction, as shown in Figure 63.

FIGURE 63
$y = \sin x, -\infty < x < \infty$

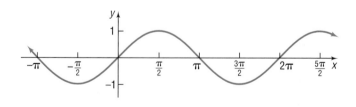

The graph of $y = \sin x$ illustrates some of the facts we already know about the sine function:

Characteristics of the
Sine Function

1. The domain is the set of all real numbers.
2. The range consists of all real numbers from -1 to 1, inclusive.
3. The sine function is an odd function, as the symmetry of the graph with respect to the origin indicates.
4. The sine function is periodic, with period 2π.
5. The x-intercepts are $\dots$, -2π, $-\pi$, 0 π, 2π, 3π, $\dots$; the y-intercept is 0.
6. The maximum value is 1 and occurs at $x = \dots$, $-3\pi/2$, $\pi/2$, $5\pi/2$, $9\pi/2$, $\dots$; the minimum value is -1 and occurs at $x = \dots$, $-\pi/2$, $3\pi/2$, $7\pi/2$, $11\pi/2$, $\dots$.

Now work Problems 1, 3, and 5.

1   The graphing techniques introduced in Chapter 2 may be used to graph functions that are transformations of the sine function. (Refer to Chapter 2, Section 2.3).

E X A M P L E   1   Graphing Transformations of $y = \sin x$ Using Shifts, Reflections, and the Like

Use the graph of $y = \sin x$ to graph $y = -\sin x + 2$.

Solution   Figure 64 illustrates the steps.

FIGURE 64

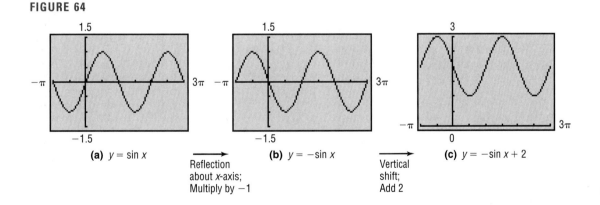

(a) $y = \sin x$

Reflection
about x-axis;
Multiply by -1

(b) $y = -\sin x$

Vertical
shift;
Add 2

(c) $y = -\sin x + 2$

E X A M P L E   2   Graphing Transformations of $y = \sin x$ Using Shifts, Reflections, and the Like

Use the graph of $y = \sin x$ to graph $y = \sin\left(x - \dfrac{\pi}{4}\right)$.

Solution Figure 65 illustrates the steps.

FIGURE 65

(a) $y = \sin x$

Horizontal shift to right $\frac{\pi}{4}$ units; Replace x by $x - \frac{\pi}{4}$

(b) $y = \sin\left(x - \frac{\pi}{4}\right)$

Now work Problem 35.

The Graph of $y = \cos x$

TABLE 8

X	Y1
0	1
1.0472	.5
1.5708	0
2.0944	-.5
3.1416	-1
4.7124	0
6.2832	1

$\frac{\pi}{3}$
$\frac{\pi}{2}$
$\frac{2\pi}{3}$
π
$\frac{3\pi}{2}$
2π

Y1 ▉cos(X)

The cosine function also has period 2π. Thus, we proceed as we did with the sine function by constructing Table 8, which lists some points on the graph of $y = \cos x$, $0 \le x \le 2\pi$. As the table shows, the graph of $y = \cos x$, $0 \le x \le 2\pi$, begins at the point $(0, 1)$. As x increases from 0 to $\pi/2$ to π, the value of y decreases from 1 to 0 to -1; as x increases from π to $3\pi/2$ to 2π, the value of y increases from -1 to 0 to 1. As before, we set the viewing rectangle as shown in Figure 66(a) and graph $y = \cos x$, $0 \le x \le 2\pi$. See Figure 66(b). Figure 66(c) shows the graph drawn by hand.

FIGURE 66
$y = \cos x, 0 \le x \le 2\pi$

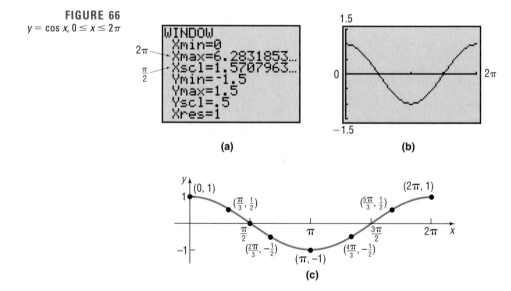

A more complete graph of $y = \cos x$ is obtained by repeating this period in each direction, as shown in Figure 67.

FIGURE 67
$y = \cos x$, $-\infty < x < \infty$

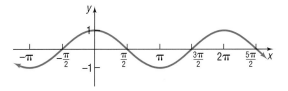

The graph of $y = \cos x$ illustrates some of the facts we already know about the cosine function:

Characteristics of the
Cosine Function

1. The domain is the set of all real numbers.
2. The range consists of all real numbers from -1 to 1, inclusive.
3. The cosine function is an even function, as the symmetry of the graph with respect to the y-axis indicates.
4. The cosine function is periodic, with period 2π.
5. The x-intercepts are . . . , $-3\pi/2$, $-\pi/2$, $\pi/2$, $3\pi/2$, $5\pi/2$, . . . ; the y-intercept is 1.
6. The maximum value is 1 and occurs at $x = $ . . . , $-2\pi, 0, 2\pi, 4\pi, 6\pi, $ . . . ; the minimum value is -1 and occurs at $x = $ . . . , $-\pi, \pi, 3\pi, 5\pi, $ . . . .

2   Again, the graphing techniques from Chapter 2 may be used to graph transformations of the cosine function.

E X A M P L E 3   Graphing Transformations of $y = \cos x$ Using Shifts, Reflections, and the Like

Use the graph of $y = \cos x$ to graph $y = 2 \cos x$

Solution   Figure 68 illustrates the graph, which is a vertical stretch of the graph of $y = \cos x$. (Multiply each y-coordinate of $y = \cos x$ by 2).

FIGURE 68
$y = 2 \cos x$

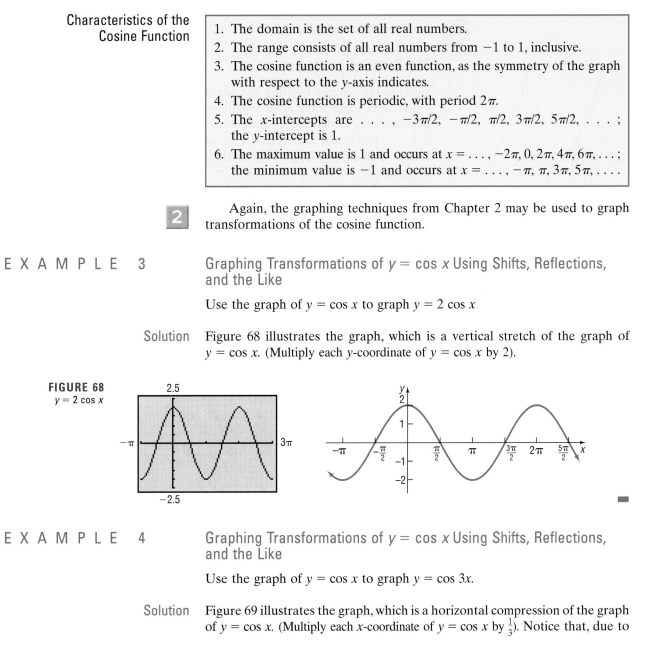

E X A M P L E 4   Graphing Transformations of $y = \cos x$ Using Shifts, Reflections, and the Like

Use the graph of $y = \cos x$ to graph $y = \cos 3x$.

Solution   Figure 69 illustrates the graph, which is a horizontal compression of the graph of $y = \cos x$. (Multiply each x-coordinate of $y = \cos x$ by $\frac{1}{3}$). Notice that, due to

Solution We start with the graph of $y = \tan x$ and shift it horizontally to the left $\pi/4$ unit. See Figure 72.

FIGURE 72

$y = \tan\left(x + \dfrac{\pi}{4}\right)$

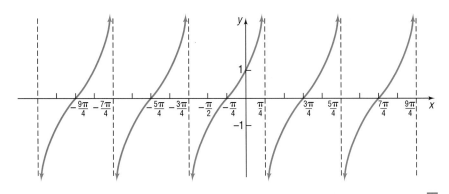

Now work Problem 51.

The Graphs of $y = \csc x$, $y = \sec x$, and $y = \cot x$

The cosecant and secant functions, sometimes referred to as **reciprocal functions,** are graphed by making use of the reciprocal identities

$$\csc x = \frac{1}{\sin x} \quad \text{and} \quad \sec x = \frac{1}{\cos x}$$

For example, the value of the cosecant function $y = \csc x$ at a given number x equals the reciprocal of the corresponding value of the sine function, provided the value of the sine function is not 0. If the value of $\sin x$ is 0, then, at such numbers x, the cosecant function is not defined. In fact, the graph of the cosecant function has vertical asymptotes at integral multiples of π. Figure 73 shows the graph drawn by hand and Figure 74 shows the graph using a graphing utility.

FIGURE 73

$y = \csc x$, $-\infty < x < \infty$, x not equal to integral multiples of π, $|y| \geq 1$

FIGURE 74

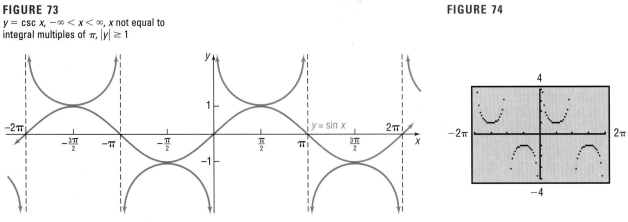

E X A M P L E   6   Graphing Transformations of $y = \csc x$ Using Shifts, Reflections, and the Like

Graph $y = 2\csc(x - \pi/2)$.

Solution   Figure 75 shows the required steps.

FIGURE 75

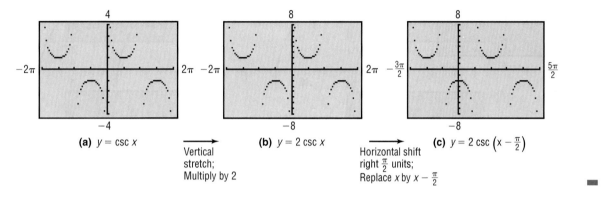

(a) $y = \csc x$ (b) $y = 2\csc x$ (c) $y = 2\csc\left(x - \frac{\pi}{2}\right)$

Vertical stretch; Multiply by 2

Horizontal shift right $\frac{\pi}{2}$ units; Replace x by $x - \frac{\pi}{2}$

Using the idea of reciprocals, we can similarly obtain the graph of $y = \sec x$. See Figure 76 and Figure 77.

FIGURE 76

$y = \sec x$, $-\infty < x < \infty$, x not equal to odd multiples of $\pi/2$, $|y| \geq 1$

FIGURE 77

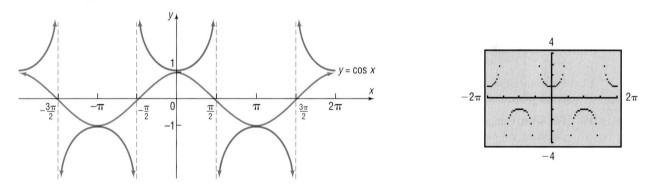

TABLE 11

x	$y = \cot x$	(x, y)
$\pi/6$	$\sqrt{3}$	$(\pi/6, \sqrt{3})$
$\pi/4$	1	$(\pi/4, 1)$
$\pi/3$	$\sqrt{3}/3$	$(\pi/3, \sqrt{3}/3)$
$\pi/2$	0	$(\pi/2, 0)$
$2\pi/3$	$-\sqrt{3}/3$	$(2\pi/3, -\sqrt{3}/3)$
$3\pi/4$	-1	$(3\pi/4, -1)$
$5\pi/6$	$-\sqrt{3}$	$(5\pi/6, -\sqrt{3})$

We obtain the graph of $y = \cot x$ as we did the graph of $y = \tan x$. The period of $y = \cot x$ is π. Because the cotangent function is not defined for integral multiples of π, we shall concentrate on the interval $(0, \pi)$. Table 11 lists some points on the graph of $y = \cot x$, $0 < x < \pi$. As x approaches 0, but remains greater than 0, the value of $\cos x$ will be close to 1 and the value of $\sin x$ will be positive and close to 0. Hence, the ratio $(\cos x)/(\sin x) = \cot x$ will be positive and large, so as x approaches 0, $\cot x$ approaches ∞ and $x = 0$ is a vertical asymptote. Similarly, as x approaches π, but remains less than π, the value of $\cos x$ will be close to -1 and the value of $\sin x$ will be positive and close to 0. Hence, the ratio $(\cos x)/(\sin x) = \cot x$ will be negative and will approach $-\infty$ as x approaches π. The line $x = \pi$ is also a vertical asymptote. Figure 78 shows the graph.

However, if we restrict the domain of $y = \cos x$ to the interval $[0, \pi]$, the restricted function

$$y = \cos x, \qquad 0 \le x \le \pi$$

is one-to-one and hence will have an inverse function.* See Figure 87.

FIGURE 87
$y = \cos x, 0 \le x \le \pi, -1 \le y \le 1$

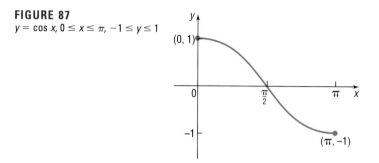

The inverse function is called the **inverse cosine** of x and is symbolized by $y = \cos^{-1} x$ (or by $y = \arccos x$). Thus,

$$y = \cos^{-1} x \quad \text{means} \quad x = \cos y \tag{2}$$
$$\text{where} \quad 0 \le y \le \pi \quad \text{and} \quad -1 \le x \le 1$$

Here, y is the angle whose cosine is x. The domain of the function $y = \cos^{-1} x$ is $-1 \le x \le 1$, and its range is $0 \le y \le \pi$. The graph of $y = \cos^{-1} x$ can be obtained by reflecting the restricted portion of the graph of $y = \cos x$ about the line $y = x$, as shown in Figure 88(a). Figure 88(b) shows the graph using a graphing utility.

FIGURE 88
$y = \cos^{-1} x, -1 \le x \le 1, 0 \le y \le \pi$

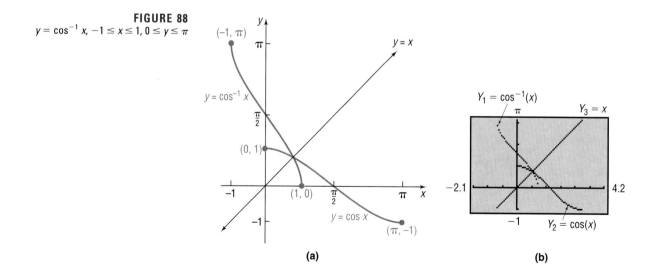

(a)

(b)

*This is the generally accepted restriction to define the inverse.

In general,

$$\cos^{-1}(\cos u) = u, \qquad \text{where} \qquad 0 \le u \le \pi$$
$$\cos(\cos^{-1} v) = v, \qquad \text{where} \qquad -1 \le v \le 1$$

E X A M P L E   5

Finding the Exact Value of an Inverse Cosine Function

Find the exact value of $\cos^{-1} 0$.

Solution   Let $\theta = \cos^{-1} 0$. We seek the angle θ, $0 \le \theta \le \pi$, whose cosine equals 0:

$$\theta = \cos^{-1} 0, \qquad 0 \le \theta \le \pi$$
$$\cos \theta = 0, \qquad 0 \le \theta \le \pi$$

From Figure 89, we see that the only angle θ within the interval $[0, \pi]$ whose cosine is 0 is $\pi/2$. (Note that $\cos 3\pi/2$ also equals 0, but $3\pi/2$ lies outside the interval $[0, \pi]$ and hence is not admissible.)

FIGURE 89

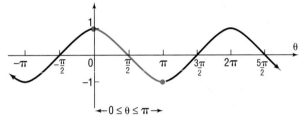

We conclude that

$$\theta = \frac{\pi}{2}$$

So

$$\cos^{-1} 0 = \frac{\pi}{2}$$

E X A M P L E   6

Finding the Exact Value of an Inverse Cosine Function

Find the exact value of $\cos^{-1}(\sqrt{2}/2)$.

Solution   Let $\theta = \cos^{-1}(\sqrt{2}/2)$. We seek the angle θ, $0 \le \theta \le \pi$, whose cosine equals $\sqrt{2}/2$:

$$\theta = \cos^{-1} \frac{\sqrt{2}}{2}, \qquad 0 \le \theta \le \pi$$
$$\cos \theta = \frac{\sqrt{2}}{2}, \qquad 0 \le \theta \le \pi$$

From Figure 90, we see that the only angle θ within the interval $[0, \pi]$ whose cosine is $\sqrt{2}/2$ is $\pi/4$.

FIGURE 90

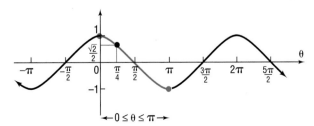

We conclude that

$$\theta = \frac{\pi}{4}$$

So

$$\cos^{-1}\frac{\sqrt{2}}{2} = \frac{\pi}{4}$$

∎

E X A M P L E   7

Finding the Exact Value of an Inverse Cosine Function

Find the exact value of $\cos^{-1}(-\frac{1}{2})$.

Solution   Let $\theta = \cos^{-1}(-\frac{1}{2})$. We seek the angle $\theta, 0 \le \theta \le \pi$, whose cosine equals $-\frac{1}{2}$:

$$\theta = \cos^{-1}(-\tfrac{1}{2}), \qquad 0 \le \theta \le \pi$$
$$\cos\theta = -\tfrac{1}{2}, \qquad 0 \le \theta \le \pi$$

(Refer to Figure 90, if necessary.) The only angle within the interval $[0, \pi]$ whose cosine is $-\frac{1}{2}$ is $2\pi/3$. Thus,

$$\theta = \frac{2\pi}{3}$$

So

$$\cos^{-1}\left(-\frac{1}{2}\right) = \frac{2\pi}{3}$$

∎

The Inverse Tangent Function

In Figure 91 we reproduce the graph of $y = \tan x$. Because every horizontal line intersects the graph infinitely many times, it follows that the tangent function is not one-to-one.

FIGURE 91
$y = \tan x, -\infty < x < \infty, x$ not equal to odd multiples of $\pi/2, -\infty < y < \infty$

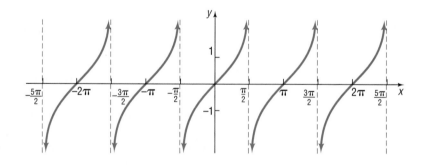

However, if we restrict the domain of $y = \tan x$ to the interval $(-\pi/2, \pi/2)$,* the restricted function

$$y = \tan x, \qquad -\frac{\pi}{2} < x < \frac{\pi}{2}$$

*This is the generally accepted restriction.

is one-to-one and hence has an inverse function. See Figure 92.

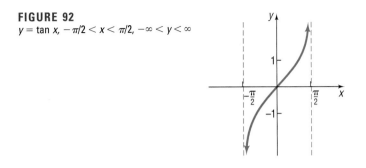

FIGURE 92
$y = \tan x, \ -\pi/2 < x < \pi/2, \ -\infty < y < \infty$

The inverse function is called the **inverse tangent** of x and is symbolized by $y = \tan^{-1} x$ (or by $y = \arctan x$). Thus,

$$y = \tan^{-1} x \quad \text{means} \quad x = \tan y \qquad (3)$$
$$\text{where} \quad -\frac{\pi}{2} < y < \frac{\pi}{2} \quad \text{and} \quad -\infty < x < \infty$$

Here, y is the angle whose tangent is x. The domain of the function $y = \tan^{-1} x$ is $-\infty < x < \infty$, and its range is $-\pi/2 < y < \pi/2$. The graph of $y = \tan^{-1} x$ can be obtained by reflecting the restricted portion of the graph of $y = \tan x$ about the line $y = x$, as shown in Figure 93(a). Figure 93(b) shows the graph using a graphing utility.

FIGURE 93
$y = \tan^{-1} x, \ -\infty < x < \infty,$
$-\pi/2 < y < \pi/2$

(a)

(b)

In general,

$$\tan^{-1}(\tan u) = u, \quad \text{where} \quad -\frac{\pi}{2} < u < \frac{\pi}{2}$$
$$\tan(\tan^{-1} v) = v, \quad \text{where} \quad -\infty < v < \infty$$

E X A M P L E   8   Finding the Exact Value of an Inverse Tangent Function

Find the exact value of $\tan^{-1} 1$.

Solution   Let $\theta = \tan^{-1} 1$. We seek the angle θ, $-\pi/2 < \theta < \pi/2$, whose tangent equals 1:

$$\theta = \tan^{-1} 1, \qquad -\frac{\pi}{2} < \theta < \frac{\pi}{2}$$

$$\tan \theta = 1, \qquad -\frac{\pi}{2} < \theta < \frac{\pi}{2}$$

Refer to Figure 92. The only angle θ within the interval $(-\pi/2, \pi/2)$ whose tangent is 1 is $\pi/4$.
  We conclude that

$$\theta = \frac{\pi}{4}$$

So

$$\tan^{-1} 1 = \frac{\pi}{4}$$

■

E X A M P L E   9   Finding the Exact Value of an Inverse Tangent Function

Find the exact value of $\tan^{-1}(-\sqrt{3})$.

Solution   Let $\theta = \tan^{-1}(-\sqrt{3})$. We seek the angle θ, $-\pi/2 < \theta < \pi/2$, whose tangent equals $-\sqrt{3}$:

$$\theta = \tan^{-1}(-\sqrt{3}), \qquad -\frac{\pi}{2} < \theta < \frac{\pi}{2}$$

$$\tan \theta = -\sqrt{3} \qquad -\frac{\pi}{2} < \theta < \frac{\pi}{2}$$

Refer again to Figure 92. The only angle θ within the interval $(-\pi/2, \pi/2)$ whose tangent is $-\sqrt{3}$ is $-\pi/3$. Thus,

$$\theta = -\frac{\pi}{3}$$

So

$$\tan^{-1}(-\sqrt{3}) = -\frac{\pi}{3}$$

■

Now work Problem 5.

E X A M P L E   10   Finding the Exact Value of Expressions Involving Inverse Trigonometric Functions

Find the exact value of $\sin[\cos^{-1}(\sqrt{3}/2)]$.

Solution   Let $\theta = \cos^{-1}(\sqrt{3}/2)$. We seek the angle θ, $0 \le \theta \le \pi$, whose cosine equals $\sqrt{3}/2$:

$$\cos \theta = \frac{\sqrt{3}}{2}, \qquad 0 \le \theta \le \pi$$

$$\theta = \frac{\pi}{6}$$

Now

$$\sin\left(\cos^{-1}\frac{\sqrt{3}}{2}\right) = \sin\theta = \sin\frac{\pi}{6} = \frac{1}{2}$$

∎

E X A M P L E   11

Finding the Exact Value of Expressions Involving Inverse Trigonometric Functions

Find the exact value of $\cos[\tan^{-1}(-1)]$.

Solution   Let $\theta = \tan^{-1}(-1)$. We seek the angle θ, $-\pi/2 < \theta < \pi/2$, whose tangent equals -1:

$$\tan\theta = -1, \qquad -\frac{\pi}{2} < \theta < \frac{\pi}{2}$$

$$\theta = -\frac{\pi}{4},$$

Now

$$\cos[\tan^{-1}(-1)] = \cos\theta = \cos\left(-\frac{\pi}{4}\right) = \frac{\sqrt{2}}{2}$$

∎

E X A M P L E   12

Finding the Exact Value of Expressions Involving Inverse Trigonometric Functions

Find the exact value of $\sec\left(\sin^{-1}\frac{1}{2}\right)$.

Solution   Let $\theta = \sin^{-1}\frac{1}{2}$. We seek the angle θ, $-\pi/2 \le \theta \le \pi/2$, whose sine equals $\frac{1}{2}$.

$$\sin\theta = \frac{1}{2}, \qquad -\frac{\pi}{2} \le \theta \le \frac{\pi}{2}$$

$$\theta = \frac{\pi}{6}$$

Now

$$\sec\left(\sin^{-1}\frac{1}{2}\right) = \sec\theta = \sec\frac{\pi}{6} = \frac{2}{\sqrt{3}} = \frac{2\sqrt{3}}{3}$$

∎

Now work Problem 25.

It is not necessary to be able to find the angle in order to solve problems like those given in Examples 10–12.

E X A M P L E   13

Finding the Exact Value of Expressions Involving Inverse Trigonometric Functions

Find the exact value of $\sin\left(\tan^{-1}\frac{1}{2}\right)$.

Solution   Let $\theta = \tan^{-1}\frac{1}{2}$. Then $\tan\theta = \frac{1}{2}$, where $-\pi/2 < \theta < \pi/2$. Because $\tan\theta > 0$, it follows that $0 < \theta < \pi/2$. Now, in Figure 94, we draw a triangle in the appropriate quadrant depicting $\tan \theta = \frac{1}{2}$. The hypotenuse of this triangle is easily found to be of length $\sqrt{5}$. Hence, the sine of θ is $1/\sqrt{5}$, so

$$\sin\left(\tan^{-1}\frac{1}{2}\right) = \sin \theta = \frac{1}{\sqrt{5}} = \frac{\sqrt{5}}{5}$$

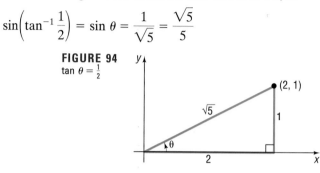

FIGURE 94
$\tan \theta = \frac{1}{2}$

E X A M P L E   14   Finding the Exact Value of Expressions Involving Inverse Trigonometric Functions

Find the exact value of $\tan(\cos^{-1}\frac{1}{3})$.

Solution Let $\theta = \cos^{-1}\frac{1}{3}$. Then $\cos \theta = \frac{1}{3}$, where $0 \le \theta \le \pi$. Because $\cos \theta > 0$, it follows that $0 \le \theta < \pi/2$. Look at the triangle in Figure 95. The side opposite angle θ is $\sqrt{8} = 2\sqrt{2}$. Thus,

$$\tan(\cos^{-1}\tfrac{1}{3}) = \tan \theta = 2\sqrt{2}$$

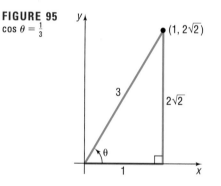

FIGURE 95
$\cos \theta = \frac{1}{3}$

E X A M P L E   15   Finding the Exact Value of Expressions Involving Inverse Trigonometric Functions

Find the exact value of $\cos[\sin^{-1}(-\frac{1}{3})]$.

Solution Let $\theta = \sin^{-1}(-\frac{1}{3})$. Then $\sin \theta = -\frac{1}{3}$ and $-\pi/2 \le \theta \le \pi/2$. Because $\sin \theta < 0$, it follows that $-\pi/2 \le \theta < 0$. Based on Figure 96, we conclude that

$$\cos\left[\sin^{-1}\left(-\frac{1}{3}\right)\right] = \cos \theta = \frac{2\sqrt{2}}{3}$$

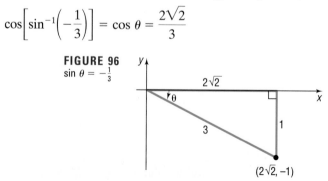

FIGURE 96
$\sin \theta = -\frac{1}{3}$

E X A M P L E   16   Finding the Exact Value of Expressions Involving Inverse Trigonometric Functions

FIGURE 97
$\cos \theta = -\frac{1}{3}$

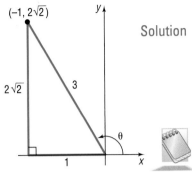

Find the exact value of $\tan[\cos^{-1}(-\frac{1}{3})]$.

Solution   Let $\theta = \cos^{-1}(-\frac{1}{3})$. Then $\cos \theta = -\frac{1}{3}$ and $0 \le \theta \le \pi$. Because $\cos \theta < 0$, it follows that $\pi/2 < \theta \le \pi$. Based on Figure 97, we conclude that

$$\tan\left[\cos^{-1}\left(-\frac{1}{3}\right)\right] = \tan \theta = \frac{2\sqrt{2}}{-1} = -2\sqrt{2}$$

 Now work Problem 39.

E X A M P L E   17   Establishing an Identity Involving Inverse Trigonometric Functions

Show that $\sin(\tan^{-1} v) = \dfrac{v}{\sqrt{1 + v^2}}$

Solution   Let $\theta = \tan^{-1} v$ so that $\tan \theta = v$, $-\pi/2 < \theta < \pi/2$. There are two possibilities: either $-\pi/2 < \theta < 0$ or $0 \le \theta < \pi/2$. If $0 \le \theta < \pi/2$, then $\tan \theta = v \ge 0$. Based on Figure 98, we conclude that

$$\sin(\tan^{-1} v) = \sin \theta = \frac{v}{\sqrt{1 + v^2}}$$

If $-\pi/2 < \theta < 0$, then $\tan \theta = v < 0$. Based on Figure 99, we conclude that

$$\sin(\tan^{-1} v) = \sin \theta = \frac{v}{\sqrt{1 + v^2}}$$

FIGURE 98
$v > 0$

FIGURE 99
$v < 0$

An alternative solution to Example 17 uses the fundamental identities. If $\theta = \tan^{-1} v$, then $\tan \theta = v$ and $-\pi/2 < \theta < \pi/2$, as before. As a result, we know that $\sec \theta > 0$. Thus,

$$\sin(\tan^{-1} v) = \sin \theta = \cos \theta \tan \theta = \frac{\tan \theta}{\sec \theta} = \frac{\tan \theta}{\sqrt{1 + \tan^2 \theta}} = \frac{v}{\sqrt{1 + v^2}}$$

$$\tan \theta = \frac{\sin \theta}{\cos \theta} \qquad \sec^2 \theta = 1 + \tan^2 \theta$$
$$\sec \theta > 0$$

Now work Problem 57.

The Remaining Inverse Trigonometric Functions

The inverse cotangent, inverse secant, and inverse cosecant functions are defined as follows:

$$y = \cot^{-1} x \quad \text{means} \quad x = \cot y \tag{4}$$
$$\text{where} \quad -\infty < x < \infty \quad \text{and} \quad 0 < y < \pi$$

$$y = \sec^{-1} x \quad \text{means} \quad x = \sec y \tag{5}$$
$$\text{where} \quad |x| \geq 1 \quad \text{and} \quad 0 \leq y \leq \pi, y \neq \frac{\pi}{2}$$

$$y = \csc^{-1} x \quad \text{means} \quad x = \csc y \tag{6}$$
$$\text{where} \quad |x| \geq 1 \quad \text{and} \quad -\frac{\pi}{2} \leq y \leq \frac{\pi}{2}, y \neq 0$$

Most calculators do not have keys for evaluating these inverse trigonometric functions. The easiest way to evaluate them is to convert to an inverse trigonometric function whose range is the same as the one to be evaluated. In this regard, notice that $y = \cot^{-1} x$ and $y = \sec^{-1} x$ (except where undefined) each have the same range as $y = \cos^{-1} x$, while $y = \csc^{-1} x$ (except where undefined) has the same range as $y = \sin^{-1} x$.

E X A M P L E 18   Approximating the Value of Inverse Trigonometric Functions

Use a calculator to approximate each expression rounded to two decimal places:

(a) $\sec^{-1} 3$ (b) $\csc^{-1}(-4)$ (c) $\cot^{-1}\frac{1}{2}$ (d) $\cot^{-1}(-2)$

Solution   First, set your calculator to radian mode.

(a) Let $\theta = \sec^{-1} 3$. Then $\sec \theta = 3$ and $0 \leq \theta \leq \pi$, $\theta \neq \pi/2$. Thus, $\cos \theta = \frac{1}{3}$ and

$$\sec^{-1} 3 = \theta = \cos^{-1} \frac{1}{3} \underset{\uparrow}{\approx} 1.23$$

Use a calculator.

(b) Let $\theta = \csc^{-1}(-4)$. Then $\csc \theta = -4$, $-\pi/2 \leq \theta \leq \pi/2$, $\theta \neq 0$. Thus, $\sin \theta = -\frac{1}{4}$ and

$$\csc^{-1}(-4) = \theta = \sin^{-1}\left(-\tfrac{1}{4}\right) \approx -0.25$$

(c) Let $\theta = \cot^{-1} \frac{1}{2}$. Then $\cot \theta = \frac{1}{2}$, $0 < \theta < \pi$. From these facts we know that θ is in quadrant I. We draw Figure 100 to help us to find $\cos \theta$. Thus, $\cos \theta = 1/\sqrt{5}$, $0 < \theta < \pi/2$, and

$$\cot^{-1}\frac{1}{2} = \theta = \cos^{-1}\frac{1}{\sqrt{5}} \approx 1.11$$

(d) Let $\theta = \cot^{-1}(-2)$. Then $\cot \theta = -2$, $0 < \theta < \pi$. From these facts we know that θ lies in quadrant II. We draw Figure 101 to help us to find $\cos \theta$. Thus, $\cos \theta = -2/\sqrt{5}$, $\pi/2 < \theta < \pi$, and

$$\cot^{-1}(-2) = \theta = \cos^{-1}\left(-\frac{2}{\sqrt{5}}\right) \approx 2.68$$

■

FIGURE 100
$\cot \theta = \frac{1}{2}, 0 < \theta < \pi$

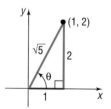

FIGURE 101
$\cot \theta = -2, 0 < \theta < \pi$

Now work Problem 67.

5.6   EXERCISES

In Problems 1–12, find the exact value of each expression.

1. $\sin^{-1} 0$
2. $\cos^{-1} 1$
3. $\sin^{-1}(-1)$
4. $\cos^{-1}(-1)$

5. $\tan^{-1} 0$
6. $\tan^{-1}(-1)$
7. $\sin^{-1} \dfrac{\sqrt{2}}{2}$
8. $\tan^{-1} \dfrac{\sqrt{3}}{3}$

9. $\tan^{-1} \sqrt{3}$
10. $\sin^{-1}\left(-\dfrac{\sqrt{3}}{2}\right)$
11. $\cos^{-1}\left(-\dfrac{\sqrt{3}}{2}\right)$
12. $\sin^{-1}\left(-\dfrac{\sqrt{2}}{2}\right)$

In Problems 13–24, use a calculator to find the value of each expression rounded to two decimal places.

13. $\sin^{-1} 0.1$
14. $\cos^{-1} 0.6$
15. $\tan^{-1} 5$
16. $\tan^{-1} 0.2$

17. $\cos^{-1}\frac{7}{8}$
18. $\sin^{-1}\frac{1}{8}$
19. $\tan^{-1}(-0.4)$
20. $\tan^{-1}(-3)$

21. $\sin^{-1}(-0.12)$
22. $\cos^{-1}(-0.44)$
23. $\cos^{-1} \dfrac{\sqrt{2}}{3}$
24. $\sin^{-1} \dfrac{\sqrt{3}}{5}$

In Problems 25–46, find the exact value of each expression.

25. $\cos\left(\sin^{-1} \dfrac{\sqrt{2}}{2}\right)$
26. $\sin\left(\cos^{-1} \dfrac{1}{2}\right)$
27. $\tan\left[\cos^{-1}\left(-\dfrac{\sqrt{3}}{2}\right)\right]$
28. $\tan\left[\sin^{-1}\left(-\dfrac{1}{2}\right)\right]$

29. $\sec\left(\cos^{-1} \dfrac{1}{2}\right)$
30. $\cot\left[\sin^{-1}\left(-\dfrac{1}{2}\right)\right]$
31. $\csc(\tan^{-1} 1)$
32. $\sec(\tan^{-1} \sqrt{3})$

33. $\sin[\tan^{-1}(-1)]$
34. $\cos\left[\sin^{-1}\left(-\dfrac{\sqrt{3}}{2}\right)\right]$
35. $\sec\left[\sin^{-1}\left(-\dfrac{1}{2}\right)\right]$
36. $\csc\left[\cos^{-1}\left(-\dfrac{\sqrt{3}}{2}\right)\right]$

37. $\tan\left(\sin^{-1} \dfrac{1}{3}\right)$
38. $\tan\left(\cos^{-1} \dfrac{1}{3}\right)$
39. $\sec\left(\tan^{-1} \dfrac{1}{2}\right)$
40. $\cos\left(\sin^{-1} \dfrac{\sqrt{2}}{3}\right)$

41. $\cot\left[\sin^{-1}\left(-\dfrac{\sqrt{2}}{3}\right)\right]$
42. $\csc[\tan^{-1}(-2)]$
43. $\sin[\tan^{-1}(-3)]$
44. $\cot\left[\cos^{-1}\left(-\dfrac{\sqrt{3}}{3}\right)\right]$

45. $\sec\left(\sin^{-1} \dfrac{2\sqrt{5}}{5}\right)$
46. $\csc\left(\tan^{-1} \dfrac{1}{2}\right)$

In Problems 47–56, use a calculator to find the value of each expression rounded to two decimal places.

47. $\sin^{-1}(\tan 0.5)$
48. $\cos^{-1}(\tan 0.4)$
49. $\tan^{-1}(\sin 0.1)$
50. $\tan^{-1}(\cos 0.2)$

51. $\cos^{-1}(\sin 1)$
52. $\tan^{-1}(\cos 1)$
53. $\sin^{-1}\left(\tan \dfrac{\pi}{8}\right)$
54. $\cos^{-1}\left(\sin \dfrac{\pi}{8}\right)$

55. $\tan^{-1}\left(\sin \dfrac{\pi}{8}\right)$
56. $\tan^{-1}\left(\cos \dfrac{\pi}{8}\right)$

57. Show that $\sec(\tan^{-1} v) = \sqrt{1 + v^2}$.
58. Show that $\tan(\sin^{-1} v) = v/\sqrt{1 - v^2}$.
59. Show that $\tan(\cos^{-1} v) = \sqrt{1 - v^2}/v$.
60. Show that $\sin(\cos^{-1} v) = \sqrt{1 - v^2}$.
61. Show that $\cos(\sin^{-1} v) = \sqrt{1 - v^2}$.
62. Show that $\cos(\tan^{-1} v) = 1/\sqrt{1 + v^2}$.
63. Show that $\sin^{-1} v + \cos^{-1} v = \pi/2$.
64. Show that $\tan^{-1} v + \cot^{-1} v = \pi/2$.
65. Show that $\tan^{-1}(1/v) = \pi/2 - \tan^{-1} v$.
66. Show that $\cot^{-1} e^v = \tan^{-1} e^{-v}$.

In Problems 67–78, use a calculator to find the value of each expression rounded to two decimal places.

67. $\sec^{-1} 4$
68. $\csc^{-1} 5$
69. $\cot^{-1} 2$
70. $\sec^{-1}(-3)$

71. $\csc^{-1}(-3)$
72. $\cot^{-1}(-\frac{1}{2})$
73. $\cot^{-1}(-\sqrt{5})$
74. $\cot^{-1}(-8.1)$

75. $\csc^{-1}(-\frac{3}{2})$
76. $\sec^{-1}(-\frac{4}{3})$
77. $\cot^{-1}(-\frac{3}{2})$
78. $\cot^{-1}(-\sqrt{10})$

79. **Being the First To See the Rising Sun** Cadillac Mountain, elevation 1530 feet, is located in Acadia National Park, Maine, and is the highest peak on the east coast of the United States. It is said that a person standing on the summit will be the first person in the United States to see the rays of the rising Sun. How much sooner would a person atop Cadillac Mountain see the first rays than a person standing below, at sea level?

[**Hint:** Consult the figure on page 468. When the person at D sees the first rays of the Sun, the person at P does not. The person at P sees the first rays of the Sun only after Earth has rotated so that

P is at location Q. Compute the length of arc s subtended by the central angle θ. Then use the fact that in 24 hours a length of 2π(3960) miles is subtended, and find the time it takes to subtend the length s.]

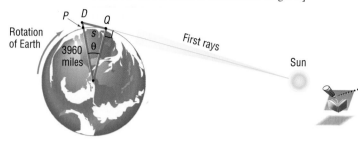

80. For what numbers x does $\sin(\sin^{-1} x) = x$?

81. For what numbers x does $\cos(\cos^{-1} x) = x$?

82. For what numbers x does $\sin^{-1}(\sin x) = x$?

83. For what numbers x does $\cos^{-1}(\cos x) = x$?

84. Graph $y = \cot^{-1} x$.

85. Graph $y = \sec^{-1} x$.

86. Graph $y = \csc^{-1} x$.

87. Explain in your own words how you would use your calculator to find the value of $\cot^{-1} 10$.

88. Consult three books on calculus and write down the definition in each of $y = \sec^{-1} x$ and $y = \csc^{-1} x$. Compare these with the definitions given in this book.

CHAPTER REVIEW

THINGS TO KNOW

Definitions

Angle in standard position	Vertex is at the origin; initial side is along the positive x-axis
Degree (1°)	$1° = \frac{1}{360}$ revolution
Radian (1)	The measure of a central angle whose rays subtend an arc whose length is the radius of the circle
Unit circle	Center at origin; radius = 1 Equation: $x^2 + y^2 = 1$
Trigonometric functions	$P = (a, b)$ is the point on the unit circle corresponding to $\theta = t$ radians;

$$\sin t = \sin \theta = b \qquad\qquad \cos t = \cos \theta = a$$

$$\tan t = \tan \theta = \frac{b}{a}, \quad a \neq 0 \qquad \cot t = \cot \theta = \frac{a}{b}, \quad b \neq 0$$

$$\csc t = \csc \theta = \frac{1}{b}, \quad b \neq 0 \qquad \sec t = \sec \theta = \frac{1}{a}, \quad a \neq 0$$

Periodic function	$f(\theta + p) = f(\theta)$, for all $\theta, p > 0$, where the smallest such p is the fundamental period
Acute angle	An angle θ whose measure is $0° < \theta < 90°$ (or $0 < \theta < \pi/2$)
Complementary angles	Two acute angles whose sum is 90° ($\pi/2$)
Cofunction	The following pairs of functions are cofunctions of each other: sine and cosine; tangent and cotangent; secant and cosecant
Reference angle of θ	The acute angle formed by the terminal side of θ and either the positive or negative x-axis

Formulas

1 revolution = 360°	
= 2π radians	
$s = r\theta$	θ is measured in radians; s is the length of arc subtended by the central angle θ of the circle of radius r.
$v = r\omega$	v is the linear speed along the circle of radius r; ω is the angular speed (measured in radians per unit time)

Table of Values

θ (RADIANS)	θ (DEGREES)	$\sin\theta$	$\cos\theta$	$\tan\theta$	$\csc\theta$	$\sec\theta$	$\cot\theta$
0	0°	0	1	0	Not defined	1	Not defined
$\pi/6$	30°	1/2	$\sqrt{3}/2$	$\sqrt{3}/3$	2	$2\sqrt{3}/3$	$\sqrt{3}$
$\pi/4$	45°	$\sqrt{2}/2$	$\sqrt{2}/2$	1	$\sqrt{2}$	$\sqrt{2}$	1
$\pi/3$	60°	$\sqrt{3}/2$	1/2	$\sqrt{3}$	$2\sqrt{3}/3$	2	$\sqrt{3}/3$
$\pi/2$	90°	1	0	Not defined	1	Not defined	0
π	180°	0	-1	0	Not defined	-1	Not defined
$3\pi/2$	270°	-1	0	Not defined	-1	Not defined	0

Identities

$$\tan\theta = \frac{\sin\theta}{\cos\theta}, \qquad \cot\theta = \frac{\cos\theta}{\sin\theta}$$

$$\cot\theta = \frac{1}{\tan\theta}, \qquad \sec\theta = \frac{1}{\cos\theta}, \qquad \csc\theta = \frac{1}{\sin\theta}$$

$$\sin^2\theta + \cos^2\theta = 1, \quad \tan^2\theta + 1 = \sec^2\theta, \quad 1 + \cot^2\theta = \csc^2\theta$$

Properties of the Trigonometric Functions

$y = \sin x$ Domain: $-\infty < x < \infty$
Range: $-1 \le y \le 1$
Periodic: period $= 2\pi$ (360°)
Odd function

$y = \cos x$ Domain: $-\infty < x < \infty$
Range: $-1 \le y \le 1$
Periodic: period $= 2\pi$ (360°)
Even function

$y = \tan x$ Domain: $-\infty < x < \infty$, except odd multiples of $\pi/2$ (90°)
Range: $-\infty < y < \infty$
Periodic: period $= \pi$ (180°)
Odd function

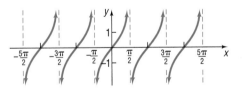

$y = \csc x$ Domain: $-\infty < x < \infty$, except integral multiples of π (180°)
Range: $|y| \ge 1$
Periodic: period $= 2\pi$ (360°)
Odd function

$y = \sec x$ Domain: $-\infty < x < \infty$, except odd multiples of $\pi/2$ (90°)
Range: $|y| \ge 1$
Periodic: period $= 2\pi$ (360°)
Even function

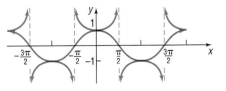

$y = \cot x$ — Domain: $-\infty < x < \infty$, except integral multiples of π (180°)
Range: $-\infty < y < \infty$
Periodic: period $= \pi$ (180°)
Odd function

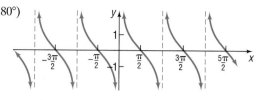

Definitions of the six inverse trigonometric functions

$y = \sin^{-1} x$ means $x = \sin y$ where $-1 \le x \le 1, -\pi/2 \le y \le \pi/2$
$y = \cos^{-1} x$ means $x = \cos y$ where $-1 \le x \le 1, 0 \le y \le \pi$
$y = \tan^{-1} x$ means $x = \tan y$ where $-\infty < x < \infty, -\pi/2 < y < \pi/2$
$y = \cot^{-1} x$ means $x = \cot y$ where $-\infty < x < \infty, 0 < y < \pi$
$y = \sec^{-1} x$ means $x = \sec y$ where $|x| \ge 1, 0 \le y \le \pi, y \ne \pi/2$
$y = \csc^{-1} x$ means $x = \csc y$ where $|x| \ge 1, -\pi/2 \le y \le \pi/2, y \ne 0$

HOW TO

Convert an angle from radian measure to degree measure

Convert an angle from degree measure to radian measure

Find the value of each of the remaining trigonometric functions if the value of one function and the quadrant of the angle are given

Use the theorem on cofunctions of complementary angles

Graph the trigonometric functions, including transformations

Use reference angles to find the value of a trigonometric function

Use a calculator to find the value of a trigonometric function

Find the exact value of certain inverse trigonometric functions

Use a calculator to find the approximate values of inverse trigonometric functions

FILL-IN-THE-BLANK ITEMS

1. Two rays drawn with a common vertex form a(n) _____. One of the rays is called the _____ _____; the other is called the _____ _____.

2. In the formula $s = r\theta$ for measuring the length s of arc along a circle of radius r, the angle θ must be measured in _____.

3. 180 degrees = _____ radians.

4. Two acute angles whose sum is a right angle are called _____.

5. The sine and _____ functions are cofunctions.

6. An angle is in _____ _____ if its vertex is at the origin and its initial side coincides with the positive x-axis.

7. The reference angle of 135° is _____.

8. The sine, cosine, cosecant, and secant functions have period _____; the tangent and cotangent functions have period _____.

9. Which of the trigonometric functions have graphs that are symmetric with respect to the y-axis?

10. Which of the trigonometric functions have graphs that are symmetric with respect to the origin?

11. The function $y = \sin^{-1} x$ has domain _____ and range _____.

12. The value of $\sin^{-1}[\cos(\pi/2)]$ is _____.

TRUE/FALSE ITEMS

T F **1.** In the formula $s = r\theta$, r is the radius of a circle and s is the arc subtended by a central angle θ, where θ is measured in degrees.

T F **2.** $|\sin \theta| \le 1$

T F **3.** $1 + \tan^2 \theta = \csc^2 \theta$

T F **4.** The only even trigonometric functions are the cosine and secant functions.

T F **5.** $\tan 62° = \cot 38°$

T F **6.** $\sin 182° = \cos 2°$

T F **7.** The graphs of $y = \tan x$, $y = \cot x$, $y = \sec x$, and $y = \csc x$ have infinitely many vertical asymptotes.

T F **8.** The domain of $y = \sin^{-1} x$ is $-\pi/2 \le x \le \pi/2$.

T F **9.** $\cos(\sin^{-1} 0) = 1$ and $\sin(\cos^{-1} 0) = 1$.

REVIEW EXERCISES

In Problems 1–4, convert each angle in degrees to radians. Express your answer as a multiple of π.

1. $135°$ **2.** $210°$ **3.** $18°$ **4.** $15°$

In Problems 5–8, convert each angle in radians to degrees.

5. $3\pi/4$ **6.** $2\pi/3$ **7.** $-5\pi/2$ **8.** $-3\pi/2$

In Problems 9–30, find the exact value of each expression. Do not use a calculator.

9. $\tan \dfrac{\pi}{4} - \sin \dfrac{\pi}{6}$

10. $\cos \dfrac{\pi}{3} + \sin \dfrac{\pi}{2}$

11. $3 \sin 45° - 4 \tan \dfrac{\pi}{6}$

12. $4 \cos 60° + 3 \tan \dfrac{\pi}{3}$

13. $6 \cos \dfrac{3\pi}{4} + 2 \tan\left(-\dfrac{\pi}{3}\right)$

14. $3 \sin \dfrac{2\pi}{3} - 4 \cos \dfrac{5\pi}{2}$

15. $\sec\left(-\dfrac{\pi}{3}\right) - \cot\left(-\dfrac{5\pi}{4}\right)$

16. $4 \csc \dfrac{3\pi}{4} - \cot\left(-\dfrac{\pi}{4}\right)$

17. $\tan \pi + \sin \pi$

18. $\cos \dfrac{\pi}{2} - \csc\left(-\dfrac{\pi}{2}\right)$

19. $\cos 180° - \tan(-45°)$

20. $\sin 270° + \cos(-180°)$

21. $\sin^2 20° + \dfrac{1}{\sec^2 20°}$

22. $\dfrac{1}{\cos^2 40°} - \dfrac{1}{\cot^2 40°}$

23. $\sec 50° \cos 50°$

24. $\tan 10° \cot 10°$

25. $\dfrac{\sin 50°}{\cos 40°}$

26. $\dfrac{\tan 20°}{\cot 70°}$

27. $\dfrac{\sin(-40°)}{\cos 50°}$

28. $\tan(-20°) \cot 20°$

29. $\sin 400° \sec(-50°)$

30. $\cot 200° \cot(-70°)$

In Problems 31–46, find the exact value of each of the remaining trigonometric functions.

31. $\sin \theta = -\dfrac{4}{5}$, $\cos \theta > 0$

32. $\cos \theta = -\dfrac{3}{5}$, $\sin \theta < 0$

33. $\tan \theta = \dfrac{12}{5}$, $\sin \theta < 0$

34. $\cot \theta = \dfrac{12}{5}$, $\cos \theta < 0$

35. $\sec \theta = -\dfrac{5}{4}$, $\tan \theta < 0$

36. $\csc \theta = -\dfrac{5}{3}$, $\cot \theta < 0$

37. $\sin \theta = \dfrac{12}{13}$, θ in quadrant II

38. $\cos \theta = -\dfrac{3}{5}$, θ in quadrant III

39. $\sin \theta = -\dfrac{5}{13}$, $3\pi/2 < \theta < 2\pi$

40. $\cos \theta = \dfrac{12}{13}$, $3\pi/2 < \theta < 2\pi$

41. $\tan \theta = \dfrac{1}{3}$, $180° < \theta < 270°$

42. $\tan \theta = -\dfrac{2}{3}$, $90° < \theta < 180°$

43. $\sec \theta = 3$, $3\pi/2 < \theta < 2\pi$

44. $\csc \theta = -4$, $\pi < \theta < 3\pi/2$

45. $\cot \theta = -2$, $\pi/2 < \theta < \pi$

46. $\tan \theta = -2$, $3\pi/2 < \theta < 2\pi$

In Problems 47–54, graph each function. Each graph should contain at least one period.

47. $y = 2 \sin 4x$ **48.** $y = -3 \cos 2x$ **49.** $y = -2 \cos\left(x + \dfrac{\pi}{2}\right)$ **50.** $y = 3 \sin(x - \pi)$

51. $y = \tan(x + \pi)$ **52.** $y = -\tan\left(x - \dfrac{\pi}{2}\right)$ **53.** $y = -2 \tan 3x$ **54.** $y = 4 \tan 2x$

In Problems 55–70, find the exact value of each expression.

55. $\sin^{-1} 1$ **56.** $\cos^{-1} 0$ **57.** $\tan^{-1} 1$ **58.** $\sin^{-1}\left(-\dfrac{1}{2}\right)$

59. $\cos^{-1}\left(-\dfrac{\sqrt{3}}{2}\right)$ **60.** $\tan^{-1}(-\sqrt{3})$ **61.** $\sin\left(\cos^{-1}\dfrac{\sqrt{2}}{2}\right)$ **62.** $\cos(\sin^{-1} 0)$

63. $\tan\left[\sin^{-1}\left(-\dfrac{\sqrt{3}}{2}\right)\right]$ **64.** $\tan\left[\cos^{-1}\left(-\dfrac{1}{2}\right)\right]$ **65.** $\sec\left(\tan^{-1}\dfrac{\sqrt{3}}{3}\right)$ **66.** $\csc\left(\sin^{-1}\dfrac{\sqrt{3}}{2}\right)$

67. $\sin\left(\tan^{-1}\dfrac{3}{4}\right)$ **68.** $\cos\left(\sin^{-1}\dfrac{3}{5}\right)$ **69.** $\tan\left[\sin^{-1}\left(-\dfrac{4}{5}\right)\right]$ **70.** $\tan\left[\cos^{-1}\left(-\dfrac{3}{5}\right)\right]$

71. Find the length of arc subtended by a central angle of 30° on a circle of radius 2 feet.

72. The minute hand of a clock is 8 inches long. How far does the tip of the minute hand move in 30 minutes? How far does it move in 20 minutes?

73. **Angular Speed of a Race Car**   A race car is driven around a circular track at a constant speed of 180 miles per hour. If the diameter of the track is $\frac{1}{2}$ mile, what is the angular speed of the car? Express your answer in revolutions per hour (which is equivalent to laps per hour).

74. **Angular Speed of a Race Car**   Repeat Problem 73 if the car goes only 150 miles per hour.

75. **Lighthouse Beacons**   The Montauk Lighthouse on Long Island has dual beams (two light sources opposite each other). Ships at sea observe a blinking light every 5 seconds. What rotation speed is required to do this?

76. **Spin Balancing Tires**   The radius of each wheel of a car is 16 inches. At how many revolutions per minute should a spin balancer be set to balance the tires at a speed of 90 miles per hour? Is the setting different for a wheel of radius 14 inches? What is this setting?

Analytic Trigonometry

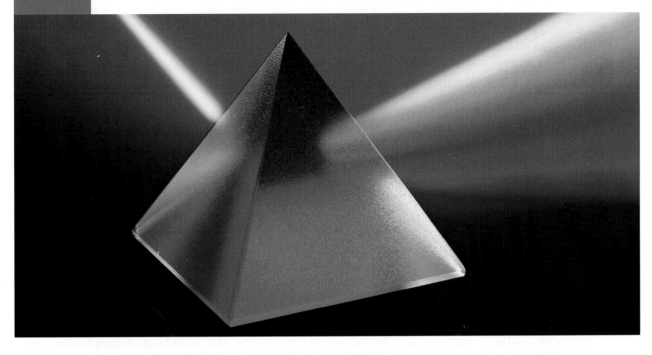

Extra credit is always welcome in school and this time your physics professor has given you a golden opportunity to shine! All you need to do is to create a display on rainbows for the college open house. You will want to use the Sullivan website at:

www.prenhall.com/sullivan

to find the information needed to complete your display.

PREPARING FOR THIS CHAPTER

Before getting started on this chapter, review the following concepts:

Fundamental Identities *(p. 426)*

Values of the Trigonometric Functions of Certain Angles *(p. 411 and p. 414)*

Solving Equations (Chapter 1, Section 1.5)

OUTLINE

THE MATHEMATICS OF RAINBOWS

Rainbows, those incredible displays of light and water, will always be a source of wonder and inspiration. From pots-o-gold to bridges for the gods, countless myths and legends have been told to explain rainbows.

In 1621, the basic Law of Refraction was proposed by the Dutch scientist Willebrord Snell. The Law of Refraction and the Law of Reflection provide the keys to the modern explanation for the formation of rainbows. Refraction is the bending of light's path as it passes from one medium into another. When this happens the different wavelengths of light separate. The algebraic expression for Snell's Law looks similar to the Law of Sines, except that the constants no longer refer to the lengths of the sides of a triangle, but instead refer to a "refractive index".

You have been asked by your physics professor to make a display on rainbows for the college open house.

1. You begin working on the project by brainstorming. Since this is the age of the World Wide Web, you decide to create an interactive display! This gets the adrenaline going. You make a web search to collect some ideas. Your web search generates thousands of sites! You select some of your favorites. At this point you want to make a list of questions that your display might answer. For example, did you ever notice that you generally will see rainbows only in the morning or afternoon? If you were to make a display of a rainbow, how would you do it?

2. Since your display is for the physics department, you want to be sure to find some examples that reveal the physical properties of rainbows. What would you add? For some examples, access *keys*.

3. To get the visitors involved in your booth, you decide to perform the following experiment: Draw a circle on a piece of paper. This will represent a water droplet. Now, with a ruler draw a line, a light ray, from the edge of your paper to your "water drop". Make a sketch of your guess of the path of the light ray as it passes through the water drop. You try this out on some of your friends to see if it will work. Do you want to use it? Why or why not?

4. Next imagine a curtain of water droplets and a light source. Try to make an illustration as to where the light source, the water curtain, the rainbow, and an observer will be located with respect to one another. For a solution check out: *Descartes and Rainbows*.

5. For the graphic of your display you decide to make a large scale illustration of the experiment of question 3. (For this exercise you will want to review *Snell's Law*.) You use colored felt tip pens, a compass, a ruler, a calculator, and a protractor to make an *exact drawing*. You want this to be right since visitors will ask you to explain your drawing to them. Make the drawing and then try to explain it to a friend. How did you do?

In this chapter we continue the derivation of identities involving the trigonometric functions. These identities play an important role in calculus, the physical and life sciences, and economics, where they are used to simplify complicated expressions.

The last section of this chapter provides techniques for solving equations that contain trigonometric functions.

6.1 | TRIGONOMETRIC IDENTITIES

1 Establish Identities

First, we review a fundamental definition:

Two functions *f* and *g* are said to be **identically equal** if

$$f(x) = g(x)$$

for every value of *x* for which both functions are defined. Such an equation is referred to as an **identity**. An equation that is not an identity is called a **conditional equation**.

For example, the following expressions are identities:

$$(x + 1)^2 = x^2 + 2x + 1 \qquad \sin^2 x + \cos^2 x = 1 \qquad \csc x = \frac{1}{\sin x}$$

The following expressions are conditional equations:

$2x + 5 = 0$ True only if $x = -\dfrac{5}{2}$.

$\sin x = 0$ True only if $x = k\pi$, k an integer.

$\sin x = \cos x$ True only if $x = \dfrac{\pi}{4} + 2k\pi$ or $x = \dfrac{5\pi}{4} + 2k\pi$, k an integer.

The following boxes summarize the trigonometric identities we have established thus far:

Quotient Identities

$$\tan \theta = \frac{\sin \theta}{\cos \theta} \qquad \cot \theta = \frac{\cos \theta}{\sin \theta}$$

Reciprocal Identities

$$\csc \theta = \frac{1}{\sin \theta} \qquad \sec \theta = \frac{1}{\cos \theta} \qquad \cot \theta = \frac{1}{\tan \theta}$$

Pythagorean Identities

$$\sin^2 \theta + \cos^2 \theta = 1 \qquad \tan^2 \theta + 1 = \sec^2 \theta$$
$$1 + \cot^2 \theta = \csc^2 \theta$$

Even-Odd Identities

$$\sin(-\theta) = -\sin\theta \qquad \cos(-\theta) = \cos\theta \qquad \tan(-\theta) = -\tan\theta$$
$$\csc(-\theta) = -\csc\theta \qquad \sec(-\theta) = \sec\theta \qquad \cot(-\theta) = -\cot\theta$$

This list of identities comprises what we shall refer to as the **basic trigonometric identities.** These identities should not merely be memorized, but should be *known* (just as you know your name rather than have it memorized). In fact, the use made of a basic identity is often a minor variation of the form listed here. For example, we might want to use $\sin^2\theta = 1 - \cos^2\theta$ instead of $\sin^2\theta + \cos^2\theta = 1$. For this reason, among others, you need to know these relationships and be quite comfortable with variations of them.

In the examples that follow, the directions will read "Establish the identity . . . ." As you will see, this is accomplished by starting with one side of the given equation (usually the one containing the more complicated expression) and, using appropriate basic identities and algebraic manipulations, arriving at the other side. The selection of an appropriate basic identity to obtain the desired result is learned only through experience and lots of practice.

EXAMPLE 1

Establishing an Identity

Establish the identity: $\sec\theta \cdot \sin\theta = \tan\theta$

Solution

We start with the left side, because it contains the more complicated expression, and apply a reciprocal identity:

$$\sec\theta \cdot \sin\theta = \frac{1}{\cos\theta} \cdot \sin\theta = \frac{\sin\theta}{\cos\theta} = \tan\theta$$

Having arrived at the right side, the identity is established. ▬

Comment   A graphing utility can be used to provide evidence of an identity. For example, if we graph $Y_1 = \sec\theta \cdot \sin\theta$ and $Y_2 = \tan\theta$, the graphs appear to be the same. This provides evidence that $Y_1 = Y_2$. However, it does not prove their equality. A graphing utility *cannot be used to establish an identity*—identities must be established algebraically.

Now work Problem 1.

EXAMPLE 2

Establishing an Identity

Establish the identity: $\sin^2(-\theta) + \cos^2(-\theta) = 1$

Solution

We begin with the left side and apply even–odd identities:

$$\sin^2(-\theta) + \cos^2(-\theta) = [\sin(-\theta)]^2 + [\cos(-\theta)]^2$$
$$= (-\sin\theta)^2 + (\cos\theta)^2$$
$$= (\sin\theta)^2 + (\cos\theta)^2$$
$$= 1$$ ▬

EXAMPLE 3

Establishing an Identity

Establish the identity: $\dfrac{\sin^2(-\theta) - \cos^2(-\theta)}{\sin(-\theta) - \cos(-\theta)} = \cos\theta - \sin\theta$

Solution We begin with two observations: The left side appears to contain the more complicated expression. Also, the left side contains expressions with the argument $-\theta$, whereas the right side contains expressions with the argument θ. We decide, therefore, to start with the left side and apply even–odd identities:

$$\frac{\sin^2(-\theta) - \cos^2(-\theta)}{\sin(-\theta) - \cos(-\theta)} = \frac{[\sin(-\theta)]^2 - [\cos(-\theta)]^2}{\sin(-\theta) - \cos(-\theta)}$$

$$= \frac{(-\sin\theta)^2 - (\cos\theta)^2}{-\sin\theta - \cos\theta} \qquad \text{Even–odd identities.}$$

$$= \frac{(\sin\theta)^2 - (\cos\theta)^2}{-\sin\theta - \cos\theta} \qquad \text{Simplify.}$$

$$= \frac{(\sin\theta - \cos\theta)(\sin\theta + \cos\theta)}{-(\sin\theta + \cos\theta)} \qquad \text{Factor.}$$

$$= \cos\theta - \sin\theta \qquad \text{Cancel and simplify.} \qquad \blacksquare$$

E X A M P L E   4   Establishing an Identity

Establish the identity: $\dfrac{1 + \tan\theta}{1 + \cot\theta} = \tan\theta$

Solution $\dfrac{1 + \tan\theta}{1 + \cot\theta} = \dfrac{1 + \tan\theta}{1 + \dfrac{1}{\tan\theta}} = \dfrac{1 + \tan\theta}{\dfrac{\tan\theta + 1}{\tan\theta}} = \dfrac{\tan\theta(1 + \tan\theta)}{\tan\theta + 1} = \tan\theta$ $\blacksquare$

 Now work Problem 9.

When sums or differences of quotients appear, it is usually best to rewrite them as a single quotient, especially if the other side of the identity consists of only one term.

E X A M P L E   5   Establishing an Identity

Establish the identity: $\dfrac{\sin\theta}{1 + \cos\theta} + \dfrac{1 + \cos\theta}{\sin\theta} = 2\csc\theta$

Solution The left side is more complicated, so we start with it and proceed to add:

$$\frac{\sin\theta}{1 + \cos\theta} + \frac{1 + \cos\theta}{\sin\theta} = \frac{\sin^2\theta + (1 + \cos\theta)^2}{(1 + \cos\theta)(\sin\theta)}$$

$$= \frac{\sin^2\theta + 1 + 2\cos\theta + \cos^2\theta}{(1 + \cos\theta)(\sin\theta)}$$

$$= \frac{(\sin^2\theta + \cos^2\theta) + 1 + 2\cos\theta}{(1 + \cos\theta)(\sin\theta)}$$

$$= \frac{2 + 2\cos\theta}{(1 + \cos\theta)(\sin\theta)}$$

$$= \frac{2(1 + \cos\theta)}{(1 + \cos\theta)(\sin\theta)}$$

$$= \frac{2}{\sin\theta}$$

$$= 2\csc\theta \qquad \blacksquare$$

E X A M P L E   6   Establishing an Identity

Establish the identity: $\dfrac{1}{\cos \theta} - \dfrac{\cos \theta}{1 + \sin \theta} = \tan \theta$

Solution $\dfrac{1}{\cos \theta} - \dfrac{\cos \theta}{1 + \sin \theta} = \dfrac{1 + \sin \theta - \cos^2 \theta}{\cos \theta(1 + \sin \theta)}$

$$= \dfrac{\sin \theta + (1 - \cos^2 \theta)}{\cos \theta(1 + \sin \theta)}$$

$$= \dfrac{\sin \theta + \sin^2 \theta}{\cos \theta(1 + \sin \theta)} \quad {\scriptstyle 1 - \cos^2 \theta = \sin^2 \theta}$$

$$= \dfrac{\sin \theta(1 + \sin \theta)}{\cos \theta(1 + \sin \theta)}$$

$$= \tan \theta \qquad\qquad\qquad ■$$

Now work Problem 23.

Sometimes, multiplying the numerator and denominator by an appropriate factor will result in a simplification.

E X A M P L E   7   Establishing an Identity

Establish the identity: $\dfrac{1 - \sin \theta}{\cos \theta} = \dfrac{\cos \theta}{1 + \sin \theta}$

Solution   We start with the left side and multiply the numerator and the denominator by $1 + \sin \theta$. (Alternatively, we could multiply the numerator and denominator of the right side by $1 - \sin \theta$.)

$$\dfrac{1 - \sin \theta}{\cos \theta} = \dfrac{1 - \sin \theta}{\cos \theta} \cdot \dfrac{1 + \sin \theta}{1 + \sin \theta}$$

$$= \dfrac{1 - \sin^2 \theta}{\cos \theta(1 + \sin \theta)}$$

$$= \dfrac{\cos^2 \theta}{\cos \theta(1 + \sin \theta)}$$

$$= \dfrac{\cos \theta}{1 + \sin \theta} \qquad\qquad ■$$

Now work Problem 35.

Although a lot of practice is the only real way to learn how to establish identities, the following guidelines should prove helpful:

Guidelines for
Establishing Identities

1. It is almost always preferable to start with the side containing the more complicated expression.
2. Rewrite sums or differences of quotients as a single quotient.
3. Sometimes rewriting one side in terms of sines and cosines only will help.
4. Always keep your goal in mind. As you manipulate one side of the expression, you must keep in mind the form of the expression on the other side.

Warning   Be careful not to handle identities to be established as if they were equations. You *cannot* establish an identity by such methods as adding the same expression to each side and obtaining a true statement. This practice is not allowed, because the original statement is precisely the one that you are trying to establish. You do not know until it has been established that it is, in fact, true.

6.1 EXERCISES

In Problems 1–80, establish each identity.

1. $\csc \theta \cdot \cos \theta = \cot \theta$

2. $\csc \theta \cdot \tan \theta = \sec \theta$

3. $1 + \tan^2(-\theta) = \sec^2 \theta$

4. $1 + \cot^2(-\theta) = \csc^2 \theta$

5. $\cos \theta(\tan \theta + \cot \theta) = \csc \theta$

6. $\sin \theta(\cot \theta + \tan \theta) = \sec \theta$

7. $\tan \theta \cot \theta - \cos^2 \theta = \sin^2 \theta$

8. $\sin \theta \csc \theta - \cos^2 \theta = \sin^2 \theta$

9. $(\sec \theta - 1)(\sec \theta + 1) = \tan^2 \theta$

10. $(\csc \theta - 1)(\csc \theta + 1) = \cot^2 \theta$

11. $(\sec \theta + \tan \theta)(\sec \theta - \tan \theta) = 1$

12. $(\csc \theta + \cot \theta)(\csc \theta - \cot \theta) = 1$

13. $\sin^2 \theta(1 + \cot^2 \theta) = 1$

14. $(1 - \sin^2 \theta)(1 + \tan^2 \theta) = 1$

15. $(\sin \theta + \cos \theta)^2 + (\sin \theta - \cos \theta)^2 = 2$

16. $\tan^2 \theta \cos^2 \theta + \cot^2 \theta \sin^2 \theta = 1$

17. $\sec^4 \theta - \sec^2 \theta = \tan^4 \theta + \tan^2 \theta$

18. $\csc^4 \theta - \csc^2 \theta = \cot^4 \theta + \cot^2 \theta$

19. $\sec \theta - \tan \theta = \dfrac{\cos \theta}{1 + \sin \theta}$

20. $\csc \theta - \cot \theta = \dfrac{\sin \theta}{1 + \cos \theta}$

21. $3 \sin^2 \theta + 4 \cos^2 \theta = 3 + \cos^2 \theta$

22. $9 \sec^2 \theta - 5 \tan^2 \theta = 5 + 4 \sec^2 \theta$

23. $1 - \dfrac{\cos^2 \theta}{1 + \sin \theta} = \sin \theta$

24. $1 - \dfrac{\sin^2 \theta}{1 - \cos \theta} = -\cos \theta$

25. $\dfrac{1 + \tan \theta}{1 - \tan \theta} = \dfrac{\cot \theta + 1}{\cot \theta - 1}$

26. $\dfrac{\csc \theta - 1}{\csc \theta + 1} = \dfrac{1 - \sin \theta}{1 + \sin \theta}$

27. $\dfrac{\sec \theta}{\csc \theta} + \dfrac{\sin \theta}{\cos \theta} = 2 \tan \theta$

28. $\dfrac{\csc \theta - 1}{\cot \theta} = \dfrac{\cot \theta}{\csc \theta + 1}$

29. $\dfrac{1 + \sin \theta}{1 - \sin \theta} = \dfrac{\csc \theta + 1}{\csc \theta - 1}$

30. $\dfrac{\cos \theta + 1}{\cos \theta - 1} = \dfrac{1 + \sec \theta}{1 - \sec \theta}$

31. $\dfrac{1 - \sin \theta}{\cos \theta} + \dfrac{\cos \theta}{1 - \sin \theta} = 2 \sec \theta$

32. $\dfrac{\cos \theta}{1 + \sin \theta} + \dfrac{1 + \sin \theta}{\cos \theta} = 2 \sec \theta$

33. $\dfrac{\sin \theta}{\sin \theta - \cos \theta} = \dfrac{1}{1 - \cot \theta}$

34. $1 - \dfrac{\sin^2 \theta}{1 + \cos \theta} = \cos \theta$

35. $\dfrac{1 - \sin \theta}{1 + \sin \theta} = (\sec \theta - \tan \theta)^2$

36. $\dfrac{1 - \cos \theta}{1 + \cos \theta} = (\csc \theta - \cot \theta)^2$

37. $\dfrac{\cos \theta}{1 - \tan \theta} + \dfrac{\sin \theta}{1 - \cot \theta} = \sin \theta + \cos \theta$

38. $\dfrac{\cot \theta}{1 - \tan \theta} + \dfrac{\tan \theta}{1 - \cot \theta} = 1 + \tan \theta + \cot \theta$

39. $\tan \theta + \dfrac{\cos \theta}{1 + \sin \theta} = \sec \theta$

40. $\dfrac{\sin \theta \cos \theta}{\cos^2 \theta - \sin^2 \theta} = \dfrac{\tan \theta}{1 - \tan^2 \theta}$

41. $\dfrac{\tan \theta + \sec \theta - 1}{\tan \theta - \sec \theta + 1} = \tan \theta + \sec \theta$

42. $\dfrac{\sin \theta - \cos \theta + 1}{\sin \theta + \cos \theta - 1} = \dfrac{\sin \theta + 1}{\cos \theta}$

43. $\dfrac{\tan \theta - \cot \theta}{\tan \theta + \cot \theta} = \sin^2 \theta - \cos^2 \theta$

44. $\dfrac{\sec \theta - \cos \theta}{\sec \theta + \cos \theta} = \dfrac{\sin^2 \theta}{1 + \cos^2 \theta}$

45. $\dfrac{\tan \theta - \cot \theta}{\tan \theta + \cot \theta} = 2 \sin^2 \theta - 1$

46. $\dfrac{\tan \theta - \cot \theta}{\tan \theta + \cot \theta} = 1 - 2 \cos^2 \theta$

47. $\dfrac{\sec \theta + \tan \theta}{\cot \theta + \cos \theta} = \tan \theta \sec \theta$

48. $\dfrac{\sec \theta}{1 + \sec \theta} = \dfrac{1 - \cos \theta}{\sin^2 \theta}$

49. $\dfrac{1 - \tan^2 \theta}{1 + \tan^2 \theta} = 2 \cos^2 \theta - 1$

50. $\dfrac{1 - \cot^2 \theta}{1 + \cot^2 \theta} = 1 - 2 \cos^2 \theta$

51. $\dfrac{\sec \theta - \csc \theta}{\sec \theta \csc \theta} = \sin \theta - \cos \theta$

52. $\dfrac{\sin^2 \theta - \tan \theta}{\cos^2 \theta - \cot \theta} = \tan^2 \theta$

53. $\sec \theta - \cos \theta = \sin \theta \tan \theta$

54. $\tan \theta + \cot \theta = \sec \theta \csc \theta$

55. $\dfrac{1}{1 - \sin \theta} + \dfrac{1}{1 + \sin \theta} = 2 \sec^2 \theta$

56. $\dfrac{1 + \sin \theta}{1 - \sin \theta} - \dfrac{1 - \sin \theta}{1 + \sin \theta} = 4 \tan \theta \sec \theta$

57. $\dfrac{\sec \theta}{1 - \sin \theta} = \dfrac{1 + \sin \theta}{\cos^3 \theta}$

58. $\dfrac{1 - \sin \theta}{1 + \sin \theta} = (\sec \theta - \tan \theta)^2$

59. $\dfrac{(\sec \theta - \tan \theta)^2 + 1}{\csc \theta (\sec \theta - \tan \theta)} = 2 \tan \theta$

60. $\dfrac{\sec^2 \theta - \tan^2 \theta + \tan \theta}{\sec \theta} = \sin \theta + \cos \theta$

61. $\dfrac{\sin \theta + \cos \theta}{\cos \theta} - \dfrac{\sin \theta - \cos \theta}{\sin \theta} = \sec \theta \csc \theta$

62. $\dfrac{\sin \theta + \cos \theta}{\sin \theta} - \dfrac{\cos \theta - \sin \theta}{\cos \theta} = \sec \theta \csc \theta$

63. $\dfrac{\sin^3 \theta + \cos^3 \theta}{\sin \theta + \cos \theta} = 1 - \sin \theta \cos \theta$

64. $\dfrac{\sin^3 \theta + \cos^3 \theta}{1 - 2 \cos^2 \theta} = \dfrac{\sec \theta - \sin \theta}{\tan \theta - 1}$

65. $\dfrac{\cos^2 \theta - \sin^2 \theta}{1 - \tan^2 \theta} = \cos^2 \theta$

66. $\dfrac{\cos \theta + \sin \theta - \sin^3 \theta}{\sin \theta} = \cot \theta + \cos^2 \theta$

67. $\dfrac{(2 \cos^2 \theta - 1)^2}{\cos^4 \theta - \sin^4 \theta} = 1 - 2 \sin^2 \theta$

68. $\dfrac{1 - 2 \cos^2 \theta}{\sin \theta \cos \theta} = \tan \theta - \cot \theta$

69. $\dfrac{1 + \sin \theta + \cos \theta}{1 + \sin \theta - \cos \theta} = \dfrac{1 + \cos \theta}{\sin \theta}$

70. $\dfrac{1 + \cos \theta + \sin \theta}{1 + \cos \theta - \sin \theta} = \sec \theta + \tan \theta$

71. $(a \sin \theta + b \cos \theta)^2 + (a \cos \theta - b \sin \theta)^2 = a^2 + b^2$

72. $(2a \sin \theta \cos \theta)^2 + a^2(\cos^2 \theta - \sin^2 \theta)^2 = a^2$

73. $\dfrac{\tan \alpha + \tan \beta}{\cot \alpha + \cot \beta} = \tan \alpha \tan \beta$

74. $(\tan \alpha + \tan \beta)(1 - \cot \alpha \cot \beta) + (\cot \alpha + \cot \beta)(1 - \tan \alpha \tan \beta) = 0$

75. $(\sin \alpha + \cos \beta)^2 + (\cos \beta + \sin \alpha)(\cos \beta - \sin \alpha) = 2 \cos \beta(\sin \alpha + \cos \beta)$

76. $(\sin \alpha - \cos \beta)^2 + (\cos \beta + \sin \alpha)(\cos \beta - \sin \alpha) = -2 \cos \beta(\sin \alpha - \cos \beta)$

77. $\ln|\sec \theta| = -\ln|\cos \theta|$

78. $\ln|\tan \theta| = \ln|\sin \theta| - \ln|\cos \theta|$

79. $\ln|1 + \cos \theta| + \ln|1 - \cos \theta| = 2 \ln|\sin \theta|$

80. $\ln|\sec \theta + \tan \theta| + \ln|\sec \theta - \tan \theta| = 0$

81. Write a few paragraphs outlining your strategy for establishing identities.

6.2 SUM AND DIFFERENCE FORMULAS

> **1** Use Sum and Difference Formulas to Find Exact Values
> **2** Use Sum and Difference Formulas to Establish Identities

In this section, we continue our derivation of trigonometric identities by obtaining formulas that involve the sum or difference of two angles, such as $\cos(\alpha + \beta)$, $\cos(\alpha - \beta)$, or $\sin(\alpha + \beta)$. These formulas are referred to as the **sum and difference formulas.** We begin with the formulas for $\cos(\alpha + \beta)$ and $\cos(\alpha - \beta)$.

Theorem   Sum and Difference Formulas for Cosines

$$\cos(\alpha + \beta) = \cos \alpha \cos \beta - \sin \alpha \sin \beta \tag{1}$$
$$\cos(\alpha - \beta) = \cos \alpha \cos \beta + \sin \alpha \sin \beta \tag{2}$$

In words, formula (1) states that the cosine of the sum of two angles equals the cosine of the first times the cosine of the second minus the sine of the first times the sine of the second.

FIGURE 1

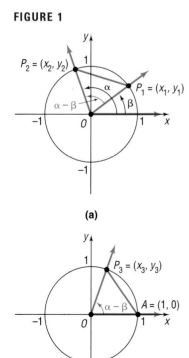

(a)

(b)

Proof We will prove formula (2) first. Although this formula is true for all numbers α and β, we shall assume in our proof that $0 < \beta < \alpha < 2\pi$. We begin with a circle with center at the origin $(0, 0)$ and radius of 1 unit (the unit circle), and we place the angles α and β in standard position, as shown in Figure 1(a). The point $P_1 = (x_1, y_1)$ lies on the terminal side of β, and the point $P_2 = (x_2, y_2)$ lies on the terminal side of α.

Now, place the angle $\alpha - \beta$ in standard position, as shown in Figure 1(b), where the point A has coordinates $(1, 0)$ and the point $P_3 = (x_3, y_3)$ is on the terminal side of the angle $\alpha - \beta$.

Looking at triangle OP_1P_2 in Figure 1(a) and triangle OAP_3 in Figure 1(b), we see that these triangles are congruent. (Do you see why? Two sides and the included angle, $\alpha - \beta$, are equal.) Hence, the unknown side of each triangle must be equal; that is,

$$d(A, P_3) = d(P_1, P_2)$$

Using the distance formula, we find that

$$\sqrt{(x_3 - 1)^2 + y_3^2} = \sqrt{(x_2 - x_1)^2 + (y_2 - y_1)^2}$$
$$(x_3 - 1)^2 + y_3^2 = (x_2 - x_1)^2 + (y_2 - y_1)^2 \quad \text{Square each side.}$$
$$x_3^2 - 2x_3 + 1 + y_3^2 = x_2^2 - 2x_1x_2 + x_1^2 + y_2^2 - 2y_1y_2 + y_1^2 \tag{3}$$

Since $P_1 = (x_1, y_1)$, $P_2 = (x_2, y_2)$, and $P_3 = (x_3, y_3)$ are points on the unit circle $x^2 + y^2 = 1$, it follows that

$$x_1^2 + y_1^2 = 1 \qquad x_2^2 + y_2^2 = 1 \qquad x_3^2 + y_3^2 = 1$$

Thus,

Theorem Sum and Difference Formulas for Sines

$$\sin(\alpha + \beta) = \sin \alpha \cos \beta + \cos \alpha \sin \beta \tag{8}$$

$$\sin(\alpha - \beta) = \sin \alpha \cos \beta - \cos \alpha \sin \beta \tag{9}$$

In words, formula (8) states that the sine of the sum of two angles equals the sine of the first times the cosine of the second plus the cosine of the first times the sine of the second.

E X A M P L E 3 **Using the Sum Formula to Find Exact Values**

Find the exact value of $\sin(7\pi/12)$.

Solution
$$\sin \frac{7\pi}{12} = \sin\left(\frac{3\pi}{12} + \frac{4\pi}{12}\right) = \sin\left(\frac{\pi}{4} + \frac{\pi}{3}\right)$$

$$= \sin \frac{\pi}{4} \cos \frac{\pi}{3} + \cos \frac{\pi}{4} \sin \frac{\pi}{3} \qquad \text{Formula (8).}$$

$$= \frac{\sqrt{2}}{2} \cdot \frac{1}{2} + \frac{\sqrt{2}}{2} \cdot \frac{\sqrt{3}}{2} = \frac{1}{4}(\sqrt{2} + \sqrt{6})$$

E X A M P L E 4 **Using the Difference Formula to Find Exact Values**

Find the exact value of $\sin 165°$.

Solution
$$\sin 165° = \sin(225° - 60°)$$

$$= \sin 225° \cos 60° - \cos 225° \sin 60° \qquad \text{Formula (9).}$$

$$= \frac{-\sqrt{2}}{2} \cdot \frac{1}{2} - \frac{-\sqrt{2}}{2} \cdot \frac{\sqrt{3}}{2} = \frac{1}{4}(\sqrt{6} - \sqrt{2})$$

Now work Problem 9.

E X A M P L E 5 **Using the Difference Formula to Find Exact Values**

Find the exact value of $\cos 80° \cos 20° + \sin 80° \sin 20°$.

Solution The form of the expression $\cos 80° \cos 20° + \sin 80° \sin 20°$ is that of the right side of the formula for $\cos(\alpha - \beta)$ with $\alpha = 80°$ and $\beta = 20°$. Thus,

$$\cos 80° \cos 20° + \sin 80° \sin 20° = \cos(80° - 20°) = \cos 60° = \frac{1}{2}$$

E X A M P L E 6 **Using the Sum Formula to Find Exact Values**

Find the exact value of $\sin \dfrac{\pi}{9} \cos \dfrac{\pi}{18} + \cos \dfrac{\pi}{9} \sin \dfrac{\pi}{18}$.

Solution We observe that the form of the given expression is that of the right side of the formula for $\sin(\alpha + \beta)$ with $\alpha = \pi/9$ and $\beta = \pi/18$. Thus,

$$\sin \frac{\pi}{9} \cos \frac{\pi}{18} + \cos \frac{\pi}{9} \sin \frac{\pi}{18} = \sin\left(\frac{\pi}{9} + \frac{\pi}{18}\right) = \sin \frac{3\pi}{18} = \sin \frac{\pi}{6} = \frac{1}{2}$$

Now work Problem 19.

E X A M P L E   7   Finding Exact Values

If it is known that $\sin \alpha = \frac{4}{5}$, $\pi/2 < \alpha < \pi$, and that $\sin \beta = -2/\sqrt{5} = -2\sqrt{5}/5$, $\pi < \beta < 3\pi/2$, find the exact value of

(a) Since $\sin \alpha = \frac{4}{5} = \frac{b}{r}$ and $\frac{6}{\pi} < \alpha < \pi$, we find $\frac{\pi}{2} = 4$ and $r = 5$.

Solution

FIGURE 2
Given $\sin \alpha = \frac{4}{5}$, $\pi/2 < \alpha < \pi$

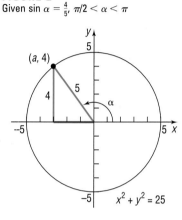

(a) See Figure 2. Since $\sin \alpha = \frac{4}{5} = \frac{b}{r}$ and $\frac{\pi}{2} < \alpha < \pi$, we find $b = 4$ and $r = 5$. Since $(a, 4)$ is on the circle $x^2 + y^2 = 25$ and is in Quadrant II, we have

$$a^2 + 4^2 = 25, \quad a < 0$$
$$a^2 = 25 - 16 = 9$$
$$a = -3$$

Thus,

$$\cos \alpha = \frac{a}{r} = -\frac{3}{5}$$

FIGURE 3
Given $\sin \beta = -2/\sqrt{5}$, $\pi < \beta < 3\pi/2$

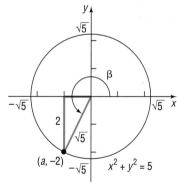

(b) See Figure 3. Since $\sin \beta = \frac{-2}{\sqrt{5}} = \frac{b}{r}$ and $\pi < \beta < \frac{3\pi}{2}$, we find $b = -2$ and $r = \sqrt{5}$. Since $(a, -2)$ is on the circle $x^2 + y^2 = 5$ and is in Quadrant III, we have

$$a^2 + (-2)^2 = 5, \quad a < 0$$
$$a^2 = 5 - 4 = 1$$
$$a = -1$$

Thus,

$$\cos \beta = \frac{a}{r} = -\frac{1}{\sqrt{5}} = -\frac{\sqrt{5}}{5}$$

(c) Using the results found in parts (a) and (b) and formula (1), we have

$$\cos(\alpha + \beta) = \cos \alpha \cos \beta - \sin \alpha \sin \beta$$
$$= -\frac{3}{5}\left(-\frac{\sqrt{5}}{5}\right) - \frac{4}{5}\left(-\frac{2\sqrt{5}}{5}\right) = \frac{11\sqrt{5}}{25}$$

(d) $$\sin(\alpha + \beta) = \sin \alpha \cos \beta + \cos \alpha \sin \beta$$
$$= \frac{4}{5}\left(-\frac{\sqrt{5}}{5}\right) + \left(-\frac{3}{5}\right)\left(-\frac{2\sqrt{5}}{5}\right) = \frac{2\sqrt{5}}{25}$$

Now work Problem 23, parts (a), (b), and (c).

Formulas for $\tan(\alpha + \beta)$ and $\tan(\alpha - \beta)$

We use the identity $\tan \theta = (\sin \theta)/(\cos \theta)$ and the sum formulas for $\sin(\alpha + \beta)$ and $\cos(\alpha + \beta)$ to derive a formula for $\tan(\alpha + \beta)$.

Proof

$$tan(\alpha + \beta) = \frac{\sin(\alpha + \beta)}{\cos(\alpha + \beta)} = \frac{\sin \alpha \cos \beta + \cos \alpha \sin \beta}{\cos \alpha \cos \beta - \sin \alpha \sin \beta}$$

Now we divide the numerator and denominator by $\cos \alpha \cos \beta$:

$$tan(\alpha + \beta) = \frac{\dfrac{\sin \alpha \cos \beta + \cos \alpha \sin \beta}{\cos \alpha \cos \beta}}{\dfrac{\cos \alpha \cos \beta - \sin \alpha \sin \beta}{\cos \alpha \cos \beta}} = \frac{\dfrac{\sin \alpha \cos \beta}{\cos \alpha \cos \beta} + \dfrac{\cos \alpha \sin \beta}{\cos \alpha \cos \beta}}{\dfrac{\cos \alpha \cos \beta}{\cos \alpha \cos \beta} - \dfrac{\sin \alpha \sin \beta}{\cos \alpha \cos \beta}}$$

$$= \frac{\dfrac{\sin \alpha}{\cos \alpha} + \dfrac{\sin \beta}{\cos \beta}}{1 - \dfrac{\sin \alpha \sin \beta}{\cos \alpha \cos \beta}} = \frac{\tan \alpha + \tan \beta}{1 - \tan \alpha \tan \beta}$$

We use the sum formula for $tan(\alpha + \beta)$ to get the difference formula:

$$tan(\alpha - \beta) = tan[\alpha + (-\beta)] = \frac{\tan \alpha + \tan(-\beta)}{1 - \tan \alpha \tan(-\beta)} = \frac{\tan \alpha - \tan \beta}{1 + \tan \alpha \tan \beta}$$

Thus, we have proved the following results:

Theorem **Sum and Difference Formulas for Tangents**

$$tan(\alpha + \beta) = \frac{\tan \alpha + \tan \beta}{1 - \tan \alpha \tan \beta} \qquad (10)$$

$$tan(\alpha - \beta) = \frac{\tan \alpha - \tan \beta}{1 + \tan \alpha \tan \beta} \qquad (11)$$

In words, formula (10) states that the tangent of the sum of two angles equals the tangent of the first plus the tangent of the second divided by 1 minus their product.

E X A M P L E   8

Establishing an Identity

2

Prove the identity: $tan(\theta + \pi) = \tan \theta$

Solution

$$tan(\theta + \pi) = \frac{\tan \theta + \tan \pi}{1 - \tan \theta \tan \pi} = \frac{\tan \theta + 0}{1 - \tan \theta \cdot 0} = \tan \theta$$

The result obtained in Example 8 verifies that the tangent function is periodic with period π, a fact we mentioned earlier.

Warning Be careful when using formulas (10) and (11). These formulas can be used only for angles α and β for which $\tan \alpha$ and $\tan \beta$ are defined, that is, all angles except odd multiples of $\pi/2$.

E X A M P L E 9   Establishing an Identity

Prove the identity: $\tan\left(\theta + \dfrac{\pi}{2}\right) = -\cot\theta$

Solution   We cannot use formula (10), since $\tan(\pi/2)$ is not defined. Instead, we proceed as follows:

$$\tan\left(\theta + \frac{\pi}{2}\right) = \frac{\sin\left(\theta + \dfrac{\pi}{2}\right)}{\cos\left(\theta + \dfrac{\pi}{2}\right)} = \frac{\sin\theta\cos\dfrac{\pi}{2} + \cos\theta\sin\dfrac{\pi}{2}}{\cos\theta\cos\dfrac{\pi}{2} - \sin\theta\sin\dfrac{\pi}{2}}$$

$$= \frac{(\sin\theta)(0) + (\cos\theta)(1)}{(\cos\theta)(0) - (\sin\theta)(1)} = \frac{\cos\theta}{-\sin\theta} = -\cot\theta \quad\blacksquare$$

E X A M P L E 10   Finding Exact Values

Find the exact value of $\sin(\cos^{-1}\frac{1}{2} + \sin^{-1}\frac{3}{5})$.

Solution   Let $\alpha = \cos^{-1}\frac{1}{2}$ and $\beta = \sin^{-1}\frac{3}{5}$. Then

$$\cos\alpha = \tfrac{1}{2}, \quad 0 \le \alpha \le \pi, \quad \text{and} \quad \sin\beta = \tfrac{3}{5}, \quad -\frac{\pi}{2} \le \beta \le \frac{\pi}{2}$$

Based on Figure 4, we obtain $\sin\alpha = \sqrt{3}/2$ and $\cos\beta = \frac{4}{5}$. Thus,

$$\sin\left(\cos^{-1}\frac{1}{2} + \sin^{-1}\frac{3}{5}\right) = \sin(\alpha + \beta) = \sin\alpha\cos\beta + \cos\alpha\sin\beta$$

$$= \frac{\sqrt{3}}{2}\cdot\frac{4}{5} + \frac{1}{2}\cdot\frac{3}{5} = \frac{4\sqrt{3} + 3}{10}$$

FIGURE 4

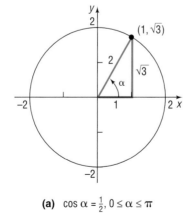

(a) $\cos\alpha = \frac{1}{2}, 0 \le \alpha \le \pi$

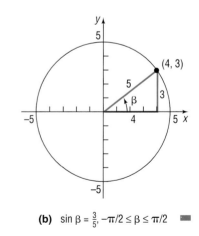

(b) $\sin\beta = \frac{3}{5}, -\pi/2 \le \beta \le \pi/2$ ∎

Now work Problem 57.

E X A M P L E 11   Writing a Trigonometric Expression as an Algebraic Expression

Write $\sin(\sin^{-1}u + \cos^{-1}v)$ as an algebraic expression containing u and v (that is, without any trigonometric functions).

71. Calculus Show that the difference quotient for $f(x) = \sin x$ is given by

$$\frac{f(x + h) - f(x)}{h} = \frac{\sin(x + h) - \sin x}{h}$$

$$= \cos x \cdot \frac{\sin h}{h} - \sin x \cdot \frac{1 - \cos h}{h}$$

72. Calculus Show that the difference quotient for $f(x) = \cos x$ is given by

$$\frac{f(x + h) - f(x)}{h} = \frac{\cos(x + h) - \cos x}{h}$$

$$= -\sin x \cdot \frac{\sin h}{h} - \cos x \cdot \frac{1 - \cos h}{h}$$

73. Show that $\sin(\sin^{-1} u + \cos^{-1} u) = 1$.

74. Show that $\cos(\sin^{-1} u + \cos^{-1} u) = 0$.

75. Explain why formula (11) cannot be used to show that

$$\tan\left(\frac{\pi}{2} - \theta\right) = \cot \theta$$

Establish this identity by using formulas (7a) and (7b).

76. If $\tan \alpha = x + 1$ and $\tan \beta = x - 1$, show that $2 \cot(\alpha - \beta) = x^2$.

77. Geometry: Angle between Two Lines Let L_1 and L_2 denote two nonvertical intersecting lines, and let θ denote the acute angle between L_1 and L_2 (see the figure). Show that

$$\tan \theta = \frac{m_2 - m_1}{1 + m_1 m_2}$$

where m_1 and m_2 are the slopes of L_1 and L_2, respectively.

[**Hint:** Use the facts that $\tan \theta_1 = m_1$ and $\tan \theta_2 = m_2$.]

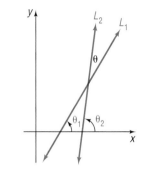

78. If $\alpha + \beta + \gamma = 180°$ and $\cot \theta = \cot \alpha + \cot \beta + \cot \gamma, 0 < \theta < 90°$, show that

$$\sin^3 \theta = \sin(\alpha - \theta) \sin(\beta - \theta) \sin(\gamma - \theta)$$

79. Discuss the following derivation:

$$\tan\left(\theta + \frac{\pi}{2}\right) = \frac{\tan \theta + \tan(\pi/2)}{1 - \tan \theta \tan(\pi/2)}$$

$$= \frac{\dfrac{\tan \theta}{\tan(\pi/2)} + 1}{\dfrac{1}{\tan(\pi/2)} - \tan \theta} = \frac{0 + 1}{0 - \tan \theta}$$

$$= \frac{1}{-\tan \theta} = -\cot \theta$$

Can you justify each step?

6.3 DOUBLE-ANGLE AND HALF-ANGLE FORMULAS

  1 Use Double-Angle Formulas to Find Exact Values
  2 Use Double-Angle and Half-Angle Formulas to Establish Identities
  3 Use Half-Angle Formulas to Find Exact Values

In this section we derive formulas for $\sin 2\theta$, $\cos 2\theta$, $\sin\left(\frac{1}{2}\theta\right)$, and $\cos\left(\frac{1}{2}\theta\right)$ in terms of $\sin \theta$ and $\cos \theta$. They are easily derived using the sum formulas.

Double-Angle Formulas

In the sum formulas for $\sin(\alpha + \beta)$ and $\cos(\alpha + \beta)$, let $\alpha = \beta = \theta$. Then

$$\sin(\alpha + \beta) = \sin \alpha \cos \beta + \cos \alpha \sin \beta$$
$$\sin(\theta + \theta) = \sin \theta \cos \theta + \cos \theta \sin \theta$$
$$\sin 2\theta = 2 \sin \theta \cos \theta \qquad (1)$$

and

$$\cos(\alpha + \beta) = \cos \alpha \cos \beta - \sin \alpha \sin \beta$$
$$\cos(\theta + \theta) = \cos \theta \cos \theta - \sin \theta \sin \theta$$
$$\cos 2\theta = \cos^2 \theta - \sin^2 \theta \qquad (2)$$

An application of the Pythagorean identity $\sin^2 \theta + \cos^2 \theta = 1$ results in two other ways to write formula (2) for $\cos 2\theta$:

$$\cos 2\theta = \cos^2 \theta - \sin^2 \theta = (1 - \sin^2 \theta) - \sin^2 \theta = 1 - 2 \sin^2 \theta$$

and

$$\cos 2\theta = \cos^2 \theta - \sin^2 \theta = \cos^2 \theta - (1 - \cos^2 \theta) = 2 \cos^2 \theta - 1$$

Thus, we have established the following **Double-Angle Formulas:**

Theorem   Double-Angle Formulas

$$\sin 2\theta = 2 \sin \theta \cos \theta \qquad (3)$$
$$\cos 2\theta = \cos^2 \theta - \sin^2 \theta \qquad (4a)$$
$$\cos 2\theta = 1 - 2 \sin^2 \theta \qquad (4b)$$
$$\cos 2\theta = 2 \cos^2 \theta - 1 \qquad (4c)$$

E X A M P L E   1

Finding Exact Values Using the Double-Angle Formula

If $\sin \theta = \frac{3}{5}$, $\pi/2 < \theta < \pi$, find the exact value of

(a) $\sin 2\theta$ (b) $\cos 2\theta$

FIGURE 5

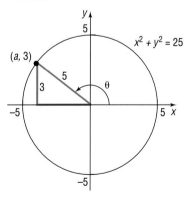

Solution   (a) Because $\sin 2\theta = 2 \sin \theta \cos \theta$ and we already know that $\sin \theta = \frac{3}{5}$, we only need to find $\cos \theta$. See Figure 5. Since $\sin \theta = \frac{3}{5} = \frac{b}{r}$, we find $b = 3$ and $r = 5$. Since $\pi/2 < \theta < \pi$, it follows that $(a, 3)$ is on the circle $x^2 + y^2 = 25$ and is in Quadrant II. Thus,

$$a^2 + 3^2 = 25, \qquad a < 0$$
$$a^2 = 25 - 9 = 16$$
$$a = -4$$

Thus, $\cos \theta = -\frac{4}{5}$. Now we use formula (3) to obtain

$$\sin 2\theta = 2 \sin \theta \cos \theta = 2(\tfrac{3}{5})(-\tfrac{4}{5}) = -\tfrac{24}{25}$$

(b) Because we are given $\sin \theta = \frac{3}{5}$, it is easiest to use formula (4b) to get $\cos 2\theta$:

$$\cos 2\theta = 1 - 2 \sin^2 \theta = 1 - 2(\tfrac{9}{25}) = 1 - \tfrac{18}{25} = \tfrac{7}{25}$$

Warning In finding $\cos 2\theta$ in Example 1(b), we chose to use a version of the Double-Angle Formula, formula (4b). Note that we are unable to use the Pythagorean Identity $\cos 2\theta = \pm\sqrt{1 - \sin^2 2\theta}$, with $\sin 2\theta = -\frac{24}{25}$, because we have no way of knowing which sign to choose.

Now work Problems 1(a) and (b).

E X A M P L E 2 Establishing Identities

2

(a) Develop a formula for $\tan 2\theta$ in terms of $\tan \theta$.

(b) Develop a formula for $\sin 3\theta$ in terms of $\sin \theta$ and $\cos \theta$.

Solution (a) In the sum formula for $\tan(\alpha + \beta)$, let $\alpha = \beta = \theta$. Then

$$\tan(\alpha + \beta) = \frac{\tan \alpha + \tan \beta}{1 - \tan \alpha \tan \beta}$$

$$\tan(\theta + \theta) = \frac{\tan \theta + \tan \theta}{1 - \tan \theta \tan \theta}$$

$$\tan 2\theta = \frac{2 \tan \theta}{1 - \tan^2 \theta} \qquad (5)$$

(b) To get a formula for $\sin 3\theta$, we use the sum formula and write 3θ as $2\theta + \theta$.

$$\sin 3\theta = \sin(2\theta + \theta) = \sin 2\theta \cos \theta + \cos 2\theta \sin \theta$$

Now use the Double-Angle Formulas to get

$$\begin{aligned}
\sin 3\theta &= (2 \sin \theta \cos \theta)(\cos \theta) + (\cos^2 \theta - \sin^2 \theta)(\sin \theta) \\
&= 2 \sin \theta \cos^2 \theta + \sin \theta \cos^2 \theta - \sin^3 \theta \\
&= 3 \sin \theta \cos^2 \theta - \sin^3 \theta
\end{aligned}$$ ∎

The formula obtained in Example 2(b) can also be written as

$$\begin{aligned}
\sin 3\theta = 3 \sin \theta \cos^2 \theta - \sin^3 \theta &= 3 \sin \theta(1 - \sin^2 \theta) - \sin^3 \theta \\
&= 3 \sin \theta - 4 \sin^3 \theta
\end{aligned}$$

That is, $\sin 3\theta$ is a third-degree polynomial in the variable $\sin \theta$. In fact, $\sin n\theta$, n a positive odd integer, can always be written as a polynomial of degree n in the variable $\sin \theta$.*

Now work Problem 47.

Other Variations of the Double-Angle Formulas

By rearranging the Double-Angle Formulas (4b) and (4c), we obtain other formulas that we will use a little later in this section.

If we solve formula (4b) for $\sin^2 \theta$, we get

$$\cos 2\theta = 1 - 2 \sin^2 \theta$$
$$2 \sin^2 \theta = 1 - \cos 2\theta$$

$$\sin^2 \theta = \frac{1 - \cos 2\theta}{2} \qquad (6)$$

*Due to the work done by P. L. Chebyshëv, these polynomials are sometimes called *Chebyshëv polynomials*.

Similarly, we can solve for $\cos^2 \theta$ in formula (4c):

$$\cos 2\theta = 2 \cos^2 \theta - 1$$
$$2 \cos^2 \theta = 1 + \cos 2\theta$$

$$\cos^2 \theta = \frac{1 + \cos 2\theta}{2} \qquad (7)$$

Formulas (6) and (7) can be used to develop a formula for $\tan^2 \theta$:

$$\tan^2 \theta = \frac{\sin^2 \theta}{\cos^2 \theta} = \frac{\dfrac{1 - \cos 2\theta}{2}}{\dfrac{1 + \cos 2\theta}{2}}$$

$$\tan^2 \theta = \frac{1 - \cos 2\theta}{1 + \cos 2\theta} \qquad (8)$$

Formulas (6) through (8) do not have to be memorized since their derivations are so straightforward.

Formulas (6) and (7) are important in calculus. The next example illustrates a problem that arises in calculus requiring the use of formula (7).

E X A M P L E   3

Establishing an Identity

Write an equivalent expression for $\cos^4 \theta$ that does not involve any powers of sine or cosine greater than 1.

Solution   The idea here is to apply formula (7) twice:

$$\cos^4 \theta = (\cos^2 \theta)^2 = \left(\frac{1 + \cos 2\theta}{2}\right)^2 \qquad \text{Formula (7).}$$

$$= \frac{1}{4}(1 + 2 \cos 2\theta + \cos^2 2\theta)$$

$$= \frac{1}{4} + \frac{1}{2} \cos 2\theta + \frac{1}{4} \cos^2 2\theta$$

$$= \frac{1}{4} + \frac{1}{2} \cos 2\theta + \frac{1}{4}\left[\frac{1 + \cos 2(2\theta)}{2}\right] \qquad \text{Formula (7).}$$

$$= \frac{1}{4} + \frac{1}{2} \cos 2\theta + \frac{1}{8}(1 + \cos 4\theta)$$

$$= \frac{3}{8} + \frac{1}{2} \cos 2\theta + \frac{1}{8} \cos 4\theta$$

Now work Problem 23.

Identities, such as the Double-Angle Formulas, can sometimes be used to rewrite expressions in a more suitable form. Let's look at an example.

E X A M P L E   4   Projectile Motion

FIGURE 6

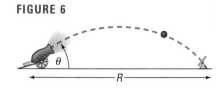

An object is propelled upward at an angle θ to the horizontal with an initial velocity of v_0 feet per second. See Figure 6. If air resistance is ignored, the horizontal distance R it travels, the **range,** is given by

$$R = \frac{1}{16} v_0^2 \sin \theta \cos \theta$$

(a) Show that $R = \dfrac{1}{32} v_0^2 \sin 2\theta$.

(b) Find the angle θ for which R is a maximum.

Solution   (a) We rewrite the given expression for the range using the Double-Angle Formula $\sin 2\theta = 2 \sin \theta \cos \theta$. Then

$$R = \frac{1}{16} v_0^2 \sin \theta \cos \theta = \frac{1}{16} v_0^2 \frac{2 \sin \theta \cos \theta}{2} = \frac{1}{32} v_0^2 \sin 2\theta$$

(b) In this form, the largest value for the range R can be easily found. For a fixed initial speed v_0, the angle θ of inclination to the horizontal determines the value of R. Since the largest value of a sine function is 1, occurring when the argument is 90°, it follows that for maximum R we must have

$$2\theta = 90°$$
$$\theta = 45°$$

An inclination to the horizontal of 45° results in maximum range.   ▬

Half-Angle Formulas

Another important use of formulas (6) through (8) is to prove the **Half-Angle Formulas.** In formulas (6) through (8), let $\theta = \alpha/2$. Then

$$\sin^2 \frac{\alpha}{2} = \frac{1 - \cos \alpha}{2} \qquad \cos^2 \frac{\alpha}{2} = \frac{1 + \cos \alpha}{2}$$

$$\tan^2 \frac{\alpha}{2} = \frac{1 - \cos \alpha}{1 + \cos \alpha} \tag{9}$$

If we solve for the trigonometric functions on the left sides of equations (9), we obtain the Half-Angle Formulas:

Theorem   Half-Angle Formulas

$$\sin \frac{\alpha}{2} = \pm\sqrt{\frac{1 - \cos \alpha}{2}} \qquad \text{(10a)}$$

$$\cos \frac{\alpha}{2} = \pm\sqrt{\frac{1 + \cos \alpha}{2}} \qquad \text{(10b)}$$

$$\tan \frac{\alpha}{2} = \pm\sqrt{\frac{1 - \cos \alpha}{1 + \cos \alpha}} \qquad \text{(10c)}$$

where the + or − sign is determined by the quadrant of the angle $\alpha/2$.

We use the Half-Angle Formulas in the next example.

E X A M P L E   5   Finding Exact Values Using Half-Angle Formulas

Find the exact value of

(a) $\cos 15°$ (b) $\sin(-15°)$

Solution   (a) Because $15° = 30°/2$, we can use the Half-Angle Formula for $\cos(\alpha/2)$ with $\alpha = 30°$. Also, because $15°$ is in quadrant I, $\cos 15° > 0$, and we choose the + sign in using formula (10b):

$$\cos 15° = \cos \frac{30°}{2} = \sqrt{\frac{1 + \cos 30°}{2}}$$

$$= \sqrt{\frac{1 + \sqrt{3}/2}{2}} = \sqrt{\frac{2 + \sqrt{3}}{4}} = \frac{\sqrt{2 + \sqrt{3}}}{2}$$

(b) We use the fact that $\sin(-15°) = -\sin 15°$ and then apply formula (10a):

$$\sin(-15°) = -\sin \frac{30°}{2} = -\sqrt{\frac{1 - \cos 30°}{2}}$$

$$= -\sqrt{\frac{1 - \sqrt{3}/2}{2}} = \sqrt{\frac{2 - \sqrt{3}}{4}} = -\frac{\sqrt{2 - \sqrt{3}}}{2}$$

It is interesting to compare the answer found in Example 5(a) with the answer to Example 2 of Section 6.2. There we calculated

$$\cos \frac{\pi}{12} = \cos 15° = \frac{1}{4}(\sqrt{6} + \sqrt{2})$$

Based on these results, we conclude that

$$\frac{1}{4}(\sqrt{6} + \sqrt{2}) \quad \text{and} \quad \frac{\sqrt{2 + \sqrt{3}}}{2}$$

In Problems 73 and 74, establish each identity.

73. $\ln|\sin\theta| = \frac{1}{2}(\ln|1 - \cos 2\theta| - \ln 2)$

74. $\ln|\cos\theta| = \frac{1}{2}(\ln|1 + \cos 2\theta| - \ln 2)$

75. Projectile Motion An object is propelled upward at an angle θ, $45° < \theta < 90°$, to the horizontal with an initial velocity of v_0 feet per second from the base of a plane that makes an angle of 45° with the horizontal. See the illustration. If air resistance is ignored, the distance R it travels up the inclined plane is given by

$$R = \frac{v_0^2 \sqrt{2}}{16} \cos\theta(\sin\theta - \cos\theta)$$

(a) Show that

$$R = \frac{v_0^2 \sqrt{2}}{32}(\sin 2\theta - \cos 2\theta - 1)$$

(b) Graph $R = R(\theta)$. (Use $v_0 = 32$ ft/sec.)

(c) What value of θ makes R the largest? (Use $v_0 = 32$ ft/sec.)

76. Sawtooth Curve An oscilloscope often displays a sawtooth curve. This curve can be approximated by sinusoidal curves of varying periods and amplitudes. A first approximation to the sawtooth curve is given by

$$y = \frac{1}{2}\sin 2\pi x + \frac{1}{4}\sin 4\pi x$$

Show that $y = \sin 2\pi x \cos^2 \pi x$.

77. Go to the library and research Chebyshëv polynomials. Write a report on your findings.

6.4 PRODUCT-TO-SUM AND SUM-TO-PRODUCT FORMULAS

1 Express Products as Sums
2 Express Sums as Products

1 The sum and difference formulas can be used to derive formulas for writing the products of sines and/or cosines as sums or differences. These identities are usually called the **Product-to-Sum Formulas.**

Theorem Product-to-Sum Formulas

$$\sin\alpha\sin\beta = \frac{1}{2}[\cos(\alpha - \beta) - \cos(\alpha + \beta)] \qquad (1)$$

$$\cos\alpha\cos\beta = \frac{1}{2}[\cos(\alpha - \beta) + \cos(\alpha + \beta)] \qquad (2)$$

$$\sin\alpha\cos\beta = \frac{1}{2}[\sin(\alpha + \beta) + \sin(\alpha - \beta)] \qquad (3)$$

MISSION POSSIBLE

How Far and How High Does a Baseball Need to Go for an Out-of-the-Park Home Run?

The sportscasters for the Cleveland Indians decided they had better be prepared for any eventuality when the new ballpark, Jacob's Field, opened in 1994. One eventuality they worried about was Jim Thome hitting a home run that went so high and so far that it left the ball park. What would they tell their listeners about the height and distance the ball went?

So they called in the Mission Possible team. The sportscasters wanted to know the distance from homeplate to the highest point of the stadium and the distance from the base of the outfield fence to the highest point of the stadium for every 5° from the 3rd base line around to the 1st base line. The problem was complicated by the fact that the distance from homeplate to the outfield fence varied from 325 feet to 410 feet. Furthermore, the height that would have to be cleared also varied, depending on where the ball was hit.

1.   First, how many distances and heights do the sportscasters want?

2.   To see one solution, use the distance from homeplate across 2nd base to the outfield fence, 410 feet. Using a transit, you find the angle of elevation from homeplate to the highest point of the stadium is 10° and the angle of elevation from the base of the outfield fence to the highest point of the stadium is 32.5°. Use this information to find the minimum distance from homeplate to the highest point of the stadium and the distance from the base of the outfield fence to the highest point of the stadium.

3.   You tell the sportscasters that if they provide you the two angles of elevation for each 5° movement around the stadium, you can solve their problem. After receiving the two angles of elevation for each 5° movement around the stadium, you decide that it would be faster to develop a general formula, so that by inputting the two angles and the distance from homeplate to the fence, the distance from homeplate to the top of the roof and the height of the roof would be computed. Let α and β denote the angles of elevation to the top of the roof from homeplate and from the base of the outfield fence and let L be the distance from homeplate to the outfield fence. What are the correct formulas?

4.   Compare your formulas with those of other groups. Are they all the same? Are they all equivalent? Which ones are simplest?

5.   Now write each correct formula in the form shown below:

$$\text{height} = \frac{L \sin \alpha \sin \beta}{\sin(\beta - \alpha)} \qquad \text{distance} = \frac{L \sin \beta}{\sin(\beta - \alpha)}$$

6.   What is the probable trajectory of a hit baseball on a windless day? How might the wind and other factors affect the path of the baseball?

7.   If a player actually did hit an out-of-the-park home run, how would the distance the ball traveled compare to the figures you have been developing for the sportscasters? What other factors will add to the distance?

8.   Could there be a longest home run? How would you measure it?

Next, replace θ by $\pi/6$ in the equation. The result is

$$\sin \frac{\pi}{6} = \frac{1}{2}$$

Thus, $\pi/6$ is a solution of the given equation. ∎

The equation given in Example 1 has other solutions besides $\theta = \pi/6$. For example, $\theta = 5\pi/6$ is also a solution, as is $\theta = 13\pi/6$. (You should check this for yourself.) In fact, the equation has an infinite number of solutions due to the periodicity of the sine function, as can be seen in Figure 7.

As before, our practice will be to solve equations, whenever possible, by finding exact solutions. In such cases, we will also verify the solution obtained by using a graphing utility. When traditional methods cannot be used, approximate solutions will be obtained using a graphing utility. The reader is encouraged to pay particular attention to the form of equations for which exact solutions are possible.

Unless otherwise indicated, in solving trigonometric equations, we need to find *all* solutions. As the next example illustrates, finding all the solutions can be accomplished by first finding solutions over an interval whose length equals the period of the function and then adding multiples of that period to the solutions found. Let's look at some examples.

FIGURE 7

E X A M P L E   2

Solving a Trigonometric Equation

Solve the equation: $\cos \theta = \frac{1}{2}$

FIGURE 8
$\cos \theta = \frac{1}{2}$

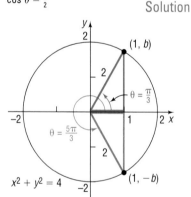

Solution The period of the cosine function is 2π. In the interval $[0, 2\pi)$, there are two angles θ for which $\cos \theta = \frac{1}{2}$: $\theta = \pi/3$ and $\theta = 5\pi/3$. See Figure 8. Because the cosine function has period 2π, all the solutions of $\cos \theta = \frac{1}{2}$ may be given by

$$\theta = \frac{\pi}{3} + 2k\pi \quad \text{or} \quad \theta = \frac{5\pi}{3} + 2k\pi \quad \text{\small k any integer.}$$

Some of the solutions are

$$\underbrace{\frac{\pi}{3}, \frac{5\pi}{3}}_{k=0}, \underbrace{\frac{7\pi}{3}, \frac{11\pi}{3}}_{k=1}, \underbrace{\frac{13\pi}{3}, \frac{17\pi}{3}}_{k=2}, \text{ and so on}$$

∎

FIGURE 9

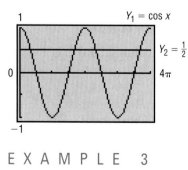

We can verify the solution by graphing $Y_1 = \cos x$ and $Y_2 = \frac{1}{2}$ to determine where the graphs intersect. (Be sure to graph in radian mode.) See Figure 9. The graph of Y_1 intersects the graph of Y_2 at $x = 1.04$ ($\approx \pi/3$), 5.23 ($\approx 5\pi/3$), 7.33 ($\approx 7\pi/3$), and 11.51 ($\approx 11\pi/3$), correct to two decimal places.

   Now work Problem 1.

In most of our work, we shall be interested only in finding solutions of trigonometric equations for $0 \leq \theta < 2\pi$.

E X A M P L E   3

Solving a Trigonometric Equation

Solve the equation: $\sin 2\theta = 1, 0 \leq \theta < 2\pi$

Solution The period of the sine function is 2π. In the interval $[0, 2\pi)$, the sine function has the value 1 only at $\pi/2$. Because the argument is 2θ in the given equation, we have

$$2\theta = \frac{\pi}{2} + 2k\pi \qquad k \text{ any integer.}$$

$$\theta = \frac{\pi}{4} + k\pi$$

In the interval $[0, 2\pi)$, the solutions of $\sin 2\theta = 1$ are $\pi/4$ $(k = 0)$ and $\pi/4 + \pi = 5\pi/4$ $(k = 1)$. Note that $k = -1$ gives $\theta = -3\pi/4$ and $k = 2$ gives $\theta = 9\pi/4$, both of which are outside $[0, 2\pi)$.

You should verify $\pi/4$ and $5\pi/4$ are solutions by graphing $Y_1 = \sin(2x)$ and $Y_2 = 1$ for $0 \le x \le 2\pi$. ∎

Warning: In solving a trigonometric equation for $\theta, 0 \le \theta < 2\pi$, in which the argument is not θ (as in Example 3), you must write down all the solutions, first, and then list those that are in the interval $[0, 2\pi)$. Otherwise, solutions may be lost. For example, in solving $\sin 2\theta = 1$, if you merely write the solution $2\theta = \pi/2$, you will find only $\theta = \pi/4$ and miss the solution $\theta = 5\pi/4$.

Now work Problem 7.

E X A M P L E 4

Solving a Trigonometric Equation

Solve the equation: $\sin 2\theta = \frac{1}{2}, 0 \le \theta < 2\pi$

FIGURE 10
$\sin 2\theta = \frac{1}{2}, 0 \le \theta < 2\pi$

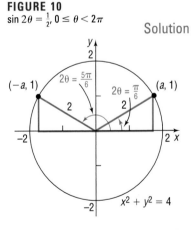

Solution The period of the sine function is 2π. In the interval $[0, 2\pi)$, the sine function has the value $\frac{1}{2}$ at $\pi/6$ and $5\pi/6$. See Figure 10. Consequently, because the argument is 2θ in the equation $\sin 2\theta = \frac{1}{2}$, we have

$$2\theta = \frac{\pi}{6} + 2k\pi \quad \text{or} \quad 2\theta = \frac{5\pi}{6} + 2k\pi \quad k \text{ any integer.}$$

$$\theta = \frac{\pi}{12} + k\pi \qquad\qquad \theta = \frac{5\pi}{12} + k\pi$$

In the interval $[0, 2\pi)$, the solutions of $\sin 2\theta = \frac{1}{2}$ are $\pi/12$, $\pi/12 + \pi = 13\pi/12$, $5\pi/12$, and $5\pi/12 + \pi = 17\pi/12$.

You should verify the solution using a graphing utility. ∎

E X A M P L E 5

Solving a Trigonometric Equation

Solve the equation: $\tan \dfrac{\theta}{4} = 1, 0 \le \theta < 2\pi$

Solution The period of the tangent function is π. In the interval $[0, \pi)$, the tangent function has the value 1 only at $\pi/4$. Because the argument is $\theta/4$ in the given equation, we have

$$\frac{\theta}{4} = \frac{\pi}{4} + k\pi \qquad k \text{ any integer.}$$

$$\theta = \pi + 4k\pi$$

In the interval $[0, 2\pi)$, $\theta = \pi$ is the only solution. You should verify the solution using a graphing utility.

EXAMPLE 6

Solving a Trigonometric Equation with a Calculator

Use a calculator to solve the equation: $\sin\theta = 0.3, 0 \le \theta < 2\pi$

Solution   To solve $\sin\theta = 0.3$ on a graphing calculator, we first choose the radian mode. See Figure 11.

FIGURE 11

```
sin⁻¹(.3)
          .304692654
```

FIGURE 12
$\sin\theta = 0.3$

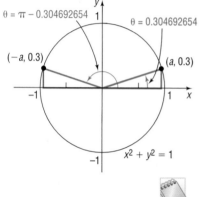

The angle 0.304692654 radian is the angle $-\pi/2 \le \theta \le \pi/2$ for which $\sin\theta = 0.3$. Another angle for which $\sin\theta = 0.3$ is $\pi - 0.304692654$. See Figure 12. The angle $\pi - 0.304692654$ is the angle in quadrant II, where $\sin\theta = 0.3$. Thus, the solutions for $\sin\theta = 0.3$, $0 \le \theta < 2\pi$, are

$$\theta = 0.304692654 \quad \text{or} \quad \theta = \pi - 0.304692654 \approx 2.8369$$

Warning   Example 6 illustrates that caution must be exercised when solving trigonometric equations on a calculator. Remember that the calculator supplies an angle only within the restrictions of the definition of the inverse trigonometric function. To find the remaining solutions, you must identify other quadrants, if any, in which the angle may be located.

Now work Problem 13.

Many trigonometric equations can be solved by applying techniques that we already know, such as applying the quadratic formula (if the equation is a second-degree polynomial) or factoring.

EXAMPLE 7

Solving a Trigonometric Equation Quadratic in Form

Solve the equation: $2\sin^2\theta - 3\sin\theta + 1 = 0, 0 \le \theta < 2\pi$

Solution   The equation we wish to solve is a quadratic equation (in $\sin\theta$) that can be factored:

$$2\sin^2\theta - 3\sin\theta + 1 = 0 \qquad 2x^2 - 3x + 1 = 0, \quad x = \sin\theta$$
$$(2\sin\theta - 1)(\sin\theta - 1) = 0 \qquad (2x - 1)(x - 1) = 0$$
$$2\sin\theta - 1 = 0 \quad \text{or} \quad \sin\theta - 1 = 0$$
$$\sin\theta = \frac{1}{2} \qquad\qquad \sin\theta = 1$$

Thus,

$$\theta = \frac{\pi}{6} \qquad \theta = \frac{5\pi}{6} \qquad \theta = \frac{\pi}{2}$$

Now work Problem 23.

When a trigonometric equation contains more than one trigonometric function, identities sometimes can be used to obtain an equivalent equation that contains only one trigonometric function.

E X A M P L E   8   Solving a Trigonometric Equation Using Identities

Solve the equation: $3 \cos \theta + 3 = 2 \sin^2 \theta, 0 \le \theta < 2\pi$

Solution   The equation in its present form contains sines and cosines. However, a form of the Pythagorean Identity can be used to transform the equation into an equivalent expression containing only cosines:

$$3 \cos \theta + 3 = 2 \sin^2 \theta$$
$$3 \cos \theta + 3 = 2(1 - \cos^2 \theta) \quad \text{sin}^2\ \theta = 1 - \cos^2 \theta$$
$$3 \cos \theta + 3 = 2 - 2 \cos^2 \theta$$
$$2 \cos^2 \theta + 3 \cos \theta + 1 = 0 \quad \text{Quadratic in } \cos \theta$$
$$(2 \cos \theta + 1)(\cos \theta + 1) = 0 \quad \text{Factor}$$
$$2 \cos \theta + 1 = 0 \quad \text{or} \quad \cos \theta + 1 = 0$$
$$\cos \theta = -\tfrac{1}{2} \qquad \cos \theta = -1$$

Thus,

$$\theta = \frac{2\pi}{3} \qquad \theta = \frac{4\pi}{3} \qquad \theta = \pi$$

Check: Graph $Y_1 = 3 \cos x + 3$ and $Y_2 = 2 \sin^2 x, 0 \le x \le 2\pi$ and find the points of intersection. How close are your approximate solutions to the exact ones found in this example? ■

E X A M P L E   9   Solving a Trigonometric Equation Using Identities

Solve the equation: $\cos 2\theta + 3 = 5 \cos \theta, 0 \le \theta < 2\pi$

Solution   First, we observe that the given equation contains two cosine functions, but with different arguments, θ and 2θ. We use the double-angle formula $\cos 2\theta = 2 \cos^2 \theta - 1$ to obtain an equivalent equation containing only $\cos \theta$:

$$\cos 2\theta + 3 = 5 \cos \theta$$
$$(2 \cos^2 \theta - 1) + 3 = 5 \cos \theta$$
$$2 \cos^2 \theta - 5 \cos \theta + 2 = 0$$
$$(\cos \theta - 2)(2 \cos \theta - 1) = 0$$
$$\cos \theta = 2 \quad \text{or} \quad \cos \theta = \tfrac{1}{2}$$

For any angle θ, $-1 \le \cos \theta \le 1$; thus, the equation $\cos \theta = 2$ has no solution. The solutions of $\cos \theta = \frac{1}{2}, 0 \le \theta < 2\pi$, are

$$\theta = \frac{\pi}{3} \qquad \theta = \frac{5\pi}{3}$$

Check: Graph $Y_1 = \cos(2x) + 3$ and $Y_2 = 5 \cos x, 0 \le x \le 2\pi$ and find the points of intersection. Compare your results with those of Example 9. ■

Now work Problem 33.

E X A M P L E   10

Solving a Trigonometric Equation Using Identities

Solve the equation: $\cos^2 \theta + \sin \theta = 2, 0 \le \theta < 2\pi$

Solution

We use a form of the Pythagorean identity:

$$\cos^2 \theta + \sin \theta = 2$$
$$(1 - \sin^2 \theta) + \sin \theta = 2 \qquad \cos^2 \theta = 1 - \sin^2 \theta$$
$$\sin^2 \theta - \sin \theta + 1 = 0$$

FIGURE 13

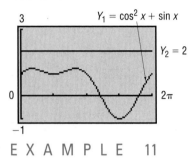

This is a quadratic equation in $\sin \theta$. The discriminant is $b^2 - 4ac = 1 - 4 = -3 < 0$. Therefore, the equation has no real solution.

Figure 13 shows the graphs of $Y_1 = \cos^2 x + \sin x$ and $Y_2 = 2$. The graphs do not intersect anywhere so the equation $Y_1 = Y_2$ has no real solution.

E X A M P L E   11

Solving a Trigonometric Equation Using Identities

Solve the equation: $\sin \theta \cos \theta = -\frac{1}{2}, 0 \le \theta < 2\pi$

Solution

The left side of the given equation is in the form of the double-angle formula $2 \sin \theta \cos \theta = \sin 2\theta$, except for a factor of 2. Thus, we multiply each side by 2:

$$\sin \theta \cos \theta = -\frac{1}{2}$$
$$2 \sin \theta \cos \theta = -1$$
$$\sin 2\theta = -1$$

The argument here is 2θ. Thus, we need to write all the solutions of this equation and then list those that are in the interval $[0, 2\pi)$.

$$2\theta = \frac{3\pi}{2} + 2k\pi \qquad \text{k any integer.}$$

$$\theta = \frac{3\pi}{4} + k\pi$$

The solutions in the interval $[0, 2\pi)$ are

$$\theta = \frac{3\pi}{4} \qquad \theta = \frac{7\pi}{4}$$

Sometimes it is necessary to square both sides of an equation in order to obtain expressions that allow the use of identities. Remember, however, that when squaring both sides extraneous solutions may be introduced. As a result, apparent solutions must be checked.

E X A M P L E   12

Other Methods for Solving a Trigonometric Equation

Solve the equation: $\sin \theta + \cos \theta = 1, 0 \le \theta < 2\pi$

Solution A

Attempts to use available identities do not lead to equations that are easy to solve. (Try it yourself.) Given the form of this equation, we decide to square each side:

$$\sin \theta + \cos \theta = 1$$
$$(\sin \theta + \cos \theta)^2 = 1$$
$$\sin^2 \theta + 2 \sin \theta \cos \theta + \cos^2 \theta = 1$$
$$2 \sin \theta \cos \theta = 0 \quad \scriptstyle{\sin^2 \theta + \cos^2 \theta = 1}$$
$$\sin \theta \cos \theta = 0$$

Thus,

$$\sin \theta = 0 \quad \text{or} \quad \cos \theta = 0$$

and the apparent solutions are

$$\theta = 0 \qquad \theta = \pi \qquad \theta = \frac{\pi}{2} \qquad \theta = \frac{3\pi}{2}$$

Because we squared both sides of the original equation, we must check these apparent solutions to see if any are extraneous:

$\theta = 0$: $\quad \sin 0 + \cos 0 = 0 + 1 = 1$ A solution.

$\theta = \pi$: $\quad \sin \pi + \cos \pi = 0 + (-1) = -1$ Not a solution.

$\theta = \dfrac{\pi}{2}$: $\quad \sin \dfrac{\pi}{2} + \cos \dfrac{\pi}{2} = 1 + 0 = 1$ A solution.

$\theta = \dfrac{3\pi}{2}$: $\quad \sin \dfrac{3\pi}{2} + \cos \dfrac{3\pi}{2} = -1 + 0 = -1$ Not a solution.

Thus, $\theta = 3\pi/2$ and $\theta = \pi$ are extraneous. The actual solutions are $\theta = 0$ and $\theta = \pi/2$. ■

We can solve the equation given in Example 12 another way.

Solution B   We start with the equation

$$\sin \theta + \cos \theta = 1$$

and divide each side by $\sqrt{2}$. (The reason for this choice will become apparent shortly.) Then

$$\frac{1}{\sqrt{2}} \sin \theta + \frac{1}{\sqrt{2}} \cos \theta = \frac{1}{\sqrt{2}}$$

The left side now resembles the formula for the sine of the sum of two angles, one of which is θ. The other angle is unknown (call it ϕ.) Then

$$\sin(\theta + \phi) = \sin \theta \cos \phi + \cos \theta \sin \phi = \frac{1}{\sqrt{2}} \tag{1}$$

where

$$\cos \phi = \frac{1}{\sqrt{2}} \qquad \sin \phi = \frac{1}{\sqrt{2}}, \qquad 0 \leq \phi < 2\pi$$

The angle ϕ is therefore $\pi/4$. As a result, equation (1) becomes

$$\sin\left(\theta + \frac{\pi}{4}\right) = \frac{1}{\sqrt{2}}$$

We solve this equation to get

$$\theta + \frac{\pi}{4} = \frac{\pi}{4} \quad \text{or} \quad \theta + \frac{\pi}{4} = \frac{3\pi}{4}$$

$$\theta = 0 \qquad\qquad \theta = \frac{\pi}{2}$$

These solutions agree with the solutions found earlier. ▬

This second method of solution can be used to solve any linear equation in the variables sin θ and cos θ.

E X A M P L E   13

Solving a Trigonometric Equation Linear in sin θ and cos θ

Solve:

$$a \sin\theta + b \cos\theta = c, \qquad 0 \le \theta < 2\pi \tag{2}$$

where a, b, and c are constants and either $a \neq 0$ or $b \neq 0$.

Solution   We divide each side of equation (2) by $\sqrt{a^2 + b^2}$. Then

$$\frac{a}{\sqrt{a^2 + b^2}} \sin\theta + \frac{b}{\sqrt{a^2 + b^2}} \cos\theta = \frac{c}{\sqrt{a^2 + b^2}} \tag{3}$$

There is a unique angle ϕ, $0 \le \phi < 2\pi$, for which

$$\cos\phi = \frac{a}{\sqrt{a^2 + b^2}} \quad \text{and} \quad \sin\phi = \frac{b}{\sqrt{a^2 + b^2}} \tag{4}$$

FIGURE 14

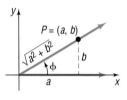

(see Figure 14). Thus, equation (3) may be written as

$$\sin\theta \cos\phi + \cos\theta \sin\phi = \frac{c}{\sqrt{a^2 + b^2}}$$

or, equivalently,

$$\sin(\theta + \phi) = \frac{c}{\sqrt{a^2 + b^2}} \tag{5}$$

where ϕ satisfies equations (4).

If $|c| > \sqrt{a^2 + b^2}$, then $\sin(\theta + \phi) > 1$ or $\sin(\theta + \phi) < -1$, and equation (5) has no solution.

If $|c| \le \sqrt{a^2 + b^2}$, then the solutions of equation (5) are

$$\theta + \phi = \sin^{-1}\frac{c}{\sqrt{a^2 + b^2}} \quad \text{or} \quad \theta + \phi = \pi - \sin^{-1}\frac{c}{\sqrt{a^2 + b^2}}$$

Because the angle ϕ is determined by equations (4), these are the solutions to equation (2). ▬

Graphing Utility Solutions

The techniques introduced in this section apply only to certain types of trigonometric equations. Solutions for other types are usually studied in calculus, using numerical methods. In the next example, we show how a graphing utility may be used to obtain solutions.

E X A M P L E   14   Solving Trigonometric Equations Using a Graphing Utility

Solve: $5 \sin x + x = 3$

Express the solution(s) correct to two decimal places.

Solution   This type of trigonometric equation cannot be solved by previous methods. A graphing utility, though, can be used here. The solution(s) of this equation are the same as the points of intersection of the graphs of $Y_1 = 5 \sin x + x$ and $Y_2 = 3$. See Figure 15.

FIGURE 15

There are three points of intersection; the x-coordinates are the solutions we seek. Using INTERSECT, we find

$$x = 0.51 \qquad x = 3.17 \qquad x = 5.71$$

correct to two decimal places.

6.5 EXERCISES

In Problems 1–12, solve each equation on the interval $0 \le \theta < 2\pi$.

1. $\sin \theta = \dfrac{1}{2}$

2. $\tan \theta = 1$

3. $\tan \theta = -\dfrac{1}{\sqrt{3}}$

4. $\cos \theta = -\dfrac{\sqrt{3}}{2}$

5. $\cos \theta = 0$

6. $\sin \theta = \dfrac{\sqrt{2}}{2}$

7. $\sin 3\theta = -1$

8. $\tan \dfrac{\theta}{2} = \sqrt{3}$

9. $\cos\left(2\theta - \dfrac{\pi}{2}\right) = -1$

10. $\sin\left(3\theta + \dfrac{\pi}{18}\right) = 1$

11. $\sec \dfrac{3\theta}{2} = -2$

12. $\cot \dfrac{2\theta}{3} = -\sqrt{3}$

In Problems 13–20, use a calculator to solve each equation on the interval $0 \le \theta < 2\pi$.

13. $\sin \theta = 0.4$

14. $\cos \theta = 0.6$

15. $\tan \theta = 5$

16. $\cot \theta = 2$

17. $\cos \theta = -0.9$

18. $\sin \theta = -0.2$

19. $\sec \theta = -4$

20. $\csc \theta = -3$

In Problems 21–50, solve each equation on the interval $0 \le \theta < 2\pi$. Verify your result using a graphing utility.

21. $2 \cos^2 \theta + \cos \theta = 0$

22. $\sin^2 \theta - 1 = 0$

23. $2 \sin^2 \theta - \sin \theta - 1 = 0$

24. $2 \cos^2 \theta + \cos \theta - 1 = 0$

25. $(\tan \theta - 1)(\sec \theta - 1) = 0$

26. $(\cot \theta + 1)(\csc \theta - \frac{1}{2}) = 0$

27. $\cos \theta = \sin \theta$

28. $\cos \theta + \sin \theta = 0$

29. $\tan \theta = 2 \sin \theta$

30. $\sin 2\theta = \cos \theta$

31. $\sin \theta = \csc \theta$

32. $\tan \theta = \cot \theta$

33. $\cos 2\theta = \cos \theta$

34. $\sin 2\theta \sin \theta = \cos \theta$

35. $\sin 2\theta + \sin 4\theta = 0$

36. $\cos 2\theta + \cos 4\theta = 0$

37. $\cos 4\theta - \cos 6\theta = 0$

38. $\sin 4\theta - \sin 6\theta = 0$

39. $1 + \sin \theta = 2 \cos^2 \theta$

40. $\sin^2 \theta = 2 \cos \theta + 2$

41. $\tan^2 \theta = \frac{3}{2} \sec \theta$

42. $\csc^2 \theta = \cot \theta + 1$

43. $3 - \sin \theta = \cos 2\theta$

44. $\cos 2\theta + 5 \cos \theta + 3 = 0$

45. $\sec^2 \theta + \tan \theta = 0$

46. $\sec \theta = \tan \theta + \cot \theta$

47. $\sin \theta - \sqrt{3} \cos \theta = 1$

48. $\sqrt{3} \sin \theta + \cos \theta = 1$

49. $\tan 2\theta + 2 \sin \theta = 0$

50. $\tan 2\theta + 2 \cos \theta = 0$

In Problems 51–56, solve each equation for $-\pi \le x \le \pi$. Express the solution(s) correct to two decimal places.

51. Solve the equation $\cos x = e^x$ by graphing $Y_1 = \cos x$ and $Y_2 = e^x$ and finding their point(s) of intersection.

52. Solve the equation $\cos x = e^x$ by graphing $Y_1 = \cos x - e^x$ and finding the x-intercept(s).

53. Solve the equation $2 \sin x = 0.7x$ by graphing $Y_1 = 2 \sin x$ and $Y_2 = 0.7x$ and finding their point(s) of intersection.

54. Solve the equation $2 \sin x = 0.7x$ by graphing $Y_1 = 2 \sin x - 0.7x$ and finding the x-intercept(s).

55. Solve the equation $\cos x = x^2$ by graphing $Y_1 = \cos x$ and $Y_2 = x^2$ and finding their point(s) of intersection.

56. Solve the equation $\cos x = x^2$ by graphing $Y_1 = \cos x - x^2$ and finding the x-intercept(s).

In Problems 57–68, use a graphing utility to solve each equation. Express the solution(s) correct to two decimal places.

57. $x + 5 \cos x = 0$

58. $x - 4 \sin x = 0$

59. $22x - 17 \sin x = 3$

60. $19x + 8 \cos x = 2$

61. $\sin x + \cos x = x$

62. $\sin x - \cos x = x$

63. $x^2 - 2 \cos x = 0$

64. $x^2 + 3 \sin x = 0$

65. $x^2 - 2 \sin 2x = 3x$

66. $x^2 = x + 3 \cos 2x$

67. $6 \sin x - e^x = 2, x > 0$

68. $4 \cos 3x - e^x = 1, x > 0$

69. Constructing a Rain Gutter A rain gutter is to be constructed of aluminum sheets 12 inches wide. After marking off a length of 4 inches from each edge, this length is bent up at an angle θ. See the illustration. The area A of the opening as a function of θ is given by

$$A = 16 \sin \theta(\cos \theta + 1), 0° < \theta < 90°$$

(a) In calculus, you will be asked to find the angle θ that maximizes A by solving the equation

$$\cos 2\theta + \cos \theta = 0, 0° < \theta < 90°$$

Solve this equation for θ by using the double-angle formula.

(b) Solve the equation for θ by writing the sum of the two cosines as a product.

(c) What is the maximum area A of the opening?

(d) Graph $A, 0° \le \theta \le 90°$, and find the angle θ that maximizes the area A. Also find the maximum area. Compare the results to the answers found earlier.

70. Projectile Motion An object is propelled upward at an angle θ, $45° < \theta < 90°$, to the horizontal with an initial velocity of v_0 feet per second from the base of a plane that makes an angle of $45°$ with the horizontal. See the illustration. If air resistance is ignored, the distance R it travels up the inclined plane is given by

$$R = \frac{v_0^2 \sqrt{2}}{32}(\sin 2\theta - \cos 2\theta - 1)$$

(a) In calculus, you will be asked to find the angle θ that maximizes R by solving the equation

$$\sin 2\theta + \cos 2\theta = 0$$

Solve this equation for θ using the method of Example 13.
(b) Solve this equation for θ by dividing each side by $\cos 2\theta$.
(c) What is the maximum distance R if $v_0 = 32$ feet per second?
(d) Graph R, $45° \le \theta \le 90°$, and find the angle θ that maximizes the distance R. Also find the maximum distance. Use $v_0 = 32$ feet per second. Compare the results with the answers found earlier.

71. Heat Transfer In the study of heat transfer, the equation $x + \tan x = 0$ occurs. Graph $Y_1 = -x$ and $Y_2 = \tan x$ for $x \ge 0$. Conclude that there are an infinite number of points of intersection of these two graphs. Now find the first two positive solutions of $x + \tan x = 0$ correct to two decimal places.

72. Carrying a Ladder around a Corner A ladder of length L is carried horizontally around a corner from a hall 3 feet wide into a hall 4 feet wide. See the illustration.
(a) Express L as a function of θ.

(b) In calculus, you will be asked to find the length of the longest ladder that can turn the corner by solving the equation

$3 \sec \theta \tan \theta - 4 \csc \theta \cot \theta = 0$,
$0° < \theta < 90°$

Solve this equation for θ.
(c) What is the length of the largest ladder that can be carried around the corner?
(d) Graph L, $0° \le \theta \le 90°$, and find the angle θ that maximizes the length L. Also find the maximum length. Compare the results with the ones found in parts (b) and (c).

73. Projectile Motion The horizontal distance a projectile will travel in the air is given by the equation

$$R = \frac{v_0^2 \sin 2\theta}{g}$$

where v_0 is the initial velocity of the projectile, θ is the angle of elevation, and g is acceleration due to gravity (9.8 m/sec^2).
(a) If you can throw a baseball with an initial speed of 34.8 m/sec, at what angle of elevation θ should you direct the throw so that the ball travels a distance of 107 m before striking the ground.
(b) Determine the maximum distance you can throw the ball.
(c) Graph R, with $v_0 = 34.8$ m/sec.
(d) Verify the results obtained in parts (a) and (b) using ZERO or ROOT.

74. Projectile Motion Refer to Problem 73.
(a) If you can throw a baseball with an initial speed of 40 m/sec, at what angle of elevation θ should you direct the throw so that the ball travels a distance of 110 m before striking the ground.
(b) Determine the maximum distance you can throw the ball.
(c) Graph R, with $v_0 = 40$ m/sec.
(d) Verify the results obtained in parts (a) and (b) using ZERO or ROOT.

*The following discussion of **Snell's Law of Refraction** (named after Willebrod Snell, 1591–1626) is needed for Problems 75–81: Light, sound, and other waves travel at different speeds, depending on the media (air, water, wood, and so on) through which they pass. Suppose that light travels from a point A in one medium, where its speed is v_1, to a point B in another medium, where its speed is v_2. Refer to the figure, where the angle θ_1 is called the **angle of incidence** and the angle θ_2 is the **angle of refraction**. Snell's Law* which can be proved using calculus, states that*

$$\frac{\sin \theta_1}{\sin \theta_2} = \frac{v_1}{v_2}$$

*The ratio v_1/v_2 is called the **index of refraction**. Some values are given in the following table.*

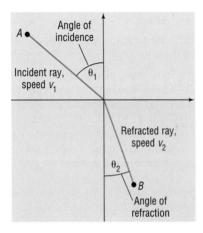

Some Indexes of Refraction	
Medium	**Index of Refraction**
Water	1.33
Ethyl alcohol	1.36
Carbon bisulfide	1.63
Air (1 atm and 20°C)	1.0003
Methylene iodide	1.74
Fused quartz	1.46
Glass, crown	1.52
Glass, dense flint	1.66
Sodium chloride	1.53

For light of wavelength 589 nanometers, measured with respect to a vacuum. The index with respect to air is negligibly different in most cases.

75. The index of refraction of light in passing from a vacuum into water is 1.33. If the angle of incidence is 40°, determine the angle of refraction.

76. The index of refraction of light in passing from a vacuum into dense glass is 1.66. If the angle of incidence is 50°, determine the angle of refraction.

77. Ptolemy, who lived in the city of Alexandria in Egypt during the second century A.D., gave the measured values in the table below for the angle of incidence θ_1 and the angle of refraction θ_2 for a light beam passing from air into water. Do these values agree with Snell's Law? If so, what index of refraction results? (These data are interesting as the oldest recorded physical measurements.)†

θ_1	θ_2
10°	7°45′
20°	15°30′
30°	22°30′
40°	29°0′
50°	35°0′
60°	40°30′
70°	45°30′
80°	50°0′

78. The speed of yellow sodium light (wavelength of 589 nanometers) in a certain liquid is measured to be 1.92×10^8 meters per second. What is the index of refraction of this liquid, with respect to air, for sodium light?‡

79. A beam of light with a wavelength of 589 nanometers traveling in air makes an angle of incidence of 40° upon a slab of transparent material, and the refracted beam makes an angle of refraction of 26°. Find the index of refraction of the material.‡

80. A light ray with a wavelength of 589 nanometers (produced by a sodium lamp) traveling through air makes an angle of incidence of 30° on a smooth, flat slab of crown glass. Find the angle of refraction.‡

81. A light beam passes from one medium to another through a thick slab of material whose index of refraction is n_2. Show that the emerging beam is parallel to the incident beam.‡

*Because this law was also deduced by René Descartes, in France it is also known as Descartes' Law.

†Adapted from Halliday and Resnick, *Physics, Parts 1 & 2,* 3rd ed. New York: Wiley, 1978, p. 953.

‡Adapted from Serway, *Physics,* 3rd ed. Philadelphia: W.B. Saunders, p. 805.

82. Explain in your own words how you would use your calculator to solve the equation $\sin x = 0.3, 0 \leq x < 2\pi$. How would you modify your approach in order to solve the equation $\cot x = 5, 0 < x < 2\pi$?

CHAPTER REVIEW

THINGS TO KNOW

Formulas

Sum and Difference Formulas

$$\cos(\alpha + \beta) = \cos \alpha \cos \beta - \sin \alpha \sin \beta$$
$$\cos(\alpha - \beta) = \cos \alpha \cos \beta + \sin \alpha \sin \beta$$
$$\sin(\alpha + \beta) = \sin \alpha \cos \beta + \cos \alpha \sin \beta$$
$$\sin(\alpha - \beta) = \sin \alpha \cos \beta - \cos \alpha \sin \beta$$
$$\tan(\alpha + \beta) = \frac{\tan \alpha + \tan \beta}{1 - \tan \alpha \tan \beta}$$
$$\tan(\alpha - \beta) = \frac{\tan \alpha - \tan \beta}{1 + \tan \alpha \tan \beta}$$

Double-Angle Formulas

$$\sin 2\theta = 2 \sin \theta \cos \theta$$
$$\cos 2\theta = \cos^2 \theta - \sin^2 \theta$$
$$\cos 2\theta = 1 - 2 \sin^2 \theta$$
$$\cos 2\theta = 2 \cos^2 \theta - 1$$
$$\tan 2\theta = \frac{2 \tan \theta}{1 - \tan^2 \theta}$$

Half-Angle Formulas

$$\sin^2 \frac{\alpha}{2} = \frac{1 - \cos \alpha}{2}$$
$$\cos^2 \frac{\alpha}{2} = \frac{1 + \cos \alpha}{2}$$
$$\tan^2 \frac{\alpha}{2} = \frac{1 - \cos \alpha}{1 + \cos \alpha}$$
$$\sin \frac{\alpha}{2} = \pm \sqrt{\frac{1 - \cos \alpha}{2}}$$
$$\cos \frac{\alpha}{2} = \pm \sqrt{\frac{1 + \cos \alpha}{2}}$$
$$\tan \frac{\alpha}{2} = \pm \sqrt{\frac{1 - \cos \alpha}{1 + \cos \alpha}} = \frac{1 - \cos \alpha}{\sin \alpha} = \frac{\sin \alpha}{1 + \cos \alpha}$$

Where the + or − sign is determined by the quadrant of the angle $\alpha/2$.

Product-to-Sum Formulas

$$\sin \alpha \sin \beta = \tfrac{1}{2}[\cos(\alpha - \beta) - \cos(\alpha + \beta)]$$
$$\cos \alpha \cos \beta = \tfrac{1}{2}[\cos(\alpha - \beta) + \cos(\alpha + \beta)]$$
$$\sin \alpha \cos \beta = \tfrac{1}{2}[\sin(\alpha + \beta) + \sin(\alpha - \beta)]$$

Sum-to-Product Formulas

$$\sin \alpha + \sin \beta = 2 \sin \frac{\alpha + \beta}{2} \cos \frac{\alpha - \beta}{2}$$
$$\sin \alpha - \sin \beta = 2 \sin \frac{\alpha - \beta}{2} \cos \frac{\alpha + \beta}{2}$$
$$\cos \alpha + \cos \beta = 2 \cos \frac{\alpha + \beta}{2} \cos \frac{\alpha - \beta}{2}$$
$$\cos \alpha - \cos \beta = -2 \sin \frac{\alpha + \beta}{2} \sin \frac{\alpha - \beta}{2}$$

HOW TO

Establish identities   Solve a trigonometric equation algebraically and graphically

CHAPTER 7

Applications of Trigonometric Functions

Alternative energy production is being studied and put into practice throughout the world. On the following page you will find your project. You will need to use the Sullivan website at:

www.prenhall.com/sullivan

to help link you to the Internet resources needed to prepare your report.

PREPARING FOR THIS CHAPTER

Before getting started on this chapter, review the following concepts:

Pythagorean Theorem *(Appendix, Section 1, p. 975)*

Right Triangle Trigonometry *(Chapter 5, Section 5.4)*

OUTLINE

Netsite: http://www.prenhall.com/sullivan/

TIDAL ENERGY: MAKING ELECTRICITY FROM OCEAN TIDES

Using tidal flows to generate electrical power seems a viable means of alternative energy production. Suppose you are the project manager for an engineering firm and the city of Hilo, HI. has contacted your firm to do a feasibility study to determine whether or not tidal energy is a practical method for generating electrical power for the city. You begin your study by doing an Internet search on "tidal energy conversion." Your plan is to model Hilo Harbor tides by a mathematical function. You know by Harmonic Analysis of the tidal oscillations that a good model is a function of the form:

$$y = A \sin(bx + c) + D \sin(ex + f)$$

The constants for A, b, c, D, e, f can be calculated using data found at the *School of Ocean and Earth Science and Technology* (SOEST). NOTE: An explanation of Harmonic Analysis can be found at: *The Harmonic Analysis Tutorial;* information on a tidal energy conversion experiment can be found at: *The TOPEX/Poseidon Project*

1. What are the advantages/disadvantages of using tidal energy to produce electrical power? (Clean energy, reduces coal burning . . . only works part time . . . etc.) What are some of the different methods available for tidal energy conversion? (Dams, floating platforms . . .)
2. From the Topex Ocean Tides graph at the SOEST website, does Hawaii seem like a good location for a tidal energy plant?
3. By looking at the SOEST Tide Chart for Hilo Harbor, can you estimate how many hours per day the power plant would produce electricity? What additional information do you need?
4. Using the data from SOEST, determine a function that models the tide at Hilo Harbor. Graph this function. How does your function compare with *the NOAA tide chart*?
5. By using a dam type of power generating plant, your plant will be able to generate 1 Kw per hour if the difference between the high and low tides is 6 meters or more. How many days per month would the plant be able to generate 1 Kw per hour at some time during the day? Use the function found in question 4.
6. If a dam type of power plant could produce electrical power for only 70 percent of the time between high and low tides, how many Kws could be generated in one day if the tides differ by more than 6 meters?
7. What other information would you need before you could determine the monthly electrical output for a Hilo Harbor Tidal Conversion Facility? If 70 percent of the time of the incoming and outgoing tides you would be able to generate 1 Kw of power per hour, how many KWs of power would the plant produce in 1 month? Use the function found in question 4.
8. Before you prepare your final report for the city of Hilo you might want to take a look at the website of the *Office of Scientific and Technical Information (OSTI), an office of the Department of Energy.*

In this chapter we use the trigonometric functions to solve applied problems. The first four sections deal with applications involving right triangles and *oblique triangles,* triangles that do not have a right angle. To solve problems involving oblique triangles, we will de-velop the Law of Sines and the Law of Cosines. We will also develop formulas for finding the area of a triangle.

The final two sections deal with applications involving the graphs of the sine and cosine functions, simple harmonic motion, and damped motion.

7.1 | SOLVING RIGHT TRIANGLES

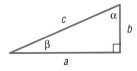

1 Solve Right Triangles

FIGURE 1

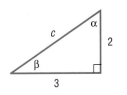

In the discussion that follows, we will always label a right triangle so that side a is opposite angle α, side b is opposite angle β, and side c is the hypotenuse, as shown in Figure 1. **To solve a right triangle** means to find the missing lengths of its sides and the measurements of its angles. We shall follow the practice of expressing the lengths of the sides rounded to two decimal places and of expressing angles in degrees rounded to one decimal place.

To solve a right triangle, we need to know either an angle (besides the 90° one) and a side or else two sides. Then we make use of the Pythagorean Theorem and the fact that the sum of the angles of a triangle is 180°. The sum of the unknown angles in a right triangle is therefore 90°. Thus, for the right triangle shown in Figure 1, we have

$$c^2 = a^2 + b^2 \qquad \alpha + \beta = 90°$$

EXAMPLE 1 — Solving a Right Triangle

FIGURE 2

Use Figure 2. If $b = 2$ and $\alpha = 40°$, find a, c, and β.

Solution

Since $\alpha = 40°$ and $\alpha + \beta = 90°$, we find that $\beta = 50°$. To find the sides a and c, we use the facts that

$$\tan 40° = \frac{a}{2} \qquad \text{and} \qquad \cos 40° = \frac{2}{c}$$

Now solve for a and c.

$$a = 2 \tan 40° \approx 1.68 \quad \text{and} \qquad c = \frac{2}{\cos 40°} \approx 2.61 \qquad \blacksquare$$

Now work Problem 1.

EXAMPLE 2 — Solving a Right Triangle

FIGURE 3

Use Figure 3. If $a = 3$ and $b = 2$, find c, α, and β.

Solution

Since $a = 3$ and $b = 2$, then by the Pythagorean Theorem we have

$$c^2 = a^2 + b^2 = 9 + 4 = 13$$
$$c = \sqrt{13} \approx 3.61$$

To find angle α, we use the fact that

$$\tan \alpha = \frac{3}{2}$$

Set the mode on your calculator to degrees. Then, rounded to one decimal place, we find $\alpha = 56.3°$. Since $\alpha + \beta = 90°$, we find that $\beta = 33.7°$. ■

Note: In subsequent examples and in the exercises that follow, unless otherwise indicated, we will measure angles in degrees and round off to one decimal place; we will round off all sides to two decimal places. To avoid round-off errors when using a calculator, we will store unrounded values in memory for use in subsequent calculations.

Now work Problem 11.

One common use for trigonometry is to measure heights and distances that are either awkward or impossible to measure by ordinary means.

E X A M P L E   3   Finding the Width of a River

A surveyor can measure the width of a river by setting up a transit* at a point C on one side of the river and taking a sighting of a point A on the other side. Refer to Figure 4. After turning through an angle of 90° at C, the surveyor walks a distance of 200 meters to point B. Using the transit at B, the angle β is measured and found to be 20°. What is the width of the river?

FIGURE 4

Solution   We seek the length of side b. We know a and β, so we use the fact that

$$\tan \beta = \frac{b}{a}$$

to get

$$\tan 20° = \frac{b}{200}$$
$$b = 200 \tan 20° \approx 72.79 \text{ meters}$$

Thus, the width of the river is approximately 73 meters, rounded to the nearest meter. ▄

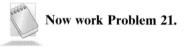

Now work Problem 21.

E X A M P L E   4   Finding the Inclination of a Mountain Trail

A straight trail with a uniform inclination leads from a hotel, elevation 8000 feet, to a scenic overlook, elevation 11,100 feet. The length of the trail is 14,100 feet. What is the inclination (grade) of the trail?

FIGURE 5

Solution   Figure 5 illustrates the situation. We seek the angle β. As the illustration shows,

$$\sin \beta = \frac{3100}{14,100}$$

Using a calculator,

$$\beta \approx 12.7°$$

The inclination (grade) of the trail is approximately 12.7°. ▄

Vertical heights can sometimes be measured using either the angle of elevation or the angle of depression. If a person is looking up at an object, the

---

*An instrument used in surveying to measure angles.

acute angle measured from the horizontal to a line-of-sight observation of the object is called the **angle of elevation.** See Figure 6(a).

If a person is standing on a cliff looking down at an object, the acute angle made by the line of sight observation of the object and the horizontal is called the **angle of depression.** See Figure 6(b).

FIGURE 6

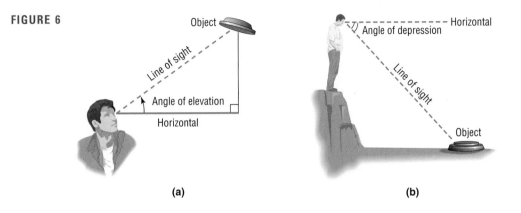

(a)   (b)

E X A M P L E   5

Finding Heights Using the Angle of Elevation

To determine the height of a radio transmission tower, a surveyor walks off a distance of 300 meters from the base of the tower. The angle of elevation is then measured and found to be 40°. If the transit is 2 meters off the ground when the sighting is taken, how high is the radio tower? See Figure 7(a).

FIGURE 7

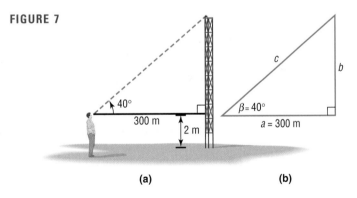

(a)   (b)

Solution   Figure 7(b) shows a triangle that replicates the illustration in Figure 7(a). To find the length b, we use the fact that $\tan \beta = b/a$. Then

$$b = a \tan \beta = 300 \tan 40° = 251.73 \text{ meters}$$

Because the transit is 2 meters high, the actual height of the tower is approximately 254 meters, rounded to the nearest meter.   ■

   Now work Problem 23.

The idea behind Example 5 can also be used to find the height of an object with a base that is not accessible to the horizontal.

E X A M P L E   6   Finding the Height of a Statue on a Building

Adorning the top of the Board of Trade building in Chicago is a statue of the Greek goddess Ceres, goddess of wheat. From street level, two observations are taken 400 feet from the center of the building. The angle of elevation to the base of the statue is found to be 45.0°; the angle of elevation to the top of the statue is 47.2°. See Figure 8(a). What is the height of the statue?

FIGURE 8

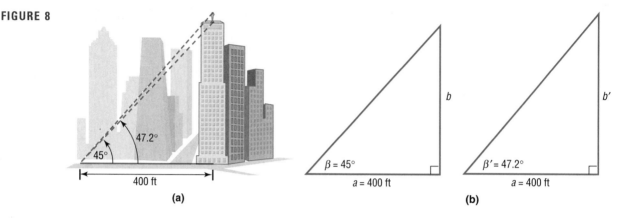

(a)　　　　　　　　　　　　　　　(b)

Solution   Figure 8(b) shows two triangles that replicate Figure 8(a). The height of the statue of Ceres will be $b' - b$. To find b and b', we refer to Figure 8(b):

$$\tan 45° = \frac{b}{400} \qquad\qquad \tan 47.2° = \frac{b'}{400}$$

$$b = 400 \tan 45° = 400 \qquad\qquad b' = 400 \tan 47.2° \approx 432$$

The height of the statue is approximately $432 - 400 = 32$ feet. ■

When it is not possible to walk off a distance from the base of the object whose height we seek, a more imaginative solution is required.

E X A M P L E   7   Finding the Height of a Mountain

To measure the height of a mountain, a surveyor takes two sightings of the peak at a distance 900 meters apart on a direct line to the mountain.* See Figure 9(a). The first observation results in an angle of elevation of 47°, whereas the second results in an angle of elevation of 35°. If the transit is 2 meters high, what is the height h of the mountain?

FIGURE 9

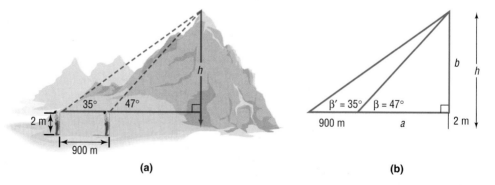

(a)　　　　　　　　　　　　　　　(b)

*For simplicity, we assume that these sightings are at the same level.

Solution Figure 9(b) shows two triangles that replicate the illustration in Figure 9(a). From the two right triangles shown, we find

$$\tan \beta' = \frac{b}{a + 900} \qquad \tan \beta = \frac{b}{a}$$

$$\tan 35° = \frac{b}{a + 900} \qquad \tan 47° = \frac{b}{a}$$

This is a system of two equations involving two variables, a and b. Because we seek b, we choose to solve the right-hand equation for a and substitute the result, $a = b/\tan 47° = b \cot 47°$, in the left-hand equation. The result is

$$\tan 35° = \frac{b}{b \cot 47° + 900}$$

$$b = (b \cot 47° + 900) \tan 35°$$

$$b = b \cot 47° \tan 35° + 900 \tan 35°$$

$$b(1 - \cot 47° \tan 35°) = 900 \tan 35°$$

$$b = \frac{900 \tan 35°}{1 - \cot 47° \tan 35°} = \frac{900 \tan 35°}{1 - \dfrac{\tan 35°}{\tan 47°}} \approx 1816$$

The height of the peak from ground level is approximately $1816 + 2 = 1818$ meters. ▬

FIGURE 10
The bearing of P from O is θ

Now work Problem 29.

In navigation, the **direction** or **bearing** from O of an object at P means the positive angle measured clockwise* from the north (N) to the ray OP. See Figure 10. Based on Figure 10, we would say that the bearing of P from O is θ degrees.

E X A M P L E   8   Finding the Bearing of an Airplane

A Boeing 777 aircraft takes off from O'Hare Airport on runway 2 LEFT, which has a bearing of $20°$. After flying for 1 mile, the pilot of the aircraft requests permission to turn $90°$ and head toward the northwest. The request is granted.

(a) What is the new bearing?
(b) After the plane goes 2 miles in this direction, what bearing should the control tower use to locate the aircraft?

FIGURE 11

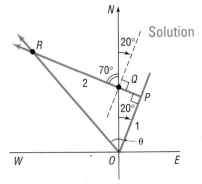

Solution (a) Figure 11 illustrates the situation. After flying 1 mile from the airport O (the control tower), the aircraft is at P. After turning $90°$ toward the northwest, we see that

$$\text{Angle } NOP = 20° \qquad \text{Angle } RQN = 90° - 20° = 70°$$

The new bearing of the aircraft is therefore

$$360° - \text{Angle } RQN = 360° - 70° = 290°$$

*Notice that this convention is just the opposite of what we are used to.

(b) After flying 2 miles at a bearing of 290°, the aircraft is at R. If $\theta =$ Angle ROP, then

$$\tan \theta = \frac{2}{1} = 2 \quad \text{so} \quad \theta = 63.4°$$

As a result,

Angle $RON = \theta - 20° = 43.4°$

The bearing of the plane from the central tower unit at O is

$$360° - \text{Angle } RON = 360° - 43.4° = 316.6°$$

E X A M P L E 9   The Gibb's Hill Lighthouse, Southampton, Bermuda

In operation since 1846, the Gibb's Hill Lighthouse stands 117 feet high on a hill 245 feet high, so its beam of light is 362 feet above sea level. A brochure states that the light can be seen on the horizon about 26 miles distant. Verify the accuracy of this statement.

Solution   Figure 12 illustrates the situation. The central angle θ obeys the equation

FIGURE 12

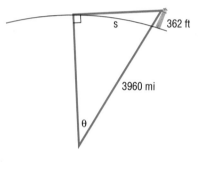

$$\cos \theta = \frac{3960}{3960 + \dfrac{362}{5280}} = 0.999982687$$

so

$$\theta = 0.00588 \text{ radian} = 0.33715° = 20.23'$$

The brochure does not indicate whether the distance is measured in nautical miles or statute miles. The distance s in nautical miles is 20.23, the measurement of θ in minutes. (Refer to Problem 86, Section 5.1). The distance s in statute miles is

$$s = r\theta = 3960(0.00588) = 23.3 \text{ miles}$$

In either case, it would seem that the brochure overstated the distance somewhat.

7.1 EXERCISES

In Problems 1–14, use the right triangle shown in the margin. Then, using the given information, solve the triangle.

1. $b = 5$, $\beta = 20°$; find a, c, and α
2. $b = 4$, $\beta = 10°$; find a, c, and α
3. $a = 6$, $\beta = 40°$; find b, c, and α
4. $a = 7$, $\beta = 50°$; find b, c, and α
5. $b = 4$, $\alpha = 10°$; find a, c, and β
6. $b = 6$, $\alpha = 20°$; find a, c, and β
7. $a = 5$, $\alpha = 25°$; find b, c, and β
8. $a = 6$, $\alpha = 40°$; find b, c, and β
9. $c = 9$, $\beta = 20°$; find b, a, and α
10. $c = 10$, $\alpha = 40°$; find b, a, and β
11. $a = 5$, $b = 3$; find c, α and β
12. $a = 2$, $b = 8$; find c, α and β
13. $a = 2$, $c = 5$; find b, α and β
14. $b = 4$, $c = 6$; find a, α and β

15. A right triangle has a hypotenuse of length 3 inches. If one angle is 35°, find the length of each leg.

16. A right triangle has a hypotenuse of length 2 centimeters. If one angle is 40°, find the length of each leg.

17. A right triangle contains a 35° angle. If one leg is of length 5 inches, what is the length of the hypotenuse?
    [**Hint:** Two answers are possible.]

18. A right triangle contains an angle of $\pi/10$ radian. If one leg is of length 3 meters, what is the length of the hypotenuse?
    [**Hint:** Two answers are possible.]

19. The hypotenuse of a right triangle is 5 inches. If one leg is 2 inches, find the degree measure of each angle.

20. The hypotenuse of a right triangle is 3 feet. If one leg is 1 foot, find the degree measure of each angle.

21. Finding the Width of a Gorge Find the distance from *A* to *C* across the gorge illustrated in the figure.

22. Finding the Distance Across a Pond Find the distance from *A* to *C* across the pond illustrated in the figure.

23. The Eiffel Tower The tallest tower built before the era of television masts, the Eiffel Tower was completed on March 31, 1889. Find the height of the Eiffel Tower (before a television mast was added to the top) using the information given in the illustration.

24. Finding the Distance of a Ship from Shore A ship, offshore from a vertical cliff known to be 100 feet in height, takes a sighting of the top of the cliff. If the angle of elevation is found to be 25°, how far offshore is the ship?

25. Suppose that you are headed toward a plateau 50 meters high. If the angle of elevation to the top of the plateau is 20°, how far are you from the base of the plateau?

26. Statue of Liberty A ship is just offshore of New York City. A sighting is taken of the Statue of Liberty, which is about 305 feet tall. If the angle of elevation to the top of the statue is 20°, how far is the ship from the base of the statue?

27. A 22 foot extension ladder leaning against a building makes a 70° angle with the ground. How far up the building does the ladder touch?

28. Finding the Height of a Building To measure the height of a building, two sightings are taken a distance of 50 feet apart. If the first angle of elevation is 40° and the second is 32°, what is the height of the building?

29. Great Pyramid of Cheops One of the original Seven Wonders of the World, the Great Pyramid of Cheops was built about 2580 B.C. Its original height was 480 feet 11 inches, but due to the loss of its topmost stones, it is now shorter.* Find the current height of the Great Pyramid, using the information given in the illustration.

30. A laser beam is to be directed through a small hole in the center of a circle of radius 10 feet. The origin of the beam is 35 feet from the circle (see the figure). At what angle of elevation should the beam be aimed to ensure that it goes through the hole?

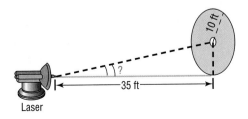

Source: Guinness Book of World Records.

31. Finding the Angle of Elevation of the Sun At 10 A.M. on April 26, 1996, a building 300 feet high casts a shadow 50 feet long. What is the angle of elevation of the Sun?

32. Mt. Rushmore To measure the height of Lincoln's caricature on Mt. Rushmore, two sightings 800 feet from the base of the mountain are taken. If the angle of elevation to the bottom of Lincoln's face is 32° and the angle of elevation to the top is 35°, what is the height of Lincoln's face?

33. Finding the Height of a Helicopter Two observers simultaneously measure the angle of elevation of a helicopter. One angle is measured as 25°, the other as 40° (see the figure). If the observers are 100 feet apart and the helicopter lies over the line joining them, how high is the helicopter?

34. Finding the Distance between Two Objects A blimp, suspended in the air at a height of 500 feet, lies directly over a line from Soldier Field to the Adler Planetarium on Lake Michigan (see the figure). If the angle of depression from the blimp to the stadium is 32° and from the blimp to the planetarium is 23°, find the distance between Soldier Field and the Adler Planetarium.

35. Finding the Length of a Guy Wire A radio transmission tower is 200 feet high. How long should a guy wire be if it is to be attached to the tower 10 feet from the top and is to make an angle of 21° with the ground?

36. Finding the Height of a Tower A guy wire 80 feet long is attached to the top of a radio transmission tower, making an angle of 25° with the ground. How high is the tower?

37. Washington Monument The angle of elevation of the Washington Monument is 35.1° at the instant it casts a shadow 789 feet long. Use this information to calculate the height of the monument.

38. Finding the Length of a Mountain Trail A straight trail with a uniform inclination of 17° leads from a hotel at an elevation of 9000 feet to a mountain lake at an elevation of 11,200 feet. What is the length of the trail?

39. Finding the Speed of a Truck A state trooper is hidden 30 feet from a highway. One second after a truck passes, the angle θ between the highway and the line of observation from the patrol car to the truck is measured. See the illustration.

(a) If the angle measures 15°, how fast is the truck traveling? Express the answer in feet per second and in miles per hour.

(b) If the angle measures 20°, how fast is the truck traveling? Express the answer in feet per second and in miles per hour.

(c) If the speed limit is 55 miles per hour and a speeding ticket is issued for speeds of 5 miles per hour or more over the limit, for what angles should the trooper issue a ticket?

40. The Gibb's Hill Lighthouse, Southampton, Bermuda In operation since 1846, the Gibb's Hill Lighthouse stands 117 feet high on a hill 245 feet high, so its beam of light is 362 feet above sea level. A brochure states that ships 40 miles away can see the light and planes flying at 10,000 feet can see it 120 miles away. Verify the accuracy of these statements. What assumption did the brochure make about the height of the ship? (See illustration, Problem 103 in Section 1 of the Appendix.)

41. Finding the Bearing of an Aircraft A DC-9 aircraft leaves Midway Airport from runway 4 RIGHT, whose bearing is 40°. After flying for

1/2 mile, the pilot requests permission to turn 90° and head toward the southeast. The permission is granted.
(a) What is the new bearing?
(b) After the airplane goes 1 mile in this direction, what bearing should the control tower use to locate the aircraft?

42. **Finding the Bearing of a Ship** A ship leaves the port of Miami with a bearing of 100° and a speed of 15 knots. After 1 hour, the ship turns 90° toward the south.
(a) What is the new bearing?
(b) After 2 hours, maintaining the same speed, what is the bearing of the ship from port?

43. **Shooting Free Throws in Basketball** The eyes of a basketball player are 6 feet above the floor. The player is at the free-throw line, which is 15 feet from the center of the basket rim (see the figure). What is the angle of elevation from the player's eyes to the center of the rim?
[**Hint:** The rim is 10 feet above the floor.]

44. **Finding the Pitch of a Roof** A carpenter is preparing to put a roof on a garage that is 20 feet by 40 feet by 20 feet. A steel support beam 46 feet in length is positioned in the center of the garage. To support the roof, another beam will be attached to the top of the center beam (see the figure). At what angle of elevation is the new beam? In other words, what is the pitch of the roof?

45. **Determining Distances At Sea** The navigator of a ship at sea spots two lighthouses she recognizes as ones 3 miles apart along a straight seashore. She determines that the angles formed between two line of sight observations of the lighthouses and the line from the ship directly to shore are 15° and 35°. See the illustration.
(a) How far is the ship from shore?
(b) How far is the ship from lighthouse A?
(c) How far is the ship from lighthouse B?

46. **Constructing a Highway** A highway whose primary directions are North-South is being constructed along the west coast of Florida. Near Naples, a bay obstructs the straight path of the road. Since the cost of a bridge is prohibitive, engineers decide to go around the bay. The illustration shows the path they decide upon and the measurements taken. What is the length of highway needed to go around the bay?

47. **Surveillance Satellites** A surveillance satellite circles Earth at a height of h miles above the surface. Suppose d is the distance, in miles, on the surface of the Earth that can be observed from the satellite. See the illustration.
(a) Find an equation that relates the central angle θ to the height h.
(b) Find an equation that relates the observable distance d and θ.
(c) Find an equation that relates d and h.

(d) If d is to be 2500 miles, how high must the satellite orbit above Earth?

(e) If the satellite orbits at a height of 300 miles, what distance d on the surface can be observed?

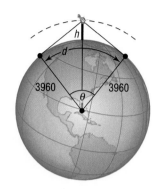

48. Compare Problem 103 in Section 1 of the Appendix with Example 9 and Problem 40 of this section. One uses the Pythagorean Theorem in its solution, while the others use trigonometry. Write a short paper that contrasts the two methods of solution. Be sure to point out the benefits as well as the shortcomings of each method. Then decide which solution you prefer. Be sure to have reasons.

7.2 THE LAW OF SINES

1. Solve SAA or ASA Triangles
2. Solve SSA Triangles

If none of the angles of a triangle is a right angle, the triangle is called **oblique.** Thus, an oblique triangle will have either three acute angles or two acute angles and one obtuse angle (an angle between 90° and 180°). See Figure 13.

FIGURE 13
Oblique triangles

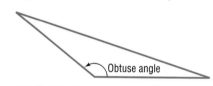

(a) All angles are acute   **(b)** Two acute angles and one obtuse angle

In the discussion that follows, we will always label an oblique triangle so that side a is opposite angle α, side b is opposite angle β, and side c is opposite angle γ, as shown in Figure 14.

FIGURE 14

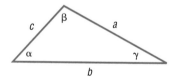

To **solve an oblique triangle** means to find the lengths of its sides and the measurements of its angles. To do this, we shall need to know the length of one side along with two other facts: either two angles, or the other two sides, or one angle and one other side.* Thus, there are four possibilities to consider:

> CASE 1: One side and two angles are known (SAA or ASA).
> CASE 2: Two sides and the angle opposite one of them are known (SSA).
> CASE 3: Two sides and the included angle are known (SAS).
> CASE 4: Three sides are known (SSS).

*Recall from plane geometry the fact that knowing three angles of a triangle determines a family of *similar triangles*—that is, triangles that have the same shape but different sizes.

Figure 15 illustrates the four cases.

FIGURE 15

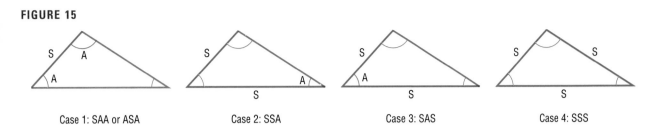

Case 1: SAA or ASA   Case 2: SSA   Case 3: SAS   Case 4: SSS

The **Law of Sines** is used to solve triangles for which Case 1 or 2 holds.

> ## Theorem   Law of Sines
>
> For a triangle with sides a, b, c and opposite angles α, β, γ, respectively,
>
> $$\frac{\sin \alpha}{a} = \frac{\sin \beta}{b} = \frac{\sin \gamma}{c} \tag{1}$$

FIGURE 16

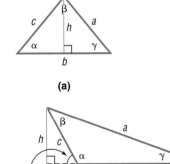

(a)

(b)

Proof To prove the Law of Sines, we construct an altitude of length h from one of the vertices of such a triangle. Figure 16(a) shows h for a triangle with three acute angles, and Figure 16(b) shows h for a triangle with an obtuse angle. In each case, the altitude is drawn from the vertex at β. Using either illustration, we have

$$\sin \gamma = \frac{h}{a}$$

from which

$$h = a \sin \gamma \tag{2}$$

From Figure 16(a), it also follows that

$$\sin \alpha = \frac{h}{c}$$

from which

$$h = c \sin \alpha \tag{3}$$

From Figure 16(b), it follows that

$$\sin \alpha = \sin(180° - \alpha) = \frac{h}{c}$$

which again gives

$$h = c \sin \alpha$$

Thus, whether the triangle has three acute angles or has two acute angles and one obtuse angle, equations (2) and (3) hold. As a result, we may equate the expressions for h in equations (2) and (3) to get

$$a \sin \gamma = c \sin \alpha$$

from which

$$\frac{\sin \alpha}{a} = \frac{\sin \gamma}{c} \qquad (4)$$

In a similar manner, by constructing the altitude h' from the vertex of angle α as shown in Figure 17, we can show that

FIGURE 17

(a)

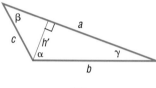

(b)

$$\sin \beta = \frac{h'}{c} \quad \text{and} \quad \sin \gamma = \frac{h'}{b}$$

Thus,

$$h' = c \sin \beta = b \sin \gamma$$

and

$$\frac{\sin \beta}{b} = \frac{\sin \gamma}{c} \qquad (5)$$

When equations (4) and (5) are combined, we have equation (1), the Law of Sines. ▬

In applying the Law of Sines to solve triangles, we use the fact that the sum of the angles of any triangle equals 180°; that is,

$$\alpha + \beta + \gamma = 180° \qquad (6)$$

1 Our first two examples show how to solve a triangle when one side and two angles are known (Case 1: SAA or ASA).

E X A M P L E   1

Using the Law of Sines to Solve a SAA Triangle

Solve the triangle: $\alpha = 40°, \beta = 60°, a = 4$

FIGURE 18

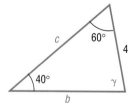

Solution   Figure 18 shows the triangle that we want to solve. The third angle γ is easily found using equation (6):

$$\alpha + \beta + \gamma = 180°$$
$$40° + 60° + \gamma = 180°$$
$$\gamma = 80°$$

Now we use the Law of Sines (twice) to find the unknown sides b and c:

$$\frac{\sin \alpha}{a} = \frac{\sin \beta}{b} \qquad \frac{\sin \alpha}{a} = \frac{\sin \gamma}{c}$$

Because $a = 4$, $\alpha = 40°$, $\beta = 60°$, and $\gamma = 80°$, we have

$$\frac{\sin 40°}{4} = \frac{\sin 60°}{b} \qquad \frac{\sin 40°}{4} = \frac{\sin 80°}{c}$$

Thus,

$$b = \frac{4 \sin 60°}{\sin 40°} \approx 5.39 \qquad c = \frac{4 \sin 80°}{\sin 40°} \approx 6.13$$

$$\uparrow \qquad\qquad\qquad \uparrow$$

From a
calculator.   From a
calculator.   ▬

Notice that in Example 1 we found b and c by working with the given side a. This is better than finding b first and working with a rounded value of b to find c.

E X A M P L E   2   Using the Law of Sines to Solve an ASA Triangle

Solve the triangle: $\alpha = 35°$, $\beta = 15°$, $c = 5$

Solution

FIGURE 19

Figure 19 illustrates the triangle that we want to solve. Because we know two angles ($\alpha = 35°$ and $\beta = 15°$), it is easy to find the third angle using equation (6):

$$\alpha + \beta + \gamma = 180°$$
$$35° + 15° + \gamma = 180°$$
$$\gamma = 130°$$

Now we know the three angles and one side ($c = 5$) of the triangle. To find the remaining two sides a and b, we use the Law of Sines (twice):

$$\frac{\sin \alpha}{a} = \frac{\sin \gamma}{c} \qquad\qquad \frac{\sin \beta}{b} = \frac{\sin \gamma}{c}$$

$$\frac{\sin 35°}{a} = \frac{\sin 130°}{5} \qquad\qquad \frac{\sin 15°}{b} = \frac{\sin 130°}{5}$$

$$a = \frac{5 \sin 35°}{\sin 130°} \approx 3.74 \qquad\qquad b = \frac{5 \sin 15°}{\sin 130°} \approx 1.69$$ ▬

Now work Problem 1.

FIGURE 20

The Ambiguous Case

Case 2 (SSA), which applies to triangles where two sides and the angle opposite one of them are known, is referred to as the **ambiguous case,** because the known information may result in one triangle, two triangles, or no triangle at all. Suppose that we are given sides a and b and angle α, as illustrated in Figure 20. The key to determining the possible triangles, if any, that may be formed from the given information, lies primarily with the height h and the fact that $h = b \sin \alpha$.

No Triangle: If $a < h = b \sin \alpha$, then clearly side a is not sufficiently long to form a triangle. See Figure 21.

One Right Triangle: If $a = h = b \sin \alpha$, then side a is just long enough to form a right triangle. See Figure 22.

FIGURE 21
$a < b \sin \alpha$

FIGURE 22
$a = b \sin \alpha$

Two Triangles: If $a < b$ and $h = b \sin \alpha < a$, then two distinct triangles can be formed from the given information. See Figure 23.

One Triangle: If $a \geq b$, then only one triangle can be formed. See Figure 24.

FIGURE 23
$b \sin \alpha < a$ and $a < b$

FIGURE 24
$a \geq b$

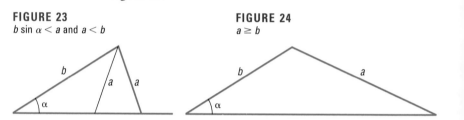

Fortunately, we do not have to rely on an illustration to draw the correct conclusion in the ambiguous case. The Law of Sines will lead us to the correct determination. Let's see how.

E X A M P L E   3

Using the Law of Sines to Solve a SSA Triangle (One Solution)

Solve the triangle: $a = 3$, $b = 2$, $\alpha = 40°$

Solution

FIGURE 25(a)

See Figure 25(a). Because $a = 3$, $b = 2$, and $\alpha = 40°$ are known, we use the Law of Sines to find the angle β:

$$\frac{\sin \alpha}{a} = \frac{\sin \beta}{b}$$

Then

$$\frac{\sin 40°}{3} = \frac{\sin \beta}{2}$$

$$\sin \beta = \frac{2 \sin 40°}{3} \approx 0.43$$

There are two angles β, $0° < \beta < 180°$, for which $\sin \beta \approx 0.43$:

$$\beta \approx 25.4° \quad \text{and} \quad \beta \approx 154.6°$$

[**Note:** Here we computed β using the stored value of $\sin \beta$. If you use the rounded value, $\sin \beta \approx 0.43$, you will obtain slightly different results.]

The second possibility is ruled out, because $\alpha = 40°$, making $\alpha + \beta \approx 194.6° > 180°$. Now, using $\beta \approx 25.4°$, we find

$$\gamma = 180° - \alpha - \beta \approx 180° - 40° - 25.4° = 114.6°$$

The third side c may now be determined using the Law of Sines:

$$\frac{\sin \gamma}{c} = \frac{\sin \alpha}{a}$$

$$\frac{\sin 114.6°}{c} = \frac{\sin 40°}{3}$$

$$c = \frac{3 \sin 114.6°}{\sin 40°} \approx 4.24$$

FIGURE 25(b)

Figure 25(b) illustrates the solved triangle.

EXAMPLE 4

Using the Law of Sines to Solve a SSA Triangle (Two Solutions)

Solve the triangle: $a = 6, b = 8, \alpha = 35°$

Solution

Because $a = 6$, $b = 8$, and $\alpha = 35°$ are known, we use the Law of Sines to find the angle β:

$$\frac{\sin \alpha}{a} = \frac{\sin \beta}{b}$$

Then

$$\frac{\sin 35°}{6} = \frac{\sin \beta}{8}$$

$$\sin \beta = \frac{8 \sin 35°}{6} \approx 0.76$$

$$\beta_1 \approx 49.9° \quad \text{or} \quad \beta_2 \approx 130.1°$$

For both possibilities we have $\alpha + \beta < 180°$. Hence, there are two triangles—one containing the angle $\beta = \beta_1 \approx 49.9°$ and the other containing the angle $\beta = \beta_2 \approx 130.1°$. The third angle γ is either

$$\gamma_1 = 180° - \alpha - \beta_1 \approx 95.1° \quad \text{or} \quad \gamma_2 = 180° - \alpha - \beta_2 \approx 14.9°$$

$$\begin{array}{cc} \uparrow & \uparrow \\ \alpha = 35° & \alpha = 35° \\ \beta_1 = 49.9° & \beta_2 = 130.1° \end{array}$$

The third side c obeys the Law of Sines, so we have

FIGURE 26

$$\frac{\sin \gamma}{c} = \frac{\sin \alpha}{a}$$

$$\frac{\sin 95.1°}{c_1} = \frac{\sin 35°}{6} \qquad \text{or} \qquad \frac{\sin 14.9°}{c_2} = \frac{\sin 35°}{6}$$

$$c_1 = \frac{6 \sin 95.1°}{\sin 35°} \approx 10.42 \qquad c_2 = \frac{6 \sin 14.9°}{\sin 35°} \approx 2.69$$

The two solved triangles are illustrated in Figure 26.

E X A M P L E   5

Using the Law of Sines to Solve a SSA Triangle (No Solution)

Solve the triangle: $a = 2, c = 1, \gamma = 50°$

Solution

Because $a = 2$, $c = 1$, and $\gamma = 50°$ are known, we use the Law of Sines to find the angle α:

$$\frac{\sin \alpha}{a} = \frac{\sin \gamma}{c}$$

$$\frac{\sin \alpha}{2} = \frac{\sin 50°}{1}$$

$$\sin \alpha = 2 \sin 50° \approx 1.53$$

FIGURE 27

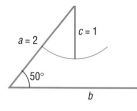

There is no angle α for which $\sin \alpha > 1$. Hence, there can be no triangle with the given measurements. Figure 27 illustrates the measurements given. Notice that, no matter how we attempt to position side c, it will never intersect side b to form a triangle. ▬

Now work Problem 17.

Applied Problems

The Law of Sines is particularly useful for solving certain applied problems.

E X A M P L E   6

Rescue at Sea

Coast Guard Station Zulu is located 120 miles due west of Station X-ray. A ship at sea sends an SOS call that is received by each station. The call to Station Zulu indicates that the location of the ship is 40° east of north; the call to Station X-ray indicates that the location of the ship is 30° west of north.

(a) How far is each station from the ship?
(b) If a helicopter capable of flying 200 miles per hour is dispatched from the nearest station to the ship, how long will it take to reach the ship?

Solution

(a) Figure 28 illustrates the situation. The angle γ is found to be

$$\gamma = 180° - 50° - 60° = 70°$$

FIGURE 28

The Law of Sines can now be used to find the two distances a and b that we seek:

$$\frac{\sin 50°}{a} = \frac{\sin 70°}{120}$$

$$a = \frac{120 \sin 50°}{\sin 70°} \approx 97.82 \text{ miles}$$

$$\frac{\sin 60°}{b} = \frac{\sin 70°}{120}$$

$$b = \frac{120 \sin 60°}{\sin 70°} \approx 110.59 \text{ miles}$$

Thus, Station Zulu is about 111 miles from the ship, and Station X-ray is about 98 miles from the ship.

(b) The time t needed for the helicopter to reach the ship from Station X-ray is found by using the formula

(Velocity, v)(Time, t) = Distance, a

Then

$$t = \frac{a}{v} = \frac{97.82}{200} \approx 0.49 \text{ hour} \approx 29 \text{ minutes}$$

It will take about 29 minutes for the helicopter to reach the ship. ▬

 Now work Problem 29.

7.2 EXERCISES

In Problems 1–8, solve each triangle.

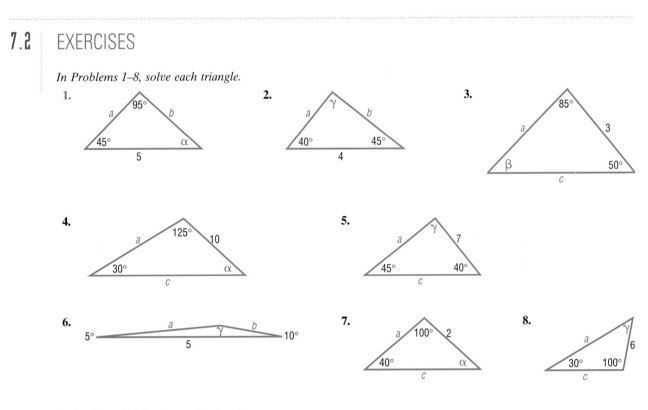

In Problems 9–16, solve each triangle.

9. $\alpha = 40°, \beta = 20°, a = 2$

10. $\alpha = 50°, \gamma = 20°, a = 3$

11. $\beta = 70°, \gamma = 10°, b = 5$

12. $\alpha = 70°, \beta = 60°, c = 4$

13. $\alpha = 110°, \gamma = 30°, c = 3$

14. $\beta = 10°, \gamma = 100°, b = 2$

15. $\alpha = 40°, \beta = 40°, c = 2$

16. $\beta = 20°, \gamma = 70°, a = 1$

In Problems 17–28, two sides and an angle are given. Determine whether the given information results in one triangle, two triangles, or no triangle at all. Solve any triangle(s) that results.

17. $a = 3, b = 2, \alpha = 50°$

18. $b = 4, c = 3, \beta = 40°$

19. $b = 5, \ c = 3, \ \beta = 100°$

20. $a = 2, c = 1, \alpha = 120°$

21. $a = 4, b = 5, \alpha = 60°$

22. $b = 2, c = 3, \beta = 40°$

23. $b = 4, c = 6, \beta = 20°$

24. $a = 3, b = 7, \alpha = 70°$

25. $a = 2, c = 1, \gamma = 100°$

26. $b = 4, c = 5, \beta = 95°$

27. $a = 2, c = 1, \gamma = 25°$

28. $b = 4, c = 5, \beta = 40°$

29. Rescue at Sea Coast Guard Station Able is located 150 miles due south of Station Baker. A ship at sea sends an SOS call that is received by each station. The call to Station Able indicates that the ship is located 35° north of east; the call to Station Baker indicates that the ship is located 30° south of east.
(a) How far is each station from the ship?
(b) If a helicopter capable of flying 200 miles per hour is dispatched from the nearest station to the ship, how long will it take to reach the ship?

30. Surveying Consult the figure. To find the distance from the house at A to the house at B, a surveyor measures the angle BAC to be 40°, then walks off a distance of 100 feet to C, and measures the angle ACB to be 50°. What is the distance from A to B?

31. Finding the Length of a Ski Lift Consult the figure. To find the length of the span of a proposed ski lift from A to B, a surveyor measures the angle DAB to be 25°, then walks off a distance of

tance of 1000 feet to C, and measures the angle ACB to be 15°. What is the distance from A to B?

32. Finding the Height of a Mountain Use the illustration in Problem 31 to find the height BD of the mountain at B.

33. Finding the Height of an Airplane An aircraft spotted by two observers who are 1000 feet apart. As the airplane passes over the line joining them, each observer takes a sighting of the angle of elevation to the plane, as indicated in the figure. How high is the airplane?

34. Finding the Height of the Bridge over the Royal Gorge The highest bridge in the world is the bridge over the Royal Gorge of the Arkansas River in Colorado.* Sightings to the same point at water level directly under the bridge are taken from each side of the 880-foot-long bridge, as indicated in the figure. How high is the bridge?

Source: Guinness Book of World Records.

35. Navigation An airplane flies from city A to city B, a distance of 150 miles, then turns through an angle of $40°$ and heads toward city C, as shown in the figure.

(a) If the distance between cities A and C is 300 miles, how far is it from city B to city C?

(b) Through what angle should the pilot turn at city C to return to City A?

36. Time Lost due to a Navigation Error In attempting to fly from city A to city B, an aircraft followed a course that was $10°$ in error, as indicated in the figure. After flying a distance of 50 miles, the pilot corrected the course by turning at point C and flying 70 miles farther. If the constant speed of the aircraft was 250 miles per hour, how much time was lost due to the error?

37. Finding the Lean of the Leaning Tower of Pisa The famous Leaning Tower of Pisa was originally 184.5 feet high.* At a distance of 123 feet from the base of the tower, the angle of eleva-

tion to the top of the tower is found to be $60°$. Find the angle CAB indicated in the figure. Also, find the perpendicular distance from C to AB.

38. Crankshafts on Cars On a certain automobile, the crankshaft is 3 inches long and the connecting rod is 9 inches long (see the figure). At the time when the angle OPA is $15°$, how far is the piston (P) from the center (O) of the crankshaft?

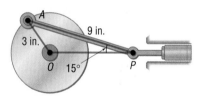

39. Constructing a Highway U.S. 41, a highway whose primary directions are north–south, is being constructed along the west coast of Florida. Near Naples, a bay obstructs the straight path of the road. Since the cost of a bridge is prohibitive, engineers decide to go around the bay. The illustration on the following page shows the path that they decide on and the measurements taken. What is the length of highway needed to go around the bay?

*In their 1986 report on the fragile seven-century-old bell tower, scientists in Pisa, Italy, said the Leaning Tower of Pisa had increased its famous lean by 1 millimeter, or 0.04 inch. That is about the annual average, although the tilting had slowed to about half that much in the previous 2 years. (*Source:* United Press International, June 29, 1986.)

Update PISA, ITALY. September, 1995. The Leaning Tower of Pisa has suddenly shifted, jeopardizing years of preservation work to stabilize it, Italian newspapers said Sunday. The tower, built on shifting subsoil between 1174 and 1350 as a belfry for the nearby cathedral, recently moved .07 inches in one night.

40. Determining Distances at Sea The navigator of a ship at sea spots two lighthouses that she knows to be 3 miles apart along a straight seashore. She determines that the angles formed between two line-of-sight observations of the lighthouses and the line from the ship directly to shore are 15° and 35°. See the illustration.
(a) How far is the ship from lighthouse *A*?
(b) How far is the ship from lighthouse *B*?
(c) How far is the ship from shore?

41. Calculating Distances at Sea The navigator of a ship at sea has the harbor in sight at which the ship is to dock. She spots a lighthouse she knows is 1 mile up the coast from the mouth of the harbor, and she measures the angle between the line-of-sight observations of the harbor and lighthouse to be 20°. With the ship heading directly toward the harbor, she repeats this measurement after 5 minutes of traveling at 12 miles per hour. If the new angle is 30°, how far is the ship from the harbor?

42. Finding Distances A forest ranger is walking on a path inclined at 5° to the horizontal directly toward a 100-foot-tall fire observation tower. The angle of elevation of the top of the tower is 40°. How far is the ranger from the tower at this time?

43. Mollweide's Formula For any triangle, **Mollweide's Formula** (named after Karl Mollweide, 1774–1825) states that

$$\frac{a + b}{c} = \frac{\cos\frac{1}{2}(\alpha - \beta)}{\sin\frac{1}{2}\gamma}$$

Derive it.
[**Hint:** Use the Law of Sines and then a sum-to-product formula.] Notice that this formula involves all six parts of a triangle. As a result, it is sometimes used to check the solution of a triangle.

44. Mollweide's Formula Another form of Mollweide's Formula is

$$\frac{a - b}{c} = \frac{\sin\frac{1}{2}(\alpha - \beta)}{\cos\frac{1}{2}\gamma}$$

Derive it.

45. For any triangle, derive the formula

$$a = b \cos \gamma + c \cos \beta$$

[**Hint:** Use the fact that $\sin \alpha = \sin(180° - \beta - \gamma)$.]

46. Law of Tangents For any triangle, derive the Law of Tangents:

$$\frac{a - b}{a + b} = \frac{\tan\frac{1}{2}(\alpha - \beta)}{\tan\frac{1}{2}(\alpha + \beta)}$$

[**Hint:** Use Mollweide's Formula.]

47. Circumscribing a Triangle Show that

$$\frac{\sin \alpha}{a} = \frac{\sin \beta}{b} = \frac{\sin \gamma}{c} = \frac{1}{2r}$$

where r is the radius of the circle circumscribing the triangle ABC whose sides are a, b, c, as shown in the figure in the margin.

[**Hint:** Draw the diameter AB'. Then β = angle ABC = angle $AB'C$ and angle $ACB' = 90°$.]

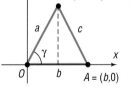**48.** Make up three problems involving oblique triangles. One should result in one triangle, the second in two triangles, and the third in no triangle.

7.3 THE LAW OF COSINES

1 Solve SAS Triangles

2 Solve SSS Triangles

In Section 7.2, we used the Law of Sines to solve Case 1 (SAA or ASA) and Case 2 (SSA) of an oblique triangle. In this section, we derive the Law of Cosines and use it to solve the remaining cases, 3 and 4.

CASE 3:	Two sides and the included angle are known (SAS).
CASE 4:	Three sides are known (SSS).

Theorem Law of Cosines

For a triangle with sides a, b, c and opposite angles α, β, γ, respectively.

$$c^2 = a^2 + b^2 - 2ab \cos \gamma \qquad (1)$$

$$b^2 = a^2 + c^2 - 2ac \cos \beta \qquad (2)$$

$$a^2 = b^2 + c^2 - 2bc \cos \alpha \qquad (3)$$

FIGURE 29

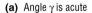

(a) Angle γ is acute

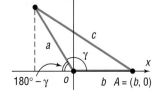

(b) Angle γ is obtuse

Proof   We will prove only formula (1) here. Formulas (2) and (3) may be proved using the same argument.

We begin by strategically placing a triangle on a rectangular coordinate system so that the vertex of angle γ is at the origin and side b lies along the positive x-axis. Regardless of whether γ is acute, as in Figure 29(a), or obtuse, as in Figure 29(b), the vertex B has coordinates $(a \cos \gamma, a \sin \gamma)$. Vertex A has coordinates $(b, 0)$.

We can now use the distance formula to compute c^2:

$$c^2 = (b - a \cos \gamma)^2 + (0 - a \sin \gamma)^2$$
$$= b^2 - 2ab \cos \gamma + a^2 \cos^2 \gamma + a^2 \sin^2 \gamma$$
$$= a^2(\cos^2 \gamma + \sin^2 \gamma) + b^2 - 2ab \cos \gamma$$
$$= a^2 + b^2 - 2ab \cos \gamma$$

Each of formulas (1), (2), and (3) may be stated in words as follows:

> ### Theorem   Law of Cosines
>
> The square of one side of a triangle equals the sum of the squares of the other two sides minus twice their product times the cosine of their included angle.

Observe that if the triangle is a right triangle (so that, say, $\gamma = 90°$), then formula (1) becomes the familiar Pythagorean Theorem: $c^2 = a^2 + b^2$. Thus, the Pythagorean Theorem is a special case of the Law of Cosines.

1 Let's see how to use the Law of Cosines to solve Case 3 (SAS), which applies to triangles for which two sides and the included angle are known.

E X A M P L E   1   **Using the Law of Cosines to Solve a SAS Triangle**

Solve the triangle: $a = 2, b = 3, \gamma = 60°$

Solution   See Figure 30. The Law of Cosines makes it easy to find the third side, c:

FIGURE 30

$$c^2 = a^2 + b^2 - 2ab \cos \gamma$$
$$= 4 + 9 - 2 \cdot 2 \cdot 3 \cdot \cos 60°$$
$$= 13 - (12 \cdot \tfrac{1}{2}) = 7$$
$$c = \sqrt{7}$$

Side c is of length $\sqrt{7}$. To find the angles α and β, we may use either the Law of Sines or the Law of Cosines. It is preferable to use the Law of Cosines, since it will lead to an equation with one solution. Using the Law of Sines would lead to an equation with two solutions that would need to be checked to determine which solution fits the given data. Thus, we choose to use formulas (2) and (3) of the Law of Cosines.

For α:

$$a^2 = b^2 + c^2 - 2bc \cos \alpha$$
$$2bc \cos \alpha = b^2 + c^2 - a^2$$
$$\cos \alpha = \frac{b^2 + c^2 - a^2}{2bc} = \frac{9 + 7 - 4}{2 \cdot 3\sqrt{7}} = \frac{12}{6\sqrt{7}} = \frac{2\sqrt{7}}{7}$$
$$\alpha \approx 40.9°$$

For β:

$$b^2 = a^2 + c^2 - 2ac \cos \beta$$
$$\cos \beta = \frac{a^2 + c^2 - b^2}{2ac} = \frac{4 + 7 - 9}{4\sqrt{7}} = \frac{1}{2\sqrt{7}} = \frac{\sqrt{7}}{14}$$
$$\beta \approx 79.1°$$

Notice that $\alpha + \beta + \gamma = 40.9° + 79.1° + 60° = 180°$, as required.

Now work Problem 1.

The next example illustrates how the Law of Cosines is used when three sides of a triangle are known, Case 4 (SSS).

E X A M P L E 2

Using the Law of Cosines to Solve a SSS Triangle

Solve the triangle: $a = 4, b = 3, c = 6$

Solution   See Figure 31. To find the angles α, β, and γ, we proceed as we did in the latter part of the solution to Example 1.

FIGURE 31

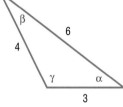

For α:

$$\cos \alpha = \frac{b^2 + c^2 - a^2}{2bc} = \frac{9 + 36 - 16}{2 \cdot 3 \cdot 6} = \frac{29}{36}$$

$$\alpha \approx 36.3°$$

For β:

$$\cos \beta = \frac{a^2 + c^2 - b^2}{2ac} = \frac{16 + 36 - 9}{2 \cdot 4 \cdot 6} = \frac{43}{48}$$

$$\beta \approx 26.4°$$

Since we know α and β,

$$\gamma = 180° - \alpha - \beta \approx 180° - 36.3° - 26.4° = 117.3°$$

Now work Problem 7.

E X A M P L E 3

Correcting a Navigational Error

A motorized sailboat leaves Naples, Florida, bound for Key West, 150 miles away. Maintaining a constant speed of 15 miles per hour, but encountering heavy cross-winds and strong currents, the crew finds, after 4 hours, that the sailboat is off course by 20°.

(a) How far is the sailboat from Key West at this time?

(b) Through what angle should the sailboat turn to correct its course?

(c) How much time has been added to the trip because of this? (Assume that the speed remains at 15 miles per hour.)

FIGURE 32

Solution   See Figure 32. With a speed of 15 miles per hour, the sailboat has gone 60 miles after 4 hours. We seek the distance x of the sailboat from Key West. We also seek the angle θ that the sailboat should turn through to correct its course.

(a) To find x, we use the Law of Cosines, since we know two sides and the included angle.

$$x^2 = 150^2 + 60^2 - 2(150)(60) \cos 20° = 9186$$
$$x = \sqrt{9186} = 95.8$$

The sailboat is about 96 miles from Key West.

(b) We now know three sides of the triangle, so we can use the Law of Cosines again to find the angle α opposite the side of length 150 miles.

$$150^2 = 96^2 + 60^2 - 2(96)(60) \cos \alpha$$
$$9684 = -11,620 \cos \alpha$$
$$\cos \alpha \approx -0.8406$$
$$\alpha \approx 147.2°$$

The sailboat should turn through an angle of

$$\theta = 180° - \alpha \approx 180° - 147.2° = 32.8°$$

The sailboat should turn through an angle of about 33° to correct its course.

(c) The total length of the trip is now $60 + 96 = 156$ miles. The extra 6 miles will only require about 0.4 hours or 24 minutes more if the speed of 15 miles per hour is maintained. ▬

Now work Problem 27.

HISTORICAL FEATURE   The Law of Sines was known vaguely, long before it was explicitly stated by Nasîr ed-dîn (about A.D. 1250). Ptolemy (about A.D. 150) was aware of it in a form using a chord function instead of the sine function. But it was first clearly stated in Europe by Regiomontanus, writing in 1464.

The Law of Cosines appears first in Euclid's *Elements* (Book II), but in a well-disguised form in which squares built on the sides of triangles are added and a rectangle representing the cosine term is subtracted. It was thus known to all mathematicians because of their familiarity with Euclid's work. An early modern form of the Law of Cosines—that for finding the angle when the sides are known—was stated by François Vièta (in 1593).

The Law of Tangents (see Problem 46 of Exercise 7.2) has become obsolete. In the past it was used in place of the Law of Cosines, because the Law of Cosines was very inconvenient for calculation with logarithms or slide rules. Mixing of addition and multiplication is now quite easy on a calculator, however, and the Law of Tangents has been shelved along with the slide rule. ▬

7.3 EXERCISES

In Problems 1–8, solve each triangle.

1.

2.

3.

4.

5.

6.

7.

8.

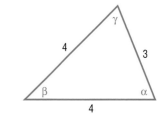

In Problems 9–24, solve each triangle.

9. $a = 3, b = 4, \gamma = 40°$

10. $a = 2, c = 1, \beta = 10°$

11. $b = 1, c = 3, \alpha = 80°$

12. $a = 6, b = 4, \gamma = 60°$

13. $a = 3, c = 2, \beta = 110°$

14. $b = 4, c = 1, \alpha = 120°$

15. $a = 2, b = 2, \gamma = 50°$

16. $a = 3, c = 2, \beta = 90°$

17. $a = 12, b = 13, c = 5$

18. $a = 4, b = 5, c = 3$

19. $a = 2, b = 2, c = 2$

20. $a = 3, b = 3, c = 2$

21. $a = 5, b = 8, c = 9$

22. $a = 4, b = 3, c = 6$

23. $a = 10, b = 8, c = 5$

24. $a = 9, b = 7, c = 10$

25. Surveying Consult the figure. To find the distance from the house at A to the house at B, a surveyor measures the angle ACB, which is found to be 70°, then walks off the distance to each house, 50 feet and 70 feet, respectively. How far apart are the houses?

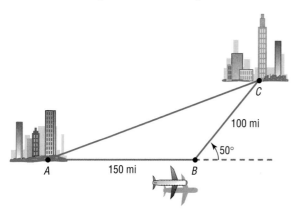

26. Navigation An airplane flies from city A to city B, a distance of 150 miles, then turns through an angle of 50° and flies to city C, a distance of 100 miles (see the figure).
(a) How far is it from city A to city C?
(b) Through what angle should the pilot turn at city C to return to city A?

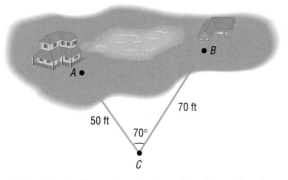

27. Revising a Flight Plan In attempting to fly from city A to city B, a distance of 330 miles, a pilot inadvertently took a course that was 10° in error, as indicated in the figure.
(a) If the aircraft maintains an average speed of 220 miles per hour and if the error in direction is discovered after 15 minutes, through what angle should the pilot turn to head toward city B?
(b) What new average speed should the pilot maintain so that the total time of the trip is 90 minutes?

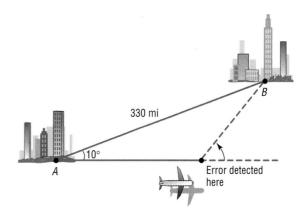

28. Avoiding a Tropical Storm A cruise ship maintains an average speed of 15 knots in going from San Juan, Puerto Rico, to Barbados, West Indies, a distance of 600 nautical miles. To avoid a tropical storm, the captain heads out of San Juan in a direction of 20° off a direct heading to Barbados. The captain maintains the 15 knot speed for 10 hours, after which time the path to Barbados becomes clear of storms.
(a) Through what angle should the captain turn to head directly to Barbados?

(b) How long will it be before the ship reaches Barbados if the same 15 knot speed is maintained?

29. **Major League Baseball Field** A Major League baseball diamond is actually a square 90 feet on a side. The pitching rubber is located 60.5 feet from home plate on a line joining home plate and second base.
    (a) How far is it from the pitching rubber to first base?
    (b) How far is it from the pitching rubber to second base?
    (c) If a pitcher faces home plate, through what angle does he need to turn to face first base?

30. **Little League Baseball Field** According to Little League baseball official regulations, the diamond is a square 60 feet on a side. The pitching rubber is located 46 feet from home plate on a line joining home plate and second base.
    (a) How far is it from the pitching rubber to first base?
    (b) How far is it from the pitching rubber to second base?
    (c) If a pitcher faces home plate, through what angle does he need to turn to face first base?

31. **Finding the Length of a Guy Wire** The height of a radio tower is 500 feet, and the ground on one side of the tower slopes upward at an angle of 10° (see the figure).
    (a) How long should a guy wire be if it is to connect to the top of the tower and be secured at a point on the sloped side 100 feet from the base of the tower?
    (b) How long should a second guy wire be if it is to connect to the middle of the tower and be secured at a point 100 feet from the base on the flat side?

32. **Finding the Length of a Guy Wire** A radio tower 500 feet high is located on the side of a hill with an inclination to the horizontal of 5° (see the figure). How long should two guy wires be if they are to connect to the top of the tower and be secured at two points 100 feet directly above and directly below the base of the tower?

33. **Wrigley Field, Home of the Chicago Cubs** The distance from home plate to dead center of Wrigley Field is 400 feet (see the figure). How far is it from dead center to third base?

34. **Little League Baseball** The distance from home plate to dead center at the Oak Lawn Little League field is 280 feet. How far is it from dead center to third base?
    [**Hint:** The distance between the bases in Little League is 60 feet.]

35. **Rods and Pistons** Rod OA (see the figure) rotates about the fixed point O so that point A travels on a circle of radius r. Connected to

point A is another rod AB of length $L > r$, and point B is connected to a piston. Show that the distance x between point O and point B is given by

$$x = r\cos\theta + \sqrt{r^2\cos^2\theta + L^2 - r^2}$$

where θ is the angle of rotation of rod OA.

36. **Geometry** Show that the length d of a chord of a circle of radius r is given by the formula

$$d = 2r\sin\frac{\theta}{2}$$

where θ is the central angle formed by the radii to the ends of the chord (see the figure). Use this result to derive the fact that $\sin\theta < \theta$, where $\theta > 0$ is measured in radians.

37. For any triangle, show that

$$\cos\frac{\gamma}{2} = \sqrt{\frac{s(s - c)}{ab}}$$

where $s = \frac{1}{2}(a + b + c)$.
[**Hint:** Use a half-angle formula and the Law of Cosines.]

38. For any triangle, show that

$$\sin\frac{\gamma}{2} = \sqrt{\frac{(s - a)(s - b)}{ab}}$$

where $s = \frac{1}{2}(a + b + c)$.

39. Use the Law of Cosines to prove the identity

$$\frac{\cos\alpha}{a} + \frac{\cos\beta}{b} + \frac{\cos\gamma}{c} = \frac{a^2 + b^2 + c^2}{2abc}$$

40. Write down your strategy for solving an oblique triangle.

7.4 THE AREA OF A TRIANGLE

1. Find the Area of SAS Triangles
2. Find the Area of SSS Triangles

In this section, we will derive several formulas for calculating the area A of a triangle. The most familiar of these is the following:

Theorem

The area A of a triangle is

$$A = \tfrac{1}{2}bh \tag{1}$$

where b is the base and h is an altitude drawn to that base.

FIGURE 33
$A = \frac{1}{2}bh$

FIGURE 34

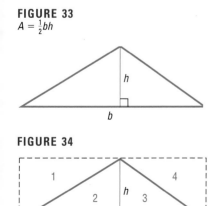

Proof The derivation of this formula is rather easy once a rectangle of base b and height h is constructed around the triangle. See Figures 33 and 34.

Triangles 1 and 2 in Figure 34 are equal in area, as are triangles 3 and 4. Consequently, the area of the triangle with base b and altitude h is exactly half the area of the rectangle, which is bh.

If the base b and altitude h to that base are known, then we can easily find the area of such a triangle using formula (1). Usually, though, the information required to use formula (1) is not given. Suppose, for example,

FIGURE 35
$h = a \sin \gamma$

that we know two sides a and b and the included angle γ (see Figure 35). Then the altitude h can be found by noting that

$$\frac{h}{a} = \sin \gamma$$

so that

$$h = a \sin \gamma$$

Using this fact in formula (1) produces

$$A = \tfrac{1}{2}bh = \tfrac{1}{2}b(a \sin \gamma) = \tfrac{1}{2}ab \sin \gamma$$

Thus, we have the formula

$$A = \tfrac{1}{2}ab \sin \gamma \qquad (2)$$

By dropping altitudes from the other two vertices of the triangle, we obtain the following corresponding formulas:

$$A = \tfrac{1}{2}bc \sin \alpha \qquad (3)$$

$$A = \tfrac{1}{2}ac \sin \beta \qquad (4)$$

It is easiest to remember these formulas using the following wording:

Theorem

The area A of a triangle equals one-half the product of two of its sides times the sine of their included angle.

E X A M P L E   1

1

Finding the Area of a SAS Triangle

Find the area A of the triangle for which $a = 8$, $b = 6$, and $\gamma = 30°$.

Solution   We use formula (2) to get

$$A = \tfrac{1}{2}ab \sin \gamma = \tfrac{1}{2} \cdot 8 \cdot 6 \sin 30° = 12$$

Now work Problem 1.

2

If the three sides of a triangle are known, another formula, called **Heron's Formula** (named after Heron of Alexandria), can be used to find the area of a triangle.

Theorem   Heron's Formula

The area A of a triangle with sides a, b, and c is

$$A = \sqrt{s(s - a)(s - b)(s - c)} \qquad (5)$$

where $s = \tfrac{1}{2}(a + b + c)$.

M I S S I O N   P O S S I B L E

Locating Lost Treasure

While scuba diving off Wreck Hill in Bermuda, a group of five entrepreneurs discovered a treasure map in a small water-tight cask on a pirate schooner that had sunk in 1747. The map directed them to an area of Bermuda now known as The Flatts, but when they got there, they realized that the most important landmark on the map was gone. They called in the Mission Possible team to help them recreate the map. They promised you 25% of whatever treasure was found.

The directions on the map read as follows:

1. From the tallest palm tree, sight the highest hill. Drop your eyes vertically until you sight the base of the hill.
2. Turn 40° clockwise from that line and walk 70 paces to the big red rock.
3. From the red rock walk 50 paces back to the sight line between the palm tree and the hill. Dig there.

The five entrepreneurs said they believed they had found the red rock and the highest hill in the vicinity, but the "tallest palm tree" had long since fallen and disintegrated. It had occurred to them that the treasure must be located on a circle with radius 50 "paces" centered around the red rock, but they had decided against digging a trench 942 feet in circumference, especially since they had no assurance that the treasure was still there. (They had decided that a "pace" must be about a yard.)

1. Determine a plan to locate the position of the lost palm tree, and write out an explanation of your procedure for the entrepreneurs.
2. Unfortunately, it turns out that the entrepreneurs had more in common with the eighteenth century pirates than you had bargained for. Once you told them the location of the lost palm tree, they tied you all to the red rock, saying they could take it from there. From the location of the palm tree, they sighted 40° counterclockwise from the rock to the hill, then ran about 50 yards to the circle they had traced about the rock and began to dig frantically. Nothing. After about an hour, they drove off shouting back at you, "25% of nothing is nothing!"
3. Fortunately, the entrepreneurs had left the shovels. After you managed to untie yourselves, you went to the correct location and found the treasure. Where was it? How far from the palm tree? Explain.
4. People who scuba dive for sunken treasure have certain legal obligations. What are they? Should you share the treasure with a lawyer, just to make sure you get to keep the rest?

Proof   The proof we shall give uses the Law of Cosines and is quite different from the proof given by Heron.

From the Law of Cosines,

$$c^2 = a^2 + b^2 - 2ab \cos \gamma$$

and the two half-angle formulas,

$$\cos^2 \frac{\gamma}{2} = \frac{1 + \cos \gamma}{2} \qquad \sin^2 \frac{\gamma}{2} = \frac{1 - \cos \gamma}{2}$$

we find

$$\cos^2 \frac{\gamma}{2} = \frac{1 + \cos \gamma}{2} = \frac{1 + \dfrac{a^2 + b^2 - c^2}{2ab}}{2}$$

$$= \frac{a^2 + 2ab + b^2 - c^2}{4ab} = \frac{(a + b)^2 - c^2}{4ab}$$

$$= \frac{(a + b - c)(a + b + c)}{4ab} = \frac{2(s - c) \cdot 2s}{4ab} = \frac{s(s - c)}{ab} \qquad (6)$$

$$\uparrow$$
$$a + b - c = a + b + c - 2c$$
$$= 2s - 2c$$

Similarly,

$$\sin^2 \frac{\gamma}{2} = \frac{(s - a)(s - b)}{ab} \qquad (7)$$

Now we use formula (2) for the area:

$$A = \frac{1}{2}ab \sin \gamma$$

$$= \frac{1}{2}ab \cdot 2 \sin \frac{\gamma}{2} \cos \frac{\gamma}{2} \qquad \sin \gamma = \sin 2\left(\frac{\gamma}{2}\right) = 2 \sin \frac{\gamma}{2} \cos \frac{\gamma}{2}$$

$$= ab \sqrt{\frac{(s - a)(s - b)}{ab}} \sqrt{\frac{s(s - c)}{ab}} \qquad \text{Use equations (6) and (7).}$$

$$= \sqrt{s(s - a)(s - b)(s - c)} \qquad \blacksquare$$

E X A M P L E   2

Finding the Area of a SSS Triangle

Find the area of a triangle whose sides are 4, 5, and 7.

Solution   We let $a = 4$, $b = 5$, and $c = 7$. Then

$$s = \tfrac{1}{2}(a + b + c) = \tfrac{1}{2}(4 + 5 + 7) = 8$$

Heron's Formula then gives the area A as

$$A = \sqrt{s(s - a)(s - b)(s - c)} = \sqrt{8 \cdot 4 \cdot 3 \cdot 1} = \sqrt{96} = 4\sqrt{6} \qquad \blacksquare$$

Now work Problem 7.

HISTORICAL FEATURE Heron's formula (also known as *Hero's Formula*) is due to Heron of Alexandria (about A.D. 75), who had, besides his mathematical talents, a good deal of engineering skills. In various temples his mechanical devices produced effects that seemed supernatural, and visitors presumably were thus influenced to generosity. Heron's book *Metrica,* on making such devices, has survived and was discovered in 1896 in the city of Constantinople.

Heron's Formula for the area of a triangle caused some mild discomfort in Greek mathematics, because a product with two factors was an area while one with three factors was a volume, but four factors seemed contradictory in Heron's time.

Karl Mollweide (1774–1875), a mathematician and astronomer, discovered the formulas named for him (see Problems 43 and 44 of Exercise 7.2). These formulas are not too important in themselves but often simplify the derivation of other formulas, as demonstrated in Historical Problems 1 and 2, which follow. ▬

HISTORICAL PROBLEMS

1. This derivation of Heron's formula uses Mollweide's formula.
   (a) Show that

   $$\frac{s}{c} = \frac{\cos(\alpha/2)\cos(\beta/2)}{\sin(\gamma/2)}$$

   where $s = \frac{1}{2}(a + b + c)$.

   Hint: Use Mollweide's formula (see Problem 43 in Exercise 7.2) and add 1 to both sides. Then use the fact that

   $$\sin\frac{\gamma}{2} = \sin\frac{180° - (\alpha + \beta)}{2} = \cos\frac{\alpha + \beta}{2}$$

   (b) Similarly, show that

   $$\frac{s}{a} = \frac{\cos(\beta/2)\cos(\gamma/2)}{\sin(\alpha/2)} \qquad \text{and} \qquad \frac{s}{b} = \frac{\cos(\alpha/2)\cos(\gamma/2)}{\sin(\beta/2)}$$

   (c) Use the results of parts (a) and (b) to show that

   $$\frac{s - a}{a} = \frac{\sin(\beta/2)\sin(\gamma/2)}{\sin(\alpha/2)} \qquad \frac{s - b}{b} = \frac{\sin(\alpha/2)\sin(\gamma/2)}{\sin(\beta/2)}$$
   $$\frac{s - c}{c} = \frac{\sin(\alpha/2)\sin(\beta/2)}{\sin(\gamma/2)}$$

   (d) Now form the product

   $$\frac{s}{c} \cdot \frac{s - a}{a} \cdot \frac{s - b}{b} \cdot \frac{s - c}{c}$$

   After cancellations, multiply each side by $cabc$, use the double-angle formulas, and use the fact that $A = \frac{1}{2}bc \sin \alpha = \frac{1}{2}ac \sin \beta$. Heron's Formula will then follow.

2. We again use Mollweide's Formula to derive some other interesting formulas. (These were derived in another way in Problems 37 and 38 of Exercise 7.3.)

(a) Add 1 to both sides of the second form of Mollweide's Formula (see Problem 44 in Exercise 7.2) and simplify to obtain

$$\frac{s - b}{c} = \frac{\sin(\alpha/2)\cos(\beta/2)}{\cos(\gamma/2)}$$

(b) Similarly, show that

$$\frac{s - c}{b} = \frac{\sin(\alpha/2)\cos(\gamma/2)}{\cos(\beta/2)}$$

(c) Use the results of part (b) and Problem 1(b) and (c) to show that

$$\cos^2\frac{\gamma}{2} = \frac{s(s - c)}{ab} \quad \text{and} \quad \sin^2\frac{\gamma}{2} = \frac{(s - a)(s - b)}{ab}$$

(d) Show that

$$\tan^2\frac{\gamma}{2} = \frac{(s - a)(s - b)}{s(s - c)}$$

3. (a) If h_1, h_2, and h_3 are the altitudes dropped from A, B, and C, respectively, in a triangle (see the figure), show that

$$\frac{1}{h_1} + \frac{1}{h_2} + \frac{1}{h_3} = \frac{s}{K}$$

where K is the area of the triangle and $s = \frac{1}{2}(a + b + c)$.

[**Hint:** $h_1 = 2K/a$.]

(b) Show that a formula for the altitude h from a vertex to the opposite side a of a triangle is

$$h = \frac{a\sin\beta\sin\gamma}{\sin\alpha}$$

4. **Inscribed Circle.** The lines that bisect each angle of a triangle meet in a single point O, and the perpendicular distance r from O to each side of the triangle is the same. The circle with center at O and radius r is called the *inscribed circle* of the triangle (see the figure).

(a) Apply Problem 3(b) to triangle OAB to show that

$$r = \frac{c\sin(\alpha/2)\sin(\beta/2)}{\cos(\gamma/2)}$$

(b) Use the results of part (a) and Problem 1(c) to show that

$$r = (s - c)\tan\frac{\gamma}{2}$$

(c) Show that

$$\cot\frac{\alpha}{2} + \cot\frac{\beta}{2} + \cot\frac{\gamma}{2} = \frac{s}{r}$$

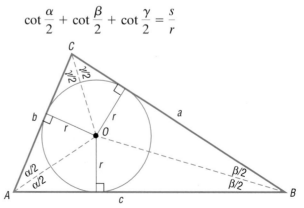

(d) Show that the area K of triangle ABC is $K = rs$. Then show that

$$r = \sqrt{\frac{(s-a)(s-b)(s-c)}{s}}$$

where $s = \frac{1}{2}(a + b + c)$.

5. Find the coordinates of the center O of the inscribed circle in terms of a, b, c, α, β, and γ. Then write the general equation of the inscribed circle.

[**Hint:** Use a system of rectangular coordinates with the vertex A at the origin and side c along the positive x-axis.] ■

7.4 EXERCISES

In Problems 1–8, find the area of each triangle. Round off answers to two decimal places.

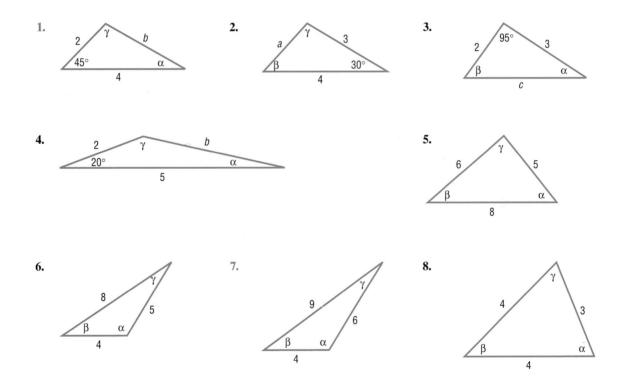

In Problems 9–24, find the area of each triangle. Round off answers to two decimal places.

9. $a = 3, b = 4, \gamma = 40°$
10. $a = 2, c = 1, \beta = 10°$
11. $b = 1, c = 3, \alpha = 80°$
12. $a = 6, b = 4, \gamma = 60°$
13. $a = 3, c = 2, \beta = 110°$
14. $b = 4, c = 1, \alpha = 120°$
15. $a = 2, b = 2, \gamma = 50°$
16. $a = 3, c = 2, \beta = 90°$
17. $a = 12, b = 13, c = 5$
18. $a = 4, b = 5, c = 3$
19. $a = 2, b = 2, c = 2$
20. $a = 3, b = 3, c = 2$
21. $a = 5, b = 8, c = 9$
22. $a = 4, b = 3, c = 6$
23. $a = 10, b = 8, c = 5$
24. $a = 9, b = 7, c = 10$

25. Cost of a Triangular Lot The dimensions of a triangular lot are 100 feet by 50 feet by 75 feet. If the price of such land is $3 per square foot, how much does the lot cost?

26. Approximating the Area of a Lake To approximate the area of a lake, a surveyor walks around the perimeter of the lake, taking the measurements shown in the illustration. Using this technique, what is the approximate area of the lake?

[**Hint:** Use the Law of Cosines on the three triangles shown and then find the sum of their areas.]

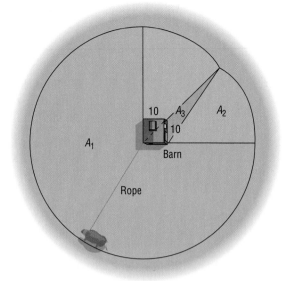

27. The Cow Problem* A cow is tethered to one corner of a square barn, 10 feet by 10 feet, with a rope 100 feet long. What is the maximum grazing area of the cow?

[**Hint:** See the illustration.]

28. Another Cow Problem If the barn in Problem 27 is rectangular, 10 feet by 20 feet, what is the maximum grazing area for the cow?

29. Area of a Triangle Prove that the area A of a triangle is given by the formula

$$A = \frac{a^2 \sin \beta \sin \gamma}{2 \sin \alpha}$$

30. Area of a Triangle Prove the two other forms of the formula given in Problem 29,

$$A = \frac{b^2 \sin \alpha \sin \gamma}{2 \sin \beta} \quad \text{and} \quad A = \frac{c^2 \sin \alpha \sin \beta}{2 \sin \gamma}$$

In Problems 31–38, use the results of Problem 29 or 30 to find the area of each triangle. Round off answers to two decimal places.

31. $\alpha = 40°, \beta = 20°, a = 2$ **32.** $\alpha = 50°, \gamma = 20°, a = 3$ **33.** $\beta = 70°, \gamma = 10°, b = 5$

34. $\alpha = 70°, \beta = 60°, c = 4$ **35.** $\alpha = 110°, \gamma = 30°, c = 3$ **36.** $\beta = 10°, \gamma = 100°, b = 2$

37. $\alpha = 40°, \beta = 40°, c = 2$ **38.** $\beta = 20°, \gamma = 70°, a = 1$

39. Geometry Consult the figure on the right, which shows a circle of radius r with center at O. Find the area A of the shaded region as a function of the central angle θ.

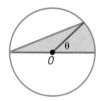

*Suggested by Professor Teddy Koukounas of SUNY at Old Westbury, who learned of it from an old farmer in Virginia. Solution provided by Professor Kathleen Miranda of SUNY at Old Westbury.

7.5   SINUSOIDAL GRAPHS; SINUSOIDAL CURVE FITTING

1. Determine the Amplitude and Period of Sinusoidal Functions
2. Find an Equation for a Sinusoidal Graph
3. Determine the Phase Shift of a Sinusoidal Function
4. Graph Sinusoidal Functions
5. Find a Sinusoidal Function from Data

Because $\sin x = \cos(x - \pi/2)$, it follows that the graph of $y = \sin x$ is the same as the graph of $y = \cos x$ after a horizontal shift of $\pi/2$ units to the right. See Figure 36.

FIGURE 36

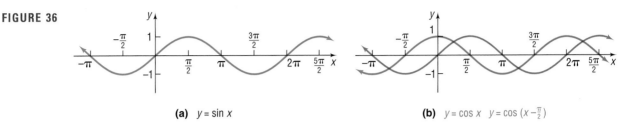

(a) $y = \sin x$ (b) $y = \cos x$ $y = \cos \left(x - \frac{\pi}{2} \right)$

Because of this, the graphs of sine functions and cosine functions are referred to as **sinusoidal graphs.**

Let's look at some general characteristics of sinusoidal graphs.

In Figure 37, we show the graph of $y = 2 \cos x$. Notice that the values of $y = 2 \cos x$ lie between -2 and 2, inclusive.

FIGURE 37
$y = 2 \cos x$

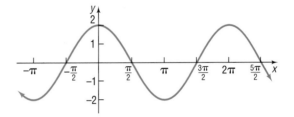

In general, the values of the functions $y = A \sin x$ and $y = A \cos x$, where $A \neq 0$, will always satisfy the inequalities.

$$-|A| \leq A \sin x \leq |A| \quad \text{and} \quad -|A| \leq A \cos x \leq |A|$$

respectively. The number $|A|$ is called the **amplitude** of $y = A \sin x$ or $y = A \cos x$. See Figure 38.

FIGURE 38

$y = A \sin x, A > 0$
Period $= 2\pi$

In Figure 39 we show the graph of $y = \cos 3x$. Notice that the period of this function is $2\pi/3$.

FIGURE 39
$y = \cos 3x$

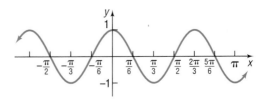

In general, if $\omega > 0$, the functions $y = \sin \omega x$ and $y = \cos \omega x$ will have period $T = 2\pi/\omega$. To see why, recall that the graph of $y = \sin \omega x$ is obtained from the graph of $y = \sin x$ by performing an appropriate horizontal compression or stretch. This horizontal compression replaces the interval $[0, 2\pi]$, which contains one period of the graph of $y = \sin x$, by the interval $[0, 2\pi/\omega]$, which contains one period of the graph of $y = \sin \omega x$. Thus, the period of the functions $y = \sin \omega x$ and $y = \cos \omega x$, $\omega > 0$, is $2\pi/\omega$. See Figure 40.

FIGURE 40

$y = A \sin \omega x, A > 0, \omega > 0$
Period $= \frac{2\pi}{\omega}$

If $\omega < 0$ in $y = \sin \omega x$ or $y = \cos \omega x$, we use the facts that

$$\sin \omega x = -\sin(-\omega x) \quad \text{and} \quad \cos \omega x = \cos(-\omega x)$$

to obtain an equivalent form in which the coefficient of x is positive.

Theorem

If $\omega > 0$, the amplitude and period of $y = A \sin \omega x$ and $y = A \cos \omega x$ are given by

$$\text{Amplitude} = |A| \qquad \text{Period} = T = \frac{2\pi}{\omega} \qquad (1)$$

E X A M P L E   1   Finding the Amplitude and Period of a Sinusoidal Function

Determine the amplitude and period of $y = 3 \sin 4x$, and graph the function.

FIGURE 41

Solution   Comparing $y = 3 \sin 4x$ to $y = A \sin \omega x$, we find that $A = 3$ and $\omega = 4$. Thus, from equation (1), the amplitude is $|A| = 3$ and the period is $T = 2\pi/\omega = 2\pi/4 = \pi/2$. We can use this information to graph $y = 3 \sin 4x$ by beginning as shown in Figure 41.

To graph $y = 3 \sin 4x$ using a graphing utility, we use the amplitude to set Ymin and Ymax and use the period to set Xmin and Xmax. See Figure 42(a). To graph $y = 3 \sin 4x$ by hand, we use the amplitude to scale the y-axis and the period to scale the x-axis. Then we fill in the graph of the sine function. See Figure 42(b).

FIGURE 42

(a)   (b)

  Now work Problem 5.

E X A M P L E   2

Finding the Amplitude and Period of a Sinusoidal Function

Determine the amplitude and period of $y = -4 \cos \pi x$, and graph the function.

Solution

FIGURE 43

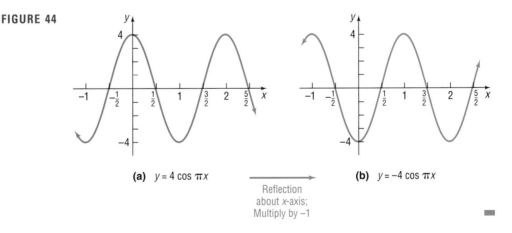

Comparing $y = -4 \cos \pi x$ with $y = A \cos \omega x$, we find that $A = -4$ and $\omega = \pi$. Thus, the amplitude is $|A| = |-4| = 4$, and the period is $T = 2\pi/\omega = 2\pi/\pi = 2$.

Figure 43 shows the graph of $y = -4 \cos \pi x$ using a graphing utility.

To graph $y = -4 \cos \pi x$ by hand, we use the amplitude to scale the y-axis, the period to scale the x-axis and fill in the graph of the cosine function, thus obtaining the graph of $y = 4 \cos \pi x$ shown in Figure 44(a). Now, since we want the graph of $y = -4 \cos \pi x$, we reflect the graph of $y = 4 \cos \pi x$ about the x-axis, as shown in Figure 44(b).

FIGURE 44

(a) $y = 4 \cos \pi x$ Reflection   **(b)** $y = -4 \cos \pi x$
                                        about x-axis;
                                        Multiply by -1

E X A M P L E   3

Finding the Amplitude and Period of a Sinusoidal Function

Determine the amplitude and period of $y = 2 \sin(-\pi x)$, and graph the function.

Solution

Since the sine function is odd, we use the equivalent form

$$y = -2 \sin \pi x$$

Comparing $y = -2 \sin \pi x$ to $y = A \sin \omega x$, we find that $A = -2$ and $\omega = \pi$. Thus, the amplitude is $|A| = 2$, and the period is $T = 2\pi/\omega = 2\pi/\pi = 2$.

FIGURE 45

Figure 45 shows the graph of $y = 2 \sin(-\pi x)$ using a graphing utility. To graph $y = -2 \sin \pi x$ by hand, we use the amplitude to scale the y-axis, the period to scale the x-axis, and fill in the graph of the sine function. Figure 46(a) shows the resulting graph of $y = 2 \sin \pi x$. To obtain the graph of $y = -2 \sin \pi x$, we reflect the graph of $y = 2 \sin \pi x$ about the x-axis, as shown in Figure 46(b).

FIGURE 46

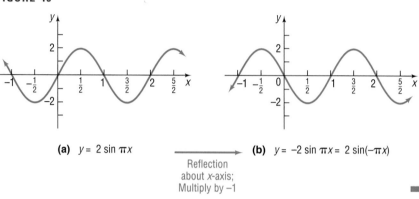

(a) $y = 2 \sin \pi x$ $\longrightarrow$ **(b)** $y = -2 \sin \pi x = 2 \sin(-\pi x)$

Reflection
about x-axis;
Multiply by -1

We also can use the ideas of amplitude and period to identify a sinusoidal function when its graph is given.

EXAMPLE 4   Finding an Equation for a Sinusoidal Graph

Find an equation for the graph shown in Figure 47.

FIGURE 47

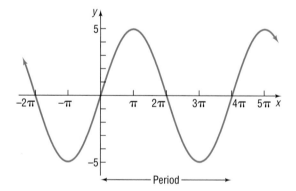

Solution   This graph can be viewed as the graph of a sine function* with amplitude $A = 5$. The period T is observed to be 4π. Thus, by equation (1),

$$T = \frac{2\pi}{\omega}$$

$$4\pi = \frac{2\pi}{\omega}$$

$$\omega = \frac{2\pi}{4\pi} = \frac{1}{2}$$

*The equation could also be viewed as a cosine function with a horizontal shift, but viewing it as a sine function is easier.

A sine function whose graph is given in Figure 47 is

$$y = A \sin \omega x = 5 \sin \frac{x}{2}$$

Check: Graph $Y_1 = 5 \sin \dfrac{x}{2}$ and compare the result with Figure 47.   ▬

E X A M P L E   5   Finding an Equation for a Sinusoidal Graph

Find an equation for the graph shown in Figure 48.

FIGURE 48

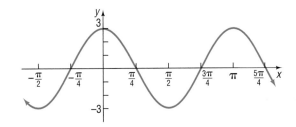

Solution   From the graph we conclude that it is easier to view the equation as a cosine function with amplitude $A = 3$ and period $T = \pi$. Thus, $2\pi/\omega = \pi$, so $\omega = 2$. A cosine function whose graph is given in Figure 48 is

$$y = A \cos \omega x = 3 \cos 2x$$

Check: Graph $Y_1 = 3 \cos 2x$ and compare the result with Figure 48.   ▬

E X A M P L E   6   Finding an Equation for a Sinusoidal Graph

Find an equation for the graph shown in Figure 49.

FIGURE 49

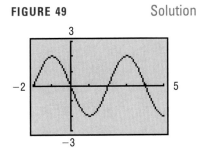

Solution   The graph is sinusoidal, with amplitude $A = 2$. The period is 4, so $2\pi/\omega = 4$ or $\omega = \pi/2$. Since the graph passes through the origin, it is easiest to view the equation as a sine function, but notice that the graph is actually the reflection of a sine function about the x-axis (since the graph is decreasing near the origin). Thus, we have

$$y = -A \sin \omega x = -2 \sin \frac{\pi}{2} x$$

Check: Graph $Y_1 = -2 \sin \dfrac{\pi}{2} x$ and compare the result with Figure 49.   ▬

Now work Problem 37.

Phase Shift

We have seen that the graph of $y = A \sin \omega x$, $\omega > 0$, has amplitude $|A|$ and period $T = 2\pi/\omega$. Thus, one period can be drawn as x varies from 0 to $2\pi/\omega$ or, equivalently, as ωx varies from 0 to 2π. See Figure 50.

We now want to discuss the graph of

$$y = A \sin(\omega x - \phi) = A \sin \omega\left(x - \frac{\phi}{\omega}\right)$$

FIGURE 50
One period of $y = A \sin \omega x$, $A > 0$, $\omega > 0$

where $\omega > 0$ and ϕ (the Greek letter phi) are real numbers. The graph will be a sine curve of amplitude $|A|$. As $\omega x - \phi$ varies from 0 to 2π, one period will be traced out. This period will begin when

$$\omega x - \phi = 0 \quad \text{or} \quad x = \frac{\phi}{\omega}$$

and will end when

$$\omega x - \phi = 2\pi \quad \text{or} \quad x = \frac{2\pi}{\omega} + \frac{\phi}{\omega}$$

FIGURE 51
One period of $y = A \sin(\omega x - \phi)$,
$A > 0$, $\omega > 0$, $\phi > 0$

See Figure 51.

Thus, we see the graph of $y = A \sin(\omega x - \phi) = A \sin \omega\left(x - \frac{\phi}{\omega}\right)$ is the same as the graph of $y = A \sin \omega x$, except that it has been shifted ϕ/ω units (to the right if $\phi > 0$ and to the left if $\phi < 0$). This number ϕ/ω is called the **phase shift** of the graph of $y = A \sin(\omega x - \phi)$.

For the graphs of $y = A \sin(\omega x - \phi)$ or $y = A \cos(\omega x - \phi)$, $\omega > 0$,

$$\text{Amplitude} = |A| \qquad \text{Period} = T = \frac{2\pi}{\omega} \qquad \text{Phase shift} = \frac{\phi}{\omega}$$

E X A M P L E   7

Finding the Amplitude, Period, and Phase Shift
of a Sinusoidal Function

Find the amplitude, period, and phase shift of $y = 3 \sin(2x - \pi)$, and graph the function.

Solution

Comparing $y = 3 \sin(2x - \pi)$ to $y = A \sin(\omega x - \phi)$, we find that $A = 3$, $\omega = 2$, and $\phi = \pi$. The graph is a sine curve with amplitude $A = 3$, period $T = 2\pi/\omega = 2\pi/2 = \pi$, and phase shift $= \dfrac{\phi}{\omega} = \dfrac{\pi}{2}$.

The graph of $y = 3\sin(2x - \pi) = 3 \sin 2\left(x - \dfrac{\pi}{2}\right)$ is obtained by using the graphing techniques discussed in Chapter 2. See Figure 52.

FIGURE 52

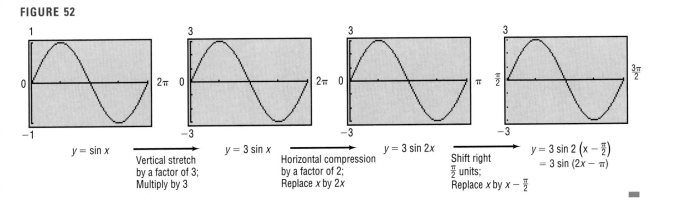

$y = \sin x$ → Vertical stretch by a factor of 3; Multiply by 3 → $y = 3 \sin x$ → Horizontal compression by a factor of 2; Replace x by $2x$ → $y = 3 \sin 2x$ → Shift right $\frac{\pi}{2}$ units; Replace x by $x - \frac{\pi}{2}$ → $y = 3 \sin 2 \left(x - \frac{\pi}{2}\right)$ $= 3 \sin (2x - \pi)$

E X A M P L E   8   Finding the Amplitude, Period, and Phase Shift of a Sinusoidal Function

Find the amplitude, period, and phase shift of $y = 2 \cos(4x + 3\pi)$, and graph the function.

Solution   Comparing $y = 2 \cos(4x + 3\pi)$ to $y = A \cos(\omega x - \phi)$, we see that $A = 2$, $\omega = 4$, and $\phi = -3\pi$. The graph is a cosine curve with amplitude $A = 2$, period $T = 2\pi/\omega = 2\pi/4 = \pi/2$, and phase shift $= \dfrac{\phi}{\omega} = \dfrac{-3\pi}{4}$. The stages to

graph $y = 2 \cos(4x + 3\pi) = 2 \cos 4\left(x + \dfrac{3\pi}{4}\right)$ are shown in Figure 53.

FIGURE 53

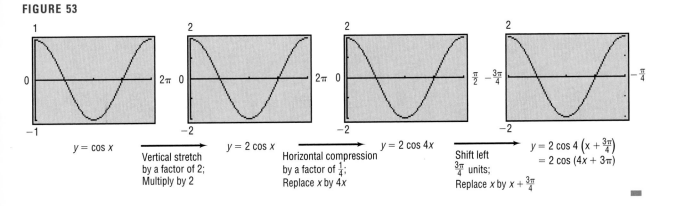

$y = \cos x$ → Vertical stretch by a factor of 2; Multiply by 2 → $y = 2 \cos x$ → Horizontal compression by a factor of $\frac{1}{4}$; Replace x by $4x$ → $y = 2 \cos 4x$ → Shift left $\frac{3\pi}{4}$ units; Replace x by $x + \frac{3\pi}{4}$ → $y = 2 \cos 4 \left(x + \frac{3\pi}{4}\right)$ $= 2 \cos (4x + 3\pi)$

Now work Problem 51.

Finding Sinusoidal Functions from Data by Hand

 Scatter diagrams of data sometimes take the form of a sinusoidal function. Let's look at an example.

The data given in Table 1 represents the average monthly temperature in Denver, Colorado. Since the data represents *average* monthly temperatures collected over so many years, the data will not vary much from year to year and so will essentially repeat itself each year. In other words, the data is periodic. Figure 54 shows the scatter diagram of this data repeated over two years, where $x = 1$ represents January, $x = 2$ represents February, and so on.

TABLE 1	

Month, x	Average Monthly Temperature, °F
January, 1	29.7
February, 2	33.4
March, 3	39.0
April, 4	48.2
May, 5	57.2
June, 6	66.9
July, 7	73.5
August, 8	71.4
September, 9	62.3
October, 10	51.4
November, 11	39.0
December, 12	31.0

Source: U.S. National Oceanic and Atmospheric Administration.

FIGURE 54

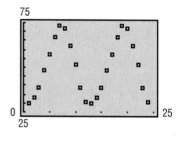

Notice the scatter diagram looks like the graph of a sinusoidal function. We choose to fit the data to a sine function of the form

$$y = A \sin(\omega x - \phi) + B$$

where A, B, ω, and ϕ are constants.

E X A M P L E   9   Finding a Sinusoidal Function from Temperature Data

By hand, fit a sine function to the data in Table 1.

FIGURE 55

Solution   We begin with a scatter diagram of the data for one year. See Figure 55. The data will be fitted to a sine function of the form

$$y = A \sin(\omega x - \phi) + B$$

STEP 1: To find the amplitude A, we compute

$$\text{Amplitude} = (\text{Largest Data Value} - \text{Smallest Data Value})/2$$
$$= (73.5 - 29.7)/2 = 21.9$$

FIGURE 56

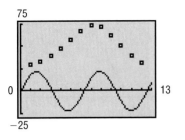

To see the remaining steps in this process, we superimpose the graph of the function $Y_1 = 21.9 \sin x$, where x represents months, on the scatter diagram. To see both, the viewing window has been adjusted. Figure 56 shows the two graphs.

To fit the data, the graph needs to be shifted vertically, shifted horizontally, and stretched horizontally.

STEP 2: We determine the vertical shift by finding the average of the highest and lowest data value.

$$\text{Vertical Shift} = (73.5 + 29.7)/2 = 51.6$$

Now we superimpose the graph of $Y_1 = 21.9 \sin x + 51.6$ on the scatter diagram. See Figure 57.

We see the graph needs to be shifted horizontally and stretched horizontally.

FIGURE 57

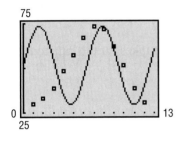

STEP 3: It is easier to find the horizontal stretch factor first. Since the temperatures repeat every 12 months, the period of the function is $T = 12$. Since $T = \dfrac{2\pi}{\omega} = 12$,

$$\omega = \frac{2\pi}{12} = \frac{\pi}{6}$$

Now we superimpose the graph of $Y_1 = 21.9 \sin \dfrac{\pi x}{6} + 51.6$ on the scatter diagram. See Figure 58.

We see the graph still needs to be shifted horizontally.

FIGURE 58

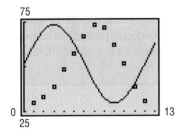

STEP 4: To determine the horizontal shift we solve the equation

$$y = 21.9 \sin\left(\frac{\pi}{6}x - \phi\right) + 51.6$$

for ϕ by letting $y = 29.7$ and $x = 1$
(the average temperature in Denver in January).*

$29.7 = 21.9 \sin\left(\dfrac{\pi}{6} \cdot 1 - \phi\right) + 51.6$	
$-21.9 = 21.9 \sin\left(\dfrac{\pi}{6} - \phi\right)$	Subtract 51.6 from both sides of the equation.
$-1 = \sin\left(\dfrac{\pi}{6} - \phi\right)$	Divide both sides of the equation by 21.9.
$-\dfrac{\pi}{2} = \left(\dfrac{\pi}{6} - \phi\right)$	Take the inverse sine of both sides of the equation.
$\phi = \dfrac{2\pi}{3}$	Solve for ϕ.

Thus,

FIGURE 59

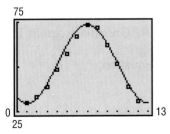

$$y = 21.9 \sin\left(\frac{\pi}{6}x - \frac{2\pi}{3}\right) + 51.6$$

The graph of $Y_1 = 21.9 \sin(\pi/6\, x - 2\pi/3) + 51.6$ and the scatter diagram of data are shown in Figure 59. ∎

*The data point selected here to find ϕ is arbitrary. Selecting a different data point will usually result in a different value for ϕ. To maintain consistency, we will always choose the data point for which y is smallest (in this case, January gives the lowest temperature).

measured as the amounts above or below the mean lower low water. The height of the water at high tide was 8.2 feet and the height of the water at low tide was −0.6 feet.
  (a) Approximately, when will the next high tide occur?
  (b) By hand, find a sinusoidal function of the form $y = A \sin(\omega x - \phi) + B$ that fits the data.
  (c) Draw a graph of the function found in (b).
  (d) Use the function found in (b) to predict the height of the water at the next high tide.

76. **Tides** Suppose the length of time between consecutive high tides is approximately 12.5 hours. According to the National Oceanic and Atmospheric Administration, on Saturday, June 28, 1997, in Juneau, AL, high tide occurred at 8:11 A.M. (08.1833 hours) and low tide occurred at 2:14 P.M. (14.2333 hours). Water heights are measured as the amounts above or below the mean lower low water. The height of the water at high tide was 13.2 feet and the height of the water at low tide was 2.2 feet.
  (a) Approximately when will the next high tide occur?
  (b) By hand, find a sinusoidal function of the form $y = A \sin(\omega x - \phi) + B$ that fits the data.
  (c) Draw a graph of the function found in (b).
  (d) Use the function found in (b) to predict the height of the water at the next high tide.

77. **Hours of Daylight** According to the *Old Farmer's Almanac,* in Miami, FL, the number of hours of sunlight on the summer solstice is 12.75 and the number of hours of sunlight on the winter solstice is 10.583.
  (a) By hand, find a sinusoidal function of the form $y = A \sin(\omega x - \phi) + B$ that fits the data.
  (b) Draw a graph of the function found in (a).
  (c) Use the function found in (a) to predict the number of hours of sunlight on April 1, the 91st day of the year.
   (d) Look up the number of hours of sunlight on April 1, 1997, in the *Old Farmer's Almanac,* and compare the actual hours of daylight to the results found in (c).

78. **Hours of Daylight** According to the *Old Farmer's Almanac,* in Detroit, MI, the number of hours of sunlight on the summer solstice is 13.65 and the number of hours of sunlight on the winter solstice is 9.067.
  (a) By hand, find a sinusoidal function of the form $y = A \sin(\omega x - \phi) + B$ that fits the data.
  (b) Draw a graph of the function found in (a).

  (c) Use the function found in (a) to predict the number of hours of sunlight on April 1, the 91st day of the year.
  (d) Look up the number of hours of sunlight on April 1, 1997, in the *Old Farmer's Almanac,* and compare the actual hours of daylight to the results found in (c).

79. **Hours of Daylight** According to the *Old Farmer's Almanac,* in Anchorage, AL, the number of hours of sunlight on the summer solstice is 16.233 and the number of hours of sunlight on the winter solstice is 5.45.
  (a) By hand, find a sinusoidal function of the form $y = A \sin(\omega x - \phi) + B$ that fits the data.
  (b) Draw a graph of the function found in (a).
  (c) Use the function found in (a) to predict the number of hours of sunlight on April 1, the 91st day of the year.
  (d) Look up the number of hours of sunlight on April 1, 1997, in the *Old Farmer's Almanac,* and compare the actual hours of daylight to the results found in (c).

80. **Hours of Daylight** According to the *Old Farmer's Almanac,* in Honolulu, HI, the numbers of hours of sunlight on the summer solstice is 12.767 and the number of hours of sunlight on the winter solstice is 10.783.
  (a) By hand, find a sinusoidal function of the form $y = A \sin(\omega x - \phi) + B$ that fits the data.
  (b) Draw a graph of the function found in (a).
  (c) Use the function found in (a) to predict the number of hours of sunlight on April 1, the 91st day of the year.
  (d) Look up the number of hours of sunlight on April 1, 1997 in the *Old Farmer's Almanac* and compare the actual hours of daylight to the results found in (c).

81. Explain how the amplitude and period of a sinusoidal graph are used to establish the scale on each coordinate axis.

82. Find an application in your major field that leads to a sinusoidal graph. Write a paper about your findings.

83. **CBL Experiment** Part 1: A student puts his/her thumb over a light probe and points it at a light source. The student then begins lifting his/her thumb from the sensor and replacing it repeatedly. Light intensity is plotted over time and the period and frequency are determined. Part 2: A light probe is pointed toward a fluorescent light bulb and its light intensity is recorded. The frequency and period of the light bulb are recorded. (Activity 15, Real-World Math with the CBL System, 1994.)

7.6 SIMPLE HARMONIC MOTION; DAMPED MOTION

> **1** Find an Equation for an Object in Simple Harmonic Motion
> **2** Analyze Simple Harmonic Motion
> **3** Analyze an Object in Damped Motion

Simple Harmonic Motion

There are many physical phenomena that can be described as simple harmonic motion. Radio and television waves, light waves, sound waves, and water waves exhibit motion that is simple harmonic.

The swinging of a pendulum, the vibrations of a tuning fork, and the bobbing of a weight attached to a coiled spring are examples of vibrational motion. In this type of motion, an object swings back and forth over the same path. In each illustration in Figure 63, the point B is the **equilibrium (rest) position** of the vibrating object. The **amplitude** of vibration is the distance from the object's rest position to its point of greatest displacement (either point A or point C in Figure 63). The **period** of a vibrating object is the time required to complete one vibration—that is, the time it takes to go from, say, point A through B to C and back to A.

FIGURE 63

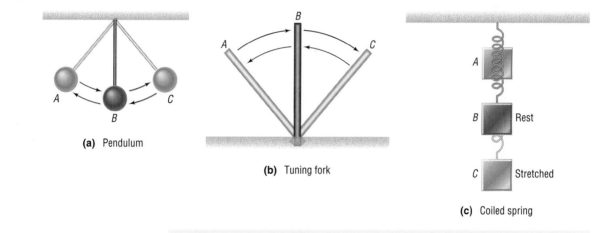

(a) Pendulum

(b) Tuning fork

(c) Coiled spring

> **Simple harmonic motion** is a special kind of vibrational motion in which the acceleration a of the object is directly proportional to the negative of its displacement d from its rest position. That is, $a = -kd$, $k > 0$.

For example, when the mass hanging from the spring in Figure 63(c) is pulled down from its rest position B to the point C, the force of the spring tries to restore the mass to its rest position. Assuming that there is no frictional force* to retard the motion, the amplitude will remain constant. The force increases in direct proportion to the distance the mass is pulled from its rest position. Since the force increases directly, the acceleration of the mass of the object must do likewise, because (by Newton's Second Law of Motion) force is directly proportional to acceleration. Thus, the acceleration of the object varies directly with its displacement, and the motion is an example of simple harmonic motion.

---

*If friction is present, the amplitude will decrease with time to 0. This type of motion is an example of **damped motion,** which is discussed later in this section.

FIGURE 64

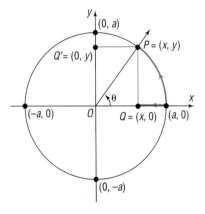

Simple harmonic motion is related to circular motion. To see this relationship, consider a circle of radius a, with center at $(0, 0)$. See Figure 64. Suppose that an object initially placed at $(a, 0)$ moves counterclockwise around the circle at constant angular speed ω. Suppose further that after time t has elapsed, the object is at the point $P = (x, y)$ on the circle. The angle θ, in radians, swept out by the ray $\overrightarrow{OP}$ in this time t is

$$\theta = \omega t$$

The coordinates of the point P at time t are

$$x = a \cos \theta = a \cos \omega t$$
$$y = a \sin \theta = a \sin \omega t$$

Corresponding to each position $P = (x, y)$ of the object moving about the circle, there is the point $Q = (x, 0)$, called the **projection of P on the x-axis.** As P moves around the circle at a constant rate, the point Q moves back and forth between the points $(a, 0)$ and $(-a, 0)$ along the x-axis with a motion that is simple harmonic. Similarly, for each point P there is a point $Q' = (0, y)$, called the **projection of P on the y-axis.** As P moves around the circle, the point Q' moves back and forth between the points $(0, a)$ and $(0, -a)$ on the y-axis with a motion that is simple harmonic. Thus, simple harmonic motion can be described as the projection of constant circular motion on a coordinate axis.

To put it another way, again consider a mass hanging from a spring where the mass is pulled down from its rest position to the point C and then released. See Figure 65(a). The graph shown in Figure 65(b) describes the distance the object is from its rest position, d, as a function of time, t, assuming no frictional force is present.

FIGURE 65

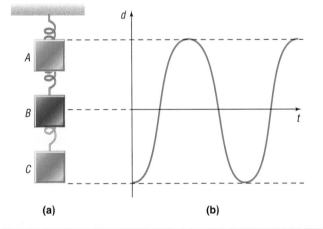

(a)   (b)

Theorem   Simple Harmonic Motion

An object that moves on a coordinate axis so that its distance d from the origin at time t is given by either

$$d = a \cos \omega t \quad \text{or} \quad d = a \sin \omega t$$

where a and $\omega > 0$ are constants, moves with simple harmonic motion. The motion has amplitude $|a|$ and period $2\pi/\omega$.

The **frequency** f of an object in simple harmonic motion is the number of oscillations per unit time. Since the period is the time required for one oscillation, it follows that frequency is the reciprocal of the period; that is,

$$f = \frac{\omega}{2\pi} \qquad \omega > 0$$

E X A M P L E   1

Finding an Equation for an Object in Harmonic Motion

Suppose that the object attached to the coiled spring in Figure 63(c) is pulled down a distance of 5 inches from its rest position and then released. If the time for one oscillation is 3 seconds, write an equation that relates the distance d of the object from its rest position after time t (in seconds). Assume no friction.

E X A

Solution

The motion of the object is simple harmonic. When the object is released ($t = 0$), the distance of the object from the rest position is -5 units (since the object was pulled down). Because $d = -5$ when $t = 0$, it is easier* to use the cosine function

FIGURE

$$d = a \cos \omega t$$

to describe the motion. Now the amplitude is $|-5| = 5$ and the period is 3. Thus,

$$a = -5 \quad \text{and} \quad \frac{2\pi}{\omega} = \text{Period} = 3 \quad \text{or} \quad \omega = \frac{2\pi}{3}$$

An equation of the motion of the object is

Rest positi

$$d = -5 \cos \frac{2\pi}{3} t$$

Note: In the solution to Example 1, we let $a = -5$, since the initial motion is down. If the initial direction was up, we would let $a = 5$.

FIGURE

20

Now work Problem 1.

0

E X A M P L E   2

Analyzing the Motion of an Object

–20

Suppose that the distance d (in meters) an object travels in time t (in seconds) satisfies the equation

FIGURE

$$d = 10 \sin 5t$$

(a) Describe the motion of the object.
(b) What is the maximum displacement from its resting position?
(c) What is the time required for one oscillation?
(d) What is the frequency?

*No phase shift is required if a cosine function is used.

(b) At what times t will the graph of V touch the graph of $y = e^{-1.9t}$? When does V touch the graph of $y = -e^{-1.9t}$?

(c) When will the voltage V be between -0.1 and 0.1 volt?

32. Graph the function $f(x) = (\sin x)/x$, $x > 0$. Based on the graph, what do you conjecture about the value of $(\sin x)/x$ for x close to 0?

33. Graph $y = x \sin x$, $y = x^2 \sin x$, and $y = x^3 \sin x$ for $x > 0$. What patterns do you observe?

34. Graph $y = \dfrac{1}{x}\sin x$, $y = \dfrac{1}{x^2}\sin x$, and $y = \dfrac{1}{x^3}\sin x$ for $x > 0$. What patterns do you observe?

35. How would you explain to a friend what simple harmonic motion is? Damped motion?

36. **CBL Experiment**   Pendulum motion is analyzed to estimate simple harmonic motion. A plot is generated with the position of the pendulum over time. The graph is used to find a sinusoidal curve of the form $y = A \cos B(x - C) + D$. Determine the amplitude, period and frequency. (Activity 16, Real-World Math with the CBL System.)

37. **CBL Experiment**   The sound from a tuning fork is collected over time. The amplitude, frequency and period of the graph are determined. A model of the form $y = A \cos B(x - C)$ is fitted to the data. (Activity 23, Real-World Math with the CBL System.)

38. **CBL Experiment**   A weight is attached to a spring and allowed to "bob" up and down. The position of the weight is recorded over time. A model of the form $f(t) = a(t) \sin (bt + c) + d$ is developed from the data. (Experiment M3, CBL System Experiment Workbook, 1994.)

CHAPTER REVIEW

THINGS TO KNOW

Formulas

Law of Sines	$\dfrac{\sin \alpha}{a} = \dfrac{\sin \beta}{b} = \dfrac{\sin \gamma}{c}$
Law of Cosines	$c^2 = a^2 + b^2 - 2ab \cos \gamma$
	$b^2 = a^2 + c^2 - 2ac \cos \beta$
	$a^2 = b^2 + c^2 - 2bc \cos \alpha$
Area of a triangle	$A = \frac{1}{2}bh$
	$A = \frac{1}{2}ab \sin \gamma$
	$A = \frac{1}{2}bc \sin \alpha$
	$A = \frac{1}{2}ac \sin \beta$
	$A = \sqrt{s(s - a)(s - b)(s - c)}$, where $s = \frac{1}{2}(a + b + c)$

Sinusoidal graphs		
	$y = A \sin \omega x$, $\omega > 0$	Period $= 2\pi/\omega$
	$y = A \cos \omega x$, $\omega > 0$	Amplitude $= |A|$
	$y = A \sin(\omega x - \phi)$	Phase shift $= \phi/\omega$
	$y = A \cos(\omega x - \phi)$	

HOW TO

Solve right triangles

Use the Law of Sines to solve a SAA, ASA, or SSA triangle

Use the Law of Cosines to solve a SAS or SSS triangle

Find the area of a triangle

Find the period and amplitude of a sinusoidal function and use them to graph the function

Find a function whose sinusoidal graph is given

Find a sinusoidal function from data by hand

Find a sinusoidal function of best fit using a graphing utility

Analyze simple harmonic motion

Analyze damped motion

FILL-IN-THE-BLANK ITEMS

1. If two sides and the angle opposite one of them are known, the Law of _____ is used to determine whether the known information results in no triangle, one triangle, or two triangles.

2. If three sides of a triangle are given, the Law of _____ is used to solve the triangle.

3. If three sides of a triangle are given, _____ Formula is used to find the area of the triangle.

4. The following function has amplitude 3 and period 2:

$y =$ _____ sin _____ x

5. The function $y = 3 \sin 6x$ has amplitude _____ and period _____.

6. The mass and damping factor of $d = 5e^{-0.5t/10} \cos\left(\sqrt{\pi^2 - \frac{(0.5)^2}{100}}\, t\right)$ are _____ and _____.

7. The motion of an object obeys the equation $d = 4 \cos 6t$. Such motion is described as _____ _____ _____.

TRUE/FALSE ITEMS

T F   **1.** An oblique triangle in which two sides and an angle are given always results in at least one triangle.

T F   **2.** Given three sides of a triangle, there is a formula for finding its area.

T F   **3.** The graphs of $y = \sin x$ and $y = \cos x$ are identical except for a horizontal shift.

T F   **4.** For $y = -2 \sin \pi x$, the amplitude is 2 and the period is $\pi/2$.

REVIEW EXERCISES

In Problems 1–4, solve each triangle.

In Problems 5–24, find the remaining angle(s) and side(s) of each triangle, if it exists. If no triangle exists, say "No triangle."

5. $\alpha = 50°, \beta = 30°, a = 1$

6. $\alpha = 10°, \gamma = 40°, c = 2$

7. $\alpha = 100°, a = 5, c = 2$

8. $a = 2, c = 5, \alpha = 60°$

9. $a = 3, c = 1, \gamma = 110°$

10. $a = 3, c = 1, \gamma = 20°$

11. $a = 3, c = 1, \beta = 100°$

12. $a = 3, b = 5, \beta = 80°$

13. $a = 2, b = 3, c = 1$

14. $a = 10, b = 7, c = 8$

15. $a = 1, b = 3, \gamma = 40°$

16. $a = 4, b = 1, \gamma = 100°$

17. $a = 5, b = 3, \alpha = 80°$

18. $a = 2, b = 3, \alpha = 20°$

19. $a = 1, b = \frac{1}{2}, c = \frac{4}{3}$

20. $a = 3, b = 2, c = 2$

21. $a = 3, \alpha = 10°, b = 4$

22. $a = 4, \alpha = 20°, \beta = 100°$

23. $c = 5, b = 4, \alpha = 70°$

24. $a = 1, b = 2, \gamma = 60°$

In Problems 25–34, find the area of each triangle.

25. $a = 2, b = 3, \gamma = 40°$

26. $b = 5, c = 4, \alpha = 20°$

27. $b = 4, c = 10, \alpha = 70°$

28. $a = 2, b = 1, \gamma = 100°$

29. $a = 4, b = 3, c = 5$

30. $a = 10, b = 7, c = 8$

31. $a = 4, b = 2, c = 5$

32. $a = 3, b = 2, c = 2$

33. $\alpha = 50°, \beta = 30°, a = 1$

34. $\alpha = 10°, \gamma = 40°, c = 3$

In Problems 35–38, determine the amplitude and period of each function without graphing.

35. $y = 4 \cos x$

36. $y = \sin 2x$

37. $y = -8 \sin \frac{\pi}{2}x$

38. $y = -2 \cos 3\pi x$

In Problems 39–46, find the amplitude, period, and phase shift of each function. Use a graphing utility to graph each function. Show at least one period.

39. $y = 4 \sin 3x$

40. $y = 2 \cos \frac{1}{3}x$

41. $y = -2 \sin\left(\frac{\pi}{2}x + \frac{1}{2}\right)$

42. $y = -6 \sin(2\pi x - 2)$

43. $y = \frac{1}{2}\sin(\frac{3}{2}x - \pi)$

44. $y = \frac{3}{2}\cos(6x + 3\pi)$

45. $y = -\frac{2}{3}\cos(\pi x - 6)$

46. $y = -7 \sin\left(\frac{\pi}{3}x + \frac{4}{3}\right)$

In Problems 47–50, find a function whose graph is given.

47.

48.

49.

50.

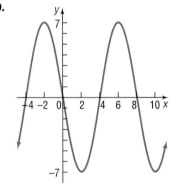

In Problems 51–54, the distance d (in feet) an object travels in time t (in seconds) is given.

(a) Describe the motion of the object.

(b) What is the maximum displacement from its resting position?

(c) What is the time required for one oscillation?

(d) What is the frequency?

51. $d = 6 \sin 2t$ **52.** $d = 2 \cos 4t$ **53.** $d = -2 \cos \pi t$ **54.** $d = -3 \sin \dfrac{\pi}{2} t$

In Problems 55–56, an object of mass m attached to a coiled spring with damping factor b is pulled down a distance a from its rest position and then released. Assume the positive direction of the motion is up and the period of the first oscillation is T.

(a) Write an equation that relates the distance d of the object from its rest position after t seconds.

(b) Graph the equation found in (b) for 5 oscillations.

55. $m = 40$ grams; $a = 15$ centimeters; $b = 0.75$ grams/second; $T = 5$ seconds

56. $m = 25$ grams; $a = 13$ centimeters; $b = 0.65$ grams/second; $T = 4$ seconds

In Problems 57–58, the distance d (in meters) of the bob of a pendulum of mass m (in kilograms) from its rest position at time t (in seconds) is given.

(a) Describe the motion of the object.

(b) What is the initial displacement of the bob? That is, what is the displacement at t = 0?

(c) Graph the motion.

(d) What is the maximum displacement after the first oscillation?

(e) What happens to the displacement of the bob as time increases without bound?

57. $d = -15e^{-0.6t/40} \cos\left(\sqrt{\left(\dfrac{2\pi}{5}\right)^2 - \dfrac{0.36}{1600}}\, t \right)$ **58.** $d = -20e^{-0.5t/60} \cos\left(\sqrt{\left(\dfrac{2\pi}{3}\right)^2 - \dfrac{0.25}{3600}}\, t \right)$

59. Alternating Voltage The electromotive force E, in volts in a certain ac circuit obeys the equation

$$E = 120 \sin 120\pi t, \ t \geq 0$$

where t is measured in seconds.

(a) What is the maximum value of E?

(b) What is the period?

(c) Graph this function over two periods.

60. Alternating Current The current I, in amperes, flowing through an ac (alternating current) circuit at time t is

$$I = 220 \sin\left(30\pi t + \dfrac{\pi}{6}\right), \qquad t \geq 0$$

(a) What is the period?

(b) What is the amplitude?

(c) What is the phase shift?

(d) Graph this function over two periods.

61. Measuring the Length of a Lake From a stationary hot air balloon 500 feet above the ground, two sightings of a lake are made (see the figure). How long is the lake?

62. Finding the Speed of a Glider From a glider 200 feet above the ground, two sightings of a stationary object directly in front are taken

1 minute apart (see the figure). What is the speed of the glider?

63. Finding the Width of a River Find the distance from *A* to *C* across the river illustrated in the figure.

64. Finding the Height of a Building Find the height of the building shown in this figure.

65. Finding the Distance to Shore The Sears Tower in Chicago is 1454 feet tall and is situated about 1 mile inland from the shore of Lake Michigan, as indicated in the figure. An observer in a pleasure boat on the lake directly in front of the Sears Tower looks at the top of the tower and measures the angle of elevation as 5°. How far offshore is the boat?

66. Finding the Grade of a Mountain Trail A straight trail with a uniform inclination leads from a hotel, elevation 5000 feet, to a lake in a valley, elevation 4100 feet. The length of the trail is 4100 feet. What is the inclination (grade) of the trail?

67. Navigation An airplane flies from city *A* to city *B*, a distance of 100 miles, then turns through an angle of 20° and heads toward city *C*, as indicated in the figure. If the distance from *A* to *C* is 300 miles, how far is it from city *B* to city *C*?

68. Correcting a Navigation Error Two cities *A* and *B* are 300 miles apart. In flying from city *A* to city *B*, a pilot inadvertently took a course that was 5° in error.

(a) If the error was discovered after flying 10 minutes at a constant speed of 420 miles per hour, through what angle should the pilot turn to correct the course? (Consult the figure below.)

(b) What new constant speed should be maintained so that no time is lost due to the error? (Assume that the speed would have been a constant 420 miles per hour if no error had occurred.)

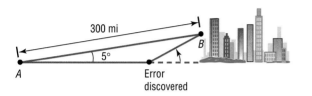

69. **Determining Distances at Sea** Rebecca, the navigator of a ship at sea, spots two lighthouses that she knows to be 2 miles apart along a straight seashore. She determines that the angles formed between two line-of-sight observations of the lighthouses and the line from the ship directly to shore are 12° and 30°. See the illustration.

(a) How far is the ship from lighthouse A?
(b) How far is the ship from lighthouse B?
(c) How far is the ship from shore?

70. **Constructing a Highway** A highway whose primary directions are north–south is being constructed along the west coast of Florida. Near Naples, a bay obstructs the straight path

of the road. Since the cost of a bridge is prohibitive, engineers decide to go around the bay. The illustration shows the path they decide on and the measurements taken. What is the length of highway needed to go around the bay?

71. **Correcting a Navigational Error** A yacht leaves St. Thomas bound for an island in the British West Indies, 200 miles away. Maintaining a constant speed of 18 miles per hour, but encountering heavy crosswinds and strong currents, the crew finds after 4 hours that the sailboat is off course by 15°.

(a) How far is the sailboat from the island at this time?
(b) Through what angle should the sailboat turn to correct its course?
(c) How much time has been added to the trip because of this? (Assume that the speed remains at 18 miles per hour.)

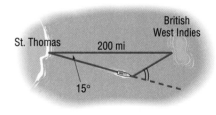

72. **Surveying** Two homes are located on opposite sides of a small hill. See the illustration. To measure the distance between them, a surveyor walks a distance of 50 feet from house A to point C, uses a transit to measure the angle ACB, which is found to be 80°, and then walks

to house B, a distance of 60 feet. How far apart are the houses?

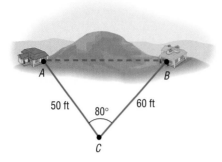

73. **Approximating the Area of a Lake** To approximate the area of a lake, Cindy walks around the perimeter of the lake, taking the measurements shown in the illustration. Using this technique, what is the approximate area of the lake?

[**Hint:** Use the Law of Cosines on the three triangles shown and then find the sum of their areas.]

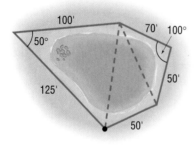

74. **Calculating the Cost of Land** The irregular parcel of land shown in the figure is being sold for $100 per square foot. What is the cost of this parcel?

75. The drive wheel of an engine is 13 inches in diameter, and the pulley on the rotary pump is 5 inches in diameter. If the shafts of the drive wheel and the pulley are 2 feet apart, what

length of belt is required to join them as shown in figure below?

76. Rework Problem 75 if the belt is crossed, as shown in the figure below.

77. **Monthly Temperature** The data below represents the average monthly temperatures for Phoenix, AZ.

Month, x	Average Monthly Temperature, °F
January, 1	51
February, 2	55
March, 3	63
April, 4	67
May, 5	77
June, 6	86
July, 7	90
August, 8	90
September, 9	84
October, 10	71
November, 11	59
December, 12	52

Source: U.S. National Oceanic and Atmospheric Administration.

(a) Using a graphing utility, draw a scatter diagram of the data for one period.
(b) By hand, find a sinusoidal function of the form $y = A \sin(\omega x - \phi) + B$ that fits the data.

(c) Draw the sinusoidal function found in (b) on the scatter diagram.

(d) Use a graphing utility to find the sinusoidal function of best fit.

(e) Draw the sinusoidal function of best fit on the scatter diagram.

78. **Monthly Temperature**   The data below represents the average monthly temperatures for Chicago, IL.

Month, x	Average Monthly Temperature, °F
January, 1	25
February, 2	28
March, 3	36
April, 4	48
May, 5	61
June, 6	72
July, 7	74
August, 8	75
September, 9	66
October, 10	55
November, 11	39
December, 12	28

Source: U.S. National Oceanic and Atmospheric Administration.

(a) Using a graphing utility, draw a scatter diagram of the data for one period.

(b) By hand, find a sinusoidal function of the form $y = A \sin(\omega x - \phi) + B$ that fits the data.

(c) Draw the sinusoidal function found in (b) on the scatter diagram.

(d) Use a graphing utility to find the sinusoidal function of best fit.

(e) Draw the sinusoidal function of best fit on the scatter diagram.

79. **Hours of Daylight**   According to the *Old Farmer's Almanac*, in Las Vegas, NV, the number of hours of sunlight on the summer solstice is 13.367 and the number of hours of sunlight on the winter solstice is 9.667.

(a) By hand, find a sinusoidal function of the form $y = A \sin(\omega x - \phi) + B$ that fits the data.

(b) Draw a graph of the function found in (a).

(c) Use the function found in (a) to predict the number of hours of sunlight on April 1, the 91st day of the year.

(d) Look up the number of hours of sunlight on April 1, 1997, in the *Old Farmer's Almanac* and compare the actual hours of daylight to the results found in (c).

80. **Hours of Daylight**   According to the *Old Farmer's Almanac*, in Seattle, WA, the number of hours of sunlight on the summer solstice is 13.967 and the number of hours of sunlight on the winter solstice is 8.417.

(a) By hand, find a sinusoidal function of the form $y = A \sin(\omega x - \phi) + B$ that fits the data.

(b) Draw a graph of the function found in (a).

(c) Use the function found in (a) to predict the number of hours of sunlight on April 1, the 91st day of the year.

(d) Look up the number of hours of sunlight on April 1, 1997, in the *Old Farmer's Almanac* and compare the actual hours of daylight to the results found in (c).

Polar Coordinates; Vectors

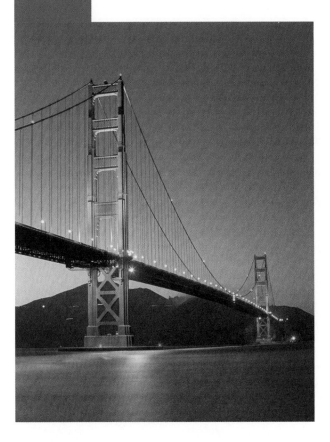

Using computer models is one way structural engineers, architects, and others try out various designs without the costs of building materials and unforeseen construction issues. On the following page you will try your hand at becoming a master bridge builder through computer modeling. The programs and other information you will need are found at:

www.prenhall.com/sullivan

Netsite: http://www.prenhall.com/sullivan/

THE GREAT BALSA BRIDGE CONTEST

Your goal is high—you intend to become The Master Balsa Bridge Builder. Over the last two years you have performed countless hours of stress testing on balsa struts to determine to strongest and lightest grains. You have performed wood joint experiments with a host of ultrastrong superglues. You have read and reread the literature on the previous contests, analyzing those bridge designs that have failed, and those that were heads above all the *competition*. You kept careful notes and have entered the data into your *bridge modeling software*. For the last few months you tested your designs on your computer. Only two survived.

You are now ready to enter into the next phase. Yours will be no *bridge disaster*. You are ready to see if your computer-generated designs match reality.

You have created a set of microtools so that your joinery will be perfect. Slowly and meticulously you build your first set of two test bridges. You need to determine how accurately your calculations match your craftsmanship. You have laminated the balsa struts, and, at last, your designs come to life! Now you place the two bridges across two chairs and begin to place weights on them . . .

1. If you think of the weights as *force vectors* acting along the struts, does it make sense that we can model the structural design for a bridge?
2. Try the *Design a Bridge* website. How did you do? Did you just overbuild everything?
3. Do you think that engineers, when they are designing a bridge, might rely a little too heavily on the computer? Do you think that might be a mistake?
4. Get out a piece of paper and make a very simple bridge design. Estimate how much weight x and y can hold. Then do a simple vector analysis using your assumptions. Are you surprised at how much your simple bridge can support?
5. Get some pieces of balsa and redo the calculations using your assumptions about the strength of the balsa. How good are you at calculating the actual values?
6. When you look carefully at a complicated bridge like the Golden Gate Bridge, how long do you think that it would take you to do all those calculations?

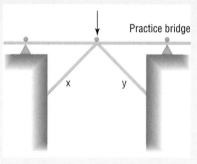

Practice bridge

This chapter is in two parts: Polar Coordinates, Sections 8.1–8.3, and Vectors, Sections 8.4–8.6. They are independent of each other and may be covered in any order.

Sections 8.1–8.3 deal with polar coordinates (an alternative to rectangular coordinates for plotting points),

graphing in polar coordinates, and finding roots of complex numbers (Demoivre's Theorem).

Sections 8.4–8.6 provide an introduction to the idea of a vector, an extremely important topic in engineering and physics.

8.1 POLAR COORDINATES

1 Plot Points Using Polar Coordinates
2 Convert between Polar Coordinates and Rectangular Coordinates

FIGURE 1

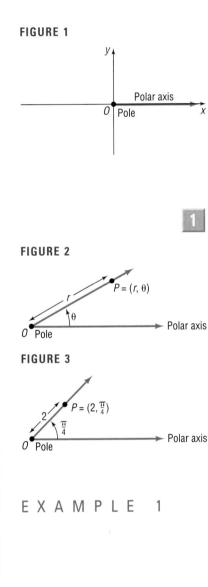

So far, we have always used a system of rectangular coordinates to plot points in the plane. Now we are ready to describe another system called *polar coordinates*. As we shall soon see, in many instances polar coordinates offer certain advantages over rectangular coordinates.

In a rectangular coordinate system, you will recall, a point in the plane is represented by an ordered pair of numbers (x, y), where x and y equal the signed distance of the point from the y-axis and x-axis, respectively. In a polar coordinate system, we select a point, called the **pole,** and then a ray with vertex at the pole, called the **polar axis.** Comparing the rectangular and polar coordinate systems, we see (in Figure 1) that the origin in rectangular coordinates coincides with the pole in polar coordinates, and the positive x-axis in rectangular coordinates coincides with the polar axis in polar coordinates.

FIGURE 2

A point P in a polar coordinate system is represented by an ordered pair of numbers (r, θ). The number r is the distance of the point from the pole, and θ is an angle (in degrees or radians) formed by the polar axis and a ray from the pole through the point. We call the ordered pair (r, θ) the **polar coordinates** of the point. See Figure 2.

As an example, suppose that the polar coordinates of a point P are $(2, \pi/4)$. We locate P by first drawing an angle of $\pi/4$ radian, placing its vertex at the pole and its initial side along the polar axis. Then we go out a distance of 2 units along the terminal side of the angle to reach the point P. See Figure 3.

FIGURE 3

Now work Problem 1.

Recall that an angle measured counterclockwise is positive, whereas one measured clockwise is negative. This convention has some interesting consequences relating to polar coordinates. Let's see what these consequences are.

E X A M P L E   1   Finding Several Polar Coordinates of a Single Point

Consider again a point P with polar coordinates $(2, \pi/4)$, as shown in Figure 4(a). Because $\pi/4$, $9\pi/4$, and $-7\pi/4$ all have the same terminal side, we also could have located this point P by using the polar coordinates $(2, 9\pi/4)$ or $(2, -7\pi/4)$, as shown in Figures 4(b) and (c).

FIGURE 4

(a)   (b)   (c)

Theorem   Conversion from Polar Coordinates to Rectangular Coordinates

If P is a point with polar coordinates (r, θ), the rectangular coordinates (x, y) of P are given by

$$x = r \cos \theta \qquad y = r \sin \theta \qquad (1)$$

Note: Most graphing calculators have the capability of converting from polar coordinates to rectangular coordinates. Consult your owner's manual to learn the proper key strokes.

FIGURE 12

Proof   Suppose that P has the polar coordinates (r, θ). We seek the rectangular coordinates (x, y) of P. Refer to Figure 12.

If $r = 0$, then, regardless of θ, the point P is the pole, for which the rectangular coordinates are $(0, 0)$. Thus, formula (1) is valid for $r = 0$.

If $r > 0$, the point P is on the terminal side of θ and $r = d(O, P) = \sqrt{x^2 + y^2}$. Thus,

$$\cos \theta = \frac{x}{r} \qquad \sin \theta = \frac{y}{r}$$

so

$$x = r \cos \theta \qquad y = r \sin \theta$$

If $r < 0$, then the point $P = (r, \theta)$ can be represented as $(-r, \pi + \theta)$, where $-r > 0$. Thus,

$$\cos(\pi + \theta) = -\cos \theta = \frac{x}{-r} \qquad \sin(\pi + \theta) = -\sin \theta = \frac{y}{-r}$$

so

$$x = r \cos \theta \qquad y = r \sin \theta \qquad \blacksquare$$

EXAMPLE 5

Converting from Polar Coordinates to Rectangular Coordinates

Find the rectangular coordinates of the points with the following polar coordinates:

(a) $(6, \pi/6)$ (b) $(-2, 5\pi/4)$ (c) $(-4, -\pi/4)$

Solution   We use formula (1): $x = r \cos \theta$ and $y = r \sin \theta$.

(a) See Figure 13(a).

$$x = r \cos \theta = 6 \cos \frac{\pi}{6} = 6 \cdot \frac{\sqrt{3}}{2} = 3\sqrt{3}$$

$$y = r \sin \theta = 6 \sin \frac{\pi}{6} = 6 \cdot \frac{1}{2} = 3$$

The rectangular coordinates of the point $(6, \pi/6)$ are $(3\sqrt{3}, 3)$.

(b) See Figure 13(b).

$$x = r \cos \theta = -2 \cos \frac{5\pi}{4} = -2\left(-\frac{\sqrt{2}}{2}\right) = \sqrt{2}$$

$$y = r \sin \theta = -2 \sin \frac{5\pi}{4} = -2\left(-\frac{\sqrt{2}}{2}\right) = \sqrt{2}$$

The rectangular coordinates of the point $(-2, 5\pi/4)$ are $(\sqrt{2}, \sqrt{2})$.

(c) See Figure 13(c).

$$x = r \cos \theta = -4 \cos\left(-\frac{\pi}{4}\right) = -4 \cdot \frac{\sqrt{2}}{2} = -2\sqrt{2}$$

$$y = r \sin \theta = -4 \sin\left(-\frac{\pi}{4}\right) = -4\left(-\frac{\sqrt{2}}{2}\right) = 2\sqrt{2}$$

The rectangular coordinates of the point $(-4, -\pi/4)$ are $(-2\sqrt{2}, 2\sqrt{2})$.

FIGURE 13

(a)   (b)   (c)

 Now work Problems 21 and 33.

To convert from rectangular coordinates (x, y) to polar coordinates (r, θ) is a little more complicated. Let's look at some examples.

E X A M P L E   6

Converting from Rectangular Coordinates to Polar Coordinates

Find polar coordinates of a point whose rectangular coordinates are $(0, 3)$.

FIGURE 14

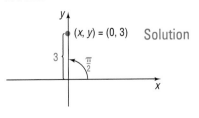

Solution   See Figure 14. The point $(0, 3)$ lies on the y-axis a distance of 3 units from the origin (pole) so $r = 3$. A ray with vertex at the pole through $(0, 3)$ forms an angle $\theta = \dfrac{\pi}{2}$ with the polar axis. Polar coordinates for this point can be given by $(3, \pi/2)$.

FIGURE 15

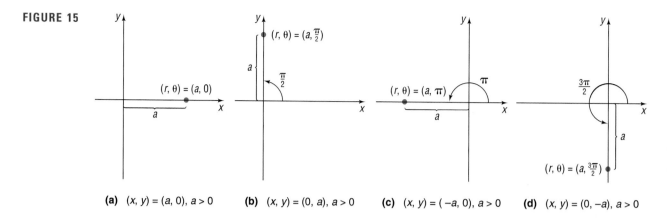

(a) $(x, y) = (a, 0), a > 0$ **(b)** $(x, y) = (0, a), a > 0$ **(c)** $(x, y) = (-a, 0), a > 0$ **(d)** $(x, y) = (0, -a), a > 0$

Figure 15 shows polar coordinates of points that lie on either the x-axis or the y-axis.

Now work Problem 37.

EXAMPLE 7

Converting from Rectangular Coordinates to Polar Coordinates

Find polar coordinates of a point whose rectangular coordinates are $(2, -2)$.

Solution

FIGURE 16

See Figure 16. The distance r from the origin to the point $(2, -2)$ is

$$r = \sqrt{x^2 + y^2} = \sqrt{(2)^2 + (-2)^2} = \sqrt{8} = 2\sqrt{2}$$

To find θ, we use the reference angle α. Then

$$\alpha = \tan^{-1}\left|\frac{y}{x}\right| = \tan^{-1}\left|\frac{-2}{2}\right| = \tan^{-1} 1 = \frac{\pi}{4}$$

Thus, $\theta = -\pi/4$, and a set of polar coordinates for this point is $(2\sqrt{2}, -\pi/4)$. Other possible representations include $(2\sqrt{2}, 7\pi/4)$ and $(-2\sqrt{2}, 3\pi/4)$. ∎

EXAMPLE 8

Converting from Rectangular Coordinates to Polar Coordinates

Find polar coordinates of a point whose rectangular coordinates are $(-1, -\sqrt{3})$.

Solution

FIGURE 17

See Figure 17. The distance r from the origin to the point $(-1, -\sqrt{3})$ is

$$r = \sqrt{x^2 + y^2} = \sqrt{(-1)^2 + (-\sqrt{3})^2} = \sqrt{4} = 2$$

To find θ, we use the reference angle α. Then

$$\alpha = \tan^{-1}\left|\frac{y}{x}\right| = \tan^{-1}\left|\frac{-\sqrt{3}}{-1}\right| = \tan^{-1}\sqrt{3} = \frac{\pi}{3}$$

Thus,

$$\theta = \pi + \alpha = \pi + \frac{\pi}{3} = \frac{4\pi}{3}$$

$(x, y) = (-1, -\sqrt{3})$

and a set of polar coordinates is $(2, 4\pi/3)$. Other possible representations include $(-2, \pi/3)$ and $(2, -2\pi/3)$. ∎

Figure 18 shows how to find polar coordinates of a point that lies in a quadrant when its rectangular coordinates (x, y) are given.

FIGURE 18

(a) $r = \sqrt{x^2 + y^2}$
$\theta = \alpha = \tan^{-1} \left|\frac{y}{x}\right|$

(b) $r = \sqrt{x^2 + y^2}$
$\theta = \pi - \alpha = \pi - \tan^{-1}\left|\frac{y}{x}\right|$

(c) $r = \sqrt{x^2 + y^2}$
$\theta = \pi + \alpha = \pi + \tan^{-1}\left|\frac{y}{x}\right|$

(d) $r = \sqrt{x^2 + y^2}$
$\theta = -\alpha = -\tan^{-1}\left|\frac{y}{x}\right|$

Now work Problem 41.

Based on the preceding discussion, we have the formulas

$$r^2 = x^2 + y^2 \qquad \tan \theta = \frac{y}{x} \qquad \text{if } x \neq 0 \qquad (2)$$

Warning: Be careful when using formula (2). Always plot the point (x, y) first, as we did in Examples 7 and 8, so that r and θ are chosen to reflect the correct quadrant. Look again at Figure 18.

Formulas (1) and (2) may also be used to transform equations.

E X A M P L E   9   Transforming Equations from Polar to Rectangular Form

Transform the equation $r = 4 \sin \theta$ from polar coordinates to rectangular coordinates, and identify the graph.

Solution   If we multiply each side by r, it will be easier to apply formulas (1) and (2):

$$r = 4 \sin \theta$$
$$r^2 = 4r \sin \theta \qquad \text{Multiply each side by } r.$$
$$x^2 + y^2 = 4y \qquad \text{Apply formulas (1) and (2).}$$

This is the equation of a circle:

$$x^2 + (y^2 - 4y) = 0$$
$$x^2 + (y^2 - 4y + 4) = 4 \qquad \text{Complete the square in } y.$$
$$x^2 + (y - 2)^2 = 4$$

Its center is at $(0, 2)$, and its radius is 2.

Now work Problem 57.

E X A M P L E 10

Transforming an Equation from Rectangular to Polar Form

Transform the equation $4xy = 9$ from rectangular coordinates to polar coordinates.

Solution   We use formula (1):

$$4xy = 9$$
$$4(r \cos \theta)(r \sin \theta) = 9 \quad \text{Formula (1).}$$
$$4r^2 \cos \theta \sin \theta = 9$$
$$2r^2 \sin 2\theta = 9 \quad \text{Double-angle formula.}$$

8.1   EXERCISES

In Problems 1–12, plot each point given in polar coordinates.

1. $(3, 90°)$ **2.** $(4, 270°)$ **3.** $(-2, 0)$ **4.** $(-3, \pi)$

5. $(6, \pi/6)$ **6.** $(5, 5\pi/3)$ **7.** $(-2, 135°)$ **8.** $(-3, 120°)$

9. $(-1, -\pi/3)$ **10.** $(-3, -3\pi/4)$ **11.** $(-2, -\pi)$ **12.** $(-3, -\pi/2)$

In Problems 13–20, plot each point given in polar coordinates, and find other polar coordinates (r, θ) of the point for which

(a) $r > 0, \quad -2\pi \leq \theta < 0$ (b) $r < 0, \quad 0 \leq \theta < 2\pi$ (c) $r > 0, \quad 2\pi \leq \theta < 4\pi$

13. $(5, 2\pi/3)$ **14.** $(4, 3\pi/4)$ **15.** $(-2, 3\pi)$ **16.** $(-3, 4\pi)$

17. $(1, \pi/2)$ **18.** $(2, \pi)$ **19.** $(-3, -\pi/4)$ **20.** $(-2, -2\pi/3)$

In Problems 21–36, polar coordinates of a point are given. Find the rectangular coordinates of each point.

21. $(3, \pi/2)$ **22.** $(4, 3\pi/2)$ **23.** $(-2, 0)$ **24.** $(-3, \pi)$

25. $(6, 150°)$ **26.** $(5, 300°)$ **27.** $(-2, 3\pi/4)$ **28.** $(-3, 2\pi/3)$

29. $(-1, -\pi/3)$ **30.** $(-3, -3\pi/4)$ **31.** $(-2, -180°)$ **32.** $(-3, -90°)$

33. $(7.5, 110°)$ **34.** $(-3.1, 182°)$ **35.** $(6.3, 3.8)$ **36.** $(8.1, 5.2)$

In Problems 37–48, the rectangular coordinates of a point are given. Find polar coordinates for each point.

37. $(3, 0)$ **38.** $(0, 2)$ **39.** $(-1, 0)$ **40.** $(0, -2)$

41. $(1, -1)$ **42.** $(-3, 3)$ **43.** $(\sqrt{3}, 1)$ **44.** $(-2, -2\sqrt{3})$

45. $(1.3, -2.1)$ **46.** $(-0.8, -2.1)$ **47.** $(8.3, 4.2)$ **48.** $(-2.3, 0.2)$

In Problems 49–56, the letters x and y represent rectangular coordinates. Write each equation using polar coordinates (r, θ).

49. $2x^2 + 2y^2 = 3$ **50.** $x^2 + y^2 = x$ **51.** $x^2 = 4y$ **52.** $y^2 = 2x$

53. $2xy = 1$ **54.** $4x^2y = 1$ **55.** $x = 4$ **56.** $y = -3$

In Problems 57–64, the letters r and θ represent polar coordinates. Write each equation using rectangular coordinates (x, y).

57. $r = \cos \theta$ **58.** $r = \sin \theta + 1$ **59.** $r^2 = \cos \theta$ **60.** $r = \sin \theta - \cos \theta$

61. $r = 2$ **62.** $r = 4$ **63.** $r = \dfrac{4}{1 - \cos \theta}$ **64.** $r = \dfrac{3}{3 - \cos \theta}$

65. Show that the formula for the distance d between two points $P_1 = (r_1, \theta_1)$ and $P_2 = (r_2, \theta_2)$ is

$$d = \sqrt{r_1^2 + r_2^2 - 2r_1 r_2 \cos(\theta_2 - \theta_1)}$$

8.2 POLAR EQUATIONS AND GRAPHS

> 1 Graph and Identify Polar Equations by Converting to Rectangular Equations
> 2 Graph Polar Equations Using a Graphing Utility
> 3 Test a Polar Equation for Symmetry
> 4 Graph Polar Equations by Hand by Plotting Points

An equation whose variables are polar coordinates is called a **polar equation**. The **graph of a polar equation** consists of all points whose polar coordinates satisfy the equation.

Just as a rectangular grid may be used to plot points given by rectangular coordinates, as in Figure 19(a), we can use a grid consisting of concentric circles (with centers at the pole) and rays (with vertices at the pole) to plot points given by polar coordinates, as shown in Figure 19(b). We shall use such **polar grids** to graph polar equations by hand.

FIGURE 19

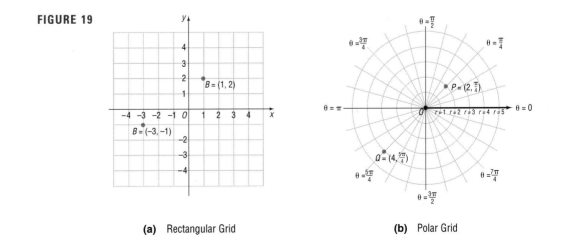

(a) Rectangular Grid

(b) Polar Grid

One method we can use to graph a polar equation by hand is to convert the equation to rectangular coordinates. In the discussion that follows, (x, y) represent the rectangular coordinates of a point P and (r, θ) represent polar coordinates of the point P.

EXAMPLE 1   Identifying and Graphing a Polar Equation by Hand (Circle)

Identify and graph the equation: $r = 3$

Solution   We convert the polar equation to a rectangular equation:

$$r = 3$$
$$r^2 = 9$$
$$x^2 + y^2 = 9$$

Thus, the graph of $r = 3$ is a circle, with center at the pole and radius 3. See Figure 20.

FIGURE 20
$r = 3$ or $x^2 + y^2 = 9$

 Now work Problem 1.

EXAMPLE 2   Identifying and Graphing a Polar Equation by Hand (Line)

Identify and graph the equation: $\theta = \pi/4$

Solution   We convert the polar equation to a rectangular equation:

$$\theta = \frac{\pi}{4}$$

$$\tan \theta = \tan \frac{\pi}{4} = 1$$

$$\frac{y}{x} = 1 \quad \text{Formula (2), Section 8.1.}$$

$$y = x$$

The graph of $\theta = \pi/4$ is a line passing through the pole making an angle of $\pi/4$ with the polar axis. See Figure 21.

FIGURE 21

$\theta = \dfrac{\pi}{4}$ or $y = x$

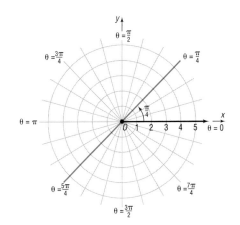

E X A M P L E   3   Identifying and Graphing a Polar Equation by Hand (Line)

Identify and graph the equation: $\theta = -\pi/6$

Solution   We convert the polar equation to a rectangular equation:

$$\theta = -\frac{\pi}{6}$$

$$\tan \theta = \tan\left(-\frac{\pi}{6}\right) = -\frac{\sqrt{3}}{3}$$

$$\frac{y}{x} = -\frac{\sqrt{3}}{3}$$

$$y = -\frac{\sqrt{3}}{3}x$$

The graph of $\theta = -\pi/6$ is a line passing through the pole making an angle of $-\pi/6$ with the polar axis. See Figure 22.

FIGURE 22

$\theta = -\dfrac{\pi}{6}$ or $y = -\dfrac{\sqrt{3}}{3}x$

 Now work Problem 3.

E X A M P L E 4

Identifying and Graphing a Polar Equation by Hand (Horizontal Line)

Identify and graph the equation sin r: $\theta = 2$

Solution

Since $y = r \sin \theta$, we can simply write the equation as

$$y = 2$$

We conclude that the graph of $r \sin \theta = 2$ is a horizontal line 2 units above the pole. See Figure 23.

FIGURE 23
$r \sin \theta = 2$ or $y = 2$

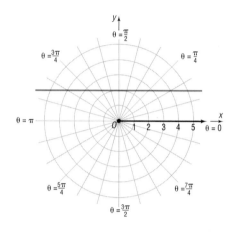

2

A second method we can use to graph a polar equation is to graph the equation using a graphing utility.

Most graphing utilities require the following steps in order to obtain the graph of an equation:

Graphing a Polar Equation Using a Graphing Utility

STEP 1: Solve the equation for r in terms of θ.
STEP 2: Select the viewing rectangle in polar mode. In addition to setting Xmin, Xmax, Xscl, and so forth, the viewing rectangle in polar mode requires setting minimum and maximum values for θ and an increment setting for θ (θ step). In addition, a square screen and radian measure should be used.
STEP 3: Enter the expression involving θ that you found in STEP 1. (Consult your manual for the correct way to enter the expression.)
STEP 4: Execute.

E X A M P L E 5

Graphing a Polar Equation Using a Graphing Utility

Use a graphing utility to graph the polar equation $r \sin \theta = 2$.

Solution

STEP 1: We solve the equation for r in terms of θ.

$$r \sin \theta = 2$$

$$r = \frac{2}{\sin \theta}$$

STEP 2: From the polar mode, select the viewing rectangle. We will use the one given next.

$$\theta\text{min} = 0$$
$$\theta\text{max} = 2\pi$$
$$\theta step = \pi/24$$
$$X\text{min} = -9$$
$$X\text{max} = 9$$
$$X scl = 1$$
$$Y\text{min} = -6$$
$$Y\text{max} = 6$$
$$Y scl = 1$$

θstep determines the number of points the graphing utility will plot. For example, if *θstep* is $\pi/24$, then the graphing utility will evaluate *r* at $\theta = 0(\theta\text{min})$, $\pi/24$, $2\pi/24$, $3\pi/24$, and so forth, up to $2\pi(\theta\text{max})$. The smaller *θstep*, the more points the graphing utility will plot. The student is encouraged to experiment with different values for *θ*min, *θ*max, and *θ*step to see how the graph is affected.

FIGURE 24

STEP 3: Enter the expression $\dfrac{2}{\sin\theta}$ after the prompt $r =$.

STEP 4: Execute.

The graph is shown in Figure 24. ■

E X A M P L E 6

Identifying and Graphing a Polar Equation (Vertical Line)

Identify and graph the equation: $r\cos\theta = -3$

Solution

Since $x = r\cos\theta$, we can simply write the equation as

$$x = -3$$

We conclude that the graph of $r\cos\theta = -3$ is a vertical line 3 units to the left of the pole. Figure 25(a) shows the graph drawn by hand. Figure 25(b) shows the graph of $r\cos\theta = -3$ using a graphing utility with $\theta\text{min} = 0$, $\theta\text{max} = 2\pi$, and $\theta\text{step} = \pi/24$.

FIGURE 25
$r\cos\theta = -3$ or $x = -3$

(a) (b)

Based on Examples 4, 5, and 6, we are led to the following results. (The proofs are left as exercises.)

Theorem

Let a be a nonzero real number, the graph of the equation

$$r \sin \theta = a$$

is a horizontal line a units above the pole if $a > 0$ and $|a|$ units below the pole if $a < 0$.

The graph of the equation

$$r \cos \theta = a$$

is a vertical line a units to the right of the pole if $a > 0$ and $|a|$ units to the left of the pole if $a < 0$.

Now work Problem 7.

E X A M P L E   7

Identifying and Graphing a Polar Equation (Circle)

Identify and graph the equation: $r = 4 \sin \theta$

Solution   To transform the equation to rectangular coordinates, we multiply each side by r:

$$r^2 = 4r \sin \theta$$

Now we use the facts that $r^2 = x^2 + y^2$ and $y = r \sin \theta$. Then

$$x^2 + y^2 = 4y$$
$$x^2 + (y^2 - 4y) = 0$$
$$x^2 + (y - 2)^2 = 4$$

This is the equation of a circle with center at $(0, 2)$ in rectangular coordinates and radius 2. Figure 26(a) shows the graph drawn by hand. Figure 26(b) shows the graph of $r = 4 \sin \theta$ using a graphing utility with $\theta\text{min} = 0$, $\theta\text{max} = 2\pi$, and $\theta\text{step} = \pi/24$.

FIGURE 26
$r = 4 \sin \theta$ or
$x^2 + (y - 2)^2 = 4$

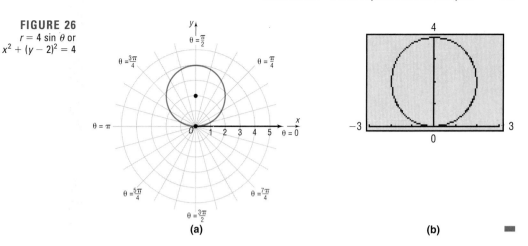

(a)   (b)

E X A M P L E   8   Identifying and Graphing a Polar Equation (Circle)

Identify and graph the equation: $r = -2 \cos \theta$

Solution   We proceed as in Example 7:

$$r^2 = -2r \cos \theta$$
$$x^2 + y^2 = -2x$$
$$x^2 + 2x + y^2 = 0$$
$$(x + 1)^2 + y^2 = 1$$

This is the equation of a circle with center at $(-1, 0)$ in rectangular coordinates and radius 1. Figure 27(a) shows the graph drawn by hand. Figure 27(b) shows the graph of $r = -2 \cos \theta$ using a graphing utility with θmin = 0, θmax = 2π, and θstep = $\pi/24$.

FIGURE 27
$r = -2 \cos \theta$ or
$(x + 1)^2 + y^2 = 1$

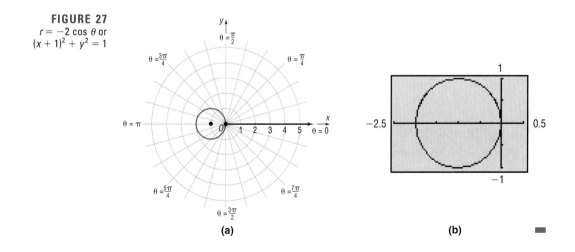

(a)   (b)

Exploration Graph $r_1 = \sin\theta$, $r_2 = 2\sin\theta$, and $r_3 = 3\sin\theta$. Do you see the pattern? Clear the screen and graph $r_1 = -\sin\theta$, $r_2 = -2\sin\theta$, and $r_3 = -3\sin\theta$. Do you see the pattern? Clear the screen and graph $r_1 = \cos\theta$, $r_2 = 2\cos\theta$, and $r_3 = 3\cos\theta$. Do you see the pattern? Clear the screen and graph $r_1 = -\cos\theta$, $r_2 = -2\cos\theta$, and $r_3 = -3\cos\theta$. Do you see the pattern?

Based on Examples 7, 8, and the Exploration we are led to the following results. (The proofs are left as exercises.)

Theorem

Let a be a positive real number. Then,

Equation   **Description**
(a) $r = 2a\sin\theta$ Circle: radius a; center at $(0, a)$ in rectangular coordinates
(b) $r = -2a\sin\theta$ Circle: radius a; center at $(0, -a)$ in rectangular coordinates
(c) $r = 2a\cos\theta$ Circle: radius a; center at $(a, 0)$ in rectangular coordinates
(d) $r = -2a\cos\theta$ Circle: radius a; center at $(-a, 0)$ in rectangular coordinates
Each circle contains the pole.

The method of converting a polar equation to an identifiable rectangular equation in order to graph it is not always helpful, nor is it always necessary. Usually, we set up a table that lists several points on the graph. By checking for symmetry, it may be possible to reduce the number of points needed to draw the graph by hand.

Symmetry

In polar coordinates the points (r, θ) and $(r, -\theta)$ are symmetric with respect to the polar axis (and to the x-axis). See Figure 28(a). The points (r, θ) and $(r, \pi - \theta)$ are symmetric with respect to the line $\theta = \pi/2$ (y-axis). See Figure 28(b). The points (r, θ) and $(-r, \theta)$ are symmetric with respect to the pole (origin). See Figure 28(c).

FIGURE 28

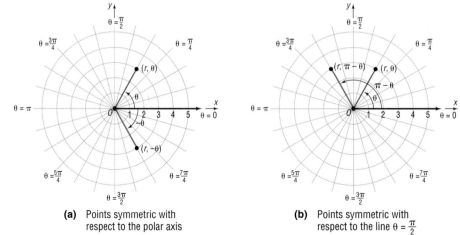

(a) Points symmetric with respect to the polar axis

(b) Points symmetric with respect to the line $\theta = \frac{\pi}{2}$

FIGURE 28
(continued)

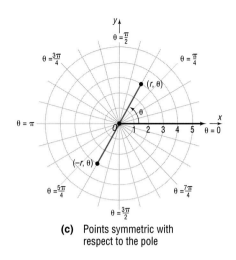

(c) Points symmetric with respect to the pole

The following tests are a consequence of these observations.

Theorem   Tests for Symmetry

Symmetry with Respect to the Polar Axis (*x*-Axis):
In a polar equation, replace θ by $-\theta$. If an equivalent equation results, the graph is symmetric with respect to the polar axis.

Symmetry with Respect to the Line $\theta = \pi/2$ (*y*-Axis):
In a polar equation, replace θ by $\pi - \theta$. If an equivalent equation results, the graph is symmetric with respect to the line $\theta = \pi/2$.

Symmetry with Respect to the Pole (Origin):
In a polar equation, replace r by $-r$. If an equivalent equation results, the graph is symmetric with respect to the pole.

The three tests for symmetry given here are *sufficient* conditions for symmetry, but they are not *necessary* conditions. That is, an equation may fail these tests and still have a graph that is symmetric with respect to the polar axis, the line $\theta = \pi/2$, or the pole. For example, the graph of $r = \sin 2\theta$ turns out to be symmetric with respect to the polar axis, the line $\theta = \pi/2$, and the pole, but all three tests given here fail. Refer to Problems 71, 72, and 73 in Exercise 8.2.

E X A M P L E   9   Graphing a Polar Equation (Cardioid)

4

Graph the equation: $r = 1 - \sin \theta$

Solution   We check for symmetry first:

Polar Axis: Replace θ by $-\theta$. The result is

$$r = 1 - \sin(-\theta) = 1 + \sin \theta$$

The test fails, so the graph may or may not be symmetric with respect to the polar axis.

The Line $\theta = \pi/2$: Replace θ by $\pi - \theta$. The result is

$$r = 1 - \sin(\pi - \theta) = 1 - (\sin \pi \cos \theta - \cos \pi \sin \theta)$$
$$= 1 - [0 \cdot \cos \theta - (-1) \sin \theta] = 1 - \sin \theta$$

Thus, the graph is symmetric with respect to the line $\theta = \pi/2$.

The Pole: Replace r by $-r$. Then the result is $-r = 1 - \sin \theta$, so $r = -1 + \sin \theta$. The test fails, so the graph may or may not be symmetric with respect to the pole.

Next, we identify points on the graph by assigning values to the angle θ and calculating the corresponding values of r. Due to the symmetry with respect to the line $\theta = \pi/2$, we only need to assign values to θ from $-\pi/2$ to $\pi/2$, as given in Table 1.

Now we plot the points (r, θ) from Table 1 and trace out the graph, beginning at the point $(2, -\pi/2)$ and ending at the point $(0, \pi/2)$. Then we reflect this portion of the graph about the line $\theta = \pi/2$ (y-axis) to obtain the complete graph.

Figure 29(a) shows the graph drawn by hand. Figure 29(b) shows the graph of $r = 1 - \sin \theta$ using a graphing utility with θmin $= 0$, θmax $= 2\pi$, and θstep $= \pi/24$.

TABLE 1	
θ	$r = 1 - \sin \theta$
$-\pi/2$	$1 + 1 = 2$
$-\pi/3$	$1 + \sqrt{3}/2 \approx 1.87$
$-\pi/6$	$1 + \frac{1}{2} = \frac{3}{2}$
0	1
$\pi/6$	$1 - \frac{1}{2} = \frac{1}{2}$
$\pi/3$	$1 - \sqrt{3}/2 \approx 0.13$
$\pi/2$	0

FIGURE 29
$r = 1 - \sin \theta$

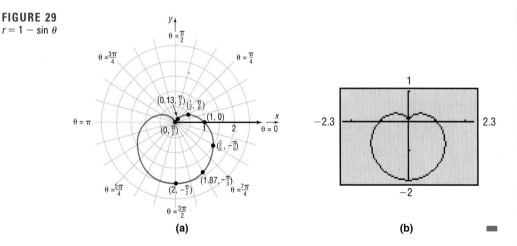

(a)   (b)

Exploration   Graph $r_1 = 1 + \sin \theta$. Clear the screen and graph $r_1 = 1 - \cos \theta$. Clear the screen and graph $r_1 = 1 + \cos \theta$. Do you see a pattern?

The curve in Figure 29 is an example of a *cardioid* (a heart-shaped curve).

Cardioids are characterized by equations of the form

$$r = a(1 + \cos\theta) \qquad r = a(1 + \sin\theta)$$
$$r = a(1 - \cos\theta) \qquad r = a(1 - \sin\theta)$$

where $a > 0$. The graph of a cardioid contains the pole.

Now work Problem 31.

E X A M P L E   10   Graphing a Polar Equation (Limaçon)

Graph the equation: $r = 3 + 2\cos\theta$

Solution   We check for symmetry first:

Polar Axis: Replace θ by $-\theta$. The result is

$$r = 3 + 2\cos(-\theta) = 3 + 2\cos\theta$$

Thus, the graph is symmetric with respect to the polar axis.

The Line $\theta = \pi/2$: Replace θ by $\pi - \theta$. The result is

$$r = 3 + 2\cos(\pi - \theta) = 3 + 2(\cos\pi\cos\theta + \sin\pi\sin\theta)$$
$$= 3 - 2\cos\theta$$

The test fails, so the graph may or may not be symmetric with respect to the line $\theta = \pi/2$.

The Pole: Replace r by $-r$. The test fails, so the graph may or may not be symmetric with respect to the pole.

Next, we identify points on the graph by assigning values to the angle θ and calculating the corresponding values of r. Due to the symmetry with respect to the polar axis, we only need to assign values to θ from 0 to π, as given in Table 2.

TABLE 2	
θ	$r = 3 + 2\cos\theta$
0	5
$\pi/6$	$3 + \sqrt{3} \approx 4.73$
$\pi/3$	4
$\pi/2$	3
$2\pi/3$	2
$5\pi/6$	$3 - \sqrt{3} \approx 1.27$
π	1

Now we plot the points (r, θ) from Table 2 and trace out the graph, beginning at the point $(5, 0)$ and ending at the point $(1, \pi)$. Then we reflect this portion of the graph about the polar axis (x-axis) to obtain the complete graph.

Figure 30(a) shows the graph drawn by hand. Figure 30(b) shows the graph of $r = 3 + 2\cos\theta$ using a graphing utility with $\theta\min = 0$, $\theta\max = 2\pi$, and $\theta\text{step} = \pi/24$.

FIGURE 30
$r = 3 + 2 \cos \theta$

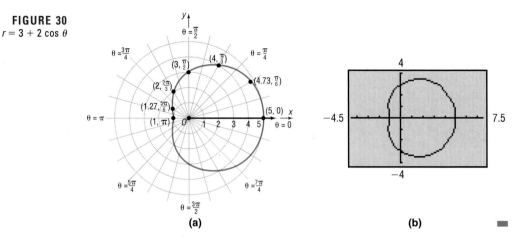

(a)

(b)

Exploration Graph $r_1 = 3 - 2 \cos \theta$. Clear the screen and graph $r_1 = 3 + 2 \sin \theta$. Clear the screen and graph $r_1 = 3 - 2 \sin \theta$. Do you see a pattern?

The curve in Figure 30 is an example of a *limaçon* (the French word for *snail*) without an inner loop.

Limaçons without an inner loop are characterized by equations of the form

$r = a + b \cos \theta$	$r = a + b \sin \theta$
$r = a - b \cos \theta$	$r = a - b \sin \theta$

where $a > 0$, $b > 0$, and $a > b$. The graph of a limaçon without an inner loop does not contain the pole.

 Now work Problem 37.

EXAMPLE 11   Graphing a Polar Equation (Limaçon with Inner Loop)

Graph the equation: $r = 1 + 2 \cos \theta$

Solution   First, we check for symmetry:

Polar Axis: Replace θ by $-\theta$. The result is

$$r = 1 + 2 \cos(-\theta) = 1 + 2 \cos \theta$$

Thus, the graph is symmetric with respect to the polar axis.

The Line $\theta = \pi/2$: Replace θ by $\pi - \theta$. The result is

$$r = 1 + 2 \cos(\pi - \theta) = 1 + 2(\cos \pi \cos \theta + \sin \pi \sin \theta)$$
$$= 1 - 2 \cos \theta$$

The test fails, so the graph may or may not be symmetric with respect to the line $\theta = \pi/2$.

The Pole: Replace r by $-r$. The test fails, so the graph may or may not be symmetric with respect to the pole.

Next, we identify points on the graph of $r = 1 + 2 \cos \theta$ by assigning values to the angle θ and calculating the corresponding values of r. Due to the symmetry with respect to the polar axis, we only need to assign values to θ from 0 to π, as given in Table 3.

Now we plot the points (r, θ) from Table 3, beginning at $(3, 0)$ and ending at $(-1, \pi)$. See Figure 31(a). Finally, we reflect this portion of the graph about the polar axis (x-axis) to obtain the complete graph. Figure 31(b) shows a complete graph drawn by hand. Figure 31(c) shows the graph of $r = 1 + 2 \cos \theta$ using a graphing utility with $\theta\min = 0$, $\theta\max = 2\pi$, and θstep $= \pi/24$.

TABLE 3	
θ	$r = 1 + 2 \cos \theta$
0	3
$\pi/6$	$1 + \sqrt{3} \approx 2.73$
$\pi/3$	2
$\pi/2$	1
$2\pi/3$	0
$5\pi/6$	$1 - \sqrt{3} \approx -0.73$
π	-1

FIGURE 31

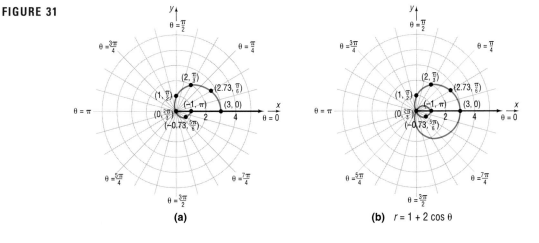

(a)   (b) $r = 1 + 2 \cos \theta$

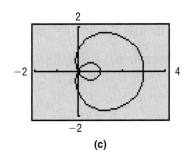

(c)

Exploration   Graph $r_1 = 1 - 2 \cos \theta$. Clear the screen and graph $r_1 = 1 + 2 \sin \theta$. Clear the screen and graph $r_1 = 1 - 2 \sin \theta$. Do you see a pattern?

The curve in Figure 31(b) is an example of a limaçon with an inner loop.

FIGURE 33

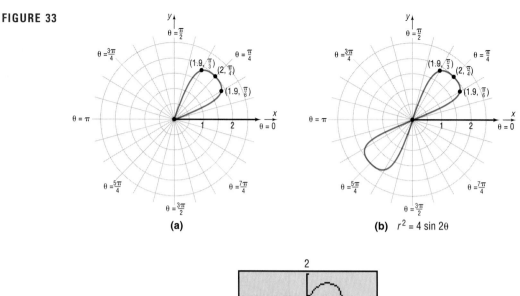

(a)

(b) $r^2 = 4 \sin 2\theta$

(c)

The curve in Figure 33(b) is an example of a *lemniscate*.

Lemniscates are characterized by equations of the form

$$r^2 = a^2 \sin 2\theta \qquad r^2 = a^2 \cos 2\theta$$

where $a \neq 0$, and have graphs that are propeller-shaped.

Now work Problem 47.

E X A M P L E   14

Graphing a Polar Equation (Spiral)

Sketch the graph of: $r = e^{\theta/5}$

Solution

The tests for symmetry with respect to the pole, the polar axis, and the line $\theta = \pi/2$ fail. Furthermore, there is no number θ for which $r = 0$. Hence, the graph does not pass through the pole. We observe that r is positive for all θ, r increases as θ increases, $r \to 0$ as $\theta \to -\infty$, and $r \to \infty$ as $\theta \to \infty$. With the help of a calculator, we obtain the values in Table 6. See Figure 34(a) for the graph drawn by hand. Figure 34(b) shows the graph of $r = e^{\theta/5}$ using a graphing utility with θmin $= -5\pi$, θmax $= 3\pi$, and θstep $= \pi/24$.

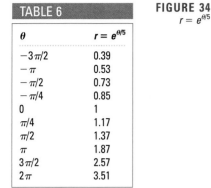

θ	$r = e^{\theta/5}$
$-3\pi/2$	0.39
$-\pi$	0.53
$-\pi/2$	0.73
$-\pi/4$	0.85
0	1
$\pi/4$	1.17
$\pi/2$	1.37
π	1.87
$3\pi/2$	2.57
2π	3.51

TABLE 6

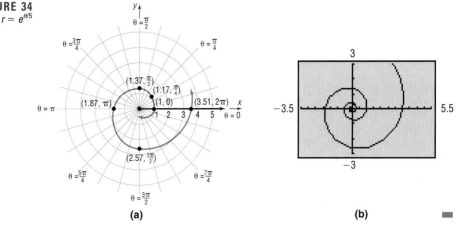

FIGURE 34
$r = e^{\theta/5}$

(a) (b)

The curve in Figure 34 is called a **logarithmic spiral,** since its equation may be written as $\theta = 5 \ln r$ and it spirals infinitely both toward the pole and away from it.

Classification of Polar Equations

The equations of some lines and circles in polar coordinates and their corresponding equations in rectangular coordinates are given in Table 7 on page 618. Also included are the names and the graphs of a few of the more frequently encountered polar equations.

Calculus Comment

For those of you who are planning to study calculus, a comment about one important role of polar equations is in order.

In rectangular coordinates, the equation $x^2 + y^2 = 1$, whose graph is the unit circle, does not define a function. In fact, on the interval $[-1, 1]$ it defines two functions,

$$y = \sqrt{1 - x^2} \quad \text{Upper semicircle.} \qquad y = -\sqrt{1 - x^2} \quad \text{Lower semicircle.}$$

In polar coordinates, the equation $r = 1$, whose graph is also the unit circle, does define a function. That is, for each choice of θ there is only one corresponding value of r, that is, $r = 1$. Since many uses of calculus require that functions be used, the opportunity to express nonfunctions in rectangular coordinates as functions in polar coordinates becomes extremely useful.

Note also that the vertical line test for functions is valid only for equations in rectangular coordinates.

HISTORICAL FEATURE Polar coordinates seem to have been invented by Jacob Bernoulli (1654–1705) about 1691, although, as with most such ideas, earlier traces of the notion exist. Early users of calculus remained committed to rectangular coordinates, and polar coordinates did not become widely used until the early

TABLE 7
Lines

Description	Line passing through the pole making an angle α with the polar axis	Vertical line	Horizontal line
Rectangular equation	$y = (\tan \alpha)x$	$x = a$	$y = b$
Polar equation	$\theta = \alpha$	$r \cos \theta = a$	$r \sin \theta = b$
Typical graph			

Circles

Description	Center at the pole, radius a	Passing through the pole, tangent to the line $\theta = \pi/2$, center on the polar axis, radius a	Passing through the pole, tangent to the polar axis, center on the line $\theta = \pi/2$, radius a
Rectangular equation	$x^2 + y^2 = a^2, a > 0$	$x^2 + y^2 = \pm 2ax, a > 0$	$x^2 + y^2 = \pm 2ay, a > 0$
Polar equation	$r = a, a > 0$	$r = \pm 2a \cos \theta, a > 0$	$r = \pm 2a \sin \theta, a > 0$
Typical graph			

Other Equations

Name	Cardioid	Limaçon without inner loop	Limaçon with inner loop
Polar equations	$r = a \pm a \cos \theta, a > 0$ $r = a \pm a \sin \theta, a > 0$	$r = a \pm b \cos \theta, 0 < b < a$ $r = a \pm b \sin \theta, 0 < b < a$	$r = a \pm b \cos \theta, 0 < a < b$ $r = a \pm b \sin \theta, 0 < a < b$
Typical graph			

Name	Lemniscate	Rose with three petals	Rose with four petals
Polar equations	$r^2 = a^2 \cos 2\theta, a > 0$ $r^2 = a^2 \sin 2\theta, a > 0$	$r = a \sin 3\theta, a > 0$ $r = a \cos 3\theta, a > 0$	$r = a \sin 2\theta, a > 0$ $r = a \cos 2\theta, a > 0$
Typical graph			

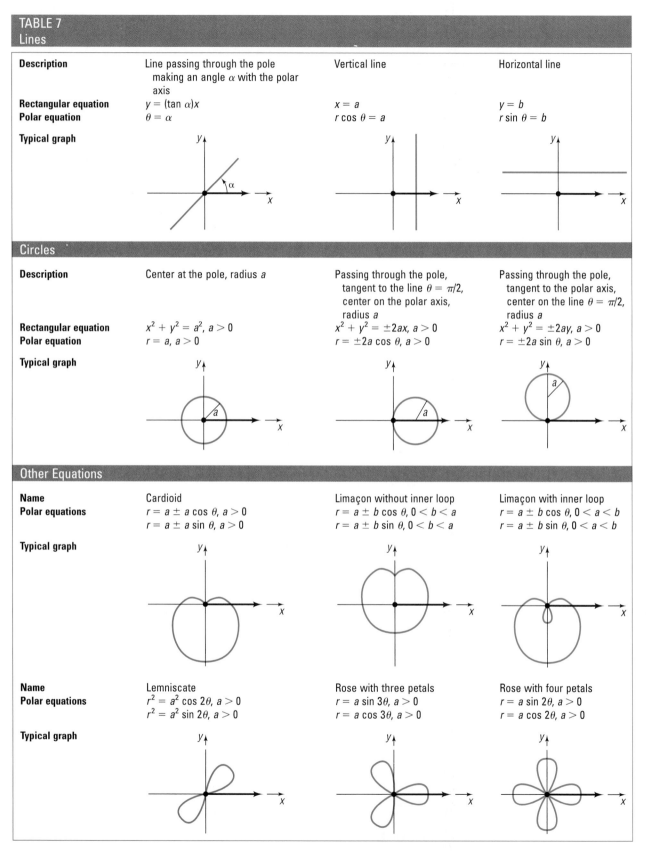

1800s. Even then, it was mostly geometers who used them for describing odd curves. Finally, about the mid-1800s, applied mathematicians realized the tremendous simplification polar coordinates make possible in the description of objects with circular or cylindrical symmetry. From then on their use became widespread.   ▬

8.2   EXERCISES

In Problems 1–16, identify each polar equation by transforming the equation to rectangular coordinates. Graph each polar equation by hand. Verify your results using a graphing utility.

1. $r = 4$
2. $r = 2$
3. $\theta = \pi/3$
4. $\theta = -\pi/4$
5. $r \sin \theta = 4$
6. $r \cos \theta = 4$
7. $r \cos \theta = -2$
8. $r \sin \theta = -2$
9. $r = 2 \cos \theta$
10. $r = 2 \sin \theta$
11. $r = -4 \sin \theta$
12. $r = -4 \cos \theta$
13. $r \sec \theta = 4$
14. $r \csc \theta = 8$
15. $r \csc \theta = -2$
16. $r \sec \theta = -4$

In Problems 17–24, match each of the graphs (A) through (H) to one of the following polar equations.

17. $r = 2$
18. $\theta = \pi/4$
19. $r = 2 \cos \theta$
20. $r \cos \theta = 2$
21. $r = 1 + \cos \theta$
22. $r = 2 \sin \theta$
23. $\theta = 3\pi/4$
24. $r \sin \theta = 2$

(A)

(B)

(C)

(D)

(E)

(F)

(G)

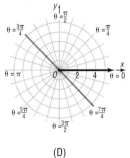

(H)

In Problems 25–30, match each of the graphs (A) through (F) to one of the following polar equations.

25. $r = 4$

26. $r = 3 \cos \theta$

27. $r = 3 \sin \theta$

28. $r \sin \theta = 3$

29. $r \cos \theta = 3$

30. $r = 2 + \sin \theta$

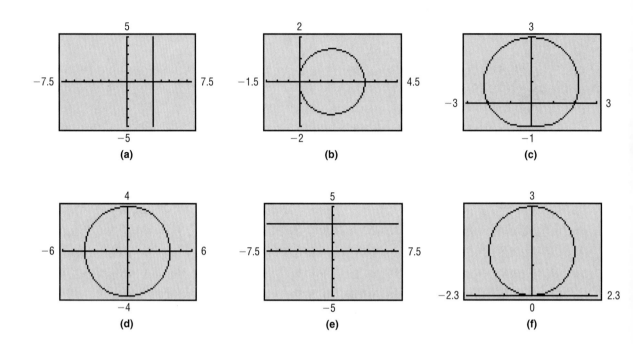

(a)   **(b)**   **(c)**

(d)   **(e)**   **(f)**

In Problems 31–54, identify and graph each polar equation by hand. Be sure to test for symmetry. Verify your results using a graphing utility.

31. $r = 2 + 2 \cos \theta$

32. $r = 1 + \sin \theta$

33. $r = 3 - 3 \sin \theta$

34. $r = 2 - 2 \cos \theta$

35. $r = 2 + \sin \theta$

36. $r = 2 - \cos \theta$

37. $r = 4 - 2 \cos \theta$

38. $r = 4 + 2 \sin \theta$

39. $r = 1 + 2 \sin \theta$

40. $r = 1 - 2 \sin \theta$

41. $r = 2 - 3 \cos \theta$

42. $r = 2 + 4 \cos \theta$

43. $r = 3 \cos 2\theta$

44. $r = 2 \sin 2\theta$

45. $r = 4 \sin 3\theta$

46. $r = 3 \cos 4\theta$

47. $r^2 = 9 \cos 2\theta$

48. $r^2 = \sin 2\theta$

49. $r = 2^\theta$

50. $r = 3^\theta$

51. $r = 1 - \cos \theta$

52. $r = 3 + \cos \theta$

53. $r = 1 - 3 \cos \theta$

54. $r = 4 \cos 3\theta$

In Problems 55–64, graph each polar equation by hand. Verify your results using a graphing utility.

55. $r = \dfrac{2}{1 - \cos \theta}$ (parabola)

56. $r = \dfrac{2}{1 - 2 \cos \theta}$ (hyperbola)

57. $r = \dfrac{1}{3 - 2 \cos \theta}$ (ellipse)

58. $r = \dfrac{1}{1 - \cos \theta}$ (parabola)

59. $r = \theta, \quad \theta \geq 0$ (spiral of Archimedes)

60. $r = \dfrac{3}{\theta}$ (reciprocal spiral)

61. $r = \csc \theta - 2, \quad 0 < \theta < \pi$ (conchoid)

62. $r = \sin \theta \tan \theta$ (cissoid)

63. $r = \tan \theta$ (kappa curve)

64. $r = \cos \dfrac{\theta}{2}$

65. Show that the graph of the equation $r \sin \theta = a$ is a horizontal line a units above the pole if $a > 0$ and $|a|$ units below the pole if $a < 0$.

66. Show that the graph of the equation $r \cos \theta = a$ is a vertical line a units to the right of the pole if $a > 0$ and $|a|$ units to the left of the pole if $a < 0$.

67. Show that the graph of the equation $r = 2a \sin \theta$, $a > 0$, is a circle of radius a with center at $(0, a)$ in rectangular coordinates.

68. Show that the graph of the equation $r = -2a \sin \theta, a > 0$, is a circle of radius a with center at $(0, -a)$ in rectangular coordinates.

69. Show that the graph of the equation $r = 2a \cos \theta$ $a > 0$, is a circle of radius a with center at $(a, 0)$ in rectangular coordinates.

70. Show that the graph of the equation $r = -2a \cos \theta, a > 0$, is a circle of radius a with center at $(-a, 0)$ in rectangular coordinates.

71. Explain why the following test for symmetry is valid: Replace r by $-r$ and θ by $-\theta$ in a polar equation. If an equivalent equation results, the graph is symmetric with respect to the line $\theta = \pi/2$ (*y*-axis).
  (a) Show that the test on page 609 fails for $r^2 = \cos \theta$, but this new test works.
  (b) Show that the test on page 609 works for $r^2 = \sin \theta$, yet this new test fails.

72. Develop a new test for symmetry with respect to the pole.
  (a) Find a polar equation for which this new test fails, yet the test on page 609 works.
  (b) Find a polar equation for which the test on page 609 fails, yet the new test works.

73. Write down two different tests for symmetry with respect to the polar axis. Find examples in which one works and the other fails. Which test do you prefer to use? Justify your position.

8.3 THE COMPLEX PLANE; DEMOIVRE'S THEOREM

  1 Convert a Complex Number from Rectangular Form to Polar Form
  2 Plot Points in the Complex Plane
  3 Find Products and Quotients of Complex Numbers in Polar Form
  4 Use DeMoivre's Theorem
  5 Find Complex Roots

When we first introduced complex numbers, we were not prepared to give a geometric interpretation of a complex number. Now we are ready. Although there are several such interpretations we could give, the one that follows is the easiest to understand.

  A complex number $z = x + yi$ can be interpreted geometrically as the point (x, y) in the *xy*-plane. Thus, each point in the plane corresponds to a complex number and, conversely, each complex number corresponds to a point in the plane. We shall refer to the collection of such points as the **complex plane.** The *x*-axis will be referred to as the **real axis,** because any point that lies on the real axis is of the form $z = x + 0i = x$, a real number. The *y*-axis is called the **imaginary axis,** because any point that lies on it is of the form $z = 0 + yi = yi$, a pure imaginary number. See Figure 35.

FIGURE 35
Complex plane

Let $z = x + yi$ be a complex number. The **magnitude** or **modulus** of z, denoted by $|z|$, is defined as the distance from the origin to the point (x, y). Thus,

$$|z| = \sqrt{x^2 + y^2} \qquad (1)$$

FIGURE 36

See Figure 36 for an illustration.

This definition for $|z|$ is consistent with the definition for the absolute value of a real number: If $z = x + yi$ is real, then $z = x + 0i$ and

$$|z| = \sqrt{x^2 + 0^2} = \sqrt{x^2} = |x|$$

Recall (Section 3.6) that if $z = x + yi$, then its **conjugate,** denoted by $\overline{z}$, is $\overline{z} = x - yi$. Because $z\overline{z} = x^2 + y^2$, it follows from equation (1) that the magnitude of z can be written as

$$|z| = \sqrt{z\overline{z}} \qquad\qquad (2)$$

Polar Form of a Complex Number

When a complex number is written in the form $z = x + yi$, we say that it is in **rectangular,** or **Cartesian, form,** because (x, y) are the rectangular coordinates of the corresponding point in the complex plane. Suppose that (r, θ) are the polar coordinates of this point. Then

$$x = r \cos \theta \qquad y = r \sin \theta \qquad\qquad (3)$$

If $r \geq 0$ and $0 \leq \theta < 2\pi$, the complex number $z = x + yi$ may be written in **polar form** as

$$z = x + yi = (r \cos \theta) + (r \sin \theta)i = r(\cos \theta + i \sin \theta) \qquad (4)$$

FIGURE 37

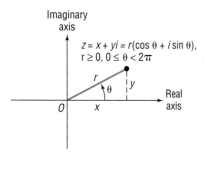

See Figure 37.

If $z = r(\cos \theta + i \sin \theta)$ is the polar form of a complex number, the angle $\theta, 0 \leq \theta < 2\pi$, is called the **argument of z.** Also, because $r \geq 0$, from equation (3) we have $r = \sqrt{x^2 + y^2}$. Thus, from equation (1) it follows that the magnitude of $z = r(\cos \theta + i \sin \theta)$ is

$$|z| = r$$

E X A M P L E   1

Plotting a Point in the Complex Plane and Writing a Complex Number in Polar Form

Plot the point corresponding to $z = \sqrt{3} - i$ in the complex plane, and write an expression for z in polar form.

Mapping Indianapolis

Supplies needed: Graph paper with squares 1/4 inch or 1 cm. or larger per side; compass, protractor, ruler; graphing calculator; map of Indianapolis (from any atlas)

Unlike many major American cities which grew up next to large bodies of water and hence were limited in their ability to spread out in all directions, Indianapolis is surrounded by land on all sides. Major roads radiate out from the city center and intersect the ring road (Rt. 465). A map of the city looks somewhat like a rectangular grid overlaying a polar coordinate system.

The Mission Possible Team has been called to Indianapolis by a car rental company that wishes to create a computer system of interactive maps to be installed in their vehicles. The customers renting their cars would be able to indicate to the computer where they would like to go and the computer screen would then show them a map giving them the quickest route from their present location to their chosen destination. Today's project represents the beginning of a study to create such a computer system.

As a first step, your team needs to set up a polar coordinate system. Using a simplified model, we designate the intersection of Meridian St. and Tenth St. as the pole. The ring road is (roughly) a circle 7 miles out from the center.

1. On graph paper sketch a large polar model of the streets of Indianapolis, using 1 square of the graph paper to represent 1 mile. The ring road would be represented by a circle with radius 7 miles. Rt. 40 is the road that intersects the ring road due east of the pole, so we will designate Rt. 40 as 0°. The other roads and their approximate angle from Rt. 40 are as follows:

Rt. 36: 30°	Rt. 65: 135°	Mann Rd.: 240°
Rt. 37: 60°	Rt. 74: 165°	Rt. 31 (Meridian St.): 270°
Rt. 31 (Meridian St.): 90°	Tenth St.: 180°	Rt. 65: 300°
Rt. 421: 105°	Rt. 70: 210°	Rt. 421: 330°

2. All but 3 of these angles are multiples of 30°. What is the greatest common factor of *all* the angles?

3. In order to establish a mathematical structure for the computer map, your team decides to create a polar equation of the petalled rose variety that would pass through all these points of intersection with the ring road and connect each to the center. What would the equation be?

4. Sketch the graph of your rose on the polar map you created for question 1. By using your graphing calculator, you can determine the order in which the petals are drawn. Number the petals on your sketch to indicate that order.

5. Find the *xy* equation determined by taking your rose equation and replacing the *r* by *y* and the θ by *x*. (Example: If your polar equation were $r = 5 \cos 7\theta$, your corresponding rectangular equation would be $y = 5 \cos 7x$. Warning: These two equations are not equivalent in the sense that they will not give the same graph.) Sketch the resulting rectangular equation in the *xy*-plane, using the domain $0 \le x \le 2\pi/3$. Label the maximum and minimum points with the numbers that correspond to the rose petals of your polar graph.

6. If your original sketch for question 1 was done on graph paper, you can now see the lines of the graph paper as representing a rectangular grid overlaying the polar graph. If not, you will need to sketch vertical and horizontal lines on your polar graph to represent the major city streets. Again, for purposes of simplifying the problem, the major streets are assumed to be occurring at 1 mile intervals.

7. Suppose that travel on the Interstate (Rt. 465, the ring road) has an average speed of 55 mph and travel on city streets has an average speed of 20 mph. Estimate the times it would take to go from the intersection of Rt. 74 and Rt. 465 to the intersection of Rt. 36 and Rt. 465 by each route. (Recall that traveling on city streets limits you to vertical and horizontal grid lines.)

8. What would be the longest time it would take to go from any intersection to another on Rt. 465, assuming you chose the shortest distance along the circle? How long would it take to go between those same two points using the city streets?

FIGURE 38

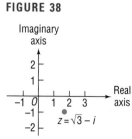

Solution The point corresponding to $z = \sqrt{3} - i$ has the rectangular coordinates $(\sqrt{3}, -1)$. The point, located in quadrant IV, is plotted in Figure 38. Because $x = \sqrt{3}$ and $y = -1$, it follows that

$$r = \sqrt{x^2 + y^2} = \sqrt{(\sqrt{3})^2 + (-1)^2} = \sqrt{4} = 2$$

and

$$\sin \theta = \frac{y}{r} = \frac{-1}{2} \qquad \cos \theta = \frac{x}{r} = \frac{\sqrt{3}}{2}, \qquad 0 \le \theta < 2\pi$$

Thus, $\theta = 11\pi/6$ and $r = 2$, so the polar form of $z = \sqrt{3} - i$ is

$$z = r(\cos \theta + i \sin \theta) = 2\left(\cos \frac{11\pi}{6} + i \sin \frac{11\pi}{6}\right)$$

Now work Problem 1.

E X A M P L E   2

Plotting a Point in the Complex Plane and Converting from Polar to Rectangular Form

Plot the point corresponding to $z = 2(\cos 30° + i \sin 30°)$ in the complex plane, and write an expression for z in rectangular form.

FIGURE 39

Solution To plot the complex number $z = 2(\cos 30° + i \sin 30°)$, we plot the point whose polar coordinates are $(r, \theta) = (2, 30°)$, as shown in Figure 39. In rectangular form,

$$z = 2(\cos 30° + i \sin 30°) = 2\left(\frac{\sqrt{3}}{2} + \frac{1}{2}i\right) = \sqrt{3} + i$$

Now work Problem 13.

3   The polar form of a complex number provides an alternative for finding products and quotients of complex numbers.

Theorem

Let $z_1 = r_1(\cos \theta_1 + i \sin \theta_1)$ and $z_2 = r_2(\cos \theta_2 + i \sin \theta_2)$ be two complex numbers. Then

$$z_1 z_2 = r_1 r_2 [\cos(\theta_1 + \theta_2) + i \sin(\theta_1 + \theta_2)] \qquad (5)$$

If $z_2 \ne 0$, then

$$\frac{z_1}{z_2} = \frac{r_1}{r_2} [\cos(\theta_1 - \theta_2) + i \sin(\theta_1 - \theta_2)] \qquad (6)$$

Proof We will prove formula (5). The proof of formula (6) is left as an exercise (see Problem 56).

$$z_1 z_2 = [r_1(\cos\theta_1 + i\sin\theta_1)][r_2(\cos\theta_2 + i\sin\theta_2)]$$
$$= r_1 r_2[(\cos\theta_1 + i\sin\theta_1)(\cos\theta_2 + i\sin\theta_2)]$$
$$= r_1 r_2[(\cos\theta_1\cos\theta_2 - \sin\theta_1\sin\theta_2) + i(\sin\theta_1\cos\theta_2 + \cos\theta_1\sin\theta_2)]$$
$$= r_1 r_2[(\cos(\theta_1 + \theta_2) + i\sin(\theta_1 + \theta_2)]$$ ▬

Because the magnitude of a complex number z is r and its argument is θ, when $z = r(\cos\theta + i\sin\theta)$, we can restate this theorem as follows:

Theorem

The magnitude of the product (quotient) of two complex numbers equals the product (quotient) of their magnitudes; the argument of the product (quotient) of two complex numbers equals the sum (difference) of their arguments.

▬

Let's look at an example of how this theorem can be used.

E X A M P L E   3

Finding Products and Quotients of Complex Numbers in Polar Form

If $z = 3(\cos 20° + i\sin 20°)$ and $w = 5(\cos 100° + \sin 100°)$, find the following (leave your answers in polar form):

(a) zw (b) z/w

Solution   (a) $zw = [3(\cos 20° + i\sin 20°)][5(\cos 100° + i\sin 100°)]$
$$= (3 \cdot 5)[\cos(20° + 100°) + i\sin(20° + 100°)]$$
$$= 15(\cos 120° + i\sin 120°)$$

(b) $\dfrac{z}{w} = \dfrac{3(\cos 20° + i\sin 20°)}{5(\cos 100° + i\sin 100°)}$

$$= \tfrac{3}{5}[\cos(20° - 100°) + i\sin(20° - 100°)]$$

$$= \tfrac{3}{5}[\cos(-80°) + i\sin(-80°)]$$

$$= \tfrac{3}{5}(\cos 280° + i\sin 280°) \quad \text{Argument must lie between 0° and 360°.}$$ ▬

Now work Problem 23.

Demoivre's Theorem

4 Demoivre's Theorem, stated by Abraham Demoivre (1667–1754) in 1730, but already known to many people by 1710, is important for the following reason: The fundamental processes of algebra are the four operations of addition, subtraction, multiplication, and division, together with powers and the extraction of roots. Demoivre's Theorem allows these latter fundamental algebraic operations to be applied to complex numbers.

$$z_k = \sqrt[n]{r}\left[\cos\left(\frac{\theta}{n} + \frac{2k\pi}{n}\right) + i\sin\left(\frac{\theta}{n} + \frac{2k\pi}{n}\right)\right] \qquad (8)$$

where $k = 0, 1, 2, \ldots, n - 1$.

Proof (Outline) We will not prove this result in its entirety. Instead, we shall show only that each z_k in equation (8) obeys the equation $z_k^n = w$ and, hence, each z_k is a complex nth root of w.

$$z_k^n = \left\{\sqrt[n]{r}\left[\cos\left(\frac{\theta}{n} + \frac{2k\pi}{n}\right) + i\sin\left(\frac{\theta}{n} + \frac{2k\pi}{n}\right)\right]\right\}^n$$

$$= (\sqrt[n]{r})^n\left[\cos n\left(\frac{\theta}{n} + \frac{2k\pi}{n}\right) + i\sin n\left(\frac{\theta}{n} + \frac{2k\pi}{n}\right)\right] \quad \text{Demoivre's Theorem.}$$

$$= r[\cos(\theta + 2k\pi) + i\sin(\theta + 2k\pi)]$$

$$= r(\cos\theta + i\sin\theta) = w$$

Thus, each z_k, $k = 0, 1, \ldots, n - 1$, is a complex nth root of w. To complete the proof, we would need to show that each z_k, $k = 0, 1, 2, \ldots, n - 1$, is, in fact, distinct and that there are no complex nth roots of w other than those given by equation (8). ∎

E X A M P L E   7

Finding Complex Cube Roots

Find the complex cube roots of $-1 + \sqrt{3}\,i$. Leave your answers in polar form, with θ in degrees.

Solution First, we express $-1 + \sqrt{3}\,i$ in polar form using degrees:

$$-1 + \sqrt{3}\,i = 2\left(-\frac{1}{2} + \frac{\sqrt{3}}{2}i\right) = 2(\cos 120° + i\sin 120°)$$

The three complex cube roots of $-1 + \sqrt{3}\,i = 2(\cos 120° + i\sin 120°)$ are

$$z_k = \sqrt[3]{2}\left[\cos\left(\frac{120°}{3} + \frac{360°k}{3}\right) + i\sin\left(\frac{120°}{3} + \frac{360°k}{3}\right)\right], \qquad k = 0, 1, 2$$

Thus,

$$z_0 = \sqrt[3]{2}(\cos 40° + i\sin 40°)$$

$$z_1 = \sqrt[3]{2}(\cos 160° + i\sin 160°)$$

$$z_2 = \sqrt[3]{2}(\cos 280° + i\sin 280°)$$

Notice that each of the three complex cube roots of $-1 + \sqrt{3}\,i$ has the same magnitude, $\sqrt[3]{2}$. This means that the points corresponding to each cube root lie the same distance from the origin; hence, the three points lie on a circle with center at the origin and radius $\sqrt[3]{2}$. Furthermore, the arguments of these cube roots are $40°, 160°$, and $280°$, the difference of consecutive pairs being $120°$. This means that the three points are equally spaced on the circle, as shown in Figure 40. These results are not coincidental. In fact, you are asked to show that these results hold for complex *n*th roots in Problems 53 through 55.

FIGURE 40

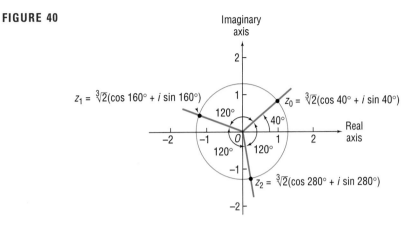

Now work Problem 43.

HISTORICAL FEATURE   The Babylonians, Greeks, and Arabs considered square roots of negative quantities to be impossible and equations with complex solutions to be unsolvable. The first hint that there was some connection between real solutions of equations and complex numbers came when Girolamo Cardano (1501–1576) and Tartaglia (1499–1557) found *real* roots of cubic equations by taking cube roots of *complex* quantities. For centuries thereafter, mathematicians worked with complex numbers without much belief in their actual existence. In 1673, John Wallis appears to have been the first to suggest the graphical representation of complex numbers, a truly significant idea that was not pursued further until about 1800. Several people, including Karl Friedrich Gauss (1777–1855), then rediscovered the idea, and the graphical representation helped to establish complex numbers as equal members of the number family. In practical applications, complex numbers have found their greatest uses in the area of alternating current, where they are a commonplace tool, and subatomic physics.

HISTORICAL PROBLEMS   1.   The quadratic formula will work perfectly well if the coefficients are complex numbers. Solve the following, using Demoivre's Theorem where necessary.

[**Hint:** The answers are "nice."]

(a) $z^2 - (2 + 5i)z - 3 + 5i = 0$ (b) $z^2 - (1 + i)z - 2 - i = 0$

8.3   EXERCISES

In Problems 1–12, plot each complex number in the complex plane and write it in polar form. Express the argument in degrees.

1. $1 + i$
2. $-1 + i$
3. $\sqrt{3} - i$
4. $1 - \sqrt{3}\,i$
5. $-3i$
6. -2
7. $4 - 4i$
8. $9\sqrt{3} + 9i$
9. $3 - 4i$
10. $2 + \sqrt{3}\,i$
11. $-2 + 3i$
12. $\sqrt{5} - i$

In Problems 13–22, write each complex number in rectangular form.

13. $2(\cos 120° + i \sin 120°)$

14. $3(\cos 210° + i \sin 210°)$

15. $4\left(\cos \dfrac{7\pi}{4} + i \sin \dfrac{7\pi}{4}\right)$

16. $2\left(\cos \dfrac{5\pi}{6} + i \sin \dfrac{5\pi}{6}\right)$

17. $3\left(\cos \dfrac{3\pi}{2} + i \sin \dfrac{3\pi}{2}\right)$

18. $4\left(\cos \dfrac{\pi}{2} + i \sin \dfrac{\pi}{2}\right)$

19. $0.2(\cos 100° + i \sin 100°)$

20. $0.4(\cos 200° + i \sin 200°)$

21. $2\left(\cos \dfrac{\pi}{18} + i \sin \dfrac{\pi}{18}\right)$

22. $3\left(\cos \dfrac{\pi}{10} + i \sin \dfrac{\pi}{10}\right)$

In Problems 23–30, find zw and z/w. Leave your answer in polar form.

23. $z = 2(\cos 40° + i \sin 40°)$
$w = 4(\cos 20° + i \sin 20°)$

24. $z = \cos 120° + i \sin 120°$
$w = \cos 100° + i \sin 100°$

25. $z = 3(\cos 130° + i \sin 130°)$
$w = 4(\cos 270° + i \sin 270°)$

26. $z = 2(\cos 80° + i \sin 80°)$
$w = 6(\cos 200° + i \sin 200°)$

27. $z = 2\left(\cos \dfrac{\pi}{8} + i \sin \dfrac{\pi}{8}\right)$
$w = 2\left(\cos \dfrac{\pi}{10} + i \sin \dfrac{\pi}{10}\right)$

28. $z = 4\left(\cos \dfrac{3\pi}{8} + i \sin \dfrac{3\pi}{8}\right)$
$w = 2\left(\cos \dfrac{9\pi}{16} + i \sin \dfrac{9\pi}{16}\right)$

29. $z = 2 + 2i$
$w = \sqrt{3} - i$

30. $z = 1 - i$
$w = 1 - \sqrt{3}\, i$

In Problems 31–42, write each expression in the standard from a + bi.

31. $[4(\cos 40° + i \sin 40°)]^3$

32. $[3(\cos 80° + i \sin 80°)]^3$

33. $\left[2\left(\cos \dfrac{\pi}{10} + i \sin \dfrac{\pi}{10}\right)\right]^5$

34. $\left[\sqrt{2}\left(\cos \dfrac{5\pi}{16} + i \sin \dfrac{5\pi}{16}\right)\right]^4$

35. $[\sqrt{3}(\cos 10° + i \sin 10°)]^6$

36. $[\tfrac{1}{2}(\cos 72° + i \sin 72°)]^5$

37. $\left[\sqrt{5}\left(\cos \dfrac{3\pi}{16} + i \sin \dfrac{3\pi}{16}\right)\right]^4$

38. $\left[\sqrt{3}\left(\cos \dfrac{5\pi}{18} + i \sin \dfrac{5\pi}{18}\right)\right]^6$

39. $(1 - i)^5$

40. $(\sqrt{3} - i)^6$

41. $(\sqrt{2} - i)^6$

42. $(1 - \sqrt{5}\, i)^8$

In Problems 43–50, find all the complex roots. Leave your answers in polar form with the argument in degrees.

43. The complex cube roots of $1 + i$

44. The complex fourth roots of $\sqrt{3} - i$

45. The complex fourth roots of $4 - 4\sqrt{3}\, i$

46. The complex cube roots of $-8 - 8i$

47. The complex fourth roots of $-16i$

48. The complex cube roots of -8

49. The complex fifth roots of i

50. The complex fifth roots of $-i$

51. Find the four complex roots of unity (1). Plot each one.

52. Find the six complex roots of unity (1). Plot each one.

53. Show that each complex nth root of a nonzero complex number w has the same magnitude.

54. Use the result of Problem 53 to draw the conclusion that each complex nth root lies on a circle with center at the origin. What is the radius of this circle?

55. Refer to Problem 54. Show that the complex nth roots of a nonzero complex number w are equally spaced on the circle.

56. Prove formula (6).

8.4 VECTORS

1. Graph Vectors
2. Find a Position Vector
3. Add and Subtract Vectors
4. Find a Scalar Product and the Magnitude of a Vector
5. Find a Unit Vector

In simple terms, a **vector** (derived from the Latin *vehere,* meaning "to carry") is a quantity that has both magnitude and direction. It is convenient to represent a vector by using an arrow. The length of the arrow represents the **magnitude** of the vector, and the arrowhead indicates the **direction** of the vector.

FIGURE 41

Many quantities in physics can be represented by vectors. For example, the velocity of an aircraft can be represented by an arrow that points in the direction of movement; the length of the arrow represents speed. Thus, if the aircraft speeds up, we lengthen the arrow; if the aircraft changes direction, we introduce an arrow in the new direction. See Figure 41. Based on this representation, it is not surprising that vectors and directed line segments are somehow related.

Directed Line Segments

If P and Q are two distinct points in the xy-plane, there is exactly one line containing both P and Q. The points on that part of the line that joins P to Q, including P and Q, form what is called the **line segment** $\overline{PQ}$. If we order the points so that they proceed from P to Q, we have a **directed line segment** from P to Q, which we denote by $\overrightarrow{PQ}$. In a directed line segment $\overrightarrow{PQ}$, we call P the **initial point** and Q the **terminal point,** as indicated in Figure 42.

FIGURE 42

The magnitude of the directed line segment $\overrightarrow{PQ}$ is the distance from the point P to the point Q; that is, it is the length of the line segment. The direction of $\overrightarrow{PQ}$ is from P to Q. If a vector $\mathbf{v}$* has the same magnitude and the same direction as the directed line segment PQ, then we write

$$\mathbf{v} = \overrightarrow{PQ}$$

---

*Boldface letters will be used to denote vectors, in order to distinguish them from numbers. For handwritten work, an arrow is placed over the letter to signify a vector.

The vector **v** whose magnitude is 0 is called the **zero vector, 0.** The zero vector is assigned no direction.

Two vectors **v** and **w** are **equal,** written

$$\mathbf{v} = \mathbf{w}$$

if they have the same magnitude and the same direction.

For example, the vectors shown in Figure 43 have the same magnitude and the same direction, so they are equal, even though they have different initial points and different terminal points. As a result, we find it useful to think of a vector simply as an arrow, keeping in mind that two arrows (vectors) are equal if they have the same direction and the same magnitude (length).

FIGURE 43

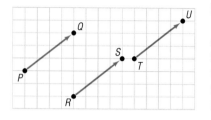

Adding Vectors

FIGURE 44

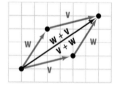

The **sum v + w** of two vectors is defined as follows: We position the vectors **v** and **w** so that the terminal point of **v** coincides with the initial point of **w**, as shown in Figure 44. The vector **v + w** is then the unique vector whose initial point coincides with the initial point of **v** and whose terminal point coincides with the terminal point of **w.**

Vector addition is **commutative.** That is, if **v** and **w** are any two vectors, then

$$\mathbf{v} + \mathbf{w} = \mathbf{w} + \mathbf{v}$$

FIGURE 45

Figure 45 illustrates this fact. (Observe that the commutative property is another way of saying that opposite sides of a parallelogram are equal and parallel.)

Vector addition is also **associative.** That is, if **u, v,** and **w** are vectors, then

$$\mathbf{u} + (\mathbf{v} + \mathbf{w}) = (\mathbf{u} + \mathbf{v}) + \mathbf{w}$$

Figure 46 illustrates the associative property for vectors.

FIGURE 46
(u + v) + w = u + (v + w)

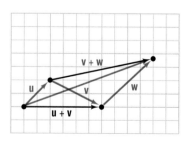

The zero vector has the property that

$$\mathbf{v} + \mathbf{0} = \mathbf{0} + \mathbf{v} = \mathbf{v}$$

FIGURE 47

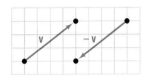

for any vector **v**.

If **v** is a vector, then −**v** is the vector having the same magnitude as **v**, but whose direction is opposite to **v**, as shown in Figure 47.

Furthermore,

$$\mathbf{v} + (-\mathbf{v}) = \mathbf{0}$$

FIGURE 48

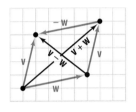

If **v** and **w** are two vectors, we define the **difference v − w** as

$$\mathbf{v} - \mathbf{w} = \mathbf{v} + (-\mathbf{w})$$

Figure 48 illustrates the relationships among **v, w, v + w,** and **v − w.**

Multiplying Vectors by Numbers

When dealing with vectors, we refer to real numbers as **scalars.** Scalars are quantities that have only magnitude. Examples from physics of scalar quantities are temperature, speed, and time. We now define how to multiply a vector by a scalar.

If α is a scalar and **v** is a vector, the **scalar product** α**v** is defined as

1. If $\alpha > 0$, the product α**v** is the vector whose magnitude is α times the magnitude of **v** and whose direction is the same as **v.**
2. If $\alpha < 0$, the product α**v** is the vector whose magnitude is $|\alpha|$ times the magnitude of **v** and whose direction is opposite that of **v.**
3. If $\alpha = 0$ or if **v** = **0,** then α**v** = **0.**

FIGURE 49

See Figure 49 for some illustrations.

For example, if **a** is the acceleration of an object of mass m due to a force **F** being exerted on it, then, by Newton's second law of motion, **F** = m**a.** Here, m**a** is the product of the scalar m and the vector **a.**

Scalar products have the following properties:

$$0\mathbf{v} = \mathbf{0} \qquad 1\mathbf{v} = \mathbf{v} \qquad -1\mathbf{v} = -\mathbf{v}$$
$$(\alpha + \beta)\mathbf{v} = \alpha\mathbf{v} + \beta\mathbf{v} \qquad \alpha(\mathbf{v} + \mathbf{w}) = \alpha\mathbf{v} + \alpha\mathbf{w}$$
$$\alpha(\beta\mathbf{v}) = (\alpha\beta)\mathbf{v}$$

E X A M P L E   1   Graphing Vectors

Use the vectors illustrated in Figure 50 to graph each expression.

FIGURE 50

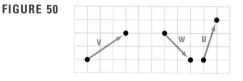

(a) $\mathbf{v} - \mathbf{w}$ (b) $2\mathbf{v} + 3\mathbf{w}$ (c) $2\mathbf{v} - \mathbf{w} + 3\mathbf{u}$

Solution   Figure 51 illustrates each graph.

FIGURE 51

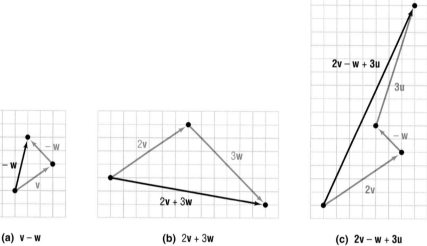

(a) $\mathbf{v} - \mathbf{w}$ (b) $2\mathbf{v} + 3\mathbf{w}$ (c) $2\mathbf{v} - \mathbf{w} + 3\mathbf{u}$

Now work Problems 1 and 3.

Magnitudes of Vectors

If $\mathbf{v}$ is a vector, we use the symbol $\|\mathbf{v}\|$ to represent the **magnitude** of $\mathbf{v}$. Since $\|\mathbf{v}\|$ equals the length of a directed line segment, it follows that $\|\mathbf{v}\|$ has the following properties:

Theorem   Properties of $\|\mathbf{v}\|$

If $\mathbf{v}$ is a vector and if α is a scalar, then

(a) $\|\mathbf{v}\| \geq 0$
(b) $\|\mathbf{v}\| = 0$ if and only if $\mathbf{v} = \mathbf{0}$
(c) $\|-\mathbf{v}\| = \|\mathbf{v}\|$
(d) $\|\alpha \mathbf{v}\| = |\alpha| \, \|\mathbf{v}\|$

Property (a) is a consequence of the fact that distance is a nonnegative number. Property (b) follows, because the length of the directed line segment $\overrightarrow{PQ}$ is positive unless P and Q are the same point, in which case the length is 0. Property (c) follows, because the length of the line segment $\overline{PQ}$ equals the length of the line segment $\overline{QP}$. Property (d) is a direct consequence of the definition of a scalar product.

A vector **u** for which $\|\mathbf{u}\| = 1$ is called a **unit vector.**

To compute the magnitude and direction of a vector, we need an algebraic way of representing vectors.

Representing Vectors in the Plane

We use a rectangular coordinate system to represent vectors in the plane. Let **i** denote a unit vector whose direction is along the positive x-axis; let **j** denote a unit vector whose direction is along the positive y-axis. If **v** is a vector with initial point at the origin O and terminal point at $P = (a, b)$, then we can represent **v** in terms of the vectors **i** and **j** as

$$\mathbf{v} = a\mathbf{i} + b\mathbf{j}$$

FIGURE 52

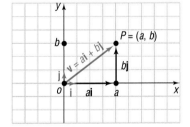

See Figure 52. The scalars a and b are called the **components** of the vector $\mathbf{v} = a\mathbf{i} + b\mathbf{j}$, with a being the component in the direction **i** and b being the component in the direction **j**.

A vector whose initial point is at the origin is called a **position vector.** The next result states that any vector whose initial point is not at the origin is equal to a unique position vector.

Theorem

Suppose that **v** is a vector with initial point $P_1 = (x_1, y_1)$, not necessarily the origin, and terminal point $P_2 = (x_2, y_2)$. If $\mathbf{v} = \overrightarrow{P_1 P_2}$, then **v** is equal to the position vector

$$\mathbf{v} = (x_2 - x_1)\mathbf{i} + (y_2 - y_1)\mathbf{j} \tag{1}$$

To see why this is true, look at Figure 53. Triangle OPA and triangle $P_1 P_2 Q$ are congruent. (Do you see why?) The line segments have the same magnitude, so $d(O, P) = d(P_1, P_2)$; and they have the same direction, so $\angle POA = \angle P_2 P_1 Q$. Since the triangles are right triangles, we have

FIGURE 53

$$\mathbf{v} = a\mathbf{i} + b\mathbf{j} = (x_2 - x_1)\mathbf{i} + (y_2 - y_1)\mathbf{j}$$

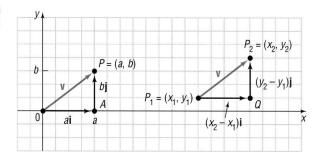

angle–side–angle. Thus, it follows that corresponding sides are equal. As a result, $x_2 - x_1 = a$ and $y_2 - y_1 = b$, so $\mathbf{v}$ may be written as

$$\mathbf{v} = a\mathbf{i} + b\mathbf{j} = (x_2 - x_1)\mathbf{i} + (y_2 - y_1)\mathbf{j} \qquad \blacksquare$$

E X A M P L E   2

2

Finding a Position Vector

Find the position vector of the vector $\mathbf{v} = \overrightarrow{P_1P_2}$ if $P_1 = (-1, 2)$ and $P_2 = (4, 6)$.

Solution

By equation (1), the position vector equal to $\mathbf{v}$ is

$$\mathbf{v} = [4 - (-1)]\mathbf{i} + (6 - 2)\mathbf{j} = 5\mathbf{i} + 4\mathbf{j}$$

FIGURE 54

See Figure 54. $\qquad \blacksquare$

Now work Problem 21.

Two position vectors $\mathbf{v}$ and $\mathbf{w}$ are equal if and only if the terminal point of $\mathbf{v}$ is the same as the terminal point of $\mathbf{w}$. This leads to the following result:

Theorem   Equality of Vectors

Two vectors $\mathbf{v}$ and $\mathbf{w}$ are equal if and only if their corresponding components are equal. That is,

If $\mathbf{v} = a_1\mathbf{i} + b_1\mathbf{j}$ and $\mathbf{w} = a_2\mathbf{i} + b_2\mathbf{j}$,
then $\mathbf{v} = \mathbf{w}$ if and only if $a_1 = a_2$ and $b_1 = b_2$.

Because of the above result, we can replace any vector (directed line segment) by a unique position vector, and vice versa. This flexibility is one of the main reasons for the wide use of vectors. Unless otherwise specified, from now on the term *vector* will mean the unique position vector equal to it.

Next, we define addition, subtraction, scalar product, and magnitude in terms of the components of a vector.

Let $\mathbf{v} = a_1\mathbf{i} + b_1\mathbf{j}$ and $\mathbf{w} = a_2\mathbf{i} + b_2\mathbf{j}$ be two vectors, and let α be a scalar. Then

$$\mathbf{v} + \mathbf{w} = (a_1 + a_2)\mathbf{i} + (b_1 + b_2)\mathbf{j} \qquad (2)$$

$$\mathbf{v} - \mathbf{w} = (a_1 - a_2)\mathbf{i} + (b_1 - b_2)\mathbf{j} \qquad (3)$$

$$\alpha\mathbf{v} = (\alpha a_1)\mathbf{i} + (\alpha b_1)\mathbf{j} \qquad (4)$$

$$\|\mathbf{v}\| = \sqrt{a_1^2 + b_1^2} \qquad (5)$$

These definitions are compatible with the geometric ones given earlier in this section. See Figure 55.

FIGURE 55

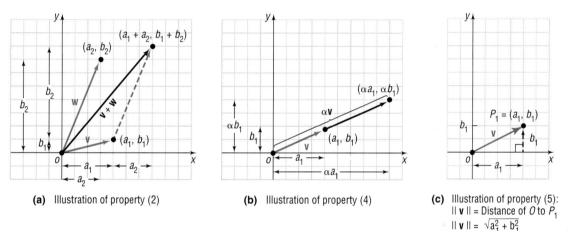

(a) Illustration of property (2)

(b) Illustration of property (4)

(c) Illustration of property (5):
$\|\mathbf{v}\| =$ Distance of O to P_1
$\|\mathbf{v}\| = \sqrt{a_1^2 + b_1^2}$

Thus, to add two vectors, simply add corresponding components. To subtract two vectors, subtract corresponding components.

E X A M P L E   3

3

Adding and Subtracting Vectors

If $\mathbf{v} = 2\mathbf{i} + 3\mathbf{j}$ and $\mathbf{w} = 3\mathbf{i} - 4\mathbf{j}$, find

(a) $\mathbf{v} + \mathbf{w}$ (b) $\mathbf{v} - \mathbf{w}$

Solution   (a) $\mathbf{v} + \mathbf{w} = (2\mathbf{i} + 3\mathbf{j}) + (3\mathbf{i} - 4\mathbf{j}) = (2 + 3)\mathbf{i} + (3 - 4)\mathbf{j}$
$= 5\mathbf{i} - \mathbf{j}$

(b) $\mathbf{v} - \mathbf{w} = (2\mathbf{i} + 3\mathbf{j}) - (3\mathbf{i} - 4\mathbf{j}) = (2 - 3)\mathbf{i} + [3 - (-4)]\mathbf{j}$
$= -\mathbf{i} + 7\mathbf{j}$

E X A M P L E   4

4

Finding Scalar Products and Magnitudes

If $\mathbf{v} = 2\mathbf{i} + 3\mathbf{j}$ and $\mathbf{w} = 3\mathbf{i} - 4\mathbf{j}$, find

(a) $3\mathbf{v}$ (b) $2\mathbf{v} - 3\mathbf{w}$ (c) $\|\mathbf{v}\|$

Solution   (a) $3\mathbf{v} = 3(2\mathbf{i} + 3\mathbf{j}) = 6\mathbf{i} + 9\mathbf{j}$

(b) $2\mathbf{v} - 3\mathbf{w} = 2(2\mathbf{i} + 3\mathbf{j}) - 3(3\mathbf{i} - 4\mathbf{j}) = 4\mathbf{i} + 6\mathbf{j} - 9\mathbf{i} + 12\mathbf{j}$
$$= -5\mathbf{i} + 18\mathbf{j}$$

(c) $\|\mathbf{v}\| = \|2\mathbf{i} + 3\mathbf{j}\| = \sqrt{2^2 + 3^2} = \sqrt{13}$ ▬

Now work Problems 27 and 33.

Recall that a unit vector **u** is one for which $\|\mathbf{u}\| = 1$. In many applications, it is useful to be able to find a unit vector **u** that has the same direction as a given vector **v**.

Theorem   Unit Vector in Direction of **v**

For any nonzero vector **v,** the vector

$$\mathbf{u} = \frac{\mathbf{v}}{\|\mathbf{v}\|}$$

is a unit vector that has the same direction as **v.**

Proof   Let $\mathbf{v} = a\mathbf{i} + b\mathbf{j}$. Then $\|\mathbf{v}\| = \sqrt{a^2 + b^2}$ and

$$\mathbf{u} = \frac{\mathbf{v}}{\|\mathbf{v}\|} = \frac{a\mathbf{i} + b\mathbf{j}}{\sqrt{a^2 + b^2}} = \frac{a}{\sqrt{a^2 + b^2}}\mathbf{i} + \frac{b}{\sqrt{a^2 + b^2}}\mathbf{j}$$

The vector **u** is in the same direction as **v,** since $\|\mathbf{v}\| > 0$. Furthermore

$$\|\mathbf{u}\| = \sqrt{\frac{a^2}{a^2 + b^2} + \frac{b^2}{a^2 + b^2}} = \sqrt{\frac{a^2 + b^2}{a^2 + b^2}} = 1$$

Thus, **u** is a unit vector in the direction of **v.**   ▬

As a consequence of this theorem, if **u** is a unit vector in the same direction as a vector **v,** then **v** may be expressed as

$$\mathbf{v} = \|\mathbf{v}\| \, \mathbf{u} \tag{6}$$

This way of expressing a vector is useful in many applications.

E X A M P L E   5   Finding a Unit Vector

5   Find a unit vector in the same direction as $\mathbf{v} = 4\mathbf{i} - 3\mathbf{j}$.

Solution   We find $\|\mathbf{v}\|$ first:

$$\|\mathbf{v}\| = \|4\mathbf{i} - 3\mathbf{j}\| = \sqrt{16 + 9} = 5$$

Now we multiply **v** by the scalar $1/\|\mathbf{v}\| = \frac{1}{5}$. The result is

$$\frac{\mathbf{v}}{\|\mathbf{v}\|} = \frac{4\mathbf{i} - 3\mathbf{j}}{5} = \frac{4}{5}\mathbf{i} - \frac{3}{5}\mathbf{j}$$

Check: This vector is, in fact, a unit vector because

$$\left(\tfrac{4}{5}\right)^2 + \left(-\tfrac{3}{5}\right)^2 = \tfrac{16}{25} + \tfrac{9}{25} = \tfrac{25}{25} = 1.$$ ▬

Now work Problem 43.

Applications

Forces provide an example of physical quantities that may be conveniently represented by vectors; two forces "combine" the way vectors "add." How do we know this? Laboratory experiments bear it out. Thus, if $\mathbf{F}_1$ and $\mathbf{F}_2$ are two forces simultaneously acting on an object, the vector sum $\mathbf{F}_1 + \mathbf{F}_2$ is the force that produces the same effect on the object as that obtained when the forces $\mathbf{F}_1$ and $\mathbf{F}_2$ act on the object. The force $\mathbf{F}_1 + \mathbf{F}_2$ is sometimes called the **resultant** of $\mathbf{F}_1$ and $\mathbf{F}_2$. See Figure 56.

FIGURE 56

Two important applications of the resultant of two vectors occur with aircraft flying in the presence of a wind and with boats cruising across a river with a current. For example, consider the velocity of wind acting on the velocity of an airplane (see Figure 57). Suppose that $\mathbf{w}$ is a vector describing the velocity of the wind; that is, $\mathbf{w}$ represents the direction and speed of the wind. If $\mathbf{v}$ is the velocity of the airplane in the absence of wind (called its **velocity relative to the air**), then $\mathbf{v} + \mathbf{w}$ is the vector equal to the actual velocity of the airplane (called its **velocity relative to the ground**).

Our next example illustrates this use of vectors in navigation.

FIGURE 57

(a) Velocity $\mathbf{w}$ of wind relative to ground

(b) Velocity $\mathbf{v}$ of airplane relative to air

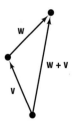

(c) Resultant $\mathbf{w} + \mathbf{v}$ equals velocity of airplane relative to the ground

EXAMPLE 6

Finding the Actual Speed of an Aircraft

A Boeing 737 aircraft maintains a constant airspeed of 500 miles per hour in the direction due south. The velocity of the jet stream is 80 miles per hour in a north-easterly direction.

(a) Find a unit vector having northeast as direction.
(b) Find a vector 80 units in magnitude having the same direction as the unit vector found in part (a).
(c) Find the actual speed of the aircraft relative to the ground.

Solution

FIGURE 58

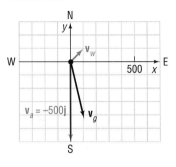

We set up a coordinate system in which north (N) is along the positive y-axis. See Figure 58. Let

$\mathbf{v}_a$ = Velocity of aircraft relative to the air = $-500\mathbf{j}$

$\mathbf{v}_g$ = Velocity of aircraft relative to ground

$\mathbf{v}_w$ = Velocity of jet stream

(a) A vector having northeast as direction is $\mathbf{i} + \mathbf{j}$. The unit vector in this direction is

$$\frac{\mathbf{i} + \mathbf{j}}{\|\mathbf{i} + \mathbf{j}\|} = \frac{\mathbf{i} + \mathbf{j}}{\sqrt{1 + 1}} = \frac{1}{\sqrt{2}}(\mathbf{i} + \mathbf{j})$$

(b) The velocity $\mathbf{v}_w$ of the jet stream is a vector with magnitude 80 in the direction of the unit vector $(1/\sqrt{2})(\mathbf{i} + \mathbf{j})$. Thus, from (6),

$$\mathbf{v}_w = 80\left[\frac{1}{\sqrt{2}}(\mathbf{i} + \mathbf{j})\right] = 40\sqrt{2}(\mathbf{i} + \mathbf{j})$$

(c) The velocity $\mathbf{v}_g$ of the aircraft relative to the ground is the resultant of the vectors $\mathbf{v}_a$ and $\mathbf{v}_w$. Thus,

$$\begin{aligned}\mathbf{v}_g &= \mathbf{v}_a + \mathbf{v}_w = -500\mathbf{j} + 40\sqrt{2}(\mathbf{i} + \mathbf{j}) \\ &= 40\sqrt{2}\mathbf{i} + (40\sqrt{2} - 500)\mathbf{j}\end{aligned}$$

The actual speed (speed relative to the ground) of the aircraft is

$$\|\mathbf{v}_g\| = \sqrt{(40\sqrt{2})^2 + (40\sqrt{2} - 500)^2} \approx 447 \text{ miles per hour} \quad \blacksquare$$

We will find the actual direction of the aircraft in Example 4 of the next section.

HISTORICAL FEATURE The history of vectors is surprisingly complicated for such a natural concept. In the xy-plane, complex numbers do a good job of imitating vectors. About 1840, mathematicians became interested in finding a system that would do for three dimensions what the complex numbers do for two dimensions. Hermann Grassmann (1809–1877), in Germany, and William Rowan Hamilton (1805–1865), in Ireland, both attempted to find solutions.

Hamilton's system was the *quaternions,* which are best thought of as a real number plus a vector, and do for four dimensions what complex numbers do for two dimensions. In this system the order of multiplication matters; that is, $\mathbf{ab} \neq \mathbf{ba}$. Hamilton spent the rest of his life working out quaternion theory and trying to get it accepted in applied mathematics, but he encountered fierce resistance due to the complicated nature of quaternion multiplication. In the work with quaternions, two products of vectors emerged, the scalar (or dot) and the vector (or cross) products.

Grassmann fared even worse than Hamilton; if people did not like Hamilton's work, at least they understood it. Grassmann's abstract style, although easily read today, was almost impenetrable during the previous century, and only a few of his ideas were appreciated. Among those few were the same scalar and vector products that Hamilton had found.

About 1880, the American physicist Josiah Willard Gibbs (1839–1903) worked out an algebra involving only the simplest concepts—the vectors and the two products. He then added some calculus, and the resulting system was simple, flexible, and well adapted to expressing a large number of physical laws. This system remains in use essentially unchanged. Hamilton's and Grassmann's more extensive systems each gave birth to much interesting mathematics, but little of this mathematics is seen at elementary levels. ■

8.4 EXERCISES

In Problems 1–8, use the vectors in the figure to the right to graph each expression.

1. **v + w**
2. **u + v**
3. **3v**
4. **4w**
5. **v − w**
6. **u − v**
7. **3v + u − 2w**
8. **2u − 3v + w**

In Problems 9–16, use the figure below to find each vector.

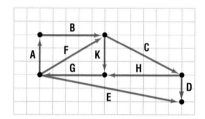

9. **x**, if **x + B = F**
10. **x**, if **x + D = E**
11. **C** in terms of **E, D,** and **F**
12. **G** in terms of **C, D, E,** and **K**
13. **E** in terms of **G, H,** and **D**
14. **E** in terms of **A, B, C,** and **D**
15. **x**, if **x = A + B + K + G**
16. **x**, if **x = A + B + C + H + G**
17. If $\|\mathbf{v}\| = 4$, what is $\|3\mathbf{v}\|$?
18. If $\|\mathbf{v}\| = 2$, what is $\|-4\mathbf{v}\|$?

In Problems 19–26, the vector **v** *has initial point P and terminal point Q. Write* **v** *in the form a**i** + b**j**, that is, find its position vector.*

19. $P = (0, 0); Q = (3, 4)$
20. $P = (0, 0); Q = (-3, -5)$
21. $P = (3, 2); Q = (5, 6)$
22. $P = (-3, 2); Q = (6, 5)$
23. $P = (-2, -1); Q = (6, -2)$
24. $P = (-1, 4); Q = (6, 2)$
25. $P = (1, 0); Q = (0, 1)$
26. $P = (1, 1); Q = (2, 2)$

In Problems 27–32, find $\|\mathbf{v}\|$.

27. $\mathbf{v} = 3\mathbf{i} - 4\mathbf{j}$
28. $\mathbf{v} = -5\mathbf{i} + 12\mathbf{j}$
29. $\mathbf{v} = \mathbf{i} - \mathbf{j}$
30. $\mathbf{v} = -\mathbf{i} - \mathbf{j}$
31. $\mathbf{v} = -2\mathbf{i} + 3\mathbf{j}$
32. $\mathbf{v} = 6\mathbf{i} + 2\mathbf{j}$

In Problems 33–38, find each quantity if $\mathbf{v} = 3\mathbf{i} - 5\mathbf{j}$ *and* $\mathbf{w} = -2\mathbf{i} + 3\mathbf{j}$.

33. $2\mathbf{v} + 3\mathbf{w}$
34. $3\mathbf{v} - 2\mathbf{w}$
35. $\|\mathbf{v} - \mathbf{w}\|$
36. $\|\mathbf{v} + \mathbf{w}\|$
37. $\|\mathbf{v}\| - \|\mathbf{w}\|$
38. $\|\mathbf{v}\| + \|\mathbf{w}\|$

In Problems 39–44, find the unit vector having the same direction as **v**.

39. $\mathbf{v} = 5\mathbf{i}$
40. $\mathbf{v} = -3\mathbf{j}$
41. $\mathbf{v} = 3\mathbf{i} - 4\mathbf{j}$
42. $\mathbf{v} = -5\mathbf{i} + 12\mathbf{j}$
43. $\mathbf{v} = \mathbf{i} - \mathbf{j}$
44. $\mathbf{v} = 2\mathbf{i} - \mathbf{j}$

45. Find a vector **v** whose magnitude is 4 and whose component in the **i** direction is twice the component in the **j** direction.

46. Find a vector **v** whose magnitude is 3 and whose component in the **i** direction is equal to the component in the **j** direction.

47. If **v** = 2**i** − **j** and **w** = x**i** + 3**j**, find all numbers x for which ‖**v** + **w**‖ = 5.

48. If P = (−3, 1) and Q = (x, 4), find all numbers x such that the vector represented by $\overrightarrow{PQ}$ has length 5.

49. **Finding Ground Speed** An airplane has an airspeed of 500 kilometers per hour in an easterly direction. If the wind velocity is 60 kilometers per hour in a northwesterly direction, find the speed of the airplane relative to the ground.

50. **Finding Airspeed** After 1 hour in the air, an airplane arrives at a point 200 miles due south of its departure point. If there was a steady wind of 30 miles per hour from the northwest during the entire flight, what was the average airspeed of the airplane?

51. **Finding Speed without Wind** An airplane travels in a northwesterly direction at a constant speed of 250 miles per hour, due to an easterly wind of 50 miles per hour. How fast would the plane have gone if there had been no wind?

52. **Finding the True Speed of a Motorboat** A small motorboat in still water maintains a speed of 10 miles per hour. In heading directly across a river (that is, perpendicular to the current) whose current is 4 miles per hour, what will be the true speed of the motorboat?

Current

Boat

Direction of boat due to current

53. Show on the graph below the force needed to prevent an object at P from moving.

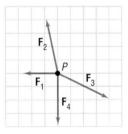

F_2

P

F_1 F_3

F_4

54. Explain in your own words what a vector is. Give an example of a vector.

55. Write a brief paragraph comparing the algebra of complex numbers and the algebra of vectors.

8.5 THE DOT PRODUCT

1 Find the Dot Product of Two Vectors
2 Find the Angle between Two Vectors
3 Find a Vector from Its Magnitude and Direction
4 Determine Whether Two Vectors Are Parallel
5 Determine Whether Two Vectors Are Orthogonal
6 Decompose a Vector into Two Orthogonal Vectors
7 Compute Work

The definition for a product of two vectors is somewhat unexpected. However, such a product has meaning in many geometric and physical applications.

If $\mathbf{v} = a_1\mathbf{i} + b_1\mathbf{j}$ and $\mathbf{w} = a_2\mathbf{i} + b_2\mathbf{j}$ are two vectors, the **dot product $\mathbf{v} \cdot \mathbf{w}$** is defined as

$$\mathbf{v} \cdot \mathbf{w} = a_1 a_2 + b_1 b_2 \qquad (1)$$

E X A M P L E   1

1

Finding Dot Products

If $\mathbf{v} = 2\mathbf{i} - 3\mathbf{j}$ and $\mathbf{w} = 5\mathbf{i} + 3\mathbf{j}$, find

(a) $\mathbf{v} \cdot \mathbf{w}$ (b) $\mathbf{w} \cdot \mathbf{v}$ (c) $\mathbf{v} \cdot \mathbf{v}$
(d) $\mathbf{w} \cdot \mathbf{w}$ (e) $\|\mathbf{v}\|$ (f) $\|\mathbf{w}\|$

Solution

(a) $\mathbf{v} \cdot \mathbf{w} = 2(5) + (-3)3 = 1$ (b) $\mathbf{w} \cdot \mathbf{v} = 5(2) + 3(-3) = 1$
(c) $\mathbf{v} \cdot \mathbf{v} = 2(2) + (-3)(-3) = 13$ (d) $\mathbf{w} \cdot \mathbf{w} = 5(5) + 3(3) = 34$
(e) $\|\mathbf{v}\| = \sqrt{2^2 + (-3)^2} = \sqrt{13}$ (f) $\|\mathbf{w}\| = \sqrt{5^2 + 3^2} = \sqrt{34}$ ■

Since the dot product $\mathbf{v} \cdot \mathbf{w}$ of two vectors $\mathbf{v}$ and $\mathbf{w}$ is a real number (scalar), we sometimes refer to it as the **scalar product.**

Properties

The results obtained in Example 1 suggest some general properties.

Theorem   Properties of Dot Product

If **u, v,** and **w** are vectors, then

Commutative Property

$$\mathbf{u} \cdot \mathbf{v} = \mathbf{v} \cdot \mathbf{u} \qquad (2)$$

Distributive Property

$$\mathbf{u} \cdot (\mathbf{v} + \mathbf{w}) = \mathbf{u} \cdot \mathbf{v} + \mathbf{u} \cdot \mathbf{w} \qquad (3)$$

$$\mathbf{v} \cdot \mathbf{v} = \|\mathbf{v}\|^2 \qquad (4)$$
$$\mathbf{0} \cdot \mathbf{v} = 0 \qquad (5)$$

Proof   We will prove properties (2) and (4) here and leave properties (3) and (5) as exercises (see Problems 29 and 30 at the end of this section).
To prove property (2), we let $\mathbf{u} = a_1\mathbf{i} + b_1\mathbf{j}$ and $\mathbf{v} = a_2\mathbf{i} + b_2\mathbf{j}.$ Then

$$\mathbf{u} \cdot \mathbf{v} = a_1 a_2 + b_1 b_2 = a_2 a_1 + b_2 b_1 = \mathbf{v} \cdot \mathbf{u}$$

To prove property (4), we let $\mathbf{v} = a\mathbf{i} + b\mathbf{j}.$ Then

$$\mathbf{v} \cdot \mathbf{v} = a^2 + b^2 = \|\mathbf{v}\|^2$$ ■

One use of the dot product is to calculate the angle between two vectors.

Angle between Vectors

FIGURE 59

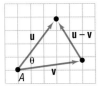

Let **u** and **v** be two vectors with the same initial point A. Then the vectors **u**, **v**, and **u** − **v** form a triangle. The angle θ at vertex A of the triangle is the **angle between the vectors u and v.** See Figure 59. We wish to find a formula for calculating the angle θ.

The sides of the triangle are of lengths $\|\mathbf{v}\|$, $\|\mathbf{u}\|$, and $\|\mathbf{u} - \mathbf{v}\|$, and θ is the included angle between the sides of length $\|\mathbf{v}\|$ and $\|\mathbf{u}\|$. The Law of Cosines (Section 7.3) can be used to find the cosine of the included angle:

$$\|\mathbf{u} - \mathbf{v}\|^2 = \|\mathbf{u}\|^2 + \|\mathbf{v}\|^2 - 2\,\|\mathbf{u}\|\,\|\mathbf{v}\|\cos\theta$$

Now we use property (4) to rewrite this equation in terms of dot products:

$$(\mathbf{u} - \mathbf{v}) \cdot (\mathbf{u} - \mathbf{v}) = \mathbf{u} \cdot \mathbf{u} + \mathbf{v} \cdot \mathbf{v} - 2\,\|\mathbf{u}\|\,\|\mathbf{v}\|\cos\theta \qquad (6)$$

and then apply the distributive property (3) twice on the left side of (6) to obtain

$$
\begin{aligned}
(\mathbf{u} - \mathbf{v}) \cdot (\mathbf{u} - \mathbf{v}) &= \mathbf{u} \cdot (\mathbf{u} - \mathbf{v}) - \mathbf{v} \cdot (\mathbf{u} - \mathbf{v}) \\
&= \mathbf{u} \cdot \mathbf{u} - \mathbf{u} \cdot \mathbf{v} - \mathbf{v} \cdot \mathbf{u} + \mathbf{v} \cdot \mathbf{v} \\
&= \mathbf{u} \cdot \mathbf{u} + \mathbf{v} \cdot \mathbf{v} - 2\,\mathbf{u} \cdot \mathbf{v} \qquad (7) \\
&\qquad\uparrow \\
&\qquad\text{Property (2).}
\end{aligned}
$$

Combining equations (6) and (7), we have

$$\mathbf{u} \cdot \mathbf{u} + \mathbf{v} \cdot \mathbf{v} - 2\,\mathbf{u} \cdot \mathbf{v} = \mathbf{u} \cdot \mathbf{u} + \mathbf{v} \cdot \mathbf{v} - 2\,\|\mathbf{u}\|\,\|\mathbf{v}\|\cos\theta$$

$$\mathbf{u} \cdot \mathbf{v} = \|\mathbf{u}\|\,\|\mathbf{v}\|\cos\theta$$

Thus, we have proved the following result:

Theorem   Angle between Vectors

If **u** and **v** are two nonzero vectors, the angle θ, $0 \le \theta \le \pi$, between **u** and **v** is determined by the formula

$$\cos\theta = \frac{\mathbf{u} \cdot \mathbf{v}}{\|\mathbf{u}\|\,\|\mathbf{v}\|} \qquad (8)$$

E X A M P L E   2   Finding the Angle θ between Two Vectors

Find the angle θ between $\mathbf{u} = 4\mathbf{i} - 3\mathbf{j}$ and $\mathbf{v} = 2\mathbf{i} + 5\mathbf{j}.$

Solution   We compute the quantities $\mathbf{u} \cdot \mathbf{v}$, $\|\mathbf{u}\|$, and $\|\mathbf{v}\|$:

$$\mathbf{u} \cdot \mathbf{v} = 4(2) + (-3)(5) = -7$$
$$\|\mathbf{u}\| = \sqrt{4^2 + (-3)^2} = 5$$
$$\|\mathbf{v}\| = \sqrt{2^2 + 5^2} = \sqrt{29}$$

By formula (8), if θ is the angle between **u** and **v**, then

$$\cos\theta = \frac{\mathbf{u} \cdot \mathbf{v}}{\|\mathbf{u}\|\,\|\mathbf{v}\|} = \frac{-7}{5\sqrt{29}} \approx -0.26$$

Using a calculator, we find that $\theta \approx 105°$. See Figure 60.

FIGURE 60

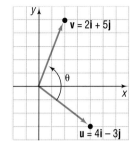

$v = 2i + 5j$

θ

$u = 4i - 3j$

Now work Problem 1.

Writing a Vector in Terms of Its Magnitude and Direction

Many applications describe a vector in terms of its magnitude and direction, rather than in terms of its components. Suppose that we are given the magnitude $\|\mathbf{v}\|$ of a nonzero vector $\mathbf{v}$ and the angle α between $\mathbf{v}$ and $\mathbf{i}$. To express $\mathbf{v}$ in terms of $\|\mathbf{v}\|$ and α, we first find the unit vector $\mathbf{u}$ having the same direction as $\mathbf{v}$:

FIGURE 61

$\mathbf{j}$ $\mathbf{u}$ $(\cos \alpha, \sin \alpha)$

α

$\mathbf{i}$ 1

$$\mathbf{u} = \frac{\mathbf{v}}{\|\mathbf{v}\|} \quad \text{or} \quad \mathbf{v} = \|\mathbf{v}\|\mathbf{u} \tag{9}$$

Look at Figure 61. The coordinates of the terminal point of $\mathbf{u}$ are $(\cos \alpha, \sin \alpha)$. Thus, $\mathbf{u} = \cos \alpha \mathbf{i} + \sin \alpha \mathbf{j}$ and, from (9),

$$\mathbf{v} = \|\mathbf{v}\|(\cos \alpha \mathbf{i} + \sin \alpha \mathbf{j}) \tag{10}$$

E X A M P L E   3

Writing a Vector When Its Magnitude and Direction Are Given

A force $\mathbf{F}$ of 5 pounds is applied in a direction that makes an angle of 30° with the positive x-axis. Express $\mathbf{F}$ in terms of $\mathbf{i}$ and $\mathbf{j}$.

Solution

The magnitude of $\mathbf{F}$ is $\|\mathbf{F}\| = 5$ and the angle between the direction of $\mathbf{F}$ and $\mathbf{i}$, the positive x-axis, is $\alpha = 30°$. Thus, by (10),

$$\mathbf{F} = \|\mathbf{F}\|(\cos \alpha \mathbf{i} + \sin \alpha \mathbf{j}) = 5(\cos 30°\mathbf{i} + \sin 30°\mathbf{j})$$
$$= 5\left(\frac{\sqrt{3}}{2}\mathbf{i} + \frac{1}{2}\mathbf{j}\right) = \frac{5}{2}(\sqrt{3}\,\mathbf{i} + \mathbf{j})$$

E X A M P L E   4

Finding the Actual Direction of an Aircraft

A Boeing 737 aircraft maintains a constant airspeed of 500 miles per hour in the direction due south. The velocity of the jet stream is 80 miles per hour in a north-easterly direction. Find the actual direction of the aircraft relative to the ground.

FIGURE 62

Solution

This is the same information given in Example 6 of Section 8.4. We repeat the figure from that example as Figure 62.

The velocity of the aircraft relative to the air is

$$\mathbf{v}_a = -500\mathbf{j}$$

The wind has magnitude 80 and direction $\alpha = 45°$. Thus, the velocity of the wind is

$$\mathbf{v}_w = 80(\cos 45°\mathbf{i} + \sin 45°\mathbf{j}) = 80\left(\frac{\sqrt{2}}{2}\mathbf{i} + \frac{\sqrt{2}}{2}\mathbf{j}\right)$$
$$= 40\sqrt{2}(\mathbf{i} + \mathbf{j})$$

The velocity of the aircraft relative to the ground is

$$\mathbf{v}_g = \mathbf{v}_a + \mathbf{v}_w = -500\mathbf{j} + 40\sqrt{2}(\mathbf{i} + \mathbf{j}) = 40\sqrt{2}\mathbf{i} + (40\sqrt{2} - 500)\mathbf{j}$$

The angle θ between $\mathbf{v}_g$ and the vector $\mathbf{v}_a = -500\mathbf{j}$ (the velocity of the aircraft relative to the air) is determined by the equation

$$\cos \theta = \frac{\mathbf{v}_g \cdot \mathbf{v}_a}{\|\mathbf{v}_g\| \ \|\mathbf{v}_a\|} = \frac{(40\sqrt{2} - 500)(-500)}{(447)(500)} \approx 0.9920$$
$$\theta \approx 7.2°$$

The direction of the aircraft relative to the ground is about 7.2° east of south. ∎

Now work Problem 19.

Parallel and Orthogonal Vectors

Two vectors $\mathbf{v}$ and $\mathbf{w}$ are said to be **parallel** if there is a nonzero scalar α so that $\mathbf{v} = \alpha\mathbf{w}$. In this case, the angle θ between $\mathbf{v}$ and $\mathbf{w}$ is 0 or π.

E X A M P L E   5   Determining Whether Vectors Are Parallel

[4]

The vectors $\mathbf{v} = 3\mathbf{i} - \mathbf{j}$ and $\mathbf{w} = 6\mathbf{i} - 2\mathbf{j}$ are parallel, since $\mathbf{v} = \frac{1}{2}\mathbf{w}$. Furthermore, since

$$\cos \theta = \frac{\mathbf{v} \cdot \mathbf{w}}{\|\mathbf{v}\| \ \|\mathbf{w}\|} = \frac{18 + 2}{\sqrt{10}\sqrt{40}} = \frac{20}{\sqrt{400}} = 1$$

the angle θ between $\mathbf{v}$ and $\mathbf{w}$ is 0. ∎

If the angle θ between two nonzero vectors $\mathbf{v}$ and $\mathbf{w}$ is $\pi/2$, the vectors $\mathbf{v}$ and $\mathbf{w}$ are called **orthogonal.***

It follows from formula (8) that if $\mathbf{v}$ and $\mathbf{w}$ are orthogonal then $\mathbf{v} \cdot \mathbf{w} = 0$, since $\cos (\pi/2) = 0$.

Orthogonal, perpendicular, and *normal* are all terms that mean "meet at a right angle." It is customary to refer to two vectors as being *orthogonal,* two lines as being *perpendicular,* and a line and a plane or a vector and a plane as being *normal.*

FIGURE 63
$\mathbf{v} \cdot \mathbf{w} = 0$; $\mathbf{v}$ is orthogonal to $\mathbf{w}$

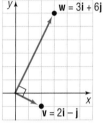

On the other hand, if $\mathbf{v} \cdot \mathbf{w} = 0$, then either $\mathbf{v} = \mathbf{0}$ or $\mathbf{w} = \mathbf{0}$ or $\cos \theta = 0$. In the latter case, $\theta = \pi/2$ and $\mathbf{v}$ and $\mathbf{w}$ are orthogonal. See Figure 63. If $\mathbf{v}$ or $\mathbf{w}$ is the zero vector, then, since the zero vector has no specific direction, we adopt the convention that the zero vector is orthogonal to every vector.

Theorem

Two vectors $\mathbf{v}$ and $\mathbf{w}$ are orthogonal if and only if

$$\mathbf{v} \cdot \mathbf{w} = 0$$

E X A M P L E   6   Determining Whether Two Vectors Are Orthogonal

FIGURE 64

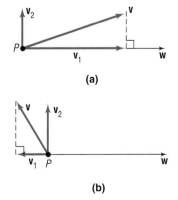

The vectors

$$\mathbf{v} = 2\mathbf{i} - \mathbf{j} \quad \text{and} \quad \mathbf{w} = 3\mathbf{i} + 6\mathbf{j}$$

are orthogonal, since

$$\mathbf{v} \cdot \mathbf{w} = 6 - 6 = 0$$

See Figure 64.

Now work Problem 11.

FIGURE 65

Projection of a Vector onto Another Vector

In many physical applications, it is necessary to find "how much" of a vector is applied in a given direction. Look at Figure 65. The force $\mathbf{F}$ due to gravity is pulling straight down (toward the center of Earth) on the block. To study the effect of gravity on the block, it is necessary to determine how much of $\mathbf{F}$ is actually pushing the block down the incline ($\mathbf{F}_1$) and how much is pressing the block against the incline ($\mathbf{F}_2$), at a right angle to the incline. Knowing the **decomposition** of $\mathbf{F}$ often will allow us to determine when friction is overcome and the block will slide down the incline.

FIGURE 66

(a)

(b)

Suppose that $\mathbf{v}$ and $\mathbf{w}$ are two nonzero vectors with the same initial point P. We seek to decompose $\mathbf{v}$ into two vectors: $\mathbf{v}_1$, which is parallel to $\mathbf{w}$, and $\mathbf{v}_2$, which is orthogonal to $\mathbf{w}$. See Figure 66(a) and 66(b). The vector $\mathbf{v}_1$ is called the **vector projection of v onto w** and is denoted by $\text{proj}_\mathbf{w} \mathbf{v}$.

The vector $\mathbf{v}_1$ is obtained as follows: From the terminal point of $\mathbf{v}$, drop a perpendicular to the line containing $\mathbf{w}$. The vector $\mathbf{v}_1$ is the vector from P to the foot of this perpendicular. The vector $\mathbf{v}_2$ is given by $\mathbf{v}_2 = \mathbf{v} - \mathbf{v}_1$. Note that $\mathbf{v} = \mathbf{v}_1 + \mathbf{v}_2$, $\mathbf{v}_1$ is parallel to $\mathbf{w}$, and $\mathbf{v}_2$ is orthogonal to $\mathbf{w}$. This is the decomposition of $\mathbf{v}$ that we wanted.

Now we seek a formula for $\mathbf{v}_1$ that is based on a knowledge of the vectors $\mathbf{v}$ and $\mathbf{w}$. Since $\mathbf{v} = \mathbf{v}_1 + \mathbf{v}_2$, we have

$$\mathbf{v} \cdot \mathbf{w} = (\mathbf{v}_1 + \mathbf{v}_2) \cdot \mathbf{w} = \mathbf{v}_1 \cdot \mathbf{w} + \mathbf{v}_2 \cdot \mathbf{w} \tag{11}$$

Since $\mathbf{v}_2$ is orthogonal to $\mathbf{w}$, we have $\mathbf{v}_2 \cdot \mathbf{w} = 0$. Since $\mathbf{v}_1$ is parallel to $\mathbf{w}$, we have $\mathbf{v}_1 = \alpha \mathbf{w}$ for some scalar α. Thus, equation (11) can be written as

$$\mathbf{v} \cdot \mathbf{w} = \alpha \mathbf{w} \cdot \mathbf{w} = \alpha \|\mathbf{w}\|^2$$

$$\alpha = \frac{\mathbf{v} \cdot \mathbf{w}}{\|\mathbf{w}\|^2}$$

Thus,

$$\mathbf{v}_1 = \alpha \mathbf{w} = \frac{\mathbf{v} \cdot \mathbf{w}}{\|\mathbf{w}\|^2}\mathbf{w}$$

Theorem

If $\mathbf{v}$ and $\mathbf{w}$ are two nonzero vectors, the vector projection of $\mathbf{v}$ onto $\mathbf{w}$ is

$$\text{proj}_{\mathbf{w}}\mathbf{v} = \frac{\mathbf{v} \cdot \mathbf{w}}{\|\mathbf{w}\|^2}\mathbf{w}$$

The decomposition of $\mathbf{v}$ into $\mathbf{v}_1$ and $\mathbf{v}_2$, where $\mathbf{v}_1$ is parallel to $\mathbf{w}$ and $\mathbf{v}_2$ is perpendicular to $\mathbf{w}$, is

$$\mathbf{v}_1 = \text{proj}_{\mathbf{w}}\mathbf{v} = \frac{\mathbf{v} \cdot \mathbf{w}}{\|\mathbf{w}\|^2}\mathbf{w} \qquad \mathbf{v}_2 = \mathbf{v} - \mathbf{v}_1 \qquad (12)$$

E X A M P L E   7   Decomposing a Vector into Two Orthogonal Vectors

FIGURE 67

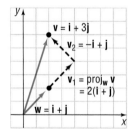

Find the vector projection of $\mathbf{v} = \mathbf{i} + 3\mathbf{j}$ onto $\mathbf{w} = \mathbf{i} + \mathbf{j}$. Decompose $\mathbf{v}$ into two vectors $\mathbf{v}_1$ and $\mathbf{v}_2$, where $\mathbf{v}_1$ is parallel to $\mathbf{w}$ and $\mathbf{v}_2$ is orthogonal to $\mathbf{w}$.

Solution   We use formulas (12).

$$\mathbf{v}_1 = \text{proj}_{\mathbf{w}}\mathbf{v} = \frac{\mathbf{v} \cdot \mathbf{w}}{\|\mathbf{w}\|^2}\mathbf{w} = \frac{1 + 3}{(\sqrt{2})^2}\mathbf{w} = 2\mathbf{w} = 2(\mathbf{i} + \mathbf{j})$$

$$\mathbf{v}_2 = \mathbf{v} - \mathbf{v}_1 = (\mathbf{i} + 3\mathbf{j}) - 2(\mathbf{i} + \mathbf{j}) = -\mathbf{i} + \mathbf{j}$$

See Figure 67.

Now work Problem 13.

Work Done by a Constant Force

7   In elementary physics, the **work** W done by a constant force $\mathbf{F}$ in moving an object from a point A to a point B is defined as

$$W = (\text{Magnitude of force})(\text{Distance}) = \|F\| \, \|\overrightarrow{AB}\|$$

(Work is commonly measured in foot-pounds or in Newton-meters.)

FIGURE 68

In this definition, it is assumed that the force **F** is applied along the line of motion. If the constant force **F** is not along the line of motion, but, instead, is at an angle θ to the direction of motion, as illustrated in Figure 68, then the **work** W **done by F** in moving an object from A to B is defined as

$$W = \mathbf{F} \cdot \overrightarrow{AB} \qquad (13)$$

This definition is compatible with the force times distance definition given above, since

$$W = (\text{Amount of force in direction of } \overrightarrow{AB})(\text{Distance})$$

$$= \|\text{proj}_{\overrightarrow{AB}}\, \mathbf{F}\| \, \|\overrightarrow{AB}\| = \frac{\mathbf{F} \cdot \overrightarrow{AB}}{\|\overrightarrow{AB}\|^2} \|\overrightarrow{AB}\| \, \|\overrightarrow{AB}\| = \mathbf{F} \cdot \overrightarrow{AB}$$

E X A M P L E   8

Computing Work

Find the work done by a force of 5 pounds acting in the direction $\mathbf{i} + \mathbf{j}$ in moving an object 1 foot from $(0, 0)$ to $(1, 0)$.

Solution

First, we must express the force **F** as a vector. The force has magnitude 5 and direction $\mathbf{i} + \mathbf{j}$. The direction of **F** therefore makes an angle of 45° with **i.** Thus, the force **F** is

$$\mathbf{F} = 5(\cos 45°\mathbf{i} + \sin 45°\mathbf{j}) = 5\left(\frac{\sqrt{2}}{2}\mathbf{i} + \frac{\sqrt{2}}{2}\mathbf{j}\right) = \frac{5\sqrt{2}}{2}(\mathbf{i} + \mathbf{j})$$

The line of motion of the object is from $A = (0, 0)$ to $B = (1, 0)$, so $\overrightarrow{AB} = \mathbf{i}.$ The work W is therefore

$$W = \mathbf{F} \cdot \overrightarrow{AB} = \frac{5\sqrt{2}}{2}(\mathbf{i} + \mathbf{j}) \cdot \mathbf{i} = \frac{5\sqrt{2}}{2} \text{ foot-pounds}$$

Now work Problem 25.

E X A M P L E   9

Computing Work

Figure 69(a) shows a girl pulling a wagon with a force of 50 pounds. How much work is done in moving the wagon 100 feet if the handle makes an angle of 30° with the ground?

FIGURE 69

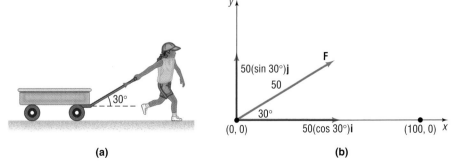

(a)

(b)

In Problems 57–64, the vector **v** *is represented by the directed line segment* $\overrightarrow{PQ}$. *Write* **v** *in the form* $a\mathbf{i} + b\mathbf{j}$ *or in the form* $\mathbf{v} = a\mathbf{i} + b\mathbf{j} + c\mathbf{k}$ *and find* $\|\mathbf{v}\|$.

57. $P = (1, -2);$ $Q = (3, -6)$ **58.** $P = (-3, 1);$ $Q = (4, -2)$ **59.** $P = (0, -2);$ $Q = (-1, 1)$
60. $P = (3, -4);$ $Q = (-2, 0)$ **61.** $P = (6, 2, 1);$ $Q = (3, 0, 2)$ **62.** $P = (4, 7, 0);$ $Q = (0, 5, 6)$
63. $P = (-1, 0, 1);$ $Q = (2, 0, 0)$ **64.** $P = (6, 2, 2);$ $Q = (2, 6, 2)$

In Problems 65–72, use the vectors $\mathbf{v} = -2\mathbf{i} + \mathbf{j}$ *and* $\mathbf{w} = 4\mathbf{i} - 3\mathbf{j}$.

65. Find $4\mathbf{v} - 3\mathbf{w}$ **66.** Find $-\mathbf{v} + 2\mathbf{w}$ **67.** Find $\|\mathbf{v}\|$
68. Find $\|\mathbf{v} + \mathbf{w}\|$ **69.** Find $\|\mathbf{v}\| + \|\mathbf{w}\|$ **70.** Find $\|2\mathbf{v}\| - 3\|\mathbf{w}\|$
71. Find a unit vector in the same direction as **v**.   **72.** Find a unit vector in the opposite direction of **w**.

In Problems 73–80, use the vectors $\mathbf{v} = 3\mathbf{i} + \mathbf{j} - 2\mathbf{k}$ *and* $\mathbf{w} = -3\mathbf{i} + 2\mathbf{j} - \mathbf{k}$.

73. Find $4\mathbf{v} - 3\mathbf{w}$ **74.** Find $-\mathbf{v} + 2\mathbf{w}$ **75.** Find $\|\mathbf{v} - \mathbf{w}\|$
76. Find $\|\mathbf{v} + \mathbf{w}\|$ **77.** Find $\|\mathbf{v}\| - \|\mathbf{w}\|$ **78.** Find $\|\mathbf{v}\| + \|\mathbf{w}\|$
79. Find a unit vector in the same direction as **v** and then in the opposite direction to **v**.
80. Find a unit vector in the same direction as **w** and then in the opposite direction to **w**.

In Problems 81–88, find the dot product $\mathbf{v} \cdot \mathbf{w}$ *and the angle between* **v** *and* **w**.

81. $\mathbf{v} = -2\mathbf{i} + \mathbf{j}, \quad \mathbf{w} = 4\mathbf{i} - 3\mathbf{j}$ **82.** $\mathbf{v} = 3\mathbf{i} - \mathbf{j}, \quad \mathbf{w} = \mathbf{i} + \mathbf{j}$ **83.** $\mathbf{v} = \mathbf{i} - 3\mathbf{j}, \quad \mathbf{w} = -\mathbf{i} + \mathbf{j}$
84. $\mathbf{v} = \mathbf{i} + 4\mathbf{j}, \quad \mathbf{w} = 3\mathbf{i} - 2\mathbf{j}$ **85.** $\mathbf{v} = \mathbf{i} + \mathbf{j} + \mathbf{k}, \quad \mathbf{w} = \mathbf{i} - \mathbf{j} + \mathbf{k}$ **86.** $\mathbf{v} = \mathbf{i} - \mathbf{j} + \mathbf{k}, \quad \mathbf{w} = 2\mathbf{i} + \mathbf{j} + \mathbf{k}$
87. $\mathbf{v} = 4\mathbf{i} - \mathbf{j} + 2\mathbf{k}, \quad \mathbf{w} = \mathbf{i} - 2\mathbf{j} - 3\mathbf{k}$ **88.** $\mathbf{v} = -\mathbf{i} - 2\mathbf{j} + 3\mathbf{k}, \quad \mathbf{w} = 5\mathbf{i} + \mathbf{j} + \mathbf{k}$

89. Find the vector projection of $\mathbf{v} = 2\mathbf{i} + 3\mathbf{j}$ onto $\mathbf{w} = 3\mathbf{i} + \mathbf{j}$.

90. Find the vector projection of $\mathbf{v} = -\mathbf{i} + 2\mathbf{j}$ onto $\mathbf{w} = 3\mathbf{i} - \mathbf{j}$.

91. Find the direction angles of the vector $\mathbf{v} = 3\mathbf{i} - 4\mathbf{j} + 2\mathbf{k}$.

92. Find the direction angles of the vector $\mathbf{v} = \mathbf{i} - \mathbf{j} + 2\mathbf{k}$.

93. **Actual Speed and Direction of a Swimmer** A swimmer can maintain a constant speed of 5 miles per hour. If the swimmer heads directly across a river that has a current moving at the rate of 2 miles per hour, what is the actual speed of the swimmer? (See the figure.) If the river is 1 mile wide, how far downstream will the swimmer end up from the point directly across the river?

Current

Swimmer's
direction

Direction
of swimmer
due to current

94. **Actual Speed and Direction of a Motorboat** A small motorboat is moving at a true speed of 11 miles per hour in a southerly direction. The current is known to be from the northeast at 3 miles per hour. What is the speed of the motorboat relative to the water? In what direction does the compass indicate that the boat is headed?

95. **Correct Direction for Crossing a Stream** A stream 1 kilometer wide has a constant current of 5 kilometers per hour. At what angle to the shore should a person head a boat that is capable of maintaining a constant speed of 15 kilometers per hour in order to reach a point directly opposite?

96. **Actual Speed and Direction of an Airplane** An airplane has an airspeed of 500 kilometers per hour in a northerly direction. The wind velocity is 60 kilometers per hour in a southeasterly direction. Find the actual speed and direction of the plane relative to the ground.

Analytic Geometry

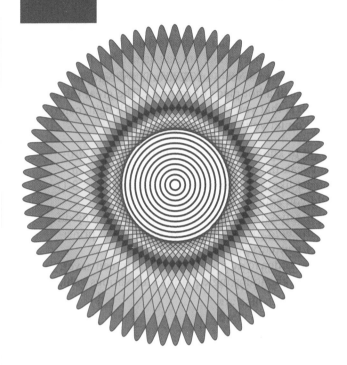

M any computer users enjoy collecting and putting on their computer the latest "screen savers". As a creative entrepreneur, you have come up with an idea for a new piece of software. You will need to use the Sullivan website at:

www.prenhall.com/sullivan

to find the computer programs and information to build your screen saver.

PREPARING FOR THIS CHAPTER

Before getting started on this chapter, review the following concepts:

Distance Formula *(p. 5)*

Completing the Square (Appendix, Section 2)

Intercepts *(pp. 22–24)*

Symmetry *(pp. 24–25)*

Circles *(pp. 29–32)*

Double-Angle and Half-Angle Formulas (Section 6.3)

Polar Coordinates (Section 8.1)

Amplitude and Period of Sinusoidal Graphs *(p. 558)*

OUTLINE

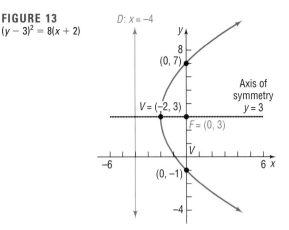

FIGURE 13
$(y - 3)^2 = 8(x + 2)$

Now work Problem 25.

E X A M P L E   8

Using a Graphing Utility to Graph a Parabola, Vertex Not at Origin

Using a graphing utility, graph the equation: $(y - 3)^2 = 8(x + 2)$

Solution   First, we must solve the equation for y.

FIGURE 14

12   $Y_1 = 3 + \sqrt{8(x + 2)}$

-11.6 15.6

-6 $Y_2 = 3 - \sqrt{8(x + 2)}$

$$(y - 3)^2 = 8(x + 2)$$
$$y - 3 = \pm\sqrt{8(x + 2)} \qquad \text{Take the square root of each side.}$$
$$y = 3 \pm \sqrt{8(x + 2)} \qquad \text{Solve for } y.$$

Figure 14 shows the graphs of the equations $Y_1 = 3 + \sqrt{8(x + 2)}$ and $Y_2 = 3 - \sqrt{8(x + 2)}$.

Now work Problem 39.

Polynomial equations define parabolas whenever they involve two variables that are quadratic in one variable and linear in the other. To discuss this type of equation, we first complete the square of the quadratic variable.

E X A M P L E   9

Discussing the Equation of a Parabola

Discuss the equation: $x^2 + 4x - 4y = 0$

Solution   Figure 15(a) shows the graph of $x^2 + 4x - 4y = 0$ using a graphing utility. We now proceed to analyze the equation.

To discuss the equation $x^2 + 4x - 4y = 0$, we complete the square involving the variable x. Thus,

$$x^2 + 4x - 4y = 0$$
$$x^2 + 4x = 4y \qquad \text{Isolate the terms involving } x \text{ on the left side.}$$
$$x^2 + 4x + 4 = 4y + 4 \qquad \text{Complete the square on the left side.}$$
$$(x + 2)^2 = 4(y + 1)$$

This equation is of the form $(x - h)^2 = 4a(y - k)$, with $h = -2$, $k = -1$, and $a = 1$. The graph is a parabola with vertex at $(h, k) = (-2, -1)$ that opens up. The focus is at $(-2, 0)$, and the directrix is the line $y = -2$. See Figure 15(b).

FIGURE 15
$x^2 + 4x - 4y = 0$

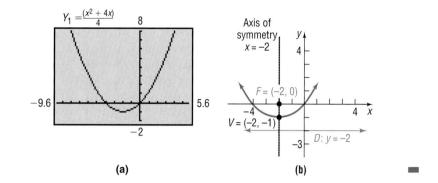

(a)

(b)

Parabolas find their way into many applications. For example, as we discussed in Section 3.1, suspension bridges have cables in the shape of a parabola. Another property of parabolas that is used in applications is their reflecting property.

Reflecting Property

FIGURE 16
Searchlight

Suppose that a mirror is shaped like a **paraboloid of revolution,** a surface formed by rotating a parabola about its axis of symmetry. If a light (or any other emitting source) is placed at the focus of the parabola, all the rays emanating from the light will reflect off the mirror in lines parallel to the axis of symmetry. This principle is used in the design of searchlights, flashlights, certain automobile headlights, and other such devices. See Figure 16.

Conversely, suppose rays of light (or other signals) emanate from a distant source so that they are essentially parallel. When these rays strike the surface of a parabolic mirror whose axis of symmetry is parallel to these rays, they are reflected to a single point at the focus. This principle is used in the design of some solar energy devices, satellite dishes, and the mirrors used in some types of telescopes. See Figure 17.

FIGURE 17

E X A M P L E   10   Satellite Dish

A satellite dish is shaped like a paraboloid of revolution. The signals that emanate from a satellite strike the surface of the dish and are reflected to a single point, where the receiver is located. If the dish is 8 feet across at its opening and is 3 feet deep at its center, at what position should the receiver be placed?

FIGURE 18

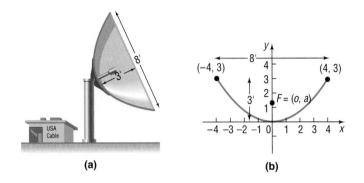

(a)   (b)

Solution   Figure 18(a) shows the satellite dish. We draw the parabola used to form the dish on a rectangular coordinate system so that the vertex of the parabola is at the origin and its focus is on the positive y-axis. See Figure 18(b). The form of the equation of the parabola is

$$x^2 = 4ay$$

and its focus is at $(0, a)$. Since $(4, 3)$ is a point on the graph, we have

$$4^2 = 4a(3)$$

$$a = \frac{4}{3}$$

The receiver should be located $1\frac{1}{3}$ feet from the base of the dish, along its axis of symmetry.   ▬

9.2   EXERCISES

In Problems 1–8, the graph of a parabola is given. Match each graph to its equation.

A. $y^2 = 4x$ B. $x^2 = 4y$ C. $y^2 = -4x$

D. $x^2 = -4y$ E. $(y - 1)^2 = 4(x - 1)$ F. $(x + 1)^2 = 4(y + 1)$

G. $(y - 1)^2 = -4(x - 1)$ H. $(x + 1)^2 = -4(y + 1)$

1.   2.   3.

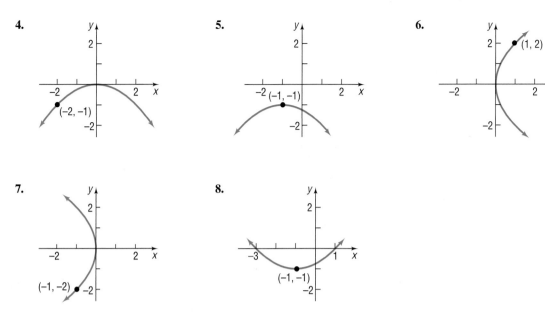

4.

(−2, −1)

5.

(−1, −1)

6.

(1, 2)

7.

(−1, −2)

8.

(−1, −1)

In Problems 9–16, the graph of a parabola is given. Match each graph to its equation.

A. $x^2 = 6y$

B. $x^2 = -6y$

C. $y^2 = 6x$

D. $y^2 = -6x$

E. $(y - 2)^2 = -6(x + 2)$

F. $(y - 2)^2 = 6(x + 2)$

G. $(x + 2)^2 = -6(y - 2)$

H. $(x + 2)^2 = 6(y - 2)$

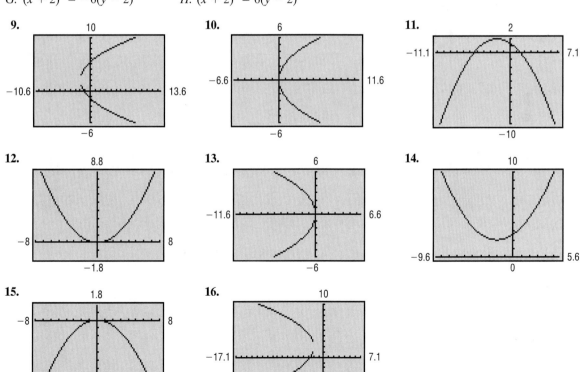

9.
10
−10.6
13.6
−6

10.
6
−6.6
11.6
−6

11.
2
−11.1
7.1
−10

12.
8.8
−8
8
−1.8

13.
6
−11.6
6.6
−6

14.
10
−9.6
5.6
0

15.
1.8
−8
8
−8.8

16.
10
−17.1
7.1
−6

In Problems 17–32, find the equation of the parabola described. Find the two points that define the latus rectum, and graph the equation by hand.

17. Focus at $(4, 0)$; vertex at $(0, 0)$

18. Focus at $(0, 2)$; vertex at $(0, 0)$

19. Focus at $(0, -3)$; vertex at $(0, 0)$

20. Focus at $(-4, 0)$; vertex at $(0, 0)$

21. Focus at $(-2, 0)$; directrix the line $x = 2$

22. Focus at $(0, -1)$; directrix the line $y = 1$

23. Directrix the line $y = -\frac{1}{2}$; vertex at $(0, 0)$

24. Directrix the line $x = -\frac{1}{2}$; vertex at $(0, 0)$

25. Vertex at $(2, -3)$; focus at $(2, -5)$

26. Vertex at $(4, -2)$; focus at $(6, -2)$

27. Vertex at $(0, 0)$; axis of symmetry the y-axis; containing the point $(2, 3)$

28. Vertex at $(0, 0)$; axis of symmetry the x-axis; containing the point $(2, 3)$

29. Focus at $(-3, 4)$; directrix the line $y = 2$

30. Focus at $(2, 4)$; directrix the line $x = -4$

31. Focus at $(-3, -2)$; directrix the line $x = 1$

32. Focus at $(-4, 4)$; directrix the line $y = -2$

In Problems 33–50, find the vertex, focus, and directrix of each parabola. Graph the equation using a graphing utility.

33. $x^2 = 4y$

34. $y^2 = 8x$

35. $y^2 = -16x$

36. $x^2 = -4y$

37. $(y - 2)^2 = 8(x + 1)$

38. $(x + 4)^2 = 16(y + 2)$

39. $(x - 3)^2 = -(y + 1)$

40. $(y + 1)^2 = -4(x - 2)$

41. $(y + 3)^2 = 8(x - 2)$

42. $(x - 2)^2 = 4(y - 3)$

43. $y^2 - 4y + 4x + 4 = 0$

44. $x^2 + 6x - 4y + 1 = 0$

45. $x^2 + 8x = 4y - 8$

46. $y^2 - 2y = 8x - 1$

47. $y^2 + 2y - x = 0$

48. $x^2 - 4x = 2y$

49. $x^2 - 4x = y + 4$

50. $y^2 + 12y = -x + 1$

In Problems 51–58, write an equation for each parabola.

51.

52.

53.

54.

55.

56.

57.

58.

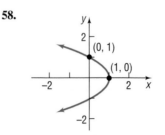

59. Satellite Dish A satellite dish is shaped like a paraboloid of revolution. The signals that emanate from a satellite strike the surface of the dish and are reflected to a single point, where the receiver is located. If the dish is 10 feet across at its opening and is 4 feet deep at is center, at what position should the receiver be placed?

60. Constructing a TV Dish A cable TV receiving dish is in the shape of a paraboloid of revolution. Find the location of the receiver, which is placed at the focus, if the dish is 6 feet across at its opening and 2 feet deep.

61. Constructing a Flashlight The reflector of a flashlight is in the shape of a paraboloid of revolution. Its diameter is 4 inches and its depth is 1 inch. How far from the vertex should the light bulb be placed so that the rays will be reflected parallel to the axis?

62. Constructing a Headlight A sealed-beam headlight is in the shape of a paraboloid of revolution. The bulb, which is placed at the focus, is 1 inch from the vertex. If the depth is to be 2 inches, what is the diameter of the headlight at its opening?

63. Suspension Bridges The cables of a suspension bridge are in the shape of a parabola, as shown in the figure. The towers supporting the cable are 600 feet apart and 80 feet high. If the cables touch the road surface midway between the towers, what is the height of the cable at a point 150 feet from the center of the bridge?

64. Suspension Bridges The cables of a suspension bridge are in the shape of a parabola. The towers supporting the cable are 400 feet apart and 100 feet high. If the cables are at a height of 10 feet midway between the towers, what is the height of the cable at a point 50 feet from the center of the bridge?

65. Searchlights A searchlight is shaped like a paraboloid of revolution. If the light source is located 2 feet from the base along the axis of symmetry and the opening is 5 feet across, how deep should the searchlight be?

66. Searchlights A searchlight is shaped like a paraboloid of revolution. If the light source is located 2 feet from the base along the axis of symmetry and the depth of the searchlight is 4 feet, what should the width of the opening be?

67. Solar Heat A mirror is shaped like a paraboloid of revolution and will be used to concentrate the rays of the sun at its focus, creating a heat source. If the mirror is 20 feet across at its opening and is 6 feet deep, where will the heat source be concentrated?

68. Reflecting Telescopes A reflecting telescope contains a mirror shaped like a paraboloid of revolution. If the mirror is 4 inches across at its opening and is 3 feet deep, where will the light collected be concentrated?

69. Parabolic Arch Bridge A bridge is built in the shape of a parabolic arch. The bridge has a span of 120 feet and a maximum height of 25 feet. See the illustration. Choose a suitable rectangular coordinate system and find the height of the arch at distances of 10, 30, and 50 feet from the center.

70. Parabolic Arch Bridge A bridge is to be built in the shape of a parabolic arch and is to have a span of 100 feet. The height of the arch a distance of 40 feet from the center is to be 10 feet. Find the height of the arch at its center.

71. Show that an equation of the form

$$Ax^2 + Ey = 0 \qquad A \neq 0, E \neq 0$$

is the equation of a parabola with vertex at $(0, 0)$ and axis of symmetry the y-axis. Find its focus and directrix.

72. Show that an equation of the form

$$Cy^2 + Dx = 0 \qquad C \neq 0, D \neq 0$$

is the equation of a parabola with vertex at $(0, 0)$ and axis of symmetry the x-axis. Find its focus and directrix.

73. Show that the graph of an equation of the form

$$Ax^2 + Dx + Ey + F = 0 \qquad A \neq 0$$

(a) Is a parabola if $E \neq 0$.
(b) Is a vertical line if $E = 0$ and $D^2 - 4AF = 0$.
(c) Is two vertical lines if $E = 0$ and $D^2 - 4AF > 0$.

(d) Contains no points if $E = 0$ and $D^2 - 4AF < 0$.

74. Show that the graph of an equation of the form

$$Cy^2 + Dx + Ey + F = 0 \qquad C \neq 0$$

(a) Is a parabola if $D \neq 0$.
(b) Is a horizontal line if $D = 0$ and $E^2 - 4CF = 0$.
(c) Is two horizontal lines if $D = 0$ and $E^2 - 4CF > 0$.
(d) Contains no points if $D = 0$ and $E^2 - 4CF < 0$.

9.3 | THE ELLIPSE

1. Find the Equation of an Ellipse
2. Graph Ellipses
3. Discuss the Equation of an Ellipse
4. Work with Ellipses with Center at (h, k)

> An **ellipse** is the collection of all points in the plane the sum of whose distances from two fixed points, called the **foci**, is a constant.

FIGURE 19

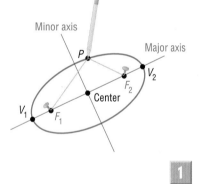

The definition actually contains within it a physical means for drawing an ellipse. Find a piece of string (the length of this string is the constant referred to in the definition). Then take two thumbtacks (the foci) and stick them on a piece of cardboard so that the distance between them is less than the length of the string. Now attach the ends of the string to the thumbtacks and, using the point of a pencil, pull the string taut. Keeping the string taut, rotate the pencil around the two thumbtacks. The pencil traces out an ellipse. as shown in Figure 19.

In Figure 19, the foci are labeled F_1 and F_2. The line containing the foci is called the **major axis.** The midpoint of the line segment joining the foci is called the **center** of the ellipse. The line through the center and perpendicular to the major axis is called the **minor axis.**

The two points of intersection of the ellipse and the major axis are the **vertices,** V_1 and V_2, of the ellipse. The distance from one vertex to the other is called the **length of the major axis.** The ellipse is symmetric with respect to its major axis and with respect to its minor axis.

With these ideas in mind, we are now ready to find the equation of an ellipse in a rectangular coordinate system. First, we place the center of the ellipse at the origin. Second, we position the ellipse so that its major axis coincides with a coordinate axis. Suppose that the major axis coincides with the x-axis, as shown in Figure 20. If c is the distance from the center to a focus, then one focus will be at $F_1 = (-c, 0)$ and the other at $F_2 = (c, 0)$. As we shall see, it is convenient to let $2a$ denote the constant distance referred to in the definition. Thus, if $P = (x, y)$ is any point on the ellipse, we have

FIGURE 20
$$d(F_1, P) + d(F_2, P) = 2a$$

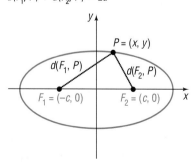

MISSION POSSIBLE

Building a Bridge over the East River

Your team is working for the transportation authority in New York City. You have been asked to study the construction plans for a new bridge over the East River in New York City. The space between supports needs to be 1050 feet; the height at the center of the arch needs to be 350 feet. One company has suggested the support be in the shape of a parabola; another company suggests a semi-ellipse. The engineering team will determine the relative strengths of the two plans; your job is to find out if there are any differences in the channel widths.

An empty tanker needs a 280-foot clearance to pass beneath the bridge. You need to find the width of the channel for each of the two different plans.

1. To determine the equation of a parabola with these characteristics, first place the parabola on coordinate axes in a convenient location and sketch it.
2. What is the equation of the parabola? (If using a decimal in the equation, you may want to carry 6 decimal places. If using a fraction, your answer will be more exact.)
3. How wide is the channel the tanker can pass through if the shape of the support is parabolic?
4. To determine the equation of a semi-ellipse with these characteristics, place the semi-ellipse on coordinate axes in a convenient location and sketch it.
5. What is the equation of the ellipse? How wide is the channel the tanker can pass through?
6. Now that you know which of the two provides the wider channel, consider some other factors. Your department is also in charge of channel depth, amount of traffic on the river, and various other factors. For example, if the river were to flood, and the water level rose by 10 feet, how would the clearances be affected? Make a decision about which plan you think would be the better one as far as your department is concerned, and explain why you think so.

In Problems 19–26, find the center, transverse axis, vertices, foci, and asymptotes. Graph each equation using a graphing utility.

19. $\dfrac{x^2}{25} - \dfrac{y^2}{9} = 1$

20. $\dfrac{y^2}{16} - \dfrac{x^2}{4} = 1$

21. $4x^2 - y^2 = 16$

22. $y^2 - 4x^2 = 16$

23. $y^2 - 9x^2 = 9$

24. $x^2 - y^2 = 4$

25. $y^2 - x^2 = 25$

26. $2x^2 - y^2 = 4$

In Problems 27–30, write an equation for each hyperbola.

27.

28.

29.

30.

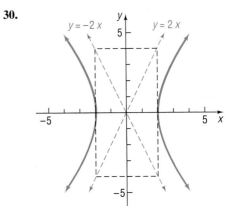

In Problems 31–38, find an equation for the hyperbola described. Graph the equation by hand.

31. Center at $(4, -1)$; focus at $(7, -1)$; vertex at $(6, -1)$

32. Center at $(-3, 1)$; focus at $(-3, 6)$; vertex at $(-3, 4)$

33. Center at $(-3, -4)$; focus at $(-3, -8)$; vertex at $(-3, -2)$

34. Center at $(1, 4)$; focus at $(-2, 4)$; vertex at $(0, 4)$

35. Foci at $(3, 7)$ and $(7, 7)$; vertex at $(6, 7)$

36. Focus at $(-4, 0)$; vertices at $(-4, 4)$ and $(-4, 2)$

37. Vertices at $(-1, -1)$ and $(3, -1)$; asymptote the line $(x - 1)/2 = (y + 1)/3$

38. Vertices at $(1, -3)$ and $(1, 1)$; asymptote the line $(x - 1)/2 = (y + 1)/3$

In Problems 39–52, find the center, transverse axis, vertices, foci, and asymptotes. Graph each equation using a graphing utility.

39. $\dfrac{(x-2)^2}{4} - \dfrac{(y+3)^2}{9} = 1$

40. $\dfrac{(y+3)^2}{4} - \dfrac{(x-2)^2}{9} = 1$

41. $(y-2)^2 - 4(x+2)^2 = 4$

42. $(x+4)^2 - 9(y-3)^2 = 9$

43. $(x+1)^2 - (y+2)^2 = 4$

44. $(y-3)^2 - (x+2)^2 = 4$

45. $x^2 - y^2 - 2x - 2y - 1 = 0$

46. $y^2 - x^2 - 4y + 4x - 1 = 0$

47. $y^2 - 4x^2 - 4y - 8x - 4 = 0$

48. $2x^2 - y^2 + 4x + 4y - 4 = 0$

49. $4x^2 - y^2 - 24x - 4y + 16 = 0$

50. $2y^2 - x^2 + 2x + 8y + 3 = 0$

51. $y^2 - 4x^2 - 16x - 2y - 19 = 0$

52. $x^2 - 3y^2 + 8x - 6y + 4 = 0$

In Problems 53–56, graph each function by hand. Verify your answer using a graphing utility.
[**Hint:** Notice that each function is "half" a hyperbola.]

53. $f(x) = \sqrt{16 + 4x^2}$

54. $f(x) = -\sqrt{9 + 9x^2}$

55. $f(x) = -\sqrt{-25 + x^2}$

56. $f(x) = \sqrt{-1 + x^2}$

57. LORAN Two LORAN stations are positioned 200 miles apart along a straight shore.
(a) A ship records a time difference of 0.00038 second between the LORAN signals. Set up an appropriate rectangular coordinate system to determine where the ship would reach shore if it were to follow the hyperbola corresponding to this time difference.
(b) If the ship wants to enter a harbor located between the two stations 20 miles from the master station, what time difference should it be looking for?
(c) If the ship is 50 miles off shore when the desired time difference is obtained, what is the exact location of the ship?
[**Note:** The speed of each radio signal is 186,000 miles per second.]

58. LORAN Two LORAN stations are positioned 100 miles apart along a straight shore.
(a) A ship records a time difference of 0.00032 second between the LORAN signals. Set up an appropriate rectangular coordinate system to determine where the ship would reach shore if it were to follow the hyperbola corresponding to this time difference.
(b) If the ship wants to enter a harbor located between the two stations 10 miles from the master station, what time difference should it be looking for?
(c) If the ship is 20 miles off shore when the desired time difference is obtained, what is the exact location of the ship?
[**Note:** The speed of each radio signal is 186,000 miles per second.]

59. Calibrating Instruments In a test of their recording devices a team of seismologists positioned two of the devices 2000 feet apart, with the device at point A to the west of the device at point B. At a point between the devices and 200 feet from point B, a small amount of explosive was detonated and a note made of the time at which the sound reached each device. A second explosion is to be carried out at a point directly north of point B.
(a) How far north should the site of the second explosion be chosen so that the measured time difference recorded by the devices for the second detonation is the same as that recorded for the first detonation?
(b) Explain why this experiment can be used to calibrate the instruments.

60. Explain in your own words the LORAN system of navigation.

61. The **eccentricity** e of a hyperbola is defined as the number c/a, where a and c are the numbers given in equation (2). Because $c > a$, it follows that $e > 1$. Describe the general shape of a hyperbola whose eccentricity is close to 1. What is the shape if e is very large?

62. A hyperbola for which $a = b$ is called an **equilateral hyperbola.** Find the eccentricity e of an equilateral hyperbola.
[**Note:** The eccentricity of a hyperbola is defined in Problem 61.]

63. Two hyperbolas that have the same set of asymptotes are called **conjugate.** Show that the hyperbolas
$$\frac{x^2}{4} - y^2 = 1 \quad \text{and} \quad y^2 - \frac{x^2}{4} = 1$$
are conjugate. Graph each hyperbola on the same set of coordinate axes.

64. Prove that the hyperbola
$$\frac{y^2}{a^2} - \frac{x^2}{b^2} = 1$$
has the two oblique asymptotes
$$y = \frac{a}{b}x \quad \text{and} \quad y = -\frac{a}{b}x$$

65. Show that the graph of an equation of the form
$$Ax^2 + Cy^2 + F = 0 \qquad A \neq 0, C \neq 0, F \neq 0$$
where A and C are of opposite sign, is a hyperbola with center at $(0, 0)$.

66. Show that the graph of an equation of the form
$$Ax^2 + Cy^2 + Dx + Ey + F = 0 \quad A \neq 0, C \neq 0$$
where A and C are of opposite sign:
(a) Is a hyperbola if $(D^2/4A) + (E^2/4C) - F \neq 0$.
(b) Is two intersecting lines if
$$(D^2/4A) + (E^2/4C) - F = 0.$$

9.5 ROTATION OF AXES; GENERAL FORM OF A CONIC

1. Identify a Conic
2. Use a Rotation of Axes to Transform Equations
3. Discuss an Equation Using a Rotation of Axes
4. Identify Conics Without a Rotation of Axes

In this section, we show that the graph of a general second-degree polynomial containing two variables x and y, that is, an equation of the form

$$Ax^2 + Bxy + Cy^2 + Dx + Ey + F = 0 \tag{1}$$

where A, B, and C are not simultaneously 0, is a conic. We shall not concern ourselves here with the degenerate cases of equation (1), such as $x^2 + y^2 = 0$, whose graph is a single point $(0, 0)$; or $x^2 + 3y^2 + 3 = 0$, whose graph contains no points; or $x^2 - 4y^2 = 0$, whose graph is two lines, $x - 2y = 0$ and $x + 2y = 0$.

We begin with the case where $B = 0$. In this case, the term containing xy is not present, so equation (1) has the form

$$Ax^2 + Cy^2 + Dx + Ey + F = 0$$

where either $A \neq 0$ or $C \neq 0$.

1   We have already discussed the procedure for identifying the graph of this kind of equation; we complete the squares of the quadratic expressions in x or y, or both. Once this has been done, the conic can be identified by comparing it to one of the forms studied in Sections 9.2 through 9.4.

In fact, though, we can identify the conic directly from the equation without completing the squares.

Theorem   Identifying Conics without Completing the Squares

Excluding degenerate cases, the equation

$$Ax^2 + Cy^2 + Dx + Ey + F = 0 \tag{2}$$

where either $A \neq 0$ or $C \neq 0$:

(a) Defines a parabola if $AC = 0$.
(b) Defines an ellipse (or a circle) if $AC > 0$.
(c) Defines a hyperbola if $AC < 0$.

Proof (a) If $AC = 0$, then either $A = 0$ or $C = 0$, but not both, so the form of equation (2) is either

$$Ax^2 + Dx + Ey + F = 0, \qquad A \neq 0$$

or

$$Cy^2 + Dx + Ey + F = 0, \qquad C \neq 0$$

Using the results of Problems 73 and 74 in Exercise 9.2, it follows that, except for the degenerate cases, the equation is a parabola.

(b) If $AC > 0$, then A and C are of the same sign. Using the results of Problems 73 and 74 in Exercise 9.3, except for the degenerate cases, the equation is an ellipse if $A \neq C$ or a circle if $A = C$.

(c) If $AC < 0$, then A and C are of opposite sign. Using the results of Problems 65 and 66 in Exercise 9.4, except for the degenerate cases, the equation is a hyperbola. ∎

We will not be concerned with the degenerate cases of equation (2). However, in practice, you should be alert to the possibility of degeneracy.

E X A M P L E 1

Identifying a Conic without Completing the Squares

Identify each equation without completing the squares.

(a) $3x^2 + 6y^2 + 6x - 12y = 0$ (b) $2x^2 - 3y^2 + 6y + 4 = 0$

(c) $y^2 - 2x + 4 = 0$

Solution (a) We compare the given equation to equation (2) and conclude that $A = 3$ and $C = 6$. Since $AC = 18 > 0$, the equation is an ellipse.

(b) Here, $A = 2$ and $C = -3$, so $AC = -6 < 0$. The equation is a hyperbola.

(c) Here, $A = 0$ and $C = 1$, so $AC = 0$. The equation is a parabola. ∎

Now work Problem 1.

Although we can now identify the type of conic represented by any equation of the form of equation (2) without completing the squares, we will still need to complete the squares if we desire additional information about a conic.

Now we turn our attention to equations of the form of equation (1), where $B \neq 0$. To discuss this case, we first need to investigate a new procedure: *rotation of axes*.

Rotation of Axes

In a **rotation of axes,** the origin remains fixed while the x-axis and y-axis are rotated through an angle θ to a new position; the new positions of the x- and y-axes are denoted by x' and y', respectively, as shown in Figure 54(a).

Now look at Figure 54(b) on page 712. There the point P has the coordinates (x, y) relative to the xy-plane, while the same point P has coordinates (x', y') relative to the $x'y'$-plane. We seek relationships that will enable us to express x and y in terms of x', y', and θ.

FIGURE 54

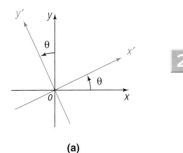

(a)

5. $3x^2 - 2y^2 + 6x + 4 = 0$

6. $4x^2 - 3y^2 - 8x + 6y + 1 = 0$

7. $2y^2 - x^2 - y + x = 0$

8. $y^2 - 8x^2 - 2x - y = 0$

9. $x^2 + y^2 - 8x + 4y = 0$

10. $2x^2 + 2y^2 - 8x + 8y = 0$

In Problems 11–20, determine the appropriate rotation formulas to use so that the new equation contains no xy-term.

11. $x^2 + 4xy + y^2 - 3 = 0$

12. $x^2 - 4xy + y^2 - 3 = 0$

13. $5x^2 + 6xy + 5y^2 - 8 = 0$

14. $3x^2 - 10xy + 3y^2 - 32 = 0$

15. $13x^2 - 6\sqrt{3}xy + 7y^2 - 16 = 0$

16. $11x^2 + 10\sqrt{3}xy + y^2 - 4 = 0$

17. $4x^2 - 4xy + y^2 - 8\sqrt{5}x - 16\sqrt{5}y = 0$

18. $x^2 + 4xy + 4y^2 + 5\sqrt{5}y + 5 = 0$

19. $25x^2 - 36xy + 40y^2 - 12\sqrt{13}x - 8\sqrt{13}y = 0$

20. $34x^2 - 24xy + 41y^2 - 25 = 0$

In Problems 21–32, graph the equation using a graphing utility. Rotate the axes so that the new equation contains no xy-term. Discuss and, by hand, graph the new equation. (Refer to Problems 11–20 for Problems 21–30).

21. $x^2 + 4xy + y^2 - 3 = 0$

22. $x^2 - 4xy + y^2 - 3 = 0$

23. $5x^2 + 6xy + 5y^2 - 8 = 0$

24. $3x^2 - 10xy + 3y^2 - 32 = 0$

25. $13x^2 - 6\sqrt{3}xy + 7y^2 - 16 = 0$

26. $11x^2 + 10\sqrt{3}xy + y^2 - 4 = 0$

27. $4x^2 - 4xy + y^2 - 8\sqrt{5}x - 16\sqrt{5}y = 0$

28. $x^2 + 4xy + 4y^2 + 5\sqrt{5}y + 5 = 0$

29. $25x^2 - 36xy + 40y^2 - 12\sqrt{13}x - 8\sqrt{13}y = 0$

30. $34x^2 - 24xy + 41y^2 - 25 = 0$

31. $16x^2 + 24xy + 9y^2 - 130x + 90y = 0$

32. $16x^2 + 24xy + 9y^2 - 60x + 80y = 0$

In Problems 33–42, identify each equation without applying a rotation of axes.

33. $x^2 + 3xy - 2y^2 + 3x + 2y + 5 = 0$

34. $2x^2 - 3xy + 4y^2 + 2x + 3y - 5 = 0$

35. $x^2 - 7xy + 3y^2 - y - 10 = 0$

36. $2x^2 - 3xy + 2y^2 - 4x - 2 = 0$

37. $9x^2 + 12xy + 4y^2 - x - y = 0$

38. $10x^2 + 12xy + 4y^2 - x - y + 10 = 0$

39. $10x^2 - 12xy + 4y^2 - x - y - 10 = 0$

40. $4x^2 + 12xy + 9y^2 - x - y = 0$

41. $3x^2 - 2xy + y^2 + 4x + 2y - 1 = 0$

42. $3x^2 + 2xy + y^2 + 4x - 2y + 10 = 0$

In Problems 43–46, apply rotation formulas (5) to

$$Ax^2 + Bxy + Cy^2 + Dx + Ey + F = 0$$

to obtain the equation

$$A'x'^2 + B'x'y' + C'y'^2 + D'x' + E'y' + F' = 0$$

43. Express A', B', C', D', E', and F' in terms of A, B, C, D, E, F, and the angle θ of rotation.

44. Show that $A + C = A' + C'$, and thus show that $A + C$ is **invariant;** that is, its value does not change under a rotation of axes.

45. Refer to Problem 44. Show that $B^2 - 4AC$ is invariant.

46. Prove that, except for degenerate cases, the equation

$$Ax^2 + Bxy + Cy^2 + Dx + Ey + F = 0$$

(a) Defines a parabola if $B^2 - 4AC = 0$.

(b) Defines an ellipse (or a circle) if $B^2 - 4AC < 0$.

(c) Defines a hyperbola if $B^2 - 4AC > 0$.

47. Use rotation formulas (5) to show that distance is invariant under a rotation of axes. That is, show that the distance from $P_1 = (x_1, y_1)$ to $P_2 = (x_2, y_2)$ in the xy-plane equals the distance from $P_1 = (x_1', y_1')$ to $P_2 = (x_2', y_2')$ in the x'y'-plane.

48. Show that the graph of the equation $x^{1/2} + y^{1/2} = a^{1/2}$ is part of the graph of a parabola.

49. Formulate a strategy for discussing and graphing an equation of the form

$$Ax^2 + Cy^2 + Dx + Ey + F = 0.$$

How does your strategy change if the equation is of the form

$$Ax^2 + Bxy + Cy^2 + Dx + Ey + F = 0?$$

9.6   POLAR EQUATIONS OF CONICS

> **1** Discuss and Graph Polar Equations of Conics
> **2** Convert a Polar Equation of a Conic to a Rectangular Equation

In Sections 9.2, 9.3, and 9.4, we gave separate definitions for the parabola, ellipse, and hyperbola based on geometric properties and the distance formula. In this section, we present an alternative definition that simultaneously defines all these conics. As we shall see, this approach is well suited to polar coordinate representation. (Refer to Section 8.1.)

Let D denote a fixed line called the **directrix**; let F denote a fixed point called the **focus**, which is not on D; and let e be a fixed positive number called the **eccentricity**. A **conic** is the set of points P in the plane such that the ratio of the distance from F to P to the distance from D to P equals e. Thus, a conic is the collection of points P for which

$$\frac{d(F, P)}{d(D, P)} = e \qquad (1)$$

If $e = 1$, the conic is a **parabola**.

If $e < 1$, the conic is an **ellipse**.

If $e > 1$, the conic is a **hyperbola**.

Observe that if $e = 1$ the definition of a parabola in equation (1) is exactly the same as the definition used earlier in Section 9.2.

In the case of an ellipse, the **major axis** is a line through the focus perpendicular to the directrix. In the case of a hyperbola, the **transverse axis** is a line through the focus perpendicular to the directrix. For both an ellipse and a hyperbola, the eccentricity e satisfies

$$e = \frac{c}{a} \qquad (2)$$

where c is the distance from the center to the focus and a is the distance from the center to a vertex.

Just as we did earlier using rectangular coordinates, we derive equations for the conics in polar coordinates by choosing a convenient position for the focus F and the directrix D. The focus F is positioned at the pole, and the directrix D is either parallel to the polar axis or perpendicular to it.

Suppose that we start with the directrix D perpendicular to the polar axis at a distance p units to the left of the pole (the focus F). See Figure 58.

FIGURE 58

If $P = (r, \theta)$ is any point on the conic, then, by equation (1),

$$\frac{d(F, P)}{d(D, P)} = e \quad \text{or} \quad d(F, P) = e \cdot d(D, P) \tag{3}$$

Now we use the point Q obtained by dropping the perpendicular from P to the polar axis to calculate $d(D, P)$:

$$d(D, P) = p + d(O, Q) = p + r \cos \theta$$

Using this expression and the fact that $d(F, P) = d(O, P) = r$ in equation (3), we get

$$\begin{aligned} d(F, P) &= e \cdot d(D, P) \\ r &= e(p + r \cos \theta) \\ r &= ep + er \cos \theta \\ r - er \cos \theta &= ep \\ r(1 - e \cos \theta) &= ep \\ r &= \frac{ep}{1 - e \cos \theta} \end{aligned}$$

Theorem **Polar Equation of a Conic; Focus at Pole; Directrix Perpendicular to Polar Axis a Distance _p_ to the Left of the Pole**

The polar equation of a conic with focus at the pole and directrix perpendicular to the polar axis at a distance p to the left of the pole is

$$r = \frac{ep}{1 - e \cos \theta} \tag{4}$$

where e is the eccentricity of the conic.

E X A M P L E   1

Discussing and Graphing the Polar Equation of a Conic

 Discuss and graph the equation: $\quad r = \dfrac{4}{2 - \cos \theta}$

Solution   Figure 59(a) shows the graph of the equation using a graphing utility in POLar mode with $\theta \min = 0$, $\theta \max = 2\pi$, and $\theta \text{step} = \pi/24$. We now proceed to discuss the equation.

The given equation is not quite in the form of equation (4), since the first term in the denominator is 2 instead of 1. Thus, we divide the numerator and denominator by 2 to obtain

$$r = \frac{2}{1 - \frac{1}{2} \cos \theta}$$

This equation is in the form of equation (4), with

$$e = \tfrac{1}{2} \quad \text{and} \quad ep = \tfrac{1}{2}p = 2$$

Thus, $e = \tfrac{1}{2}$ and $p = 4$. We conclude that the conic is an ellipse, since $e = \tfrac{1}{2} < 1$. One focus is at the pole, and the directrix is perpendicular to the polar axis, a distance of 4 units to the left of the pole. It follows that the major axis is along the polar axis. To find the vertices, we let $\theta = 0$ and $\theta = \pi$. Thus, the vertices of the ellipse are $(4, 0)$ and $(\tfrac{4}{3}, \pi)$. At the midpoint of the vertices, we locate the center of the ellipse at $(\tfrac{4}{3}, 0)$. [Do you see why? The vertices $(4, 0)$ and $(\tfrac{4}{3}, \pi)$ in polar coordinates are $(4, 0)$ and $(-\tfrac{4}{3}, 0)$ in rectangular coordinates. The midpoint in rectangular coordinates is $(\tfrac{4}{3}, 0)$, which is also $(\tfrac{4}{3}, 0)$ in polar coordinates.] Thus, $a =$ distance from the center to a vertex $= \tfrac{8}{3}$. Using $a = \tfrac{8}{3}$ and $e = \tfrac{1}{2}$ in equation (2), $e = c/a$, we find $c = \tfrac{4}{3}$. Finally, using $a = \tfrac{8}{3}$ and $c = \tfrac{4}{3}$ in $b^2 = a^2 - c^2$, we have

$$b^2 = a^2 - c^2 = \frac{64}{9} - \frac{16}{9} = \frac{48}{9}$$

$$b = \frac{4\sqrt{3}}{3}$$

Figure 59(b) shows the graph drawn by hand.

FIGURE 59

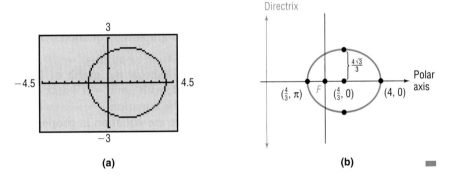

(a)   (b)

Exploration: Graph $r_1 = 4/(2 + \cos\theta)$ and compare the result with Figure 59(a). What do you conclude? Clear the screen and graph $r_1 = 4/(2 - \sin\theta)$ and then $r_1 = 4/(2 + \sin\theta)$. Compare each of these graphs with Figure 59(a). What do you conclude?

 Now work Problem 5.

Equation (4) was obtained under the assumption that the directrix was perpendicular to the polar axis at a distance p units to the left of the pole. A similar derivation (see Problem 37), in which the directrix is perpendicular to the polar axis at a distance p units to the right of the pole, results in the equation

$$r = \frac{ep}{1 + e\cos\theta}$$

$$x = f(t) \qquad y = g(t)$$

where t is in I, are called **parametric equations** of the curve. The variable t is called a **parameter.**

1   Parametric equations are particularly useful in describing movement along a curve. Suppose that a curve is defined by the parametric equations

$$x = f(t) \qquad y = g(t)$$

where f and g are each defined over some interval I. For a given value of t in I, we can find the value of $x = f(t)$ and $y = g(t)$, thus obtaining a point (x, y) on the curve. In fact, as t varies over the interval I in some order, say from left to right, successive values of t give rise to a directed movement along the curve. That is, as t varies over I in some order, the curve is traced out in a certain direction by the corresponding succession of points (x, y). Let's look at an example.

E X A M P L E   1   Discussing a Curve Defined by Parametric Equations

Discuss the curve defined by the parametric equations

$$x = 3t^2 \qquad y = 2t, \qquad -2 \le t \le 2 \qquad\qquad (1)$$

Solution   For each number t, $-2 \le t \le 2$, there corresponds a number x and a number y. For example, when $t = -2$, then $x = 12$ and $y = -4$. When $t = 0$, then $x = 0$ and $y = 0$. Indeed, we can set up a table listing various choices of the parameter t and the corresponding values for x and y, as shown in Table 6. Plotting these points and connecting them with a smooth curve leads to Figure 62.

TABLE 6

t	x	y	(x, y)
-2	12	-4	$(12, -4)$
-1	3	-2	$(3, -2)$
0	0	0	$(0, 0)$
1	3	2	$(3, 2)$
2	12	4	$(12, 4)$

FIGURE 62

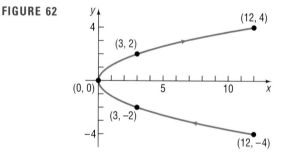

Notice the arrows on the curve in Figure 62. They indicate the direction, or **orientation,** of the curve for increasing values of the parameter t.

2   Most graphing utilities have the capability of graphing parametric equations. The following steps are usually required in order to obtain the graph of parametric equations. Check your owner's manual to see how yours works.

Graphing Parametric Equations
Using a Graphing Utility

STEP 1: Set the mode to PARametric. Enter $x(t)$ and $y(t)$.
STEP 2: Select the viewing rectangle. In addition to setting Xmin, Xmax, Xscl, and so on, the viewing rectangle in parametric mode requires setting minimum and maximum values for the parameter t and an increment setting for t (Tstep).
STEP 3: Execute.

E X A M P L E   2   Graphing a Curve Defined by Parametric Equations Using a Graphing Utility

Graph the curve defined by the parametric equations

$$x = 3t^2 \qquad y = 2t \qquad -2 \le t \le 2$$

Solution   STEP 1: Enter the equations $x(t) = 3t^2, y(t) = 2t$ with the graphing utility in PARametric mode.

STEP 2: Select the viewing rectangle. The interval I is $-2 \le t \le 2$, so we select the following square viewing rectangle:

$$T\text{min} = -2$$
$$T\text{max} = 2$$
$$T\text{step} = 0.1$$
$$X\text{min} = 0$$
$$X\text{max} = 15$$
$$X\text{scl} = 1$$
$$Y\text{min} = -5$$
$$Y\text{max} = 5$$
$$Y\text{scl} = 1$$

FIGURE 63

We choose $T\text{min} = -2$ and $T\text{max} = 2$ because $-2 \le t \le 2$. Finally, the choice for Tstep will determine the number of points the graphing utility will plot. For example, with Tstep at 0.1, the graphing utility will evaluate x and y at $t = -2, -1.9, -1.8$, and so on. The smaller the Tstep, the more points the graphing utility will plot. The reader is encouraged to experiment with different values of Tstep to see how the graph is affected.

STEP 3: Execute. Notice the direction the graph is drawn. This direction shows the orientation of the curve.

The graph shown in Figure 63 is complete.   ■

Now work Problem 1.

Exploration   Graph the following parametric equations using a graphing utility with $X\text{min} = 0$, $X\text{max} = 15$, $Y\text{min} = -5$, $Y\text{max} = 5$, and $T\text{step} = 0.1$:

1. $x = \dfrac{3t^2}{4}, y = t, -4 \le t \le 4$

2. $x = 3t^2 + 12t + 12, y = 2t + 4, -4 \le t \le 0$

3. $x = 3t^{2/3}, y = 2\sqrt[3]{t}, -8 \le t \le 8$

Compare these graphs to the graph in Figure 63. Conclude that parametric equations defining a curve are not unique; that is, different parametric equations can represent the same graph.   ■

3   The curve given in Example 1 and 2 should be familiar. To identify it accurately, we find the corresponding rectangular equation by eliminating the parameter t from the parametric equations (1) given in Example 1,

$$x = 3t^2 \qquad y = 2t, \qquad -2 \le t \le 2$$

Noting that we can readily solve for t in $y = 2t$, obtaining $t = y/2$, we substitute this expression in the other equation:

$$x = 3t^2 \underset{\underset{t = \frac{y}{2}}{\uparrow}}{=} 3\left(\frac{y}{2}\right)^2 = \frac{3y^2}{4}$$

This equation, $x = 3y^2/4$, is the equation of a parabola with vertex at $(0, 0)$ and axis of symmetry along the x-axis.

Exploration   In   FUNCtion   mode   graph $x = \dfrac{3y^2}{4}\left(Y_1 = \sqrt{\dfrac{4x}{3}}\right.$ and $Y_2 = -\sqrt{\dfrac{4x}{3}}\bigg)$ with Xmin $= 0$, Xmax $= 15$, Ymin $= -5$, Ymax $= 5$. Compare this graph with Figure 63. Why do the graphs differ?   ▬

Note that the parameterized curve defined by equation (1) and shown in Figure 62 (or 63) is only a part of the parabola $x = 3y^2/4$. Thus, the graph of the rectangular equation obtained by eliminating the parameter will, in general, contain more points than the original parameterized curve. Care must therefore be taken when a parameterized curve is sketched by hand after eliminating the parameter. Even so, the process of eliminating the parameter t of a parameterized curve in order to identify it accurately is sometimes a better approach than merely plotting points. However, the elimination process sometimes requires a little ingenuity.

E X A M P L E   3

Finding the Rectangular Equation of a Curve Defined Parametrically

Find the rectangular equation of the curve whose parametric equations are

$$x = a \cos t \qquad y = a \sin t$$

where $a > 0$ is a constant. Graph this curve, indicating its orientation.

Solution   The presence of sines and cosines in the parametric equations suggests that we use a Pythagorean identity. In fact, since

$$\cos t = \frac{x}{a} \qquad \sin t = \frac{y}{a}$$

we find that

$$\cos^2 t + \sin^2 t = 1$$
$$\left(\frac{x}{a}\right)^2 + \left(\frac{y}{a}\right)^2 = 1$$
$$x^2 + y^2 = a^2$$

FIGURE 64

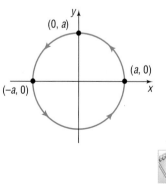

Thus, the curve is a circle with center at $(0, 0)$ and radius a. As the parameter t increases, say from $t = 0$ [the point $(a, 0)$] to $t = \pi/2$ [the point $(0, a)$] to $t = \pi$ [the point $(-a, 0)$], we see that the corresponding points are traced in a counterclockwise direction around the circle. Hence, the orientation is as indicated in Figure 64.   ▬

Now work Problem 13.

Let's discuss the curve in Example 3 further. The domain of each parametric equation is $-\infty < t < \infty$. Thus, the graph in Figure 64 is actually being repeated each time t increases by 2π. If we wanted the curve to consist of exactly 1 revolution in the counterclockwise direction, we could write

$$x = a \cos t \qquad y = a \sin t, \qquad 0 \le t \le 2\pi$$

This curve starts at $t = 0$ [the point $(a, 0)$] and, proceeding counterclockwise around the circle, ends at $t = 2\pi$ [also the point $(a, 0)$].

If we wanted the curve to consist of exactly three revolutions in the counterclockwise direction, we could write

$$x = a \cos t \qquad y = a \sin t, \qquad -2\pi \le t \le 4\pi$$

or

$$x = a \cos t \qquad y = a \sin t, \qquad 0 \le t \le 6\pi$$

or

$$x = a \cos t \qquad y = a \sin t, \qquad 2\pi \le t \le 8\pi$$

If we wanted the curve to consist of the upper semicircle of radius a with a counterclockwise orientation, we could write

$$x = a \cos t \qquad y = a \sin t, \qquad 0 \le t \le \pi$$

See Figure 65.

If we wanted the curve to consist of the left semicircle of radius a with a clockwise orientation, we could write

$$x = -a \sin t \qquad y = -a \cos t, \qquad 0 \le t \le \pi$$

See Figure 66.

FIGURE 65

FIGURE 66

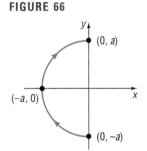

Seeing the Concept (a) Graph $x = \cos t$, $y = \sin t$ for $0 \le t \le 2\pi$. Compare to Figure 64. Graph $x = \cos t$, $y = \sin t$ for $0 \le t \le \pi$. Compare to Figure 65. Graph $x = -\sin t$, $y = -\cos t$ for $0 \le t \le \pi$. Compare to Figure 66. ▬

Time as a Parameter: Projectile Motion; Simulated Motion

4

If we think of the parameter t as time, then the parametric equations $x = f(t)$ and $y = g(t)$ of a curve C specify how the x- and y-coordinates of a moving point vary with time.

For example, we can use parametric equations to describe the motion of an object, sometimes referred to as **curvilinear motion.** Using parametric equations, we can specify not only where the object travels, that is, its location (x, y), but also when it gets there, that is, the time t.

When an object is propelled upward at an inclination θ to the horizontal with initial speed v_0, the resulting motion is called **projectile motion.** See Figure 67(a).

In calculus it is shown that the parametric equations of the path of a projectile fired at an inclination θ to the horizontal, with an initial speed v_0, from a height h above the horizontal are

$$x = (v_0 \cos \theta)t \qquad y = -\frac{1}{2}gt^2 + (v_0 \sin \theta)t + h \qquad (2)$$

where t is the time and g is the constant acceleration due to gravity (approximately 32 ft/sec/sec or 9.8 m/sec/sec). See Figure 67(b).

FIGURE 67

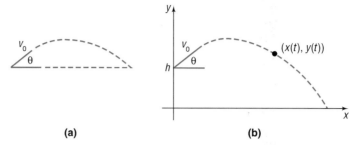

(a)　　　　　　　　　　　　(b)

E X A M P L E   4

FIGURE 68

Projectile Motion

Suppose Jim hit a golf ball with an initial velocity of 150 feet per second at an angle of 30° to the horizontal. See Figure 68.

(a) Find parametric equations that describe the position of the ball as a function of time.

(b) How long is the golf ball in the air?

(c) When is the ball at its maximum height? Determine the maximum height of the ball.

(d) Determine the distance the ball traveled.

(e) Using a graphing utility, simulate the motion of the golfball by simultaneously graphing the equations found in (a).

Solution　(a) We have $v_0 = 150$, $\theta = 30°$, $h = 0$ (the ball is on the ground) and $g = 32$ (since units are in feet and seconds). Substituting these values into equations (2), we find

$$x = (150 \cos 30°)t = 75\sqrt{3}\,t$$

$$y = \frac{-1}{2}(32)t^2 + (150 \sin 30°)t + 0 = -16t^2 + (150 \sin 30°)t$$

$$= -16t^2 + 75t$$

(b) To determine the length of time the ball is in the air, we solve the equation $y = 0$:

$$-16t^2 + 75t = 0$$

$$t(-16t + 75) = 0$$

$$t = 0 \text{ seconds or } t = \frac{75}{16} = 4.6875 \text{ seconds}$$

The ball will strike the ground after 4.6875 seconds.

(c) Notice the height y of the ball is a quadratic function of t. Thus, the maximum height of the ball can be found by determining the vertex of $y = -16t^2 + 75t$. The value of t at the vertex is

$$t = \frac{-b}{2a} = \frac{-75}{-32} = 2.34375 \text{ seconds}$$

The ball is at its maximum height after 2.34375 seconds. The maximum height of the ball is found by evaluating the function y at $t = 2.34375$ seconds.

$$\text{Maximum height} = -16(2.34375)^2 + (75)2.34375 \approx 87.89 \text{ feet}$$

FIGURE 69

(d) Since the ball is in the air for 4.6875 seconds, the horizontal distance the ball travels is

$$x = (75\sqrt{3})4.6875 \approx 608.92 \text{ feet}$$

(e) We enter the equations from (a) into a graphing utility with Tmin $= 0$, Tmax $= 4.7$ and Tstep $= 0.1$. See Figure 69. ▬

Exploration   Simulate the motion of a ball thrown straight up with an initial speed of 100 feet per second from a height of 5 feet above the ground. Use PARametric mode with Tmin $= 0$, Tmax $= 6.5$, Tstep $= 0.1$, Xmin $= 0$, Xmax $= 5$, Ymin $= 0$, and Ymax $= 180$. What happens to the speed at which the graph is drawn as the ball goes up and then comes back down? How do you interpret this physically? Repeat the experiment using other values for Tstep. How does this affect the experiment?

Hint: In the projectile motion equations, let $\theta = 90°$, $v_0 = 100$, $h = 5$, and $g = 32$.) We use $x = 3$ instead of $x = 0$ to see the vertical motion better.

Result   See Figure 70. In Figure 70(a) the ball is going up. In Figure 70(b), the ball is near its highest point. Finally, in Figure 70(c), the ball is coming back down.

FIGURE 70

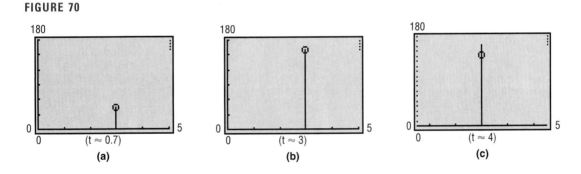

Notice that, as the ball goes up, its speed decreases, until at the highest point it is zero. Then the speed increases as the ball comes back down. ▬

A graphing utility can be used to simulate other kinds of motion as well. Let's work again Example 5 from Section 1.6.

E X A M P L E   5

Simulating Motion

Tanya, who is a long distance runner, runs at an average velocity of 8 miles per hour. Two hours after Tanya leaves your house, you leave in your Honda and follow the same route. If your average velocity is 40 miles per hour, how long will it be before you catch up to Tanya? See Figure 71. Use a simulation of the two motions to answer the question.

FIGURE 71

Solution We begin by setting up two sets of parametric equations: one to describe Tanya's motion, the other to describe the motion of the Honda. We choose time $t = 0$ to be when Tanya leaves the house. We choose $y_1 = 2$ as Tanya's path and $y_2 = 4$ as the parallel path of the Honda. The horizontal distances traversed in time t are:

$$\text{Tanya: } x_1 = 8t \qquad \text{Honda: } x_2 = 40(t - 2)$$

The Honda catches up to Tanya when $x_1 = x_2$.

$$8t = 40(t - 2)$$
$$8t = 40t - 80$$
$$-32t = -80$$
$$t = \frac{-80}{-32} = 2.5$$

The Honda catches up to Tanya after 2.5 hours.

In PARametric mode with Tstep $= 0.01$, we simultaneously graph

$$\text{Tanya: } x_1 = 8t \qquad \text{Honda: } x_2 = 40(t - 2)$$
$$y_1 = 2 \qquad\qquad y_2 = 4$$

for $0 \le t \le 3$.

Figure 72 shows the relative position of Tanya and the Honda for $t = 0$, $t = 2$, $t = 2.25$, $t = 2.5$, and $t = 2.75$.

FIGURE 72

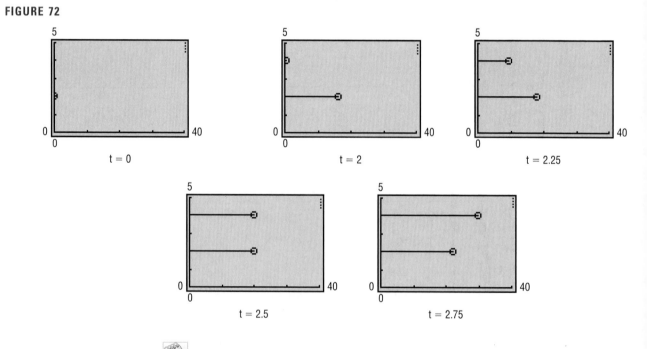

$t = 0$ $\qquad$ $t = 2$ $\qquad$ $t = 2.25$

$t = 2.5$ $\qquad$ $t = 2.75$

Now work Problem 27.

Finding Parametric Equations

We now take up the question of how to find parametric equations of a given curve.

5

If a curve is defined by the equation $y = f(x)$, where f is a function, one way of finding parametric equations is simply to let $x = t$. Then $y = f(t)$. Thus,

$$x = t \qquad y = f(t) \qquad t \text{ in the domain of } f$$

are parametric equations of the curve.

EXAMPLE 6

Finding Parametric Equations for a Curve Defined by a Rectangular Equation

Find parametric equations for the equation $y = x^2 - 4$.

Solution Let $x = t$. Then the parametric equations are

$$x = t \qquad y = t^2 - 4, \qquad -\infty < t < \infty$$

Another less obvious approach to Example 6 is to let $x = t^3$. Then the parametric equations become

$$x = t^3 \qquad y = t^6 - 4, \qquad -\infty < t < \infty$$

Care must be taken when using this approach, since the substitution for x must be a function that allows x to take on all the values stipulated by the domain of f. Thus, for example, letting $x = t^2$ so that $y = t^4 - 4$ does not result in equivalent parametric equations for $y = x^2 - 4$, since only points for which $x \geq 0$ are obtained.

EXAMPLE 7

Finding Parametric Equations for an Object in Motion

Find parametric equations for the ellipse

$$x^2 + \frac{y^2}{9} = 1$$

where the parameter t is time (in seconds) and

(a) The motion around the ellipse is clockwise, begins at the point $(0, 3)$, and requires 1 second for a complete revolution.

(b) The motion around the ellipse is counterclockwise, begins at the point $(1, 0)$, and requires 2 seconds for a complete revolution.

Solution (a) Since the motion begins at the point $(0, 3)$, we want $x = 0$ and $y = 3$ when $t = 0$. Furthermore, since the given equation is an ellipse, we begin by letting

$$x = \sin \omega t \qquad \frac{y}{3} = \cos \omega t$$

for some constant ω. These parametric equations clearly satisfy the equation. Furthermore, with this choice, when $t = 0$, we have $x = 0$ and $y = 3$. For the motion to be clockwise, the motion will have to begin with the value of x increasing and y decreasing as t increases. Thus, $\omega > 0$. Finally, since 1 revolution requires 1 second, the period $2\pi/\omega = 1$, so $\omega = 2\pi$. Thus, parametric equations that satisfy the conditions stipulated are

$$x = \sin 2\pi t \qquad y = 3 \cos 2\pi t, \qquad 0 \leq t \leq 1 \qquad (3)$$

(b) Since the motion begins at the point (1, 0), we want $x = 1$ and $y = 0$ when $t = 0$. Furthermore, since the given equation is an ellipse, we begin by letting

$$x = \cos \omega t \qquad \frac{y}{3} = \sin \omega t$$

for some constant ω. These parametric equations clearly satisfy the equation. Furthermore, with this choice, when $t = 0$, we have $x = 1$ and $y = 0$. For the motion to be counterclockwise, the motion will have to begin with the value of x decreasing and y increasing as t increases. Thus, $\omega > 0$. Finally, since 1 revolution requires 2 seconds, the period is $2\pi/\omega = 2$, so $\omega = \pi$. Thus, the parametric equations that satisfy the conditions stipulated are

$$x = \cos \pi t \qquad y = 3 \sin \pi t, \qquad 0 \le t \le 2 \tag{4}$$

∎

Either of equations (3) or (4) can serve as parametric equations for the ellipse $x^2 + y^2/9 = 1$ given in Example 6. The direction of the motion, the beginning point, and the time for 1 revolution merely serve to help arrive at a particular parametric representation.

Now work Problem 35.

The Cycloid

Suppose that a circle radius a rolls along a horizontal line without slipping. As the circle rolls along the line, a point P on the circle will trace out a curve called a **cycloid** (see Figure 73). We now seek parametric equations* for a cycloid.

We begin with a circle of radius a and take the fixed line on which the circle rolls as the x-axis. Let the origin be one of the points at which the point P comes in contact with the x-axis. Figure 73 illustrates the position of this

FIGURE 73

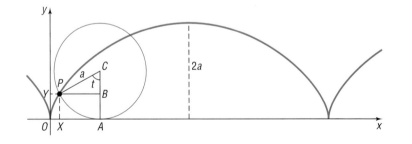

*Any attempt to derive the rectangular equation of a cycloid would soon demonstrate how complicated the task is.

point P after the circle has rolled somewhat. The angle t (in radians) measures the angle through which the circle has rolled.

Since we require no slippage, it follows that

Arc $AP = d(O, A)$

Therefore,

$at = d(O, A)$

The x-coordinate of the point P is

$d(O, X) = d(O, A) - d(X, A) = at - a \sin t = a(t - \sin t)$

The y-coordinate of the point P is equal to

$d(O, Y) = d(A, C) - d(B, C) = a - a \cos t = a(1 - \cos t)$

Thus, the parametric equations of the cycloid are

$$x = a(t - \sin t) \qquad y = a(1 - \cos t) \tag{5}$$

Exploration Graph $x = t - \sin t$, $y = 1 - \cos t$, $0 \le t \le 3\pi$, using your graphing utility with Tstep $= \pi/36$ and a square screen. Compare your results with Figure 73. ▬

Applications to Mechanics

If a is negative in equation (5), we obtain an inverted cycloid, as shown in Figure 74(a). The inverted cycloid occurs as a result of some remarkable applications in the field of mechanics. We shall mention two of them: the *brachistochrone* and the *tautochrone.**

The **brachistochrone** is the curve of quickest descent. If a particle is constrained to follow some path from one point A to a lower point B (not on the same vertical line) and is acted on only by gravity, the time needed to make the descent is least if the path is an inverted cycloid. See Figure 74(b). This remarkable discovery, which is attributed to many famous mathemati-

---

*In Greek, *brachistochrone* means "the shortest time," and *tautochrone* means "equal time."

FIGURE 74

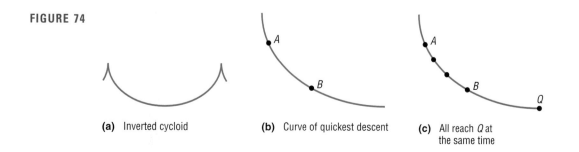

(a) Inverted cycloid

(b) Curve of quickest descent

(c) All reach Q at the same time

FIGURE 75
A flexible pendulum constrained by cycloids swings in a cycloid.

cians (including Johann Bernoulli and Blaise Pascal), was a significant step in creating the branch of mathematics known as the *calculus of variations*.

To define the **tautochrone**, let Q be the lowest point on an inverted cycloid. If several particles placed at various positions on an inverted cycloid simultaneously begin to slide down the cycloid, they will reach the point Q at the same time, as indicated in Figure 74(c). The tautochrone property of the cycloid was used by Christiaan Huygens (1629–1695), the Dutch mathematician, physicist, and astronomer, to construct a pendulum clock with a bob that swings along a cycloid (see Figure 75). In Huygen's clock, the bob was made to swing along a cycloid by suspending the bob on a thin wire constrained by two plates shaped like cycloids. In a clock of this design, the period of the pendulum is independent of its amplitude.

9.7 EXERCISES

In Problems 1–20, graph the curve whose parametric equations are given using a graphing utility. Find the rectangular equation of each curve. Graph the curve by hand and show its orientation.

1. $x = 3t + 2$, $y = t + 1$; $0 \le t \le 4$

2. $x = t - 3$, $y = 2t + 4$; $0 \le t \le 2$

3. $x = t + 2$, $y = \sqrt{t}$; $t \ge 0$

4. $x = \sqrt{2t}$, $y = 4t$; $t \ge 0$

5. $x = t^2 + 4$, $y = t^2 - 4$; $-\infty < t < \infty$

6. $x = \sqrt{t} + 4$, $y = \sqrt{t} - 4$; $t \ge 0$

7. $x = 3t^2$, $y = t + 1$; $-\infty < t < \infty$

8. $x = 2t - 4$, $y = 4t^2$; $-\infty < t < \infty$

9. $x = 2e^t$, $y = 1 + e^t$; $t \ge 0$

10. $x = e^t$, $y = e^{-t}$; $t \ge 0$

11. $x = \sqrt{t}$, $y = t^{3/2}$; $t \ge 0$

12. $x = t^{3/2} + 1$, $y = \sqrt{t}$; $t \ge 0$

13. $x = 2 \cos t$, $y = 3 \sin t$; $0 \le t \le 2\pi$

14. $x = 2 \cos t$, $y = 3 \sin t$; $0 \le t \le \pi$

15. $x = 2 \cos t$, $y = 3 \sin t$; $-\pi \le t \le 0$

16. $x = 2 \cos t$, $y = \sin t$; $0 \le t \le \pi/2$

17. $x = \sec t$, $y = \tan t$; $0 \le t \le \pi/4$

18. $x = \csc t$, $y = \cot t$; $\pi/4 \le t \le \pi/2$

19. $x = \sin^2 t$, $y = \cos^2 t$; $0 \le t \le 2\pi$

20. $x = t^2$, $y = \ln t$; $t > 0$

21. Projectile Motion Bob throws a ball straight up with an initial speed of 50 feet per second from a height of 6 feet.
(a) Find parametric equations that describe the motion of the ball as a function of time.
(b) How long is the ball in the air?
(c) When is the ball at its maximum height? Determine the maximum height of the ball.
(d) Simulate the motion of the ball by graphing the equations found in (a).

22. Projectile Motion Alice throws a ball straight up with an initial speed of 40 feet per second from a height of 5 feet.
(a) Find parametric equations that describe the motion of the ball as a function of time.
(b) How long is the ball in the air?
(c) When is the ball at its maximum height? Determine the maximum height of the ball.

(d) Simulate the motion of the ball by graphing the equations found in (a).

23. Catching a Train Bill's train leaves at 8:06 A.M. and accelerates at the rate of 2 meters/second/second. Bill arrives at the train station 5 seconds after the train had left. Bill can run 5 meters/second. Will Bill catch the train?
(a) Find parametric equations that describe the motion of the train and Bill as a function of time.
(b) Determine algebraically whether Bill will catch the train. If so, when?
(c) Simulate the motion of the train and Bill by simultaneously graphing the equations found in (a).

24. Catching a Bus Jodi's bus leaves at 5:03 P.M. and accelerates at the rate of 3 meters/second/second. Jodi arrives at the bus station 2 seconds after the bus had left. Jodi can run 5 meters/second. Will Jodi catch the bus?

(a) Find parametric equations that describe the motion of the bus and Jodi as a function of time.

(b) Determine algebraically whether Jodi will catch the bus. If so, when?

(c) Simulate the motion of the bus and Jodi by simultaneously graphing the equations found in (a).

25. **Projectile Motion** Nolan Ryan throws a baseball with an initial speed of 145 feet per second at an angle of 20° to the horizontal. The ball leaves Nolan Ryan's hand at a height of 5 feet.

(a) Find parametric equations that describe the position of the ball as a function of time.

(b) How long is the ball in the air?

(c) When is the ball at its maximum height? Determine the maximum height of the ball.

(d) Determine the distance the ball traveled.

(e) Using a graphing utility, simultaneously graph the equations found in (a).

26. **Projectile Motion** Mark McGuire hit a baseball with an initial speed of 180 feet per second at an angle of 40° to the horizontal. The ball was hit at a height of 3 feet off the ground.

(a) Find parametric equations that describe the position of the ball as a function of time.

(b) How long is the ball in the air?

(c) When is the ball at its maximum height? Determine the maximum height of the ball.

(d) Determine the distance the ball traveled.

(e) Using a graphing utility, simultaneously graph the equations found in (a).

27. **Projectile Motion** Suppose Adam throws a tennis ball off a cliff 300 meters high with an initial speed of 40 meters per second at an angle of 45° to the horizontal.

(a) Find parametric equations that describe the position of the ball as a function of time.

(b) How long is the ball in the air?

(c) When is the ball at its maximum height? Determine the maximum height of the ball.

(d) Determine the distance the ball traveled.

(e) Using a graphing utility, simultaneously graph the equations found in (a).

28. **Projectile Motion** Suppose Adam throws a tennis ball off a cliff 300 meters high with an initial speed of 40 meters per second at an angle of 45° to the horizontal on the moon (gravity on the Moon is 1/6 of that on earth).

(a) Find parametric equations that describe the position of the ball as a function of time.

(b) How long is the ball in the air?

(c) When is the ball at its maximum height? Determine the maximum height of the ball.

(d) Determine the distance the ball traveled.

(e) Using a graphing utility, simultaneously graph the equations found in (a).

29. **Uniform Motion** A Toyota Paseo (traveling East at 40 mph) and Pontiac Bonneville (traveling North at 30 mph) are heading toward the same intersection. The Paseo is 5 miles from the intersection when the Bonneville is 4 miles from the intersection. See the figure.

(a) Find a formula for the distance between the cars as a function of time.

(b) Graph the function in (a) using a graphing utility.

(c) What is the minimum distance between the cars? When are the cars closest?

(d) Find parametric equations that describe the motion of the Paseo and Bonneville.

(e) Simulate the motion of the cars by simultaneously graphing the equations found in (d).

30. **Uniform Motion** A Cesna (heading South at 120 mph) and a Boeing 747 (heading West at 600 mph) are flying toward each other at the same altitude. The Cesna is 100 miles from the point the flight patterns intersect and the 747 is 550 miles from this intersection point. See the figure.

(a) Find a formula for the distance between the planes as a function of time.

(b) Graph the function in (a) using a graphing utility.

(c) What is the minimum distance between the planes? When are the planes closest?

(d) Find parametric equations that describe the motion of the Cesna and 747.

(e) Simulate the motion of the planes by simultaneously graphing the equations found in (d).

In Problems 31–34, find two different parametric equations for each rectangular equation.

31. $y = x^3$
32. $y = x^4 + 1$
33. $x = y^{3/2}$
34. $x = \sqrt{y}$

In Problems 35–38, find parametric equations for an object that moves along the ellipse $x^2/4 + y^2/9 = 1$ with the motion described.

35. The motion begins at $(2, 0)$, is clockwise, and requires 2 seconds for a complete revolution.

36. The motion begins at $(0, 3)$, is clockwise, and requires 1 second for a complete revolution.

37. The motion begins at $(0, 3)$, is counterclockwise, and requires 1 second for a complete revolution.

38. The motion begins at $(2, 0)$, is counterclockwise, and requires 3 seconds for a complete revolution.

In Problems 39 and 40, the parametric equations of four curves are given. Graph each of them by hand, indicating the orientation.

39. C_1: $x = t$, $y = t^2$; $-4 \le t \le 4$
$\quad C_2$: $x = \cos t$, $y = 1 - \sin^2 t$; $0 \le t \le \pi$
$\quad C_3$: $x = e^t$, $y = e^{2t}$; $0 \le t \le \ln 4$
$\quad C_4$: $x = \sqrt{t}, y = t$; $0 \le t \le 16$

40. C_1: $x = t$, $y = \sqrt{1 - t^2}$; $-1 \le t \le 1$
$\quad C_2$: $x = \sin t$, $y = \cos t$; $0 \le t \le 2\pi$
$\quad C_3$: $x = \cos t$, $y = \sin t$; $0 \le t \le 2\pi$
$\quad C_4$: $x = \sqrt{1 - t^2}$, $y = t$; $-1 \le t \le 1$

41. Show that the parametric equations for a line passing through the points (x_1, y_1) and (x_2, y_2) are

$$x = (x_2 - x_1)t + x_1 \quad y = (y_2 - y_1)t + y_1,$$
$$-\infty < t < \infty$$

What is the orientation of this line?

42. Projectile Motion The position of a projectile fired with an initial velocity v_0 feet per second and at an angle θ to the horizontal at the end of t seconds is given by the parametric equations

$$x = (v_0 \cos \theta)t \quad y = (v_0 \sin \theta)t - 16t^2$$

See the following illustration.

(a) Obtain the rectangular equation of the trajectory and identify the curve.

(b) Show that the projectile hits the ground $(y = 0)$ when $t = \frac{1}{16} v_0 \sin \theta$.

(c) How far has the projectile traveled (horizontally) when it strikes the ground? In other words, find the **range R.**

(d) Find the time t when $x = y$. Then find the horizontal distance x and the vertical distance y traveled by the projectile in this time. Then compute $\sqrt{x^2 + y^2}$. This is the distance R, the range, that the projectile travels up a plane inclined at $45°$ to the horizontal $(x = y)$. See the following illustration. (See also Problem 75 in Exercise 6.3.)

In Problems 43–46, use a graphing utility to graph the curve defined by the given parametric equations.

43. $x = t \sin t, \quad y = t \cos t$

44. $x = \sin t + \cos t, \quad y = \sin t - \cos t$

45. $x = 4 \sin t - 2 \sin 2t$
$y = 4 \cos t - 2 \cos 2t$

46. $x = 4 \sin t + 2 \sin 2t$
$y = 4 \cos t + 2 \cos 2t$

47. The Hypocycloid The hypocycloid is a curve defined by the parametric equations

$$x(t) = \cos^3 t \qquad y(t) = \sin^3 t, \qquad 0 \le t \le 2\pi$$

(a) Graph the hypocycloid using a graphing utility.

(b) Find rectangular equations of the hypocycloid.

48. In Problem 47, we graphed the hypocycloid. Now graph the rectangular equations of the hypocycloid. Did you obtain a complete graph? If not, experiment until you do.

49. Look up the curves called *hypocycloid* and *epicycloid*. Write a report on what you find. Be sure to draw comparisons with the cycloid.

CHAPTER REVIEW

THINGS TO KNOW

Equations

Parabola	See Tables 1 and 2.
Ellipse	See Table 3.
Hyperbola	See Table 4.

General equation of a conic

$$Ax^2 + Bxy + Cy^2 + Dx + Ey + F = 0$$

Parabola if $B^2 - 4AC = 0$
Ellipse (or circle) if $B^2 - 4AC < 0$
Hyperbola if $B^2 - 4AC > 0$

Conic in polar coordinates

$$\frac{d(F, P)}{d(P, D)} = e$$

Parabola if $e = 1$
Ellipse if $e < 1$
Hyperbola if $e > 1$

Polar equations of a conic   See Table 5.

Plane curve $(x, y) = (f(t), g(t))$, t a parameter

Definitions

Parabola   Set of points P in the plane for which $d(F, P) = d(P, D)$, where F is the focus and D is the directrix

Ellipse   Set of points P in the plane, the sum of whose distances from two fixed points (the foci) is a constant

Hyperbola   Set of points P in the plane, the difference of whose distances from two fixed points (the foci) is a constant

Formulas

Rotation formulas

$$x = x' \cos\theta - y' \sin\theta$$
$$y = x' \sin\theta + y' \cos\theta$$

Angle θ of rotation that eliminates the $x'y'$-term

$$\cot 2\theta = \frac{A - C}{B} \qquad 0 < \theta < \pi/2$$

HOW TO

Find the vertex, focus, and directrix of a parabola given its equation

Graph a parabola given its equation (by hand and by using a graphing utility)

Find an equation of a parabola given certain information about the parabola

Find the center, foci, and vertices of an ellipse given its equation

Graph an ellipse given its equation (by hand and by using a graphing utility)

Find an equation of an ellipse given certain information about the ellipse

Find the center, foci, vertices, and asymptotes of a hyperbola given its equation

Graph a hyperbola given its equation (by hand and by using a graphing utility)

Find an equation of a hyperbola given certain information about the hyperbola

Identify conics without completing the square

Identify conics without a rotation of axes

Use rotation formulas to transform second-degree equations so that no xy-term is present

Identify and graph (by hand and by using a graphing utility) conics given by a polar equation

Graph parametric equations by hand and by using a graphing utility

Find the rectangular equation given the parametric equations

Simulate motion problems

FILL-IN-THE-BLANK ITEMS

1. A(n) _____ is the collection of all points in the plane such that the distance from each point to a fixed point equals its distance to a fixed line.

2. A(n) _____ is the collection of all points in the plane the sum of whose distances from two fixed points is a constant.

3. A(n) _____ is the collection of all points in the plane the difference of whose distances from two fixed points is a constant.

4. For an ellipse, the foci lie on the _____ axis; for a hyperbola, the foci lie on the _____ axis.

5. For the ellipse $(x^2/9) + (y^2/16) = 1$, the major axis is along the _____.

6. The equations of the asymptotes of the hyperbola $(y^2/9) - (x^2/4) = 1$ are _____ and _____.

7. To transform the equation

$$Ax^2 + Bxy + Cy^2 + Dx + Ey + F = 0, \qquad B \neq 0$$

into one in x' and y' without an $x'y'$-term, rotate the axes through an acute angle θ that satisfies the equation _____.

8. The polar equation

$$r = \frac{8}{4 - 2\sin\theta}$$

is a conic whose eccentricity is _____. It is a(n) _____ whose directrix is _____ to the polar axis at a distance _____ units _____ the pole.

9. The parametric equations $x = 2\sin t$ and $y = 3\cos t$ represent a(n) _____.

TRUE/FALSE ITEMS

T F   **1.** On a parabola, the distance from any point to the focus equals the distance from that point to the directrix.

T F   **2.** The foci of an ellipse lie on its minor axis.

T F   **3.** The foci of a hyperbola lie on its transverse axis.

T F   **4.** Hyperbolas always have asymptotes, and ellipses never have asymptotes.

T F   **5.** A hyperbola never intersects its conjugate axis.

T F   **6.** A hyperbola always intersects its transverse axis.

T F   **7.** The equation $ax^2 + 6y^2 - 12y = 0$ defines an ellipse if $a > 0$.

T F   **8.** The equation $3x^2 + bxy + 12y^2 = 10$ defines a parabola if $b = -12$.

T F   **9.** If (r, θ) are polar coordinates, the equation $r = 2/(2 + 3 \sin \theta)$ defines a hyperbola.

T F   **10.** Parametric equations defining a curve are unique.

REVIEW EXERCISES

In Problems 1–20, identify each equation. If it is a parabola, give its vertex, focus, and directrix; if it is an ellipse, give its center, vertices, and foci; if it is a hyperbola, give its center, vertices, foci, and asymptotes.

1. $y^2 = -16x$

2. $16x^2 = y$

3. $\dfrac{x^2}{25} - y^2 = 1$

4. $\dfrac{y^2}{25} - x^2 = 1$

5. $\dfrac{y^2}{25} + \dfrac{x^2}{16} = 1$

6. $\dfrac{x^2}{9} + \dfrac{y^2}{16} = 1$

7. $x^2 + 4y = 4$

8. $3y^2 - x^2 = 9$

9. $4x^2 - y^2 = 8$

10. $9x^2 + 4y^2 = 36$

11. $x^2 - 4x = 2y$

12. $2y^2 - 4y = x - 2$

13. $y^2 - 4y - 4x^2 + 8x = 4$

14. $4x^2 + y^2 + 8x - 4y + 4 = 0$

15. $4x^2 + 9y^2 - 16x - 18y = 11$

16. $4x^2 + 9y^2 - 16x + 18y = 11$

17. $4x^2 - 16x + 16y + 32 = 0$

18. $4y^2 + 3x - 16y + 19 = 0$

19. $9x^2 + 4y^2 - 18x + 8y = 23$

20. $x^2 - y^2 - 2x - 2y = 1$

In Problems 21–36, obtain an equation of the conic described. Graph the equation by hand.

21. Parabola; focus at $(-2, 0)$; directrix the line $x = 2$

22. Ellipse; center at $(0, 0)$; focus at $(0, 3)$; vertex at $(0, 5)$

23. Hyperbola; center at $(0, 0)$; focus at $(0, 4)$; vertex at $(0, -2)$

24. Parabola; vertex at $(0, 0)$; directrix the line $y = -3$

25. Ellipse; foci at $(-3, 0)$ and $(3, 0)$; vertex at $(4, 0)$

26. Hyperbola; vertices at $(-2, 0)$ and $(2, 0)$; focus at $(4, 0)$

27. Parabola; vertex at $(2, -3)$; focus at $(2, -4)$

28. Ellipse; center at $(-1, 2)$; focus at $(0, 2)$; vertex at $(2, 2)$

29. Hyperbola; center at $(-2, -3)$; focus at $(-4, -3)$; vertex at $(-3, -3)$

30. Parabola; focus at $(3, 6)$; directrix the line $y = 8$

31. Ellipse; foci at $(-4, 2)$ and $(-4, 8)$; vertex at $(-4, 10)$

32. Hyperbola; vertices at $(-3, 3)$ and $(5, 3)$; focus at $(7, 3)$

33. Center at $(-1, 2)$; $a = 3$; $c = 4$; transverse axis parallel to the x-axis

34. Center at $(4, -2)$; $a = 1$; $c = 4$; transverse axis parallel to the y-axis

35. Vertices at $(0, 1)$ and $(6, 1)$; asymptote the line $3y + 2x - 9 = 0$

36. Vertices at $(4, 0)$ and $(4, 4)$; asymptote the line $y + 2x - 10 = 0$

In Problems 37–46, identify each conic without completing the squares and without applying a rotation of axes.

37. $y^2 + 4x + 3y - 8 = 0$

38. $2x^2 - y + 8x = 0$

39. $x^2 + 2y^2 + 4x - 8y + 2 = 0$

40. $x^2 - 8y^2 - x - 2y = 0$

41. $9x^2 - 12xy + 4y^2 + 8x + 12y = 0$

42. $4x^2 + 4xy + y^2 - 8\sqrt{5}x + 16\sqrt{5}y = 0$

43. $4x^2 + 10xy + 4y^2 - 9 = 0$

44. $4x^2 - 10xy + 4y^2 - 9 = 0$

45. $x^2 - 2xy + 3y^2 + 2x + 4y - 1 = 0$

46. $4x^2 + 12xy - 10y^2 + x + y - 10 = 0$

In Problems 47–52, rotate the axes so that the new equation contains no xy-term. Discuss and graph the new equation by hand. Verify your result using a graphing utility.

47. $2x^2 + 5xy + 2y^2 - \frac{9}{2} = 0$

48. $2x^2 - 5xy + 2y^2 - \frac{9}{2} = 0$

49. $6x^2 + 4xy + 9y^2 - 20 = 0$

50. $x^2 + 4xy + 4y^2 + 16\sqrt{5}x - 8\sqrt{5}y = 0$

51. $4x^2 - 12xy + 9y^2 + 12x + 8y = 0$

52. $9x^2 - 24xy + 16y^2 + 80x + 60y = 0$

In Problems 53–58, identify the conic that each polar equation represents, and graph it by hand. Verify your result using a graphing utility.

53. $r = \dfrac{4}{1 - \cos\theta}$

54. $r = \dfrac{6}{1 + \sin\theta}$

55. $r = \dfrac{6}{2 - \sin\theta}$

56. $r = \dfrac{2}{3 + 2\cos\theta}$

57. $r = \dfrac{8}{4 + 8\cos\theta}$

58. $r = \dfrac{10}{5 + 20\sin\theta}$

In Problems 59–62, convert each polar equation to a rectangular equation.

59. $r = \dfrac{4}{1 - \cos\theta}$

60. $r = \dfrac{6}{2 - \sin\theta}$

61. $r = \dfrac{8}{4 + 8\cos\theta}$

62. $r = \dfrac{2}{3 + 2\cos\theta}$

In Problems 63–68, graph the curve whose parametric equations are given using a graphing utility. Find the rectangular equation of each curve. Graph the curve by hand and show its orientation.

63. $x = 4t - 2$, $y = 1 - t$; $-\infty < t < \infty$

64. $x = 2t^2 + 6$, $y = 5 - t$; $-\infty < t < \infty$

65. $x = 3\sin t$, $y = 4\cos t + 2$; $0 \le t \le 2\pi$

66. $x = \ln t$, $y = t^3$; $t > 0$

67. $x = \sec^2 t$, $y = \tan^2 t$; $0 \le t \le \pi/4$

68. $x = t^{3/2}$, $y = 2t + 4$; $t \ge 0$

69. Find an equation of the hyperbola whose foci are the vertices of the ellipse $4x^2 + 9y^2 = 36$ and whose vertices are the foci of this ellipse.

70. Find an equation of the ellipse whose foci are the vertices of the hyperbola $x^2 - 4y^2 = 16$ and whose vertices are the foci of this hyperbola.

71. Describe the collection of points in a plane so that the distance from each point to the point $(3, 0)$ is three-fourths of its distance from the line $x = \frac{16}{3}$.

72. Describe the collection of points in a plane so that the distance from each point to the point $(5, 0)$ is five-fourths of its distance from the line $x = \frac{16}{5}$.

73. Mirrors A mirror is shaped like a paraboloid of revolution. If a light source is located 1 foot from the base along the axis of symmetry and the opening is 2 feet across, how deep should the mirror be?

74. Parabolic Arch Bridge A bridge is built in the shape of a parabolic arch. The bridge has a span of 60 feet and a maximum height of 20 feet. Find the height of the arch at distances of 5, 10, and 20 feet from the center.

75. Semielliptical Arch Bridge A bridge is built in the shape of a semielliptical arch. The bridge has a span of 60 feet and a maximum height of 20 feet. Find the height of the arch at distances of 5, 10, and 20 feet from the center.

76. Whispering Galleries The figure shows the specifications for an elliptical ceiling in a hall designed to be a whispering gallery. Where in the hall are the foci located?

77. LORAN Two LORAN stations are positioned 150 miles apart along a straight shore.
(a) A ship records a time difference of 0.00032 second between the LORAN signals. Set up an appropriate rectangular coordinate system to determine where the ship would reach shore if it were to follow the hyperbola corresponding to this time difference.
(b) If the ship wants to enter a harbor located between the two stations 15 miles from the master station, what time difference should it be looking for?
(c) If the ship is 20 miles offshore when the desired time difference is obtained, what is the exact location of the ship?

[**Note:** The speed of each radio signal is 186,000 miles per second.]

78. Uniform Motion Mary's train leaves at 7:15 A.M. and accelerates at the rate of 3 meters/second/second. Mary arrives at the train station 2 seconds after the train had left. Mary can run 6 meters/second. Will Mary catch the train?

(a) Find parametric equations that describe the motion of the train and Mary as a function of time.

(b) Determine algebraically whether Mary will catch the train. If so, when?

(c) Simulate the motion of the train and Mary by simultaneously graphing the equations found in (a).

79. Projectile Motion Drew Bledsoe throws a football with an initial speed of 100 feet per second at an angle of 35° to the horizontal. The ball leaves Drew Bledsoe's hand at a height of 6 feet.

(a) Find parametric equations that describe the position of the ball as a function of time.

(b) How long is the ball in the air?

(c) When is the ball at its maximum height? Determine the maximum height of the ball.

(d) Determine the distance the ball traveled.

(e) Using a graphing utility, simultaneously graph the equations found in (a).

80. Formulate a strategy for discussing and graphing an equation of the form
$$Ax^2 + Bxy + Cy^2 + Dx + Ey + F = 0.$$

Systems of Equations and Inequalities

Internet usage and the security of the information passing through the Internet is a major concern for corporations and other organizations. On the following page, you and your friends decide to start an encryption software company. You will need to use the Sullivan website at:

www.prenhall.com/sullivan

in order to find the resources available to build your encryption program.

PREPARING FOR THIS CHAPTER

Before getting started on this chapter, review the following concepts:

For Sections 10.1, 10.2, 10.7, 10.8: Lines *(Section 1.3)*

For Section 10.5: Identity *(p. 475)*

Proper and Improper Rational Functions *(p. 233)*

Fundamental Theorem of Algebra *(p. 274)*

For Section 10.6: Circles *(pp. 29–32)*

Parabolas *(pp. 667–675)*

Ellipses *(pp. 680–689)*

Hyperbolas *(pp. 694–704)*

For Section 10.7: Inequalities *(Section 1.7)*

OUTLINE

The manipulations just performed on the augmented matrix are called **row operations.** There are three basic row operations:

Row Operations

1. Interchange any two rows.
2. Replace a row by a nonzero multiple of that row.
3. Replace a row by the sum of that row and a constant multiple of some other row.

These three row operations correspond to the three rules given earlier for obtaining an equivalent system of equations. Thus, when a row operation is performed on a matrix, the resulting matrix represents a system of equations equivalent to the system represented by the original matrix.

For example, consider the augmented matrix

$$\begin{bmatrix} 1 & 2 & | & 3 \\ 4 & -1 & | & 2 \end{bmatrix}$$

Suppose we want to apply a row operation to this matrix that results in a matrix whose entry in row 2, column 1 is a 0. The row operation to use is

Multiply each entry in row 1 by -4 and add the result
to the corresponding entries in row 2.   (2)

If we use R_2 to represent the new entries in row 2 and we use r_1 and r_2 to represent the original entries in rows 1 and 2, respectively, then we can represent the row operation in statement (2) by

$$R_2 = -4r_1 + r_2$$

Then

$$\begin{bmatrix} 1 & 2 & | & 3 \\ 4 & -1 & | & 2 \end{bmatrix} \xrightarrow[\underset{R_2 = -4r_1 + r_2}{\uparrow}]{} \begin{bmatrix} 1 & 2 & | & 3 \\ -4(1) + 4 & -4(2) + (-1) & | & -4(3) + 2 \end{bmatrix} = \begin{bmatrix} 1 & 2 & | & 3 \\ 0 & -9 & | & -10 \end{bmatrix}$$

As desired, we now have the entry 0 in row 2, column 1.

E X A M P L E   4   Applying a Row Operation to an Augmented Matrix

Apply the row operation $R_2 = -3r_1 + r_2$ to the augmented matrix

$$\begin{bmatrix} 1 & -2 & | & 2 \\ 3 & -5 & | & 9 \end{bmatrix}$$

Solution   The row operation $R_2 = -3r_1 + r_2$ tells us that the entries in row 2 are to be replaced by the entries obtained after multiplying each entry in row 1 by -3 and adding the result to the corresponding entries in row 2. Thus,

$$\begin{bmatrix} 1 & -2 & | & 2 \\ 3 & -5 & | & 9 \end{bmatrix} \xrightarrow[\underset{R_2 = -3r_1 + r_2}{\uparrow}]{} \begin{bmatrix} 1 & -2 & | & 2 \\ -3(1) + 3 & (-3)(-2) + (-5) & | & -3(2) + 9 \end{bmatrix} = \begin{bmatrix} 1 & -2 & | & 2 \\ 0 & 1 & | & 3 \end{bmatrix}$$

◼

E X A M P L E   5

Finding a Particular Row Operation

Using the matrix

$$\begin{bmatrix} 1 & -2 & | & 2 \\ 0 & 1 & | & 3 \end{bmatrix}$$

find a row operation that will result in a matrix with a 0 in row 1, column 2.

Solution   We want a 0 in row 1, column 2. This result can be accomplished by multiplying row 2 by 2 and adding the result to row 1. That is, we apply the row operation $R_1 = 2r_2 + r_1$:

$$\begin{bmatrix} 1 & -2 & | & 2 \\ 0 & 1 & | & 3 \end{bmatrix} \underset{\underset{R_1 = 2r_2 + r_1}{\uparrow}}{\rightarrow} \begin{bmatrix} 2(0) + 1 & 2(1) + (-2) & | & 2(3) + 2 \\ 0 & 1 & | & 3 \end{bmatrix} = \begin{bmatrix} 1 & 0 & | & 8 \\ 0 & 1 & | & 3 \end{bmatrix}. \blacksquare$$

A word about the notation we have introduced. A row operation such as $R_1 = 2r_2 + r_1$ changes the entries in row 1. Note also that to change the entries in a given row we multiply the entries in some other row by an appropriate number and add the results to the original entries of the row to be changed.

Now let's see how we use row operations to solve a system of linear equations.

E X A M P L E   6

Solving a System of Linear Equations Using Matrices

Solve: $\begin{cases} 4x + 3y = 11 & \text{(1)} \\ x - 3y = -1 & \text{(2)} \end{cases}$

Solution   First, we write the augmented matrix that represents this system:

$$\begin{bmatrix} 4 & 3 & | & 11 \\ 1 & -3 & | & -1 \end{bmatrix}$$

The first step requires getting the entry 1 in row 1, column 1. An interchange of rows 1 and 2 is the easiest way to do this:

$$\begin{bmatrix} 1 & -3 & | & -1 \\ 4 & 3 & | & 11 \end{bmatrix}$$

Next, we want a 0 under the entry 1 in column 1. We use the row operation $R_2 = -4r_1 + r_2$:

$$\begin{bmatrix} 1 & -3 & | & -1 \\ 4 & 3 & | & 11 \end{bmatrix} \underset{\underset{R_2 = -4r_1 + r_2}{\uparrow}}{\rightarrow} \begin{bmatrix} 1 & -3 & | & -1 \\ 0 & 15 & | & 15 \end{bmatrix}$$

Now we want the entry 1 in row 2, column 2. We use $R_2 = \frac{1}{15}r_2$:

$$\begin{bmatrix} 1 & -3 & | & -1 \\ 0 & 15 & | & 15 \end{bmatrix} \underset{\underset{R_2 = \frac{1}{15}r_2}{\uparrow}}{\rightarrow} \begin{bmatrix} 1 & -3 & | & -1 \\ 0 & 1 & | & 1 \end{bmatrix}$$

or, equivalently,

$$\begin{cases} x = \frac{2}{11}z + \frac{7}{11} & \text{(1)} \\ y = \frac{1}{11}z - \frac{2}{11} & \text{(2)} \end{cases}$$

where z can be any real number.

Now work Problem 53.

E X A M P L E   9   Solving a System of Linear Equations Using Matrices

$$\text{Solve: } \begin{cases} x + y + z = 6 \\ 2x - y - z = 3 \\ x + 2y + 2z = 0 \end{cases}$$

Solution   The augmented matrix is

$$\begin{bmatrix} 1 & 1 & 1 & | & 6 \\ 2 & -1 & -1 & | & 3 \\ 1 & 2 & 2 & | & 0 \end{bmatrix} \rightarrow \begin{bmatrix} 1 & 1 & 1 & | & 6 \\ 0 & -3 & -3 & | & -9 \\ 0 & 1 & 1 & | & -6 \end{bmatrix} \rightarrow \begin{bmatrix} 1 & 1 & 1 & | & 6 \\ 0 & 1 & 1 & | & -6 \\ 0 & -3 & -3 & | & -9 \end{bmatrix} \rightarrow \begin{bmatrix} 1 & 1 & 1 & | & 6 \\ 0 & 1 & 1 & | & -6 \\ 0 & 0 & 0 & | & -27 \end{bmatrix}$$

$$\underset{\substack{R_2 = -2r_1 + r_2 \\ R_3 = -1r_1 + r_3}}{\uparrow} \qquad \underset{\text{Interchange rows 2 and 3.}}{\uparrow} \qquad \underset{R_3 = 3r_2 + r_3.}{\uparrow}$$

This matrix is in echelon form. The bottom row is equivalent to the equation

$$0x + 0y + 0z = -27$$

which has no solution. Hence, the original system is inconsistent.   ∎

Now work Problems 23 and 29.

The matrix method is especially effective for systems of equations for which the number of equations and the number of variables are unequal. Here, too, such a system is either inconsistent or consistent. If it is consistent, it will have either exactly one solution or infinitely many solutions.

Let's look at a system of four equations containing three variables.

E X A M P L E   10   Solving a System of Linear Equations Using Matrices

$$\text{Solve: } \begin{cases} x - 2y + z = 0 & \text{(1)} \\ 2x + 2y - 3z = -3 & \text{(2)} \\ y - z = -1 & \text{(3)} \\ -x + 4y + 2z = 13 & \text{(4)} \end{cases}$$

Solution   The augmented matrix is

$$\begin{bmatrix} 1 & -2 & 1 & | & 0 \\ 2 & 2 & -3 & | & -3 \\ 0 & 1 & -1 & | & -1 \\ -1 & 4 & 2 & | & 13 \end{bmatrix} \rightarrow \begin{bmatrix} 1 & -2 & 1 & | & 0 \\ 0 & 6 & -5 & | & -3 \\ 0 & 1 & -1 & | & -1 \\ 0 & 2 & 3 & | & 13 \end{bmatrix} \rightarrow \begin{bmatrix} 1 & -2 & 1 & | & 0 \\ 0 & 1 & -1 & | & -1 \\ 0 & 6 & -5 & | & -3 \\ 0 & 2 & 3 & | & 13 \end{bmatrix}$$

$$\underset{\substack{R_2 = -2r_1 + r_2 \\ R_4 = r_1 + r_4}}{\uparrow} \qquad \underset{\text{Interchange rows 2 and 3.}}{\uparrow}$$

$$\rightarrow \begin{bmatrix} 1 & -2 & 1 & \bigm| & 0 \\ 0 & 1 & -1 & \bigm| & -1 \\ 0 & 0 & 1 & \bigm| & 3 \\ 0 & 0 & 5 & \bigm| & 15 \end{bmatrix} \rightarrow \begin{bmatrix} 1 & -2 & 1 & \bigm| & 0 \\ 0 & 1 & -1 & \bigm| & -1 \\ 0 & 0 & 1 & \bigm| & 3 \\ 0 & 0 & 0 & \bigm| & 0 \end{bmatrix}$$

$\uparrow$
$R_3 = -6r_2 + r_3$
$R_4 = -2r_2 + r_4$

$\uparrow$
$R_4 = -5r_3 + r_4$

We could stop here, since the matrix is in echelon form, and back-substitute $z = 3$ to find x and y. Or we can continue to obtain the reduced echelon form:

$$\rightarrow \begin{bmatrix} 1 & 0 & -1 & \bigm| & -2 \\ 0 & 1 & -1 & \bigm| & -1 \\ 0 & 0 & 1 & \bigm| & 3 \\ 0 & 0 & 0 & \bigm| & 0 \end{bmatrix} \rightarrow \begin{bmatrix} 1 & 0 & 0 & \bigm| & 1 \\ 0 & 1 & 0 & \bigm| & 2 \\ 0 & 0 & 1 & \bigm| & 3 \\ 0 & 0 & 0 & \bigm| & 0 \end{bmatrix}$$

$\uparrow$
$R_1 = 2r_2 + r_1$

$\uparrow$
$R_1 = r_3 + r_1$
$R_2 = r_3 + r_2$

The matrix is now in reduced echelon form, and we can see that the solution is $x = 1, y = 2, z = 3$. ∎

E X A M P L E   11   **Mixing Acids**

A chemistry laboratory has three containers of nitric acid, HNO_3. One container holds a solution with a concentration of 10% HNO_3, the second holds 20% HNO_3, and the third holds 40% HNO_3. How many liters of each solution should be mixed to obtain 100 liters of a solution whose concentration is 25% HNO_3?

Solution   Let x, y, and z represent the number of liters of 10%, 20%, and 40% concentrations of HNO_3, respectively. We want 100 liters in all, and the concentration of HNO_3 from each solution must sum to 25% of 100 liters. Thus, we find that

$$\begin{cases} x + y + z = 100 \\ 0.10x + 0.20y + 0.40z = 0.25(100) \end{cases}$$

Now, the augmented matrix is

$$\begin{bmatrix} 1 & 1 & 1 & \bigm| & 100 \\ 0.10 & 0.20 & 0.40 & \bigm| & 25 \end{bmatrix} \rightarrow \begin{bmatrix} 1 & 1 & 1 & \bigm| & 100 \\ 0 & 0.10 & 0.30 & \bigm| & 15 \end{bmatrix}$$

$\uparrow$
$R_2 = -0.10r_1 + r_2$

$$\rightarrow \begin{bmatrix} 1 & 1 & 1 & \bigm| & 100 \\ 0 & 1 & 3 & \bigm| & 150 \end{bmatrix} \rightarrow \begin{bmatrix} 1 & 0 & -2 & \bigm| & -50 \\ 0 & 1 & 3 & \bigm| & 150 \end{bmatrix}$$

$\uparrow$
$R_2 = 10r_2$

$\uparrow$
$R_1 = -1r_2 + r_1$

The matrix is now in reduced echelon form. The final matrix represents the system

$$\begin{cases} x - 2z = -50 & \text{(1)} \\ y + 3z = 150 & \text{(2)} \end{cases}$$

In Problems 11–20, use a graphing utility to perform, in order (a), followed by (b), followed by (c) on the given augmented matrix.

11. $\begin{bmatrix} 1 & -3 & -5 & | & -2 \\ 2 & -5 & -4 & | & 5 \\ -3 & 5 & 4 & | & 6 \end{bmatrix}$ (a) $R_2 = -2r_1 + r_2$ (b) $R_3 = 3r_1 + r_3$ (c) $R_3 = 4r_2 + r_3$

12. $\begin{bmatrix} 1 & -3 & -3 & | & -3 \\ 2 & -5 & 2 & | & -4 \\ -3 & 2 & 4 & | & 6 \end{bmatrix}$ (a) $R_2 = -2r_1 + r_2$ (b) $R_3 = 3r_1 + r_3$ (c) $R_3 = 7r_2 + r_3$

13. $\begin{bmatrix} 1 & -3 & 4 & | & 3 \\ 2 & -5 & 6 & | & 6 \\ -3 & 3 & 4 & | & 6 \end{bmatrix}$ (a) $R_2 = -2r_1 + r_2$ (b) $R_3 = 3r_1 + r_3$ (c) $R_3 = 6r_2 + r_3$

14. $\begin{bmatrix} 1 & -3 & 3 & | & -5 \\ 2 & -5 & -3 & | & -5 \\ -3 & -2 & 4 & | & 6 \end{bmatrix}$ (a) $R_2 = -2r_1 + r_2$ (b) $R_3 = 3r_1 + r_3$ (c) $R_3 = 11r_2 + r_3$

15. $\begin{bmatrix} 1 & -3 & 2 & | & -6 \\ 2 & -5 & 3 & | & -4 \\ -3 & -6 & 4 & | & 6 \end{bmatrix}$ (a) $R_2 = -2r_1 + r_2$ (b) $R_3 = 3r_1 + r_3$ (c) $R_3 = 15r_2 + r_3$

16. $\begin{bmatrix} 1 & -3 & -4 & | & -6 \\ 2 & -5 & 6 & | & -6 \\ -3 & 1 & 4 & | & 6 \end{bmatrix}$ (a) $R_2 = -2r_1 + r_2$ (b) $R_3 = 3r_1 + r_3$ (c) $R_3 = 8r_2 + r_3$

17. $\begin{bmatrix} 1 & -3 & 1 & | & -2 \\ 2 & -5 & 6 & | & -2 \\ -3 & 1 & 4 & | & 6 \end{bmatrix}$ (a) $R_2 = -2r_1 + r_2$ (b) $R_3 = 3r_1 + r_3$ (c) $R_3 = 8r_2 + r_3$

18. $\begin{bmatrix} 1 & -3 & -1 & | & 2 \\ 2 & -5 & 2 & | & 6 \\ -3 & -6 & 4 & | & 6 \end{bmatrix}$ (a) $R_2 = -2r_1 + r_2$ (b) $R_3 = 3r_1 + r_3$ (c) $R_3 = 15r_2 + r_3$

19. $\begin{bmatrix} 1 & -3 & -2 & | & 3 \\ 2 & -5 & 2 & | & -1 \\ -3 & -2 & 4 & | & 6 \end{bmatrix}$ (a) $R_2 = -2r_1 + r_2$ (b) $R_3 = 3r_1 + r_3$ (c) $R_3 = 11r_2 + r_3$

20. $\begin{bmatrix} 1 & -3 & 5 & | & -3 \\ 2 & -5 & 1 & | & -4 \\ -3 & 3 & 4 & | & 6 \end{bmatrix}$ (a) $R_2 = -2r_1 + r_2$ (b) $R_3 = 3r_1 + r_3$ (c) $R_3 = 6r_2 + r_3$

In Problems 21–30, the reduced echelon form of a system of linear equations is given. Write the system of equations corresponding to the given matrix. Use x, y, or x, y, z or x_1, x_2, x_3, x_4 as variables. Determine whether the system is consistent or inconsistent. If it is consistent, give the solution.

21. $\begin{bmatrix} 1 & 0 & | & 5 \\ 0 & 1 & | & -1 \end{bmatrix}$

22. $\begin{bmatrix} 1 & 0 & | & -4 \\ 0 & 1 & | & 0 \end{bmatrix}$

23. $\begin{bmatrix} 1 & 0 & 0 & | & 1 \\ 0 & 1 & 0 & | & 2 \\ 0 & 0 & 0 & | & 3 \end{bmatrix}$

24. $\begin{bmatrix} 1 & 0 & 0 & | & 0 \\ 0 & 1 & 0 & | & 0 \\ 0 & 0 & 0 & | & 2 \end{bmatrix}$

25. $\begin{bmatrix} 1 & 0 & 2 & | & -1 \\ 0 & 1 & -4 & | & -2 \\ 0 & 0 & 0 & | & 0 \end{bmatrix}$

26. $\begin{bmatrix} 1 & 0 & 4 & | & 4 \\ 0 & 1 & 3 & | & 2 \\ 0 & 0 & 0 & | & 0 \end{bmatrix}$

27. $\begin{bmatrix} 1 & 0 & 0 & 0 & | & 1 \\ 0 & 1 & 0 & 1 & | & 2 \\ 0 & 0 & 1 & 2 & | & 3 \end{bmatrix}$

28. $\begin{bmatrix} 1 & 0 & 0 & 0 & | & 1 \\ 0 & 1 & 0 & 2 & | & 2 \\ 0 & 0 & 1 & 3 & | & 0 \end{bmatrix}$

29. $\begin{bmatrix} 1 & 0 & 0 & 4 & | & 2 \\ 0 & 1 & 1 & 3 & | & 3 \\ 0 & 0 & 0 & 0 & | & 0 \end{bmatrix}$

30. $\begin{bmatrix} 1 & 0 & 0 & 0 & | & 1 \\ 0 & 1 & 0 & 0 & | & 2 \\ 0 & 0 & 1 & 2 & | & 3 \end{bmatrix}$

In Problems 31–72, solve each system of equations using matrices (row operations). If the system has no solution, say it is inconsistent.

31. $\begin{cases} x + y = 8 \\ x - y = 4 \end{cases}$

32. $\begin{cases} x + 2y = 5 \\ x + y = 3 \end{cases}$

33. $\begin{cases} x - 5y = -13 \\ 3x + 2y = 12 \end{cases}$

34. $\begin{cases} x + 3y = 5 \\ 2x - 3y = -8 \end{cases}$

35. $\begin{cases} 3x - 6y = 24 \\ 5x + 4y = 12 \end{cases}$

36. $\begin{cases} 2x + 4y = 16 \\ 3x - 5y = -9 \end{cases}$

37. $\begin{cases} 2x + y = 1 \\ 4x + 2y = 6 \end{cases}$

38. $\begin{cases} x - y = 5 \\ -3x + 3y = 2 \end{cases}$

39. $\begin{cases} 2x - 4y = -2 \\ 3x + 2y = 3 \end{cases}$

40. $\begin{cases} 3x + 3y = 3 \\ 4x + 2y = \frac{8}{3} \end{cases}$

41. $\begin{cases} x + 2y = 4 \\ 2x + 4y = 8 \end{cases}$

42. $\begin{cases} 3x - y = 7 \\ 9x - 3y = 21 \end{cases}$

43. $\begin{cases} 2x + 3y = 6 \\ x - y = \frac{1}{2} \end{cases}$

44. $\begin{cases} \frac{1}{2}x + y = -2 \\ x - 2y = 8 \end{cases}$

45. $\begin{cases} 3x - 5y = 3 \\ 15x + 5y = 21 \end{cases}$

46. $\begin{cases} 2x - y = -1 \\ x + \frac{1}{2}y = \frac{3}{2} \end{cases}$

47. $\begin{cases} x - y = 6 \\ 2x - 3z = 16 \\ 2y + z = 4 \end{cases}$

48. $\begin{cases} 2x + y = -4 \\ -2y + 4z = 0 \\ 3x - 2z = -11 \end{cases}$

49. $\begin{cases} x - 2y + 3z = 7 \\ 2x + y + z = 4 \\ -3x + 2y - 2z = -10 \end{cases}$

50. $\begin{cases} 2x + y - 3z = 0 \\ -2x + 2y + z = -7 \\ 3x - 4y - 3z = 7 \end{cases}$

51. $\begin{cases} 2x - 2y - 2z = 2 \\ 2x + 3y + z = 2 \\ 3x + 2y = 0 \end{cases}$

52. $\begin{cases} 2x - 3y - z = 0 \\ -x + 2y + z = 5 \\ 3x - 4y - z = 1 \end{cases}$

53. $\begin{cases} -x + y + z = -1 \\ -x + 2y - 3z = -4 \\ 3x - 2y - 7z = 0 \end{cases}$

54. $\begin{cases} 2x - 3y - z = 0 \\ 3x + 2y + 2z = 2 \\ x + 5y + 3z = 2 \end{cases}$

55. $\begin{cases} 2x - 2y + 3z = 6 \\ 4x - 3y + 2z = 0 \\ -2x + 3y - 7z = 1 \end{cases}$

56. $\begin{cases} 3x - 2y + 2z = 6 \\ 7x - 3y + 2z = -1 \\ 2x - 3y + 4z = 0 \end{cases}$

57. $\begin{cases} x + y - z = 6 \\ 3x - 2y + z = -5 \\ x + 3y - 2z = 14 \end{cases}$

58. $\begin{cases} x - y + z = -4 \\ 2x - 3y + 4z = -15 \\ 5x + y - 2z = 12 \end{cases}$

59. $\begin{cases} x + 2y - z = -3 \\ 2x - 4y + z = -7 \\ -2x + 2y - 3z = 4 \end{cases}$

60. $\begin{cases} x + 4y - 3z = -8 \\ 3x - y + 3z = 12 \\ x + y + 6z = 1 \end{cases}$

61. $\begin{cases} 3x + y - z = \frac{2}{3} \\ 2x - y + z = 1 \\ 4x + 2y = \frac{8}{3} \end{cases}$

62. $\begin{cases} x + y = 1 \\ 2x - y + z = 1 \\ x + 2y + z = \frac{8}{3} \end{cases}$

63. $\begin{cases} x + y + z + w = 4 \\ 2x - y + z = 0 \\ 3x + 2y + z - w = 6 \\ x - 2y - 2z + 2w = -1 \end{cases}$

64. $\begin{cases} x + y + z + w = 4 \\ -x + 2y + z = 0 \\ 2x + 3y + z - w = 6 \\ -2x + y - 2z + 2w = -1 \end{cases}$

65. $\begin{cases} x + 2y + z = 1 \\ 2x - y + 2z = 2 \\ 3x + y + 3z = 3 \end{cases}$

66. $\begin{cases} x + 2y - z = 3 \\ 2x - y + 2z = 6 \\ x - 3y + 3z = 4 \end{cases}$

67. $\begin{cases} x - y + z = 5 \\ 3x + 2y - 2z = 0 \end{cases}$

68. $\begin{cases} 2x + y - z = 4 \\ -x + y + 3z = 1 \end{cases}$

69. $\begin{cases} 2x + 3y - z = 3 \\ x - y - z = 0 \\ -x + y + z = 0 \\ x + y + 3z = 5 \end{cases}$

70. $\begin{cases} x - 3y + z = 1 \\ 2x - y - 4z = 0 \\ x - 3y + 2z = 1 \\ x - 2y = 5 \end{cases}$

71. $\begin{cases} 4x + y + z - w = 4 \\ x - y + 2z + 3w = 3 \end{cases}$

72. $\begin{cases} -4x + y = 5 \\ 2x - y + z - w = 5 \\ z + w = 4 \end{cases}$

Equations (3) and (4) lead us to the following result, called **Cramer's Rule:**

Theorem Cramer's Rule for Two Equations Containing Two Variables

The solution to the system of equations

$$\begin{cases} ax + by = s & (1) \\ cx + dy = t & (2) \end{cases} \qquad (5)$$

is given by

$$x = \frac{\begin{vmatrix} s & b \\ t & d \end{vmatrix}}{\begin{vmatrix} a & b \\ c & d \end{vmatrix}}, \qquad y = \frac{\begin{vmatrix} a & s \\ c & t \end{vmatrix}}{\begin{vmatrix} a & b \\ c & d \end{vmatrix}} \qquad (6)$$

provided that

$$D = \begin{vmatrix} a & b \\ c & d \end{vmatrix} = ad - bc \neq 0$$

In the derivation given for Cramer's Rule above, we assumed that none of the numbers a, b, c, and d were 0. In Problem 58 at the end of this section you will be asked to complete the proof under the less stringent conditions that $D = ad - bc \neq 0$.

Now look carefully at the pattern in Cramer's Rule. The denominator in the solution (6) is the determinant of the coefficients of the variables:

$$\begin{cases} ax + by = s \\ cx + dy = t \end{cases} \qquad D = \begin{vmatrix} a & b \\ c & d \end{vmatrix}$$

In the solution for x, the numerator is the determinant, denoted by D_x, formed by replacing the entries in the first column (the coefficients of x) in D by the constants on the right side of the equal sign:

$$D_x = \begin{vmatrix} s & b \\ t & d \end{vmatrix}$$

In the solution for y, the numerator is the determinant, denoted by D_y, formed by replacing the entries in the second column (the coefficients of y) in D by the constants on the right side of the equal sign:

$$D_y = \begin{vmatrix} a & s \\ c & t \end{vmatrix}$$

Cramer's Rule then states that, if $D \neq 0$,

$$x = \frac{D_x}{D}, \qquad y = \frac{D_y}{D} \qquad (7)$$

E X A M P L E   3   Solving a System of Linear Equations Using Determinants

Use Cramer's Rule, if applicable, to solve the system

$$\begin{cases} 3x - 2y = 4 & (1) \\ 6x + y = 13 & (2) \end{cases}$$

Algebraic Solution The determinant D of the coefficients of the variables is

$$D = \begin{vmatrix} 3 & -2 \\ 6 & 1 \end{vmatrix} = (3)(1) - (6)(-2) = 15$$

Because $D \neq 0$, Cramer's Rule (7) can be used:

$$x = \frac{D_x}{D} = \frac{\begin{vmatrix} 4 & -2 \\ 13 & 1 \end{vmatrix}}{15} = \frac{30}{15} = 2, \qquad y = \frac{D_y}{D} = \frac{\begin{vmatrix} 3 & 4 \\ 6 & 13 \end{vmatrix}}{15} = \frac{15}{15} = 1$$

The solution is $x = 2$, $y = 1$.

Graphing Solution We enter the coefficient matrix into our graphing utility and call it A. We enter the matrices D_x and D_y into our graphing utility and call them B and C, respectively. Evaluate det $[A]$. Since det $[A] \neq 0$, we can use Cramer's Rule. Finally, we find x by calculating det$[B]$/det$[A]$ and y by calculating det$[C]$/det$[A]$. The results are shown in Figure 10.

The solution is $x = 2$, $y = 1$. ▬

FIGURE 10

```
det [A]
            15
det [B]/det [A]
            2
det [C]/det [A]
            1
```

If, in attempting to use Cramer's Rule, the determinant D of the coefficients of the variables is found to equal 0 (so that Cramer's Rule is not applicable), then the system either is inconsistent or has infinitely many solutions. (Refer to Problem 90 in Exercise 10.2.)

Now work Problem 11.

3 by 3 Determinants

In order to use Cramer's Rule to solve a system of three equations containing three variables, we need to define a 3 by 3 determinant.

A **3 by 3 determinant** is symbolized by

$$\begin{vmatrix} a_{11} & a_{12} & a_{13} \\ a_{21} & a_{22} & a_{23} \\ a_{31} & a_{32} & a_{33} \end{vmatrix} \tag{8}$$

in which $a_{11}, a_{12}, \ldots$ are real numbers.

As with matrices, we use a double subscript to identify an entry by indicating its row and column numbers. For example, the entry a_{23} is in row 2, column 3.

The value of a 3 by 3 determinant may be defined in terms of 2 by 2 determinants by the following formula:

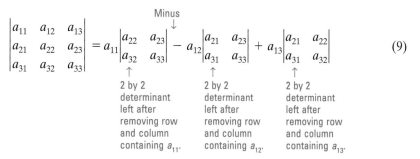

$$\begin{vmatrix} a_{11} & a_{12} & a_{13} \\ a_{21} & a_{22} & a_{23} \\ a_{31} & a_{32} & a_{33} \end{vmatrix} = a_{11} \begin{vmatrix} a_{22} & a_{23} \\ a_{32} & a_{33} \end{vmatrix} \overset{\text{Minus}}{-} a_{12} \begin{vmatrix} a_{21} & a_{23} \\ a_{31} & a_{33} \end{vmatrix} + a_{13} \begin{vmatrix} a_{21} & a_{22} \\ a_{31} & a_{32} \end{vmatrix} \tag{9}$$

2 by 2 determinant left after removing row and column containing a_{11}.

2 by 2 determinant left after removing row and column containing a_{12}.

2 by 2 determinant left after removing row and column containing a_{13}.

Be sure to take note of the minus sign that appears with the second term—it's easy to forget it! Formula (9) is best remembered by noting that each entry in row 1 is multiplied by the 2 by 2 determinant that remains after the row and column containing the entry have been removed, as follows:

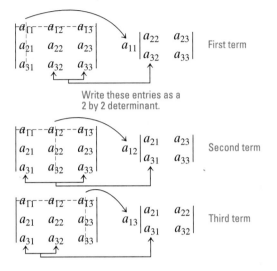

First term

Write these entries as a
2 by 2 determinant.

Second term

Third term

Now insert the minus sign before the middle expression and add

$$
\begin{vmatrix} a_{11} & a_{12} & a_{13} \\ a_{21} & a_{22} & a_{23} \\ a_{31} & a_{32} & a_{33} \end{vmatrix} = a_{11} \begin{vmatrix} a_{22} & a_{23} \\ a_{32} & a_{33} \end{vmatrix} \overset{\text{Minus}}{\underset{\downarrow}{-}} a_{12} \begin{vmatrix} a_{21} & a_{23} \\ a_{31} & a_{33} \end{vmatrix} + a_{13} \begin{vmatrix} a_{21} & a_{22} \\ a_{31} & a_{32} \end{vmatrix}
$$

Formula (9) exhibits one way to find the value of a 3 by 3 determinant, *by expanding across row 1.* In fact, the expansion can take place across any row or down any column. The terms to be added or subtracted consist of the row (or column) entry times the value of the 2 by 2 determinant that remains after removing the row and column entry. The value of the determinant is found by adding or subtracting the terms according to the following scheme:

$$
\begin{array}{ccc}
+ & - & + \\
- & + & - \\
+ & - & +
\end{array}
$$

For example, if we choose to expand down column 2, we obtain

$$
\begin{vmatrix} a_{11} & a_{12} & a_{13} \\ a_{21} & a_{22} & a_{23} \\ a_{31} & a_{32} & a_{33} \end{vmatrix} = - a_{12} \begin{vmatrix} a_{21} & a_{23} \\ a_{31} & a_{33} \end{vmatrix} + a_{22} \begin{vmatrix} a_{11} & a_{13} \\ a_{31} & a_{33} \end{vmatrix} - a_{32} \begin{vmatrix} a_{11} & a_{13} \\ a_{21} & a_{23} \end{vmatrix}
$$

Expand down column 2 $(-, +, -)$.

If we choose to expand across row 3, we obtain

$$
\begin{vmatrix} a_{11} & a_{12} & a_{13} \\ a_{21} & a_{22} & a_{23} \\ a_{31} & a_{32} & a_{33} \end{vmatrix} = a_{31} \begin{vmatrix} a_{12} & a_{13} \\ a_{22} & a_{23} \end{vmatrix} - a_{32} \begin{vmatrix} a_{11} & a_{13} \\ a_{21} & a_{23} \end{vmatrix} + a_{33} \begin{vmatrix} a_{11} & a_{12} \\ a_{21} & a_{22} \end{vmatrix}
$$

Expand across row 3 $(+, -, +)$.

MISSION POSSIBLE

Keeping Secrets

Your group has discovered that a rival consulting firm has been intercepting and reading the reports you are sending one another. You need a way to encode your final decisions so that this rival firm does not take business away from you by stealing your ideas. You have decided to find a way to encode your most important messages to each other so that prying eyes are not able to read them.

Your method of coding begins with the simple one used by children, that is, assigning to each letter of the alphabet the number which represents its position in the order. For example, A = 1, B = 2, ..., M = 13, N = 14, ..., Z = 26. A space would be represented by 27; a period by 0. Then the message is translated into 2 × 1 matrices as in the example which follows.

Suppose you wish to send the message: *Choose Dealer C.* You would write this out first as letters in groups of two, then as matrices.

$$\begin{matrix} Ch & oo & se & D & ea & le & r & C \\ \begin{bmatrix} 3 \\ 8 \end{bmatrix} & \begin{bmatrix} 15 \\ 15 \end{bmatrix} & \begin{bmatrix} 19 \\ 5 \end{bmatrix} & \begin{bmatrix} 27 \\ 4 \end{bmatrix} & \begin{bmatrix} 5 \\ 1 \end{bmatrix} & \begin{bmatrix} 12 \\ 5 \end{bmatrix} & \begin{bmatrix} 18 \\ 27 \end{bmatrix} & \begin{bmatrix} 3 \\ 0 \end{bmatrix} \end{matrix}$$

Next you would use a coding matrix, a 2 × 2 square matrix with numbers chosen so that the inverse exists. We'll use $\begin{bmatrix} 6 & 4 \\ -2 & 7 \end{bmatrix}$ as our coding matrix. To encode a message, you multiply the coding matrix times each matrix of the message, as follows:.

$$\begin{bmatrix} 6 & 4 \\ -2 & 7 \end{bmatrix} \cdot \begin{bmatrix} 3 \\ 8 \end{bmatrix} = \begin{bmatrix} 50 \\ 50 \end{bmatrix}, \begin{bmatrix} 6 & 4 \\ -2 & 7 \end{bmatrix} \cdot \begin{bmatrix} 15 \\ 15 \end{bmatrix} = \begin{bmatrix} 150 \\ 75 \end{bmatrix}, \text{ and so on}$$

The resulting set of matrices would be sent along to your partner as a sequence of numbers: 50, 50, 150, 75, and so on. Your partner would use the inverse of the encoding matrix to decipher your message. Obviously you would have to guard the encoding matrix and its inverse very carefully!

1. Find the inverse of the coding matrix above. Make sure it is correct by multiplying the two encoded matrices times your matrix to see if you can recover the original two matrices.

2. Next, create your own message (not too long!) and encrypt it using the coding matrix above. Copy the resulting code onto a separate piece of paper as a sequence of numbers.

3. Exchange coded messages with another group.

4. Use the inverse matrix to decode the message sent to you by the other group.

5. Cryptography is a science practiced by governments at war, or businesses which are competitive, in situations where secrecy is crucial. It is especially important in transferring valuable information, such as credit card numbers, over the Internet. If your consulting firm is serious about keeping secrets, you may need to upgrade your coding procedures. For example, the message *Choose Dealer C*, given above, could be broken up into four 4 × 1 matrices. Then the coding matrix would have to be 4 × 4 with an inverse to match. A government might use a 20 × 20 coding matrix to decrease the probability that it could be deciphered. Create your own 4 × 4 coding matrix, find its inverse, and encode a message of your choosing in 4 ×1 matrices. Then pass the code and the key to a neighboring group for decoding.

It can be shown that the value of a determinant does not depend on the choice of the row or column used in the expansion.

E X A M P L E   4

Evaluating a 3 × 3 Determinant

Find the value of the 3 by 3 determinant: $\begin{vmatrix} 3 & 4 & -1 \\ 4 & 6 & 2 \\ 8 & -2 & 3 \end{vmatrix}$

Solution   We choose to expand across row 1.

Remember the minus sign.
↓

$$\begin{vmatrix} 3 & 4 & -1 \\ 4 & 6 & 2 \\ 8 & -2 & 3 \end{vmatrix} = 3\begin{vmatrix} 6 & 2 \\ -2 & 3 \end{vmatrix} - 4\begin{vmatrix} 4 & 2 \\ 8 & 3 \end{vmatrix} + (-1)\begin{vmatrix} 4 & 6 \\ 8 & -2 \end{vmatrix}$$

$$= 3(18 + 4) - 4(12 - 16) + (-1)(-8 - 48)$$
$$= 3(22) - 4(-4) + (-1)(-56)$$
$$= 66 + 16 + 56 = 138$$

We could also find the value of the 3 by 3 determinant in Example 4 by expanding down column 3 (the signs are +, −, +):

$$\begin{vmatrix} 3 & 4 & -1 \\ 4 & 6 & 2 \\ 8 & -2 & 3 \end{vmatrix} = (-1)\begin{vmatrix} 4 & 6 \\ 8 & -2 \end{vmatrix} - 2\begin{vmatrix} 3 & 4 \\ 8 & -2 \end{vmatrix} + 3\begin{vmatrix} 3 & 4 \\ 4 & 6 \end{vmatrix}$$

$$= -1(-8 - 48) - 2(-6 - 32) + 3(18 - 16)$$
$$= 56 + 76 + 6 = 138$$

Evaluating 3 × 3 determinants on a graphing utility follows the same procedure as evaluating 2 × 2 determinants.

Now work Problem 7.

Systems of Three Equations Containing Three Variables

Consider the following system of three equations containing three variables:

$$\begin{cases} a_{11}x + a_{12}y + a_{13}z = c_1 \\ a_{21}x + a_{22}y + a_{23}z = c_2 \\ a_{31}x + a_{32}y + a_{33}z = c_3 \end{cases} \qquad (10)$$

It can be shown that if the determinant D of the coefficients of the variables is not 0, that is, if

$$D = \begin{vmatrix} a_{11} & a_{12} & a_{13} \\ a_{21} & a_{22} & a_{23} \\ a_{31} & a_{32} & a_{33} \end{vmatrix} \neq 0$$

then the unique solution of system (10) is given by

Cramer's Rule for Three Equations Containing Three Variables

$$x = \frac{D_x}{D}, \qquad y = \frac{D_y}{D}, \qquad z = \frac{D_z}{D}$$

where

$$D_x = \begin{vmatrix} c_1 & a_{12} & a_{13} \\ c_2 & a_{22} & a_{23} \\ c_3 & a_{32} & a_{33} \end{vmatrix} \qquad D_y = \begin{vmatrix} a_{11} & c_1 & a_{13} \\ a_{21} & c_2 & a_{23} \\ a_{31} & c_3 & a_{33} \end{vmatrix} \qquad D_z = \begin{vmatrix} a_{11} & a_{12} & c_1 \\ a_{21} & a_{22} & c_2 \\ a_{31} & a_{32} & c_3 \end{vmatrix}$$

The similarity of this pattern and the pattern observed earlier for a system of two equations containing two variables should be apparent.

E X A M P L E 5   **Using Cramer's Rule**

Use Cramer's Rule, if applicable, to solve the following system:

$$\begin{cases} 2x + y - z = 3 & \text{(1)} \\ -x + 2y + 4z = -3 & \text{(2)} \\ x - 2y - 3z = 4 & \text{(3)} \end{cases}$$

Solution   The value of the determinant D of the coefficients of the variables is

$$D = \begin{vmatrix} 2 & 1 & -1 \\ -1 & 2 & 4 \\ 1 & -2 & -3 \end{vmatrix} = 2\begin{vmatrix} 2 & 4 \\ -2 & -3 \end{vmatrix} - 1\begin{vmatrix} -1 & 4 \\ 1 & -3 \end{vmatrix} + (-1)\begin{vmatrix} -1 & 2 \\ 1 & -2 \end{vmatrix}$$
$$= 2(2) - 1(-1) + (-1)(0)$$
$$= 4 + 1 = 5$$

Because $D \neq 0$, we proceed to find the values of D_x, D_y, and D_z:

$$D_x = \begin{vmatrix} 3 & 1 & -1 \\ -3 & 2 & 4 \\ 4 & -2 & -3 \end{vmatrix} = 3\begin{vmatrix} 2 & 4 \\ -2 & -3 \end{vmatrix} - 1\begin{vmatrix} -3 & 4 \\ 4 & -3 \end{vmatrix} + (-1)\begin{vmatrix} -3 & 2 \\ 4 & -2 \end{vmatrix}$$
$$= 3(2) - 1(-7) + (-1)(-2) = 15$$

$$D_y = \begin{vmatrix} 2 & 3 & -1 \\ -1 & -3 & 4 \\ 1 & 4 & -3 \end{vmatrix} = 2\begin{vmatrix} -3 & 4 \\ 4 & -3 \end{vmatrix} - 3\begin{vmatrix} -1 & 4 \\ 1 & -3 \end{vmatrix} + (-1)\begin{vmatrix} -1 & -3 \\ 1 & 4 \end{vmatrix}$$
$$= 2(-7) - 3(-1) + (-1)(-1)$$
$$= -14 + 3 + 1 = -10$$

$$D_z = \begin{vmatrix} 2 & 1 & 3 \\ -1 & 2 & -3 \\ 1 & -2 & 4 \end{vmatrix} = 2\begin{vmatrix} 2 & -3 \\ -2 & 4 \end{vmatrix} - 1\begin{vmatrix} -1 & -3 \\ 1 & 4 \end{vmatrix} + 3\begin{vmatrix} -1 & 2 \\ 1 & -2 \end{vmatrix}$$
$$= 2(2) - 1(-1) + 3(0) = 5$$

As a result,

$$x = \frac{D_x}{D} = \frac{15}{5} = 3, \qquad y = \frac{D_y}{D} = \frac{-10}{5} = -2, \qquad z = \frac{D_z}{D} = \frac{5}{5} = 1$$

The solution is $x = 3, y = -2, z = 1$. ∎

If the determinant of the coefficients of the variables of a system of three linear equations containing three variables is 0, then Cramer's Rule is not applicable. In such a case, the system either is inconsistent or has infinitely many solutions.

Solving systems of three equations containing three variables using Cramer's Rule on a graphing utility follows the same procedure as that for solving systems of two equations containing two variables.

Now work Problem 29.

More about Determinants

Determinants have several properties that are sometimes helpful for obtaining their value. We list some of them here.

> **Theorem**
>
> The value of a determinant changes sign if any two rows (or any two columns) are interchanged. (11)

Proof for 2 by 2 Determinants

$$\begin{vmatrix} a & b \\ c & d \end{vmatrix} = ad - bc \quad \text{and} \quad \begin{vmatrix} c & d \\ a & b \end{vmatrix} = bc - ad = -(ad - bc)$$

E X A M P L E 6   Demonstrating a Theorem (11)

$$\begin{vmatrix} 3 & 4 \\ 1 & 2 \end{vmatrix} = 6 - 4 = 2 \qquad \begin{vmatrix} 1 & 2 \\ 3 & 4 \end{vmatrix} = 4 - 6 = -2$$

> **Theorem**
>
> If all the entries in any row (or any column) equal 0, the value of the determinant is 0. (12)

Proof Merely expand across the row (or down the column) containing the 0's.

> **Theorem**
>
> If any two rows (or any two columns) of a determinant have corresponding entries that are equal, the value of the determinant is 0. (13)

You are asked to prove this result for a 3 by 3 determinant in which the entries in column 1 equal the entries in column 3 in Problem 61 at the end of this section.

E X A M P L E   7   Demonstrating a Theorem (13)

$$\begin{vmatrix} 1 & 2 & 3 \\ 1 & 2 & 3 \\ 4 & 5 & 6 \end{vmatrix} = 1\begin{vmatrix} 2 & 3 \\ 5 & 6 \end{vmatrix} - 2\begin{vmatrix} 1 & 3 \\ 4 & 6 \end{vmatrix} + 3\begin{vmatrix} 1 & 2 \\ 4 & 5 \end{vmatrix}$$

$$= 1(-3) - 2(-6) + 3(-3)$$
$$= -3 + 12 - 9 = 0$$

■

Theorem

If any row (or any column) of a determinant is multiplied by a nonzero number k, the value of the determinant is also changed by a factor of k.

(14)

■

You are asked to prove this result for a 3 by 3 determinant using row 2 in Problem 60 at the end of this section.

E X A M P L E   8   Demonstrating a Theorem (14)

$$\begin{vmatrix} 1 & 2 \\ 4 & 6 \end{vmatrix} = 6 - 8 = -2$$

$$\begin{vmatrix} k & 2k \\ 4 & 6 \end{vmatrix} = 6k - 8k = -2k = k(-2) = k\begin{vmatrix} 1 & 2 \\ 4 & 6 \end{vmatrix}$$

■

Theorem

If the entries of any row (or any column) of a determinant are multiplied by a nonzero number k and the result is added to the corresponding entries of another row (or column), the value of the determinant remains unchanged.

(15)

■

In Problem 62 at the end of this section, you are asked to prove this result for a 3 by 3 determinant using rows 1 and 2.

E X A M P L E   9   Demonstrating a Theorem (15)

$$\begin{vmatrix} 3 & 4 \\ 5 & 2 \end{vmatrix} \underset{\uparrow}{=} \begin{vmatrix} -7 & 0 \\ 5 & 2 \end{vmatrix} = -14$$

Multiply row 2 by −2 and add to row 1.

■

10.3 EXERCISES

In Problems 1–10, find the value of each determinant (a) by hand; (b) using a graphing utility.

1. $\begin{vmatrix} 3 & 1 \\ 4 & 2 \end{vmatrix}$
  2. $\begin{vmatrix} 6 & 1 \\ 5 & 2 \end{vmatrix}$
  3. $\begin{vmatrix} 6 & 4 \\ -1 & 3 \end{vmatrix}$
  4. $\begin{vmatrix} 8 & -3 \\ 4 & 2 \end{vmatrix}$
  5. $\begin{vmatrix} -3 & -1 \\ 4 & 2 \end{vmatrix}$

6. $\begin{vmatrix} -4 & 2 \\ -5 & 3 \end{vmatrix}$
  7. $\begin{vmatrix} 3 & 4 & 2 \\ 1 & -1 & 5 \\ 1 & 2 & -2 \end{vmatrix}$
  8. $\begin{vmatrix} 1 & 3 & -2 \\ 6 & 1 & -5 \\ 8 & 2 & 3 \end{vmatrix}$
  9. $\begin{vmatrix} 4 & -1 & 2 \\ 6 & -1 & 0 \\ 1 & -3 & 4 \end{vmatrix}$
  10. $\begin{vmatrix} 3 & -9 & 4 \\ 1 & 4 & 0 \\ 8 & -3 & 1 \end{vmatrix}$

In Problems 11–38, solve each system of equations using Cramer's Rule if it is applicable (a) by hand; (b) using a graphing utility. If Cramer's Rule is not applicable, say so.

11. $\begin{cases} x + y = 8 \\ x - y = 4 \end{cases}$
  12. $\begin{cases} x + 2y = 5 \\ x - y = 3 \end{cases}$
  13. $\begin{cases} 5x - y = 13 \\ 2x + 3y = 12 \end{cases}$
  14. $\begin{cases} x + 3y = 5 \\ 2x - 3y = -8 \end{cases}$

15. $\begin{cases} 3x = 24 \\ x + 2y = 0 \end{cases}$
  16. $\begin{cases} 4x + 5y = -3 \\ -2y = -4 \end{cases}$
  17. $\begin{cases} 3x - 6y = 24 \\ 5x + 4y = 12 \end{cases}$
  18. $\begin{cases} 2x + 4y = 16 \\ 3x - 5y = -9 \end{cases}$

19. $\begin{cases} 3x - 2y = 4 \\ 6x - 4y = 0 \end{cases}$
  20. $\begin{cases} -x + 2y = 5 \\ 4x - 8y = 6 \end{cases}$
  21. $\begin{cases} 2x - 4y = -2 \\ 3x + 2y = 3 \end{cases}$
  22. $\begin{cases} 3x + 3y = 3 \\ 4x + 2y = \frac{8}{3} \end{cases}$

23. $\begin{cases} 2x - 3y = -1 \\ 10x + 10y = 5 \end{cases}$
  24. $\begin{cases} 3x - 2y = 0 \\ 5x + 10y = 4 \end{cases}$
  25. $\begin{cases} 2x + 3y = 6 \\ x - y = \frac{1}{2} \end{cases}$
  26. $\begin{cases} \frac{1}{2}x + y = -2 \\ x - 2y = 8 \end{cases}$

27. $\begin{cases} 3x - 5y = 3 \\ 15x + 5y = 21 \end{cases}$
  28. $\begin{cases} 2x - y = -1 \\ x + \frac{1}{2}y = \frac{3}{2} \end{cases}$
  29. $\begin{cases} x + y - z = 6 \\ 3x - 2y + z = -5 \\ x + 3y - 2z = 14 \end{cases}$

30. $\begin{cases} x - y + z = -4 \\ 2x - 3y + 4z = -15 \\ 5x + y - 2z = 12 \end{cases}$
  31. $\begin{cases} x + 2y - z = -3 \\ 2x - 4y + z = -7 \\ -2x + 2y - 3z = 4 \end{cases}$
  32. $\begin{cases} x + 4y - 3z = -8 \\ 3x - y + 3z = 12 \\ x + y + 6z = 1 \end{cases}$

33. $\begin{cases} x - 2y + 3z = 1 \\ 3x + y - 2z = 0 \\ 2x - 4y + 6z = 2 \end{cases}$
  34. $\begin{cases} x - y + 2z = 5 \\ 3x + 2y = 4 \\ -2x + 2y - 4z = -10 \end{cases}$
  35. $\begin{cases} x + 2y - z = 0 \\ 2x - 4y + z = 0 \\ -2x + 2y - 3z = 0 \end{cases}$

36. $\begin{cases} x + 4y - 3z = 0 \\ 3x - y + 3z = 0 \\ x + y + 6z = 0 \end{cases}$
  37. $\begin{cases} x - 2y + 3z = 0 \\ 3x + y - 2z = 0 \\ 2x - 4y + 6z = 0 \end{cases}$
  38. $\begin{cases} x - y + 2z = 0 \\ 3x + 2y = 0 \\ -2x + 2y - 4z = 0 \end{cases}$

39. Solve $\begin{cases} \dfrac{1}{x} + \dfrac{1}{y} = 8 \\ \dfrac{3}{x} - \dfrac{5}{y} = 0 \end{cases}$
  40. Solve $\begin{cases} \dfrac{4}{x} - \dfrac{3}{y} = 0 \\ \dfrac{6}{x} + \dfrac{3}{2y} = 2 \end{cases}$

[**Hint:** Let $u = 1/x$ and $v = 1/y$ and solve for u and v.]

In Problems 41–46, solve for x.

41. $\begin{vmatrix} x & x \\ 4 & 3 \end{vmatrix} = 5$
  42. $\begin{vmatrix} x & 1 \\ 3 & x \end{vmatrix} = -2$
  43. $\begin{vmatrix} x & 1 & 1 \\ 4 & 3 & 2 \\ -1 & 2 & 5 \end{vmatrix} = 2$

44. $\begin{vmatrix} 3 & 2 & 4 \\ 1 & x & 5 \\ 0 & 1 & -2 \end{vmatrix} = 0$
  45. $\begin{vmatrix} x & 2 & 3 \\ 1 & x & 0 \\ 6 & 1 & -2 \end{vmatrix} = 7$
  46. $\begin{vmatrix} x & 1 & 2 \\ 1 & x & 3 \\ 0 & 1 & 2 \end{vmatrix} = -4x$

In Problems 47–54, use properties of determinants to find the value of each determinant if it is known that

$$\begin{vmatrix} x & y & z \\ u & v & w \\ 1 & 2 & 3 \end{vmatrix} = 4$$

47. $\begin{vmatrix} 1 & 2 & 3 \\ u & v & w \\ x & y & z \end{vmatrix}$

48. $\begin{vmatrix} x & y & z \\ u & v & w \\ 2 & 4 & 6 \end{vmatrix}$

49. $\begin{vmatrix} x & y & z \\ -3 & -6 & -9 \\ u & v & w \end{vmatrix}$

50. $\begin{vmatrix} 1 & 2 & 3 \\ x-u & y-v & z-w \\ u & v & w \end{vmatrix}$

51. $\begin{vmatrix} 1 & 2 & 3 \\ x-3 & y-6 & z-9 \\ 2u & 2v & 2w \end{vmatrix}$

52. $\begin{vmatrix} x & y & z-x \\ u & v & w-u \\ 1 & 2 & 2 \end{vmatrix}$

53. $\begin{vmatrix} 1 & 2 & 3 \\ 2x & 2y & 2z \\ u-1 & v-2 & w-3 \end{vmatrix}$

54. $\begin{vmatrix} x+3 & y+6 & z+9 \\ 3u-1 & 3v-2 & 3w-3 \\ 1 & 2 & 3 \end{vmatrix}$

55. **Geometry: Equation of a Line** An equation of the line containing the two points (x_1, y_1) and (x_2, y_2) may be expressed as the determinant

$$\begin{vmatrix} x & y & 1 \\ x_1 & y_1 & 1 \\ x_2 & y_2 & 1 \end{vmatrix} = 0$$

Prove this result by expanding the determinant and comparing the result to the two-point form of the equation of a line.

56. **Geometry: Collinear Points** Using the result obtained in Problem 55, show that three distinct points (x_1, y_1), (x_2, y_2), and (x_3, y_3) are collinear (lie on the same line) if and only if

$$\begin{vmatrix} x_1 & y_1 & 1 \\ x_2 & y_2 & 1 \\ x_3 & y_3 & 1 \end{vmatrix} = 0$$

57. Show that $\begin{vmatrix} x^2 & x & 1 \\ y^2 & y & 1 \\ z^2 & z & 1 \end{vmatrix} = (y-z)(x-y)(x-z)$.

58. Complete the proof of Cramer's Rule for two equations containing two variables.

[**Hint:** In system (5), page 780, if $a = 0$, then $b \neq 0$ and $c \neq 0$, since $D = -bc \neq 0$. Now show that equations (6) provide a solution of the system when $a = 0$. There are then three remaining cases: $b = 0$, $c = 0$, and $d = 0$.]

59. Interchange columns 1 and 3 of a 3 by 3 determinant. Show that the value of the new determinant is -1 times the value of the original determinant.

60. Multiply each entry in row 2 of a 3 by 3 determinant by the number k, $k \neq 0$. Show that the value of the new determinant is k times the value of the original determinant.

61. Prove that a 3 by 3 determinant in which the entries in column 1 equal those in column 3 has the value 0.

62. Prove that, if row 2 of a 3 by 3 determinant is multiplied by k, $k \neq 0$, and the result is added to the entries in row 1, then there is no change in the value of the determinant.

10.4 MATRIX ALGEBRA

1 Work with Equality and Addition of Matrices

2 Know Properties of Matrices

3 Know How to Multiply Matrices

4 Find the Inverse of a Matrix

5 Solve Systems of Equations Using Inverse Matrices

In Section 10.2, we defined a matrix as an array of real numbers and used an augmented matrix to represent a system of linear equations. There is, however, a branch of mathematics, called **linear algebra,** that deals with matrices in such a way that an algebra of matrices is permitted. In this section, we provide a survey of how this **matrix algebra** is developed.

Before getting started, we restate the definition of a matrix.

A **matrix** is defined as a rectangular array of numbers:

$$
\begin{array}{c}
 \quad \text{Column 1} \quad \text{Column 2} \qquad \text{Column } j \qquad \text{Column } n \\
\begin{array}{l}
\text{Row 1} \\
\text{Row 2} \\
\vdots \\
\text{Row } i \\
\vdots \\
\text{Row } m
\end{array}
\begin{bmatrix}
a_{11} & a_{12} & \cdots & a_{1j} & \cdots & a_{1n} \\
a_{21} & a_{22} & \cdots & a_{2j} & \cdots & a_{2n} \\
\vdots & \vdots & & \vdots & & \vdots \\
a_{i1} & a_{i2} & \cdots & a_{ij} & \cdots & a_{in} \\
\vdots & \vdots & & \vdots & & \vdots \\
a_{m1} & a_{m2} & \cdots & a_{mj} & \cdots & a_{mn}
\end{bmatrix}
\end{array}
$$

Each number a_{ij} of the matrix has two indices: the **row index** i and the **column index** j. The matrix shown above has m rows and n columns. The $m \cdot n$ numbers a_{ij} are usually referred to as the **entries** of the matrix.

Let's begin with an example that illustrates how matrices can be used to conveniently represent an array of information.

EXAMPLE 1 Arranging Data in a Matrix

In a survey of 900 people, the following information was obtained:

200 males	Thought federal defense spending was too high
150 males	Thought federal defense spending was too low
45 males	Had no opinion
315 females	Thought federal defense spending was too high
125 females	Thought federal defense spending was too low
65 females	Had no opinion

We can arrange the above data in a rectangular array as follows:

	Too High	Too Low	No Opinion
Male	200	150	45
Female	315	125	65

or as

$$
\begin{bmatrix}
200 & 150 & 45 \\
315 & 125 & 65
\end{bmatrix}
$$

This matrix has two rows (representing males and females) and three columns (representing "too high," "too low," and "no opinion"). ■

The matrix we developed in Example 1 has 2 rows and 3 columns. In general, a matrix with m rows and n columns is called an m **by** n **matrix.** Thus, the matrix we developed in Example 1 is a 2 by 3 matrix. Notice that an m by n matrix will contain $m \cdot n$ entries.

If an m by n matrix has the same number of rows as columns, that is, if $m = n$, then the matrix is referred to as a **square matrix.**

E X A M P L E   2

Examples of Matrices

(a) $\begin{bmatrix} 5 & 0 \\ -6 & 1 \end{bmatrix}$ A 2 by 2 square matrix   (b) $\begin{bmatrix} 1 & 0 & 3 \end{bmatrix}$ A 1 by 3 matrix

(c) $\begin{bmatrix} 6 & -2 & 4 \\ 4 & 3 & 5 \\ 8 & 0 & 1 \end{bmatrix}$ A 3 by 3 square matrix   ∎

Equality and Addition of Matrices

We begin our discussion of matrix algebra by first defining what is meant by two matrices being equal and then defining the operations of addition and subtraction. It is important to note that these definitions require each matrix to have the same number of rows *and* the same number of columns as a prerequisite for equality and for addition and subtraction.

We usually represent matrices by capital letters, such as A, B, C, and so on.

Two m by n matrices A and B are said to be **equal,** written as

$$A = B$$

provided each entry a_{ij} in A is equal to the corresponding entry b_{ij} in B.

For example,

$$\begin{bmatrix} 2 & 1 \\ 0.5 & -1 \end{bmatrix} = \begin{bmatrix} \sqrt{4} & 1 \\ \frac{1}{2} & -1 \end{bmatrix} \quad \text{and} \quad \begin{bmatrix} 3 & 2 & 1 \\ 0 & 1 & -2 \end{bmatrix} = \begin{bmatrix} \sqrt{9} & \sqrt{4} & 1 \\ 0 & 1 & \sqrt[3]{-8} \end{bmatrix}$$

$$\begin{bmatrix} 4 & 1 \\ 6 & 1 \end{bmatrix} \neq \begin{bmatrix} 4 & 0 \\ 6 & 1 \end{bmatrix}$$ Because the entries in row 1, column 2 are not equal.

$$\begin{bmatrix} 4 & 1 & 2 \\ 6 & 1 & 2 \end{bmatrix} \neq \begin{bmatrix} 4 & 1 & 2 & 3 \\ 6 & 1 & 2 & 4 \end{bmatrix}$$ Because the matrix on the left is 2 by 3 and the matrix on the right is 2 by 4.

If each of A and B is an m by n matrix (and each therefore contains $m \cdot n$ entries), the statement $A = B$ actually represents a system of $m \cdot n$ ordinary equations. We will make use of this fact a little later.

Suppose A and B represent two m by n matrices. We define their **sum** $A + B$ to be the m by n matrix formed by adding the corresponding entries a_{ij} of A and b_{ij} of B. The **difference $A - B$** is defined as the m by n matrix formed by subtracting the entries b_{ij} in B from the corresponding entries a_{ij} in A. Addition and subtraction of matrices are allowed only for matrices having the same number m of rows and the same number n of columns. Thus, for example, a 2 by 3 matrix and a 2 by 4 matrix cannot be added or subtracted.

The advent of graphing utilities has made the sometimes tedious process of matrix algebra easy. Let's compare how a graphing utility adds and subtracts matrices with doing it by hand.

E X A M P L E   3   Adding and Subtracting Matrices

Suppose that

$$A = \begin{bmatrix} 2 & 4 & 8 & -3 \\ 0 & 1 & 2 & 3 \end{bmatrix} \quad \text{and} \quad B = \begin{bmatrix} -3 & 4 & 0 & 1 \\ 6 & 8 & 2 & 0 \end{bmatrix}.$$

Find   (a) $A + B$ (b) $A - B$

Graphing Solution   Enter the matrices into a graphing utility. Name them $[A]$ and $[B]$. Figure 11 shows the results of adding and subtracting $[A]$ and $[B]$.

Algebraic Solution   (a) $A + B = \begin{bmatrix} 2 & 4 & 8 & -3 \\ 0 & 1 & 2 & 3 \end{bmatrix} + \begin{bmatrix} -3 & 4 & 0 & 1 \\ 6 & 8 & 2 & 0 \end{bmatrix}$

FIGURE 11

$$= \begin{bmatrix} 2 + (-3) & 4 + 4 & 8 + 0 & -3 + 1 \\ 0 + 6 & 1 + 8 & 2 + 2 & 3 + 0 \end{bmatrix} \quad \text{Add corresponding entries.}$$

$$= \begin{bmatrix} -1 & 8 & 8 & -2 \\ 6 & 9 & 4 & 3 \end{bmatrix}$$

(b) $A - B = \begin{bmatrix} 2 & 4 & 8 & -3 \\ 0 & 1 & 2 & 3 \end{bmatrix} - \begin{bmatrix} -3 & 4 & 0 & 1 \\ 6 & 8 & 2 & 0 \end{bmatrix}$

$$= \begin{bmatrix} 2 - (-3) & 4 - 4 & 8 - 0 & -3 - 1 \\ 0 - 6 & 1 - 8 & 2 - 2 & 3 - 0 \end{bmatrix} \quad \text{Subtract corresponding entries.}$$

$$= \begin{bmatrix} 5 & 0 & 8 & -4 \\ -6 & -7 & 0 & 3 \end{bmatrix}$$

Now work Problem 1.

2   Many of the algebraic properties of sums of real numbers are also true for sums of matrices. Suppose that A, B, and C are m by n matrices. Then matrix addition is **commutative.** That is,

---

Commutative Property

$$A + B = B + A$$

Matrix addition is also **associative.** That is,

Associative Property

$$(A + B) + C = A + (B + C)$$

Although we shall not prove these results, the proofs, as the following example illustrates, are based on the commutative and associative properties for real numbers.

E X A M P L E   4   Demonstrating the Commutative Property

$$\begin{bmatrix} 2 & 3 & -1 \\ 4 & 0 & 7 \end{bmatrix} + \begin{bmatrix} -1 & 2 & 1 \\ 5 & -3 & 4 \end{bmatrix} = \begin{bmatrix} 2 + (-1) & 3 + 2 & -1 + 1 \\ 4 + 5 & 0 + (-3) & 7 + 4 \end{bmatrix}$$

$$= \begin{bmatrix} -1+2 & 2+3 & 1+(-1) \\ 5+4 & -3+0 & 4+7 \end{bmatrix}$$

$$= \begin{bmatrix} -1 & 2 & 1 \\ 5 & -3 & 4 \end{bmatrix} + \begin{bmatrix} 2 & 3 & -1 \\ 4 & 0 & 7 \end{bmatrix} \quad \blacksquare$$

A matrix whose entries are all equal to 0 is called a **zero matrix.** Each of the following matrices is a zero matrix:

$\begin{bmatrix} 0 & 0 \\ 0 & 0 \end{bmatrix}$ 2 by 2 square zero matrix. $\begin{bmatrix} 0 & 0 & 0 \\ 0 & 0 & 0 \end{bmatrix}$ 2 by 3 zero matrix. $\begin{bmatrix} 0 & 0 & 0 \end{bmatrix}$ 1 by 3 zero matrix.

Zero matrices have properties similar to the real number 0. Thus, if A is an m by n matrix and 0 is an m by n zero matrix, then

$$A + 0 = A$$

In other words, the zero matrix is the additive identity in matrix algebra.

We also can multiply a matrix by a real number. If k is a real number and A is an m by n matrix, the matrix kA is the m by n matrix formed by multiplying each entry a_{ij} in A by k. The number k is sometimes referred to as a **scalar,** and the matrix kA is called a **scalar multiple** of A.

E X A M P L E   5   Operations Using Matrices

Suppose that

$$A = \begin{bmatrix} 3 & 1 & 5 \\ -2 & 0 & 6 \end{bmatrix} \qquad B = \begin{bmatrix} 4 & 1 & 0 \\ 8 & 1 & -3 \end{bmatrix} \qquad C = \begin{bmatrix} 9 & 0 \\ -3 & 6 \end{bmatrix}$$

Find   (a) $4A$ (b) $\frac{1}{3}C$ (c) $3A - 2B$

Graphing Solution   Enter the matrices $[A]$, $[B]$, and $[C]$ into a graphing utility. Figure 12 shows the required computations.

FIGURE 12

(a)   (b)   (c)

Algebraic Solution   (a) $4A = 4\begin{bmatrix} 3 & 1 & 5 \\ -2 & 0 & 6 \end{bmatrix} = \begin{bmatrix} 4\cdot3 & 4\cdot1 & 4\cdot5 \\ 4(-2) & 4\cdot0 & 4\cdot6 \end{bmatrix} = \begin{bmatrix} 12 & 4 & 20 \\ -8 & 0 & 24 \end{bmatrix}$

(b) $\frac{1}{3}C = \frac{1}{3}\begin{bmatrix} 9 & 0 \\ -3 & 6 \end{bmatrix} = \begin{bmatrix} \frac{1}{3}\cdot9 & \frac{1}{3}\cdot0 \\ \frac{1}{3}(-3) & \frac{1}{3}\cdot6 \end{bmatrix} = \begin{bmatrix} 3 & 0 \\ -1 & 2 \end{bmatrix}$

(c) $3A - 2B = 3\begin{bmatrix} 3 & 1 & 5 \\ -2 & 0 & 6 \end{bmatrix} - 2\begin{bmatrix} 4 & 1 & 0 \\ 8 & 1 & -3 \end{bmatrix}$

$= \begin{bmatrix} 3 \cdot 3 & 3 \cdot 1 & 3 \cdot 5 \\ 3(-2) & 3 \cdot 0 & 3 \cdot 6 \end{bmatrix} - \begin{bmatrix} 2 \cdot 4 & 2 \cdot 1 & 2 \cdot 0 \\ 2 \cdot 8 & 2 \cdot 1 & 2(-3) \end{bmatrix}$

$= \begin{bmatrix} 9 & 3 & 15 \\ -6 & 0 & 18 \end{bmatrix} - \begin{bmatrix} 8 & 2 & 0 \\ 16 & 2 & -6 \end{bmatrix}$

$= \begin{bmatrix} 9 - 8 & 3 - 2 & 15 - 0 \\ -6 - 16 & 0 - 2 & 18 - (-6) \end{bmatrix}$

$= \begin{bmatrix} 1 & 1 & 15 \\ -22 & -2 & 24 \end{bmatrix}$ ∎

Now work Problem 5.

We list next some of the algebraic properties of scalar multiplication. Let h and k be real numbers, and let A and B be m by n matrices. Then

---

Properties of Scalar Multiplication

$$k(hA) = (kh)A$$
$$(k + h)A = kA + hA$$
$$k(A + B) = kA + kB$$

The proofs of these properties are based on properties of real numbers. For example, if A and B are 2 by 2 matrices, then

$$k(A + B) = k\left(\begin{bmatrix} a_{11} & a_{12} \\ a_{21} & a_{22} \end{bmatrix} + \begin{bmatrix} b_{11} & b_{12} \\ b_{21} & b_{22} \end{bmatrix}\right) = k\begin{bmatrix} a_{11} + b_{11} & a_{12} + b_{12} \\ a_{21} + b_{21} & a_{22} + b_{22} \end{bmatrix}$$

$$= \begin{bmatrix} k(a_{11} + b_{11}) & k(a_{12} + b_{12}) \\ k(a_{21} + b_{21}) & k(a_{22} + b_{22}) \end{bmatrix} = \begin{bmatrix} ka_{11} + kb_{11} & ka_{12} + kb_{12} \\ ka_{21} + kb_{21} & ka_{22} + kb_{22} \end{bmatrix}$$

$$= \begin{bmatrix} ka_{11} & ka_{12} \\ ka_{21} & ka_{22} \end{bmatrix} + \begin{bmatrix} kb_{11} & kb_{12} \\ kb_{21} & kb_{22} \end{bmatrix} = k\begin{bmatrix} a_{11} & a_{12} \\ a_{21} & a_{22} \end{bmatrix} + k\begin{bmatrix} b_{11} & b_{12} \\ b_{21} & b_{22} \end{bmatrix} = kA + kB$$

Multiplication of Matrices

Unlike the straightforward definition for adding two matrices, the definition for multiplying two matrices is not what we might expect. In preparation for this definition, we need the following definitions:

A **row vector** R is a 1 by n matrix

$$R = [r_1 \quad r_2 \quad \cdots \quad r_n]$$

A **column vector** C is an n by 1 matrix

$$C = \begin{bmatrix} c_1 \\ c_2 \\ \vdots \\ c_n \end{bmatrix}$$

The **product** RC of R times C is defined as the number

$$RC = [r_1 \ r_2 \ \cdots \ r_n] \begin{bmatrix} c_1 \\ c_2 \\ \vdots \\ c_n \end{bmatrix} = r_1 c_1 + r_2 c_2 + \cdots + r_n c_n$$

Notice that a row vector and a column vector can be multiplied only if they contain the same number of entries.

E X A M P L E   6   The Product of a Row Vector by a Column Vector

If $R = [3 \ \ -5 \ \ 2]$ and $C = \begin{bmatrix} 3 \\ 4 \\ -5 \end{bmatrix}$, then

$$RC = [3 \ \ -5 \ \ 2] \begin{bmatrix} 3 \\ 4 \\ -5 \end{bmatrix} = 3 \cdot 3 + (-5)4 + 2(-5)$$

$$= 9 - 20 - 10 = -21 \qquad \blacksquare$$

Let's look at an application of the product of a row vector by a column vector.

E X A M P L E   7   Using Matrices to Compute Revenue

A clothing store sells men's shirts for $25, silk ties for $8, and wool suits for $300. Last month, the store had sales consisting of 100 shirts, 200 ties, and 50 suits. What was the total revenue due to these sales?

Solution   We set up a row vector R to represent the prices of each item and a column vector C to represent the corresponding number of items sold.
Then

$$\begin{array}{cc} \text{Prices} & \text{Number} \\ \text{Shirts Ties Suits} & \text{sold} \end{array}$$

$$R = [25 \ \ 8 \ \ 300] \qquad C = \begin{bmatrix} 100 \\ 200 \\ 50 \end{bmatrix} \begin{array}{l} \text{Shirts} \\ \text{Ties} \\ \text{Suits} \end{array}$$

The total revenue obtained is the product RC. That is,

$$RC = [25 \ \ 8 \ \ 300] \begin{bmatrix} 100 \\ 200 \\ 50 \end{bmatrix}$$

$$= 25 \cdot 100 + 8 \cdot 200 + 300 \cdot 50 = \$19{,}100$$

Shirt revenue   Tie revenue   Suit revenue   Total revenue $\blacksquare$

The definition for multiplying two matrices is based on the definition of a row vector times a column vector.

Let *A* denote an *m* by *r* matrix and let *B* denote an *r* by *n* matrix. The **product** *AB* is defined as the *m* by *n* matrix whose entry in row *i*, column *j* is the product of the *i*th row of *A* and the *j*th column of *B*.

The definition of the product *AB* of two matrices *A* and *B*, in this order, requires that the number of columns of *A* equal the number of rows of *B*; otherwise, no product is defined:

An example will help clarify the definition.

E X A M P L E 8   Multiplying Two Matrices

Find the product *AB* if

$$A = \begin{bmatrix} 2 & 4 & -1 \\ 5 & 8 & 0 \end{bmatrix} \quad \text{and} \quad B = \begin{bmatrix} 2 & 5 & 1 & 4 \\ 4 & 8 & 0 & 6 \\ -3 & 1 & -2 & -1 \end{bmatrix}$$

Solution   First, we note that *A* is 2 by 3 and *B* is 3 by 4, so the product *AB* is defined and will be a 2 by 4 matrix.

Graphing Solution   Enter the matrices *A* and *B* into a graphing utility. Figure 13 shows the product *AB*.

FIGURE 13

```
[A] [B]
 [[23 41 4 33]
  [42 89 5 68]]
```

Algebraic Solution   Suppose we want the entry in row 2, column 3 of *AB*. To find it, we find the product of the row vector from row 2 of *A* and the column vector from column 3 of *B:*

Column 3 of *B*

$$\begin{array}{c} \text{Row 2 of } A \\ \begin{bmatrix} 5 & 8 & 0 \end{bmatrix} \end{array} \begin{bmatrix} 1 \\ 0 \\ -2 \end{bmatrix} = 5 \cdot 1 + 8 \cdot 0 + 0(-2) = 5$$

So far, we have

$$AB = \begin{bmatrix} \underline{\quad} & \underline{\quad} & \overset{\text{Column 3}}{\underset{\downarrow}{5}} & \underline{\quad} \\ \underline{\quad} & \underline{\quad} & \underline{\quad} & \underline{\quad} \end{bmatrix} \leftarrow \text{Row 2}$$

Now, to find the entry in row 1, column 4 of AB, we find the product of row 1 of A and column 4 of B:

Column 4 of B

Row 1 of A

$$[2 \quad 4 \quad -1]\begin{bmatrix} 4 \\ 6 \\ -1 \end{bmatrix} = 2 \cdot 4 + 4 \cdot 6 + (-1)(-1) = 33$$

Continuing in this fashion, we find AB:

$$AB = \begin{bmatrix} 2 & 4 & -1 \\ 5 & 8 & 0 \end{bmatrix}\begin{bmatrix} 2 & 5 & 1 & 4 \\ 4 & 8 & 0 & 6 \\ -3 & 1 & -2 & -1 \end{bmatrix}$$

$$= \begin{bmatrix} \text{Row 1 of } A & \text{Row 1 of } A & \text{Row 1 of } A & \text{Row 1 of } A \\ \text{times} & \text{times} & \text{times} & \text{times} \\ \text{column 1 of } B & \text{column 2 of } B & \text{column 3 of } B & \text{column 4 of } B \\ & & & \\ \text{Row 2 of } A & \text{Row 2 of } A & \text{Row 2 of } A & \text{Row 2 of } A \\ \text{times} & \text{times} & \text{times} & \text{times} \\ \text{column 1 of } B & \text{column 2 of } B & \text{column 3 of } B & \text{column 4 of } B \end{bmatrix}$$

$$= \begin{bmatrix} 2 \cdot 2 + 4 \cdot 4 + (-1)(-3) & 2 \cdot 5 + 4 \cdot 8 + (-1)1 & 2 \cdot 1 + 4 \cdot 0 + (-1)(-2) & 33 \text{ (from earlier)} \\ 5 \cdot 2 + 8 \cdot 4 + 0(-3) & 5 \cdot 5 + 8 \cdot 8 + 0 \cdot 1 & 5 \text{ (from earlier)} & 5 \cdot 4 + 8 \cdot 6 + 0(-1) \end{bmatrix}$$

$$= \begin{bmatrix} 23 & 41 & 4 & 33 \\ 42 & 89 & 5 & 68 \end{bmatrix}$$

Now work Problem 17.

Notice that for the matrices given in Example 8 the product BA is not defined, because B is 3 by 4 and A is 2 by 3. Try calculating BA on a graphing utility. What do you notice?

Another result that can occur when multiplying two matrices is illustrated in the next example.*

E X A M P L E   9

Multiplying Two Matrices

If

$$A = \begin{bmatrix} 2 & 1 & 3 \\ 1 & -1 & 0 \end{bmatrix} \quad \text{and} \quad B = \begin{bmatrix} 1 & 0 \\ 2 & 1 \\ 3 & 2 \end{bmatrix}$$

find   (a) AB (b) BA

Solution   (a) $AB = \begin{bmatrix} 2 & 1 & 3 \\ 1 & -1 & 0 \end{bmatrix}\begin{bmatrix} 1 & 0 \\ 2 & 1 \\ 3 & 2 \end{bmatrix} = \begin{bmatrix} 13 & 7 \\ -1 & -1 \end{bmatrix}$

                                2 by 3   3 by 2   2 by 2

*For most of the examples that follow, we will multiply matrices by hand. You should verify each result using a graphing utility.

(b) $BA = \begin{bmatrix} 1 & 0 \\ 2 & 1 \\ 3 & 2 \end{bmatrix} \begin{bmatrix} 2 & 1 & 3 \\ 1 & -1 & 0 \end{bmatrix} = \begin{bmatrix} 2 & 1 & 3 \\ 5 & 1 & 6 \\ 8 & 1 & 9 \end{bmatrix}$

         3 by 2   2 by 3   3 by 3   ■

Notice in Example 9 that AB is 2 by 2 and BA is 3 by 3. Thus, it is possible for both AB and BA to be defined, yet be unequal. In fact, even if A and B are both n by n matrices, so that AB and BA are each defined and n by n, AB and BA will nearly always be unequal.

E X A M P L E   10   Multiplying Two Square Matrices

If

$$A = \begin{bmatrix} 2 & 1 \\ 0 & 4 \end{bmatrix} \quad \text{and} \quad B = \begin{bmatrix} -3 & 1 \\ 1 & 2 \end{bmatrix}$$

find   (a) AB (b) BA

Solution   (a) $AB = \begin{bmatrix} 2 & 1 \\ 0 & 4 \end{bmatrix} \begin{bmatrix} -3 & 1 \\ 1 & 2 \end{bmatrix}$

$= \begin{bmatrix} 2(-3) + 1 \cdot 1 & 2 \cdot 1 + 1 \cdot 2 \\ 0(-3) + 4 \cdot 1 & 0 \cdot 1 + 4 \cdot 2 \end{bmatrix} = \begin{bmatrix} -5 & 4 \\ 4 & 8 \end{bmatrix}$

(b) $BA = \begin{bmatrix} -3 & 1 \\ 1 & 2 \end{bmatrix} \begin{bmatrix} 2 & 1 \\ 0 & 4 \end{bmatrix}$

$= \begin{bmatrix} (-3)2 + 1 \cdot 0 & (-3)1 + 1 \cdot 4 \\ 1 \cdot 2 + 2 \cdot 0 & 1 \cdot 1 + 2 \cdot 4 \end{bmatrix} = \begin{bmatrix} -6 & 1 \\ 2 & 9 \end{bmatrix}$ ■

The preceding examples demonstrate that an important property of real numbers, the commutative property of multiplication, is not shared by matrices. Thus, in general:

Theorem

Matrix multiplication is not commutative.

■

Now work Problems 7 and 9.

Next we give two of the properties of real numbers that are shared by matrices. Assuming each product and sum is defined, we have:

Associative Property

$$A(BC) = (AB)C$$

Distributive Property

$$A(B + C) = AB + AC$$

The Identity Matrix

For an n by n square matrix, the entries located in row i, column i, $1 \leq i \leq n$, are called the **diagonal entries.** An n by n square matrix whose diagonal entries are 1's, while all other entries are 0's, is called the **identity matrix** I_n. For example,

$$I_2 = \begin{bmatrix} 1 & 0 \\ 0 & 1 \end{bmatrix} \quad I_3 = \begin{bmatrix} 1 & 0 & 0 \\ 0 & 1 & 0 \\ 0 & 0 & 1 \end{bmatrix}$$

and so on.

E X A M P L E   11   Multiplication with an Identity Matrix

Let

$$A = \begin{bmatrix} -1 & 2 & 0 \\ 0 & 1 & 3 \end{bmatrix} \quad \text{and} \quad B = \begin{bmatrix} 3 & 2 \\ 4 & 6 \\ 5 & 2 \end{bmatrix}$$

Find   (a) AI_3 (b) I_2A (c) BI_2

Solution   (a) $AI_3 = \begin{bmatrix} -1 & 2 & 0 \\ 0 & 1 & 3 \end{bmatrix} \begin{bmatrix} 1 & 0 & 0 \\ 0 & 1 & 0 \\ 0 & 0 & 1 \end{bmatrix} = \begin{bmatrix} -1 & 2 & 0 \\ 0 & 1 & 3 \end{bmatrix} = A$

(b) $I_2A = \begin{bmatrix} 1 & 0 \\ 0 & 1 \end{bmatrix} \begin{bmatrix} -1 & 2 & 0 \\ 0 & 1 & 3 \end{bmatrix} = \begin{bmatrix} -1 & 2 & 0 \\ 0 & 1 & 3 \end{bmatrix} = A$

(c) $BI_2 = \begin{bmatrix} 3 & 2 \\ 4 & 6 \\ 5 & 2 \end{bmatrix} \begin{bmatrix} 1 & 0 \\ 0 & 1 \end{bmatrix} = \begin{bmatrix} 3 & 2 \\ 4 & 6 \\ 5 & 2 \end{bmatrix} = B$

Example 11 demonstrates the following property:

Identity Property

If A is an m by n matrix, then

$$I_m A = A \quad \text{and} \quad AI_n = A$$

If A is an n by n square matrix, then $AI_n = I_n A = A$.

Thus, an identity matrix has properties analogous to those of the real number 1. In other words, the identity matrix is a multiplicative identity in matrix algebra.

4 The Inverse of a Matrix

Let A be a square n by n matrix. If there exists an n by n matrix A^{-1}, read "A inverse," for which

$$AA^{-1} = A^{-1}A = I_n$$

then A^{-1} is called the **inverse** of the matrix A.

As we shall soon see, not every square matrix has an inverse. When a matrix A does have an inverse A^{-1}, then A is said to be **nonsingular.** If a matrix A has no inverse, it is called **singular.**

E X A M P L E   12   Multiplying a Matrix by Its Inverse

Show that the inverse of

$$A = \begin{bmatrix} 3 & 1 \\ 2 & 1 \end{bmatrix} \quad \text{is} \quad A^{-1} = \begin{bmatrix} 1 & -1 \\ -2 & 3 \end{bmatrix}$$

Solution   We need to show that $AA^{-1} = A^{-1}A = I_2$.

$$AA^{-1} = \begin{bmatrix} 3 & 1 \\ 2 & 1 \end{bmatrix}\begin{bmatrix} 1 & -1 \\ -2 & 3 \end{bmatrix} = \begin{bmatrix} 3 \cdot 1 + 1(-2) & 3(-1) + 1 \cdot 3 \\ 2 \cdot 1 + 1(-2) & 2(-1) + 1 \cdot 3 \end{bmatrix}$$

$$= \begin{bmatrix} 1 & 0 \\ 0 & 1 \end{bmatrix} = I_2$$

$$A^{-1}A = \begin{bmatrix} 1 & -1 \\ -2 & 3 \end{bmatrix}\begin{bmatrix} 3 & 1 \\ 2 & 1 \end{bmatrix} = \begin{bmatrix} 3 - 2 & 1 - 1 \\ -6 + 6 & -2 + 3 \end{bmatrix} = \begin{bmatrix} 1 & 0 \\ 0 & 1 \end{bmatrix} = I_2$$

We now show one way to find the inverse of

$$A = \begin{bmatrix} 3 & 1 \\ 2 & 1 \end{bmatrix}$$

Suppose that A^{-1} is given by

$$A^{-1} = \begin{bmatrix} x & y \\ z & w \end{bmatrix} \tag{1}$$

where x, y, z, and w are four variables. Based on the definition of an inverse, if, indeed, A has an inverse, we have

$$AA^{-1} = I_2$$

$$\begin{bmatrix} 3 & 1 \\ 2 & 1 \end{bmatrix}\begin{bmatrix} x & y \\ z & w \end{bmatrix} = \begin{bmatrix} 1 & 0 \\ 0 & 1 \end{bmatrix}$$

$$\begin{bmatrix} 3x + z & 3y + w \\ 2x + z & 2y + w \end{bmatrix} = \begin{bmatrix} 1 & 0 \\ 0 & 1 \end{bmatrix}$$

Because corresponding entries must be equal, it follows that this matrix equation is equivalent to four ordinary equations:

$$\begin{cases} 3x + z = 1 \\ 2x + z = 0 \end{cases} \qquad \begin{cases} 3y + w = 0 \\ 2y + w = 1 \end{cases}$$

The augmented matrix of each system is

$$\begin{bmatrix} 3 & 1 & \vert & 1 \\ 2 & 1 & \vert & 0 \end{bmatrix} \qquad \begin{bmatrix} 3 & 1 & \vert & 0 \\ 2 & 1 & \vert & 1 \end{bmatrix} \tag{2}$$

The usual procedure would be to transform each augmented matrix into reduced echelon form. Notice, though, that the left sides of the augmented matrices are equal, so the same row operations (see Section 10.2) can be used to reduce each one. Thus, we find it more efficient to combine the two augmented matrices (2) into a single matrix, as shown next, and then transform it into reduced echelon form:

$$\begin{bmatrix} 3 & 1 & \vert & 1 & 0 \\ 2 & 1 & \vert & 0 & 1 \end{bmatrix}$$

Now we attempt to transform the left side into an identity matrix:

$$\begin{bmatrix} 3 & 1 & \vert & 1 & 0 \\ 2 & 1 & \vert & 0 & 1 \end{bmatrix} \xrightarrow[\substack{\uparrow \\ R_1 = -1r_2 + r_1}]{} \begin{bmatrix} 1 & 0 & \vert & 1 & -1 \\ 2 & 1 & \vert & 0 & 1 \end{bmatrix}$$

$$\xrightarrow[\substack{\uparrow \\ R_2 = -2r_1 + r_2}]{} \begin{bmatrix} 1 & 0 & \vert & 1 & -1 \\ 0 & 1 & \vert & -2 & 3 \end{bmatrix} \tag{3}$$

Matrix (3) is in reduced echelon form. Now we reverse the earlier step of combining the two augmented matrices in (2) and write the single matrix (3) as two augmented matrices:

$$\begin{bmatrix} 1 & 0 & \vert & 1 \\ 0 & 1 & \vert & -2 \end{bmatrix} \text{ and } \begin{bmatrix} 1 & 0 & \vert & -1 \\ 0 & 1 & \vert & 3 \end{bmatrix}$$

We conclude from these matrices that $x = 1$, $z = -2$, and $y = -1$, $w = 3$. Substituting these values into matrix (1), we find that

$$A^{-1} = \begin{bmatrix} 1 & -1 \\ -2 & 3 \end{bmatrix}$$

Notice in display (3) that the 2 by 2 matrix to the right of the vertical bar is, in fact, the inverse of A. Also notice that the identity matrix I_2 is the matrix that appears to the left of the vertical bar. These observations and the procedures followed above will work in general.

Procedure for Finding the Inverse of a Nonsingular Matrix

> To find the inverse of an n by n nonsingular matrix A, proceed as follows:
>
> STEP 1: Form the matrix $[A|I_n]$.
>
> STEP 2: Transform the matrix $[A|I_n]$ into reduced echelon form.
>
> STEP 3: The reduced echelon form of $[A|I_n]$ will contain the identity matrix I_n on the left of the vertical bar; the n by n matrix on the right of the vertical bar is the inverse of A.

In other words, if A is nonsingular, we begin with the matrix $[A|I_n]$ and, after transforming it into reduced echelon form, we end up with the matrix $[I_n|A^{-1}]$.

Let's look at another example.

E X A M P L E 13 Finding the Inverse of a Matrix

The matrix

$$A = \begin{bmatrix} 1 & 1 & 0 \\ -1 & 3 & 4 \\ 0 & 4 & 3 \end{bmatrix}$$

is nonsingular. Find its inverse.

Graphing Solution Enter the matrix A into a graphing utility. Figure 14 shows A^{-1}.

FIGURE 14

```
[A]-1
[[1.75  .75  -1]
 [-.75  -.75  1 ]
 [1     1    -1]]
```

Algebraic Solution First, we form the matrix

$$[A|I_3] = \begin{bmatrix} 1 & 1 & 0 & | & 1 & 0 & 0 \\ -1 & 3 & 4 & | & 0 & 1 & 0 \\ 0 & 4 & 3 & | & 0 & 0 & 1 \end{bmatrix}$$

Next, we use row operations to transform $[A|I_3]$ into reduced echelon form:

$$\begin{bmatrix} 1 & 1 & 0 & | & 1 & 0 & 0 \\ -1 & 3 & 4 & | & 0 & 1 & 0 \\ 0 & 4 & 3 & | & 0 & 0 & 1 \end{bmatrix} \rightarrow \begin{bmatrix} 1 & 1 & 0 & | & 1 & 0 & 0 \\ 0 & 4 & 4 & | & 1 & 1 & 0 \\ 0 & 4 & 3 & | & 0 & 0 & 1 \end{bmatrix} \rightarrow \begin{bmatrix} 1 & 1 & 0 & | & 1 & 0 & 0 \\ 0 & 1 & 1 & | & \frac{1}{4} & \frac{1}{4} & 0 \\ 0 & 4 & 3 & | & 0 & 0 & 1 \end{bmatrix}$$

$$\uparrow \\ R_2 = r_1 + r_2 \qquad\qquad\qquad\qquad \uparrow \\ R_2 = \tfrac{1}{4} r_2$$

$$\rightarrow \begin{bmatrix} 1 & 0 & -1 & | & \frac{3}{4} & -\frac{1}{4} & 0 \\ 0 & 1 & 1 & | & \frac{1}{4} & \frac{1}{4} & 0 \\ 0 & 0 & -1 & | & -1 & -1 & 1 \end{bmatrix} \rightarrow \begin{bmatrix} 1 & 0 & -1 & | & \frac{3}{4} & -\frac{1}{4} & 0 \\ 0 & 1 & 1 & | & \frac{1}{4} & \frac{1}{4} & 0 \\ 0 & 0 & 1 & | & 1 & 1 & -1 \end{bmatrix}$$

$$\uparrow \\ R_1 = -1r_2 + r_1 \qquad\qquad\qquad\qquad \uparrow \\ R_3 = -4r_2 + r_3 \qquad\qquad\qquad\qquad R_3 = -1r_3$$

$$\rightarrow \begin{bmatrix} 1 & 0 & 0 & | & \frac{7}{4} & \frac{3}{4} & -1 \\ 0 & 1 & 0 & | & -\frac{3}{4} & -\frac{3}{4} & 1 \\ 0 & 0 & 1 & | & 1 & 1 & -1 \end{bmatrix}$$

$$\uparrow \\ R_1 = r_3 + r_1 \\ R_2 = -1r_3 + r_2$$

The matrix $[A|I_3]$ is now in reduced echelon form, and the identity matrix I_3 is on the left of the vertical bar. Hence, the inverse of A is

$$A^{-1} = \begin{bmatrix} \frac{7}{4} & \frac{3}{4} & -1 \\ -\frac{3}{4} & -\frac{3}{4} & 1 \\ 1 & 1 & -1 \end{bmatrix}$$

You can (and should) verify that this is the correct inverse by showing that $AA^{-1} = A^{-1}A = I_3$.

Now work Problem 23.

If transforming the matrix $[A|I_n]$ into reduced echelon form does not result in the identity matrix I_n to the left of the vertical bar, then A has no inverse. The next example demonstrates such a matrix.

E X A M P L E   14   Showing a Matrix Has No Inverse

Show that the following matrix has no inverse.

$$A = \begin{bmatrix} 4 & 6 \\ 2 & 3 \end{bmatrix}$$

Graphing Solution   Enter the matrix A. Figure 15 shows the result when we try to find its inverse. The ERRor comes about because A is singular.

FIGURE 15

```
ERR:SINGULAR MAT
1∎Goto
2:Quit
```

Algebraic Solution   Proceeding as in Example 13, we form the matrix

$$[A|I_2] = \begin{bmatrix} 4 & 6 & | & 1 & 0 \\ 2 & 3 & | & 0 & 1 \end{bmatrix}$$

Then we use row operations to transform $[A \mid I_2]$ into reduced echelon form:

$$[A|I_2] = \begin{bmatrix} 4 & 6 & | & 1 & 0 \\ 2 & 3 & | & 0 & 1 \end{bmatrix} \rightarrow \begin{bmatrix} 1 & \frac{3}{2} & | & \frac{1}{4} & 0 \\ 2 & 3 & | & 0 & 1 \end{bmatrix} \rightarrow \begin{bmatrix} 1 & \frac{3}{2} & | & \frac{1}{4} & 0 \\ 0 & 0 & | & -\frac{1}{2} & 1 \end{bmatrix}$$

$$\underset{R_1 = \frac{1}{4} r_1}{\uparrow} \qquad\qquad \underset{R_2 = -2r_1 + r_2}{\uparrow}$$

The matrix $[A|I_2]$ is sufficiently reduced for us to see the identity matrix cannot appear to the left of the vertical bar. We conclude that A has no inverse. ■

Now work Problem 51.

Solving Systems of Linear Equations

5 Inverse matrices can be used to solve systems of equations in which the number of equations is the same as the number of variables.

E X A M P L E   15   Using the Inverse Matrix to Solve a System of Linear Equations

Solve the system of equations: $\begin{cases} x + y = 3 \\ -x + 3y + 4z = -3 \\ 4y + 3z = 2 \end{cases}$

Solution   If we let

$$A = \begin{bmatrix} 1 & 1 & 0 \\ -1 & 3 & 4 \\ 0 & 4 & 3 \end{bmatrix}, \quad X = \begin{bmatrix} x \\ y \\ z \end{bmatrix}, \quad \text{and} \quad B = \begin{bmatrix} 3 \\ -3 \\ 2 \end{bmatrix}$$

then the original system of equations can be written compactly as the matrix equation

$$AX = B \qquad\qquad (4)$$

We know from Example 13 that the matrix A has the inverse A^{-1}, so we multiply each side of equation (4) by A^{-1}:

$$AX = B$$
$$A^{-1}(AX) = A^{-1}B$$
$$(A^{-1}A)X = A^{-1}B \qquad \text{Associative property of multiplication.}$$
$$I_3X = A^{-1}B \qquad \text{Definition of inverse matrix.}$$
$$X = A^{-1}B \qquad \text{Property of identity matrix.} \qquad\qquad (5)$$

Now we use (5) to find $X = \begin{bmatrix} x \\ y \\ z \end{bmatrix}$.

Graphing Solution Enter the matrices A and B into a graphing utility. Figure 16 shows the solution to the system of equations.

FIGURE 16

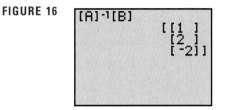

Algebraic Solution $$X = \begin{bmatrix} x \\ y \\ z \end{bmatrix} = A^{-1}B = \begin{bmatrix} \frac{7}{4} & \frac{3}{4} & -1 \\ -\frac{3}{4} & -\frac{3}{4} & 1 \\ 1 & 1 & -1 \end{bmatrix} \begin{bmatrix} 3 \\ -3 \\ 2 \end{bmatrix} = \begin{bmatrix} 1 \\ 2 \\ -2 \end{bmatrix}$$

Example 13

Thus, $x = 1, y = 2, z = -2$. ∎

The method used in Example 15 to solve a system of equations is particularly useful when it is necessary to solve several systems of equations in which the constants appearing to the right of the equal signs change, while the coefficients of the variables on the left side remain the same. See Problems 31–50 for some illustrations.

HISTORICAL FEATURE Matrices were invented in 1857 by Arthur Cayley (1821–1895) as a way of efficiently computing the result of substituting one linear system into another (see Historical Problem 2). The resulting system had incredible richness, in the sense that a very wide variety of mathematical systems could be mimicked by the matrices. Cayley and his friend J.J. Sylvester (1814–1897) spent much of the rest of their lives elaborating the theory. The torch was then passed to G. Frobenius (1849–1917), whose deep investigations established a central place for matrices in modern mathematics. In 1924, rather to the surprise of physicists, it was found that matrices (with complex numbers in them) were exactly the right tool for describing the behavior of atomic systems. Today, matrices are used in a wide variety of applications. ∎

HISTORICAL PROBLEMS 1. *Matrices and Complex Numbers* Frobenius emphasized in his research how matrices could be used to mimic other mathematical systems. Here, we mimic the behavior of complex numbers using matrices. Mathematicians call such a relationship an *isomorphism*.

Complex number $\longleftrightarrow$ Matrix

$$a + bi \longleftrightarrow \begin{bmatrix} a & b \\ -b & a \end{bmatrix}$$

Note that the complex number can be read off the top line of the matrix. Thus,

$$2 + 3i \longleftrightarrow \begin{bmatrix} 2 & 3 \\ -3 & 2 \end{bmatrix} \quad \text{and} \quad \begin{bmatrix} 4 & -2 \\ 2 & 4 \end{bmatrix} \longleftrightarrow 4 - 2i$$

(a) Find the matrices corresponding to $2 - 5i$ and $1 + 3i$.

(b) Multiply the two matrices.

(c) Find the corresponding complex number for the matrix found in part (b).

(d) Multiply $2 - 5i$ by $1 + 3i$. The result should be the same as that found in part (c).

The process also works for addition and subtraction. Try it for yourself.

2. *Cayley's Definition of Matrix Multiplication* Cayley invented matrix multiplication to simplify the following problem:

$$\begin{cases} u = ar + bs \\ v = cr + ds \end{cases} \qquad \begin{cases} x = ku + lv \\ y = mu + nv \end{cases}$$

(a) Find x and y in terms of r and s by substituting u and v from the first system of equations into the second system of equations.

(b) Use the result of part (a) to find the 2 by 2 matrix A in

$$\begin{bmatrix} x \\ y \end{bmatrix} = A \begin{bmatrix} r \\ s \end{bmatrix}$$

(c) Now look at the following way to do it. Write the equations in matrix form

$$\begin{bmatrix} u \\ v \end{bmatrix} = \begin{bmatrix} a & b \\ c & d \end{bmatrix} \begin{bmatrix} r \\ s \end{bmatrix} \qquad \begin{bmatrix} x \\ y \end{bmatrix} = \begin{bmatrix} k & l \\ m & n \end{bmatrix} \begin{bmatrix} u \\ v \end{bmatrix}$$

So

$$\begin{bmatrix} x \\ y \end{bmatrix} = \begin{bmatrix} k & l \\ m & n \end{bmatrix} \begin{bmatrix} a & b \\ c & d \end{bmatrix} \begin{bmatrix} r \\ s \end{bmatrix}$$

Do you see how Cayley defined matrix multiplication?

10.4 EXERCISES

In Problems 1–16, use the following matrices to compute the given expression: (a) by hand; (b) using a graphing utility.

$$A = \begin{bmatrix} 0 & 3 & -5 \\ 1 & 2 & 6 \end{bmatrix} \qquad B = \begin{bmatrix} 4 & 1 & 0 \\ -2 & 3 & -2 \end{bmatrix} \qquad C = \begin{bmatrix} 4 & 1 \\ 6 & 2 \\ -2 & 3 \end{bmatrix}$$

1. $A + B$	**2.** $A - B$	**3.** $4A$	**4.** $-3B$
5. $3A - 2B$	**6.** $2A + 4B$	**7.** AC	**8.** BC
9. CA	**10.** CB	**11.** $C(A + B)$	**12.** $(A + B)C$
13. $AC - 3I_2$	**14.** $CA + 5I_3$	**15.** $CA - CB$	**16.** $AC + BC$

In Problems 17–20, compute each product: (a) by hand; (b) using a graphing utility.

17. $\begin{bmatrix} 2 & -2 \\ 1 & 0 \end{bmatrix} \begin{bmatrix} 2 & 1 & 4 & 6 \\ 3 & -1 & 3 & 2 \end{bmatrix}$

18. $\begin{bmatrix} 4 & 1 \\ 2 & 1 \end{bmatrix} \begin{bmatrix} -6 & 6 & 1 & 0 \\ 2 & 5 & 4 & -1 \end{bmatrix}$

19. $\begin{bmatrix} 1 & 0 & 1 \\ 2 & 4 & 1 \\ 3 & 6 & 1 \end{bmatrix} \begin{bmatrix} 1 & 3 \\ 6 & 2 \\ 8 & -1 \end{bmatrix}$

20. $\begin{bmatrix} 4 & -2 & 3 \\ 0 & 1 & 2 \\ -1 & 0 & 1 \end{bmatrix} \begin{bmatrix} 2 & 6 \\ 1 & -1 \\ 0 & 2 \end{bmatrix}$

In Problems 21–30, each matrix is nonsingular. Find the inverse of each matrix. Be sure to check your answer using a graphing utility (when possible).

21. $\begin{bmatrix} 2 & 1 \\ 1 & 1 \end{bmatrix}$ **22.** $\begin{bmatrix} 3 & -1 \\ -2 & 1 \end{bmatrix}$ **23.** $\begin{bmatrix} 6 & 5 \\ 2 & 2 \end{bmatrix}$ **24.** $\begin{bmatrix} -4 & 1 \\ 6 & -2 \end{bmatrix}$

25. $\begin{bmatrix} 2 & 1 \\ a & a \end{bmatrix}, \quad a \neq 0$ **26.** $\begin{bmatrix} b & 3 \\ b & 2 \end{bmatrix}, \quad b \neq 0$ **27.** $\begin{bmatrix} 1 & -1 & 1 \\ 0 & -2 & 1 \\ -2 & -3 & 0 \end{bmatrix}$ **28.** $\begin{bmatrix} 1 & 0 & 2 \\ -1 & 2 & 3 \\ 1 & -1 & 0 \end{bmatrix}$

29. $\begin{bmatrix} 1 & 1 & 1 \\ 3 & 2 & -1 \\ 3 & 1 & 2 \end{bmatrix}$ **30.** $\begin{bmatrix} 3 & 3 & 1 \\ 1 & 2 & 1 \\ 2 & -1 & 1 \end{bmatrix}$

In Problems 31–50, use the inverses found in Problems 21–30 to solve each system of equations by hand.

31. $\begin{cases} 2x + y = 8 \\ x + y = 5 \end{cases}$ **32.** $\begin{cases} 3x - y = 8 \\ -2x + y = 4 \end{cases}$ **33.** $\begin{cases} 2x + y = 0 \\ x + y = 5 \end{cases}$

34. $\begin{cases} 3x - y = 4 \\ -2x + y = 5 \end{cases}$ **35.** $\begin{cases} 6x + 5y = 7 \\ 2x + 2y = 2 \end{cases}$ **36.** $\begin{cases} -4x + y = 0 \\ 6x - 2y = 14 \end{cases}$

37. $\begin{cases} 6x + 5y = 13 \\ 2x + 2y = 5 \end{cases}$ **38.** $\begin{cases} -4x + y = 5 \\ 6x - 2y = -9 \end{cases}$ **39.** $\begin{cases} 2x + y = -3 \\ ax + ay = -a \end{cases} \quad a \neq 0$

40. $\begin{cases} bx + 3y = 2b + 3 \\ bx + 2y = 2b + 2 \end{cases} \quad b \neq 0$ **41.** $\begin{cases} 2x + y = 7/a \\ ax + ay = 5 \end{cases} \quad a \neq 0$ **42.** $\begin{cases} bx + 3y = 14 \\ bx + 2y = 10 \end{cases} \quad b \neq 0$

43. $\begin{cases} x - y + z = 0 \\ -2y + z = -1 \\ -2x - 3y = -5 \end{cases}$ **44.** $\begin{cases} x + 2z = 6 \\ -x + 2y + 3z = -5 \\ x - y = 6 \end{cases}$ **45.** $\begin{cases} x - y + z = 2 \\ -2y + z = 2 \\ -2x - 3y = \frac{1}{2} \end{cases}$

46. $\begin{cases} x + 2z = 2 \\ -x + 2y + 3z = -\frac{3}{2} \\ x - y = 2 \end{cases}$ **47.** $\begin{cases} x + y + z = 9 \\ 3x + 2y - z = 8 \\ 3x + y + 2z = 1 \end{cases}$ **48.** $\begin{cases} 3x + 3y + z = 8 \\ x + 2y + z = 5 \\ 2x - y + z = 4 \end{cases}$

49. $\begin{cases} x + y + z = 2 \\ 3x + 2y - z = \frac{7}{3} \\ 3x + y + 2z = \frac{10}{3} \end{cases}$ **50.** $\begin{cases} 3x + 3y + z = 1 \\ x + 2y + z = 0 \\ 2x - y + z = 4 \end{cases}$

In Problems 51–56, by hand, show that each matrix has no inverse.

51. $\begin{bmatrix} 4 & 2 \\ 2 & 1 \end{bmatrix}$ **52.** $\begin{bmatrix} -3 & \frac{1}{2} \\ 6 & -1 \end{bmatrix}$ **53.** $\begin{bmatrix} 15 & 3 \\ 10 & 2 \end{bmatrix}$

54. $\begin{bmatrix} -3 & 0 \\ 4 & 0 \end{bmatrix}$ **55.** $\begin{bmatrix} -3 & 1 & -1 \\ 1 & -4 & -7 \\ 1 & 2 & 5 \end{bmatrix}$ **56.** $\begin{bmatrix} 1 & 1 & -3 \\ 2 & -4 & 1 \\ -5 & 7 & 1 \end{bmatrix}$

In Problems 57–60, use a graphing utility to find the inverse, if it exists, of each matrix. Round answers to two decimal places.

57. $\begin{bmatrix} 25 & 61 & -12 \\ 18 & -2 & 4 \\ 8 & 35 & 21 \end{bmatrix}$ **58.** $\begin{bmatrix} 18 & -3 & 4 \\ 6 & -20 & 14 \\ 10 & 25 & -15 \end{bmatrix}$

59. $\begin{bmatrix} 44 & 21 & 18 & 6 \\ -2 & 10 & 15 & 5 \\ 21 & 12 & -12 & 4 \\ -8 & -16 & 4 & 9 \end{bmatrix}$ **60.** $\begin{bmatrix} 16 & 22 & -3 & 5 \\ 21 & -17 & 4 & 8 \\ 2 & 8 & 27 & 20 \\ 5 & 15 & -3 & -10 \end{bmatrix}$

In Problems 61–64, use the idea behind Example 15 with a graphing utility to solve the following systems of equations. Round answers to two decimal places.

61. $\begin{cases} 25x + 61y - 12z = 10 \\ 18x - 12y + 7z = -9 \\ 3x + 4y - z = 12 \end{cases}$

62. $\begin{cases} 25x + 61y - 12z = 15 \\ 18x - 12y + 7z = -3 \\ 3x + 4y - z = 12 \end{cases}$

63. $\begin{cases} 25x + 61y - 12z = 21 \\ 18x - 12y + 7z = 7 \\ 3x + 4y - z = -2 \end{cases}$

64. $\begin{cases} 25x + 61y - 12z = 25 \\ 18x - 12y + 7z = 10 \\ 3x + 4y - z = -4 \end{cases}$

65. **Computing the Cost of Production**   The Acme Steel Company is a producer of stainless steel and aluminum containers. On a certain day, the following stainless steel containers were manufactured: 500 with 10-gallon capacity, 350 with 5-gallon capacity, and 400 with 1-gallon capacity. On the same day, the following aluminum containers were manufactured: 700 with 10-gallon capacity, 500 with 5-gallon capacity, and 850 with 1-gallon capacity.
   (a) Find a 2 by 3 matrix representing the above data. Find a 3 by 2 matrix to represent the same data.
   (b) If the amount of material used in the 10-gallon containers is 15 pounds, the amount used in the 5-gallon containers is 8 pounds, and the amount used in the 1-gallon containers is 3 pounds, find a 3 by 1 matrix representing the amount of material.
   (c) Multiply the 2 by 3 matrix found in part (a) and the 3 by 1 matrix found in part (b) to get a 2 by 1 matrix showing the day's usage of material.
   (d) If stainless steel costs Acme $0.10 per pound and aluminum costs $0.05 per pound, find a 1 by 2 matrix representing cost.
   (e) Multiply the matrices found in parts (c) and (d) to determine what the total cost of the day's production was.

66. **Computing Profit**   Rizza Ford has two locations, one in the city and the other in the suburbs. In January, the city location sold 400 sub-

compacts, 250 intermediate-size cars, and 50 station wagons; in February, it sold 350 subcompacts, 100 intermediates, and 30 station wagons. At the suburban location in January, 450 subcompacts, 200 intermediates, and 140 station wagons were sold. In February, the suburban location sold 350 subcompacts, 300 intermediates, and 100 station wagons.
   (a) Find 2 by 3 matrices that summarize the sales data for each location for January and February (one matrix for each month).
   (b) Use matrix addition to obtain total sales for the two-month period.
   (c) The profit on each kind of car is $100 per subcompact, $150 per intermediate, and $200 per station wagon. Find a 3 by 1 matrix representing this profit.
   (d) Multiply the matrices found in parts (b) and (c) to get a 2 by 1 matrix showing the profit at each location.

67. Consider the 2 by 2 square matrix

$$A = \begin{bmatrix} a & b \\ c & d \end{bmatrix}$$

If $D = ad - bc \neq 0$, show that A is nonsingular and that

$$A^{-1} = \frac{1}{D} \begin{bmatrix} d & -b \\ -c & a \end{bmatrix}$$

68. Make up a situation different from any found in the text that can be represented by a matrix.

10.5 PARTIAL FRACTION DECOMPOSITION

1. Decompose P/Q, Where Q Has Only Nonrepeated Linear Factors
2. Decompose P/Q, Where Q Has Repeated Linear Factors
3. Decompose P/Q, Where Q Has Only Nonrepeated Irreducible Quadratic Factors
4. Decompose P/Q, Where Q Has Repeated Irreducible Quadratic Factors

Consider the problem of adding two fractions

$$\frac{3}{x + 4} \quad \text{and} \quad \frac{2}{x - 3}$$

The result is

$$\frac{3}{x + 4} + \frac{2}{x - 3} = \frac{3(x - 3) + 2(x + 4)}{(x + 4)(x - 3)} = \frac{5x - 1}{x^2 + x - 12}$$

The reverse procedure, of starting with the rational expression $(5x - 1)/(x^2 + x - 12)$ and writing it as the sum (or difference) of the two simpler fractions $3/(x + 4)$ and $2/(x - 3)$ is referred to as **partial fraction decomposition,** and the two simpler fractions are called **partial fractions.** Decomposing a rational expression into a sum of partial fractions is important in solving certain types of calculus problems. This section presents a systematic way to decompose rational expressions.

We begin by recalling that a rational expression is the ratio of two polynomials, say, P and $Q \neq 0$, that have no common factors. Recall also that a rational expression P/Q is called **proper** if the degree of the polynomial in the numerator is less than the degree of the polynomial in the denominator. Otherwise, the rational expression is termed **improper.**

Because any improper rational expression can be reduced by long division to a mixed form consisting of the sum of a polynomial and a proper rational expression, we shall restrict the discussion that follows to proper rational expressions.

The partial fraction decomposition of the rational expression P/Q depends on the factors of the denominator Q. Recall (from Section 3.5) that any polynomial whose coefficients are real numbers can be factored (over the real numbers) into products of linear and/or irreducible quadratic factors. Thus, the denominator Q of the rational expression P/Q will contain only factors of one or both of the following types:

1. *Linear factors* of the form $x - a$, where a is a real number.
2. *Irreducible quadratic factors* of the form $ax^2 + bx + c$, where a, b, and c are real numbers, $a \neq 0$, and $b^2 - 4ac < 0$ (which guarantees that $ax^2 + bx + c$ cannot be written as the product of two linear factors with real coefficients).

As it turns out, there are four cases to be examined. We begin with the case for which Q has only nonrepeated linear factors.

CASE 1: Q has only nonrepeated linear factors.

Under the assumption that Q has only nonrepeated linear factors, the polynomial Q has the form

$$Q(x) = (x - a_1)(x - a_2) \cdot \ldots \cdot (x - a_n)$$

where none of the numbers $a_1, a_2, \ldots, a_n$ are equal. In this case, the partial fraction decomposition of P/Q is of the form

$$\frac{P(x)}{Q(x)} = \frac{A_1}{x - a_1} + \frac{A_2}{x - a_2} + \cdots + \frac{A_n}{x - a_n} \qquad (1)$$

where the numbers $A_1, A_2, \ldots, A_n$ are to be determined.

We show how to find these numbers in the example that follows.

E X A M P L E 1   Nonrepeated Linear Factors

Write the partial fraction decomposition of $\dfrac{x}{x^2 - 5x + 6}$.

Solution   First, we factor the denominator,

$$x^2 - 5x + 6 = (x - 2)(x - 3)$$

and conclude that the denominator contains only nonrepeated linear factors. Then we decompose the rational expression according to equation (1):

$$\frac{x}{x^2 - 5x + 6} = \frac{A}{x - 2} + \frac{B}{x - 3} \qquad (2)$$

where A and B are to be determined. To find A and B, we clear the fractions by multiplying each side by $(x - 2)(x - 3) = x^2 - 5x + 6$. The result is

$$x = A(x - 3) + B(x - 2) \qquad (3)$$

or

$$x = (A + B)x + (-3A - 2B)$$

This equation is an identity in x. Thus, we may equate the coefficients of like powers of x to get

$$\begin{cases} 1 = A + B \\ 0 = -3A - 2B \end{cases}$$
$\quad$ Equate coefficients of x: $1x = (A + B)x$.
$\quad$ Equate coefficients of x^0, the constants:
$\quad$ $0x^0 = (-3A - 2B)x^0$.

This system of two equations containing two variables, A and B, can be solved using whatever method you wish. Solving it, we get

$$A = -2 \qquad B = 3$$

Thus, from equation (2), the partial fraction decomposition is

$$\frac{x}{x^2 - 5x + 6} = \frac{-2}{x - 2} + \frac{3}{x - 3}$$

Check: The decomposition can be checked by adding the fractions:

$$\frac{-2}{x - 2} + \frac{3}{x - 3} = \frac{-2(x - 3) + 3(x - 2)}{(x - 2)(x - 3)} = \frac{x}{(x - 2)(x - 3)}$$

$$= \frac{x}{x^2 - 5x + 6}$$

Now work Problem 9.

The numbers to be found in the partial fraction decomposition can some-times be found more readily by using suitable choices for x (which may in-clude complex numbers) in the identity obtained after fractions have been cleared. In Example 1, the identity after clearing fractions, equation (3), is

$$x = A(x - 3) + B(x - 2)$$

If we let $x = 2$ in this expression, the term containing B drops out, leaving $2 = A(-1)$, or $A = -2$. Similarly, if we let $x = 3$, the term containing A drops out, leaving $3 = B$. Thus, as before $A = -2$ and $B = 3$.

We use this method in the next example.

2   **CASE 2:** Q has repeated linear factors.

If the polynomial Q has a repeated factor, say, $(x - a)^n$, $n \geq 2$ an in-teger, then, in the partial fraction decomposition of P/Q, we allow for the terms

$$\frac{A_1}{x - a} + \frac{A_2}{(x - a)^2} + \cdots + \frac{A_n}{(x - a)^n}$$

where the numbers $A_1, A_2, \ldots, A_n$ are to be determined.

E X A M P L E 2

Repeated Linear Factors

Write the partial fraction decomposition of $\dfrac{x + 2}{x^3 - 2x^2 + x}$.

Solution   First, we factor the denominator,

$$x^3 - 2x^2 + x = x(x^2 - 2x + 1) = x(x - 1)^2$$

and find that the denominator has the nonrepeated linear factor x and the repeated linear factor $(x - 1)^2$. By Case 1, we must allow for the term A/x in the decomposition; and, by Case 2, we must allow for the terms $B/(x - 1) + C/(x - 1)^2$ in the decomposition.

Thus, we write

$$\frac{x + 2}{x^3 - 2x^2 + x} = \frac{A}{x} + \frac{B}{x - 1} + \frac{C}{(x - 1)^2} \tag{4}$$

Again, we clear fractions by multiplying each side by $x^3 - 2x^2 + x = x(x - 1)^2$. The result is the identity

$$x + 2 = A(x - 1)^2 + Bx(x - 1) + Cx \tag{5}$$

If we let $x = 0$ in this expression, the terms containing B and C drop out leaving $2 = A(-1)^2$, or $A = 2$. Similarly, if we let $x = 1$, the terms containing A and B drop out, leaving $3 = C$. Thus, equation (5) becomes

$$x + 2 = 2(x - 1)^2 + Bx(x - 1) + 3x$$

Now, let $x = 2$ (any choice other than 0 or 1 will work as well). The result is

$$4 = 2(1)^2 + B(2)(1) + 3(2)$$
$$2B = 4 - 2 - 6 = -4$$
$$B = -2$$

Thus, we have $A = 2$, $B = -2$, and $C = 3$.
From equation (4), the partial fraction decomposition is

$$\frac{x + 2}{x^3 - 2x^2 + x} = \frac{2}{x} + \frac{-2}{x - 1} + \frac{3}{(x - 1)^2}$$

∎

E X A M P L E   3

Repeated Linear Factors

Write the partial fraction decomposition of: $\dfrac{x^3 - 8}{x^2(x - 1)^3}$

Solution   The denominator contains the repeated linear factor x^2 and the repeated linear factor $(x - 1)^3$. Thus, the partial fraction decomposition takes the form

$$\frac{x^3 - 8}{x^2(x - 1)^3} = \frac{A}{x} + \frac{B}{x^2} + \frac{C}{x - 1} + \frac{D}{(x - 1)^2} + \frac{E}{(x - 1)^3} \tag{6}$$

As before, we clear fractions and obtain the identity

$$x^3 - 8 = Ax(x - 1)^3 + B(x - 1)^3 + Cx^2(x - 1)^2 + Dx^2(x - 1) + Ex^2 \tag{7}$$

Let $x = 0$. (Do you see why this choice was made?) Then,

$$-8 = B(-1)$$
$$B = 8$$

Now let $x = 1$ in equation (7). Then

$$-7 = E$$

Use $B = 8$ and $E = -7$ in equation (7) and collect like terms:

$$x^3 - 8 = Ax(x - 1)^3 + 8(x - 1)^3$$
$$+ Cx^2(x - 1)^2 + Dx^2(x - 1) - 7x^2$$
$$x^3 - 8 - 8(x^3 - 3x^2 + 3x - 1) + 7x^2 = Ax(x - 1)^3 + Cx^2(x - 1)^2 + Dx^2(x - 1)$$
$$-7x^3 + 31x^2 - 24x = x(x - 1)[A(x - 1)^2 + Cx(x - 1) + Dx]$$
$$x(x - 1)(-7x + 24) = x(x - 1)[A(x - 1)^2 + Cx(x - 1) + Dx]$$
$$-7x + 24 = A(x - 1)^2 + Cx(x - 1) + Dx \tag{8}$$

We now work with equation (8). Let $x = 0$. Then

$$24 = A$$

Now, let $x = 1$ in equation (8). Then,

$$17 = D$$

Use $A = 24$ and $D = 17$ in equation (8) and collect like terms:

$$-7x + 24 = 24(x - 1)^2 + Cx(x - 1) + 17x$$
$$-24x^2 + 48x - 24 - 17x - 7x + 24 = Cx(x - 1)$$
$$-24x^2 + 24x = Cx(x - 1)$$
$$-24x(x - 1) = Cx(x - 1)$$
$$-24 = C$$

We now know all the numbers A, B, C, D, and E, so, from equation (6), we have the decomposition

$$\frac{x^3 - 8}{x^2(x - 1)^3} = \frac{24}{x} + \frac{8}{x^2} + \frac{-24}{x - 1} + \frac{17}{(x - 1)^2} + \frac{-7}{(x - 1)^3}$$

∎

The method employed in Example 3, although somewhat tedious, is still preferable to solving the system of five equations containing five variables that the expansion of equation (6) leads to.

Now work Problem 15.

The final two cases involve irreducible quadratic factors. As mentioned in Section 3.5, a quadratic factor is irreducible if it cannot be factored into linear factors with real coefficients. A quadratic expression $ax^2 + bx + c$ is irreducible whenever $b^2 - 4ac < 0$. For example, $x^2 + x + 1$ and $x^2 + 4$ are irreducible.

> **CASE 3:** Q contains a nonrepeated irreducible quadratic factor.
>
> If Q contains a nonrepeated irreducible quadratic factor of the form $ax^2 + bx + c$, then, in partial fraction decomposition of P/Q, allow for the term
>
> $$\boxed{\frac{Ax + B}{ax^2 + bx + c}}$$
>
> where the numbers A and B are to be determined.

EXAMPLE 4   Nonrepeated Irreducible Quadratic Factor

Write the partial fraction decomposition of $\dfrac{3x - 5}{x^3 - 1}$.

Solution   We factor the denominator,

$$x^3 - 1 = (x - 1)(x^2 + x + 1)$$

and find that is has a nonrepeated linear factor $x - 1$ and a nonrepeated irreducible quadratic factor $x^2 + x + 1$. Thus, we allow for the term $A/(x - 1)$ by Case 1, and we allow for the term $(Bx + C)/(x^2 + x + 1)$ by Case 3. Hence, we write

$$\frac{3x - 5}{x^3 - 1} = \frac{A}{x - 1} + \frac{Bx + C}{x^2 + x + 1} \tag{9}$$

We clear fractions by multiplying each side of equation (9) by $x^3 - 1 = (x - 1)(x^2 + x + 1)$ to obtain the identity

$$3x - 5 = A(x^2 + x + 1) + (Bx + C)(x - 1) \tag{10}$$

10.6 SYSTEMS OF NONLINEAR EQUATIONS

> **1** Solve a System of Nonlinear Equations Using Substitution
> **2** Solve a System of Nonlinear Equations Using Elimination

1 There is no general methodology for solving a system of nonlinear equations by hand. There are times when substitution is best; other times, elimination is best; and there are times when neither of these methods works. Experience and a certain degree of imagination are your allies here.

Before we begin, two comments are in order:

1. If the system contains two variables, then graph them. By graphing each equation in the system, we can get an idea of how many solutions a system has and approximately where they are located.
2. Extraneous solutions can creep in when solving nonlinear systems algebraically, so it is imperative that all apparent solutions be checked.

E X A M P L E 1   Solving a System of Nonlinear Equations

Solve the following system of equations:

$$\begin{cases} 3x - y = -2 & \text{(1) A line.} \\ 2x^2 - y = 0 & \text{(2) A parabola.} \end{cases}$$

Graphing Solution   We use a graphing utility to graph $Y_1 = 3x + 2$ and $Y_2 = 2x^2$. From Figure 17, we see that the system apparently has two solutions. Using INTERSECT, the solutions to the system of equations are $(-0.5, 0.5)$ and $(2, 8)$.

FIGURE 17

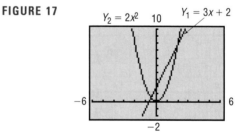

Algebraic Solution
Using Substitution

First, we notice that the system contains two variables and that we know how to graph each equation by hand. In Figure 18, we see that the system apparently has two solutions.

We will use substitution to solve the system. Equation (1) is easily solved for y:

$$3x - y = -2$$
$$y = 3x + 2$$

FIGURE 18

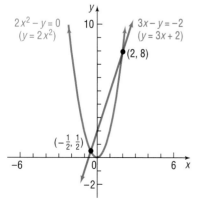

We substitute this expression for y in equation (2). The result is an equation containing just the variable x, which we can then solve:

$$2x^2 - y = 0$$
$$2x^2 - (3x + 2) = 0$$
$$2x^2 - 3x - 2 = 0$$
$$(2x + 1)(x - 2) = 0$$
$$2x + 1 = 0 \quad \text{or} \quad x - 2 = 0$$
$$x = -\frac{1}{2} \qquad\qquad x = 2$$

Using these values for x in $y = 3x + 2$, we find

$$y = 3\left(-\frac{1}{2}\right) + 2 = \frac{1}{2} \quad \text{or} \quad y = 3(2) + 2 = 8$$

The apparent solutions are $x = -\frac{1}{2}$, $y = \frac{1}{2}$ and $x = 2$, $y = 8$.

Check: For $x = -\frac{1}{2}$, $y = \frac{1}{2}$:

$$\begin{cases} 2\left(-\frac{1}{2}\right)^2 - \frac{1}{2} = 2\left(\frac{1}{4}\right) - \frac{1}{2} = 0 & \text{(1)} \\ 3\left(-\frac{1}{2}\right) - \frac{1}{2} = -\frac{3}{2} - \frac{1}{2} = -2 & \text{(2)} \end{cases}$$

For $x = 2$, $y = 8$:

$$\begin{cases} 3(2) - 8 = \quad 6 - 8 = -2 & \text{(1)} \\ 2(2)^2 - 8 = 2(4) - 8 = \quad 0 & \text{(2)} \end{cases}$$

Each solution checks. Now we know the graphs in Figure 18 intersect at $\left(-\frac{1}{2}, \frac{1}{2}\right)$ and at $(2, 8)$. ∎

Now work Problem 11.

Our next example illustrates how the method of elimination works for nonlinear systems.

E X A M P L E   2

Solving a System of Nonlinear Equations

Solve: $\begin{cases} x^2 + y^2 = 13 & \text{(1) A circle.} \\ x^2 - y = 7 & \text{(2) A parabola.} \end{cases}$

Graphing Solution   We use a graphing utility to graph $x^2 + y^2 = 13$ and $x^2 - y = 7$. (Remember that to graph $x^2 + y^2 = 13$ requires two functions, $Y_1 = \sqrt{13 - x^2}$ and $Y_2 = -\sqrt{13 - x^2}$, and a square screen.) From Figure 19 we see that the system apparently has four solutions. Using INTERSECT, the solutions to the system of equations are $(-3, 2)$, $(3, 2)$, $(-2, -3)$, and $(2, -3)$.

Algebraic Solution   First, we graph each equation, as shown in Figure 20. Based on the graph,
Using Elimination   we expect four solutions. By subtracting equation (2) from equation (1), the variable x is eliminated, leaving

$$y^2 + y = 6$$

FIGURE 19

FIGURE 20

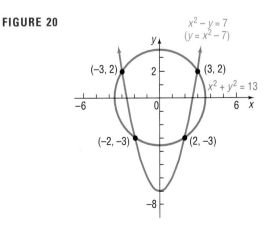

Using INTERSECT, the points of intersection are $(-1, 1)$ and $(0, 1)$. Because x cannot be 0, the value $x = 0$ is extraneous, and we discard it. Thus, the only solution is $x = -1$ and $y = 1$.

Algebraic Solution
Using Elimination

First, we multiply equation (2) by x to eliminate the fraction. The result is an equivalent system because x cannot be 0 [look at equation (2) to see why]:

$$\begin{cases} x^2 + x + y^2 - 3y + 2 = 0 & (1) \\ x^2 + x + y^2 - y = 0 & (2) \end{cases}$$

Now subtract equation (2) from equation (1) to eliminate x. The result is

$$-2y + 2 = 0$$
$$y = 1$$

To find x, we back-substitute $y = 1$ in equation (1):

$$x^2 + x + 1 - 3 + 2 = 0$$
$$x^2 + x = 0$$
$$x(x + 1) = 0$$
$$x = 0 \quad \text{or} \quad x = -1$$

Because x cannot be 0, the value $x = 0$ is extraneous, and we discard it. Thus, the solution is $x = -1, y = 1$.

Check: We now check $x = -1, y = 1$:

$$\begin{cases} (-1)^2 + (-1) + 1^2 - 3(1) + 2 = 1 - 1 + 1 - 3 + 2 = 0 & (1) \\ -1 + 1 + \dfrac{1^2 - 1}{-1} = 0 + \dfrac{0}{-1} = 0 & (2) \end{cases}$$

Thus, the only solution to the system is $x = -1, y = 1$. ∎

Now work Problems 25 and 49.

E X A M P L E   5

Solving a System of Nonlinear Equations

Solve: $\begin{cases} x^2 - y^2 = 4 & \text{(1) A hyperbola.} \\ y = x^2 & \text{(2) A parabola.} \end{cases}$

Graphing Solution

We graph $Y_1 = x^2$ and $x^2 - y^2 = 4$ in Figure 23(a). You will need to graph $x^2 - y^2 = 4$ as two functions:

$$Y_2 = \sqrt{x^2 - 4} \quad \text{and} \quad Y_3 = -\sqrt{x^2 - 4}$$

From Figure 23(a) we see that the graphs of these two equations do not intersect. Thus, the system is inconsistent.

FIGURE 23(a)

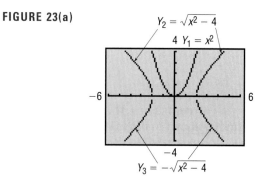

$Y_2 = \sqrt{x^2 - 4}$
$4 \; Y_1 = x^2$
$Y_3 = -\sqrt{x^2 - 4}$

Algebraic Solution Either substitution or elimination can be used here. We use substitution and replace x^2 by y in equation (1). The result is

$$y - y^2 = 4$$
$$y^2 - y + 4 = 0$$

This is a quadratic equation whose discriminant is $1 - 4 \cdot 4 = -15 < 0$. Thus, the equation has no real solutions, and hence, the system is inconsistent. The graphs of these two equations will not intersect. See Figure 23(b).

FIGURE 23(b)

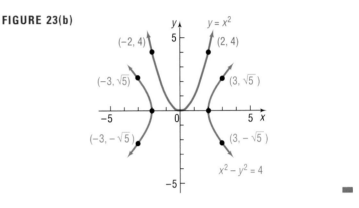

The following examples illustrate two of the more imaginative ways to solve systems of nonlinear equations algebraically.

E X A M P L E 6   Solving a System of Nonlinear Equations

Solve: $\begin{cases} 4x^2 - 9xy - 28y^2 = 0 & (1) \\ 16x^2 - 4xy = 16 & (2) \end{cases}$

Graphing Solution Instead of solving for y by completing the square, we notice that equation (1) can be factored:

$$4x^2 - 9xy - 28y^2 = 0$$
$$(4x + 7y)(x - 4y) = 0$$

This results in the two equations:

$$4x + 7y = 0 \qquad \text{or} \qquad x - 4y = 0$$
$$Y_1 = -\frac{4}{7}x \qquad\qquad Y_2 = \frac{1}{4}x$$

We graph these two equations and equation (2): $Y_3 = \dfrac{4(x^2 - 1)}{x}$. See Figure 24. Using INTERSECT, we find the intersection points to be $(-0.93, 0.53)$, $(-1.03, -0.25)$, $(0.93, -0.53)$, and $(1.03, 0.25)$, each correct to two decimal places.

FIGURE 24

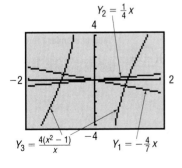

Algebraic Solution We take note of the fact that equation (1) can be factored:

$$4x^2 - 9xy - 28y^2 = 0$$
$$(4x + 7y)(x - 4y) = 0$$

From this system, we get $x^2 = mx + 4 - 2m$ or $x^2 - mx + (2m - 4) = 0$. By using the quadratic formula, we get

$$x = \frac{m \pm \sqrt{m^2 - 4(2m - 4)}}{2}$$

To obtain a unique solution for x, the two roots must be equal; in other words, the discriminant $m^2 - 4(2m - 4)$ must be 0. Complete the work to get m, and write an equation of the tangent line.

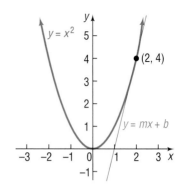

In Problems 88–94, use Descartes' method from Problem 87 to find the equation of the line tangent to each graph at the given point.

88. $x^2 + y^2 = 10$; at $(1, 3)$

89. $y = x^2 + 2$; at $(1, 3)$

90. $x^2 + y = 5$; at $(-2, 1)$

91. $2x^2 + 3y^2 = 14$; at $(1, 2)$

92. $3x^2 + y^2 = 7$; at $(-1, 2)$

93. $x^2 - y^2 = 3$; at $(2, 1)$

94. $2y^2 - x^2 = 14$; at $(2, 3)$

95. If r_1 and r_2 are two solutions of a quadratic equation $ax^2 + bx + c = 0$, then it can be shown that

$$r_1 + r_2 = -\frac{b}{a} \quad \text{and} \quad r_1 r_2 = \frac{c}{a}$$

Solve this system of equations for r_1 and r_2.

 96. A circle and a line intersect at most twice. A circle and a parabola intersect at most four times. Deduce that a circle and the graph of a polynomial of degree 3 intersect at most six times. What do you conjecture about a polynomial of degree 4? What about a polynomial of degree n? Can you explain your conclusions using an algebraic argument?

97. Suppose that you are the manager of a sheet metal shop. A customer asks you to manufacture 10,000 boxes, each box being open on top. The boxes are required to have a square base and a 9-cubic foot capacity. You construct the boxes by cutting a square out from each corner of a square piece of sheet metal and folding along the edges.
(a) What are the dimensions of the square to be cut if the area of the square piece of sheet metal is 100 square feet?
(b) Could you make the box using a smaller piece of sheet metal? Make a list of the dimensions of the box for various pieces of sheet metal.

10.7 SYSTEMS OF INEQUALITIES

1 Graph an Inequality by Hand
2 Graph an Inequality Using a Graphing Utility
3 Graph a System of Linear Inequalities

In Chapter 1, we discussed inequalities in one variable. In this section, we discuss inequalities in two variables. Samples are given in Example 1.

EXAMPLE 1   Samples of Inequalities in Two Variables

(a) $3x + y - 6 < 0$ (b) $x^2 + y^2 < 4$ (c) $y^2 \le x$

1

An inequality in two variables x and y is **satisfied** by an ordered pair (a, b) if, when x is replaced by a and y by b, a true statement results. A **graph of an inequality in two variables** x and y consists of all points (x, y) whose coordinates satisfy the inequality.

Let's look at an example.

E X A M P L E   2   Graphing a Linear Inequality

Graph the linear inequality: $3x + y - 6 \leq 0$

Solution   We begin with the associated problem of the graph of the linear equation

$$3x + y - 6 = 0$$

formed by replacing (for now) the $\leq$ symbol with an $=$ sign. The graph of the linear equation is a line. See Figure 26(a). This line is part of the graph of the inequality we seek because the inequality is nonstrict. (Do you see why? We are seeking points for which $3x + y - 6$ is less than *or equal to* 0.)

Now, let's test a few randomly selected points to see whether they belong to the graph of the inequality

	$3x + y - 6$	**Conclusion**
$(4, -1)$	$3(4) + (-1) - 6 = 5 > 0$	Does not belong to graph
$(5, 5)$	$3(5) + 5 - 6 = 14 > 0$	Does not belong to graph
$(-1, 2)$	$3(-1) + 2 - 6 = -7 < 0$	Belongs to graph
$(-2, -2)$	$3(-2) + (-2) - 6 = -14 < 0$	Belongs to graph

Look again at Figure 26(a). Notice that the two points that belong to the graph both lie on the same side of the line, and the two points that do not belong to the graph lie on the opposite side. As it turns out, this is always the case. Thus, the graph we seek consists of all points that lie on the same side of the line as do $(-1, 2)$ and $(-2, -2)$. The graph we seek is the shaded region in Figure 26(b).

FIGURE 26

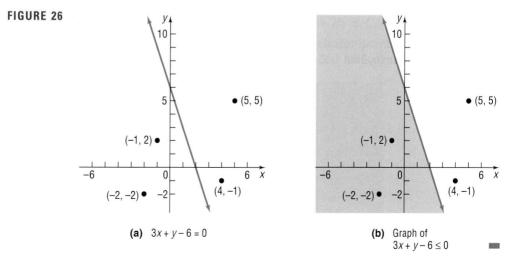

(a) $3x + y - 6 = 0$

(b) Graph of $3x + y - 6 \leq 0$

other for which $Ax + By > C$. Because of this, for linear inequalities, only one test point is required.

EXAMPLE 5   Graphing Linear Inequalities

Graph: (a) $y < 2$ (b) $y \geq 2x$

Solution   (a) The graph of the equation $y = 2$ is a horizontal line and is not part of the graph of the inequality. Since $(0,0)$ satisfies the inequality, the graph consists of the half-plane below the line $y = 2$. See Figure 31.

(b) The graph of the equation $y = 2x$ is a line and is part of the graph of the inequality. Using $(3, 0)$ as a test point, we find it does not satisfy the inequality $[0 < 2 \cdot 3]$. Thus, points in the half-plane on the opposite side of $y = 2x$ satisfy the inequality. See Figure 32.

FIGURE 31

FIGURE 32

 Now work Problem 3.

Systems of Inequalities in Two Variables

3 The **graph of a system of inequalities** in two variables x and y is the set of all points (x, y) that simultaneously satisfy *each* of the inequalities in the system. Thus, the graph of a system of inequalities can be obtained by graphing each inequality individually and then determining where, if at all, they intersect.

EXAMPLE 6   Graphing a System of Linear Inequalities Using a Graphing Utility

Graph the system: $\begin{cases} x + y \geq 2 \\ 2x - y \leq 4 \end{cases}$

Solution   First, we graph the lines $x + y = 2$ $(Y_1 = -x + 2)$ and $2x - y = 4$ $(Y_2 = 2x - 4)$. See Figure 33.

FIGURE 33

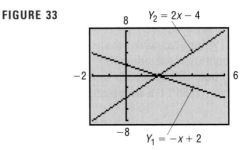

$Y_2 = 2x - 4$

$Y_1 = -x + 2$

Notice the graphs divide the viewing window into four regions. We select a test point for each region and determine whether the point makes *both* inequalities true. We choose to test $(0, 0), (2, 3), (4, 0)$, and $(2, -2)$. Figure 34(a) shows that $(2, 3)$ is the only point for which both inequalities are true. Thus, we obtain the graph shown in Figure 34(b).

FIGURE 34

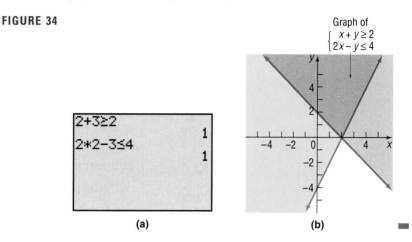

(a)   (b)

Notice in Figure 34(b) that the darker region, the graph of the system of inequalities, is the intersection of the graphs of the single inequalities $x + y \geq 2$ and $2x - y \leq 4$. Obtaining Figure 34(b) by this method is sometimes faster than using test points.

E X A M P L E 7

Graphing a System of Linear Inequalities

Graph the system: $\begin{cases} x + y \leq 2 \\ x + y \geq 0 \end{cases}$

Solution   See Figure 35. The overlapping, darker shaded region between the two boundary lines is the graph of the system.

FIGURE 35 $x + y = 0$ $x + y = 2$

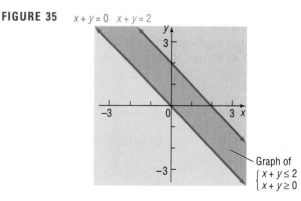

Graph of
$\begin{cases} x + y \leq 2 \\ x + y \geq 0 \end{cases}$

Now work Problem 25.

E X A M P L E 8

Graphing a System of Linear Inequalities

Graph the system: $\begin{cases} 2x - y \geq 2 \\ 2x - y \geq 0 \end{cases}$

FIGURE 41

EXAMPLE 12   Financial Planning

A retired couple has up to $25,000 to invest. As their financial adviser, you recommend that they place at least $15,000 in Treasury bills yielding 6% and at most $5000 in corporate bonds yielding 9%.

(a) Using x to denote the amount of money invested in Treasury bills and y the amount invested in corporate bonds, write a system of linear inequalities that describes the possible amounts of each investment.

(b) Graph the system.

Solution   (a) The system of linear inequalities is

$$\begin{cases} x \geq 0 & \text{x and y are nonnegative variables since they represent money invested.} \\ y \geq 0 & \\ x + y \leq 25,000 & \text{The total of the two investments, $x + y$, cannot exceed \$25,000.} \\ x \geq 15,000 & \text{At least \$15,000 in Treasury bills.} \\ y \leq 5000 & \text{At most \$5000 in corporate bonds.} \end{cases}$$

(b) See the shaded region in Figure 42. Note that the inequalities $x \geq 0$ and $y \geq 0$ again require that the graph of the system be in quadrant I.

FIGURE 42

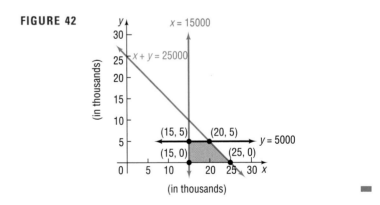

The graph of the system of linear inequalities in Figure 42 is said to be **bounded,** because it can be contained within some circle of sufficiently large radius. A graph that cannot be contained in any circle is said to be **unbounded.** For example, the graph of the system of linear inequalities in Figure 41 is unbounded, since it extends indefinitely in a particular direction.

Notice in Figures 41 and 42 that those points belonging to the graph that are also points of intersection of boundary lines have been plotted. Such points are referred to as **vertices** or **corner points** of the graph. Thus, the system graphed in Figure 41 has three corner points: (0, 4), (1, 2), and (3, 0). The system graphed in Figure 42 has four corner points: (15, 0), (25, 0), (20, 5), and (15, 5).

These ideas will be used in the next section in developing a method for solving linear programming problems, an important application of linear inequalities.

Now work Problem 33.

10.7 EXERCISES

In Problems 1–12, graph each inequality.

1. $x \geq 0$ **2.** $y \geq 0$ **3.** $x \geq 4$ **4.** $y \leq 2$

5. $2x + y \geq 6$ **6.** $3x + 2y \leq 6$ **7.** $x^2 + y^2 > 1$ **8.** $x^2 + y^2 \leq 9$

9. $y \leq x^2 - 1$ **10.** $y > x^2 + 2$ **11.** $xy \geq 4$ **12.** $xy \leq 1$

In Problems 13–30, graph each system of inequalities.

13. $\begin{cases} x + y \leq 2 \\ 2x + y \geq 4 \end{cases}$
14. $\begin{cases} 3x - y \geq 6 \\ x + 2y \leq 2 \end{cases}$
15. $\begin{cases} 2x - y \leq 4 \\ 3x + 2y \geq -6 \end{cases}$
16. $\begin{cases} 4x - 5y \leq 0 \\ 2x - y \geq 2 \end{cases}$

17. $\begin{cases} 2x - 3y \leq 0 \\ 3x + 2y \leq 6 \end{cases}$
18. $\begin{cases} 4x - y \geq 2 \\ x + 2y \geq 2 \end{cases}$
19. $\begin{cases} x^2 + y^2 \leq 9 \\ x + y \geq 3 \end{cases}$
20. $\begin{cases} x^2 + y^2 \geq 9 \\ x + y \leq 3 \end{cases}$

21. $\begin{cases} y \geq x^2 - 4 \\ y \leq x - 2 \end{cases}$
22. $\begin{cases} y^2 \leq x \\ y \geq x \end{cases}$
23. $\begin{cases} xy \geq 4 \\ y \geq x^2 + 1 \end{cases}$
24. $\begin{cases} y + x^2 \leq 1 \\ y \geq x^2 - 1 \end{cases}$

25. $\begin{cases} x - 2y \leq 6 \\ 2x - 4y \geq 0 \end{cases}$
26. $\begin{cases} x + 4y \leq 8 \\ x + 4y \geq 4 \end{cases}$
27. $\begin{cases} 2x + y \geq -2 \\ 2x + y \geq 2 \end{cases}$
28. $\begin{cases} x - 4y \leq 4 \\ x - 4y \geq 0 \end{cases}$

29. $\begin{cases} 2x + 3y \geq 6 \\ 2x + 3y \leq 0 \end{cases}$
30. $\begin{cases} 2x + y \geq 0 \\ 2x + y \geq 2 \end{cases}$

In Problems 31–40, graph each system of linear inequalities. Tell whether the graph is bounded or unbounded, and label the corner points.

31. $\begin{cases} x \geq 0 \\ y \geq 0 \\ 2x + y \leq 6 \\ x + 2y \leq 6 \end{cases}$
32. $\begin{cases} x \geq 0 \\ y \geq 0 \\ x + y \geq 4 \\ 2x + 3y \geq 6 \end{cases}$
33. $\begin{cases} x \geq 0 \\ y \geq 0 \\ x + y \geq 2 \\ 2x + y \geq 4 \end{cases}$
34. $\begin{cases} x \geq 0 \\ y \geq 0 \\ 3x + y \leq 6 \\ 2x + y \leq 2 \end{cases}$
35. $\begin{cases} x \geq 0 \\ y \geq 0 \\ x + y \geq 2 \\ 2x + 3y \leq 12 \\ 3x + y \leq 12 \end{cases}$

36. $\begin{cases} x \geq 0 \\ y \geq 0 \\ x + y \geq 2 \\ x + y \leq 10 \\ 2x + y \leq 3 \end{cases}$
37. $\begin{cases} x \geq 0 \\ y \geq 0 \\ x + y \geq 2 \\ x + y \leq 8 \\ 2x + y \leq 10 \end{cases}$
38. $\begin{cases} x \geq 0 \\ y \geq 0 \\ x + y \geq 2 \\ x + y \leq 8 \\ x + 2y \geq 1 \end{cases}$
39. $\begin{cases} x \geq 0 \\ y \geq 0 \\ x + 2y \geq 1 \\ x + 2y \leq 10 \end{cases}$
40. $\begin{cases} x \geq 0 \\ y \geq 0 \\ x + 2y \geq 1 \\ x + 2y \leq 10 \\ x + y \geq 2 \\ x + y \leq 8 \end{cases}$

Similarly, the supply of 120 pounds, or 1920 ounces, of special-blend coffee leads to the constraint

$12x + 8y \le 1920$ Special-blend coffee constraint.

The linear programming problem may be stated as

Maximize $P = 0.3x + 0.4y$

subject to the constraints

$x \ge 0, \quad y \ge 0, \quad 4x + 8y \le 1600, \quad 12x + 8y \le 1920$

The graph of the constraints (the feasible points) is illustrated in Figure 45. We list the corner points and evaluate the objective function at each one. In Table 3, we can see that the maximum profit, $84, is achieved with 40 packages of the low-grade mixture and 180 packages of the high-grade mixture.

FIGURE 45

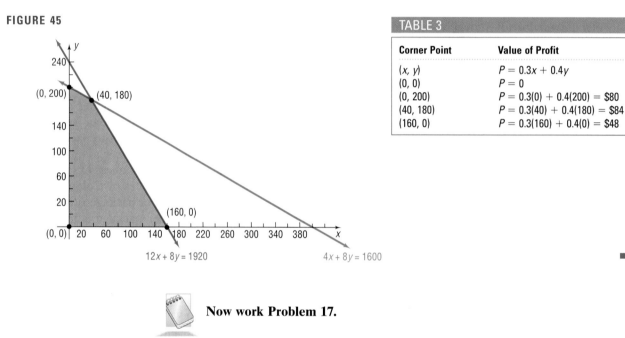

TABLE 3

Corner Point	Value of Profit
(x, y)	$P = 0.3x + 0.4y$
(0, 0)	$P = 0$
(0, 200)	$P = 0.3(0) + 0.4(200) = \$80$
(40, 180)	$P = 0.3(40) + 0.4(180) = \$84$
(160, 0)	$P = 0.3(160) + 0.4(0) = \$48$

$12x + 8y = 1920$ $4x + 8y = 1600$

Now work Problem 17.

10.8 EXERCISES

In Problems 1–6, find the maximum and minimum value of the given objective function.

The figure illustrates the graph of the feasible points of a linear programming problem.

1. $z = x + y$ **2.** $z = 2x + 3y$ **3.** $z = x + 10y$

4. $z = 10x + y$ **5.** $z = 5x + 7y$ **6.** $z = 7x + 5y$

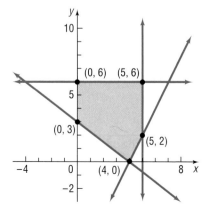

In Problems 7–16, solve each linear programming problem.

7. Maximize $z = 2x + y$ subject to $x \geq 0$, $y \geq 0$, $x + y \leq 6$, $x + y \geq 1$

8. Maximize $z = x + 3y$ subject to $x \geq 0$, $y \geq 0$, $x + y \geq 3$, $x \leq 5, y \leq 7$

9. Minimize $z = 2x + 5y$ subject to $x \geq 0$, $y \geq 0$, $x + y \geq 2$, $x \leq 5, y \leq 3$

10. Minimize $z = 3x + 4y$ subject to $x \geq 0$, $y \geq 0$, $2x + 3y \geq 6$, $x + y \leq 8$

11. Maximize $z = 3x + 5y$ subject to $x \geq 0$, $y \geq 0$, $x + y \geq 2$, $2x + 3y \leq 12$, $3x + 2y \leq 12$

12. Maximize $z = 5x + 3y$ subject to $x \geq 0$, $y \geq 0$, $x + y \geq 2$, $x + y \leq 8$, $2x + y \leq 10$

13. Minimize $z = 5x + 4y$ subject to $x \geq 0$, $y \geq 0$, $x + y \geq 2$, $2x + 3y \leq 12$, $3x + y \leq 12$

14. Minimize $z = 2x + 3y$ subject to $x \geq 0$, $y \geq 0$, $x + y \geq 3$, $x + y \leq 9$, $x + 3y \geq 6$

15. Maximize $z = 5x + 2y$ subject to $x \geq 0$, $y \geq 0$, $x + y \leq 10$, $2x + y \geq 10$, $x + 2y \geq 10$

16. Maximize $z = 2x + 4y$ subject to $x \geq 0$, $y \geq 0$, $2x + y \geq 4$, $x + y \leq 9$

17. Maximizing Profit A manufacturer of skis produces two types: downhill and cross-country. Use the following table to determine how many of each kind of ski should be produced to achieve a maximum profit. What is the maximum profit? What would the maximum profit be if the maximum time available for manufacturing is increased to 48 hours?

	Downhill	Cross-Country	Maximum Time Available
Manufacturing time per ski	2 hours	1 hour	40 hours
Finishing time per ski	1 hour	1 hour	32 hours
Profit per ski	$70	$50	

18. Farm Management A farmer has 70 acres of land available for planting either soybeans or wheat. The cost of preparing the soil, the workdays required, and the expected profit per acre planted for each type of crop are given in the following table

	Soybeans	Wheat
Preparation cost per acre	$60	$30
Workdays required per acre	3	4
Profit per acre	$180	$100

The farmer cannot spend more than $1800 in preparation costs nor more than a total of 120 workdays. How many acres of each crop should be planted in order to maximize the profit? What is the maximum profit? What is the maximum profit if the farmer is willing to spend no more than $2400 on preparation?

19. Farm Management A small farm in Illinois has 100 acres of land available on which to grow corn and soybeans. The following table shows the cultivation cost per acre, the labor cost per acre, and the expected profit per acre. The column on the right shows the amount of money available for each of these expenses. Find the number of acres of each crop that should be planted in order to maximize profit.

	Soybeans	Corn	Money Available
Cultivation cost per acre	$40	$60	$1800
Labor cost per acre	$60	$60	$2400
Profit per acre	$200	$250	

20. Dietary Requirements A certain diet requires at least 60 units of carbohydrates, 45 units of protein, and 30 units of fat each day. Each ounce of Supplement A provides 5 units of carbohydrates, 3 units of protein, and 4 units of fat. Each ounce of Supplement B provides 2 units of carbohydrates, 2 units of protein, and 1 unit of fat. If Supplement A costs $1.50 per ounce and Supplement B costs $1.00 per ounce, how many ounces of each supplement should be taken daily to minimize the cost of the diet?

21. Production Scheduling In a factory, machine 1 produces 8″ plyers at the rate of 60 units per hour, and 6″ plyers at the rate of 70 units per hour. Machine 2 produces 8″ plyers at the rate of 40 units per hour and 6″ plyers at the rate of 20 units per hour. It costs $50 per hour to operate machine 1, while machine 2 costs $30 per hour to operate. The production schedule requires that at least 240 units of 8″ plyers and at least 140 units of 6″ plyers must be produced during each 10 hour day. Which combination of machines will cost the least money to operate?

In Problems 57–66, write the partial fraction decomposition of each rational expression.

57. $\dfrac{6}{x(x-4)}$

58. $\dfrac{x}{(x+2)(x-3)}$

59. $\dfrac{x-4}{x^2(x-1)}$

60. $\dfrac{2x-6}{(x-2)^2(x-1)}$

61. $\dfrac{x}{(x^2+9)(x+1)}$

62. $\dfrac{3x}{(x-2)(x^2+1)}$

63. $\dfrac{x^3}{(x^2+4)^2}$

64. $\dfrac{x^3+1}{(x^2+16)^2}$

65. $\dfrac{x^2}{(x^2+1)(x^2-1)}$

66. $\dfrac{4}{(x^2+4)(x^2-1)}$

In Problems 67–76, solve each system of equations algebraically. Verify your result using a graphing utility.

67. $\begin{cases} 2x+y+3=0 \\ x^2+y^2=5 \end{cases}$

68. $\begin{cases} x^2+y^2=16 \\ 2x-y^2=-8 \end{cases}$

69. $\begin{cases} 2xy+y^2=10 \\ 3y^2-xy=2 \end{cases}$

70. $\begin{cases} 3x^2-y^2=1 \\ 7x^2-2y^2-5=0 \end{cases}$

71. $\begin{cases} x^2+y^2=6y \\ x^2=3y \end{cases}$

72. $\begin{cases} 2x^2+y^2=9 \\ x^2+y^2=9 \end{cases}$

73. $\begin{cases} 3x^2+4xy+5y^2=8 \\ x^2+3xy+2y^2=0 \end{cases}$

74. $\begin{cases} 3x^2+2xy-2y^2=6 \\ xy-2y^2+4=0 \end{cases}$

75. $\begin{cases} x^2-3x+y^2+y=-2 \\ \dfrac{x^2-x}{y}+y+1=0 \end{cases}$

76. $\begin{cases} x^2+x+y^2=y+2 \\ x+1=\dfrac{2-y}{x} \end{cases}$

In Problems 77–82, graph each system of inequalities. Tell whether the graph is bounded or unbounded, and label the corner points.

77. $\begin{cases} -2x+y\le 2 \\ x+y\ge 2 \end{cases}$

78. $\begin{cases} x-2y\le 6 \\ 2x+y\ge 2 \end{cases}$

79. $\begin{cases} x\ge 0 \\ y\ge 0 \\ x+y\le 4 \\ 2x+3y\le 6 \end{cases}$

80. $\begin{cases} x\ge 0 \\ y\ge 0 \\ 3x+y\ge 6 \\ 2x+y\ge 2 \end{cases}$

81. $\begin{cases} x\ge 0 \\ y\ge 0 \\ 2x+y\le 8 \\ x+2y\ge 2 \end{cases}$

82. $\begin{cases} x\ge 0 \\ y\ge 0 \\ 3x+y\le 9 \\ 2x+3y\ge 6 \end{cases}$

In Problems 83–86, graph each system of inequalities.

83. $\begin{cases} x^2+y^2\le 16 \\ x+y\ge 2 \end{cases}$

84. $\begin{cases} y^2\le x-1 \\ x-y\le 3 \end{cases}$

85. $\begin{cases} y\le x^2 \\ xy\le 4 \end{cases}$

86. $\begin{cases} x^2+y^2\ge 1 \\ x^2+y^2\le 4 \end{cases}$

In Problems 87–92, solve each linear programming problem.

87. Maximize $z=3x+4y$ subject to $x\ge 0$, $y\ge 0$, $3x+2y\ge 6$, $x+y\le 8$

88. Maximize $z=2x+4y$ subject to $x\ge 0$, $y\ge 0$, $x+y\le 6$, $x\ge 2$

89. Minimize $z=3x+5y$ subject to $x\ge 0$, $y\ge 0$, $x+y\ge 1$, $3x+2y\le 12$, $x+3y\le 12$

90. Minimize $z=3x+y$ subject to $x\ge 0$, $y\ge 0$, $x\le 8$, $y\le 6$, $2x+y\ge 4$

91. Maximize $z=5x+4y$ subject to $x\ge 0$, $y\ge 0$, $x+2y\ge 2$, $3x+4y\le 12$, $y\ge x$

92. Maximize $z=4x+5y$ subject to $x\ge 0$, $y\ge 0$, $2x+3y\ge 6$, $x\ge y$, $2x+y\le 12$

93. Find A such that the system of equations has infinitely many solutions.

$$\begin{cases} 2x+5y=5 \\ 4x+10y=A \end{cases}$$

94. Find A such that the system in Problem 93 is inconsistent.

95. Curve Fitting Find the quadratic function $y=ax^2+bx+c$ that passes through the three points $(0,1)$, $(1,0)$, and $(-2,1)$.

96. Curve Fitting Find the general equation of the circle that passes through the three points $(0,1)$, $(1,0)$, and $(-2,1)$.

[**Hint:** The general equation of a circle is $x^2+y^2+Dx+Ey+F=0$.]

97. Blending Coffee A coffee distributor is blending a new coffee that will cost $3.90 per pound. It will consist of a blend of $3.00 per pound coffee and $6.00 per pound coffee. What amounts of each type of coffee should be mixed to achieve the desired blend?

[**Hint:** Assume that the weight of the blended coffee is 100 pounds.]

$3.00/lb $3.90/lb $6.00/lb

98. **Farming** A 1000 acre farm in Illinois is used to raise corn and soy beans. The cost per acre for raising corn is $65 and the cost per acre for soy beans is $45. If $54,325 has been budgeted for costs and all the acreage is to be used, how many acres should be allocated for each crop?

99. **Cookie Orders** A cookie company makes three kinds of cookies, oatmeal raisin, chocolate chip, and shortbread, packaged in small, medium, and large boxes. The small box contains 1 dozen oatmeal raisin and 1 dozen chocolate chip; the medium box has 2 dozen oatmeal raisin, 1 dozen chocolate chip, and 1 dozen shortbread; the large box contains 2 dozen oatmeal raisin, 2 dozen chocolate chip, and 3 dozen shortbread. If you require exactly 15 dozen oatmeal raisin, 10 dozen chocolate chip, and 11 dozen shortbread, how many of each size box should you buy?

100. **Mixed Nuts** A store that specializes in selling nuts has 72 pounds of cashews and 120 pounds of peanuts available. These are to be mixed in 12-ounce packages as follows: a lower-priced package containing 8 ounces of peanuts and 4 ounces of cashews and a quality package containing 6 ounces of peanuts and 6 ounces of cashews.
    (a) Use x to denote the number of lower-priced packages and use y to denote the number of quality packages. Write a system of linear inequalities that describes the possible number of each kind of package.
    (b) Graph the system and label the corner points.

101. A small rectangular lot has a perimeter of 68 feet. If its diagonal is 26 feet, what are the dimensions of the lot?

102. The area of a rectangular window is 4 square feet. If the diagonal measures $2\sqrt{2}$ feet, what are the dimensions of the window?

103. **Geometry** A certain right triangle has a perimeter of 14 inches. If the hypotenuse is 6 inches long, what are the lengths of the legs?

104. **Geometry** A certain isosceles triangle has a perimeter of 18 inches. If the altitude is 6 inches, what is the length of the base?

105. **Building a Fence** How much fence is required to enclose 5000 square feet by two squares whose sides are in the ratio of 1:2?

106. **Mixing Acids** A chemistry laboratory has three containers of hydrochloric acid, HCl. One container holds a solution with a concentration of 10% HCl, the second holds 25% HCl, and the third holds 40% HCl. How many liters of each should be mixed to obtain 100 liters of a solution with a concentration of 30% HCl? Construct a table showing some of the possible combinations.

107. **Calculating Allowances** Katy, Mike, Danny, and Colleen agreed to do yard work at home for $45 to be split among them. After they finished, their father determined that Mike deserves twice what Katy gets, Katy and Colleen deserve the same amount, and Danny deserves half of what Katy gets. How much does each one receive?

108. **Finding the Speed of the Jet Stream** On a flight between Midway Airport in Chicago and Ft. Lauderdale, Florida, a Boeing 737 jet maintains an airspeed of 475 miles per hour. If the trip from Chicago to Ft. Lauderdale takes 2 hours, 30 minutes and the return flight takes 2 hours, 50 minutes, what is the speed of the jet stream? (Assume the speed of the jet stream remains constant at the various altitudes of the plane.)

109. **Constant Rate Jobs** If Bruce and Bryce work together for 1 hour and 20 minutes, they will finish a certain job. If Bryce and Marty work together for 1 hour and 36 minutes, the same job can be finished. If Marty and Bruce work together, they can complete this job in 2 hours and 40 minutes. How long will it take each of them working alone to finish the job?

110. **Maximizing Profit on Figurines** A factory manufactures two kinds of ceramic figurines: a dancing girl and a mermaid, each requiring three processes—molding, painting, and glazing. The daily labor available for molding is no more than 90 work-hours, labor available for painting does not exceed 120 work-hours, and labor available for glazing is no more than 60 work-hours. The dancing girl requires 3 work-hours for molding, 6 work-hours for painting, and 2 work-hours for glazing. The mermaid requires 3 work-hours for molding, 4 work-hours

for painting, and 3 work-hours for glazing. If the profit on each figurine is $25 for dancing girls and $30 for mermaids, how many of each should be produced each day to maximize profit? If management decides to produce the number of each figurine that maximizes profit, determine which of these processes has excess work-hours assigned to it.

111. Minimizing Production Cost A factory produces gasoline engines and diesel engines. Each week the factory is obligated to deliver at least 20 gasoline engines and at least 15 diesel engines. Due to physical limitations, however, the factory cannot make more than 60 gasoline en-

gines nor more than 40 diesel engines. Finally, to prevent layoffs, a total of at least 50 engines must be produced. If gasoline engines cost $450 each to produce and diesel engines cost $550 each to produce, how many of each should be produced per week to minimize the cost? What is the excess capacity of the factory; that is, how many of each kind of engine are being produced in excess of the number the factory is obligated to deliver?

112. Describe four ways of solving a system of three linear equations containing three variables. Which method do you prefer? Why?

Sequences; Induction; Counting; Probability

Bicycle riding has grown in popularity in the past few years. With the increase of riders also comes an increase in bicycle accidents. On the following page you are placed in the position of being a law student and having to argue a case regarding helmet safety. You will need to use the Sullivan website at:

www.prenhall.com/sullivan

to find the relevant data upon which to base your case.

PREPARING FOR THIS CHAPTER

Before getting started on this chapter, review the following concepts:

For Section 11.3: Compound Interest (*Section 4.6*)

OUTLINE

Graphing utilities can be used to generate recursively defined sequences when the nth term depends upon the term preceding it.

EXAMPLE 7 Using a Graphing Utility to Write the Terms of a Recursively Defined Sequence

Use a graphing utility to write down the first five terms of the following recursively defined sequence

$$s_1 = 1, \qquad s_n = 4s_{n-1}$$

Solution First, put the graphing utility into sequence mode and dot mode. Using $Y = $, enter the recursive formula into the graphing utility. Next, set up the viewing window to generate the desired sequence. Finally, graph the recursion relation and use TRACE to determine the terms in the sequence. For example, in Figure 7 we see the fourth term of the sequence is 64.

Note: If your graphing utility is capable of generating tables, the TABLE feature displays the terms in the sequence. See Table 2.

FIGURE 7

EXAMPLE 8 Writing the Terms of a Recursively Defined Sequence

Write down the first five terms of the following recursively defined sequence.

$$u_1 = 1, \qquad u_2 = 1, \qquad u_{n+2} = u_n + u_{n+1}$$

Solution We are given the first two terms. To get the third term requires that we know each of the previous two terms. Thus,

$$u_1 = 1$$
$$u_2 = 1$$
$$u_3 = u_1 + u_2 = 2$$
$$u_4 = u_2 + u_3 = 1 + 2 = 3$$
$$u_5 = u_3 + u_4 = 2 + 3 = 5$$

The sequence defined in Example 8 is called a **Fibonacci sequence,** and the terms of this sequence are called **Fibonacci numbers.** These numbers appear in a wide variety of applications (see Problems 67 and 68).

E X A M P L E   9   Writing the Terms of a Recursively Defined Sequence

Write down the first five terms of the following recursively defined sequence.

$$f_1 = 1, \qquad f_{n+1} = (n + 1)f_n$$

Solution   Here

$$f_1 = 1$$
$$f_2 = 2f_1 = 2 \cdot 1 = 2$$
$$f_3 = 3f_2 = 3 \cdot 2 = 6$$
$$f_4 = 4f_3 = 4 \cdot 6 = 24$$
$$f_5 = 5f_4 = 5 \cdot 24 = 120$$

■

You should recognize the nth term of the sequence in Example 9 as $n!$.

Now work Problems 21 and 29.

Adding the First n Terms of a Sequence; Summation Notation

It is often important to be able to find the sum of the first n terms of a sequence $\{a_n\}$, namely,

$$a_1 + a_2 + a_3 + \cdots + a_n \tag{1}$$

Rather than write down all these terms, we introduce a more concise way to express the sum, called **summation notation.** Using summation notation, we would write the sum (1) as

$$a_1 + a_2 + a_3 + \cdots + a_n = \sum_{k=1}^{n} a_k$$

The symbol Σ (a stylized version of the Greek letter sigma, which is an S in our alphabet) is simply an instruction to sum, or add up, the terms. The integer k is called the **index** of the sum; it tells you where to start the sum and where to end it. Therefore, the expression

$$\sum_{k=1}^{n} a_k \tag{2}$$

is an instruction to add the terms a_k of the sequence $\{a_n\}$ from $k = 1$ through $k = n$. We read expression (2) as "the sum of a_k from $k = 1$ to $k = n$."

Next, we list some properties of sequences using summation notation.

Theorem   Properties of Sequences

If $\{a_n\}$ and $\{b_n\}$ are two sequences and c is a real number, then

1. $\displaystyle\sum_{k=1}^{n} c = c \cdot n$

2. $\displaystyle\sum_{k=1}^{n} ca_k = c \sum_{k=1}^{n} a_k$

3. $\displaystyle\sum_{k=1}^{n} (a_k + b_k) = \sum_{k=1}^{n} a_k + \sum_{k=1}^{n} b_k$

4. $\displaystyle\sum_{k=1}^{n} (a_k - b_k) = \sum_{k=1}^{n} a_k - \sum_{k=1}^{n} b_k$

5. $\displaystyle\sum_{k=1}^{n} a_k = \sum_{k=1}^{j} a_k + \sum_{k=j+1}^{n} a_k,$ when $1 < j < n$

Although we shall not prove these properties, the proofs are based on properties of real numbers.

EXAMPLE 10   Finding the Sum of a Sequence

Find the sum of each sequence.

(a) $\displaystyle\sum_{k=1}^{5} 3k$ (b) $\displaystyle\sum_{k=1}^{4} (k^2 - 7k + 2)$

Solution   (a) $\displaystyle\sum_{k=1}^{5} 3k = 3\sum_{k=1}^{5} k = 3(1 + 2 + 3 + 4 + 5) = 3(15) = 45$

↑
Property 2

(b) $\displaystyle\sum_{k=1}^{4} (k^2 - 7k + 2) = \sum_{k=1}^{4} k^2 - \sum_{k=1}^{4} 7k + \sum_{k=1}^{4} 2$ Properties 3, 4.

$\displaystyle = \sum_{k=1}^{4} k^2 - 7\sum_{k=1}^{4} k + \sum_{k=1}^{4} 2$ Property 2.

$= (1^2 + 2^2 + 3^2 + 4^2) - 7(1 + 2 + 3 + 4) + 2(4)$ Property 1.

$= (1 + 4 + 9 + 16) - 7(10) + 8$

$= 30 - 70 + 8$

$= -32$

EXAMPLE 11   Using a Graphing Utility to Find the Sum of a Sequence

Using a graphing utility, find the sum of each sequence.

(a) $\displaystyle\sum_{k=1}^{5} 3k$ (b) $\displaystyle\sum_{k=1}^{4} (k^2 - 7k + 2)$

Solution   (a) Figure 8 shows the solution using a TI-83 graphing calculator.

FIGURE 8

Thus, $\displaystyle\sum_{k=1}^{5} 3k = 45$.

(b) Figure 9 shows the solution using a TI-83 graphing calculator.

FIGURE 9

Thus, $\displaystyle\sum_{k=1}^{4} (k^2 - 7k + 2) = -32$.

Now work Problem 39.

E X A M P L E   12

Expanding Summation Notation

Write out each sum.

(a) $\displaystyle\sum_{k=1}^{n} \frac{1}{k}$ (b) $\displaystyle\sum_{k=1}^{n} k!$

Solution   (a) $\displaystyle\sum_{k=1}^{n} \frac{1}{k} = \frac{1}{1} + \frac{1}{2} + \frac{1}{3} + \cdots + \frac{1}{n}$ (b) $\displaystyle\sum_{k=1}^{n} k! = 1! + 2! + \cdots + n!$

E X A M P L E   13

Writing a Sum in Summation Notation

Express each sum using summation notation.

(a) $1^2 + 2^2 + 3^2 + \cdots + 9^2$ (b) $1 + \dfrac{1}{2} + \dfrac{1}{4} + \dfrac{1}{8} + \cdots + \dfrac{1}{2^{n-1}}$

Solution   (a) The sum $1^2 + 2^2 + 3^2 + \cdots + 9^2$ has 9 terms, each of the form k^2 and starts at $k = 1$ and ends at $k = 9$. Thus,

$$1^2 + 2^2 + 3^2 + \cdots + 9^2 = \sum_{k=1}^{9} k^2$$

(b) The sum

$$1 + \frac{1}{2} + \frac{1}{4} + \frac{1}{8} + \cdots + \frac{1}{2^{n-1}}$$

has n terms, each of the form $1/2^{k-1}$ and starts at $k = 1$ and ends at $k = n$. Thus,

$$1 + \frac{1}{2} + \frac{1}{4} + \frac{1}{8} + \cdots + \frac{1}{2^{n-1}} = \sum_{k=1}^{n} \frac{1}{2^{k-1}}$$

The index of summation need not always begin at 1 or end at n; for example,

$$\sum_{k=0}^{n-1} \frac{1}{2^k} = 1 + \frac{1}{2} + \frac{1}{4} + \cdots + \frac{1}{2^{n-1}}$$

Letters other than k may be used as the index. For example,

$$\sum_{j=1}^{n} j! \quad \text{and} \quad \sum_{i=1}^{n} i!$$

each represent the same sum as the one given in Example 12(b).

Now work Problems 49 and 59.

11.1 EXERCISES

In Problems 1–12, write down the first five terms of each sequence by hand. Verify your results using a graphing utility.

1. $\{n\}$

2. $\{n^2 + 1\}$

3. $\left\{\dfrac{n}{n + 2}\right\}$

4. $\left\{\dfrac{2n + 1}{2n}\right\}$

5. $\{(-1)^{n+1}n^2\}$

6. $\left\{(-1)^{n-1}\left(\dfrac{n}{2n - 1}\right)\right\}$

7. $\left\{\dfrac{2^n}{3^n + 1}\right\}$

8. $\left\{\left(\dfrac{4}{3}\right)^n\right\}$

9. $\left\{\dfrac{(-1)^n}{(n + 1)(n + 2)}\right\}$

10. $\left\{\dfrac{3^n}{n}\right\}$

11. $\left\{\dfrac{n}{e^n}\right\}$

12. $\left\{\dfrac{n^2}{2^n}\right\}$

In Problems 13–20, the given pattern continues. Write down the nth term of each sequence suggested by the pattern.

13. $\dfrac{1}{2}, \dfrac{2}{3}, \dfrac{3}{4}, \dfrac{4}{5}, \ldots$

14. $\dfrac{1}{1 \cdot 2}, \dfrac{1}{2 \cdot 3}, \dfrac{1}{3 \cdot 4}, \dfrac{1}{4 \cdot 5}, \ldots$

15. $1, \dfrac{1}{2}, \dfrac{1}{4}, \dfrac{1}{8}, \ldots$

16. $\dfrac{2}{3}, \dfrac{4}{9}, \dfrac{8}{27}, \dfrac{16}{81}, \ldots$

17. $1, -1, 1, -1, 1, -1, \ldots$

18. $1, \dfrac{1}{2}, 3, \dfrac{1}{4}, 5, \dfrac{1}{6}, 7, \dfrac{1}{8}, \ldots$

19. $1, -2, 3, -4, 5, -6, \ldots$

20. $2, -4, 6, -8, 10, \ldots$

In Problems 21–34, a sequence is defined recursively. Write the first five terms by hand. When possible, use a graphing utility to verify your results.

21. $a_1 = 2; a_{n+1} = 3 + a_n$

22. $a_1 = 3; a_{n+1} = 4 - a_n$

23. $a_1 = -2; a_{n+1} = n + a_n$

24. $a_1 = 1; a_{n+1} = n - a_n$

25. $a_1 = 5; a_{n+1} = 2a_n$

26. $a_1 = 2; a_{n+1} = -a_n$

27. $a_1 = 3; a_{n+1} = \dfrac{a_n}{n}$

28. $a_1 = -2; a_{n+1} = n + 3a_n$

29. $a_1 = 1; a_2 = 2; a_{n+2} = a_n a_{n+1}$

30. $a_1 = -1; a_2 = 1; a_{n+2} = a_{n+1} + na_n$

31. $a_1 = A; a_{n+1} = a_n + d$

32. $a_1 = A; a_{n+1} = ra_n, r \neq 0$

33. $a_1 = \sqrt{2}; a_{n+1} = \sqrt{2 + a_n}$

34. $a_1 = \sqrt{2}; a_{n+1} = \sqrt{a_n/2}$

In Problems 35–46, find the sum of each sequence. Verify your results using a graphing utility.

35. $\displaystyle\sum_{k=1}^{10} 5$

36. $\displaystyle\sum_{k=1}^{20} 8$

37. $\displaystyle\sum_{k=1}^{6} k$

38. $\displaystyle\sum_{k=1}^{4} (-k)$

39. $\displaystyle\sum_{k=1}^{5} (5k + 3)$

40. $\displaystyle\sum_{k=1}^{6} (3k - 7)$

41. $\displaystyle\sum_{k=1}^{3} (k^2 + 4)$

42. $\displaystyle\sum_{k=0}^{4} (k^2 - 4)$

43. $\displaystyle\sum_{k=1}^{6} (-1)^k 2^k$

44. $\displaystyle\sum_{k=1}^{4} (-1)^k 3^k$

45. $\displaystyle\sum_{k=1}^{4} (k^3 - 1)$

46. $\displaystyle\sum_{k=0}^{3} (k^3 + 2)$

In Problems 47–56, write out each sum.

47. $\displaystyle\sum_{k=1}^{n}(k+2)$

48. $\displaystyle\sum_{k=1}^{n}(2k+1)$

49. $\displaystyle\sum_{k=1}^{n}\frac{k^2}{2}$

50. $\displaystyle\sum_{k=1}^{n}(k+1)^2$

51. $\displaystyle\sum_{k=0}^{n}\frac{1}{3^k}$

52. $\displaystyle\sum_{k=0}^{n}\left(\frac{3}{2}\right)^k$

53. $\displaystyle\sum_{k=0}^{n-1}\frac{1}{3^{k+1}}$

54. $\displaystyle\sum_{k=0}^{n-1}(2k+1)$

55. $\displaystyle\sum_{k=2}^{n}(-1)^k\ln k$

56. $\displaystyle\sum_{k=3}^{n}(-1)^{k+1}2^k$

In Problems 57–66, express each sum using summation notation.

57. $1+2+3+\cdots+20$

58. $1^3+2^3+3^3+\cdots+8^3$

59. $\dfrac{1}{2}+\dfrac{2}{3}+\dfrac{3}{4}+\cdots+\dfrac{13}{13+1}$

60. $1+3+5+7+\cdots+[2(12)-1]$

61. $1-\dfrac{1}{3}+\dfrac{1}{9}-\dfrac{1}{27}+\cdots+(-1)^6\left(\dfrac{1}{3^6}\right)$

62. $\dfrac{2}{3}-\dfrac{4}{9}+\dfrac{8}{27}-\cdots+(-1)^{11+1}\left(\dfrac{2}{3}\right)^{11}$

63. $3+\dfrac{3^2}{2}+\dfrac{3^3}{3}+\cdots+\dfrac{3^n}{n}$

64. $\dfrac{1}{e}+\dfrac{2}{e^2}+\dfrac{3}{e^3}+\cdots+\dfrac{n}{e^n}$

65. $a+(a+d)+(a+2d)+\cdots+(a+nd)$

66. $a+ar+ar^2+\cdots+ar^{n-1}$

67. Growth of a Rabbit Colony A colony of rabbits begins with one pair of mature rabbits, which will produce a pair of offspring (one male, one female) each month. Assume that all rabbits mature in 1 month and produce a pair of offspring (one male, one female) after 2 months. If no rabbits ever die, how many pairs of mature rabbits are there after 7 months?

[**Hint:** A Fibonacci sequence models this colony. Do you see why?]

1 mature pair

1 mature pair

2 mature pairs

3 mature pairs

68. Fibonacci Sequence Let

$$u_n=\frac{(1+\sqrt5)^n-(1-\sqrt5)^n}{2^n\sqrt5}$$

define the *n*th term of a sequence.
(a) Show that $u_1=1$ and $u_2=1$.
(b) Show that $u_{n+2}=u_{n+1}+u_n$.
(c) Draw the conclusion that $\{u_n\}$ is a Fibonacci sequence.

In Problems 69 and 70, we use the fact that in some programming languages it is possible to have a function subroutine include a call to itself.

69. Programming Exercise Write a program that accepts an integer as input and prints the number and its factorial. Use a recursively defined function, that is, use a function subroutine that calls itself.

70. Programming Exercise Write a program that accepts a positive integer N as input and outputs the Nth Fibonacci number. Use a recursively defined subroutine.

71. Pascal's Triangle Divide the triangular array shown (called Pascal's triangle) using diagonal lines as indicated. Find the sum of the numbers in each of these diagonal rows. Do you recognize this sequence?

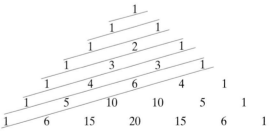

72. Fibonacci Sequence Use the result of Problem 68 to do the following problems:

(a) Write the first 10 terms of the Fibonacci sequence.

The **amount of an annuity** is the sum of all deposits made plus all interest paid.

Suppose the interest an account earns is i percent per payment period (expressed as a decimal). For example, if an account pays 12% compounded monthly (12 times a year) then $i = 0.12/12 = 0.01$. If an account pays 8% compounded quarterly (4 times a year) then $i = 0.08/4 = 0.02$. To develop a formula for the amount of an annuity, suppose P is deposited each payment period for n payment periods in an account that earns i percent per payment period. When the last deposit is made at the nth payment period, the first deposit of P has earned interest compounded for $n - 1$ payment periods, the second deposit of P has earned interest compounded for $n - 2$ payment periods, and so on. Table 3 shows the value of each deposit after n deposits have been made.

TABLE 3

Deposit	1	2	3	$n - 1$	n
Amount	$P(1 + i)^{n-1}$	$P(1 + i)^{n-2}$	$P(1 + i)^{n-3}$	$P(1 + i)$	P

The amount A of the annuity is the sum of the amounts shown in Table 3, namely,

$$A = P(1 + i)^{n-1} + P(1 + i)^{n-2} + \cdots + P(1 + i) + P$$
$$= P[1 + (1 + i) + \cdots + (1 + i)^{n-1}]$$

The expression in brackets is the sum of a geometric sequence with n terms and a common ratio of $(1 + i)$. As a result,

$$A = P[1 + (1 + i) + \cdots + (1 + i)^{n-2} + (1 + i)^{n-1}]$$
$$= P\frac{1 - (1 + i)^n}{1 - (1 + i)} = P\frac{1 - (1 + i)^n}{-i} = P\frac{(1 + i)^n - 1}{i}$$

We have established the following result.

Theorem Amount of an Annuity

If P represents the deposit in dollars made at each payment period for an annuity at i percent interest per payment period, the amount A of the annuity after n payment periods is

$$A = P\frac{(1 + i)^n - 1}{i} \tag{6}$$

E X A M P L E 7   Determining the Amount of an Annuity

To save for retirement, Brett decides to place $2000 into an Individual Retirement Account (IRA) each year for the next 30 years. What will the value of the IRA be when Bill retires in 30 years if the rate of return of the IRA is assumed to be 10% per annum compounded annually?

Solution   This is an ordinary annuity with 30 annual deposits of $P = \$2000$. The rate of interest per payment period is $i = 0.10/1 = 0.10$. The number of payment periods is $n = 30$. The amount A of the annuity in 30 years is

$$A = 2000\left[\frac{(1 + 0.10)^{30} - 1}{0.10}\right] = \$2000(164.49402) = \$328,988.05$$

EXAMPLE 8

To save for his daughter's college education, Mr. McGowen decides to put \$50 aside every month in a credit union account paying 10% interest compounded monthly. If he begins this savings program when his daughter is 3 years old, how much will he have saved by the time his daughter is 18 years old?

Solution   When his daughter is 18 years old, Mr. McGowen will have made his 180th payment (15 years $\times$ 12 payments per year). This is an annuity with $P = \$50$, $n = 180$, and $i = \frac{0.10}{12}$. The amount A saved is

$$A = 50\left[\frac{\left(1 + \dfrac{0.10}{12}\right)^{180} - 1}{\dfrac{0.10}{12}}\right] = \$50(414.4703) = \$20,723.52$$

Geometric Series

An infinite sum of the form

$$a + ar + ar^2 + \cdots + ar^{n-1} + \cdots$$

with first term a and common ratio r, is called an **infinite geometric series** and is denoted by

$$\sum_{k=1}^{\infty} ar^{k-1}$$

Based on formula (3), the sum S_n of the first n terms of a geometric series is

$$S_n = a\,\frac{1 - r^n}{1 - r} = \frac{a}{1 - r} - \frac{ar^n}{1 - r} \tag{7}$$

If this finite sum S_n approaches a number L as $n \to \infty$, then we call L the **sum of the infinite geometric series,** and we write

$$L = \sum_{k=1}^{\infty} ar^{k-1}$$

Theorem   Sum of an Infinite Geometric Series

If $|r| < 1$, the sum of the infinite geometric series $\displaystyle\sum_{k=1}^{\infty} ar^{k-1}$ is

$$\sum_{k=1}^{\infty} ar^{k-1} = \frac{a}{1 - r} \tag{8}$$

(book IX, 35, 36), where, like all of Euclid's algebra, it is presented in a geometric form.

Investigations of other kinds of sequences began in the 1500s, when algebra became sufficiently developed to handle the more complicated problems. The development of calculus in the 1600s added a powerful new tool, especially for finding the sum of infinite series, and the subject continues to flourish today.

HISTORICAL PROBLEMS

1. *Arithmetic sequence problem from the Rhind papyrus (statement modified slightly for clarity)* One hundred loaves of bread are to be divided among five people so that the amounts they receive form an arithmetic sequence. The first two together receive one-seventh of what the last three receive. How many does each receive? [*Partial answer:* First person receives $1\frac{2}{3}$ loaves.]

2. The following old English children's rhyme resembles one of the Rhind papyrus problems:

> As I was going to St. Ives
> I met a man with seven wives
> Each wife had seven sacks
> Each sack had seven cats
> Each cat had seven kits [kittens]
> Kits, cats, sacks, wives
> How many were going to St. Ives?

(a) Assuming that the speaker and the cat fanciers met by traveling in opposite directions, what is the answer?

(b) How many kittens are being transported?

(c) Kits, cats, sacks, wives; how many?

[**Hint:** It is easier to include the man, find the sum with the formula, and then subtract 1 for the man.]

11.3 EXERCISES

In Problems 1–10, a geometric sequence is given. Find the common ratio and write out the first four terms.

1. $\{3^n\}$

2. $\{(-5)^n\}$

3. $\left\{-3\left(\frac{1}{2}\right)^n\right\}$

4. $\left\{\left(\frac{5}{2}\right)^n\right\}$

5. $\left\{\frac{2^{n-1}}{4}\right\}$

6. $\left\{\frac{3^n}{9}\right\}$

7. $\{2^{n/3}\}$

8. $\{3^{2n}\}$

9. $\left\{\frac{3^{n-1}}{2^n}\right\}$

10. $\left\{\frac{2^n}{3^{n-1}}\right\}$

In Problems 11–24, determine whether the given sequence is arithmetic, geometric, or neither. If the sequence is arithmetic, find the common difference; if it is geometric, find the common ratio.

11. $\{n+2\}$

12. $\{2n-5\}$

13. $\{4n^2\}$

14. $\{5n^2+1\}$

15. $\{3-\frac{2}{3}n\}$

16. $\{8-\frac{3}{4}n\}$

17. $1, 3, 6, 10, \ldots$

18. $2, 4, 6, 8, \ldots$

19. $\{(\frac{2}{3})^n\}$

20. $\{(\frac{5}{4})^n\}$

21. $-1, -2, -4, -8, \ldots$

22. $1, 1, 2, 3, 5, 8, \ldots$

23. $\{3^{n/2}\}$

24. $\{(-1)^n\}$

In Problems 25–32, find the fifth term and the nth term of the geometric sequence whose initial term a and common ratio r are given.

25. $a = 2; r = 3$

26. $a = -2; r = 4$

27. $a = 5; r = -1$

28. $a = 6; r = -2$

29. $a = 0; r = \frac{1}{2}$

30. $a = 1; r = -\frac{1}{3}$

31. $a = \sqrt{2}; r = \sqrt{2}$

32. $a = 0; r = 1/\pi$

In Problems 33–38, find the indicated term of each geometric sequence.

33. 7th term of $1, \frac{1}{2}, \frac{1}{4}, \ldots$

34. 8th term of $1, 3, 9, \ldots$

35. 9th term of $1, -1, 1, \ldots$

36. 10th term of $-1, 2, -4, \ldots$

37. 8th term of $0.4, 0.04, 0.004, \ldots$

38. 7th term of $0.1, 1.0, 10.0, \ldots$

In Problems 39–44, find the sum.

39. $\dfrac{1}{4} + \dfrac{2}{4} + \dfrac{2^2}{4} + \dfrac{2^3}{4} + \cdots + \dfrac{2^{n-1}}{4}$

40. $\dfrac{3}{9} + \dfrac{3^2}{9} + \dfrac{3^3}{9} + \cdots + \dfrac{3^n}{9}$

41. $\displaystyle\sum_{k=1}^{n} \left(\tfrac{2}{3}\right)^k$

42. $\displaystyle\sum_{k=1}^{n} 4 \cdot 3^{k-1}$

43. $-1 - 2 - 4 - 8 - \cdots - (2^{n-1})$

44. $2 + \dfrac{6}{5} + \dfrac{18}{25} + \cdots + 2\left(\dfrac{3}{5}\right)^n$

For Problems 45–50, use a graphing utility to find the sum of each geometric sequence.

45. $\dfrac{1}{4} + \dfrac{2}{4} + \dfrac{2^2}{4} + \dfrac{2^3}{4} + \cdots + \dfrac{2^{14}}{4}$

46. $\dfrac{3}{9} + \dfrac{3^2}{9} + \dfrac{3^3}{9} + \cdots + \dfrac{3^{15}}{9}$

47. $\displaystyle\sum_{n=1}^{15} \left(\tfrac{2}{3}\right)^n$

48. $\displaystyle\sum_{n=1}^{15} 4 \cdot 3^{n-1}$

49. $-1 - 2 - 4 - 8 - \cdots - 2^{14}$

50. $2 + \dfrac{6}{5} + \dfrac{18}{25} + \cdots + 2\left(\dfrac{3}{5}\right)^{15}$

In Problems 51–60, find the sum of each infinite geometric series.

51. $1 + \dfrac{1}{3} + \dfrac{1}{9} + \cdots$

52. $2 + \dfrac{4}{3} + \dfrac{8}{9} + \cdots$

53. $8 + 4 + 2 + \cdots$

54. $6 + 2 + \dfrac{2}{3} + \cdots$

55. $2 - \dfrac{1}{2} + \dfrac{1}{8} - \dfrac{1}{32} + \cdots$

56. $1 - \dfrac{3}{4} + \dfrac{9}{16} - \dfrac{27}{64} + \cdots$

57. $\displaystyle\sum_{k=1}^{\infty} 5\left(\tfrac{1}{4}\right)^{k-1}$

58. $\displaystyle\sum_{k=1}^{\infty} 8\left(\tfrac{1}{3}\right)^{k-1}$

59. $\displaystyle\sum_{k=1}^{\infty} 6\left(-\tfrac{2}{3}\right)^{k-1}$

60. $\displaystyle\sum_{k=1}^{\infty} 4\left(-\tfrac{1}{2}\right)^{k-1}$

61. Find x so that x, $x + 2$, and $x + 3$ are terms of a geometric sequence.

62. Find x so that $x - 1$, x, and $x + 2$ are terms of a geometric sequence.

63. **Retirement** Christine contributes $100 each month to her 401(k). What will be the value of Christine's 401(k) in 30 years if the per annum rate of return is assumed to be 12% compounded monthly?

64. **Saving for a Home** Jolene wants to purchase a new home. Suppose she invests $400 per month into a mutual fund. If the per annum rate of return of the mutual fund is assumed to be 10% compounded monthly, how much will Jolene have for a down payment after 3 years?

65. **Tax Sheltered Annuity** Don contributes $500 at the end of each quarter to a Tax Sheltered Annuity (TSA). What will the value of the TSA be in 20 years if the per annum rate of return is assumed to be 8% compounded quarterly?

66. **Retirement** Ray, planning on retiring in 15 years, contributes $1000 to an Individual Retirement Account (IRA) semiannually. What

will the value of the IRA be when Ray retires if the per annum rate of return is assumed to be 10% compounded semiannually?

67. **Sinking Fund** Scott and Alice want to purchase a vacation home in 10 years and have $50,000 for a down payment. How much should they place in a savings account each month if the per annum rate of return is assumed to be 6% compounded monthly?

68. **Sinking Fund** For a child born in 1996, a 4-year college education at a public university is projected to be $150,000. Assuming an 8% per annum rate of return compounded monthly how much must be contributed to a college fund every month in order to have $150,000 in 18 years when the child begins college.

69. **Pendulum Swings** Initially, a pendulum swings through an arc of 2 feet. On each successive swing, the length of arc is 0.9 of the previous length.
   (a) What is the length of arc after 10 swings?
   (b) On which swing is the length of arc first less than 1 foot?
   (c) After 15 swings, what total length will the pendulum have swung?
   (d) When it stops, what total length will the pendulum have swung?

where the blanks are numbers to be found. This is, in fact, the case, as we shall see shortly.

First, we need to introduce a symbol.

The Symbol $\dbinom{n}{j}$

1 We define the symbol $\dbinom{n}{j}$, read "n taken j at a time," as follows:

If j and n are integers with $0 \leq j \leq n$, the **symbol** $\dbinom{n}{j}$ is defined as

$$\binom{n}{j} = \frac{n!}{j!(n-j)!} \tag{1}$$

Comment: On a graphing calculator, the symbol $\dbinom{n}{j}$ may be denoted by the key $\boxed{\text{nCr}}$.

E X A M P L E   1   Evaluating $\dbinom{n}{j}$

Find:

(a) $\dbinom{3}{1}$ (b) $\dbinom{4}{2}$ (c) $\dbinom{8}{7}$ (d) $\dbinom{65}{15}$

Solution (a) $\dbinom{3}{1} = \dfrac{3!}{1!(3-1)!} = \dfrac{3!}{1!2!} = \dfrac{3 \cdot 2 \cdot 1}{1(2 \cdot 1)} = \dfrac{6}{2} = 3$

FIGURE 15

```
65 nCr 15
   2.073746998E14
```

(b) $\dbinom{4}{2} = \dfrac{4!}{2!(4-2)!} = \dfrac{4!}{2!2!} = \dfrac{4 \cdot 3 \cdot 2 \cdot 1}{(2 \cdot 1)(2 \cdot 1)} = \dfrac{24}{4} = 6$

(c) $\dbinom{8}{7} = \dfrac{8!}{7!(8-7)!} = \dfrac{8!}{7!1!} = \dfrac{8 \cdot 7!}{7! \cdot 1!} = \dfrac{8}{1} = 8$

$\uparrow$
$8! = 8 \cdot 7!$

(d) Figure 15 shows the solution using a TI-83 graphing calculator.

Thus, $\dbinom{65}{15} = 2.073746998 \times 10^{14}$. ∎

Now work Problem 1.

Two useful formulas involving the symbol $\dbinom{n}{j}$ are

$$\binom{n}{0} = 1 \quad \text{and} \quad \binom{n}{n} = 1$$

M I S S I O N   P O S S I B L E

Getting the Most Out of Your Contract

You and your band have just signed a recording contract with the nationally acclaimed recording studio, NASHBURG TENS. They've promised $200,000 a year for six years, plus a choice of one of the following four options:

(a) a bonus of $10,000 per year
(b) an annual increase of 4.5% per year (starting after the first year)
(c) an annual increase of 6% per year (starting after the second year)
(d) an annual increase of $9,500 per year (starting after the first year)

You have to tell them today which option you want to take. Your agent is out of town and out of cellular phone range, so you'll have to do the math yourselves.

1. For each option, find out what your payment would be every year for the six years. Round to the nearest dollar.
2. Identify which options are examples of arithmetic or geometric sequences.
3. For each option, find out how much the recording studio will have paid in total over the whole six years. Do you have a formula that will shorten your work?
4. Which option pays your band the most over all? Are there any considerations or circumstances that would make other options better choices even though they pay less money? Are there advantages to being paid more at the beginning of the contract? What are they?

thermore, because a set is a collection, the order in which the elements are listed is immaterial. Thus, {1, 2, 3}, {1, 3, 2}, {2, 1, 3}, and so on, all represent the same set.

E X A M P L E   1

Writing the Elements of a Set

Write the set consisting of the possible outcomes from tossing a coin twice. Use H for "heads" and T for "tails."

Solution   In tossing a coin twice, we can get heads each time, HH; or heads the first time and tails the second, HT; or tails the first time and heads the second, TH; or tails each time, TT. Because no other possibilities exist, the set of outcomes is

{HH, HT, TH, TT}   ∎

If two sets A and B have precisely the same elements, then we say that A and B are **equal** and write $A = B$.

If each element of a set A is also an element of a set B, then we say that A is a **subset** of B and write $A \subseteq B$.

If $A \subseteq B$ and $A \neq B$, then we say that A is a **proper subset** of B and write $A \subset B$.

Thus, if $A \subseteq B$, every element in set A is also in set B, but B may or may not have additional elements. If $A \subset B$, every element in A is also in B, and B has at least one element not found in A.

Finally, we agree that the empty set is a subset of every set; that is,

$$\varnothing \subseteq A \qquad \text{for any set } A$$

E X A M P L E   2

Finding All the Subsets of a Set

Write down all the subsets of the set {a, b, c}.

Solution   To organize our work, we write down all the subsets with no elements, then those with one element, then those with two elements, and finally those with three elements. These will give us all the subsets. Do you see why?

0 Elements	1 Element	2 Elements	3 Elements
∅	{a}, {b}, {c}	{a, b}, {b, c}, {a, c}	{a, b, c}

∎

Now work Problem 21.

> If A and B are sets, the **intersection** of A with B, denoted $A \cap B$, is the set consisting of elements that belong to both A and B. The **union** of A with B, denoted $A \cup B$, is the set consisting of elements that belong to *either A or B,* or both.

E X A M P L E 3

Finding the Intersection and Union of Sets

2

Let $A = \{1, 3, 5, 8\}$, $B = \{3, 5, 7\}$, and $C = \{2, 4, 6, 8\}$. Find:

(a) $A \cap B$ (b) $A \cup B$ (c) $B \cap (A \cup C)$

Solution

(a) $A \cap B = \{1, 3, 5, 8\} \cap \{3, 5, 7\} = \{3, 5\}$
(b) $A \cup B = \{1, 3, 5, 8\} \cup \{3, 5, 7\} = \{1, 3, 5, 7, 8\}$
(c) $B \cap (A \cup C) = \{3, 5, 7\} \cap [\{1, 3, 5, 8\} \cup \{2, 4, 6, 8\}]$
                 $= \{3, 5, 7\} \cap \{1, 2, 3, 4, 5, 6, 8\} = \{3, 5\}$ ▬

Now work Problem 5.

3

Usually, in working with sets, we designate a **universal set,** the set consisting of all the elements we wish to consider. Once a universal set has been designated, we can consider elements of the universal set not found in a given set.

> If A is a set, the **complement** of A, denoted A', is the set consisting of all the elements not in A.

E X A M P L E 4

Finding the Complement of a Set

If the universal set is $U = \{1, 2, 3, 4, 5, 6, 7, 8, 9\}$, and if $A = \{1, 3, 5, 7, 9\}$, then $A' = \{2, 4, 6, 8\}$.   ▬

Notice that: $A \cup A' = U$ and $A \cap A' = \varnothing$.

FIGURE 17

Now work Problem 13.

It is often helpful to draw pictures of sets. Such pictures, called **Venn diagrams,** represent sets as circles enclosed in a rectangle, which represents the universal set. Such diagrams often help us to visualize various relationships among sets. See Figure 17.

If we know that $A \subseteq B$, we might use the Venn diagram in Figure 18(a). If we know that A and B have no elements in common, that is, if $A \cap B = \varnothing$, we might use the Venn diagram in Figure 18(b).

FIGURE 18

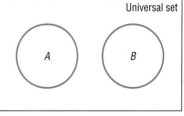

(a) $A \subseteq B$ **(b)** $A \cap B = \varnothing$

Figures 19(a), 19(b), and 19(c) use Venn diagrams to illustrate the definitions of intersection, union, and complement, respectively.

FIGURE 19

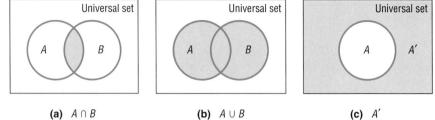

(a) $A \cap B$ **(b)** $A \cup B$ **(c)** A'

Counting

As you count the number of students in a classroom or the number of pennies in your pocket, what you are really doing is matching, on a one-to-one basis, each object to be counted with the counting numbers $1, 2, 3, \ldots, n$, for some number n. If a set A matched up in this fashion with the set $\{1, 2, \ldots, 25\}$, you would conclude that there are 25 elements in the set A. We use the notation $n(A) = 25$ to indicate that there are 25 elements in the set A.

Because the empty set has no elements, we write

$$n(\varnothing) = 0$$

If the number of elements in a set is a nonnegative integer, we say the set is **finite.** Otherwise, it is **infinite.** We shall concern ourselves only with finite sets.

From Example 2, we can see that a set with 3 elements has $2^3 = 8$ subsets. In fact, it can be shown that a set with n elements has exactly 2^n subsets. This fact has an important application to computers, which we take up at the end of this section.

E X A M P L E 5

Analyzing Survey Data

In a survey of 100 college students, 35 were registered in College Algebra, 52 were registered in Introduction to Computer Science, and 18 were in both courses. How many were registered in neither course?

FIGURE 20

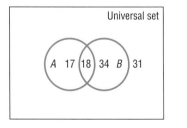

Solution First, we let $A = $ Set of students in College Algebra
$B = $ Set of students in Introduction to Computer Science
Then the given information tells us that

$$n(A) = 35 \qquad n(B) = 52 \qquad n(A \cap B) = 18$$

Refer to Figure 20. Do you see how the numerical entries were determined? Based on the diagram, we conclude that $17 + 18 + 34 = 69$ students were registered in at least one of the two courses. Since 100 students were surveyed, it follows that $100 - 69 = 31$ were registered in neither course. ▬

Now work Problem 35.

The conclusions drawn in Example 5 lead us to formulate a general counting formula. If we count the elements in each of two sets A and B, we necessarily count twice any elements that are in both A and B, that is, those elements in $A \cap B$. Thus, to count correctly the elements that are in A or B, that is, to find $n(A \cup B)$, we need to subtract those in $A \cap B$ from $n(A) + n(B)$.

> **Theorem**   Counting Formula
>
> If A and B are finite sets, then
>
> $$n(A \cup B) = n(A) + n(B) - n(A \cap B) \qquad (1)$$

A special case of the counting formula (1) occurs if A and B have no elements in common. In this case, $A \cap B = \varnothing$ so that $n(A \cap B) = 0$.

> **Theorem**   Addition Principle of Counting
>
> If two sets A and B have no elements in common, then
>
> $$n(A \cup B) = n(A) + n(B) \qquad (2)$$

E X A M P L E   6   Counting the Number of Possible Codes

A certain code is to consist of either a letter of the alphabet or a digit, but not both. How many codes are possible?

Solution   Let the sets A and B be defined as

A = Set of letters in the alphabet
B = Set of digits $\{0, 1, 2, \ldots, 9\}$

Then

$$n(A) = 26 \qquad n(B) = 10$$

Because letters and digits are different, $A \cap B = \varnothing$. The number of ways either a letter or a digit can be chosen is, therefore,

$$n(A \cup B) = n(A) + n(B) = 26 + 10 = 36$$

Application to Computers

Information stored in a computer may be thought of as a series of switches, which are either on or off and are denoted by either the number 0 (off) or the number 1 (on). These numbers are the binary digits, or **bits.** A **register** holds a certain fixed number of bits. For example, the Z-80 microprocessor has 8-bit registers; the PDP-11 minicomputer has 16-bit registers, and the IBM-370 computer has 32-bit registers. Thus, a Z-80 register may hold an entry that looks like this: 01111001 (8 bits). We wish to find out how many different representations are possible in a given register.

TABLE 4

a	b	c	Subset
0	0	0	∅
1	0	0	{a}
0	1	0	{b}
0	0	1	{c}
1	1	0	{a, b}
0	1	1	{b, c}
1	0	1	{a, c}
1	1	1	{a, b, c}

We proceed in steps, first looking at a hypothetical 3-bit register. Look again at the solution to Example 2 and arrange all the subsets of $\{a, b, c\}$ as shown in Table 4. As the table illustrates, the number of subsets of a set with 3 elements equals the number of different representations in a 3-bit register. A set with n elements has 2^n subsets; thus, an n-bit register has 2^n representations. So an 8-bit register can hold $2^8 = 256$ different symbols, a 16-bit register can hold $2^{16} = 65,536$ different symbols, and a 32-bit register can hold $2^{32} \approx 4.3 \times 10^9$ different symbols.

11.6 EXERCISES

In Problems 1–10, use $A = \{1, 3, 5, 7, 9\}$, $B = \{1, 5, 6, 7\}$, and $C = \{1, 2, 4, 6, 8, 9\}$ to find each set.

1. $A \cup B$ **2.** $A \cup C$ **3.** $A \cap B$ **4.** $A \cap C$

5. $(A \cup B) \cap C$ **6.** $(A \cap C) \cup (B \cap C)$ **7.** $(A \cap B) \cup C$ **8.** $(A \cup B) \cup C$

9. $(A \cup C) \cap (B \cup C)$ **10.** $(A \cap B) \cap C$

In Problems 11–20, use $U = $ Universal set $= \{0, 1, 2, 3, 4, 5, 6, 7, 8, 9\}$, $A = \{1, 3, 4, 5, 9\}$, $B = \{2, 4, 6, 7, 8\}$, and $C = \{1, 3, 4, 6\}$ to find each set.

11. A' **12.** C' **13.** $(A \cap B)'$ **14.** $(B \cup C)'$ **15.** $A' \cup B'$

16. $B' \cap C'$ **17.** $(A \cap C')'$ **18.** $(B' \cup C)'$ **19.** $(A \cup B \cup C)'$ **20.** $(A \cap B \cap C)'$

21. Write down all the subsets of $\{a, b, c, d\}$.

22. Write down all the subsets of $\{a, b, c, d, e\}$.

23. If $n(A) = 15$, $n(B) = 20$, and $n(A \cap B) = 10$, find $n(A \cup B)$.

24. If $n(A) = 20$, $n(B) = 40$, and $n(A \cup B) = 35$, find $n(A \cap B)$.

25. If $n(A \cup B) = 50$, $n(A \cap B) = 10$, and $n(B) = 20$, find $n(A)$.

26. If $n(A \cup B) = 60$, $n(A \cap B) = 40$, and $n(A) = n(B)$, find $n(A)$.

In Problems 27–34, use the information given in the figure.

27. How many are in set A?

28. How many are in set B?

29. How many are in A or B?

30. How many are in A and B?

31. How many are in A but not C?

32. How many are not in A?

33. How many are in A and B and C?

34. How many are in A or B or C?

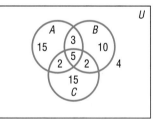

35. **Analyzing Survey Data** In a consumer survey of 500 people, 200 indicated that they would be buying a major appliance within the next month; 150 indicated they would buy a car, and 25 said they would purchase both a major appliance and a car. How many will purchase neither? How many will purchase only a car?

36. **Analyzing Survey Data** In a student survey, 200 indicated that they would attend Summer Session I and 150 indicated Summer Session II. If 75 students plan to attend both summer sessions and 275 indicated that they would attend

neither session, how many students participated in the survey?

37. **Analyzing Survey Data** In a survey of 100 investors in the stock market,

50 owned shares in IBM

40 owned shares in AT&T

45 owned shares in GE

20 owned shares in both IBM and GE

15 owned shares in both AT&T and GE

20 owned shares in both IBM and AT&T

5 owned shares in all three

(a) How many of the investors surveyed did not have shares in any of the three companies?
(b) How many owned just IBM shares?
(c) How many owned just GE shares?
(d) How many owned neither IBM nor GE?
(e) How many owned either IBM or AT&T but no GE?

38. Classifying Blood Types Human blood is classified as either Rh+ or Rh−. Blood is also classified by type: A, if it contains an A antigen; B, if it contains a B antigen; AB, if it contains both A and B antigens; and O, if it contains neither antigen. Draw a Venn diagram illustrating the various blood types. Based on this classification, how many different kinds of blood are there?

39. Make up a problem different from any found in the text that requires the addition principle of counting to solve. Give it to a friend to solve and critique.

40. Investigate the notion of counting as it relates to infinite sets. Write an essay on your findings.

11.7 PERMUTATIONS AND COMBINATIONS

> 1 Solve Counting Problems Using the Multiplication Principle
> 2 Solve Counting Problems Using Permutations
> 3 Solve Counting Problems Using Combinations

Counting plays a major role in many diverse areas, such as probability, statistics, and computer science. In this section we shall look at special types of counting problems and develop general formulas for solving them.

We begin with an example that will demonstrate a general counting principle.

E X A M P L E   1   Counting the Number of Possible Meals

The fixed-price dinner at a restaurant provides the following choices:

Appetizer: soup or salad
Entree: baked chicken, broiled beef patty, baby beef liver, or roast beef au jus
Dessert: ice cream or cheese cake

How many different meals can be ordered?

Solution   Ordering such a meal requires three separate decisions:

| **Choose an Appetizer** | **Choose an Entree** | **Choose a Dessert** |
| 2 choices | 4 choices | 2 choices |

Look at the **tree diagram** in Figure 21. We see that, for each choice of appetizer, there are 4 choices of entrees. And for each of these $2 \cdot 4 = 8$ choices, there are 2 choices for dessert. Thus, there are a total of

$$2 \cdot 4 \cdot 2 = 16$$

different meals that can be ordered.

FIGURE 21

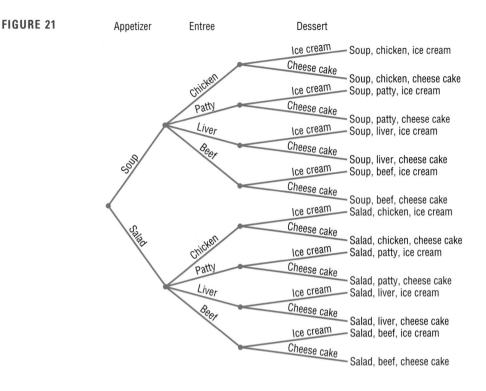

Example 1 illustrates a general counting principle.

Theorem   Multiplication Principle of Counting

If a task consists of a sequence of choices in which there are p selections for the first choice, q selections for the second choice, r selections for the third choice, and so on, then the task of making these selections can be done in

$$p \cdot q \cdot r \cdot \ldots$$

different ways.

E X A M P L E   2

Counting Airport Codes

The International Airline Transportation Association (IATA) assigns three-letter codes to represent airport locations. For example, JFK represents Kennedy International in New York. How many different airport codes are possible?

Solution   The task consists of making three selections. Each selection requires choosing a letter of the alphabet (26 choices). Thus, by the multiplication principle, there are

$$26 \cdot 26 \cdot 26 = 17{,}576$$

different airport codes.

In Example 2, we were allowed to repeat a letter. For example, a valid airport code is FLL (Ft. Lauderdale International Airport), in which the letter L appears twice. In the next example, such repetition is not allowed.

E X A M P L E 3   Counting without Repetition

Suppose that we wish to establish a three-letter code using any of the 26 letters of the alphabet, but we require that no letter be used more than once. How many different three-letter codes are there?

Solution   The task consists of making three selections. The first selection requires choosing from 26 letters. Because no letter can be used more than once, the second selection requires choosing from 25 letters. The third selection requires choosing from 24 letters. (Do you see why?) By the multiplication principle, there are

$$26 \cdot 25 \cdot 24 = 15{,}600$$

different three-letter codes with no letter repeated.

Now work Problems 25 and 29.

Example 3 illustrates a type of counting problem referred to as a *permutation*.

> A **permutation** is an ordered arrangement of n distinct objects without repetitions. The symbol $P(n, r)$ represents the number of permutations of n distinct objects, taken r at a time, where $r \leq n$.

For example, the question posed in Example 3 asks for the number of ways the 26 letters of the alphabet can be arranged using three nonrepeated letters. The answer is

$$P(26, 3) = 26 \cdot 25 \cdot 24 = 15{,}600$$

To arrive at a formula for $P(n, r)$, we note that the task of obtaining an ordered arrangement of n objects in which only $r \leq n$ of them are used, without repeating any of them, requires making r selections. For the first selection, there are n choices; for the second selection, there are $n - 1$ choices; for the third selection, there are $n - 2$ choices; . . . ; for the rth selection, there are $n - (r - 1)$ choices. By the multiplication principle, we have

$$
\begin{array}{cccc}
\text{1st} & \text{2nd} & \text{3rd} & r\text{th}
\end{array}
$$
$$P(n, r) = n \cdot (n - 1) \cdot (n - 2) \cdot \ldots \cdot [n - (r - 1)]$$
$$= n \cdot (n - 1) \cdot (n - 2) \cdot \ldots \cdot (n - r + 1)$$

This formula for $P(n, r)$ can be compactly written using factorial notation.*

$$P(n, r) = n \cdot (n - 1) \cdot (n - 2) \cdot \ldots \cdot (n - r + 1)$$
$$= n \cdot (n - 1) \cdot (n - 2) \cdot \ldots \cdot (n - r + 1) \cdot \frac{(n - r) \cdot \ldots \cdot 3 \cdot 2 \cdot 1}{(n - r) \cdot \ldots \cdot 3 \cdot 2 \cdot 1} = \frac{n!}{(n - r)!}$$

*Recall that $0! = 1, 1! = 1, 2! = 2 \cdot 1, \cdots, n! = n(n - 1) \cdot \ldots \cdot 3 \cdot 2 \cdot 1$.

The form of the answer to Example 10 is suggestive of a general result. Had the letters in REARRANGE each been different, there would have been $P(9, 9) = 9!$ possible words formed. This is the numerator of the answer. The presence of 3 R's, 2 A's, and 2 E's reduces the number of different words, as the entries in the denominator illustrate. We are led to the following result:

Theorem Permutations with Repetition

The number of permutations of n objects of which n_1 are of one kind, n_2 are of a second kind, . . . , and n_k are of a kth kind is given by

$$\frac{n!}{n_1! \cdot n_2! \cdot \ldots \cdot n_k!} \tag{3}$$

where $n = n_1 + n_2 + \cdots + n_k$.

E X A M P L E   11   Arranging Flags

How many different vertical arrangements are there of 8 flags if 4 are white, 3 are blue, and 1 is red?

Solution   We seek the number of permutations of 8 objects, of which 4 are of one kind, 3 of a second kind, and 1 of a third kind. Using formula (3), we find that there are

$$\frac{8!}{4! \cdot 3! \cdot 1!} = \frac{8 \cdot 7 \cdot 6 \cdot 5 \cdot 4!}{4! \cdot 3! \cdot 1!} = 280 \text{ different arrangements:}$$

Now work Problem 53.

11.7   EXERCISES

In Problems 1–8, find the value of each permutation. Verify your results using a graphing calculator.

1. $P(6, 2)$ 2. $P(7, 2)$ 3. $P(5, 5)$ 4. $P(4, 4)$
5. $P(8, 0)$ 6. $P(9, 0)$ 7. $P(8, 3)$ 8. $P(8, 5)$

In Problems 9–16, use formula (2) to find the value of each combination. Verify your results using a graphing calculator.

9. $C(8, 2)$ 10. $C(8, 6)$ 11. $C(6, 4)$ 12. $C(6, 2)$
13. $C(15, 15)$ 14. $C(18, 1)$ 15. $C(26, 13)$ 16. $C(18, 9)$

17. List all the permutations of 5 objects a, b, c, d, and e taken 3 at a time. What is $P(5, 3)$?

18. List all the permutations of 5 objects a, b, c, d, and e taken 2 at a time. What is $P(5, 2)$?

19. List all the permutations of 4 objects 1, 2, 3, and 4 taken 3 at a time. What is $P(4, 3)$?

20. List all the permutations of 6 objects, 1, 2, 3, 4, 5, and 6 taken 3 at a time. What is $P(6, 3)$?

21. List all the combinations of the 5 objects a, b, c, d, and e taken 3 at a time. What is $C(5, 3)$?

22. List all the combinations of the 5 objects a, b, c, d, and e taken 2 at a time. What is $C(5, 2)$?

23. List all the combinations of the 4 objects 1, 2, 3, and 4 taken 3 at a time. What is $C(4, 3)$?

24. List all the combinations of the 6 objects 1, 2, 3, 4, 5, and 6 taken 3 at a time. What is $C(6, 3)$?

25. A man has 5 shirts and 3 ties. How many different shirt and tie combinations can he wear?

26. A woman has 3 blouses and 5 skirts. How many different outfits can she wear?

27. **Forming Codes** How many two-letter codes can be formed using the letters A, B, C, and D? Repeated letters are allowed.

28. **Forming Codes** How many two-letter codes can be formed using the letters A, B, C, D, and E? Repeated letters are allowed.

29. **Forming Numbers** How many three-digit numbers can be formed using the digits 0 and 1? Repeated digits are allowed.

30. **Forming Numbers** How many three-digit numbers can be formed using the digits 0, 1, 2, 3, 4, 5, 6, 7, 8, and 9? Repeated digits are allowed.

31. In how many ways can 4 people be lined up?

32. In how many ways can 5 different boxes be stacked?

33. **Forming Codes** How many different three-letter codes are there if only the letters A, B, C, D, and E can be used and no letter can be used more than once?

34. **Forming Codes** How many different four-letter codes are there if only the letters A, B, C, D, E, and F can be used and no letter can be used more than once?

35. **Arranging Letters** How many arrangements are there of the letters in the word MONEY?

36. **Arranging Digits** How many arrangements are there of the digits in the number 51,342?

37. **Establishing Committees** In how many ways can a committee of 4 students be formed from a pool of 7 students?

38. **Establishing Committees** In how many ways can a committee of 3 professors be formed from a department having 8 professors?

39. **Possible Answers on a True/False Test** How many arrangements of answers are possible for a true/false test with 10 questions?

40. **Possible Answers on a Multiple-choice Test** How many arrangements of answers are possible in a multiple-choice test with 5 questions, each of which has 4 possible answers?

41. How many four-digit numbers can be formed using the digits 0, 1, 2, 3, 4, 5, 6, 7, 8, and 9 if the first digit cannot be 0? Repeated digits are allowed.

42. How many five-digit numbers can be formed using the digits 0, 1, 2, 3, 4, 5, 6, 7, 8, and 9 if the first digit cannot be 0 or 1? Repeated digits are allowed.

43. **Arranging Books** Five different mathematics books are to be arranged on a student's desk. How many arrangements are possible?

44. **Forming License Plate Numbers** How many different license plate numbers can be made us-

ing 2 letters followed by 4 digits selected from the digits 0 through 9, if
(a) Letters and digits may be repeated?
(b) Letters may be repeated, but digits are not repeated?
(c) Neither letters nor digits may be repeated?

45. **Stock Portfolios** As a financial planner, you are asked to select one stock each from the following groups: 8 DOW stocks, 15 NASDAQ stocks, and 4 global stocks. How many different portfolios are possible?

46. **Combination Locks** A combination lock has 50 numbers on it. To open it, you turn to a number, then rotate clockwise to a second number, and then counterclockwise to the third number. How many different lock combinations are there?

FIGURE 27

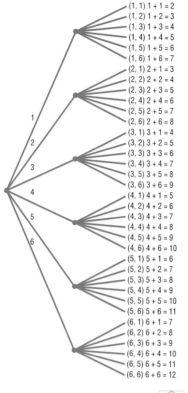

Solution   We begin by constructing a probability model for the experiment. The tree diagram in Figure 27 will help us to see all the possibilities.

Because the dice are fair, no one of the 36 possible outcomes in the sample space S is more likely to occur than any other. Thus, we have equally likely outcomes with $n(S) = 36$.

(a) The event E, "the dice total 7 or 11," consists of the outcomes

$$E = \{(1, 6), (2, 5), (3, 4), (4, 3), (5, 2), (6, 1), (5, 6), (6, 5)\}$$

Because $n(E) = 8$, we have

$$P(E) = \frac{n(E)}{n(S)} = \frac{8}{36} = \frac{2}{9} \approx 0.222$$

(b) The event F, "the dice total 2, 3, or 12," consists of the outcomes

$$F = \{(1, 1), (1, 2), (2, 1), (6, 6)\}$$

Because $n(F) = 4$,

$$P(F) = \frac{n(F)}{n(S)} = \frac{4}{36} = \frac{1}{9} \approx 0.111$$

(c) The number of possibilities that require that you throw again is $36 - n(E) - n(F) = 36 - 8 - 4 = 24$. Thus, the probability that another throw is required is

$$\frac{24}{36} = \frac{2}{3} \approx 0.667$$

∎

Now work Problem 29.

Applications Involving Permutations and Combinations

E X A M P L E   9   Computing Probabilities

Because of a mistake in packaging, 5 defective phones were packaged with 15 good ones. All phones look alike and have equal probability of being chosen. Three phones are selected.

(a) What is the probability that all 3 are defective?
(b) What is the probability that exactly 2 are defective?
(c) What is the probability that at least 2 are defective?

Solution   The sample space S consists of the number of ways 3 objects can be selected from 20 objects, that is, the number of combinations of 20 things taken 3 at a time.

$$n(S) = C(20, 3) = \frac{20!}{17! \cdot 3!} = \frac{20 \cdot 19 \cdot 18}{6} = 1140$$

Each of these outcomes is equally likely to occur.

(a) If E is the event "3 are defective," then the number of elements in E is the number of ways the 3 defective phones can be chosen from the 5 defective phones: $C(5, 3) = 10$. Thus, the probability of E is

$$P(E) = \frac{n(E)}{n(S)} = \frac{10}{1140} \approx 0.0088$$

(b) If F is the event "exactly 2 are defective" and 3 phones are selected, then the number of elements in F is the number of ways to select 2 defective phones from the 5 defective phones and 1 good phone from the 15 good ones. The first of these can be done in $C(5, 2)$ ways and the second in $C(15, 1)$ ways. By the multiplication principle, the event F can occur in

$$C(5, 2) \cdot C(15, 1) = \frac{5!}{3! \cdot 2!} \cdot \frac{15!}{14! \cdot 1!} = 10 \cdot 15 = 150 \text{ ways}$$

The probability of F is therefore

$$P(F) = \frac{n(F)}{n(S)} = \frac{150}{1140} \approx 0.1316$$

(c) The event G, "at least two are defective," when 3 are chosen, is equivalent to requiring that either exactly 2 defective are chosen or exactly 3 defective are chosen. That is, $G = E \cup F$. Since E and F are mutually exclusive (it is not possible to select 2 defective phones and, at the same time, select 3 defective phones), we find

$$P(G) = P(E) + P(F) \approx 0.0088 + 0.1316 = 0.1404 \qquad \blacksquare$$

Now work Problem 43.

E X A M P L E 10

Tossing a Coin

A fair coin is tossed 6 times.

(a) What is the probability of obtaining exactly 5 heads and one tail?
(b) What is the probability of obtaining between 4 and 6 heads, inclusive?

Solution

The number of elements in the sample space S is found using the Multiplication Principle. Each toss results in a head (H) or a tail (T). Since the coin is tossed 6 times, we have

$$n(S) = \underbrace{2 \cdot 2 \cdot \ldots \cdot 2}_{6 \text{ tosses}} = 2^6 = 64$$

The outcomes are equally likely since the coin is fair.

(a) Any sequence that contains 5 heads and 1 tail is determined once the position of the 5 heads (or 1 tail) is known. The number of ways we can position 5 heads in a sequence of 6 slots is $C(6, 5) = 6$. The probability of the event E—exactly 5 heads and one tail—is

$$P(E) = \frac{n(E)}{n(S)} = \frac{C(6, 5)}{2^6} = \frac{6}{64} \approx 0.0938$$

(b) Let F be the event: between 4 and 6 heads, inclusive. To obtain between 4 and 6 heads is equivalent to the event: either 4 heads or 5 heads or 6 heads. Since each of these is mutually exclusive (it is impossible to obtain both 4 heads and 5 heads when tossing a coin 6 times), we have

$$P(F) = P(4 \text{ heads or 5 heads or 6 heads})$$
$$= P(4 \text{ heads}) + P(5 \text{ heads}) + P(6 \text{ heads})$$

The probabilities on the right are obtained as in part (a). Thus

$$P(F) = \frac{C(6, 4)}{2^6} + \frac{C(6, 5)}{2^6} + \frac{C(6, 6)}{2^6} = \frac{15}{64} + \frac{6}{64} + \frac{1}{64} = \frac{22}{64} \approx 0.3438 \qquad \blacksquare$$

(b) French and Spanish books must alternate in the grouping, beginning with a French book?

91. Telephone Numbers Using the digits 0, 1, 2, ..., 9, how many 7-digit numbers can be formed if the first digit cannot be 0 or 9 and if the last digit is greater than or equal to 2 and less than or equal to 3? Repeated digits are allowed.

92. Home Choices A contractor constructs homes with 5 different choices of exterior finish, 3 different roof arrangements, and 4 different window designs. How many different types of homes can be built?

93. License Plate Possibilities A license plate consists of 1 letter, excluding O and I, followed by a 4-digit number that cannot have a 0 in the lead position. How many different plates are possible?

94. Using the digits 0 and 1, how many different numbers consisting of 8 digits can be formed?

95. Forming Different Words How many different words can be formed using all the letters in the word MISSING?

96. Arranging Flags How many different vertical arrangements are there of 10 flags, if 4 are white, 3 are blue, 2 are green, and 1 is red?

97. Forming Committees A group of 9 people is going to be formed into committees of 4, 3, and 2 people. How many committees can be formed if
(a) A person can serve on any number of committees?
(b) No person can serve on more than one committee?

98. Forming Committees A group consists of 5 men and 8 women. A committee of 4 is to be formed from this group, and policy dictates that at least 1 woman be on this committee.
(a) How many committees can be formed that contain exactly 1 man?
(b) How many committees can be formed that contain exactly 2 women?
(c) How many committees can be formed that contain at least 1 man?

99. From a box containing three 40-watt bulbs, six 60-watt bulbs, and eleven 75-watt bulbs, a bulb is drawn at random. What is the probability that the bulb is 40 watts? What is the probability that it is not a 75 watt bulb?

100. You have four $1 bills, three $5 bills, and two $10 bills in your wallet. If you pick a bill at random, what is the probability it will be a $1 bill?

101. Each of the letters in the word ROSE is written on an index card and the cards are then shuffled. What is the probability that, when the cards are dealt out, they spell the word ROSE?

102. Each of the numbers, 1, 2, ..., 100 is written on an index card and the cards are then shuffled. If a card is selected at random, what is the probability that the number on the card is divisible by 5? What is the probability that the card selected either is a 1 or names a prime number?

103. Computing Probabilities Because of a mistake in packaging, a case of 12 bottles of red wine contained 5 Merlot and 7 Cabernet, each without labels. All the bottles look alike and have equal probability of being chosen. Three bottles are selected.
(a) What is the probability all 3 are Merlot?
(b) What is the probability exactly 2 are Merlot?
(c) What is the probability none is a Merlot?

104. Tossing a Coin A fair coin is tossed 10 times.
(a) What is the probability of obtaining exactly 5 heads?
(b) What is the probability of obtaining all heads?

105. Constructing a Brick Staircase A brick staircase has a total of 25 steps. The bottom step requires 80 bricks. Each successive step requires three less bricks than the prior step.
(a) How many bricks are required for the top step?
(b) How many bricks are required to build the staircase?

106. Creating a Floor Design A mosaic tile floor is designed in the shape of a trapezoid 30 feet wide at the base and 15 feet wide at the top. See Figure 11, page 870. The tiles, 12 inches by 12 inches, are to be placed so that each successive row contains one less tile than the row below. How many tiles will be required?

107. Retirement Planning Chris gets paid once a month and contributes $200 each pay period into his 401(k). If Chris plans on retiring in 20 years, what will the value of his 401(k) be if the per annum rate of return of the 401(k) is 10% compounded monthly?

108. Retirement Planning Jacky contributes $500 every quarter into an IRA. If Jacky plans on retiring in 30 years, what will the value of the IRA be if the per annum rate of return of the IRA is 8% compounded quarterly?

109. Bouncing Balls A ball is dropped from a height of 20 feet. Each time it strikes the ground, it bounces up to $\frac{3}{4}$ of the previous height.
  (a) What height will the ball bounce up to after it strikes the ground for the third time?
  (b) What is its height after it strikes the ground for the nth time?

  (c) How many times does the ball need to strike the ground before its height is less than 6 inches?
  (d) What total distance does the ball travel before it stops bouncing?

110. Salary Increases Your friend has just been hired at an annual salary of $20,000. If she expects to receive annual increases of 4%, what will her salary be as she begins her fifth year?

111. At the Milex tune-up and brake repair shop, the manager has found that a car will require a tune-up with a probability of 0.6, a brake job with a probability of 0.1, and both with a probability of 0.02.
  (a) What is the probability that a car requires either a tune-up or a brake job?
  (b) What is the probability that a car requires a tune-up but not a brake job?
  (c) What is the probability that a car requires neither type of repair?

A Preview of Calculus: The Limit and the Derivative of a Function

Population growth has always been a concern of the human race. It has long been contended that if the population continues to grow, the Earth's limited resources will not be capable of supporting the expanding population. Thomas Malthus is frequently cited for his extreme views on population growth. On the following page you will test Malthus' theories by modeling world population. You will need to use the Sullivan website at:

www.prenhall.com/Sullivan

to help link you to the Internet resources needed to produce your model.

PREPARING FOR THIS CHAPTER

Before getting started on this chapter, review the following concepts:

OUTLINE

THOMAS MALTHUS AND POPULATION GROWTH

In response to the poverty and misery among the lower classes, Thomas Malthus wrote *A Summary View of the Principle of Population* in 1830. In this article, Malthus contended "population, when unchecked, increases in a geometrical progression of such a nature as to double itself every twenty-five years." Malthus also felt the rate of increase in the output of agriculture "would fall short, beyond all comparison, of the natural increases of population in a geometric progression." This shortfall would cause population growth to subside.

Population growth has long been a concern. Living on a planet with limited resources necessarily implies the planet has an upper bound at which it can sustain a population. Malthus felt the misery of his time was a direct result of geometric increases in population combined with arithmetic increases in agricultural output. Of course, we now know that agricultural output has increased faster than arithmetically so Malthus' doomsday predictions did not come to fruition. However, do populations grow geometrically as Malthus proposed? Geometric growth would imply $\dfrac{P_t}{P_{t-1}} = r + 1$, where r is the growth rate.

1. Using *world population data* and a graphing utility find the logistic growth function of best fit treating the year as the independent variable. Let $t = 0$ represent 1950, $t = 1$ represent 1951, etc. until you have entered all the years and the corresponding population up to the current year.
2. Graph $Y_1 = f(t)$, where $f(t)$ represents the logistic growth function of best fit found in question 1.
3. Determine the instantaneous rate of growth of population in 1960 using the numerical derivative function on your graphing utility.
4. Use the result from question 3 to predict the population in 1961. What was the actual population in 1961?
5. Determine the instantaneous growth of population in 1970, 1980 and 1990. What is happening to the instantaneous growth rate as time passes? Is Malthus' contention of a geometric growth rate accurate?
6. Using the numerical derivative function on your graphing utility, graph $Y_2 = f'(t)$, where $f'(t)$ represents the first derivative of $f(t)$ with respect to time. Y_2 is the growth rate of the population at any time t.
7. Using the MAXIMUM function on your graphing utility, determine the year in which the growth rate of the population is largest. What is happening to the growth rate in the years following the maximum? Find this point on the graph of $Y_1 = f(t)$.
8. Evaluate $\lim\limits_{t \to \infty} f(t)$. This limiting value is the carrying capacity of Earth. What is the carrying capacity of Earth?
9. What do you think will happen if the population of the Earth exceeds the carrying capacity? Do you think that agricultural output will continue to increase at the same rate as population growth? What effect will urban sprawl have on agricultural output?

One of the questions that motivated the development of calculus was a geometry problem, the Tangent Problem. This problem asks, "What is the slope of the tangent line to the graph of a function f at a point P on its graph?" See the illustration.

Various solutions to the Tangent Problem have been given, beginning with the Greeks about 2000 years ago. Much later, Rene Descartes (1596–1650) and Pierre de Fermat (1601–1665), working independently, provided solutions to the Tangent Problem. Their work inspired Gottfried Wilhelm von Leibniz (1646–1716), who shares with Sir Isaac Newton (1642–1727) the honor of founding the calculus.

12.1   FINDING LIMITS USING TABLES AND GRAPHS

> **1**   Find a Limit Using Tables
>
> **2**   Find a Limit Using a Graph

The idea of the limit of a function is what connects algebra and geometry to the mathematics of calculus. In working with the limit of a function, we encounter notation of the form

$$\lim_{x \to c} f(x) = N$$

This is read as "the limit of $f(x)$ as x approaches c equals the number N." Here f is a function defined on some open interval containing the number c; f need not be defined at c, however.

We may describe the meaning of $\lim_{x \to c} f(x) = N$ as follows:

> For all x approximately equal to c, with $x \neq c$, the corresponding value $f(x)$ is approximately equal to N.

Another description of $\lim_{x \to c} f(x) = N$ is

> As x gets closer to c, but remains unequal to c, the corresponding values of $f(x)$ get closer to N.

1   Tables, either generated on a graphing utility or computed with the help of a calculator, are useful for finding limits.

EXAMPLE 1   Finding a Limit Using Tables

Find: $\lim_{x \to 3} 5x^2$

Solution   Here $f(x) = 5x^2$ and $c = 3$. We set up a TABLE with $Y_1 = 5x^2$, choosing values of x closer to $c = 3$ using numbers less than 3 and greater than 3. See Table 1.

From Table 1, we infer that as x gets closer to 3, the value of $f(x) = 5x^2$ gets closer to 45. That is,

$$\lim_{x \to 3} 5x^2 = 45$$

TABLE 1

X	Y1
2.99	44.701
2.999	44.97
2.9999	44.997
3.0001	45.003
3.001	45.03
3.01	45.301

Y1■5X²

When choosing the values of x in a table, the number to start with and the increment is arbitrary. However, they should be chosen so that the table makes it clear what the corresponding values of f are getting close to.

   Now work Problem 1.

EXAMPLE 2   Finding a Limit Using Tables

Find: (a) $\lim\limits_{x \to 2} \dfrac{x^2 - 4}{x - 2}$ (b) $\lim\limits_{x \to 2} (x + 2)$

Solution   (a) Here $f(x) = \dfrac{x^2 - 4}{x - 2}$ and $c = 2$. Notice that the domain of f is $\{x | x \ne 2\}$, so f is not defined at $x = 2$. See Table 2. We infer that as x gets closer to 2, the value of $f(x) = \dfrac{x^2 - 4}{x - 2}$ gets closer to 4. That is,

$$\lim\limits_{x \to 2} \dfrac{x^2 - 4}{x - 2} = 4$$

TABLE 2

X	Y1
1.99	3.99
1.999	3.999
1.9999	3.9999
2.0001	4.0001
2.001	4.001
2.01	4.01

Y1 ☐ (X²-4)/(X-2)

(b) Here $g(x) = x + 2$ and $c = 2$. The domain of g is all real numbers. See Table 3. We infer that as x gets closer to 2, the value of $g(x)$ gets closer to 4. That is,

$$\lim\limits_{x \to 2} (x + 2) = 4$$

TABLE 3

X	Y1
1.99	3.99
1.999	3.999
1.9999	3.9999
2.0001	4.0001
2.001	4.001
2.01	4.01

Y1 ☐ X+2

The conclusion that $\lim\limits_{x \to 2} (x + 2) = 4$ could have been obtained without the use of Table 3; as x gets closer to 2, it follows that $x + 2$ will get closer to 4.

Also, for part (a), you are right if you make the observation that, since $x \ne 2$, then

$$f(x) = \dfrac{x^2 - 4}{x - 2} = \dfrac{(x - 2)(x + 2)}{x - 2} = x + 2 \qquad x \ne 2$$

Now it is easy to conclude that $\lim\limits_{x \to 2} \dfrac{x^2 - 4}{x - 2} = \lim\limits_{x \to 2} (x + 2) = 4$.

Let's look at an example for which the factoring technique used above does not work.

EXAMPLE 3   Finding a Limit Using a Table

Find: $\lim\limits_{x \to 0} \dfrac{\sin x}{x}$

Solution   First, we observe that the domain of the function $f(x) = \dfrac{\sin x}{x}$ is $\{x | x \ne 0\}$. We create Table 4, where x is measured in radians. We infer from Table 4 that $\lim\limits_{x \to 0} \dfrac{\sin x}{x} = 1$.

TABLE 4

X	Y1
-.05	.99958
-.02	.99993
-.01	.99998
.01	.99998
.02	.99993
.05	.99958

Y1 ☐ sin(X)/X

The graph of a function f can also be of help in finding limits. See Figure 1.

FIGURE 1

$f(c) = N; \lim_{x \to c} f(c) = N$

(a)

$f(c) \neq N; \lim_{x \to c} f(c) = N$

(b)

$f(c)$ not defined; $\lim_{x \to c} f(c) = N$

(c)

In each graph, notice that as x gets closer to c the values of f get closer to the number N. We conclude that

$$\lim_{x \to c} f(x) = N$$

This is the conclusion regardless of the value of f at c. In Figure 1(a), $f(c) = N$ and in Figure 1(b), $f(c) \neq N$. Figure 1(c) illustrates that $\lim_{x \to c} f(x) = N$ even if f is not defined at c.

E X A M P L E   4

Finding a Limit by Graphing

Find: $\lim_{x \to 2} f(x)$ if $f(x) = \begin{cases} 3x - 2 & \text{if } x \neq 2 \\ 3 & \text{if } x = 2 \end{cases}$

Solution

The function f is a piecewise-defined function. Its graph is shown in Figure 2(a) drawn by hand. Figure 2(b) shows the graph using a graphing utility.

FIGURE 2

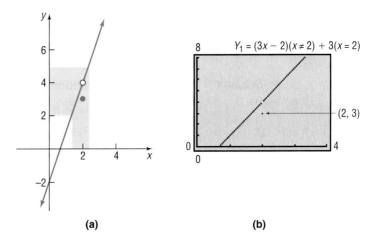

(a)

(b)

Upon ZOOMing-IN around $x = 2$ and TRACEing, we conclude that $\lim_{x \to 2} f(x) = 4$. See Figure 3.

FIGURE 3

Tangent Problem

Find the slope of the tangent line to the graph of a function $y = f(x)$ at a point P on its graph. See Figure 13.

FIGURE 13

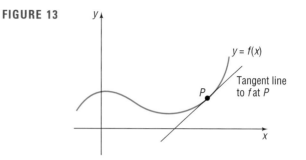

We first need to define what we mean by a *tangent* line. In high school geometry, the tangent line to a circle is defined as the line that intersects the graph in exactly one point. Look at Figure 14. Notice that the tangent line just *touches* the graph of the circle.

This definition, however, does not work in general. Look at Figure 15. The lines L_1 and L_2 only intersect the graph in one point P, but none of them touches the graph at P. Additionally, the tangent line L_T shown in Figure 16 touches the graph of f at P, but also intersects the graph elsewhere. So how should we define the tangent line to the graph of f at a point P?

FIGURE 14   **FIGURE 15**   **FIGURE 16**

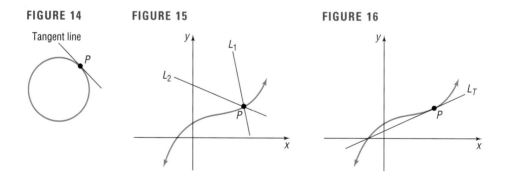

1 The tangent line L_T to the graph of a function $y = f(x)$ at a point P necessarily contains the point P. Because of the point–slope form of the equation of a line, it remains to find the slope m_{tan} of the tangent line.

Suppose that the coordinates of the point P are $(c, f(c))$. Locate another point $Q = (x, f(x))$ on the graph of f. The line containing P and Q is the secant line. (Refer to Section 2.2.) The slope m_{sec} of the secant line is

$$m_{\text{sec}} = \frac{f(x) - f(c)}{x - c}$$

MISSION POSSIBLE

Finding the Profit Maximizing Level of Output

The owners of Two for the Road Bicycle Shop are concerned they are not maximizing their profits. They hire the Mission Possible Team to determine the profit maximizing level of sales for their shop. Your team is provided with the following monthly data regarding number of bicycles produced and sold, total cost of production, and total revenue.

Number of Bicycles, x	Total Cost of Production, C	Total Revenue, R
0	24,000	0
25	27,750	28,000
60	31,500	45,000
102	35,250	53,400
150	39,000	59,160
190	42,750	62,360
223	46,500	61,485
249	50,250	59,875

1. Calculate the profit (total revenue − total cost) for 0, 25, 60, 102, 150, 190, 223, and 249 bicycles. What is the profit maximizing level of output?

2. The owners want a more specific result. Draw a scatter diagram of x versus C and x versus R on the same viewing window treating number of bicycles as the independent variable. Be sure to use different plotting symbols for cost and revenue.

3. To estimate a cost function, use a graphing utility to find the cubic function of best fit treating total cost as the dependent variable and number of bicycles as the independent variable.

4. To estimate a revenue function, use a graphing utility to find the quadratic function of best fit treating total revenue as the dependent variable and number of bicycles as the independent variable.

5. Find the profit function, $P(x) = R(x) - C(x)$ where $R(x)$ is the revenue function found in (4) and $C(x)$ is the cost function found in (3).

6. Graph $Y_1 = R(x)$, $Y_2 = C(x)$ and $Y_3 = P(x)$ on the same viewing window. Using the MAXIMUM feature of a graphing utility, determine the profit maximizing level of output. What is the maximum profit? How does this compare with your answer in (1)?

7. Using the numerical derivative function on your graphing utility, graph $Y_4 = P'(x)$ where $P'(x)$ represents the first derivative of profit with respect to x.

8. Using ROOT (or ZERO) determine the value of x where $P'(x) = 0$. Compare this result to your result to (6). What do you conclude?

9. Using the numerical derivative function on your graphing utility, graph $Y_5 = R'(x)$ where $R'(x)$ represents the first derivative of revenue with respect to x. Economists call $R'(x)$ the marginal revenue function. Using the numerical derivative function on your graphing utility, graph $Y_6 = C'(x)$ where $C'(x)$ represents the marginal cost function.

10. Using the INTERSECT feature on your graphing utility, determine the value of x where marginal revenue equals marginal cost. Compare this result to your result from (8). What do you conclude?

11. Interpret marginal revenue and marginal cost in the context of the derivative. Use this interpretation to explain to the owners why the level of output where marginal revenue equals marginal cost is the profit maximizing level of output.

Real Numbers

Real numbers are represented by symbols such as

$$25, \quad 0, \quad -3, \quad \tfrac{1}{2}, \quad -\tfrac{5}{4}, \quad 0.125, \quad \sqrt{2}, \quad \pi, \quad \sqrt[3]{-2}, \quad 0.666\ldots$$

The set of **counting numbers,** or **natural numbers,** is the set $\{1, 2, 3, 4, \ldots\}$. (The three dots, called an **ellipsis,** indicate that the pattern continues indefinitely.) The set of **integers** is the set $\{\ldots, -3, -2, -1, 0, 1, 2, 3, \ldots\}$. A **rational number** is a number that can be expressed as a *quotient* a/b of two integers, where the integer b cannot be 0. Examples of rational numbers are $\tfrac{3}{4}, \tfrac{5}{2}, \tfrac{0}{4}$, and $-\tfrac{2}{3}$. Since $a/1 = a$ for any integer a, every integer is also a rational number. Real numbers that are not rational are called **irrational.** Examples of irrational numbers are $\sqrt{2}$ and π (the Greek letter pi), which equals the constant ratio of the circumference to the diameter of a circle. See Figure 1.

Real numbers can be represented as **decimals.** Rational real numbers have decimal representations that either **terminate** or are nonterminating with **repeating** blocks of digits. For example, $\tfrac{3}{4} = 0.75$, which terminates; and $\tfrac{2}{3} = 0.666\ldots$, in which the digit 6 repeats indefinitely. Irrational real numbers have decimal representations that neither repeat nor terminate. For example, $\sqrt{2} = 1.414213\ldots$ and $\pi = 3.14159\ldots$. In practice, irrational numbers are generally represented by approximations. We use the symbol $\approx$ (read as "approximately equal to") to write $\sqrt{2} \approx 1.4142$ and $\pi \approx 3.1416$.

Often, letters are used to represent numbers. If the letter used is to represent *any* number from a given set of numbers, it is referred to as a **variable.** A **constant** is either a fixed number, such as 5, $\sqrt{2}$, and so on, or a letter that represents a fixed (possibly unspecified) number. In general, we will follow the practice of using letters near the beginning of the alphabet, such as a, b, and c, for constants and using those near the end, such as x, y, and z, as variables.

In working with expressions or formulas involving variables, the variables may only be allowed to take on values from a certain set of numbers, called the **domain of the variable.** For example, in the expression $1/x$, the variable x cannot take on the value 0, since division by 0 is not allowed.

It can be shown that there is a one-to-one correspondence between real numbers and points on a line. That is, every real number corresponds to a point on the line and, conversely, each point on the line has a unique real number associated with it. We establish this correspondence of real numbers with points on a line in the following manner.

We start with a line that is, for convenience, drawn horizontally. Pick a point on the line and label it O, for **origin.** Then pick another point some fixed distance to the right of O and label it U, for **unit.** The fixed distance, which may be 1 inch, 1 centimeter, 1 light-year, or any unit distance, determines the **scale.** We associate the real number 0 with the origin O and the number 1 with the point U. Refer to Figure 2. The point to the right of U that is twice as far from O as U is associated with the number 2. The point to the right of U that is three times as far from O as U is associated with the number 3. The point midway between O and U is assigned the number 0.5, or $\tfrac{1}{2}$. Corresponding points to the left of the origin O are assigned the numbers $-\tfrac{1}{2}$, -1, -2, -3, and so on. The real number x associated with a point P is called the **coordinate** of P, and the line whose points have been assigned coordinates is called the **real number line.** Notice in Figure 2 that we placed

FIGURE 1

$$\pi = \frac{C}{d}$$

FIGURE 2
Real number line

an arrowhead on the right end of the line to indicate the direction in which the assigned numbers increase. Figure 2 also shows the points associated with the irrational numbers $\sqrt{2}$ and π.

The real number line divides the real numbers into three classes: the **negative real numbers** are the coordinates of points to the left of the origin O; the real number **zero** is the coordinate of the origin O; the **positive real numbers** are the coordinates of points to the right of the origin O.

Let a and b be two real numbers. If the difference $a - b$ is positive, then we say that a is **greater than** b and write $a > b$. Alternatively, if $a - b$ is positive, we can also say that b is **less than** a and write $b < a$. Thus, $a > b$ and $b < a$ are equivalent statements.

On the real number line, if $a > b$, the point with coordinate a is to the right of the point with coordinate b. For example, $0 > -1$, $\pi > 3$, and $\sqrt{2} < 2$. Furthermore,

$$a > 0 \text{ is equivalent to } a \text{ is positive}$$
$$a < 0 \text{ is equivalent to } a \text{ is negative}$$

If the difference $a - b$ of two real numbers is positive or 0, that is, if $a > b$ or $a = b$, then we say that a is **greater than or equal to** b and write $a \geq b$. Alternatively, if $a \geq b$, we can also say that b is **less than or equal to** a and write $b \leq a$.

Statements of the form $a < b$ or $b > a$ are called **strict inequalities;** statements of the form $a \leq b$ or $b \geq a$ are called **nonstrict inequalities.** The symbols $>, <, \geq$, and $\leq$ are called **inequality signs.**

If x is a real number and $x \geq 0$, then x is either positive or 0. As a result, we describe the inequality $x \geq 0$ by saying that x is nonnegative.

Inequalities are useful in representing certain subsets of real numbers. In so doing, though, other variations of the inequality notation may be used.

E X A M P L E   2

FIGURE 3
$x > 4$

FIGURE 4
$x > 4$ and $x \leq 6$

Graphing Inequalities

(a) In the inequality $x > 4$, x is any number greater than 4. In Figure 3, we use a left parenthesis to indicate that the number 4 is not part of the graph.

(b) In the inequality $4 < x \leq 6$, x is any number between 4 and 6, including 6 but excluding 4. In Figure 4, we use a right bracket to indicate that 6 is part of the graph. ▬

Now work Problem 15 (in the exercise set at the end of this section).

Let a and b represent two real numbers with $a < b$: A **closed interval,** denoted by **[a, b],** consists of all real numbers x for which $a \leq x \leq b$. An **open interval,** denoted by **(a, b),** consists of all real numbers x for which $a < x < b$. The **half-open,** or **half-closed intervals** are **(a, b],** consisting of all real numbers x for which $a < x \leq b$, and **[a, b),** consisting of all real numbers x for which $a \leq x < b$. In each of these definitions, a is called the **left end point** and b the **right end point** of the interval. Figure 5 illustrates each type of interval.

The converse of the Pythagorean Theorem is also true.

> **Converse of the Pythagorean Theorem**
>
> In a triangle, if the square of the length of one side equals the sum of the squares of the lengths of the other two sides, then the triangle is a right triangle. The 90° angle is opposite the longest side.

E X A M P L E   10   **Verifying That a Triangle Is a Right Triangle**

Show that a triangle whose sides are of lengths 5, 12, and 13 is a right triangle. Identify the hypotenuse.

Solution   We square the lengths of the sides:

FIGURE 10

$$5^2 = 25, \quad 12^2 = 144, \quad 13^2 = 169$$

Notice that the sum of the first two squares (25 and 144) equals the third square (169). Hence, the triangle is a right triangle. The longest side, 13, is the hypotenuse. See Figure 10. ∎

Now work Problem 79.

Geometry Formulas

Certain formulas from geometry are useful in solving algebra problems. We list some of these formulas next.

For a rectangle of length l and width w,

> Area $= lw$ Perimeter $= 2l + 2w$

For a triangle with base b and altitude h,

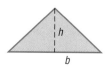

> Area $= \dfrac{1}{2}bh$

For a circle of radius r (diameter $d = 2r$),

> Area $= \pi r^2$ Circumference $= 2\pi r = \pi d$

For a rectangular box of length l, width w, and height h,

> Volume $= lwh$

1 | EXERCISES

In Problems 1–10, replace the question mark by $<$, $>$, or $=$, whichever is correct.

1. $\frac{1}{2}$? 0   **2.** 5 ? 6   **3.** -1 ? -2 **4.** -3 ? $-\frac{5}{2}$ **5.** π ? 3.14

6. $\sqrt{2}$? 1.41   **7.** $\frac{1}{2}$? 0.5   **8.** $\frac{1}{3}$? 0.33   **9.** $\frac{2}{3}$? 0.67   **10.** $\frac{1}{4}$? 0.25

11. On the real number line, label the points with coordinates 0, 1, -1, $\frac{5}{2}$, -2.5, $\frac{3}{4}$, and 0.25.

12. Repeat Problem 11 for the coordinates 0, -2, 2, -1.5, $\frac{3}{2}$, $\frac{1}{3}$, and $\frac{2}{3}$.

In Problems 13–20, write each statement as an inequality.

13. x is positive   **14.** z is negative

15. x is less than 2   **16.** y is greater than -5

17. x is less than or equal to 1   **18.** x is greater than or equal to 2

19. x is less than 5 and x is greater than 2

20. y is less than or equal to 2 and y is greater than 0

In Problems 21–24, write each inequality using interval notation, and illustrate each inequality using the real number line.

21. $0 \le x \le 4$ **22.** $-1 < x < 5$ **23.** $4 \le x < 6$ **24.** $-2 < x \le 0$

In Problems 25–28, write each interval as an inequality involving x, and illustrate each inequality using the real number line.

25. $[2, 5]$ **26.** $(1, 2)$ **27.** $[4, \infty)$ **28.** $(-\infty, 2]$

In Problems 29–32, find the value of each expression if $x = 2$ and $y = -3$.

29. $|x + y|$ **30.** $|x - y|$ **31.** $|x| + |y|$ **32.** $|x| - |y|$

In Problems 33–52, simplify each expression.

33. 3^0 **34.** 3^2 **35.** 4^{-2} **36.** $(-3)^2$

37. $\left(\frac{2}{3}\right)^2$ **38.** $\left(\frac{-4}{5}\right)^3$ **39.** $3^{-6} \cdot 3^4$ **40.** $4^{-2} \cdot 4^3$

41. $\left(\frac{2}{3}\right)^{-2}$ **42.** $\left(\frac{3}{2}\right)^{-3}$ **43.** $\dfrac{2^3 \cdot 3^2}{2 \cdot 3^{-2}}$ **44.** $\dfrac{3^{-2} \cdot 5^3}{3 \cdot 5}$

45. $9^{3/2}$ **46.** $16^{3/4}$ **47.** $(-8)^{4/3}$ **48.** $(-27)^{2/3}$

49. $\sqrt{32}$ **50.** $\sqrt[3]{24}$ **51.** $\sqrt[3]{-\frac{8}{27}}$ **52.** $\sqrt[4]{\frac{4}{9}}$

In Problems 53–68, simplify each expression so that all exponents are positive. Whenever an exponent is negative or 0, we assume that the base does not equal 0.

53. $x^0 y^2$ **54.** $x^{-1} y$ **55.** $x^{-2} y$ **56.** $x^4 y^0$

57. $\dfrac{x^{-2} y^3}{xy^4}$ **58.** $\dfrac{x^{-2} y}{xy^2}$ **59.** $\left(\dfrac{4x}{5y}\right)^{-2}$ **60.** $(xy)^{-2}$

61. $\dfrac{x^{-1} y^{-2} z}{x^2 y z^3}$ **62.** $\dfrac{3x^{-2} y z^2}{x^4 y^{-3} z}$ **63.** $\dfrac{(-2)^3 x^4 (yz)^2}{3^2 xy^3 z^4}$ **64.** $\dfrac{4x^{-2}(yz)^{-1}}{(-5)^2 x^4 y^2 z^{-2}}$

65. $\dfrac{x^{-2}}{\dfrac{1}{x^2} + \dfrac{1}{y^2}}$ **66.** $\dfrac{x^{-1} + y^{-1}}{x^{-1} - y^{-1}}$ **67.** $\left(\dfrac{3x^{-1}}{4y^{-1}}\right)^{-2}$ **68.** $\left(\dfrac{5x^{-2}}{6y^{-2}}\right)^{-3}$

2 | EXERCISES

In Problems 1–6, tell what number should be added to complete the square of each expression.

1. $x^2 - 4x$

2. $x^2 - 2x$

3. $x^2 + \frac{1}{2}x$

4. $x^2 - \frac{1}{3}x$

5. $x^2 - \frac{2}{3}x$

6. $x^2 - \frac{2}{5}x$

In Problems 7–12, complete the squares of x and y in each equation.

7. $x^2 + y^2 - 4x + 4y - 1 = 0$

8. $x^2 + y^2 + 4x + 4y - 8 = 0$

9. $x^2 + y^2 + 6x - 2y + 1 = 0$

10. $x^2 + y^2 - 8x + 2y + 1 = 0$

11. $x^2 + y^2 + x - y - \frac{1}{2} = 0$

12. $x^2 + y^2 - x + y - \frac{3}{2} = 0$

In Problems 13–18, solve each equation by completing the square.

13. $x^2 + 4x - 21 = 0$

14. $x^2 - 6x = 13$

15. $x^2 - \frac{1}{2}x = \frac{3}{16}$

16. $x^2 + \frac{2}{3}x = \frac{1}{3}$

17. $3x^2 + x - \frac{1}{2} = 0$

18. $2x^2 - 3x = 1$

In Problems 19–30, find the real solutions, if any, of each equation. Use the quadratic formula.

19. $x^2 - 4x + 2 = 0$

20. $x^2 + 4x + 2 = 0$

21. $x^2 - 4x - 1 = 0$

22. $x^2 + 6x + 1 = 0$

23. $2x^2 - 5x + 3 = 0$

24. $2x^2 + 5x + 3 = 0$

25. $4y^2 - y + 2 = 0$

26. $4t^2 + t + 1 = 0$

27. $4x^2 = 1 - 2x$

28. $2x^2 = 1 - 2x$

29. $9t^2 - 6t + 1 = 0$

30. $4u^2 - 6u + 9 = 0$

3 | SYNTHETIC DIVISION

1 Use Synthetic Division

To find the quotient as well as the remainder when a polynomial function f of degree 1 or higher is divided by $g(x) = x - c$, a shortened version of long division, called **synthetic division,** makes the task simpler.

To see how synthetic division works, we will use long division to divide the polynomial $f(x) = 2x^3 - x^2 + 3$ by $g(x) = x - 3$.

$$
\begin{array}{r}
2x^2 + 5x + 15 \\
x - 3 \overline{)\, 2x^3 - x^2 \qquad + 3} \\
\underline{2x^3 - 6x^2} \\
5x^2 \\
\underline{5x^2 - 15x} \\
15x + 3 \\
\underline{15x - 45} \\
48
\end{array}
$$

The process of synthetic division arises from rewriting the long division in a more compact form, using simpler notation. For example, in the long division above, the terms in color are not really necessary because they are identical to the terms directly above them. With these terms removed, we have

$$
\begin{array}{r}
2x^2 + 5x + 15 \\
x - 3 \overline{)\, 2x^3 - x^2 \qquad + 3} \\
\underline{- 6x^2} \\
5x^2 \\
\underline{- 15x} \\
15x \\
\underline{- 45} \\
48
\end{array}
$$

Most of the x's that appear in this process can also be removed, provided we are careful about positioning each coefficient. In this regard, we will need to use 0 as the coefficient of x in the dividend, because that power of x is missing. Now we have

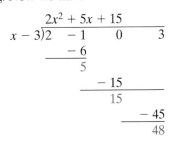

We can make this display more compact by moving the lines up until the numbers in color align horizontally:

$$
\begin{array}{r}
2x^2 + 5x + 15 \quad \text{Row 1.}\\
x - 3\overline{)2 \quad -1 \quad\quad 0 \quad\quad 3} \quad \text{Row 2.}\\
-6 \quad -15 \quad -45 \quad \text{Row 3.}\\
\bigcirc \quad 5 \quad\quad 15 \quad\quad 48 \quad \text{Row 4.}
\end{array}
$$

Now, if we place the leading coefficient of the quotient (2) in the circled position, the first three numbers in row 4 are precisely the coefficients of the quotient, and the last number in row 4 is the remainder. Thus, row 1 is not really needed, so we can compress the process to three rows, where the bottom row contains the coefficients of both the quotient and the remainder:

$$
\begin{array}{r}
x - 3\overline{)2 \quad -1 \quad\quad 0 \quad\quad 3} \quad \text{Row 1.}\\
-6 \quad -15 \quad -45 \quad \text{Row 2 (subtract).}\\
\hline
2 \quad\quad 5 \quad\quad 15 \quad\quad 48 \quad \text{Row 3.}
\end{array}
$$

Recall that the entries in row 3 are obtained by subtracting the entries in row 2 from those in row 1. Rather than subtracting the entries in row 2, we can change the sign of each entry and add. With this modification, our display will look like this:

$$
\begin{array}{r}
x - 3\overline{)2 \quad -1 \quad\quad 0 \quad\quad 3} \quad \text{Row 1.}\\
6 \quad\quad 15 \quad\quad 45 \quad \text{Row 2 (add).}\\
\hline
2 \quad\quad 5 \quad\quad 15 \quad\quad 48 \quad \text{Row 3.}
\end{array}
$$

Notice that the entries in row 2 are three times the prior entries in row 3. Our last modification to the display replaces the $x - 3$ by 3. The entries in row 3 give the quotient and the remainder as shown next.

Quotient   Remainder

$2x^2 + 5x + 15$ $R = 48$

Let's go through another example step by step.

E X A M P L E   1   **Using Synthetic Division to Find the Quotient and Remainder**

Use synthetic division to find the quotient and remainder when

$$f(x) = 3x^4 + 8x^2 - 7x + 4 \quad \text{is divided by} \quad g(x) = x - 1$$

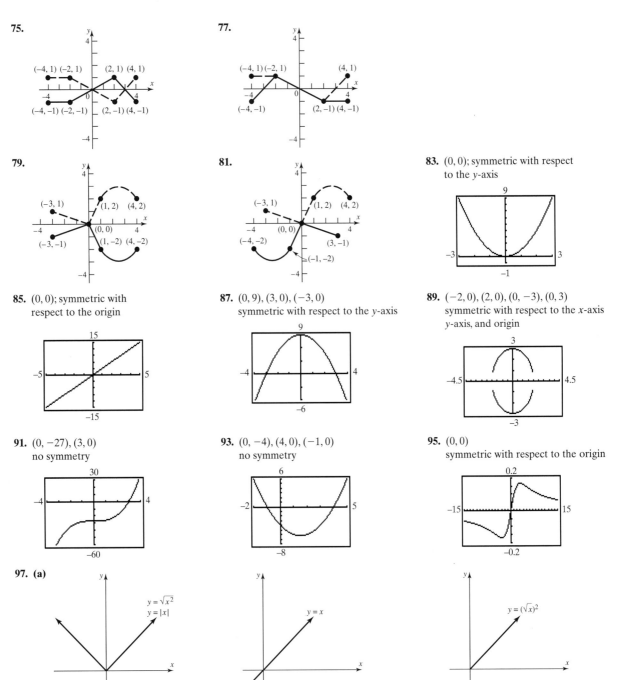

75.

77.

79.

81.

83. $(0, 0)$; symmetric with respect to the y-axis

85. $(0, 0)$; symmetric with respect to the origin

87. $(0, 9), (3, 0), (-3, 0)$ symmetric with respect to the y-axis

89. $(-2, 0), (2, 0), (0, -3), (0, 3)$ symmetric with respect to the x-axis y-axis, and origin

91. $(0, -27), (3, 0)$ no symmetry

93. $(0, -4), (4, 0), (-1, 0)$ no symmetry

95. $(0, 0)$ symmetric with respect to the origin

97. (a)

(b) Since $\sqrt{x^2} = |x|$, for all x, the graphs of $y = \sqrt{x^2}$ and $y = |x|$ are the same.   **(c)** For $y = (\sqrt{x})^2$, the domain of the variable x is $x \geq 0$; for $y = x$, the domain of the variable x is all real numbers. Thus, $(\sqrt{x})^2 = x$ only for $x \geq 0$.   **(d)** For $y = \sqrt{x^2}$, the range of the variable y is $y \geq 0$; for $y = x$, the range of the variable y is all real numbers. Also, $\sqrt{x^2} = |x|$, which equals x only if $x \geq 0$.

1.3 Exercises

1. (a) $\dfrac{1}{2}$ **(b)** For every 2 unit change in x, y will change 1 unit; if x increases by 2 units, y will increase by 1 unit.

3. (a) -1 **(b)** If x increases by 1 unit, y will decrease by 1 unit.

5. Slope $= -\dfrac{3}{2}$ **7.** Slope $= -\dfrac{1}{2}$ **9.** Slope $= 0$

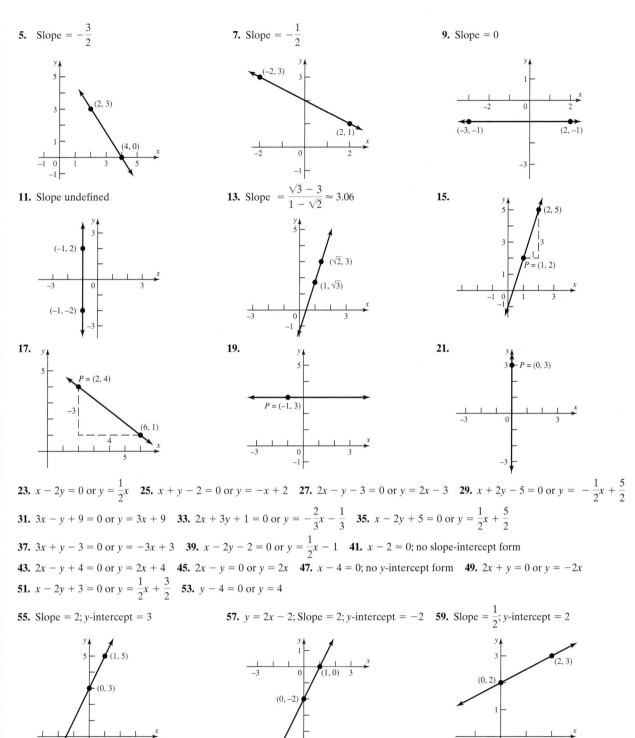

11. Slope undefined   **13.** Slope $= \dfrac{\sqrt{3} - 3}{1 - \sqrt{2}} \approx 3.06$ **15.**

17.   **19.**   **21.**

23. $x - 2y = 0$ or $y = \dfrac{1}{2}x$ **25.** $x + y - 2 = 0$ or $y = -x + 2$ **27.** $2x - y - 3 = 0$ or $y = 2x - 3$ **29.** $x + 2y - 5 = 0$ or $y = -\dfrac{1}{2}x + \dfrac{5}{2}$

31. $3x - y + 9 = 0$ or $y = 3x + 9$ **33.** $2x + 3y + 1 = 0$ or $y = -\dfrac{2}{3}x - \dfrac{1}{3}$ **35.** $x - 2y + 5 = 0$ or $y = \dfrac{1}{2}x + \dfrac{5}{2}$

37. $3x + y - 3 = 0$ or $y = -3x + 3$ **39.** $x - 2y - 2 = 0$ or $y = \dfrac{1}{2}x - 1$ **41.** $x - 2 = 0$; no slope-intercept form

43. $2x - y + 4 = 0$ or $y = 2x + 4$ **45.** $2x - y = 0$ or $y = 2x$ **47.** $x - 4 = 0$; no y-intercept form   **49.** $2x + y = 0$ or $y = -2x$

51. $x - 2y + 3 = 0$ or $y = \dfrac{1}{2}x + \dfrac{3}{2}$ **53.** $y - 4 = 0$ or $y = 4$

55. Slope $= 2$; y-intercept $= 3$ **57.** $y = 2x - 2$; Slope $= 2$; y-intercept $= -2$ **59.** Slope $= \dfrac{1}{2}$; y-intercept $= 2$

61. $y = -\dfrac{1}{2}x + 2$; Slope $= -\dfrac{1}{2}$

y-intercept $= 2$

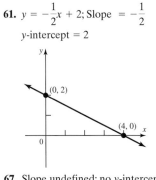

63. $y = \dfrac{2}{3}x - 2$; Slope $= \dfrac{2}{3}$

y-intercept $= -2$

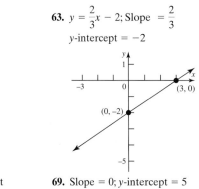

65. $y = -x + 1$; Slope $= -1$

y-intercept $= 1$

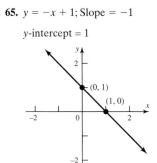

67. Slope undefined; no y-intercept

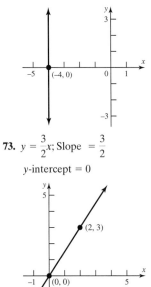

69. Slope $= 0$; y-intercept $= 5$

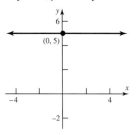

71. $y = x$; Slope $= 1$; y-intercept $= 0$

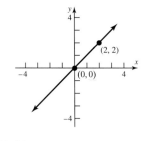

73. $y = \dfrac{3}{2}x$; Slope $= \dfrac{3}{2}$

y-intercept $= 0$

75. $y = 0$

79. **(d)**

77. **(b)**

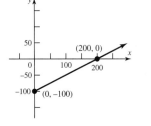

81. $x - y + 2 = 0$ or $y = x + 2$ **83.** $x + 3y - 3 = 0$ or $y = -\dfrac{1}{3}x + 1$ **85.** $2x + 3y = 0$ or $y = -\dfrac{2}{3}x$ **87.** $°C = \dfrac{5}{9}(°F - 32)$; approx. 21°C

89. (a) $P = 0.5x - 100$ **(b)** \$400 **(c)** \$2400 **91.** $C = 0.10819x + 9.06$; $C = \$41.52$; $C = \$106.43$

93. (a)
$$x^2 + (mx + b)^2 = r^2$$
$$(1 + m^2)x^2 + 2mbx + b^2 - r^2 = 0$$
One solution if and only if discriminate $= 0$
$$(2mb)^2 - 4(1 + m^2)(b^2 - r^2) = 0$$
$$-4b^2 + 4r^2 + 4m^2r^2 = 0$$
$$r^2(1 + m^2) = b^2$$

(b) $x = \dfrac{-2mb}{2(1 + m^2)} = \dfrac{-2mb}{2b^2/r^2} = \dfrac{-r^2m}{b}$

$y = m\left(\dfrac{-r^2m}{b}\right) + b = \dfrac{-r^2m^2}{b} + b$

$= \dfrac{-r^2m^2 + b^2}{b} = \dfrac{r^2}{b}$

(c) Slope of tangent line $= m$
Slope of line joining center
to point of tangency

$= \dfrac{r^2/b}{-r^2m/b} = -\dfrac{1}{m}$

95. $\sqrt{2}x + 4y - 11\sqrt{2} + 12 = 0$ **97.** $x + 5y + 13 = 0$ **99.** All have the same slope, 2; the lines are parallel.   **101.** $y = 2$

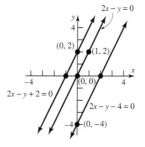

1.4 Exercises

1. Linear relation   **3.** Linear relation   **5.** Nonlinear relation

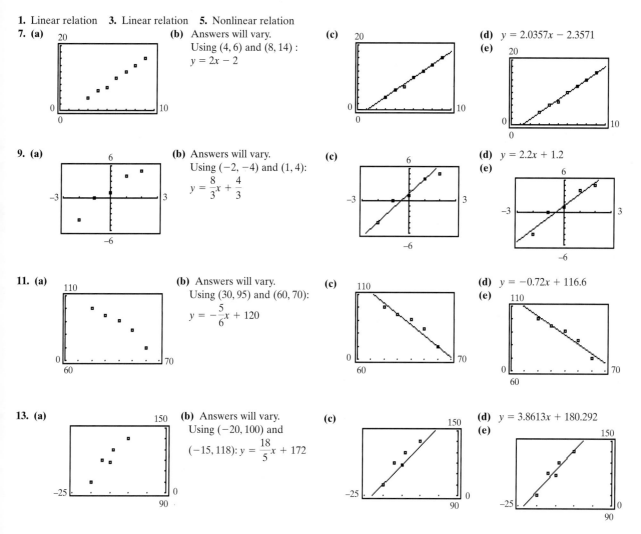

7. (a)
(b) Answers will vary.
Using $(4, 6)$ and $(8, 14)$:
$y = 2x - 2$
(c)
(d) $y = 2.0357x - 2.3571$
(e)

9. (a)
(b) Answers will vary.
Using $(-2, -4)$ and $(1, 4)$:
$y = \dfrac{8}{3}x + \dfrac{4}{3}$
(c)
(d) $y = 2.2x + 1.2$
(e)

11. (a)
(b) Answers will vary.
Using $(30, 95)$ and $(60, 70)$:
$y = -\dfrac{5}{6}x + 120$
(c)
(d) $y = -0.72x + 116.6$
(e)

13. (a)
(b) Answers will vary.
Using $(-20, 100)$ and
$(-15, 118)$: $y = \dfrac{18}{5}x + 172$
(c)
(d) $y = 3.8613x + 180.292$
(e)

15. (a)

17. (a)

(19) (a)

(b) $C = 0.755I + 0.6266$

(c) As disposable income increases by \$1, consumption increases by about \$0.755

(d) \$32,337

(b) $y = -0.8265x + 70.3903$

(c) As speed increases by 1 mph, mpg decreases by 0.8265.

(d) 20 mpg

(b) $NIBT = 0.0842$ sales $- 88.5776$

(c) As sales increase by \$1, NIBT increases by \$0.0842

(d) \$118.3 billion

21. (a) $y = 1.13x - 2841.69$ **(b)** If per capita disposable income increases \$1, per capita personal consumption increases \$1.13.
(c) \$14,095.88 **23. (a)** $y = 323.42x - 62,080.03$ **(b)** If average fuel consumption per car increases by 1 gallon, pollutant emissions of carbon monoxide increase 323.42 thousand tons. **(c)** 101,247 thousand tons.

1.5 Exercises

1. 7 **3.** -3 **5.** 4 **7.** $\dfrac{5}{4}$ **9.** -1 **11.** -18 **13.** -3 **15.** -16 **17.** 0.5 **19.** 2 **21.** 2 **23.** -1 **25.** 3 **27.** No real solution

29. $\{0, 9\}$ **31.** $\{0, 9\}$ **33.** No real solution **35.** 2 **37.** 21 **39.** $\{-2, 2\}$ **41.** $-\dfrac{20}{39}$ **43.** 6 **45.** $\{-3, 3\}$ **47.** $\{-4, 1\}$ **49.** $\left\{-1, \dfrac{3}{2}\right\}$

51. $\{-4, 4\}$ **53.** 2 **55.** $\{-12, 12\}$ **57.** $\left\{-\dfrac{36}{5}, \dfrac{24}{5}\right\}$ **59.** No real solution **61.** $\{-2, 2\}$ **63.** $\{-1, 3\}$ **65.** $\{-2, -1, 0, 1\}$

67. $\{0, 4\}$ **69.** $\{-6, 2\}$ **71.** $\left\{-\dfrac{1}{2}, 3\right\}$ **73.** $\{3, 4\}$ **75.** $\dfrac{3}{2}$ **77.** $\left\{-\dfrac{2}{3}, \dfrac{3}{2}\right\}$ **79.** $\left\{-\dfrac{3}{4}, 2\right\}$ **81.** $x = \dfrac{b + c}{a}$ **83.** $x = \dfrac{abc}{a + b}$ **85.** $x = a^2$

87. No real solution **89.** Repeated real solution **91.** Two unequal real solutions **93.** $\{0.58, 3.41\}; \{2 - \sqrt{2}, 2 + \sqrt{2}\}$

95. $\{-2.80, 1.07\}; \dfrac{-\sqrt{3} - \sqrt{15}}{2}, \dfrac{-\sqrt{3} + \sqrt{15}}{2}$ **97.** $\{-0.85, 1.17\}; \left\{\dfrac{1 \pm \sqrt{1 + 4\pi^2}}{2\pi}\right\}$ **99.** $(-8.15, -0.22); \left\{\dfrac{-8\pi \pm \sqrt{64\pi^2 - 12\sqrt{29}}}{6}\right\}$

101. $\{-2.64, 2.64\}; \{-\sqrt{7}, \sqrt{7}\}$ **103.** $0.25, \dfrac{1}{4}$ **105.** $\{-0.60, 2.50\}; \left\{-\dfrac{3}{5}, \dfrac{5}{2}\right\}$ **107.** $\{-0.50, 0.66\}; \left\{-\dfrac{1}{2}, \dfrac{2}{3}\right\}$

109. $\{-1.70, 0.29\}; \left\{\dfrac{-\sqrt{2} + 2}{2}, \dfrac{-\sqrt{2} - 2}{2}\right\}$ **111.** $\{-2.56, 1.56\}; \left\{\dfrac{-1 - \sqrt{17}}{2}, \dfrac{-1 + \sqrt{17}}{2}\right\}$ **113.** 3.00 **115.** 4.50 **117.** $0.31, 12.30$

119. $1.00, 23.00$ **121.** $R = \dfrac{R_1 R_2}{R_1 + R_2}$ **123.** $R = \dfrac{mv^2}{F}$ **125.** $r = \dfrac{S - a}{S}$

1.6 Exercises

1. $A = \pi r^2$; r = Radius, A = Area **3.** $A = s^2$, A = Area, s = Length of a side **5.** $F = ma$; F = Force, m = Mass, a = Acceleration
7. $W = Fd$; W = Work, F = Force, d = Distance **9.** $C = 150x$; C = Total cost, x = number of dishwashers
11. \$11,000 will be invested in bonds and \$9000 in CD's. **13.** David will receive \$400,000, Paige \$300,000, and Dan \$200,000.
15. The regular hourly rate is \$8.50. **17.** The Bears got 5 touchdowns. **19.** The length is 19ft; the width is 11 ft.
21. (a) The dimensions are 10 ft by 5 ft. **(b)** The area is 50 sq ft. **(c)** The dimensions are 7.5 ft by 7.5 ft. **(d)** The area is 56.25 sq ft.
23. Invest \$31,250 in bonds and \$18,750 in CD's. **25.** \$11,600 was loaned out at 8%. **27.** The dimensions are 11 ft by 13 ft.
29. The dimensions are 5 m by 8 m. **31.** The dimensions of the sheet metal should be 4 ft by 4 ft.
33. (a) The ball strikes the ground after 6 seconds. **(b)** The ball passes the top of the building on its way down after 5 seconds.
35. The original price was \$147,058.82; purchasing the model saves \$22,058.82. **37.** The bookstore paid \$44.80.
39. Working together, it takes 12 min. **41.** Brooke needs a score of 85.
43. The defensive back catches up to the tight end at the tight end's 45 yard line.
45. The border will be 2.56 ft wide. **47.** The dimensions of the reduced candy bar are 11.55 cm by 6.55 cm by 3 cm.
49. The border will be 2.71 ft wide. **51.** The current is 2.286 mi/hr. **53.** Mix 40 pounds of cashews with the peanuts.
55. Mike passes Dan $\dfrac{1}{3}$ mile from the start, 2 minutes from the time Mike started to race. **57.** The rescue craft reaches the ship in 2 hrs.
59. Start the auxiliary pump at 9:45 AM. **61.** The tub will fill in 1 hr. **63.** The most you can invest in the CD is \$66,667.
65. The average speed is 49.5 mi/hr. **67.** Set the original price at \$40. At 50% off, there will be no profit at all.

1.7 Exercises

1. (a) $0 < 2$ **(b)** $-2 < 0$ **(c)** $9 < 15$ **(d)** $-6 > -10$ **3. (a)** $2x - 2 < -1$ **(b)** $2x - 4 < -3$ **(c)** $6x + 3 < 6$ **(d)** $-4x - 2 > -4$
5. $<$ **7.** $>$ **9.** $>$ **11.** $>$
13. $\{x \mid x < 4\}$ or $(-\infty, 4)$ **15.** $\{x \mid x \geq -1\}$ or $[-1, \infty)$ **17.** $\{x \mid x > 3\}$ or $(3, \infty)$ **19.** $\{x \mid x \geq 2\}$ or $[2, \infty)$

21. $\{x \mid x > -7\}$ or $(-7, \infty)$ **23.** $\left\{x \mid x \leq \frac{2}{3}\right\}$ or $\left(-\infty, \frac{2}{3}\right]$ **25.** $\{x \mid x < -20\}$ or $(-\infty, -20)$ **27.** $\left\{x \mid x \geq \frac{4}{3}\right\}$ or $\left[\frac{4}{3}, \infty\right)$

29. $\{x \mid 3 \leq x \leq 5\}$ or $[3, 5]$ **31.** $\left\{x \mid \frac{2}{3} \leq x \leq 3\right\}$ or $\left[\frac{2}{3}, 3\right]$ **33.** $\left\{x \mid -\frac{11}{2} < x < \frac{1}{2}\right\}$ or $\left(-\frac{11}{2}, \frac{1}{2}\right)$ **35.** $\{x \mid -6 < x < 0\}$ or $(-6, 0)$

37. $\{x \mid x < -5\}$ or $(-\infty, -5)$ **39.** $\{x \mid x \geq -1\}$ or $[-1, \infty)$ **41.** $\left\{x \mid \frac{1}{2} \leq x < \frac{5}{4}\right\}$ or $\left[\frac{1}{2}, \frac{5}{4}\right)$ **43.** $\left\{x \mid x < -\frac{1}{2}\right\}$ or $\left(-\infty, -\frac{1}{2}\right)$

45. $\{x \mid x > 5\}$ or $(5, \infty)$ **47.** $\{x \mid x > 3\}$ or $(3, \infty)$

49. $(-4, 4)$ **51.** $(-\infty, -4)$ or $(4, \infty)$ **53.** $(1, 3)$ **55.** $\left[-\frac{2}{3}, 2\right]$

57. $(-\infty, 1]$ or $[5, \infty)$ **59.** $\left(-1, \frac{3}{2}\right)$ **61.** $(-\infty, -1)$ or $(2, \infty)$ **63.** $(-\infty, -3)$ or $(-3, \infty)$

65. $(-\infty, \infty)$ **67.** $(0.49, 0.51)$

69. $|x - 2| < \frac{1}{2}$; $\{x \mid \frac{3}{2} < x < \frac{5}{2}\}$ **71.** $|x + 3| > 2$; $\{x \mid x < -5 \text{ or } x > -1\}$ **73.** $21 < \text{age} < 30$
75. $|x - 98.6| \geq 1.5$; $\{x \mid x \leq 97.1°F \text{ or } x \geq 100.1°F\}$
77. (a) Male ≥ 73.4 **(b)** Female ≥ 79.7 **(c)** A female can expect to live at least 6.3 years longer.
79. The agent's commission ranges from \$45,000 to \$95,000, inclusive. As a percent of selling price, the commission ranges from 5% to 8.6%, inclusive.   **81.** The amount withheld varies from \$70.62 to \$84.62, inclusive.
83. The usage varies from 675.48 kWhr to 2500.88 kWhr, inclusive.   **85.** The dealer's cost varies from \$7457.63 to \$7857.14, inclusive.
87. $74 \leq$ Fifth test score < 124 **89.** The amount of gasoline ranged from 12 to 20 gal, inclusive.   **91.** For t between 2 and 3 seconds
93. Between 10 and 50 watches must be sold; $10 \leq x \leq 50$
95. $\dfrac{a + b}{2} - a = \dfrac{a + b - 2a}{2} = \dfrac{b - a}{2} > 0$; therefore, $a < \dfrac{a + b}{2}$

$b - \dfrac{a + b}{2} = \dfrac{2b - a - b}{2} = \dfrac{b - a}{2} > 0$; therefore, $b > \dfrac{a + b}{2}$

97. $(\sqrt{ab})^2 - a^2 = ab - a^2 = a(b - a) > 0$; thus, $(\sqrt{ab})^2 > a^2$ and $\sqrt{ab} > a$
$b^2 - (\sqrt{ab})^2 = b^2 - ab = b(b - a) > 0$; thus, $b^2 > (\sqrt{ab})^2$ and $b > \sqrt{ab}$

33.

35.

37.

39.

41.

43.

45.

47.

49.

51.

53.

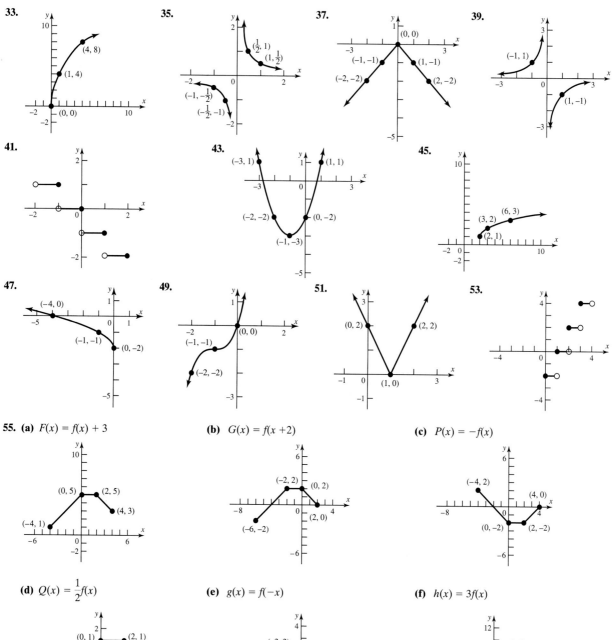

55. (a) $F(x) = f(x) + 3$

(b) $G(x) = f(x + 2)$

(c) $P(x) = -f(x)$

(d) $Q(x) = \dfrac{1}{2}f(x)$

(e) $g(x) = f(-x)$

(f) $h(x) = 3f(x)$

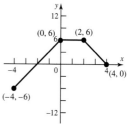

57. (a) $F(x) = f(x) + 3$ **(b)** $G(x) = f(x + 2)$ **(c)** $P(x) = -f(x)$

(d) $Q(x) = \frac{1}{2}f(x)$ **(e)** $g(x) = f(-x)$ **(f)** $h(x) = 3f(x)$

59. (a) $F(x) = f(x) + 3$ **(b)** $G(x) = f(x + 2)$ **(c)** $P(x) = -f(x)$

(d) $Q(x) = \frac{1}{2}f(x)$ **(e)** $g(x) = f(-x)$ **(f)** $h(x) = 3f(x)$

61. (a) $y = |x + 1|$ **(b)** $y = |4 - x^2|$ **(c)** $y = |x^3 + x|$

(d) Any part of the graph of $y = f(x)$ that lies below the x-axis is reflected about the x-axis to obtain the graph of $y = |f(x)|$

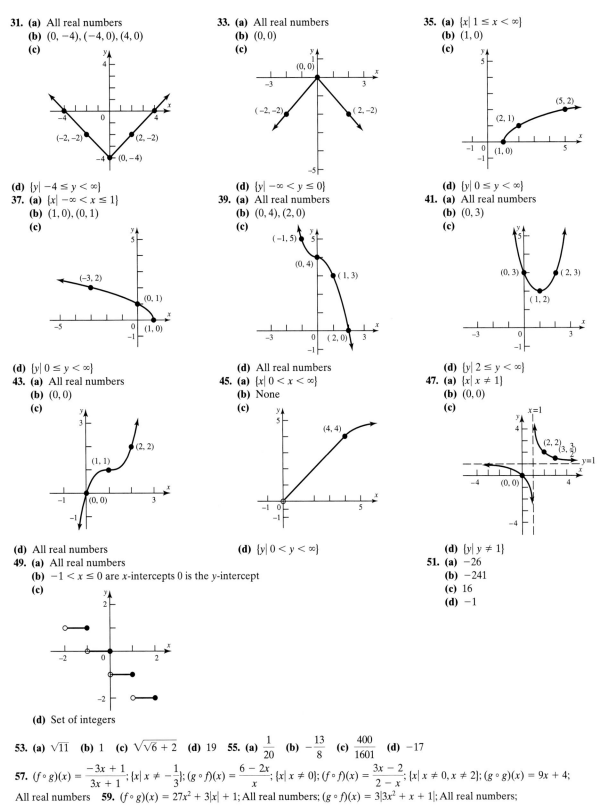

31. (a) All real numbers
(b) $(0, -4), (-4, 0), (4, 0)$
(c)
(d) $\{y|\ -4 \le y < \infty\}$
37. (a) $\{x|\ -\infty < x \le 1\}$
(b) $(1, 0), (0, 1)$
(c)
(d) $\{y|\ 0 \le y < \infty\}$
43. (a) All real numbers
(b) $(0, 0)$
(c)
(d) All real numbers
49. (a) All real numbers
(b) $-1 < x \le 0$ are x-intercepts 0 is the y-intercept
(c)
(d) Set of integers

33. (a) All real numbers
(b) $(0, 0)$
(c)
(d) $\{y|\ -\infty < y \le 0\}$
39. (a) All real numbers
(b) $(0, 4), (2, 0)$
(c)
(d) All real numbers
45. (a) $\{x|\ 0 < x < \infty\}$
(b) None
(c)
(d) $\{y|\ 0 < y < \infty\}$

35. (a) $\{x|\ 1 \le x < \infty\}$
(b) $(1, 0)$
(c)
(d) $\{y|\ 0 \le y < \infty\}$
41. (a) All real numbers
(b) $(0, 3)$
(c)
(d) $\{y|\ 2 \le y < \infty\}$
47. (a) $\{x|\ x \ne 1\}$
(b) $(0, 0)$
(c)
(d) $\{y|\ y \ne 1\}$
51. (a) -26
(b) -241
(c) 16
(d) -1

53. (a) $\sqrt{11}$ **(b)** 1 **(c)** $\sqrt{\sqrt{6} + 2}$ **(d)** 19 **55. (a)** $\dfrac{1}{20}$ **(b)** $-\dfrac{13}{8}$ **(c)** $\dfrac{400}{1601}$ **(d)** -17

57. $(f \circ g)(x) = \dfrac{-3x + 1}{3x + 1}; \{x|\ x \ne -\dfrac{1}{3}\}; (g \circ f)(x) = \dfrac{6 - 2x}{x}; \{x|\ x \ne 0\}; (f \circ f)(x) = \dfrac{3x - 2}{2 - x}; \{x|\ x \ne 0, x \ne 2\}; (g \circ g)(x) = 9x + 4;$
All real numbers   **59.** $(f \circ g)(x) = 27x^2 + 3|x| + 1;$ All real numbers; $(g \circ f)(x) = 3|3x^2 + x + 1|;$ All real numbers;
$(f \circ f)(x) = 3(3x^2 + x + 1)^2 + 3x^2 + x + 2;$ All real numbers; $(g \circ g)(x) = 9|x|;$ All real numbers

61. $(f \circ g)(x) = \dfrac{1 + x}{1 - x}$; $\{x \mid x \neq 0, x \neq 1\}$; $(g \circ f)(x) = \dfrac{x - 1}{x + 1}$; $\{x \mid x \neq -1, x \neq 1\}$; $(f \circ f)(x) = x$; $\{x \mid x \neq 1\}$; $(g \circ g)(x) = x$; $\{x \mid x \neq 0\}$

63. (a)

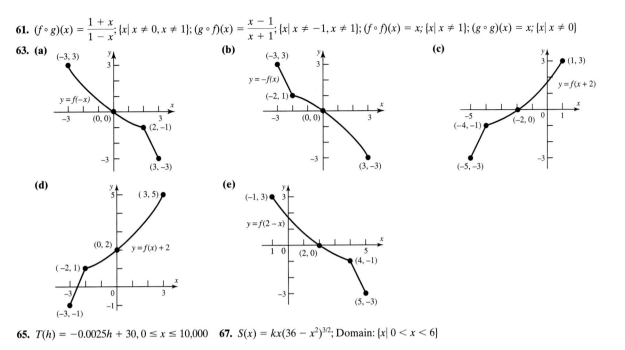

65. $T(h) = -0.0025h + 30, 0 \leq x \leq 10,000$ **67.** $S(x) = kx(36 - x^2)^{3/2}$; Domain: $\{x \mid 0 < x < 6\}$

CHAPTER 3   *3.1 Exercises*

1. D   **3.** A   **5.** B   **7.** E   **9.** D   **11.** B

13. $f(x) = \dfrac{1}{4}(x - 0)^2 + 0$

15. $f(x) = \dfrac{1}{4}(x - 0)^2 - 2$

17. $f(x) = \dfrac{1}{4}(x - 0)^2 + 2$

19. $f(x) = -\dfrac{1}{4}(x - 0)^2 + 1$

21. $f(x) = (x + 2)^2 - 2$

23. $f(x) = 2(x - 1)^2 - 1$

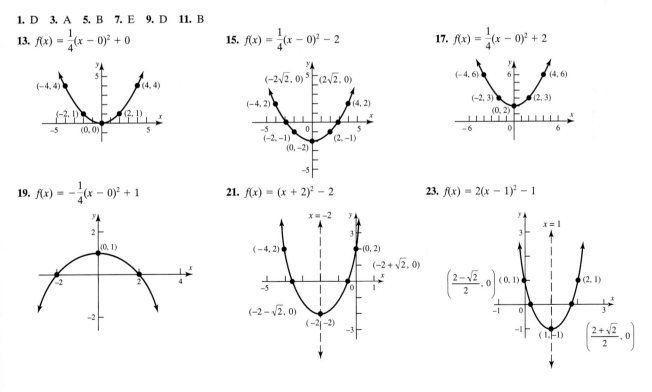

25. $f(x) = -(x + 1)^2 + 1$

27. $f(x) = \frac{1}{2}(x + 1)^2 - \frac{3}{2}$

29. $f(x) = (x + 1)^2 - 9$

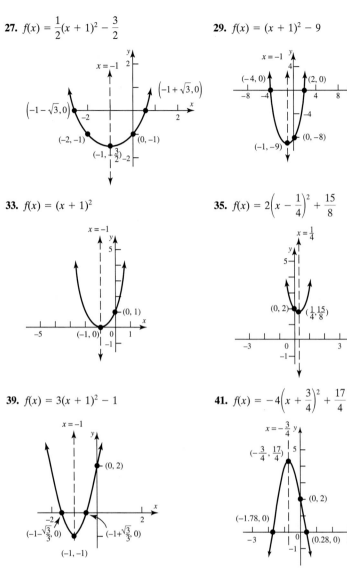

31. $f(x) = -\left(x + \frac{3}{2}\right)^2 + \frac{25}{4}$

33. $f(x) = (x + 1)^2$

35. $f(x) = 2\left(x - \frac{1}{4}\right)^2 + \frac{15}{8}$

37. $f(x) = -2\left(x - \frac{1}{2}\right)^2 - \frac{5}{2}$

39. $f(x) = 3(x + 1)^2 - 1$

41. $f(x) = -4\left(x + \frac{3}{4}\right)^2 + \frac{17}{4}$

43. Minimum value; -21 **45.** Maximum value; 21   **47.** Maximum value; 13

49. Opens up; vertex at $(-1, f(-1))$; axis of symmetry $x = -1$

51. Each parabola opens up and passes through $(0, 1)$. Each one has the same shape.   **53.** Price: $500; maximum revenue: $1,000,000

55. 10,000 ft²; 100 ft by 100 ft **57.** 2,000,000 m² **59.** 4,166,666.7 m² **61. (a)** **(b)** 219.53 ft

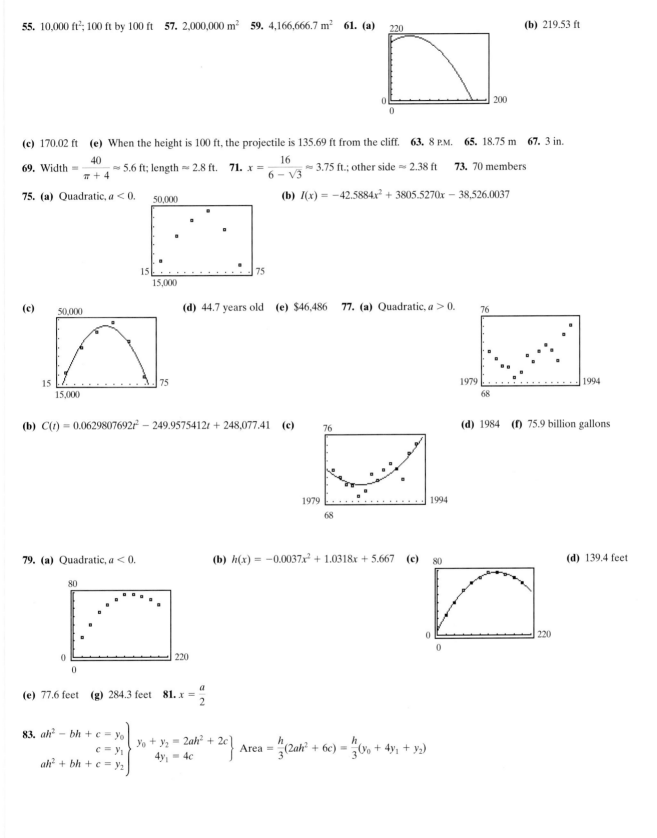

(c) 170.02 ft **(e)** When the height is 100 ft, the projectile is 135.69 ft from the cliff. **63.** 8 P.M. **65.** 18.75 m **67.** 3 in.

69. Width $= \dfrac{40}{\pi + 4} \approx 5.6$ ft; length ≈ 2.8 ft. **71.** $x = \dfrac{16}{6 - \sqrt{3}} \approx 3.75$ ft.; other side ≈ 2.38 ft **73.** 70 members

75. (a) Quadratic, $a < 0$. **(b)** $I(x) = -42.5884x^2 + 3805.5270x - 38,526.0037$

(c) **(d)** 44.7 years old **(e)** \$46,486 **77. (a)** Quadratic, $a > 0$.

(b) $C(t) = 0.0629807692t^2 - 249.9575412t + 248,077.41$ **(c)** **(d)** 1984 **(f)** 75.9 billion gallons

79. (a) Quadratic, $a < 0$. **(b)** $h(x) = -0.0037x^2 + 1.0318x + 5.667$ **(c)** **(d)** 139.4 feet

(e) 77.6 feet **(g)** 284.3 feet **81.** $x = \dfrac{a}{2}$

83. $\left.\begin{array}{r} ah^2 - bh + c = y_0 \\ c = y_1 \\ ah^2 + bh + c = y_2 \end{array}\right\}$ $\left.\begin{array}{r} y_0 + y_2 = 2ah^2 + 2c \\ 4y_1 = 4c \end{array}\right\}$ Area $= \dfrac{h}{3}(2ah^2 + 6c) = \dfrac{h}{3}(y_0 + 4y_1 + y_2)$

3.5 Exercises

1. No; $f(2) = 8$ **3.** Yes; $f(2) = 0$ **5.** Yes; $f(-3) = 0$ **7.** No; $f(-4) = 1$ **9.** Yes; $f\left(\dfrac{1}{2}\right) = 0$ **11.** 7; 3 or 1 positive; 2 or 0 negative

13. 6; 2 or 0 positive; 2 or 0 negative   **15.** 3; 2 or 0 positive; 1 negative   **17.** 4; 2 or 0 positive; 2 or 0 negative

19. 5; 0 positive; 3 or 1 negative   **21.** 6; 1 positive; 1 negative   **23.** $\pm 1, \pm\dfrac{1}{3}$ **25.** $\pm 1, \pm 3$ **27.** $\pm 1, \pm 2, \pm\dfrac{1}{4}, \pm\dfrac{1}{2}$ **29.** $\pm\dfrac{1}{3}, \pm\dfrac{2}{3}, \pm 1, \pm 2$

31. $\pm\dfrac{1}{2}, \pm 1, \pm 2, \pm 4$ **33.** $\pm\dfrac{1}{6}, \pm\dfrac{1}{3}, \pm\dfrac{1}{2}, \pm\dfrac{2}{3}, \pm 1, \pm 2$ **35.** -1 and 1 **37.** -3 and 7 **39.** -5 and 2

41. $-3, -1, 2; f(x) = (x + 3)(x + 1)(x - 2)$ **43.** $\dfrac{1}{2}; f(x) = 2\left(x - \dfrac{1}{2}\right)(x^2 + 1)$ **45.** $-1, 1; f(x) = (x + 1)(x - 1)(x^2 + 2)$

47. $-\dfrac{1}{2}, \dfrac{1}{2}; f(x) = 4\left(x + \dfrac{1}{2}\right)\left(x - \dfrac{1}{2}\right)(x^2 + 2)$ **49.** $-2, -1, 1, 1; f(x) = (x + 2)(x + 1)(x - 1)^2$

51. $-0.70, 0.70, 2; f(x) = 4\left(x + \dfrac{\sqrt{2}}{2}\right)\left(x - \dfrac{\sqrt{2}}{2}\right)(x - 2)\left(x^2 + \dfrac{1}{2}\right)$ **53.** $-5.90, -0.30, 3.00$ **55.** $-3.80, 4.50$ **57.** $-3.42, 0.31, 12.30$

59. $-43.50, 1.00, 23.00$ **61.** $\{-1, 2\}$ **63.** $\left\{\dfrac{2}{3}, -1 + \sqrt{2}, -1 - \sqrt{2}\right\}$ **65.** $\left\{\dfrac{1}{3}, \sqrt{5}, -\sqrt{5}\right\}$ **67.** $\{-3, -2\}$ **69.** $-\dfrac{1}{3}$

71. Approximately 4 or 8 Cavaliers can be produced at an average cost of $10,500.   **73.** $k = 5$ **75.** -7
77. No (use the Rational Zeros Theorem)   **79.** No (use the Rational Zeros Theorem)   **81.** 7 in.
83. All the potential rational zeros are integers. Hence, r either is an integer or is not a rational root (and is therefore irrational).
85.

$$y^3 + by^2 + cy + d = 0$$
$$\left(x - \dfrac{b}{3}\right)^3 + b\left(x - \dfrac{b}{3}\right)^2 + c\left(x - \dfrac{b}{3}\right) + d = 0$$
$$x^3 - \dfrac{3b}{3}x^2 + 3\left(\dfrac{b^2}{9}\right)x - \dfrac{b^3}{27} + b\left(x^2 - \dfrac{2bx}{3} + \dfrac{b^2}{9}\right) + cx - \dfrac{bc}{3} + d = 0$$
$$x^3 - \dfrac{b^2}{3}x + cx - \dfrac{b^3}{27} + \dfrac{b^3}{9} - \dfrac{bc}{3} + d = 0$$
$$x^3 + \left(c - \dfrac{b^2}{3}\right)x + \left(\dfrac{2b^3}{27} - \dfrac{bc}{3} + d\right) = 0$$

87. $K = \dfrac{-p}{3H}$

$$H^3 + \left(\dfrac{-p}{3H}\right)^3 = -q$$
$$H^6 + qH^3 - \dfrac{p^3}{27} = 0$$
$$H^3 = \dfrac{-q \pm \sqrt{q^2 + \dfrac{4p^3}{27}}}{2} \quad \text{Choose + sign}$$
$$H = \sqrt[3]{\dfrac{-q}{2} + \sqrt{\dfrac{q^2}{4} + \dfrac{p^3}{27}}}$$

89. $x = H + K$; now use results from Problems 87 and 88   **91.** $p = 3, q = -14; x = \sqrt[3]{7 + 5\sqrt{2}} + \sqrt[3]{7 - 5\sqrt{2}}$ **93.** 8

3.6 Exercises

1. $8 + 5i$ **3.** $-7 + 6i$ **5.** $-6 - 11i$ **7.** $6 - 18i$ **9.** $6 + 4i$ **11.** $10 - 5i$ **13.** 37 **15.** $\dfrac{6}{5} + \dfrac{8}{5}i$ **17.** $1 - 2i$ **19.** $\dfrac{5}{2} - \dfrac{7}{2}i$

21. $-\dfrac{1}{2} + \dfrac{\sqrt{3}}{2}i$ **23.** $2i$ **25.** $-i$ **27.** i **29.** -6 **31.** $-10i$ **33.** $-2 + 2i$ **35.** 0 **37.** 0 **39.** $2i$ **41.** $5i$ **43.** $5i$ **45.** $\{-2i, 2i\}$

47. $\{-4, 4\}$ **49.** $\{3 - 2i, 3 + 2i\}$ **51.** $\{3 - i, 3 + i\}$ **53.** $\left\{\dfrac{1}{4} - \dfrac{1}{4}i, \dfrac{1}{4} + \dfrac{1}{4}i\right\}$ **55.** $\left\{-\dfrac{1}{5} - \dfrac{2}{5}i, -\dfrac{1}{5} + \dfrac{2}{5}i\right\}$ **57.** $\left\{-\dfrac{1}{2} - \dfrac{\sqrt{3}}{2}i, -\dfrac{1}{2} + \dfrac{\sqrt{3}}{2}i\right\}$

59. $\{2, -1 - \sqrt{3}i, -1 + \sqrt{3}i\}$ **61.** $\{-2, 2, -2i, 2i\}$ **63.** $\{-3i, -2i, 2i, 3i\}$ **65.** Two complex solutions.   **67.** Two unequal real solutions.
69. A repeated real solution.   **71.** $2 - 3i$ **73.** 6   **75.** 25   **77.** $z + \bar{z} = (a + bi) + (a - bi) = 2a; z - \bar{z} = (a + bi) - (a - bi) = 2bi$
79. $z + w = (a + bi) + (c + di) = \overline{(a + c) + (b + d)i} = (a + c) - (b + d)i = (a - bi) + (c - di) = \bar{z} + \bar{w}$

3.7 Exercises

1. $4 + i$ **3.** $-i, 1 - i$ **5.** $-i, -2i$ **7.** $-i$ **9.** $2 - i, -3 + i$ **11.** $f(x) = x^4 - 14x^3 + 77x^2 - 200x + 208; a = 1$

13. $f(x) = x^5 - 4x^4 + 7x^3 - 8x^2 + 6x - 4; a = 1$ **15.** $f(x) = x^4 - 6x^3 + 10x^2 - 6x + 9; a = 1$ **17.** $-2i, 4$ **19.** $2i, -3, \dfrac{1}{2}$

21. $3 + 2i, -2, 5$ **23.** $4i, -\sqrt{11}, \sqrt{11}, -\dfrac{2}{3}$ **25.** $1, -\dfrac{1}{2} - \dfrac{\sqrt{3}}{2}i, -\dfrac{1}{2} + \dfrac{\sqrt{3}}{2}i$ **27.** $2, 3 - 2i, 3 + 2i$ **29.** $-i, i, -2i, 2i$ **31.** $-5i, 5i, -3, 1$

33. $-4, \dfrac{1}{3}, 2 - 3i, 2 + 3i$ **35.** Zeros that are complex numbers must occur in conjugate pairs; or a polynomial with real coefficients of odd degree must have at least one real zero.   **37.** If the remaining zero were a complex number, then its conjugate would also be a zero.

3.8 Exercises

1. $\{x| -2 < x < 5\}$ 3. $\{x| -\infty < x < 0 \text{ or } 4 < x < \infty\}$ 5. $\{x| -3 < x < 3\}$ 7. $\{x| -\infty < x < -4 \text{ or } 3 < x < \infty\}$ 9. $\{x| -\frac{1}{2} < x < 3\}$

11. $\{x| -\infty < x < -1 \text{ or } 8 < x < \infty\}$ 13. No real solution 15. $\{x| -\infty < x < -\frac{2}{3} \text{ or } \frac{3}{2} < x < \infty\}$ 17. $\{x| 1 < x < \infty\}$

19. $\{x| -\infty < x < 1 \text{ or } 2 < x < 3\}$ 21. $\{x| -1 < x < 0 \text{ or } 3 < x < \infty\}$ 23. $\{x| -\infty < x < -1 \text{ or } 1 < x < \infty\}$ 25. $\{x| 1 < x < \infty\}$

27. $\{x| -\infty < x < -1 \text{ or } 1 < x < \infty\}$ 29. $\{x| -1 < x < 8\}$ 31. $\{x| 2.14 \leq x < \infty\}$ 33. $\{x| -\infty < x < -2 \text{ or } 2 < x < \infty\}$

35. $\{x| -1 \leq x \leq -0.23 \text{ or } 4.23 \leq x < \infty\}$ 37. $\{x| -\infty < x < -1 \text{ or } 1 < x < \infty\}$ 39. $\{x| -\infty < x < -1 \text{ or } 0 < x < 1\}$

41. $\{x| -\infty < x < -1 \text{ or } 1 < x < \infty\}$ 43. $\{x| -\infty < x < -\frac{2}{3} \text{ or } 0 < x < \frac{3}{2}\}$ 45. $\{x| -\infty < x < 2\}$ 47. $\{x| -2 < x \leq 9\}$

49. $\{x| -\infty < x < 2 \text{ or } 3 < x < 5\}$ 51. $\{x| -\infty < x < -3 \text{ or } -1 < x < 1 \text{ or } 2 < x < \infty\}$

53. $\{x| -\infty < x < -5 \text{ or } -4 < x < -3 \text{ or } 1 < x < \infty\}$ 55. $\{x| -\infty < x \leq -0.5 \text{ or } 1 \leq x < 4\}$ 57. $\{x| -3.30 < x < -3 \text{ or } 0.30 < x < \infty\}$

59. $\{x| 4 < x < \infty\}$ 61. $\{x| -\infty < x \leq -4 \text{ or } 4 \leq x < \infty\}$ 63. $\{x| -\infty < x < -4 \text{ or } 2 \leq x < \infty\}$

65. **(a)** The ball is more than 96 feet above the ground for time t between 2 and 3 seconds, $2 < t < 3$. **(b)**

(c) 100 feet **(d)** 2.5 seconds 67. **(a)** For a profit of at least \$50, between 8 and 32 watches must be sold, $8 \leq x \leq 32$.

(b) **(c)** \$2000 **(d)** 100 **(e)** **(f)** \$80 **(g)** 20

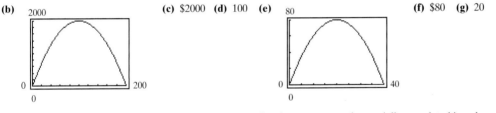

69. Chevy can produce at most 8 Cavaliers in a day, assuming that cars cannot be partially completed in a day.

71. **(a)**

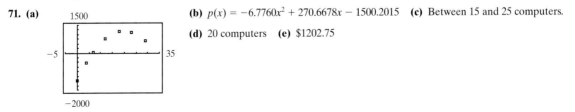

(b) $p(x) = -6.7760x^2 + 270.6678x - 1500.2015$ **(c)** Between 15 and 25 computers.

(d) 20 computers **(e)** \$1202.75

73. $b - a = (\sqrt{b} - \sqrt{a})(\sqrt{b} + \sqrt{a})$; since $a \geq 0$ and $b \geq 0$, then $\sqrt{a} \geq 0$ and $\sqrt{b} \geq 0$ so that $\sqrt{b} + \sqrt{a} \geq 0$; thus, $b - a \geq 0$ is equivalent to $\sqrt{b} - \sqrt{a} \geq 0$ and $a \leq b$ is equivalent to $\sqrt{a} \leq \sqrt{b}$

Fill-in-the-Blank Items

1. parabola; vertex 2. Remainder; Dividend 3. $f(c)$ 4. $f(c) = 0$ 5. zero 6. three; one; two; no 7. $\pm 1, \pm\frac{1}{2}$ 8. $y = 1$
9. $x = -1$ 10. $3 - 4i$

True/False Items

1. F 2. F 3. T 4. T 5. T

25.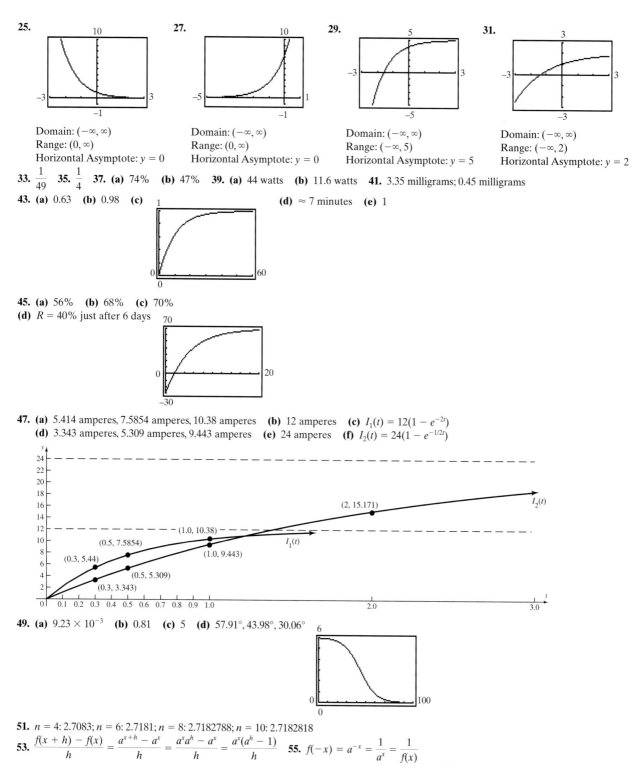

Domain: $(-\infty, \infty)$
Range: $(0, \infty)$
Horizontal Asymptote: $y = 0$

27.

Domain: $(-\infty, \infty)$
Range: $(0, \infty)$
Horizontal Asymptote: $y = 0$

29.

Domain: $(-\infty, \infty)$
Range: $(-\infty, 5)$
Horizontal Asymptote: $y = 5$

31.

Domain: $(-\infty, \infty)$
Range: $(-\infty, 2)$
Horizontal Asymptote: $y = 2$

33. $\dfrac{1}{49}$ **35.** $\dfrac{1}{4}$ **37. (a)** 74% **(b)** 47% **39. (a)** 44 watts **(b)** 11.6 watts **41.** 3.35 milligrams; 0.45 milligrams

43. (a) 0.63 **(b)** 0.98 **(c)**   **(d)** ≈ 7 minutes **(e)** 1

45. (a) 56% **(b)** 68% **(c)** 70%
(d) $R = 40\%$ just after 6 days

47. (a) 5.414 amperes, 7.5854 amperes, 10.38 amperes **(b)** 12 amperes **(c)** $I_1(t) = 12(1 - e^{-2t})$
(d) 3.343 amperes, 5.309 amperes, 9.443 amperes **(e)** 24 amperes **(f)** $I_2(t) = 24(1 - e^{-1/2t})$

49. (a) 9.23×10^{-3} **(b)** 0.81 **(c)** 5 **(d)** $57.91°, 43.98°, 30.06°$

51. $n = 4: 2.7083; n = 6: 2.7181; n = 8: 2.7182788; n = 10: 2.7182818$

53. $\dfrac{f(x + h) - f(x)}{h} = \dfrac{a^{x+h} - a^x}{h} = \dfrac{a^x a^h - a^x}{h} = \dfrac{a^x(a^h - 1)}{h}$ **55.** $f(-x) = a^{-x} = \dfrac{1}{a^x} = \dfrac{1}{f(x)}$

57. (a) $\sinh(-x) = \dfrac{1}{2}(e^{-x} - e^{x}) = -\dfrac{1}{2}(e^{x} - e^{-x}) = -\sinh x$ **(b)**

59. $f(1) = 5, f(2) = 17, f(3) = 257, f(4) = 65{,}537, \ f(5) = 4{,}294{,}967{,}297 = 641 \times 6{,}700{,}417$

4.3 Exercises

1. $2 = \log_3 9$ **3.** $2 = \log_a 1.6$ **5.** $2 = \log_{1.1} M$ **7.** $x = \log_2 7.2$ **9.** $\sqrt{2} = \log_x \pi$ **11.** $x = \ln 8$ **13.** $2^3 = 8$ **15.** $a^6 = 3$ **17.** $3^x = 2$

19. $2^{1.3} = M$ **21.** $(\sqrt{2})^x = \pi$ **23.** $e^x = 4$ **25.** 0 **27.** 2 **29.** -4 **31.** $\dfrac{1}{2}$ **33.** 4 **35.** $\dfrac{1}{2}$ **37.** $\{x \mid x < 3\}$ **39.** All real numbers except 0

41. $\{x \mid x < -2 \text{ or } x > 3\}$ **43.** $\{x \mid x > 0, x \neq 1\}$ **45.** $\{x \mid x < -1 \text{ or } x > 0\}$ **47.** 0.511 **49.** 30.099 **51.** $\sqrt{2}$ **53.** B **55.** D **57.** A

59. E **61.** C **63.** A **65.** D

67.

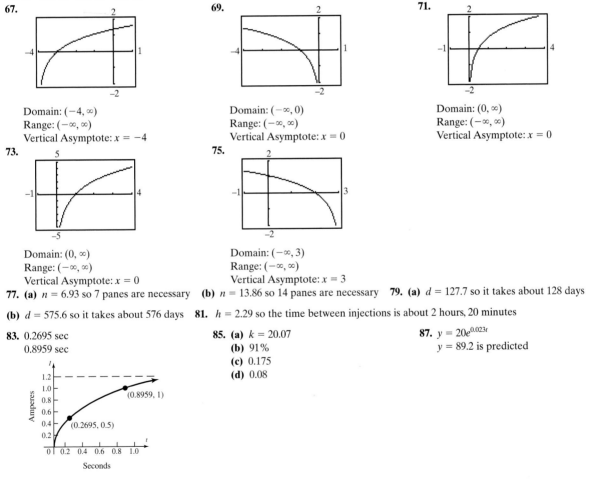

Domain: $(-4, \infty)$
Range: $(-\infty, \infty)$
Vertical Asymptote: $x = -4$

69.

Domain: $(-\infty, 0)$
Range: $(-\infty, \infty)$
Vertical Asymptote: $x = 0$

71.

Domain: $(0, \infty)$
Range: $(-\infty, \infty)$
Vertical Asymptote: $x = 0$

73.

Domain: $(0, \infty)$
Range: $(-\infty, \infty)$
Vertical Asymptote: $x = 0$

75.

Domain: $(-\infty, 3)$
Range: $(-\infty, \infty)$
Vertical Asymptote: $x = 3$

77. (a) $n = 6.93$ so 7 panes are necessary **(b)** $n = 13.86$ so 14 panes are necessary **79. (a)** $d = 127.7$ so it takes about 128 days

(b) $d = 575.6$ so it takes about 576 days **81.** $h = 2.29$ so the time between injections is about 2 hours, 20 minutes

83. 0.2695 sec
0.8959 sec

85. (a) $k = 20.07$
(b) 91%
(c) 0.175
(d) 0.08

87. $y = 20e^{0.023t}$
$y = 89.2$ is predicted

5. (a) 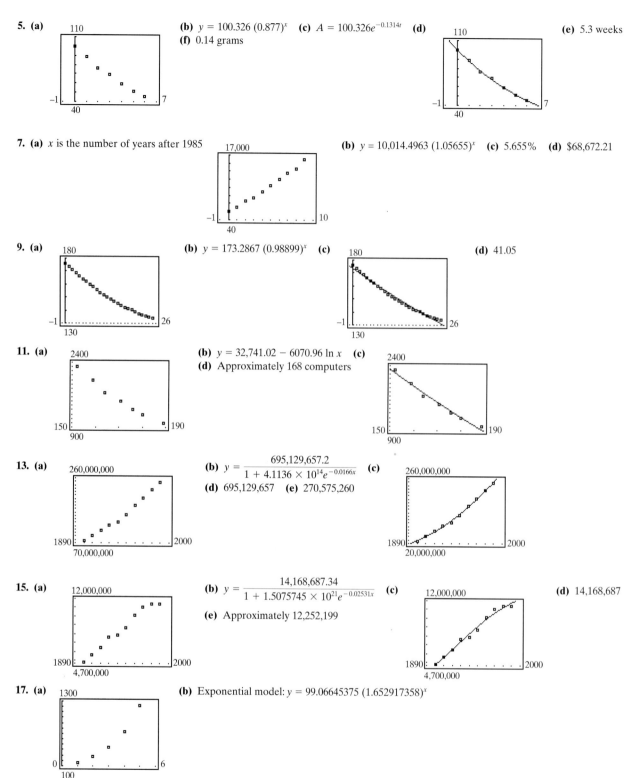 **(b)** $y = 100.326 \, (0.877)^x$ **(c)** $A = 100.326e^{-0.1314t}$ **(d)** **(e)** 5.3 weeks
(f) 0.14 grams

7. (a) x is the number of years after 1985 **(b)** $y = 10{,}014.4963 \, (1.05655)^x$ **(c)** 5.655% **(d)** $68,672.21

9. (a) **(b)** $y = 173.2867 \, (0.98899)^x$ **(c)** **(d)** 41.05

11. (a) **(b)** $y = 32{,}741.02 - 6070.96 \ln x$ **(c)**
(d) Approximately 168 computers

13. (a) **(b)** $y = \dfrac{695{,}129{,}657.2}{1 + 4.1136 \times 10^{14}e^{-0.0166x}}$ **(c)**
(d) 695,129,657 **(e)** 270,575,260

15. (a) **(b)** $y = \dfrac{14{,}168{,}687.34}{1 + 1.5075745 \times 10^{21}e^{-0.02531x}}$ **(c)** **(d)** 14,168,687
(e) Approximately 12,252,199

17. (a) **(b)** Exponential model: $y = 99.06645375 \, (1.652917358)^x$

19. (a)

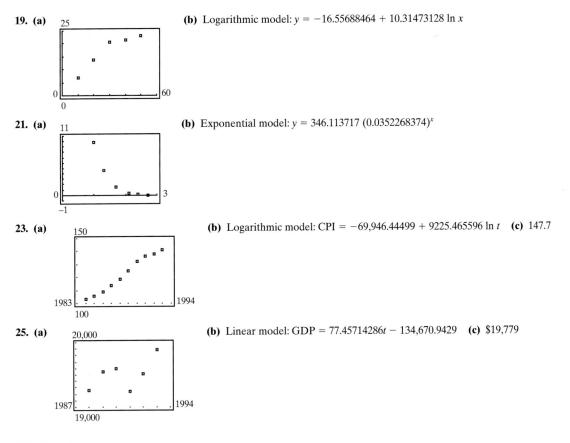

(b) Logarithmic model: $y = -16.55688464 + 10.31473128 \ln x$

21. (a)

(b) Exponential model: $y = 346.113717 \,(0.0352268374)^x$

23. (a)

(b) Logarithmic model: CPI $= -69{,}946.44499 + 9225.465596 \ln t$ **(c)** 147.7

25. (a)

(b) Linear model: GDP $= 77.45714286t - 134{,}670.9429$ **(c)** \$19,779

4.9 Exercises

1. 70 decibels   **3.** 111.76 decibels   **5.** 10 watts/m^2 **7.** 4.0 on the Richter scale
9. 125,892.54 mm; the Mexico City earthquake was 15.85 times as intense as the one in San Francisco.

Fill-in-the-Blank Items

1. one-to-one   **2.** $y = x$ **3.** $(0, 1)$ and $(1, a)$ **4.** 1   **5.** 4   **6.** sum   **7.** 1   **8.** 7   **9.** $x > 0$ **10.** $(1, 0)$ and $(a, 1)$ **11.** 1   **12.** 7

True/False Items

1. F   **2.** T   **3.** T   **4.** T   **5.** F   **6.** F   **7.** T   **8.** F   **9.** F   **10.** T

Review Exercises

1. $f^{-1}(x) = \dfrac{2x + 3}{5x - 2}$; $f(f^{-1}(x)) = \dfrac{2\left(\dfrac{2x + 3}{5x - 2}\right) + 3}{5\left(\dfrac{2x + 3}{5x - 2}\right) - 2} = x$; $f^{-1}(f(x)) = \dfrac{2\left(\dfrac{2x + 3}{5x - 2}\right) + 3}{5\left(\dfrac{2x + 3}{5x - 2}\right) - 2} = x$;

Domain f = Range f^{-1} = All real numbers except $\dfrac{2}{5}$; Range f = Domain f^{-1} = All real numbers except $\dfrac{2}{5}$

3. $f^{-1}(x) = \dfrac{x+1}{x}$; $f(f^{-1}(x)) = \dfrac{1}{\dfrac{x+1}{x} - 1} = x$; $f^{-1}(f(x)) = \dfrac{\dfrac{1}{x-1} + 1}{\dfrac{1}{x-1}} = x$;

Domain f = Range f^{-1} = All real numbers except 1; Range f = Domain f^{-1} = All real numbers except 0

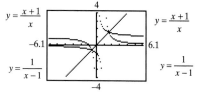

5. $f^{-1}(x) = \dfrac{27}{x^3}$; $f(f^{-1}(x)) = \dfrac{3}{(27/x^3)^{1/3}} = x$; $f^{-1}(f(x)) = \dfrac{27}{(3/x^{1/3})^3} = x$;

Domain f = Range f^{-1} = All real numbers except 0; Range f = Domain f^{-1} = All real numbers except 0

7. -3 **9.** $\sqrt{2}$ **11.** 0.4   **13.** $\dfrac{25}{4}\log_4 x$ **15.** $\ln\left[\dfrac{1}{(x+1)^2}\right] = -2\ln(x+1)$ **17.** $\log\left(\dfrac{4x^3}{[(x+3)(x-2)]^{1/2}}\right)$ **19.** $y = Ce^{2x^2}$

21. $y = (Ce^{3x^2})^2$ **23.** $y = \sqrt{e^{x+C} + 9}$ **25.** $y = \ln(x^2 + 4) - C$

27.

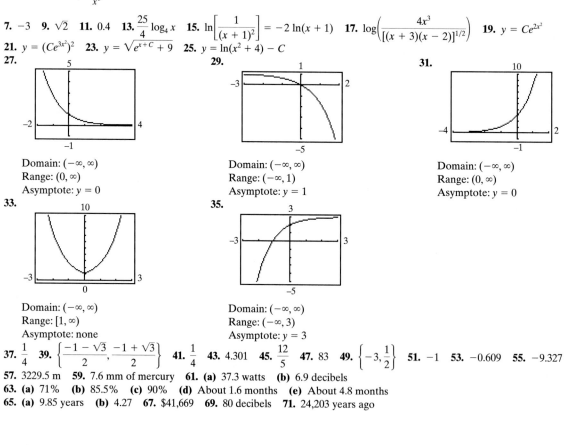

Domain: $(-\infty, \infty)$
Range: $(0, \infty)$
Asymptote: $y = 0$

29.

Domain: $(-\infty, \infty)$
Range: $(-\infty, 1)$
Asymptote: $y = 1$

31.

Domain: $(-\infty, \infty)$
Range: $(0, \infty)$
Asymptote: $y = 0$

33.

Domain: $(-\infty, \infty)$
Range: $[1, \infty)$
Asymptote: none

35.

Domain: $(-\infty, \infty)$
Range: $(-\infty, 3)$
Asymptote: $y = 3$

37. $\dfrac{1}{4}$ **39.** $\left\{\dfrac{-1 - \sqrt{3}}{2}, \dfrac{-1 + \sqrt{3}}{2}\right\}$ **41.** $\dfrac{1}{4}$ **43.** 4.301   **45.** $\dfrac{12}{5}$ **47.** 83   **49.** $\left\{-3, \dfrac{1}{2}\right\}$ **51.** -1 **53.** -0.609 **55.** -9.327

57. 3229.5 m   **59.** 7.6 mm of mercury   **61. (a)** 37.3 watts   **(b)** 6.9 decibels
63. (a) 71%   **(b)** 85.5%   **(c)** 90%   **(d)** About 1.6 months   **(e)** About 4.8 months
65. (a) 9.85 years   **(b)** 4.27   **67.** $41,669   **69.** 80 decibels   **71.** 24,203 years ago

73. (a)

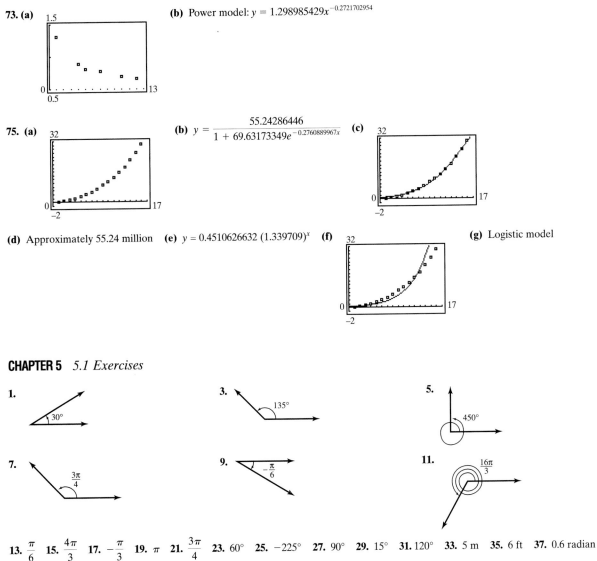

(b) Power model: $y = 1.298985429x^{-0.2721702954}$

75. (a)

(b) $y = \dfrac{55.24286446}{1 + 69.63173349e^{-0.2760889967x}}$

(c)

(d) Approximately 55.24 million   **(e)** $y = 0.4510626632\,(1.339709)^x$ **(f)**

(g) Logistic model

CHAPTER 5 *5.1 Exercises*

1. 30°

3. 135°

5. 450°

7. $\dfrac{3\pi}{4}$

9. $-\dfrac{\pi}{6}$

11. $\dfrac{16\pi}{3}$

13. $\dfrac{\pi}{6}$ **15.** $\dfrac{4\pi}{3}$ **17.** $-\dfrac{\pi}{3}$ **19.** π **21.** $\dfrac{3\pi}{4}$ **23.** 60° **25.** −225° **27.** 90° **29.** 15° **31.** 120° **33.** 5 m **35.** 6 ft **37.** 0.6 radian

39. $\dfrac{\pi}{3} \approx 1.047$ in. **41.** 0.30 **43.** −0.70 **45.** 2.18 **47.** 5.93 **49.** 179.91° **51.** 587.28° **53.** 114.59° **55.** 362.11° **57.** 40.17°

59. 1.03° **61.** 9.15° **63.** 40°19′12″ **65.** 18°15′18″ **67.** 19°59′24″ **69.** $3\pi \approx 9.4248$ in.; $5\pi \approx 15.7080$ in.

71. $\omega = \dfrac{1}{60}$ radian/sec; $v = \dfrac{1}{12}$ cm/sec **73.** 452.5 rpm **75.** 37.7 in. **77.** 2292 mph **79.** $\dfrac{3}{4}$ rpm **81.** 2.86 mph

83. 31.47 revolutions/minute **85.** 1037 mi/hr

5.2 Exercises

1. $\sin\theta = \dfrac{4}{5}$; $\cos\theta = -\dfrac{3}{5}$; $\tan\theta = -\dfrac{4}{3}$; $\csc\theta = \dfrac{5}{4}$; $\sec\theta = -\dfrac{5}{3}$; $\cot\theta = -\dfrac{3}{4}$

3. $\sin\theta = -\dfrac{3\sqrt{13}}{13}$; $\cos\theta = \dfrac{2\sqrt{13}}{13}$; $\tan\theta = -\dfrac{3}{2}$; $\csc\theta = -\dfrac{\sqrt{13}}{3}$; $\sec\theta = \dfrac{\sqrt{13}}{2}$; $\cot\theta = -\dfrac{2}{3}$

5. $\sin\theta = -\dfrac{\sqrt{2}}{2}$; $\cos\theta = -\dfrac{\sqrt{2}}{2}$; $\tan\theta = 1$; $\csc\theta = -\sqrt{2}$; $\sec\theta = -\sqrt{2}$; $\cot\theta = 1$

7. $\sin\theta = -\dfrac{2\sqrt{13}}{13}$; $\cos\theta = -\dfrac{3\sqrt{13}}{13}$; $\tan\theta = \dfrac{2}{3}$; $\csc\theta = -\dfrac{\sqrt{13}}{2}$; $\sec\theta = -\dfrac{\sqrt{13}}{3}$; $\cot\theta = \dfrac{3}{2}$

6.3 Exercises

1. (a) $\dfrac{24}{25}$ (b) $\dfrac{7}{25}$ (c) $\dfrac{\sqrt{10}}{10}$ (d) $\dfrac{3\sqrt{10}}{10}$ **3.** (a) $\dfrac{24}{25}$ (b) $-\dfrac{7}{25}$ (c) $\dfrac{2\sqrt5}{5}$ (d) $-\dfrac{\sqrt5}{5}$ **5.** (a) $-\dfrac{2\sqrt2}{3}$ (b) $\dfrac{1}{3}$ (c) $\sqrt{\dfrac{3+\sqrt6}{6}}$

(d) $\sqrt{\dfrac{3-\sqrt6}{6}}$ **7.** (a) $\dfrac{4\sqrt2}{9}$ (b) $-\dfrac{7}{9}$ (c) $\dfrac{\sqrt3}{3}$ (d) $\dfrac{\sqrt6}{3}$ **9.** (a) $-\dfrac{4}{5}$ (b) $\dfrac{3}{5}$ (c) $\sqrt{\dfrac{5+2\sqrt5}{10}}$ (d) $\sqrt{\dfrac{5-2\sqrt5}{10}}$ **11.** (a) $-\dfrac{3}{5}$

(b) $-\dfrac{4}{5}$ (c) $\dfrac{1}{2}\sqrt{\dfrac{10-\sqrt{10}}{5}}$ (d) $-\dfrac{1}{2}\sqrt{\dfrac{10+\sqrt{10}}{5}}$ **13.** $\dfrac{\sqrt{2-\sqrt2}}{2}$ **15.** $1-\sqrt2$ **17.** $-\dfrac{\sqrt{2+\sqrt3}}{2}$ **19.** $\dfrac{2}{\sqrt{2+\sqrt2}}=(2-\sqrt2)\sqrt{2+\sqrt2}$

21. $-\dfrac{\sqrt{2-\sqrt2}}{2}$ **23.** $\sin^4\theta=(\sin^2\theta)^2=\left(\dfrac{1-\cos 2\theta}{2}\right)^2=\dfrac{1}{4}(1-2\cos 2\theta+\cos^2\theta)=\dfrac{1}{4}-\dfrac{1}{2}\cos 2\theta+\dfrac{1}{4}\cos^2 2\theta$

$=\dfrac{1}{4}-\dfrac{1}{2}\cos 2\theta+\dfrac{1}{4}\left(\dfrac{1+\cos 4\theta}{2}\right)=\dfrac{1}{4}-\dfrac{1}{2}\cos 2\theta+\dfrac{1}{8}+\dfrac{1}{8}\cos 4\theta=\dfrac{3}{8}-\dfrac{1}{2}\cos 2\theta+\dfrac{1}{8}\cos 4\theta$

25. $\sin 4\theta=\sin 2(2\theta)=2\sin 2\theta\cos 2\theta=(4\sin\theta\cos\theta)(1-2\sin^2\theta)=4\sin\theta\cos\theta-8\sin^3\theta\cos\theta=(\cos\theta)(4\sin\theta-8\sin^3\theta)$

27. $16\sin^5\theta-20\sin^3\theta+5\sin\theta$ **29.** $\cos^4\theta-\sin^4\theta=(\cos^2\theta+\sin^2\theta)(\cos^2\theta-\sin^2\theta)=\cos 2\theta$

31. $\cot 2\theta=\dfrac{1}{\tan 2\theta}=\dfrac{1}{\dfrac{2\tan\theta}{1-\tan^2\theta}}=\dfrac{1-\tan^2\theta}{2\tan\theta}=\dfrac{1-\dfrac{1}{\cot^2\theta}}{2\left(\dfrac{1}{\cot\theta}\right)}=\dfrac{\dfrac{\cot^2\theta-1}{\cot^2\theta}}{\dfrac{2}{\cot\theta}}=\dfrac{\cot^2\theta-1}{\cot^2\theta}\cdot\dfrac{\cot\theta}{2}=\dfrac{\cot^2\theta-1}{2\cot\theta}$

33. $\sec 2\theta=\dfrac{1}{\cos 2\theta}=\dfrac{1}{2\cos^2\theta-1}=\dfrac{1}{\dfrac{2}{\sec^2\theta}-1}=\dfrac{1}{\dfrac{2-\sec^2\theta}{\sec^2\theta}}=\dfrac{\sec^2\theta}{2-\sec^2\theta}$ **35.** $\cos^2 2\theta-\sin^2 2\theta=\cos[2(2\theta)]=\cos 4\theta$

37. $\dfrac{\cos 2\theta}{1+\sin 2\theta}=\dfrac{\cos^2\theta-\sin^2\theta}{1+2\sin\theta\cos\theta}=\dfrac{(\cos\theta-\sin\theta)(\cos\theta+\sin\theta)}{\sin^2+\cos^2\theta+2\sin\theta\cos\theta}=\dfrac{(\cos\theta-\sin\theta)(\cos\theta+\sin\theta)}{(\sin\theta+\cos\theta)(\sin\theta+\cos\theta)}=\dfrac{\cos\theta-\sin\theta}{\cos\theta+\sin\theta}$

$=\dfrac{\dfrac{\cos\theta-\sin\theta}{\sin\theta}}{\dfrac{\cos\theta+\sin\theta}{\sin\theta}}=\dfrac{\dfrac{\cos\theta}{\sin\theta}-\dfrac{\sin\theta}{\sin\theta}}{\dfrac{\cos\theta}{\sin\theta}+\dfrac{\sin\theta}{\sin\theta}}=\dfrac{\cot\theta-1}{\cos\theta+1}$ **39.** $\sec^2\dfrac{\theta}{2}=\dfrac{1}{\cos^2\left(\dfrac{\theta}{2}\right)}=\dfrac{1}{\dfrac{1+\cos\theta}{2}}=\dfrac{2}{1+\cos\theta}$

41. $\cot^2\dfrac{\theta}{2}=\dfrac{1}{\tan^2\left(\dfrac{\theta}{2}\right)}=\dfrac{1}{\dfrac{1-\cos\theta}{1+\cos\theta}}=\dfrac{1+\cos\theta}{1-\cos\theta}=\dfrac{1+\dfrac{1}{\sec\theta}}{1-\dfrac{1}{\sec\theta}}=\dfrac{\dfrac{\sec\theta+1}{\sec\theta}}{\dfrac{\sec\theta-1}{\sec\theta}}=\dfrac{\sec\theta+1}{\sec\theta}\cdot\dfrac{\sec\theta}{\sec\theta-1}=\dfrac{\sec\theta+1}{\sec\theta-1}$

43. $\dfrac{1-\tan^2\left(\dfrac{\theta}{2}\right)}{1+\tan^2\left(\dfrac{\theta}{2}\right)}=\dfrac{1-\dfrac{1-\cos\theta}{1+\cos\theta}}{1+\dfrac{1-\cos\theta}{1+\cos\theta}}=\dfrac{\dfrac{1+\cos\theta-(1-\cos\theta)}{1+\cos\theta}}{\dfrac{1+\cos\theta+1-\cos\theta}{1+\cos\theta}}=\dfrac{\dfrac{2\cos\theta}{1+\cos\theta}}{\dfrac{2}{1+\cos\theta}}=\dfrac{2\cos\theta}{1+\cos\theta}\cdot\dfrac{1+\cos\theta}{2}=\cos\theta$

45. $\dfrac{\sin 3\theta}{\sin\theta}-\dfrac{\cos 3\theta}{\cos\theta}=\dfrac{\sin 3\theta\cos\theta-\cos 3\theta\sin\theta}{\sin\theta\cos\theta}=\dfrac{\sin(3\theta-\theta)}{\dfrac{1}{2}(2\sin\theta\cos\theta)}=\dfrac{2\sin 2\theta}{\sin 2\theta}=2$

47. $\tan 3\theta=\tan(\theta+2\theta)=\dfrac{\tan\theta+\tan 2\theta}{1-\tan\theta\tan 2\theta}=\dfrac{\tan\theta+\dfrac{2\tan\theta}{1-\tan^2\theta}}{1-\dfrac{\tan\theta(2\tan\theta)}{1-\tan^2\theta}}=\dfrac{\tan\theta-\tan^3\theta+2\tan\theta}{1-\tan^2\theta-2\tan^2\theta}=\dfrac{3\tan\theta-\tan^3\theta}{1-3\tan^2\theta}$

49. $\dfrac{\sqrt3}{2}$ **51.** $\dfrac{7}{25}$ **53.** $\dfrac{24}{7}$ **55.** $\dfrac{24}{25}$ **57.** $\dfrac{1}{5}$ **59.** $\dfrac{25}{7}$ **61.** 4 **63.** $-\dfrac{1}{4}$ **65.** $A=\dfrac{1}{2}h\left(\dfrac{1}{2}\text{base}\right)+\dfrac{1}{2}h\left(\dfrac{1}{2}\text{base}\right)=s\cos\dfrac{\theta}{2}\cdot s\sin\dfrac{\theta}{2}=\dfrac{1}{2}s^2\sin\theta$

67.

69. $\sin\dfrac{\pi}{24}=\dfrac{\sqrt2}{4}\sqrt{4-\sqrt6-\sqrt2};\ \cos\dfrac{\pi}{24}=\dfrac{\sqrt2}{4}\sqrt{4+\sqrt6+\sqrt2}$

71. $\sin^3\theta + \sin^3(\theta + 120°) + \sin^3(\theta + 240°) = \sin^3\theta + (\sin\theta\cos 120° + \cos\theta\sin 120°)^3 + (\sin\theta\cos 240° + \cos\theta\sin 240°)^3$

$$= \sin^3\theta + \left(-\frac{1}{2}\sin\theta + \frac{\sqrt{3}}{2}\cos\theta\right)^3 + \left(-\frac{1}{2}\sin\theta - \frac{\sqrt{3}}{2}\cos\theta\right)^3$$

$$= \sin^3\theta + \frac{1}{8}(3\sqrt{3}\cos^3\theta - 9\cos^2\theta\sin\theta + 3\sqrt{3}\cos\theta\sin^2\theta - \sin^3\theta) - \frac{1}{8}(\sin^3\theta + 3\sqrt{3}\sin^2\theta\cos\theta + 9\sin\theta\cos^2\theta + 3\sqrt{3}\cos^3\theta)$$

$$= \frac{3}{4}\sin^3\theta - \frac{9}{4}\cos^2\theta\sin\theta = \frac{3}{4}[\sin^3\theta - 3\sin\theta(1 - \sin^2\theta)] = \frac{3}{4}(4\sin^3\theta - 3\sin\theta) = -\frac{3}{4}\sin 3\theta$$

73. $\frac{1}{2}(\ln|1 - \cos 2\theta| - \ln 2) = \ln\left(\frac{|1 - \cos 2\theta|}{2}\right)^{1/2} = \ln|\sin^2\theta|^{1/2} = \ln|\sin\theta|$

75. (a) $R = \frac{v_0^2\sqrt{2}}{16}(\sin\theta\cos\theta - \cos^2\theta) = \frac{v_0^2\sqrt{2}}{16}\left(\frac{1}{2}\sin 2\theta - \frac{1 + \cos 2\theta}{2}\right)$ **(b)**

$$= \frac{v_0^2\sqrt{2}}{32}(\sin 2\theta - \cos 2\theta - 1)$$

(c) $\theta = 67.5°$ makes R largest

45°   90°

6.4 Exercises

1. $\frac{1}{2}(\cos 2\theta - \cos 6\theta)$ **3.** $\frac{1}{2}(\sin 6\theta + \sin 2\theta)$ **5.** $\frac{1}{2}(\cos 8\theta + \cos 2\theta)$ **7.** $\frac{1}{2}(\cos\theta - \cos 3\theta)$ **9.** $\frac{1}{2}(\sin 2\theta + \sin\theta)$ **11.** $2\sin\theta\cos 3\theta$

13. $2\cos 3\theta\cos\theta$ **15.** $2\sin 2\theta\cos\theta$ **17.** $2\sin\theta\sin\left(\frac{\theta}{2}\right)$ **19.** $\frac{\sin\theta + \sin 3\theta}{2\sin 2\theta} = \frac{2\sin 2\theta\cos(-\theta)}{2\sin 2\theta} = \cos(-\theta) = \cos\theta$

21. $\frac{\sin 4\theta + \sin 2\theta}{\cos 4\theta + \cos 2\theta} = \frac{2\sin 3\theta\cos\theta}{2\cos 3\theta\cos\theta} = \frac{\sin 3\theta}{\cos 3\theta} = \tan 3\theta$ **23.** $\frac{\cos\theta - \cos 3\theta}{\sin\theta + \sin 3\theta} = \frac{-2\sin 2\theta\sin(-\theta)}{2\sin 2\theta\cos(-\theta)} = \frac{\sin\theta}{\cos\theta} = \tan\theta$

25. $\sin\theta(\sin\theta + \sin 3\theta) = \sin\theta[2\sin 2\theta\cos(-\theta)] = 2\sin 2\theta\sin\theta\cos\theta = \cos\theta(2\sin 2\theta\sin\theta) = \cos\theta\left[2 \cdot \frac{1}{2}(\cos\theta - \cos 3\theta)\right]$

$= \cos\theta(\cos\theta - \cos 3\theta)$

27. $\frac{\sin 4\theta + \sin 8\theta}{\cos 4\theta + \cos 8\theta} = \frac{2\sin 6\theta\cos(-2\theta)}{2\cos 6\theta\cos(-2\theta)} = \frac{\sin 6\theta}{\cos 6\theta} = \tan 6\theta$

29. $\frac{\sin 4\theta + \sin 8\theta}{\sin 4\theta - \sin 8\theta} = \frac{2\sin 6\theta\cos(-2\theta)}{2\sin(-2\theta)\cos 6\theta} = \frac{\sin 6\theta}{\cos 6\theta} \cdot \frac{\cos 2\theta}{-\sin 2\theta} = \tan 6\theta(-\cot 2\theta) = -\frac{\tan 6\theta}{\tan 2\theta}$

31. $\frac{\sin\alpha + \sin\beta}{\sin\alpha - \sin\beta} = \frac{2\sin\dfrac{\alpha + \beta}{2}\cos\dfrac{\alpha - \beta}{2}}{2\sin\dfrac{\alpha - \beta}{2}\cos\dfrac{\alpha + \beta}{2}} = \frac{\sin\dfrac{\alpha + \beta}{2}}{\cos\dfrac{\alpha + \beta}{2}} \cdot \frac{\cos\dfrac{\alpha - \beta}{2}}{\sin\dfrac{\alpha - \beta}{2}} = \tan\frac{\alpha + \beta}{2}\cot\frac{\alpha - \beta}{2}$

33. $\frac{\sin\alpha + \sin\beta}{\cos\alpha + \cos\beta} = \frac{2\sin\dfrac{\alpha + \beta}{2}\cos\dfrac{\alpha - \beta}{2}}{2\cos\dfrac{\alpha + \beta}{2}\cos\dfrac{\alpha - \beta}{2}} = \frac{\sin\dfrac{\alpha + \beta}{2}}{\cos\dfrac{\alpha + \beta}{2}} = \tan\frac{\alpha + \beta}{2}$

35. $1 + \cos 2\theta + \cos 4\theta + \cos 6\theta = (1 + \cos 6\theta) + (\cos 2\theta + \cos 4\theta) = 2\cos^2 3\theta + 2\cos 3\theta\cos(-\theta) = 2\cos 3\theta(\cos 3\theta + \cos\theta)$

$= 2\cos 3\theta(2\cos 2\theta\cos\theta) = 4\cos\theta\cos 2\theta\cos 3\theta$

37. (a) $y = 2\sin 2061\pi t\cos 357\pi t$ **(b)**   **(c)** $y_{\max} = 2$

39. $\sin 2\alpha + \sin 2\beta + \sin 2\gamma = 2\sin(\alpha + \beta)\cos(\alpha - \beta) + \sin 2\gamma = 2\sin(\alpha + \beta)\cos(\alpha - \beta) + 2\sin\gamma\cos\gamma$

$= 2\sin(\pi - \gamma)\cos(\alpha - \beta) + 2\sin\gamma\cos\gamma = 2\sin\gamma\cos(\alpha - \beta) + 2\sin\gamma\cos\gamma = 2\sin\gamma[\cos(\alpha - \beta) + \cos\gamma]$

$= 2\sin\gamma\left(2\cos\frac{\alpha - \beta + \gamma}{2}\cos\frac{\alpha - \beta - \gamma}{2}\right) = 4\sin\gamma\cos\frac{\pi - 2\beta}{2}\cos\frac{2\alpha - \pi}{2} = 4\sin\gamma\cos\left(\frac{\pi}{2} - \beta\right)\cos\left(\alpha - \frac{\pi}{2}\right)$

$= 4\sin\gamma\sin\beta\sin\alpha$

41.
$$\sin(\alpha - \beta) = \sin \alpha \cos \beta - \cos \alpha \sin \beta$$
$$\sin(\alpha + \beta) = \sin \alpha \cos \beta + \cos \alpha \sin \beta$$
$$\sin(\alpha - \beta) + \sin(\alpha + \beta) = 2 \sin \alpha \cos \beta$$
$$\sin \alpha \cos \beta = \frac{1}{2}[\sin(\alpha + \beta) + \sin(\alpha - \beta)]$$

43. $2 \cos \dfrac{\alpha + \beta}{2} \cos \dfrac{\alpha - \beta}{2} = 2 \cdot \dfrac{1}{2}\left[\cos\left(\dfrac{\alpha + \beta}{2} + \dfrac{\alpha - \beta}{2}\right) + \cos\left(\dfrac{\alpha + \beta}{2} - \dfrac{\alpha - \beta}{2}\right)\right] = \cos \dfrac{2\alpha}{2} + \cos \dfrac{2\beta}{2} = \cos \alpha + \cos \beta$

6.5 Exercises

1. $\dfrac{\pi}{6}, \dfrac{5\pi}{6}$ **3.** $\dfrac{5\pi}{6}, \dfrac{11\pi}{6}$ **5.** $\dfrac{\pi}{2}, \dfrac{3\pi}{2}$ **7.** $\dfrac{\pi}{2}, \dfrac{7\pi}{6}, \dfrac{11\pi}{6}$ **9.** $\dfrac{3\pi}{4}, \dfrac{7\pi}{4}$ **11.** $\dfrac{4\pi}{9}, \dfrac{8\pi}{9}, \dfrac{16\pi}{9}$ **13.** $0.4115168, 2.73007581$

15. $1.3734008, 4.5149934$ **17.** $2.6905658, 3.5926195$ **19.** $1.8234766, 4.4597087$ **21.** $\dfrac{\pi}{2}, \dfrac{2\pi}{3}, \dfrac{4\pi}{3}, \dfrac{3\pi}{2}$

23. $\dfrac{\pi}{2}, \dfrac{7\pi}{6}, \dfrac{11\pi}{6}$ **25.** $0, \dfrac{\pi}{4}, \dfrac{5\pi}{4}$ **27.** $\dfrac{\pi}{4}, \dfrac{5\pi}{4}$ **29.** $0, \dfrac{\pi}{3}, \pi, \dfrac{5\pi}{3}$ **31.** $\dfrac{\pi}{2}, \dfrac{3\pi}{2}$ **33.** $0, \dfrac{2\pi}{3}, \dfrac{4\pi}{3}$ **35.** $0, \dfrac{\pi}{3}, \dfrac{2\pi}{3}, \dfrac{\pi}{2}, \dfrac{4\pi}{3}, \pi, \dfrac{5\pi}{3}, \dfrac{3\pi}{2}$

37. $0, \dfrac{\pi}{5}, \dfrac{2\pi}{5}, \dfrac{3\pi}{5}, \dfrac{4\pi}{5}, \pi, \dfrac{6\pi}{5}, \dfrac{7\pi}{5}, \dfrac{8\pi}{5}, \dfrac{9\pi}{5}$ **39.** $\dfrac{\pi}{6}, \dfrac{5\pi}{6}, \dfrac{3\pi}{2}$ **41.** $\dfrac{\pi}{3}, \dfrac{5\pi}{3}$ **43.** No real solutions **45.** No real solutions **47.** $\dfrac{\pi}{2}, \dfrac{7\pi}{6}$

49. $0, \dfrac{\pi}{3}, \pi, \dfrac{5\pi}{3}$

51.

0, −1.29

53.

−2.24, 0, 2.24

55.

−0.82, 0.82

57. $-1.30, 1.97, 3.83$ **59.** 0.52 **61.** 1.25 **63.** $-1.02, 1.02$ **65.** $0, 2.14$ **67.** $0.76, 1.34$ **69. (a)** $60°$ **(b)** $60°$ **(c)** $A(60°) = 12\sqrt{3}$ sq in.
(d) θmax $= 60°$
Maximum Area $= 20.78$ sq in.
71. $2.02, 4.91$ **73. (a)** $29.99°$ **(b)**

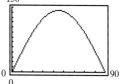

(d) 123.6 meters **75.** $28.9°$ **77.** Yes; it varies from 1.27 to 1.34 **79.** 1.47

81. If θ is the original angle of incidence and ϕ is the angle of refraction, then $\dfrac{\sin \theta}{\sin \phi} = n_2$. The angle of incidence of the emerging

beam is also ϕ, and the index of refraction is $\dfrac{1}{n_2}$. Thus, θ is the angle of refraction of the emerging beam.

Fill-in-the-Blank Items

1. identity; conditional **2.** $-$ **3.** $+$ **4.** $\sin^2 \theta$; $2 \cos^2 \theta$; $2 \sin^2 \theta$ **5.** $1 - \cos \alpha$

True/False Items

1. T **2.** F **3.** T **4.** F **5.** F **6.** F

Review Exercises

1. $\tan \theta \cot \theta - \sin^2 \theta = 1 - \sin^2 \theta = \cos^2 \theta$ **3.** $\cos^2 \theta(1 + \tan^2 \theta) = \cos^2 \theta \sec^2 \theta = 1$
5. $4 \cos^2 \theta + 3 \sin^2 \theta = \cos^2 \theta + 3(\cos^2 \theta + \sin^2 \theta) = 3 + \cos^2 \theta$
7. $\dfrac{1 - \cos \theta}{\sin \theta} + \dfrac{\sin \theta}{1 - \cos \theta} = \dfrac{(1 - \cos \theta)^2 + \sin^2 \theta}{\sin \theta(1 - \cos \theta)} = \dfrac{1 - 2 \cos \theta + \cos^2 \theta + \sin^2 \theta}{\sin \theta(1 - \cos \theta)} = \dfrac{2(1 - \cos \theta)}{\sin \theta(1 - \cos \theta)} = 2 \csc \theta$

9. $\dfrac{\cos\theta}{\cos\theta - \sin\theta} = \dfrac{\dfrac{\cos\theta}{\cos\theta}}{\dfrac{\cos\theta - \sin\theta}{\cos\theta}} = \dfrac{1}{1 - \dfrac{\sin\theta}{\cos\theta}} = \dfrac{1}{1 - \tan\theta}$

11. $\dfrac{\csc\theta}{1 + \csc\theta} = \dfrac{\dfrac{1}{\sin\theta}}{1 + \dfrac{1}{\sin\theta}} = \dfrac{1}{1 + \sin\theta} = \dfrac{1}{1 + \sin\theta} \cdot \dfrac{1 - \sin\theta}{1 - \sin\theta} = \dfrac{1 - \sin\theta}{1 - \sin^2\theta} = \dfrac{1 - \sin\theta}{\cos^2\theta}$

13. $\csc\theta - \sin\theta = \dfrac{1}{\sin\theta} - \sin\theta = \dfrac{1 - \sin^2\theta}{\sin\theta} = \dfrac{\cos^2\theta}{\sin\theta} = \cos\theta \cdot \dfrac{\cos\theta}{\sin\theta} = \cos\theta\cot\theta$

15. $\dfrac{1 - \sin\theta}{\sec\theta} = \cos\theta(1 - \sin\theta) \cdot \dfrac{1 + \sin\theta}{1 + \sin\theta} = \dfrac{\cos\theta(1 - \sin^2\theta)}{1 + \sin\theta} = \dfrac{\cos^3\theta}{1 + \sin\theta}$

17. $\cot\theta - \tan\theta = \dfrac{\cos\theta}{\sin\theta} - \dfrac{\sin\theta}{\cos\theta} = \dfrac{\cos^2\theta - \sin^2\theta}{\sin\theta\cos\theta} = \dfrac{1 - 2\sin^2\theta}{\sin\theta\cos\theta}$

19. $\dfrac{\cos(\alpha + \beta)}{\cos\alpha\sin\beta} = \dfrac{\cos\alpha\cos\beta - \sin\alpha\sin\beta}{\cos\alpha\sin\beta} = \dfrac{\cos\alpha\cos\beta}{\cos\alpha\sin\beta} - \dfrac{\sin\alpha\sin\beta}{\cos\alpha\sin\beta} = \cot\beta - \tan\alpha$

21. $\dfrac{\cos(\alpha - \beta)}{\cos\alpha\cos\beta} = \dfrac{\cos\alpha\cos\beta + \sin\alpha\sin\beta}{\cos\alpha\cos\beta} = \dfrac{\cos\alpha\cos\beta}{\cos\alpha\cos\beta} + \dfrac{\sin\alpha\sin\beta}{\cos\alpha\cos\beta} = 1 + \tan\alpha\tan\beta$

23. $(1 + \cos\theta)\left(\tan\dfrac{\theta}{2}\right) = \left(2\cos^2\dfrac{\theta}{2}\right)\dfrac{\sin\left(\dfrac{\theta}{2}\right)}{\cos\left(\dfrac{\theta}{2}\right)} = 2\sin\dfrac{\theta}{2}\cos\dfrac{\theta}{2} = \sin\theta$

25. $2\cot\theta\cot 2\theta = 2\left(\dfrac{\cos\theta}{\sin\theta}\right)\left(\dfrac{\cos 2\theta}{\sin 2\theta}\right) = \dfrac{2\cos\theta(\cos^2\theta - \sin^2\theta)}{2\sin^2\theta\cos\theta} = \dfrac{\cos^2\theta - \sin^2\theta}{\sin^2\theta} = \cot^2\theta - 1$

27. $1 - 8\sin^2\theta\cos^2\theta = 1 - 2(2\sin\theta\cos\theta)^2 = 1 - 2\sin^2 2\theta = \cos 4\theta$ **29.** $\dfrac{\sin 2\theta + \sin 4\theta}{\cos 2\theta + \cos 4\theta} = \dfrac{2\sin 3\theta\cos(-\theta)}{2\cos 3\theta\cos(-\theta)} = \tan 3\theta$

31. $\dfrac{\cos 2\theta - \cos 4\theta}{\cos 2\theta + \cos 4\theta} - \tan\theta\tan 3\theta = \dfrac{-2\sin 3\theta\sin(-\theta)}{2\cos 3\theta\cos(-\theta)} - \tan\theta\tan 3\theta = \tan 3\theta\tan\theta - \tan\theta\tan 3\theta = 0$ **33.** $\dfrac{1}{4}(\sqrt{6} - \sqrt{2})$

35. $\dfrac{1}{4}(\sqrt{6} - \sqrt{2})$ **37.** $\dfrac{1}{2}$ **39.** $\sqrt{\dfrac{2 - \sqrt{2}}{2 + \sqrt{2}}} = \sqrt{2} - 1$ **41. (a)** $-\dfrac{33}{65}$ **(b)** $-\dfrac{56}{65}$ **(c)** $-\dfrac{63}{65}$ **(d)** $\dfrac{33}{56}$ **(e)** $\dfrac{24}{25}$ **(f)** $\dfrac{119}{169}$ **(g)** $\dfrac{5\sqrt{26}}{26}$

(h) $\dfrac{2\sqrt{5}}{5}$ **43. (a)** $-\dfrac{16}{65}$ **(b)** $-\dfrac{63}{65}$ **(c)** $-\dfrac{56}{65}$ **(d)** $\dfrac{16}{63}$ **(e)** $\dfrac{24}{25}$ **(f)** $\dfrac{119}{169}$ **(g)** $\dfrac{\sqrt{26}}{26}$ **(h)** $-\dfrac{\sqrt{10}}{10}$ **45. (a)** $-\dfrac{63}{65}$ **(b)** $\dfrac{16}{65}$ **(c)** $\dfrac{33}{65}$

(d) $-\dfrac{63}{16}$ **(e)** $\dfrac{24}{25}$ **(f)** $-\dfrac{119}{169}$ **(g)** $\dfrac{2\sqrt{13}}{13}$ **(h)** $-\dfrac{\sqrt{10}}{10}$ **47. (a)** $\dfrac{(-\sqrt{3} - 2\sqrt{2})}{6}$ **(b)** $\dfrac{(1 - 2\sqrt{6})}{6}$ **(c)** $\dfrac{(-\sqrt{3} + 2\sqrt{2})}{6}$

(d) $\dfrac{(-\sqrt{3} - 2\sqrt{2})}{(1 - 2\sqrt{6})} = \dfrac{(8\sqrt{2} + 9\sqrt{3})}{23}$ **(e)** $-\dfrac{\sqrt{3}}{2}$ **(f)** $-\dfrac{7}{9}$ **(g)** $\dfrac{\sqrt{3}}{3}$ **(h)** $\dfrac{\sqrt{3}}{2}$ **49. (a)** 1   **(b)** 0   **(c)** $-\dfrac{1}{9}$ **(d)** Not defined

(e) $\dfrac{4\sqrt{5}}{9}$ **(f)** $-\dfrac{1}{9}$ **(g)** $\dfrac{\sqrt{30}}{6}$ **(h)** $-\dfrac{\sqrt{6}\sqrt{3} - \sqrt{5}}{6}$ **51.** $\dfrac{\pi}{3}, \dfrac{5\pi}{3}$ **53.** $\dfrac{3\pi}{4}, \dfrac{5\pi}{4}$ **55.** $\dfrac{3\pi}{4}, \dfrac{7\pi}{4}$ **57.** $0, \dfrac{\pi}{2}, \pi, \dfrac{3\pi}{2}$

59. $1.1197695, \pi - 1.1197695$ **61.** $0, \pi$ **63.** $0, \dfrac{2\pi}{3}, \pi, \dfrac{4\pi}{3}$ **65.** $0, \dfrac{\pi}{6}, \dfrac{5\pi}{6}$ **67.** $\dfrac{\pi}{6}, \dfrac{\pi}{2}, \dfrac{5\pi}{6}$ **69.** $\dfrac{\pi}{2}, \pi$ **71.** 1.11   **73.** 0.86   **75.** 2.21

CHAPTER 7   *7.1 Exercises*

1. $a \approx 13.74, c \approx 14.62, \alpha = 70°$ **3.** $b \approx 5.03, c \approx 7.83, \alpha = 50°$ **5.** $a \approx 0.71, c \approx 4.06, \beta = 80°$ **7.** $b \approx 10.72, c \approx 11.83, \beta = 65°$
9. $b \approx 3.08, a \approx 8.46, \alpha = 70°$ **11.** $c \approx 5.83, \alpha \approx 59.0°, \beta \approx 31.0°$ **13.** $b \approx 4.58, \alpha \approx 23.6°, \beta \approx 66.4°$ **15.** 1.72 in., 2.46 in.
17. 6.10 in. or 8.72 in.   **19.** 23.6° and 66.4°   **21.** 70.02 ft   **23.** 985.91 ft   **25.** 137 m   **27.** 20.67 ft   **29.** 449.36 ft   **31.** 80.5°   **33.** 30 ft
35. 530 ft   **37.** 555 ft   **39. (a)** 112 ft/sec or 76.3 mph   **(b)** 82.41 ft/sec or 56.2 mph   **(c)** under 18.8°   **41. (a)** 130°   **(b)** 103.4°
43. 14.9°   **45. (a)** 3.1 mi   **(b)** 3.2 mi   **(c)** 3.8 mi   **47. (a)** $\cos\dfrac{\theta}{2} = \dfrac{3960}{3960 + h}$ **(b)** $d = 3960\,\theta$ **(c)** $\cos\dfrac{d}{7920} = \dfrac{3960}{3960 + h}$

(d) 206 mi   **(e)** 2990 miles

7.2 Exercises

1. $a = 3.23, b = 3.55, \alpha = 40°$ **3.** $a = 3.25, c = 4.23, \beta = 45°$ **5.** $\gamma = 95°, c = 9.86, a = 6.36$ **7.** $\alpha = 40°, a = 2, c = 3.06$
9. $\gamma = 120°, b = 1.06, c = 2.69$ **11.** $\alpha = 100°, a = 5.24, c = 0.92$ **13.** $\beta = 40°, a = 5.64, b = 3.86$ **15.** $\gamma = 100°, a = 1.31, b = 1.31$
17. One triangle; $\beta = 30.7°, \gamma = 99.3°, c = 3.86$ **19.** One triangle; $\gamma = 36.2°, \alpha = 43.8°, a = 3.51$ **21.** No triangle

8.6 Exercises

1. All points of the form $(x, 0, z)$. **3.** All points of the form $(x, 3, z)$. **5.** All points of the form $(-4, y, z)$. **7.** All points of the form $(1, 2, z)$.
9. $\sqrt{21}$ **11.** $\sqrt{33}$ **13.** $\sqrt{26}$ **15.** $(2, 0, 0)$; $(2, 1, 0)$; $(0, 1, 0)$; $(2, 0, 3)$; $(0, 1, 3)$; $(0, 0, 3)$ **17.** $(1, 4, 3)$; $(3, 2, 3)$; $(3, 4, 3)$; $(3, 2, 5)$; $(1, 4, 5)$; $(1, 2, 5)$
19. $(-1, 2, 2)$; $(4, 0, 2)$; $(4, 2, 2)$; $(-1, 2, 5)$; $(4, 0, 5)$; $(-1, 0, 5)$ **21.** $\mathbf{v} = 3\mathbf{i} + 4\mathbf{j} - \mathbf{k}$ **23.** $2\mathbf{v} = 2\mathbf{i} + 4\mathbf{j} + \mathbf{k}$ **25.** $\mathbf{v} = 8\mathbf{i} - \mathbf{j} + 5\mathbf{k}$ **27.** 7 **29.** $\sqrt{3}$
31. $\sqrt{22}$ **33.** $2\mathbf{v} + 3\mathbf{w} = -\mathbf{j} - 2\mathbf{k}$ **35.** $\sqrt{105}$ **37.** $\sqrt{38} - \sqrt{17}$ **39.** $\dfrac{\mathbf{v}}{\|\mathbf{v}\|} = \mathbf{i}$ **41.** $\dfrac{\mathbf{v}}{\|\mathbf{v}\|} = \dfrac{3}{7}\mathbf{i} - \dfrac{6}{7}\mathbf{j} - \dfrac{2}{7}\mathbf{k}$ **43.** $\dfrac{\mathbf{v}}{\|\mathbf{v}\|} = \dfrac{\sqrt{3}}{3}\mathbf{i} + \dfrac{\sqrt{3}}{3}\mathbf{j} + \dfrac{\sqrt{3}}{3}\mathbf{k}$
45. $\mathbf{v} \cdot \mathbf{w} = 0$; $\theta = 90°$ **47.** $\mathbf{v} \cdot \mathbf{w} = -2$, $\theta \approx 100°$ **49.** $\mathbf{v} \cdot \mathbf{w} = 0$; $\theta = 90°$ **51.** $\mathbf{v} \cdot \mathbf{w} = 52$; $\theta = 0°$ **53.** $\alpha \approx 65°$; $\beta \approx 149°$; $\gamma \approx 107°$
55. $\alpha \approx 55°$; $\beta \approx 125°$; $\gamma \approx 125°$ **57.** $\alpha = 45°$; $\beta = 45°$, $\gamma = 90°$ **59.** $\alpha \approx 61°$; $\beta \approx 144°$; $\gamma \approx 71°$
61. $d(P_0, P) = \sqrt{(x - x_0)^2 + (y - y_0)^2 + (z - z_0)^2}$ **63.** $(x - 1)^2 + (y - 2)^2 + (z - 2)^2 = 4$ **65.** Center: $(-1, 1, 0)$; Radius: 2
$\qquad R = \sqrt{(x - x_0)^2 + (y - y_0)^2 + (z - z_0)^2}$
$\qquad R^2 = (x - x_0)^2 + (y - y_0)^2 + (z - z_0)^2$
67. Center: $(2, -2, 1)$; Radius: 3 **69.** Center: $(2, 0, -1)$; Radius: $\dfrac{3\sqrt{2}}{2}$ **71.** $\dfrac{6\sqrt{5}}{5}$ Joules **73.** 9

Fill-in-the-Blank Items

1. pole; polar axis **2.** -2 **3.** $r = 2 \cos \theta$ **4.** the polar axis (x-axis) **5.** magnitude or modulus; argument **6.** unit **7.** 0

True/False Items

1. F **2.** T **3.** F **4.** T **5.** T **6.** T **7.** T **8.** T

Review Exercises

1. $\left(\dfrac{3\sqrt{3}}{2}, \dfrac{3}{2}\right)$ **3.** $(1, \sqrt{3})$ **5.** $(0, 3)$

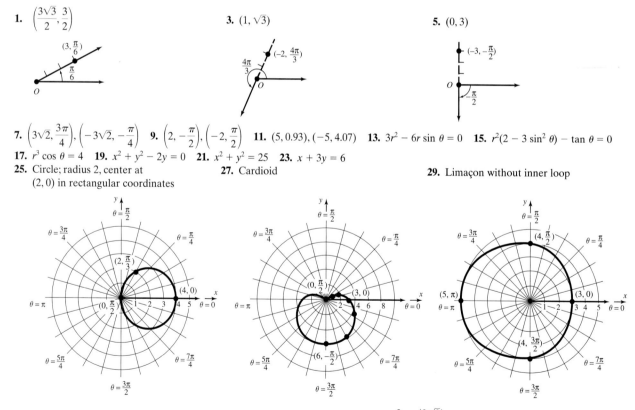

7. $\left(3\sqrt{2}, \dfrac{3\pi}{4}\right), \left(-3\sqrt{2}, -\dfrac{\pi}{4}\right)$ **9.** $\left(2, -\dfrac{\pi}{2}\right), \left(-2, \dfrac{\pi}{2}\right)$ **11.** $(5, 0.93), (-5, 4.07)$ **13.** $3r^2 - 6r \sin \theta = 0$ **15.** $r^2(2 - 3 \sin^2 \theta) - \tan \theta = 0$
17. $r^3 \cos \theta = 4$ **19.** $x^2 + y^2 - 2y = 0$ **21.** $x^2 + y^2 = 25$ **23.** $x + 3y = 6$
25. Circle; radius 2, center at $(2, 0)$ in rectangular coordinates **27.** Cardioid **29.** Limaçon without inner loop

31. $\sqrt{2}(\cos 225° + i \sin 225°)$ **33.** $5(\cos 323.1° + i \sin 323.1°)$ **35.** $-\sqrt{3} + i$ **37.** $-\dfrac{3}{2} + \left(\dfrac{3\sqrt{3}}{2}\right)i$ **39.** $0.098 - 0.017i$

41. $zw = \cos 130° + i \sin 130°$; $\dfrac{z}{w} = \cos 30° + i \sin 30°$ **43.** $zw = 6(\cos 0 + i \sin 0)$; $\dfrac{z}{w} = \dfrac{3}{2}\left(\cos\dfrac{8\pi}{5} + i \sin\dfrac{8\pi}{5}\right)$

45. $zw = 5(\cos 5° + i \sin 5°)$; $\dfrac{z}{w} = 5(\cos 15° + i \sin 15°)$ **47.** $\dfrac{27}{2} + \dfrac{27\sqrt{3}}{2}i$ **49.** $4i$ **51.** 64 **53.** $-527 - 336i$

55. $3, 3(\cos 120° + i \sin 120°), 3(\cos 240° + i \sin 240°)$ **57.** $\mathbf{v} = 2\mathbf{i} - 4\mathbf{j}$; $\|\mathbf{v}\| = 2\sqrt{5}$ **59.** $\mathbf{v} = -\mathbf{i} + 3\mathbf{j}$; $\|\mathbf{v}\| = \sqrt{10}$

61. $\mathbf{v} = -3\mathbf{i} - 2\mathbf{j} + \mathbf{k}$; $|\mathbf{v}| = \sqrt{14}$ **63.** $\mathbf{v} = 3\mathbf{i} - \mathbf{k}$; $|\mathbf{v}| = \sqrt{10}$ **65.** $-20\mathbf{i} + 13\mathbf{j}$ **67.** $\sqrt{5}$ **69.** $\sqrt{5} + 5 \approx 7.24$ **71.** $\dfrac{-2\sqrt{5}}{5}\mathbf{i} + \dfrac{\sqrt{5}}{5}\mathbf{j}$

73. $21\mathbf{i} - 2\mathbf{j} - 5\mathbf{k}$ **75.** $\sqrt{38}$ **77.** 0 **79.** $\dfrac{3\sqrt{14}}{14}\mathbf{i} + \dfrac{\sqrt{14}}{14}\mathbf{j} - \dfrac{\sqrt{14}}{7}\mathbf{k}$; $-\dfrac{3\sqrt{14}}{14}\mathbf{i} - \dfrac{\sqrt{14}}{14}\mathbf{j} + \dfrac{\sqrt{14}}{7}\mathbf{k}$ **81.** $\mathbf{v} \cdot \mathbf{w} = -11$; $\theta \approx 169.7°$

83. $\mathbf{v} \cdot \mathbf{w} = -4$; $\theta \approx 153.4°$ **85.** $\mathbf{v} \cdot \mathbf{w} = 1$; $\theta \approx 70.5$ **87.** $\mathbf{v} \cdot \mathbf{w} = 0$; $\theta = 90°$ **89.** $\text{proj}_w\,\mathbf{v} = \dfrac{9}{10}(3\mathbf{i} + \mathbf{j})$ **91.** $\alpha \approx 56.1°$; $\beta \approx 138°$; $\gamma \approx 68.2°$

93. $\sqrt{29} \approx 5.39$ mph; 0.4 mi **95.** At an angle of 70.5° to the shore

CHAPTER 9 *9.2 Exercises*

1. B **3.** E **5.** H **7.** C **9.** F **11.** G **13.** D **15.** B
17. $y^2 = 16x$ **19.** $x^2 = -12y$ **21.** $y^2 = -8x$

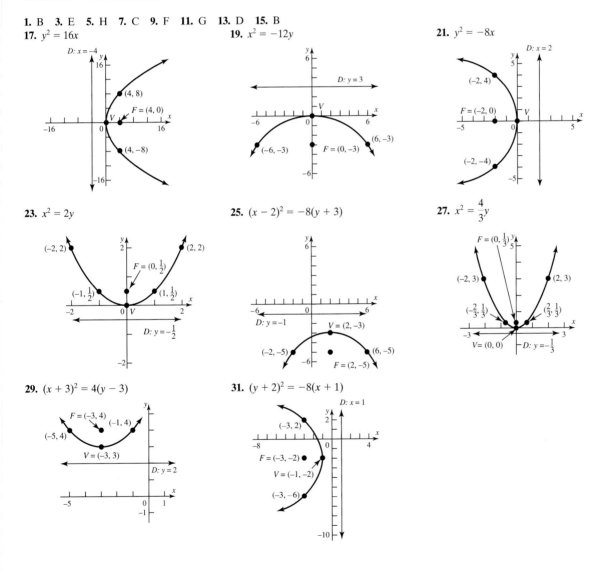

23. $x^2 = 2y$ **25.** $(x - 2)^2 = -8(y + 3)$ **27.** $x^2 = \dfrac{4}{3}y$

29. $(x + 3)^2 = 4(y - 3)$ **31.** $(y + 2)^2 = -8(x + 1)$

25.

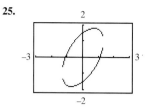

$\theta = 60°$ (see Problem 15)

$$\frac{x'^2}{4} + y'^2 = 1$$

Ellipse
Center at $(0, 0)$
Major axis is the x'-axis.
Vertices at $(\pm 2, 0)$

27.

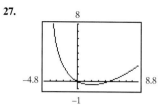

$\theta \approx 63°$ (see Problem 17)

$$y'^2 = 8x'$$

Parabola
Vertex at $(0, 0)$
Focus at $(2, 0)$

29.

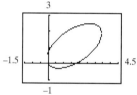

$\theta \approx 34°$ (see Problem 19)

$$\frac{(x' - 2)^2}{4} + y'^2 = 1$$

Ellipse
Center at $(2, 0)$
Major axis is the x'-axis
Vertices at $(4, 0)$ and $(0, 0)$

31.

$$\cot 2\theta = \frac{7}{24}; \theta \approx 37°$$

$$(x' - 1)^2 = -6\left(y' - \frac{1}{6}\right)$$

Parabola

Vertex at $\left(1, \frac{1}{6}\right)$

Focus at $\left(1, -\frac{4}{3}\right)$

33. Hyperbola **35.** Hyperbola **37.** Parabola **39.** Ellipse **41.** Ellipse
43. Refer to equation (6):
$A' = A \cos^2 \theta + B \sin \theta \cos \theta + C \sin^2 \theta$
$B' = B(\cos^2 \theta - \sin^2 \theta) + 2(C - A)(\sin \theta \cos \theta)$
$C' = A \sin^2 \theta - B \sin \theta \cos \theta + C \cos^2 \theta$
$D' = D \cos \theta + E \sin \theta$
$E' = -D \sin \theta + E \cos \theta$
$F' = F$
45. Use Problem 43 to find $B'^2 - 4A'C'$. After much cancellation, $B'^2 - 4A'C' = B^2 - 4AC$.
47. Use formulas (5) and find $d^2 = (x_2 - x_1)^2 + (y_2 - y_1)^2$. After simplifying, $(x_2 - x_1)^2 + (y_2 - y_1)^2 = (x'_2 - x'_1)^2 + (y'_2 - y'_1)^2$.

Exercises 9.6

1. Parabola; directrix is perpendicular to the polar axis 1 unit to the right of the pole.

3. Hyperbola; directrix is parallel to the polar axis $\frac{4}{3}$ units below the pole.

5. Ellipse; directrix is perpendicular to the polar axis $\frac{3}{2}$ units to the left of the pole.

7.

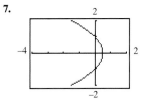

Parabola; directrix is perpendicular to the polar axis 1 unit to the right of the pole.

9.

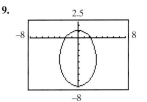

Ellipse; directrix is parallel to the polar axis $\frac{8}{3}$ units above the pole.

11.

Hyperbola; directrix is perpendicular to the polar axis $\frac{3}{2}$ units to the left of the pole.

13.

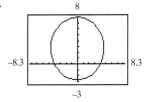

Ellipse; directrix is parallel to the polar axis 8 units below the pole; vertices are at $\left(8, \frac{\pi}{2}\right)$ and $\left(\frac{8}{3}, \frac{3\pi}{2}\right)$.

(d)

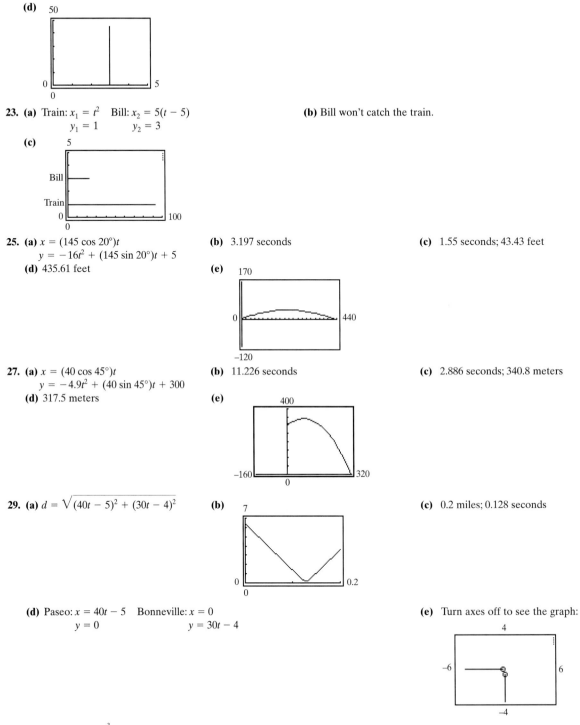

23. (a) Train: $x_1 = t^2$ Bill: $x_2 = 5(t - 5)$
            $y_1 = 1$ $y_2 = 3$

(b) Bill won't catch the train.

(c)

25. (a) $x = (145 \cos 20°)t$
            $y = -16t^2 + (145 \sin 20°)t + 5$

(b) 3.197 seconds

(c) 1.55 seconds; 43.43 feet

(d) 435.61 feet

(e)

27. (a) $x = (40 \cos 45°)t$
            $y = -4.9t^2 + (40 \sin 45°)t + 300$

(b) 11.226 seconds

(c) 2.886 seconds; 340.8 meters

(d) 317.5 meters

(e)

29. (a) $d = \sqrt{(40t - 5)^2 + (30t - 4)^2}$

(b)

(c) 0.2 miles; 0.128 seconds

(d) Paseo: $x = 40t - 5$ Bonneville: $x = 0$
            $y = 0$ $y = 30t - 4$

(e) Turn axes off to see the graph:

31. $x = t, y = t^3; x = \sqrt[3]{t}, y = t$ **33.** $x = t, y = t^{2/3}; x = t^{3/2}, y = t$ **35.** $x = 2 \cos \pi t, y = -3 \sin \pi t, 0 \le t \le 2$
37. $x = -2 \sin 2\pi t, y = 3 \cos 2\pi t, 0 \le t \le 1$

39.

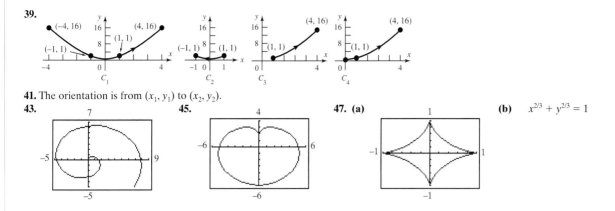

41. The orientation is from (x_1, y_1) to (x_2, y_2).

43.

45.

47. (a)

(b) $x^{2/3} + y^{2/3} = 1$

Fill-in-the-Blank Items

1. parabola **2.** ellipse **3.** hyperbola **4.** major; transverse **5.** y-axis **6.** $\dfrac{y}{3} = \dfrac{x}{2}; \dfrac{y}{3} = -\dfrac{x}{2}$ **7.** $\cot 2\theta = \dfrac{A - C}{B}$

8. $\dfrac{1}{2}$; ellipse; parallel; 4; below **9.** ellipse

True/False Items

1. T **2.** F **3.** T **4.** T **5.** T **6.** T **7.** T **8.** T **9.** T **10.** F

Review Exercises

1. Parabola; vertex $(0, 0)$, focus $(-4, 0)$, directrix $x = 4$

3. Hyperbola; center $(0, 0)$, vertices $(5, 0)$ and $(-5, 0)$, foci $(\sqrt{26}, 0)$ and $(-\sqrt{26}, 0)$, asymptotes $y = \dfrac{1}{5}x$ and $y = -\dfrac{1}{5}x$

5. Ellipse; center $(0, 0)$, vertices $(0, 5)$ and $(0, -5)$, foci $(0, 3)$ and $(0, -3)$

7. $x^2 = -4(y - 1)$: Parabola; vertex $(0, 1)$, focus $(0, 0)$, directrix $y = 2$

9. $\dfrac{x^2}{2} - \dfrac{y^2}{8} = 1$: Hyperbola; center $(0, 0)$, vertices $(\sqrt{2}, 0)$ and $(-\sqrt{2}, 0)$, foci $(\sqrt{10}, 0)$ and $(-\sqrt{10}, 0)$, asymptotes $y = 2x$ and $y = -2x$

11. $(x - 2)^2 = 2(y + 2)$: Parabola; vertex $(2, -2)$, focus $\left(2, -\dfrac{3}{2}\right)$, directrix $y = -\dfrac{5}{2}$

13. $\dfrac{(y - 2)^2}{4} - (x - 1)^2 = 1$: Hyperbola; center $(1, 2)$, vertices $(1, 4)$ and $(1, 0)$, foci $(1, 2 + \sqrt{5})$ and $(1, 2 - \sqrt{5})$,

asymptotes $y - 2 = \pm 2(x - 1)$

15. $\dfrac{(x - 2)^2}{9} + \dfrac{(y - 1)^2}{4} = 1$: Ellipse; center $(2, 1)$, vertices $(5, 1)$ and $(-1, 1)$, foci $(2 + \sqrt{5}, 1)$ and $(2 - \sqrt{5}, 1)$

17. $(x - 2)^2 = -4(y + 1)$: Parabola; vertex $(2, -1)$, focus $(2, -2)$, directrix $y = 0$

19. $\dfrac{(x - 1)^2}{4} + \dfrac{(y + 1)^2}{9} = 1$: Ellipse; center $(1, -1)$, vertices $(1, 2)$ and $(1, -4)$, foci $(1, -1 + \sqrt{5})$ and $(1, -1 - \sqrt{5})$

CHAPTER 10 *10.1 Exercises*

1. $2(2) - (-1) = 5$ and $5(2) + 2(-1) = 8$ **3.** $3(2) - 4\left(\dfrac{1}{2}\right) = 4$ and $\dfrac{1}{2}(2) - 3\left(\dfrac{1}{2}\right) = -\dfrac{1}{2}$ **5.** $2^2 - 1^2 = 3$ and $(2)(1) = 2$

7. $\dfrac{0}{1+0} + 3(2) = 6$ and $0 + 9(2)^2 = 36$ **9.** $3(1) + 3(-1) + 2(2) = 4$, $1 - (-1) - 2 = 0$, and $2(-1) - 3(2) = -8$ **11.** $x = 6, y = 2$

13. $x = 3, y = 2$ **15.** $x = 8, y = -4$ **17.** $x = \dfrac{1}{3}, y = -\dfrac{1}{6}$ **19.** Inconsistent **21.** $x = 1, y = 2$ **23.** $x = 4 - 2y$, y is any real number

25. $x = 1, y = 1$ **27.** $x = \dfrac{3}{2}, y = 1$ **29.** $x = 4, y = 3$ **31.** $x = \dfrac{4}{3}, y = \dfrac{1}{5}$ **33.** $x = 8, y = 2, z = 0$ **35.** $x = 2, y = -1, z = 1$

37. Inconsistent **39.** $x = 5z - 2, y = 4z - 3$ where z is any real number, or $x = \dfrac{5}{4}y + \dfrac{7}{4}, z = \dfrac{1}{4}y + \dfrac{3}{4}$ where y is any real number, or

$y = \dfrac{4}{5}x - \dfrac{7}{5}, z = \dfrac{1}{5}x + \dfrac{2}{5}$ where x is any real number **41.** Inconsistent **43.** $x = 1, y = 3, z = -2$ **45.** $x = -3, y = \dfrac{1}{2}, z = 1$

47. $x = \dfrac{1}{5}, y = \dfrac{1}{3}$ **49.** $x = 48.15, y = 15.18$ **51.** $x = -21.47, y = 16.12$ **53.** $x = 0.26, y = 0.06$ **55.** 20 and 61

57. Length 30 ft; width 15 ft **59.** Cheeseburger \$1.55; shake \$0.85 **61.** 22.5 lb

63. Average wind speed 25 mph; average airspeed 175 mph **65.** 80 \$25 sets and 120 \$45 sets **67.** \$5.56 **69.** 12, 15, 21

71. $I_1 = \dfrac{10}{71}, I_2 = \dfrac{65}{71}, I_3 = \dfrac{55}{71}$ **73.** 100 orchestra, 210 main, and 190 balcony seats **79.** $b = -\dfrac{1}{2}; c = \dfrac{3}{2}$ **81.** $a = \dfrac{4}{3}, b = -\dfrac{5}{3}, c = 1$

83. $x = \dfrac{b_1 - b_2}{m_2 - m_1}, y = \dfrac{m_2 b_1 - m_1 b_2}{m_2 - m_1}$ **85.** $y = mx + b, x$ is any real number

10.2 Exercises

1. $\begin{bmatrix} 1 & -5 & | & 5 \\ 4 & 3 & | & 6 \end{bmatrix}$ **3.** $\begin{bmatrix} 2 & 3 & | & 6 \\ 4 & -6 & | & -2 \end{bmatrix}$ **5.** $\begin{bmatrix} 0.01 & -0.03 & | & 0.06 \\ 0.13 & 0.10 & | & 0.20 \end{bmatrix}$ **7.** $\begin{bmatrix} 1 & -1 & 1 & | & 10 \\ 3 & 2 & 0 & | & 5 \\ 1 & 1 & 2 & | & 2 \end{bmatrix}$ **9.** $\begin{bmatrix} 1 & 1 & -1 & | & 2 \\ 3 & -2 & 0 & | & 2 \end{bmatrix}$

11.
```
[A]
[[1  -3  -5  -2]
 [0   1   6   9]
 [0   0  13  36]]
```

13.
```
[A]
[[1  -3   4   3 ]
 [0   1  -2   0 ]
 [0   0   4  15]]
```

15.
```
[A]
[[1  -3   2   -6 ]
 [0   1  -1    8 ]
 [0   0  -5  108]]
```

17.
```
[A]
[[1  -3   1   -2]
 [0   1   4    2]
 [0   0  39   16]]
```

19.
```
[A]
[[1  -3  -2   3 ]
 [0   1   6  -7 ]
 [0   0  64  -62]]
```

21. $\begin{cases} x = 5 \\ y = -1 \end{cases}$

consistent; $x = 5, y = -1$

23. $\begin{cases} x = 1 \\ y = 2 \\ 0 = 3 \end{cases}$

inconsistent

25. $\begin{cases} x + 2z = -1 \\ y - 4z = -2 \\ 0 = 0 \end{cases}$

consistent;
$x = -1 - 2z$,
$y = -2 + 4z$,
z any real number

27. $\begin{cases} x_1 = 1 \\ x_2 + x_4 = 2 \\ x_3 + 2x_4 = 3 \end{cases}$

consistent;
$x_1 = 1, x_2 = 2 - x_4$,
$x_3 = 3 - 2x_4$,
x_4 any real number

29. $\begin{cases} x_1 + 4x_4 = 2 \\ x_2 + x_3 + 3x_4 = 3 \\ 0 = 0 \end{cases}$

consistent;
$x_1 = 2 - 4x_4$,
$x_2 = 3 - x_3 - 3x_4$,
x_3, x_4 any real numbers

31. $x = 6, y = 2$ **33.** $x = 2, y = 3$ **35.** $x = 4, y = -2$ **37.** Inconsistent **39.** $x = \dfrac{1}{2}, y = \dfrac{3}{4}$ **41.** $x = 4 - 2y$, y is any real number

43. $x = \dfrac{3}{2}, y = 1$ **45.** $x = \dfrac{4}{3}, y = \dfrac{1}{5}$ **47.** $x = 8, y = 2, z = 0$ **49.** $x = 2, y = -1, z = 1$ **51.** Inconsistent

53. $x = 5z - 2, y = 4z - 3$, where z is any real number, or $x = \dfrac{5}{4}y + \dfrac{7}{4}, z = \dfrac{1}{4}y + \dfrac{3}{4}$, where y is any real number, or

$y = \dfrac{4}{5}x - \dfrac{7}{5}, z = \dfrac{1}{5}x + \dfrac{2}{5}$, where x is any real number **55.** Inconsistent **57.** $x = 1, y = 3, z = -2$ **59.** $x = -3, y = \dfrac{1}{2}, z = 1$

61. $x = \dfrac{1}{3}, y = \dfrac{2}{3}, z = 1$ **63.** $x = 1, y = 2, z = 0, w = 1$ **65.** $y = 0, z = 1 - x$, x is any real number

67. $x = 2, y = z - 3$, z is any real number **69.** $x = \dfrac{13}{9}, y = \dfrac{7}{18}, z = \dfrac{19}{18}$

71. $x = \dfrac{7}{5} - \dfrac{3}{5}z - \dfrac{2}{5}w$, $y = -\dfrac{8}{5} + \dfrac{7}{5}z + \dfrac{13}{5}w$, where z and w are any real numbers   **73.** $y = -2x^2 + x + 3$ **75.** $f(x) = 3x^3 - 4x^2 + 5$

77. $x =$ liters of 15% H_2SO_4, $y =$ liters of 25% H_2SO_4, $z =$ liters of 50% H_2SO_4: $\begin{cases} x = \frac{5}{2}z - 150 \\ y = 250 - \frac{7}{2}z \end{cases}$

15%	25%	50%	40%
0	40	60	100
10	26	64	100
20	12	68	100

79. If $x =$ Price of hamburgers, $y =$ Price of fries, $z =$ Price of colas, then $x = 2.75 - z$, $y = 0.68 + \dfrac{1}{3}z$, z any real number.

There is not sufficient information:

x	$2.15	$2.00	$1.85
y	$0.88	$0.93	$0.98
z	$0.60	$0.75	$0.90

81. **(a)** **Amount Invested At**

7%	9%	11%
0	10,000	10,000
1000	8000	11,000
2000	6000	12,000
3000	4000	13,000
4000	2000	14,000
5000	0	15,000

(b) **Amount Invested At**

7%	9%	11%
12,500	12,500	0
14,500	8500	2000
16,500	4500	4000
18,750	0	6250

(c) All the money invested 7% provides $2100, more than what is required.

83. $I_1 = \dfrac{4}{15}$, $I_2 = \dfrac{8}{15}$, $I_3 = \dfrac{4}{5}$ **85.** $I_1 = 3.5$, $I_2 = 2.5$, $I_3 = 1$

89. If $a_1 \neq 0$,

$$\begin{bmatrix} a_1 & b_1 & | & c_1 \\ a_2 & b_2 & | & c_2 \end{bmatrix} \rightarrow \begin{bmatrix} 1 & \frac{b_1}{a_1} & | & \frac{c_1}{a_1} \\ a_2 & b_2 & | & c_2 \end{bmatrix} \rightarrow \begin{bmatrix} 1 & \frac{b_1}{a_1} & | & \frac{c_1}{a_1} \\ 0 & -\frac{a_2b_1}{a_1} + b_2 & | & -\frac{a_2c_1}{a_1} + c_2 \end{bmatrix} \rightarrow \begin{bmatrix} 1 & \frac{b_1}{a_1} & | & \frac{c_1}{a_1} \\ 0 & \frac{-a_2b_1 + b_2a_1}{a_1} & | & \frac{-a_2c_1 + c_2a_1}{a_1} \end{bmatrix}$$

$$\rightarrow \begin{bmatrix} 1 & \frac{b_1}{a_1} & | & \frac{c_1}{a_1} \\ 0 & 1 & | & \frac{-a_2c_1 + c_2a_1}{a_1} \cdot \frac{a_1}{-a_2b_1 + b_2a_1} \end{bmatrix} \rightarrow \begin{bmatrix} 1 & \frac{b_1}{a_1} & | & \frac{c_1}{a_1} \\ 0 & 1 & | & \frac{-a_2c_1 + c_2a_1}{-a_2b_1 + b_2a_1} \end{bmatrix} \rightarrow \begin{bmatrix} 1 & 0 & | & \frac{-b_1c_2 + b_2c_1}{-a_2b_1 + b_2a_1} \\ 0 & 1 & | & \frac{-a_2c_1 + c_2a_1}{-a_2b_1 + b_2a_1} \end{bmatrix}$$

$x = \dfrac{1}{a_1b_2 - a_2b_1}(c_1b_2 - c_2b_1) = \dfrac{1}{D}(c_1b_2 - c_2b_1)$, $y = \dfrac{1}{a_1b_2 - a_2b_1}(a_1c_2 - a_2c_1) = \dfrac{1}{D}(a_1c_2 - a_2c_1)$

If $a_1 = 0$, then $a_2 \neq 0$, $b_1 \neq 0$, and

$$\begin{bmatrix} 0 & b_1 & | & c_1 \\ a_2 & b_2 & | & c_2 \end{bmatrix} \rightarrow \begin{bmatrix} a_2 & b_2 & | & c_2 \\ 0 & b_1 & | & c_1 \end{bmatrix} \rightarrow \begin{bmatrix} 1 & \frac{b_2}{a_2} & | & \frac{c_2}{a_2} \\ 0 & b_1 & | & c_1 \end{bmatrix} \rightarrow \begin{bmatrix} 1 & \frac{b_2}{a_2} & | & \frac{c_2}{a_2} \\ 0 & 1 & | & \frac{c_1}{b_1} \end{bmatrix} \rightarrow \begin{bmatrix} 1 & 0 & | & \frac{c_2}{a_2} - \frac{b_2c_1}{a_2b_1} = \frac{c_1b_2 - c_2b_1}{-a_2b_1} \\ 0 & 1 & | & \frac{c_1}{b_1} = \frac{-a_2c_1}{-a_2b_1} \end{bmatrix}$$

10.3 Exercises

1. 2   **3.** 22   **5.** -2 **7.** 10   **9.** -26 **11.** $x = 6$, $y = 2$ **13.** $x = 3$, $y = 2$ **15.** $x = 8$, $y = -4$ **17.** $x = 4$, $y = -2$ **19.** Not applicable

21. $x = \dfrac{1}{2}$, $y = \dfrac{3}{4}$ **23.** $x = \dfrac{1}{10}$, $y = \dfrac{2}{5}$ **25.** $x = \dfrac{3}{2}$, $y = 1$ **27.** $x = \dfrac{4}{3}$, $y = \dfrac{1}{5}$ **29.** $x = 1$, $y = 3$, $z = -2$ **31.** $x = -3$, $y = \dfrac{1}{2}$, $z = 1$

33. Not applicable   **35.** $x = 0$, $y = 0$, $z = 0$ **37.** Not applicable   **39.** $x = \dfrac{1}{5}$, $y = \dfrac{1}{3}$ **41.** -5 **43.** $\dfrac{13}{11}$ **45.** 0 or -9 **47.** -4 **49.** 12

51. 8   **53.** 8

55. $(y_1 - y_2)x - (x_1 - x_2)y + (x_1y_2 - x_2y_1) = 0$

$(y_1 - y_2)x + (x_2 - x_1)y = x_2y_1 - x_1y_2$

$(x_2 - x_1)y - (x_2 - x_1)y_1 = (y_2 - y_1)x + x_2y_1 - x_1y_2 - (x_2 - x_1)y_1$

$(x_2 - x_1)(y - y_1) = (y_2 - y_1)x - (y_2 - y_1)x_1$

$y - y_1 = \dfrac{y_2 - y_1}{x_2 - x_1}(x - x_1)$

13.

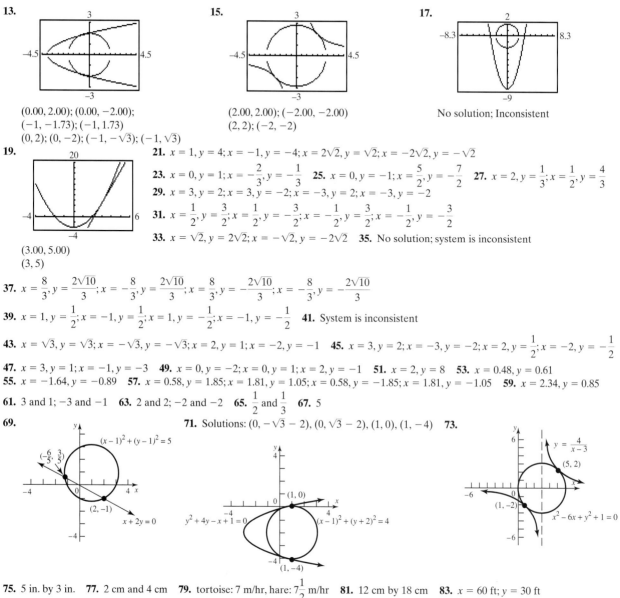

(0.00, 2.00); (0.00, −2.00);
(−1, −1.73); (−1, 1.73)
(0, 2); (0, −2); (−1, −√3); (−1, √3)

15.

(2.00, 2.00); (−2.00, −2.00)
(2, 2); (−2, −2)

17.

No solution; Inconsistent

19.

(3.00, 5.00)
(3, 5)

21. $x = 1, y = 4; x = -1, y = -4; x = 2\sqrt{2}, y = \sqrt{2}; x = -2\sqrt{2}, y = -\sqrt{2}$

23. $x = 0, y = 1; x = -\dfrac{2}{3}, y = -\dfrac{1}{3}$ **25.** $x = 0, y = -1; x = \dfrac{5}{2}, y = -\dfrac{7}{2}$ **27.** $x = 2, y = \dfrac{1}{3}; x = \dfrac{1}{2}, y = \dfrac{4}{3}$

29. $x = 3, y = 2; x = 3, y = -2; x = -3, y = 2; x = -3, y = -2$

31. $x = \dfrac{1}{2}, y = \dfrac{3}{2}; x = \dfrac{1}{2}, y = -\dfrac{3}{2}; x = -\dfrac{1}{2}, y = \dfrac{3}{2}; x = -\dfrac{1}{2}, y = -\dfrac{3}{2}$

33. $x = \sqrt{2}, y = 2\sqrt{2}; x = -\sqrt{2}, y = -2\sqrt{2}$ **35.** No solution; system is inconsistent

37. $x = \dfrac{8}{3}, y = \dfrac{2\sqrt{10}}{3}; x = -\dfrac{8}{3}, y = \dfrac{2\sqrt{10}}{3}; x = \dfrac{8}{3}, y = -\dfrac{2\sqrt{10}}{3}; x = -\dfrac{8}{3}, y = -\dfrac{2\sqrt{10}}{3}$

39. $x = 1, y = \dfrac{1}{2}; x = -1, y = \dfrac{1}{2}; x = 1, y = -\dfrac{1}{2}; x = -1, y = -\dfrac{1}{2}$ **41.** System is inconsistent

43. $x = \sqrt{3}, y = \sqrt{3}; x = -\sqrt{3}, y = -\sqrt{3}; x = 2, y = 1; x = -2, y = -1$ **45.** $x = 3, y = 2; x = -3, y = -2; x = 2, y = \dfrac{1}{2}; x = -2, y = -\dfrac{1}{2}$

47. $x = 3, y = 1; x = -1, y = -3$ **49.** $x = 0, y = -2; x = 0, y = 1; x = 2, y = -1$ **51.** $x = 2, y = 8$ **53.** $x = 0.48, y = 0.61$

55. $x = -1.64, y = -0.89$ **57.** $x = 0.58, y = 1.85; x = 1.81, y = 1.05; x = 0.58, y = -1.85; x = 1.81, y = -1.05$ **59.** $x = 2.34, y = 0.85$

61. 3 and 1; −3 and −1   **63.** 2 and 2; −2 and −2   **65.** $\dfrac{1}{2}$ and $\dfrac{1}{3}$ **67.** 5

69.

71. Solutions: $(0, -\sqrt{3} - 2), (0, \sqrt{3} - 2), (1, 0), (1, -4)$ **73.**

75. 5 in. by 3 in.   **77.** 2 cm and 4 cm   **79.** tortoise: 7 m/hr, hare: $7\dfrac{1}{2}$ m/hr   **81.** 12 cm by 18 cm   **83.** $x = 60$ ft; $y = 30$ ft

85. $l = \dfrac{P + \sqrt{P^2 - 16A}}{4}; w = \dfrac{P - \sqrt{P^2 - 16A}}{4}$ **87.** $y = 4x - 4$ **89.** $y = 2x + 1$ **91.** $y = -\dfrac{1}{3}x + \dfrac{7}{3}$ **93.** $y = 2x - 3$

95. $r_1 = \dfrac{-b + \sqrt{b^2 - 4ac}}{2a}; r_2 = \dfrac{-b - \sqrt{b^2 - 4ac}}{2a}$

10.7 Exercises

1.

3.

5.

7.

9.

11.

13.

15.

17.

19.

21.

23.

25.

27.

29. No solution

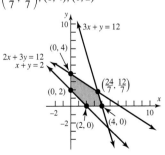

31. Bounded; corner points

$(0, 0), (3, 0), (2, 2), (0, 3)$

33. Unbounded; corner points

$(2, 0), (0, 4)$

35. Bounded; corner points $(2, 0), (4, 0),$

$\left(\dfrac{24}{7}, \dfrac{12}{7}\right), (0, 4), (0, 2)$

83.

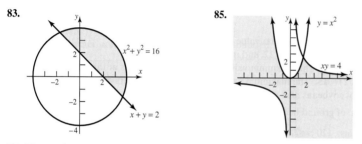

85.

87. The maximum value is 32 when $x = 0$ and $y = 8$ **89.** The minimum value is 3 when $x = 1$ and $y = 0$

91. The maximum value is $\dfrac{108}{7}$ when $x = \dfrac{12}{7}$ and $y = \dfrac{12}{7}$ **93.** 10   **95.** $y = -\dfrac{1}{3}x^2 - \dfrac{2}{3}x + 1$

97. 70 lbs of \$3 coffee and 30 lbs of \$6 coffee   **99.** 1 small, 5 medium, 2 large   **101.** 24 ft by 10 ft   **103.** $4 + \sqrt{2}$ in. and $4 - \sqrt{2}$ in.
105. $120\sqrt{10}$ ft   **107.** Katy gets \$10, Mike gets \$20, Danny gets \$5, Colleen gets \$10   **109.** Katy: 4 hr; Mike: 2hr; Danny: 8 hr
111. 35 gasoline engines, 15 diesel engines; 15 gasoline engines, 0 diesel engines

CHAPTER 11   *11.1 Exercises*

1. 1, 2, 3, 4, 5   **3.** $\dfrac{1}{3}, \dfrac{2}{4} = \dfrac{1}{2}, \dfrac{3}{5}, \dfrac{4}{6} = \dfrac{2}{3}, \dfrac{5}{7}$ **5.** $1, -4, 9, -16, 25$ **7.** $\dfrac{1}{2}, \dfrac{2}{5}, \dfrac{2}{7}, \dfrac{8}{41}, \dfrac{8}{61}$ **9.** $-\dfrac{1}{6}, \dfrac{1}{12}, -\dfrac{1}{20}, \dfrac{1}{30}, -\dfrac{1}{42}$

11. $\dfrac{1}{e}, \dfrac{2}{e^2}, \dfrac{3}{e^3}, \dfrac{4}{e^4}, \dfrac{5}{e^5}$ **13.** $\dfrac{n}{(n+1)}$ **15.** $\dfrac{1}{2^{n-1}}$ **17.** $(-1)^{n+1}$ **19.** $(-1)^{n+1}n$ **21.** $a_1 = 2, a_2 = 5, a_3 = 8, a_4 = 11, a_5 = 14$

23. $a_1 = -2, a_2 = -1, a_3 = 1, a_4 = 4, a_5 = 8$ **25.** $a_1 = 5, a_2 = 10, a_3 = 20, a_4 = 40, a_5 = 80$ **27.** $a_1 = 3, a_2 = 3, a_3 = \dfrac{3}{2}, a_4 = \dfrac{1}{2}, a_5 = \dfrac{1}{8}$

29. $a_1 = 1, a_2 = 2, a_3 = 2, a_4 = 4, a_5 = 8$ **31.** $a_1 = A, a_2 = A + d, a_3 = A + 2d, a_4 = A + 3d, a_5 = A + 4d$

33. $a_1 = \sqrt{2}, a_2 = \sqrt{2 + \sqrt{2}}, a_3 = \sqrt{2 + \sqrt{2 + \sqrt{2}}}, a_4 = \sqrt{2 + \sqrt{2 + \sqrt{2 + \sqrt{2}}}}, a_5 = \sqrt{2 + \sqrt{2 + \sqrt{2 + \sqrt{2 + \sqrt{2}}}}}$

35. 50   **37.** 21   **39.** 90   **41.** 26   **43.** 10   **45.** 96   **47.** $3 + 4 + \cdots + (n+2)$ **49.** $\dfrac{1}{2} + 2 + \dfrac{9}{2} + \cdots + \dfrac{n^2}{2}$ **51.** $1 + \dfrac{1}{3} + \dfrac{1}{9} + \cdots + \dfrac{1}{3^n}$

53. $\dfrac{1}{3} + \dfrac{1}{9} + \cdots + \dfrac{1}{3^n}$ **55.** $\ln 2 - \ln 3 + \ln 4 - \cdots + (-1)^n \ln n$ **57.** $\displaystyle\sum_{k=1}^{20} k$ **59.** $\displaystyle\sum_{k=1}^{13} \dfrac{k}{k+1}$ **61.** $\displaystyle\sum_{k=0}^{6}(-1)^k\left(\dfrac{1}{3^k}\right)$ **63.** $\displaystyle\sum_{k=1}^{n} \dfrac{3^k}{k}$

65. $\displaystyle\sum_{k=0}^{n}(a + kd)$ **67.** 21   **71.** A Fibonacci sequence

11.2 Exercises

1. $d = 1$; 5, 6, 7, 8   **3.** $d = 2$; $-3, -1, 1, 3$ **5.** $d = -2$; 4, 2, 0, -2 **7.** $d = -\dfrac{1}{3}$; $\dfrac{1}{6}, -\dfrac{1}{6}, -\dfrac{1}{2}, -\dfrac{5}{6}$ **9.** $d = \ln 3$; $\ln 3, 2 \ln 3, 3 \ln 3, 4 \ln 3$

11. $a_5 = 14$; $a_n = 3n - 1$ **13.** $a_5 = -7$; $a_n = 8 - 3n$ **15.** $a_5 = 2$; $a_n = \dfrac{1}{2}(n - 1)$ **17.** $a_5 = 5\sqrt{2}$; $a_n = \sqrt{2}n$ **19.** $a_{12} = 24$ **21.** $a_{10} = -26$

23. $a_8 = a + 7b$ **25.** $a_1 = -13$; $d = 3$; $a_{n+1} = a_n + 3$ **27.** $a_1 = -53, d = 6$; $a_{n+1} = a_n + 6$ **29.** $a_1 = 28$; $d = -2$; $a_{n+1} = a_n - 2$

31. $a_1 = 25$; $d = -2$; $a_{n+1} = a_n - 2$ **33.** n^2 **35.** $\dfrac{n}{2}(9 + 5n)$ **37.** 1260   **39.** 324

41.
```
sum(seq(3.45n+4.
12,n,1,20,1))
          806.9
```

43.
```
sum(seq(2.4n+.4,
n,1,15,1))
             294
```

45.
```
sum(seq(2.58n+2.
32,n,1,25,1))
           896.5
```

47. $-\dfrac{3}{2}$ **49.** 1185 seats   **51.** 210 of (1st colors name) and 190 (2nd colors name)

11.3 Exercises

1. $r = 3; 3, 9, 27, 81$ **3.** $r = \dfrac{1}{2}; -\dfrac{3}{2}, -\dfrac{3}{4}, -\dfrac{3}{8}, -\dfrac{3}{16}$ **5.** $r = 2; \dfrac{1}{4}, \dfrac{1}{2}, 1, 2$ **7.** $r = 2^{1/3}; 2^{1/3}, 2^{2/3}, 2, 2^{4/3}$ **9.** $r = \dfrac{3}{2}; \dfrac{1}{2}, \dfrac{3}{4}, \dfrac{9}{8}, \dfrac{27}{16}$

11. Arithmetic; $d = 1$ **13.** Neither **15.** Arithmetic; $d = -\dfrac{2}{3}$ **17.** Neither **19.** Geometric; $r = \dfrac{2}{3}$ **21.** Geometric; $r = 2$

23. Geometric; $r = 3^{1/2}$ **25.** $a_5 = 162; a_n = 2 \cdot 3^{n-1}$ **27.** $a_5 = 5; a_n = 5 \cdot (-1)^{n-1}$ **29.** $a_5 = 0; a_n = 0$ **31.** $a_5 = 4\sqrt{2}; a_n = (\sqrt{2})^n$

33. $a_7 = \dfrac{1}{64}$ **35.** $a_9 = 1$ **37.** $a_8 = 0.00000004$ **39.** $-\dfrac{1}{4}(1 - 2^n)$ **41.** $2\left[1 - \left(\dfrac{2}{3}\right)^n\right]$ **43.** $1 - 2^n$

45.
```
(1/4)sum(seq(2^n
,n,0,14,1))
        8191.75
```
47.
```
sum(seq((2/3)^n,
n,1,15,1))
      1.995432683
```
49.
```
-1sum(seq(2^n,n,
0,14,1))
         -32767
```

51. $\dfrac{3}{2}$ **53.** 16 **55.** $\dfrac{8}{5}$ **57.** $\dfrac{20}{3}$ **59.** $\dfrac{18}{5}$ **61.** -4 **63.** \$349,496.41 **65.** \$96,885.98 **67.** \$305.10

69. (a) 0.775 ft **(b)** 8th **(c)** 15.88 ft **(d)** 20 ft **71.** \$21,879.11
73. Option 2 results in the most: \$16,038,304; Option 1 results in the least: \$14,700,000 **75.** 1.845×10^{19} **79.** 3, 5, 9, 24, 73
81. A: \$25,250 per year in 5th year, \$112,742 total; B: \$24,761 per year in 5th year, \$116,801 total

11.4 Exercises

1. (I) $n = 1: 2 \cdot 1 = 2$ and $1(1 + 1) = 2$
   (II) If $2 + 4 + 6 + \cdots + 2k = k(k + 1)$, then $2 + 4 + 6 + \cdots + 2k + 2(k + 1) = (2 + 4 + 6 + \cdots + 2k) + 2(k + 1)$
   $= k(k + 1) + 2(k + 1) = k^2 + 3k + 2 = (k + 1)(k + 2)$.

3. (I) $n = 1: 1 + 2 = 3$ and $\dfrac{1}{2}(1)(1 + 5) = \dfrac{1}{2}(6) = 3$

   (II) If $3 + 4 + 5 + \cdots + (k + 2) = \dfrac{1}{2}k(k + 5)$, then $3 + 4 + 5 + \cdots + (k + 2) + [(k + 1) + 2]$

   $= [3 + 4 + 5 + \cdots + (k + 2)] + (k + 3) = \dfrac{1}{2}k(k + 5) + k + 3 = \dfrac{1}{2}(k^2 + 7k + 6) = \dfrac{1}{2}(k + 1)(k + 6)$

5. (I) $n = 1: 3 \cdot 1 - 1 = 2$ and $\dfrac{1}{2}(1)[3(1) + 1] = \dfrac{1}{2}(4) = 2$

   (II) If $2 + 5 + 8 + \cdots + (3k - 1) = \dfrac{1}{2}k(3k + 1)$, then $2 + 5 + 8 + \cdots + (3k - 1) + [3(k + 1) - 1]$

   $= [2 + 5 + 8 + \cdots + (3k - 1)] + 3k + 2 = \dfrac{1}{2}k(3k + 1) + (3k + 2) = \dfrac{1}{2}(3k^2 + 7k + 4) = \dfrac{1}{2}(k + 1)(3k + 4)$

7. (I) $n = 1: 2^{1-1} = 1$ and $2^1 - 1 = 1$
   (II) If $1 + 2 + 2^2 + \cdots + 2^{k-1} = 2^k - 1$, then $1 + 2 + 2^2 + \cdots + 2^{k-1} + 2^{(k+1)-1} = (1 + 2 + 2^2 + \cdots + 2^{k-1}) + 2^k$
   $= 2^k - 1 + 2^k = 2(2^k) - 1 = 2^{k+1} - 1$.

9. (I) $n = 1: 4^{1-1} = 1$ and $\dfrac{1}{3}(4^1 - 1) = \dfrac{1}{3}(3) = 1$

   (II) If $1 + 4 + 4^2 + \cdots + 4^{k-1} = \dfrac{1}{3}(4^k - 1)$, then $1 + 4 + 4^2 + \cdots + 4^{k-1} + 4^{(k+1)-1} = (1 + 4 + 4^2 + \cdots + 4^{k-1}) + 4^k$

   $= \dfrac{1}{3}(4^k - 1) + 4^k = \dfrac{1}{3}[4^k - 1 + 3(4^k)] = \dfrac{1}{3}[4(4^k) - 1] = \dfrac{1}{3}(4^{k+1} - 1)$.

11. (I) $n = 1: \dfrac{1}{1 \cdot 2} = \dfrac{1}{2}$ and $\dfrac{1}{1 + 1} = \dfrac{1}{2}$

   (II) If $\dfrac{1}{1 \cdot 2} + \dfrac{1}{2 \cdot 3} + \dfrac{1}{3 \cdot 4} + \cdots + \dfrac{1}{k(k + 1)} = \dfrac{k}{k + 1}$, then $\dfrac{1}{1 \cdot 2} + \dfrac{1}{2 \cdot 3} + \dfrac{1}{3 \cdot 4} + \cdots + \dfrac{1}{k(k + 1)} + \dfrac{1}{(k + 1)[(k + 1) + 1]}$

   $= \left[\dfrac{1}{1 \cdot 2} + \dfrac{1}{2 \cdot 3} + \dfrac{1}{3 \cdot 4} + \cdots + \dfrac{1}{k(k + 1)}\right] + \dfrac{1}{(k + 1)(k + 2)} = \dfrac{k}{k + 1} + \dfrac{1}{(k + 1)(k + 2)} = \dfrac{k + 1}{k + 2}$.

13. (I) $n = 1: 1^2 = 1$ and $\dfrac{1}{6} \cdot 1 \cdot 2 \cdot 3 = 1$

   (II) If $1^2 + 2^2 + 3^2 + \cdots + k^2 = \dfrac{1}{6}k(k + 1)(2k + 1)$, then $1^2 + 2^2 + 3^2 + \cdots + k^2 + (k + 1)^2$

   $= (1^2 + 2^2 + 3^2 + \cdots + k^2) + (k + 1)^2 = \dfrac{1}{6}k(k + 1)(2k + 1) + (k + 1)^2 = \dfrac{1}{6}(2k^3 + 9k^2 + 13k + 6) = \dfrac{1}{6}(k + 1)(k + 2)(2k + 3)$.

Review Exercises

1. 120   **3.** 10   **5.** 336   **7.** 56   **9.** $-\dfrac{4}{3}, \dfrac{5}{4}, -\dfrac{6}{5}, \dfrac{7}{6}, -\dfrac{8}{7}$ **11.** $2, 1, \dfrac{8}{9}, 1, \dfrac{32}{25}$ **13.** $3, 2, \dfrac{4}{3}, \dfrac{8}{9}, \dfrac{16}{27}$ **15.** $2, 0, 2, 0, 2$

17. Arithmetic; $d = 1; \dfrac{n}{2}(n + 11)$ **19.** Neither   **21.** Geometric; $r = 8; \dfrac{8}{7}(8^n - 1)$ **23.** Arithmetic ; $d = 4; 2n(n - 1)$

25. Geometric; $r = \dfrac{1}{2}; 6\left[1 - \left(\dfrac{1}{2}\right)^n\right]$ **27.** Neither   **29.** 115   **31.** 75   **33.** 0.49977   **35.** 35   **37.** $\dfrac{1}{10^{10}}$ **39.** $9\sqrt{2}$ **41.** $5n - 4$

43. $n - 10$ **45.** $\dfrac{9}{2}$ **47.** $\dfrac{4}{3}$ **49.** 8

51. (I) $n = 1: 3 \cdot 1 = 3$ and $\dfrac{3 \cdot 1}{2}(2) = 3$

(II) If $3 + 6 + 9 + \cdots + 3k = \dfrac{3k}{2}(k + 1)$, then $3 + 6 + 9 + \cdots + 3k + 3(k + 1) = (3 + 6 + 9 + \cdots + 3k) + (3k + 3)$

$= \dfrac{3k}{2}(k + 1) + (3k + 3) = \dfrac{3k^2}{2} + \dfrac{9k}{2} + \dfrac{6}{2} = \dfrac{3}{2}(k^2 + 3k + 2) = \dfrac{3}{2}(k + 1)(k + 2).$

53. (I) $n = 1: 2 \cdot 3^{1-1} = 2$ and $3^1 - 1 = 2$
(II) If $2 + 6 + 18 + \cdots + 2 \cdot 3^{k-1} = 3^k - 1$, then $2 + 6 + 18 + \cdots + 2 \cdot 3^{k-1} + 2 \cdot 3^{(k+1)-1}$
$= (2 + 6 + 18 + \cdots + 2 \cdot 3^{k-1}) + 2 \cdot 3^k = 3^k - 1 + 2 \cdot 3^k = 3 \cdot 3^k - 1 = 3^{k+1} - 1.$

55. (I) $n = 1: 1^2 = 1$ and $\dfrac{1}{2}(6 - 3 - 1) = \dfrac{1}{2}(2) = 1$

(II) If $1^2 + 4^2 + 7^2 + \cdots + (3k - 2)^2 = \dfrac{1}{2}k(6k^2 - 3k - 1)$, then

$1^2 + 4^2 + 7^2 + \cdots + (3k - 2)^2 + [3(k + 1) - 2]^2 = [1^2 + 4^2 + 7^2 + \cdots + (3k - 2)^2] + (3k + 1)^2 = \dfrac{1}{2}k(6k^2 - 3k - 1) + (3k + 1)^2$

$= \dfrac{1}{2}(6k^3 + 15k^2 + 11k + 2) = \dfrac{1}{2}(k + 1)(6k^2 + 9k + 2) = \dfrac{1}{2}(k + 1)[6(k + 1)^2 - 3(k + 1) - 1].$

57. $x^5 + 10x^4 + 40x^3 + 80x^2 + 80x + 32$ **59.** $32x^5 + 240x^4 + 720x^3 + 1080x^2 + 810x + 243$ **61.** 144   **63.** 84   **65.** $\{1, 3, 5, 6, 7, 8\}$
67. $\{3, 7\}$ **69.** $\{1, 2, 4, 6, 8, 9\}$ **71.** $\{1, 2, 4, 5, 6, 9\}$ **73.** 17   **75.** 29   **77.** 7   **79.** 25   **81.** 60   **83.** 128   **85.** 3024   **87.** 70   **89.** 91
91. 1,600,000   **93.** 216,000   **95.** 1260   **97.** (a) 381,024   (b) 1260   **99.** $\dfrac{3}{20}, \dfrac{9}{20}$ **101.** $\dfrac{1}{24}$ **103.** (a) 0.045   (b) 0.318   (c) 0.159

105. (a) 8   (b) 1100   **107.** \$151,873.77   **109.** (a) $\left(\dfrac{3}{4}\right)^3 \cdot 20 = \dfrac{135}{16}$ ft   (b) $20\left(\dfrac{3}{4}\right)^n$ ft   (c) after the 13th time   (d) 140 ft
111. (a) 0.68   (b) 0.58   (c) 0.32

CHAPTER 12

12.1 Exercises

1. 32   **3.** 1   **5.** 4   **7.** 2   **9.** 0   **11.** 3   **13.** 4   **15.** Does not exist   **17.** 13   **19.** -3 **21.** 6   **23.** 1   **25.** 1   **27.** -1

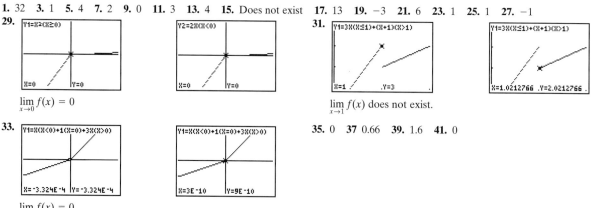

29.
$\lim\limits_{x \to 0} f(x) = 0$

31.
$\lim\limits_{x \to 1} f(x)$ does not exist.

33.
$\lim\limits_{x \to 0} f(x) = 0$

35. 0   **37** 0.66   **39.** 1.6   **41.** 0

12.2 Exercises

1. 5 **3.** 4 **5.** 8 **7.** 8 **9.** -1 **11.** 8 **13.** 3 **15** -1 **17.** 32 **19.** 2 **21.** $\dfrac{7}{6}$ **23.** 3 **25.** 0 **27.** $\dfrac{2}{3}$ **29.** $\dfrac{8}{5}$ **31.** 0 **33.** 5 **37.** 0
39. 0 **41.** -1 **43.** 1 **45** 1

12.3 Exercises

1. Continuous for all real numbers. **3.** Continuous for all real numbers. **5.** Continuous for all real numbers.
7. Continuous for all real numbers except $x = \dfrac{k\pi}{2}$, k is an odd integer. **9.** Continuous for all real numbers except $x = -2$ and $x = 2$.

11. Continous for all positive real numbers except $x = 1$. **13.** 5 **15.** 7 **17.** 1 **19.** 4 **21.** 0 **23.** $\dfrac{3}{2}$ **25.** Continuous

27. Continuous **29.** Not continuous **31.** Not continuous **33.** Not continuous **35.** Continuous **37.** Not continuous
39. Continuous **41.** $\{x \mid -8 \le x < -6 \text{ or } -6 < x < 4 \text{ or } 4 < x \le 6\}$ **43.** $-8, -5, -3$ **45.** $f(-8) = 0; f(-4) = 2$ **47.** ∞ **49.** 2 **51.** 1
53. Limit exists; 0 **55.** No **57.** Yes **59** No
61. Discontinuous at $x = -1$ and $x = 1$.

$\displaystyle\lim_{x\to 1} R(x) = \frac{1}{2}$: Hole at $\left(1, \frac{1}{2}\right)$

$\displaystyle\lim_{x\to -1^-} R(x) = -\infty$; $\displaystyle\lim_{x\to -1^+} R(x) = \infty$: Vertical Asymptote at $x = -1$

63. Discontinuous at $x = -1$ and $x = 1$.

$\displaystyle\lim_{x\to -1} R(x) = \frac{1}{2}$: Hole at $\left(-1, \frac{1}{2}\right)$

$\displaystyle\lim_{x\to 1^-} R(x) = -\infty$; $\displaystyle\lim_{x\to 1^+} = \infty$: Vertical asympotote at $x = 1$

65. $x = -\sqrt[3]{2}$: Asymptote; $x = 1$: Hole **67.** $x = -3$: Asymptote; $x = 2$: Hole **69.** $x = -\sqrt[3]{2}$: Asymptote; $x = -1$: Hole

12.4 Exercises

1. $m_{\tan} = 3$ **3.** $m_{\tan} = -2$ **5.** $m_{\tan} = 12$ **7.** $m_{\tan} = 5$

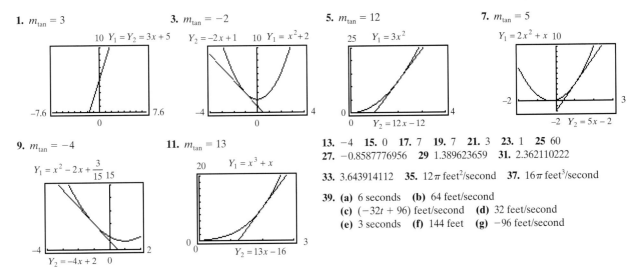

9. $m_{\tan} = -4$ **11.** $m_{\tan} = 13$

13. -4 **15.** 0 **17.** 7 **19.** 7 **21.** 3 **23.** 1 **25** 60
27. -0.8587776956 **29** 1.389623659 **31.** 2.362110222

33. 3.643914112 **35.** 12π feet^2/second **37.** 16π feet^3/second

39. (a) 6 seconds **(b)** 64 feet/second
(c) $(-32t + 96)$ feet/second **(d)** 32 feet/second
(e) 3 seconds **(f)** 144 feet **(g)** -96 feet/second

41. (a) $-23\dfrac{1}{3}$ feet/second **(b)** -21 feet/second **(c)** -18 feet/second **(d)** $-2.631t^2 - 10.269t + 999.933$ **(e)** -15.531 feet/second

Conics

Parabola

$$y^2 = 4ax \qquad y^2 = -4ax \qquad x^2 = 4ay \qquad x^2 = -4ay$$

Ellipse

$$\frac{x^2}{a^2} + \frac{y^2}{b^2} = 1, \quad c^2 = a^2 - b^2 \qquad\qquad \frac{x^2}{b^2} + \frac{y^2}{a^2} = 1, \quad c^2 = a^2 - b^2$$

Hyperbola

$$\frac{x^2}{a^2} - \frac{y^2}{b^2} = 1, \quad c^2 = a^2 + b^2 \qquad\qquad \frac{y^2}{a^2} - \frac{x^2}{b^2} = 1, \quad c^2 = a^2 + b^2$$

$$\text{Asymptotes:} \quad y = \frac{b}{a}x, \quad y = -\frac{b}{a}x \qquad\qquad \text{Asymptotes:} \quad y = \frac{a}{b}x, \quad y = -\frac{a}{b}x$$

Properties of Logarithms

$$\log_a MN = \log_a M + \log_a N$$

$$\log_a\left(\frac{M}{N}\right) = \log_a M - \log_a N$$

$$\log_a M^r = r \log^a M$$

$$\log_a M = \frac{\log M}{\log a} = \frac{\ln M}{\ln a}$$

Binomial Theorem

$$(a + b)^n = a^n + \binom{n}{1}ba^{n-1} + \binom{n}{2}b^2a^{n-2}$$

$$+ \cdots + \binom{n}{1}b^{n-1}a + b^n$$

Arithmetic Sequence

$$a + (a + d) + (a + 2d) + \cdots + [a + (n - 1)d]$$

$$= na + \frac{n(n - 1)}{2}d$$

Permutations/Combinations

$$0! = 1 \qquad 1! = 1$$

$$n! = n(n - 1) \cdot \ldots \cdot (3)(2)(1)$$

$$P(n, r) = \frac{n!}{(n - r)!}$$

$$C(n, r) = \binom{n}{r} = \frac{n!}{(n - r)!r!}$$

Geometric Sequence

$$a + ar + ar^2 + \cdots + ar^{n-1} = a\frac{1 - r^n}{1 - r}$$

Geometric Series

$$\text{If } |r| < 1, a + ar + ar^2 + \cdots = \sum_{k=1}^{\infty} ar^{k-1}$$

$$= \frac{a}{1 - r}$$